PATTY'S INDUSTRIAL HYGIENE AND TOXICOLOGY

Volume I
GENERAL PRINCIPLES

Volume II
TOXICOLOGY

Volume III
THEORY AND RATIONALE
OF INDUSTRIAL HYGIENE
PRACTICE

PATTY'S INDUSTRIAL HYGIENE AND TOXICOLOGY

Volume III
THEORY AND RATIONALE OF
INDUSTRIAL HYGIENE PRACTICE

SECOND EDITION
3A, THE WORK ENVIRONMENT

LEWIS J. CRALLEY, PH.D.
LESTER V. CRALLEY, PH.D.
Editors

Contributors

L. R. Allen
E. W. Arp, Jr.
K. A. Busch
K. J. Caplan
W. C. Cooper
L. J. Cralley
L. V. Cralley

P. M. Eller
P. E. Enterline
J. E. Fielding
R. L. Fischoff
F. G. Freiberger
R. L. Harris, Jr.
B. J. Held

M. J. Kotowski
N. A. Leidel
J. R. Lynch
M. H. Munsch
R. S. Parkinson
D. J. Paustenbach
C. Silberstein

A WILEY-INTERSCIENCE PUBLICATION
JOHN WILEY & SONS, New York · Chichester · Brisbane · Toronto · Singapore

Library of Congress Cataloging in Publication Data:
(Revised for volume 3)
Main entry under title:

Theory and rationale of industrial hygiene practice.

  At head of title: Patty's Industrial hygiene and
toxicology, volume III.
  "A Wiley-Interscience publication."
  Includes index.
  Contents: 3A. The work environment—3B. Biological
responses.
  1. Industrial hygiene—Collected works. 2. Industrial
toxicology—Collected works. I. Cralley, Lewis J.,
1911–   . II. Cralley, Lester Vincent, 1911–
III. Patty's Industrial
hygiene and toxicology.

RC967.T48 1985   613.6′2   84-25727
ISBN 0-471-86137-5

Printed in the United States of America

10 9 8 7 6 5 4 3 2 1

Contributors

LARAINE R. ALLEN, J.D., Reed, Smith, Shaw and McClay, Pittsburgh, Pennsylvania

EARL W. ARP, JR., Ph.D., Ashland Oil, Inc., Ashland, Kentucky

KENNETH A. BUSCH, Senior Advisor for Statistics, Division of Standards Development and Technology Transfer, NIOSH, Cincinnati, Ohio

KNOWLTON J. CAPLAN, Industrial Health Engineering Associates, Inc., Hopkins, Minnesota

W. CLARK COOPER, M.D., Occupational Medical Consultant, Lafayette, California

LESTER V. CRALLEY, Ph.D., 1453 Banyan Drive, Fallbrook, California

LEWIS J. CRALLEY, Ph.D., 7126 Golden Gate Drive, Cincinnati, Ohio

PETER M. ELLER, Ph.D., Office of Director, Division of Physical Sciences and Engineering, NIOSH, Cincinnati, Ohio

PHILIP E. ENTERLINE, Ph.D., Professor and Chairman, Department of Biostatistics, Graduate School of Public Health, University of Pittsburgh, Pittsburgh, Pennsylvania

JONATHAN E. FIELDING, M.D., Professor of Pediatrics and Public Health, Co-Director, Center for Health Enhancement, University of California, Los Angeles, California

ROBERT L. FISCHOFF, Manager, Industrial Hygiene Programs/Europe, Middle East, Africa, International Business Machines Corporation, Paris, France

FRED G. FREIBERGER, Safety Engineering/Emergency Control Program Administrator, International Business Machines Corporation, Bethesda, Maryland

ROBERT L. HARRIS, JR., Ph.D., Director, Occupational Health Studies Group, University of North Carolina, Chapel Hill, North Carolina

BRUCE J. HELD, Leader, Safety Science Group, Lawrence Livermore Laboratory, University of California, Livermore, California

MATTHIAS J. KOTOWSKI, Lawrence Livermore Laboratory, University of California, Livermore, California

NELSON A. LEIDEL, Sc.D., Senior Science Advisor, NIOSH, Atlanta, Georgia

JEREMIAH R. LYNCH, Manager, Industrial Hygiene, Medical and Environmental Affairs Department, Exxon Chemical Company, East Millstone, New Jersey

MARTHA HARTLE MUNSCH, J. D., Reed, Smith, Shaw and McClay, Pittsburgh, Pennsylvania

REBECCA S. PARKINSON, Program Manager, Employee Health Promotion, American Telephone and Telegraph Company, New York, New York

DENNIS J. PAUSTENBACH, Ph.D., Manager, Industrial and Environmental Toxicology, Syntex Corporation, Palo Alto, California

CAROL SILBERSTEIN, R.N., Ph.D., Director, Occupational Health Nursing, Associate Professor of Nursing, University of Cincinnati, Cincinnati, Ohio

Preface

The scope and depth in the development of industrial hygiene as a science has continued at an accelerated pace over the past thirty years. Prior to this period, that is, the first half of the twentieth century, the concepts of industrial hygiene began to develop with emphasis on preventive aspects of maintaining a healthful work environment. This encompassed research into obtaining basic knowledge and techniques for the recognition, evaluation, and control of health hazards in the work environment. These programs undertaken by government, industry, academia, insurance carriers, associations, foundations, and labor organizations developed the basic knowledge and techniques needed by the participating multidisciplinary professions to define full-concept programs in preventing occupational disease arising from the workplace. In the United States, these along with the enactment of the Occupational Safety and Health Act of 1970 gave support to the establishment of industrial hygiene as a science and supported its development in the fullest concept in assuring a healthful working environment.

An important approach in the early development of industrial hygiene was the emphasis placed on quantitative excellence wherever this program quality was encountered. This led to the current procedures and techniques used in establishing threshold limit values, now designated in government standards as personal exposure limits, the assessment of exposure levels, and the development of control procedures needed to keep hazardous exposures within safe limits.

An exceedingly high level of professionalism has always existed in the science of industrial hygiene. This is attested by the acceptance by management of control programs as an integral part of industrial processes and of industrial hygienists as important members of the management team. This close interrelationship is essential in the emergence and application of high technology in industry.

It is timely and important that these aspects be strengthened and that the theoretical basis and rationale of industrial hygiene practice be examined thoroughly and continually to restate fundamental facts and to direct attention to areas of weakness before they influence and become incorporated into acceptable practices by virtue of precedence.

LEWIS J. CRALLEY, PH.D.
LESTER V. CRALLEY, PH.D.

February 1985
Cincinnati, Ohio
Fallbrook, California

Notation

The subject areas covered in this volume are based on information and interpretation of regulations available in 1983. The practice of industrial hygiene necessitates a continuing updating in these areas.

Advancing industrial technology, including the more recent high technology, is associated with a number of newer types of health stresses with the increasing potential for synergistic interactions. This, along with similar health stresses associated with an ever-expanding life-style, often makes it difficult to distinguish health problems of the workplace from those of off-the-job activities.

The complexity of industrial technology, its application in industrial production, and the associated health stresses at the worksite are such that persons with expertise in the related industrial hygiene and allied specialties are needed for the recognition, evaluation, and control of these stresses. Similarly, persons with the proper expertise are needed in the interpretation of information presented in this volume and in the extrapolation of these data to specific situations.

Contents

PATTY'S INDUSTRIAL HYGIENE AND TOXICOLOGY

Volume I
GENERAL PRINCIPLES

Volume II
TOXICOLOGY

Volume III
THEORY AND RATIONALE
OF INDUSTRIAL HYGIENE
PRACTICE

CHAPTER ONE

Rationale

LEWIS J. CRALLEY, Ph.D., and
LESTER V. CRALLEY, Ph.D.

1 BACKGROUND

The emergence of industrial hygiene as a science has followed a predictable pattern. Whenever a gap of knowledge exists and an urgent need arises for such knowledge, dedicated people will gain the knowledge.

The harmful effects from exposures to toxic substances in the workplace, producing disease and death among workers, have been known well for over two thousand years. Knowledge on the toxicity of materials encountered in industry and means for control of exposure to toxic materials as well as techniques for its measurement and evaluation, however, were not available during the earlier period of industrial development. With few exceptions, the earliest attention given to worker health was in applying the knowledge at hand, which concerned primarily the recognition and treatment of illnesses associated with the job.

However, the devotion of prime attention to the preventive aspects of worker health maintenance through controlling job-associated health hazards became quite evident if the best interest of the worker was to be served in preventing occupational diseases. Not until around the turn of the twentieth century, though, did major effort begin to be directed toward the recognition, measurement, evaluation, and control of workplace environmental health stresses in the prevention of occupational diseases.

The aim of this chapter is not to document or present chronologically the major past contributors to worker health and their relevant works or the events and episodes that gave urgency to the development of industrial hygiene as a science. Rather, the purpose of the chapter is to place in perspective the many factors involved in relating environmental stresses to health and the rationale upon which the practice of industrial hygiene is based, including the recognition, measurement, evaluation, and control of workplace stresses, the biological responses to these stresses, the body defense mechanisms involved, and their interrelationships.

The individual chapters in Volume III, A and B, cover these aspects in detail.

Similarly, it is not the intent of this volume, A and B, to present procedures, instrumental or otherwise, for measuring airborne levels of exposure to chemical and other stress agents. This aspect is covered in detail in Patty's Industrial Hygiene, Volume I, *General Principles*. Rather, attention in this volume, A and B, is devoted to other aspects of airborne exposure such as representative and adequate sampling, variations in exposure levels, exposure durations, and the like.

2 INSEPARABILITY OF ENVIRONMENT AND HEALTH

Knowledge is being constantly developed on the ecological balance that exists between the earth's natural environmental forces and the existing biological species and how the effects of changes in either may affect the other. In the earth's early history this balance was maintained by the natural interrelationships of stresses and accommodations between the environment and the existing biological species at that site. This system related as well to the ecological balance within the species, both plant and animal.

Studies of past catastrophic events such as the ice age have shown the effects that changes in this balance can have on the existing biological species. The forces that brought on the demise of the dinosaurs that lived during the Cretaceous and Mesozoic periods are uncertain. Most probably major geological events were involved. Studies have also shown that in the earth's past history other animal species, as well, have originated and disappeared.

The human species, however, has been an exception to the ecological balance that existed in the earth's earlier history between the natural environment and the evolving biological species. The human ability to think, create, and change the natural environment has brought on changes, above and beyond those of the existing natural forces and environment, that have an ever increasing impact upon the previous overall ecology balance.

The capacity of humans to alter the environment to serve their benefits is beyond the bounds of anticipation. In early human history these efforts predictably addressed themselves to a better means of survival, that is, food, shelter, and protection. As these efforts progressed, emphasis was placed on gaining knowledge concerning factors affecting human health and well-being. Thus evolved the medical sciences, including public health. In some instances, these efforts resulted in the intervention of the ecological balance in the control of disease. In other situations the environment may have been altered to make necessary resources available as in the building of dams for flood control and for developing hydroelectric power. This type of alteration of the localized natural environment and the associated ecological systems may have an impact in developing additional stresses in readjusting the existing ecological balance.

Of more recent impact on health has been the stresses in living and on health brought on by activities associated with personal gratifications such as life-styles as well as those associated with an ever more complex and advancing technology in all of the sciences.

The quality of indoor environment is receiving increasing attention in relation to good health. This applies to the study and control of factors giving rise to psychological stresses associated with living or working in enclosed spaces, as well as air pollution arising from life-styles and building designs and materials. Examples of the latter include airborne contamination from smoking, insulating materials, and fabrics as well as location of housing sites.

Thus the advantages associated with changes in the environment for human benefit and improving the essential quality of living must be at the same time equated with their cost-effectiveness as well as their potential to deteriorate the environment and present a new scheme of stresses.

That humans, for optimal health, must exist in harmony with the surrounding 24-hour daily environment and its stresses is self-evident.

To better understand the significance of the on-the-job environmental health hazards, an overview of the 24-hour daily stress patterns of workers will be helpful, for this permits a perspective in which the overall component stresses are related to the whole of the worker's health.

Our habitat, the earth and its flora and fauna, is in reality a chemical one, that is, an entity that can be described in terms of an almost infinite number of related elements and compounds, the habitat in which the species was derived and in which a sort of symbiosis exists that supports the survival of the individual species.

The intricacy of this relationship is illustrated in the presently recognized essential trace elements: copper, chromium, fluorine, iodine, molybdenum, manganese, nickel, selenium, silicon, vanadium, and zinc. All are toxic when ingested in excess and all are listed in the standards relating to permissible exposure limits in the working environment. Some forms of several of these trace elements are classified as carcinogens. It is most revealing that certain trace elements essential for survival are under some circumstances capable of destruction. Thus it is a question of how much.

The environment is both friendly and hostile. The friendly milieu provides the components necessary for survival: oxygen, food, and water. On the other hand, the hostile environment constitutes a stress in which survival is constantly challenged.

Although numerous factors are obviously involved in the optimal health of an individual, stresses arising out of the overall environment, that is, the workplace, life-style, and off-the-job activities, are dominant. Thus the stresses over the 24-hour period have an overall impact on an individual's health. Any activity over the same period of time that can be stress relieving will have a beneficial effect in helping the body to adjust to the remaining insults of the day.

The inseparability of the environment and its relation to health is presented graphically in Figure 1.1.

2.1 Environmental Health Stresses

An environmental health stress may be thought of as any agent in the environment capable of significantly diminishing a sense of well-being, causing severe discomfort, interfering with proper body organ functions, or causing disease. These stresses may be chemical, physical, biological, or psychological in nature. They may arise from natural or created sources.

2.1.1 Macrocosmic Sources

Macrocosmic sources of environmental stress agents are those emanating from the solar system or from extensive geographical regions and capable of affecting large geographical areas such as the quality of the air, soil, or water.

Examples of natural sources of these stress agents include ultraviolet, thermal, and other radiations from the sun, volcanic eruptions that release huge quantities of gases and particulates into the upper atmosphere, the changing of the upper air jetstream and other factors that influence the climate, and the movement of the earth's surface plate structure resulting in earthquakes, and tidal waves.

Examples of created macrocosmic stresses include the potential interference of the ozone layer of the upper atmosphere that shields the earth from excessive ultraviolet radiation through the release of large quantities of some organic compounds into the air, the excessive burning of fossil fuels that increase the carbon dioxide level and temperature of the atmosphere, or excessive industrial, community, or life-style pollution over a prolonged period that may affect the quality of the air, water, or soil.

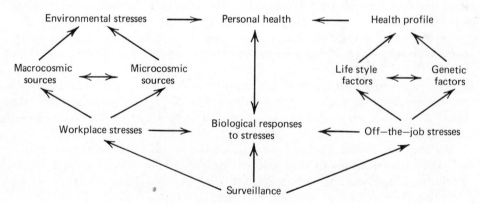

Figure 1.1. Inseparability of environment and health.

These stresses may act directly upon the individual as in excessive exposure to ultraviolet and thermal radiation and toxic materials, or indirectly by influencing the earth's climate—sunshine, rain, and temperature—thus affecting vegetation and habitability.

2.1.2   Microcosmic Sources

Microcosmic sources of stress agents are those emanating from localized areas and generally affecting the same region. These are most commonly at the community or regional community level and may also include the home and work environment.

Examples of natural sources of microcosmic stress agents are pollen, which gives rise to sensitization, allergy, and hay fever; water pollution from ground sources having a high mineral or salt content; and the release of methane, radon, sulfur gases and other contaminants from underground and surface areas.

Created sources of health stress agents at the community level are, however, by far the most common source. These include noise from everyday activities as in lawn mowing, motorcycle and truck street traffic, and loud music; air pollution from motor vehicle exhaust, release of industrial emissions into the air, emissions from refuse and garbage land fills and from toxic waste disposal sites, spraying of crops, and life-styles; water pollution through the release of contaminants from home, community and industrial activities into the waste water and seepage into the ground water from garbage and land fills and from toxic waste disposal sites.

These stress agents may have a direct response such as the effects of noise on hearing and toxic exposures on health, or an indirect response such as acid smog and rain affecting vegetation and soil quality downwind from the source.

2.1.3   Life-Style

The life-style of an individual, including habits, nutrition, off-the-job activities, recreation, exercise, and rest, may have beneficial effects as well as innate and insidious stresses that exert a profound influence on that person's health.

Extensive knowledge is being developed constantly on the influence of habits such as excessive smoking, alcohol consumption, and the use of drugs on health. These may have additive, accumulative, and synergistic actions or may exert superimposed responses on other exposures in the workplace. These agents can be an overwhelming cause of off-the-job respiratory, cardiovascular, renal, and other diseases and may create a grave health problem in an individual previous to other exposures encountered in the workplace.

Knowledge is also emerging on the deleterious effects of excessive smoking, alcohol consumption, and drug use by women during pregnancy on the health of their offspring, including malformation and improper functioning of body organs and systems, abnormal weight, and so forth.

The benefits to health of a good, adequately balanced nutrition, that is, vitamins, minerals, and other essential food intake, are gaining increased attention in relation to general fitness, weight control, prevention of disease, supporting the natural body defense mechanisms, recovery from exposure to environmental stresses, and aging. Conversely, malnutrition and obesity are the cause of many illnesses and may have a synergistic effect on exposure to other stresses.

Recreational activities are important aspects of good health practices, releasing tension brought on through both off- and on-the-job stresses. Conversely, many recreational activities may be harmful, such as listening to loud music at home or at discos, which may lead to a loss in hearing, frequent engagements in events and schedules that may interfere with the body's internal rhythmic functions, failing to observe needed precautions while using chemicals in hobby activities, and pursuing activities to the point of exhaustion.

Both exercise and rest are important activities in maintaining good health. Exercise helps maintain proper muscular tone as well as weight control. The exercise should, however, be designed for specific purposes, maintained on a regular basis, and structured to accommodate the body physique and health profile of an individual. Otherwise, more harm than benefit may result. Rest provides time for the body to recuperate from physical and psychological stresses.

The hours between shifts and during weekends provide time for the body to excrete many agents absorbed during the workshift. Work patterns that disturb this recovery period, as in moonlighting, may have an even increased deleterious effect if similar stresses are encountered on the second job. Likewise, smoking, alcohol consumption, and use of drugs may impair the body's proper recuperation from previous stresses.

2.1.4 Off-the-Job Stresses

The worker may encounter a host of innate stresses outside the workplace. These may be of a physical or psychological nature and usually include the eight to ten hour period between the end of the shift and retiring and on weekends. Most any of the off-the-job activities if performed in excess and without regard to necessary precautions are capable of producing stress and injury. Hobby and recreational activities likely account for a good portion of this time. More recently, increasing periods of time are spent sedentarily watching television during which eyestrain, mental and physical fatigue, and general inactivity occur—poor preparation for a refreshing night's sleep.

In hobby activities the participant often encounters many of the environmental exposures experienced on the job, such as in soldering, welding, cleaning, gluing, woodworking, grinding, sanding, and painting, during which the related exposures are well controlled. The home hobbyist, however, will not likely have available the protective devices needed such as local exhaust ventilation, protective clothing, goggles, and respirators, or protective creams. The same hobbyist often does not read or observe the precautionary labels, takes short cuts to save time, and fails to use the basic principles of keeping toxic materials from the skin, thoroughly washing after skin contact, not smoking or eating while working with these materials, and keeping upwind of harmful materials encountered as in insecticide and other spraying in gardening.

The activities previous to and supporting the sleep environment may not augment sleep. Thus the body may be fatigued or otherwise stressed prior to entering the workplace and endanger the worker or those around that person.

The off-the-job gamut of health stresses is wide and formidable. To deal with these stresses satisfactorily requires that a degree of accommodation be reached based on judgment, feasibility, personal option, objectives, and other factors. It is evident that although the components of the total environmental health stresses must be considered on an individual basis, each component cannot stand alone and apart from the others.

2.1.5 Workplace Stresses

The workplace may be a most important source of health stresses if the operations have not been studied thoroughly and the associated health hazards eliminated or the necessary controls incorporated into plant processes.

This was evidenced during the earlier industrial growth when little information was available on the methodology for identifying, measuring, evaluating, and controlling work related stresses. During this period workplace exposures were often severe, leading to a high incidence of disease and related deaths.

At the turn of this century, and especially since the 1950s, management, unions, government, academic, and other groups have taken an increased interest in worker health and in the control of exposures to stresses in the workplace. This, along with the setting of standards of exposure limits, has created an urgency for a broad support of functions and knowledge to prevent on-the-job illnesses.

2.1.6 Biological Response to Environmental Health Stresses

The human body consists of a number of discrete organs and systems derived from the embryonic state, encased in a dermal sheath, and developed to perform specific functions necessary for the functioning of the body as an integral unit. These organs are interdependent so that a malfunction in one may affect the functioning of many others. As an example, a malfunction in the alveoli which hinders the passage of oxygen into the blood transport system may have a direct effect on other organs through their diminished oxygen supply. Similarly, any hormonal imbalance or enzyme aberration may affect the functioning of many other body organs. Once invasion has occurred, toxic agents may selectively target one or more of the organs. Each organ has a varying degree for resisting stress, adjusting to injury, and propensity for repair.

Although the organ structures and functions of the several animal species have many similarities, with some more approximate than others to that of the human body, care has to be taken in extrapolating research data on any one animal species to that of the human body. Similarly, research data obtained within a smaller frame of reference, such as may be done through cell cultures, has to be cautiously interpreted with regard to the whole integrated human body organ system.

2.1.7 Body Protective Mechanisms Against Environmental Stresses

The human body in coexisting with the hostile stresses of the external environment has established a formidable scheme of protection against many of these stresses. This is accomplished in a remarkable manner by the ectodermal and endodermal barriers preventing absorption through inhalation, skin contact, and ingestion, and supported by the backup mesodermal and biotransformation mechanisms once invasion has occurred. These external and internal protective mechanisms against invasion, however, are not absolute and can be overwhelmed by a stress agent to the extent that they are ineffective, with resultant disease and death. Also, these protective mechanisms may become impaired in various degrees from insults associated with life-styles and other daily activities.

In studying the effects of specific stresses on health it is important to be aware of the body protective mechanisms involved. Suitable control of a specific stress should suppliment the body's protective response to that stress.

2.1.8 Coaptation of Health and Environmental Stresses

To survive, the human body must live in balance with the surrounding environment and its concomitant stresses. Since these stresses, singularly or combined, are not constant in value even for short periods of time or over limited geographical areas, the body must have built-in mechanisms for adjusting to differing levels of stresses through adaptation, acclimatization, acclimation, and other regulating systems. There is a limit, however, to which the body can protect itself against these stresses without a breakdown occurring in these protective systems.

Consequently, if humans are to have freedom of geographic movement and of living in an excessively hostile environment, some kind of an additional accommodation must be reached with the environmental insults that are not satisfactorily handled by or overwhelm the related body protective mechanisms. Exceptions include limited living and working in confined spaces where the environment is under absolute control against outside catastrophic stresses as in space travel and underwater activities.

Examples of this coaptive relationship include exposure to ultraviolet radiation from the sun. It is obvious that avoidance of all ultraviolet radiation from this source is impracticable. In addition to the ozone layer of the atmosphere shielding the earth from major levels of ultraviolet radiation emitted by the sun and the body's own protective mechanisms such as pigmentation, further accommodation is reached through the use of skin barriers, eye protection through use of sun glasses, and a managed limitation to exposure.

In the case of allergens in foods, cosmetics, air, fabrics, drugs, and so on, discovery of the offending agents and avoidance of them constitute a prime approach, though medical management is highly important.

In high altitudes where the oxygen content of the air is diminished, the body acclimatizes by increasing the number of red blood cells and hemoglobin that carry oxygen to the tissues. Further accommodation may be provided through the use of supplemental oxygen supply.

In exposures to excessively high and low temperatures, the body has a number of regulating systems to keep the body temperature within normal limits. This in turn permits living in a wide range of environmental temperatures. Further accommodation to extremes in environmental temperatures may be provided through special clothing, protective equipment, and living and working in climate controlled buildings.

Where excessive exposures to environmental stresses exist in the workplace, emphasis is placed on their elimination or lowering the stress levels through recognized control procedures to the point that the body defense mechanisms can adequately prevent injury to health, though some situations may recommend the use of personal protective equipment or other control strategies.

There is a limit to what can be done to alter environmental stresses from natural macrocosmic sources. Thus, these become ubiquitous background stresses upon which other exposures from microcosmic sources are added including community and industrial pollution, off-the-job activities, life-styles, and on-the-job activities.

Rationally, then, it is primarily the created stresses from a few macrocosmic but predominantly microcosmic sources that are amenable to control and must be kept within acceptable limits to permit the human body to reasonably exist in an increasingly complex industrial technological age.

3 INDUSTRIAL TECHNOLOGICAL ADVANCES

In the early history of civilization, the many activities associated with living were at the tribal level where emphasis was placed upon survival, that is, procuring adequate food, clothing, protection, and shelter. Though the tribes were nomadic, they were undoubtedly aware that they must live in accommodation with the environment. This would have been evidenced through the appropriate use of clothing, safe use of fire, and observing climatic patterns.

It was inevitable that the nomadic way of life would give way in most instances to a more settled life-style in which food, shelter, and protection could be more dependable, based upon individual effort and ingenuity. During this period the accommodations to the natural elements were made easier through a more permanent shelter and pattern of living.

The next advancement in industrialization came through the realization that increased production could be attained through specialization of work pursuits wherein a designated work group devoted its entire effort to the making of a single product as clothing, pottery, and tools, or the growing of foods, and in which each group shared their commodity in exchange for the commodity of other groups. This was the beginning of the cottage industries that related primarily to the community level of trade. Even at that level of production many of the health stresses associated with the different pursuits were intensified over that period of nomadic living where every person was a sort of jack-of-all-trades. This was especially true where the operations tended to be restricted to confined and crowded spaces.

As means of communications and transportation improved, trade increased between adjoining communities, and the search continued for better ways of producing commodities at an increased volume with less manpower. Similar production operations tended to expand and be concentrated within the same housing structure. This led to an increase in the health stresses of the whole workforce in instances where the stress agents were cumulative in intensity and response.

This industrialization trend was continued and intensified with the advent of the steam engine, which gave rise to an industrial revolution involving larger factories and newer production techniques along with increased associated health risks. Where in the past, exposures to health risks in any one workplace were dominantly to a few stress agents, the newer industrialization and technology led to a more complex pattern of exposures.

Since the 1950s technological advances and their application to production have expanded at an ever increasing pace. The application in industry of knowledge gained through space technology research has rapidly expanded into the newer electronic high-technology age.

The advent of this high technology and its application to production is having its effect upon both the nature of employment and the concomitant health stresses. While in the past workers needed only special instructions to perform most job operations—and this will continue for some time—the move into higher technology has created a demand for highly trained employees for many job positions, and this trend will increase dramatically. Computers, word processors, video display terminals, radar, and other electronic equipment are becoming commonplace in industry. Robots are rapidly taking over many repetitive operations.

Similarly, the nature and extent of associated health stresses are becoming more complex with the advent of the high technology industry.

The urgent need for knowledge concerning the effects of exposures to health stresses associated with an ever expanding industrial technology, along with the methodology for their recognition, evaluation, and control, gave rise to the science of industrial hygiene. This science must keep attuned to the ever changing applications of technology in industry.

4 EMERGENCE OF INDUSTRIAL HYGIENE AS A SCIENCE

Science may be defined as an organized body of knowledge and facts established through research, observations, and hypothesis. As such, a science may be basic as relating to the physical sciences, or applied in the sense that the principles of other sciences are brought to bear in developing facts and knowledge in a specific area of application.

The lack of knowledge during the early history of industrial development on the effects of health stresses associated with industrial operations and how these could be controlled, with the often concomitant massive exposures to harmful materials in the workplace, led to many serious episodes of illnesses and deaths among workers. An example is the high incidence of silicosis and silicotuberculosis that existed at the turn of the century in the workers of the hard rock mines, granite industry, and tunnel operations where the dust had a high free silica content.

During this early period the major effort on behalf of the workers was to apply the knowledge at hand, which related primarily to the recognition and treatment of occupational illnesses.

It was not until around the turn of this century that specific attention began to be devoted to the preventive aspects of industrial illnesses. Scientists, including engineers, chemists, and physicists, began to apply their knowledge and expertise toward the development of methods and procedures for identifying, measuring, and controlling exposures to harmful airborne dusts and chemicals in the workplace. At that time there were no recognized procedures for accomplishing these objectives.

After trying various potential procedures, the impingement method was judged the most adaptable one at that time for collecting many of the airborne contaminants such as particulates, mists, some fumes, and gases. The light field dust counting technique was developed for enumerating levels of dust in the air; and conventional analytical methods available at that time were adapted for measuring levels of chemicals in these samples.

Even during this early period of developing airborne sample collection and analytical procedures, these scientists realized that the exposure patterns that existed were more complex and complicated than the instrumentation for sample collection and analysis would define. These scientists also knew of many of the deficiencies associated with the proposed procedures being developed. They were aware that the data being collected represented only a segment of the overall exposure pattern. They believed, however, that this segment value could be used as an index which would represent the overall exposure pattern so long as the production techniques and other operational factors remained the same. It must be remembered that at that time information was not available on respiratory deposition and dust size. It was imperative to them, and rightly so, that some method, with whatever deficiencies that may have been incumbent, be developed for indexing airborne levels of contamination in the workplace both for estimating levels of exposure and for use as a benchmark in determining degrees of air quality improvement after controls had been established.

Since the earliest instrumentation for collecting airborne contaminant samples was nonportable, such samples represented general room levels of exposure depending upon where the samples were taken.

The above procedures for airborne sampling and analysis in the workroom, as primitive as they may seem in comparison to those methods available today, served that period of time well. They accounted for the drastic reduction in massive exposures that existed in many work sites and were the methods and procedures upon which future refinements would be made.

These scientists showed that it was feasible to lower the massive workplace airborne contamination that often existed at that time, and by relating the data to the health profile of the worker, observed that lowering of the exposure level also lowered the incidence of the associated disease. Thus began the first field studies that were to have a profound influence on the collection of relevant data upon which to base permissible limits and in developing the rationale upon which the practice of industrial hygiene is predicated. The latter was to be further substantiated through laboratory and clinical research.

The development of the hand-operated midget impinger pump during the 1930s was a decided improvement over the standard impinger pump, since it was portable and permitted taking airborne samples closer to the worker using impingement, fritted glass bubbler, and other collecting techniques. The newer exposure data showed that the worker often had a higher level of exposure than that indicated by the general room airborne level.

Other instrumentation, such as the electrostatic precipitator and evacuated containers, came into use during this period. In the late 1940s the filter paper method for collecting some airborne particulate samples came into use. This procedure was found to be superior in many instances to the impingement method. This method did not fracture particulates or disperse conglomerates, which often accompanied impingement collection, and permitted direct gravimetric measurement of the sample.

Data also began to emerge on the importance of particulate size in relation to retention and deposition in the respiratory tract. The electron microscope made it possible to study the characteristics and response of submicron particle sizes.

Also, a dramatic increase began in toxicologic and epidemiological studies by government, industry, educational institutions, and foundations directed to obtaining data upon which to base exposure standards as well as good industrial hygiene practices.

Another major advancement in developing better methodology for studying occupational diseases occurred at midcentury. Toxicologic and other studies had revealed that lowering the exposure and extending the period of exposure changed the dose–response pattern of many toxic agents. As an example, a high airborne level of exposure to lead produces an acute response over a relatively short period of time. In contrast, lowering the exposure level of the same agent and extending the exposure time shows a different dose–response pattern, a chronic form of lead poisoning. Thus in studying the effects of exposure to health stress agents it is important to obtain relevant dose–response data over an extended period of exposure time. One method of obtaining relevant health profile data on workers is through medical records and special studies. Another and more recent method is through the study of causes of death on death certificates through Social Security, management retirement system, and Union records. These studies have revealed that a lifetime of work exposure to an agent or an extended latency period of twenty or more years from time of initial exposure may be necessary to fully define the wide range of dose–response relationships. This may be especially true for carcinogenic and other long-term types of exposure response.

A more recent advancement relates to chronobiology, the study relating to the body's internal biological rhythms and their effects on organ functions, etc. The workweek schedule can have a direct effect on these rhythmic patterns and health. Also, there is some evidence that the rate of absorption of toxic materials and reaction to stress may relate in some way to an individual's chronobiology.

A surge of improved and sophisticated techniques took place in the 1960s for quantifying worker exposure to health stress agents in the workplace. This applied both to sample procurement and analytical techniques in which much lower levels of exposure to specific agents could be determined.

The establishment of professional associations to support the interests and growth of the profession has played an important role in developing industrial hygiene as a science. In the United States, the American Public Health Association in the 1930s had a section on industrial hygiene that supported the early growth of the profession. The American Conference of Governmental Industrial Hygienists was organized in 1938. The American Industrial Hygiene Association was organized in 1939. The American Board of Industrial Hygiene was created and held its first meeting in 1960. The Board certifies in the comprehensive practice of industrial hygiene as well as in six additional specialties. Industrial hygienists certified by the Board have the status of Diplomates and as such are eligible for membership in the American Academy of Industrial Hygiene.

The American Industrial Hygiene Association has established a laboratory accreditation program with the objective of assisting those laboratories performing industrial hygiene analysis in achieving and maintaining the highest level of professional performance.

The American Industrial Hygiene Foundation was established in 1979 under the auspices of the American Industrial Hygiene Association. The functions of the Foundation are carried out by an independent Board of Trustees. The Foundation provides fellowships to worthy industrial hygiene graduate students, encourages qualified science students to enter the industrial hygiene profession, and entices major universities to establish industrial hygiene graduate programs.

In the United States, a number of occupational health regulations were established in the early 1900s with emphasis, in several, on listing limits of exposure to a relatively few agents. These regulations were effective at the local, state, and federal levels depending on governmental jurisdiction.

The Social Security Act of 1935 and the Walsh-Healy Act of 1936 had an immense impact in giving increased stability, incentive, and expanded concepts in the practice of industrial hygiene. These acts stimulated industry into incorporating industrial hygiene programs as an integral part of management. They also stimulated broad base programs in industry, foundations, educational institutions, insurance carriers, unions, and government into the cause, recognition, and control of occupational diseases. These acts established the philosophy that the worker had a right to earn a living without endangerment to health, and were the forerunners for the passage of the Occupational Safety and Health Act of 1970.

The passage of the latter act, enacted for the purpose of assuring "so far as possible every man and woman in the nation safe and healthful working conditions," had a very broad bearing on the further development and practice in the industrial hygiene profession. The above regulations and Acts greatly supported the establishment of industrial hygiene as a science. It has been necessary to expand the profession in all of its concepts and technical aspects to meet these responsibilities.

Other industrialized countries have had a similar experience in the professional recognition and growth of the science relating to the recognition, measurement, evaluation, and control of work related health stresses.

4.1 Definition of Industrial Hygiene

The American Industrial Hygiene Association defines industrial hygiene as "that science and art devoted to the recognition, evaluation, and control of those environmental factors or stresses, arising in or from the workplace, which may cause sickness, impaired health and well-being, or significant discomfort and inefficiency among workers or among the citizens of a community."

By definition, industrial hygiene is an applied science encompassing the application of knowledge from a multidisciplinary profession including the sciences of chemistry, engineering, biology, mathematics, medicine, physics, toxicology, and other specialties. Industrial hygiene meets the criteria for the definition as a science since it brings together in context and practice an organized body of knowledge necessary for the recognition, evaluation, and control of health stresses in the work environment.

In the early 1900s, the major thrust in the control of workplace health stresses was directed toward those areas in industry having massive exposures to highly toxic materials. The professional talents of the engineer, chemist, physician, physcist, and statistician were those largely used in these programs.

As industrial technology advanced, the complexity of the worker exposure also increased along with the professional talents needed to study the effects, recognition, evaluation, and control of the newer health stress agents. This is especially true with the advent of high technology in the electronic and allied industries. Stress factors such as improper lighting and contrast, glare, posture, need for strict attention for a given task, fatigue, ability to concentrate, tension, and many others arise in the operation of computers, word processors, video display terminals, and radar equipment, which are becoming commonplace in industry. Thus, concerns for the health of the employees above and beyond that of toxicity response arise. The study and control of these newer stress agents point to the need for the occupational health nurse, psychologist, human factors engineer, and others to join the professional team studying the effects and control of the ever widening list of health stress agents in the workplace.

In the early practice of occupational health nursing, emphasis was placed around such activities as the emergency treatment of traumatic injuries stipulated in written orders of a physician and in obtaining records information relating to physical examinations and the like. With the advanced training of the occupational health nurse, this limited role was found to be wasteful of professional talent and resources. The occupational health nurse is often the first interface between the worker and pending health problems and is in a position to gain information on situations and health stresses both on and off-the-job that may, unless addressed, lead to a more serious response. The occupational health nurse has increasingly become a member of the multidisciplinary team needed in the recognition of job associated health stresses.

Similarly, the psychologist, in the study of the effects of strain, tension, and similar factors, and the human factors engineer in designing machines, tools, and equipment to meet the requirements of the worker, are examples of other professionals joining the multidisciplinary team studying the effects and control of the ever increasingly complex health stresses associated with advanced industrial technology.

The complexity of this multidisciplinary profession needed for carrying out the related responsibilities of the Occupational Safety and Health Act of 1970 is further illustrated in the more than 20 technical committees of the American Industrial Hygiene Association and the seven different areas of certification by the American Board of Industrial Hygiene.

4.2   Rationale of Industrial Hygiene Practice

The practice of industrial hygiene is based upon the following tenets:

Environmental health stresses in the workplace can be quantitatively measured and expressed in terms that relate to the degree of stress.

Stresses in the workplace, in main, show a dose–response relationship. The dose can be expressed as a value integrating the concentration and the time duration of the exposure to the stress agent. In general, as the dose increases the severity of the response also increases. As the dose decreases the biological response decreases and may at some time in dose value exhibit a different kind of response, chronic versus acute, depending on the time duration of the stress even though the total stress expressed as a dose–response value may be the same.

The human body has an intricate mechanization of protection, both in preventing the invasion of hostile stresses into the body and in dealing with stress agents once invasion has occurred. For most stress agents there is some point above zero level of exposure which the body can thus tolerate over a working lifetime without injury to health. Levels of exposure to specific stress agents should be kept within prescribed safe limits. Regardless of their type, all exposures in the workplace should be kept within the limits attainable through good industrial hygiene and work practices. Some stress agents, though, may have a biological response at such a low level of exposure that a minimal level of exposure should be observed. Examples of such agents include those that have hypersensitive, hypersusceptible and genotoxic properties.

The elimination of health hazards through process design should be the first objective in maintaining a healthful workplace. Where this is not feasible, recognized engineering methods should be used to keep exposures within acceptable limits. In some instances, supplemental programs such as the use of personal protective equipment and other control strategies have application.

Surveillance of both the work environment and the worker should be maintained to assure a healthful workplace.

4.3   Elements of an Industrial Hygiene Program

The purpose of an industrial hygiene program is to assure a healthful workplace for employees. It should include all the functions needed in the recognition, evaluation, and control of occupational health hazards associated with production, office and other work. This requires a comprehensive program designed around the nature of the operations, documented to preserve a sound retrospective record and executed in a professional manner.

The basic components of a comprehensive program include the following:

1.   An ongoing data collection system that provides the essential functions for identifying and assessing the level of health hazards in the workplace.

2.   Participation in the periodic review of worker exposure and health records to detect the emergence of health stresses in the workplace.

3.   Participation in research, including toxicological and epidemiological studies designed to generate data useful in establishing safe levels of exposure.

4.   A data storage system that permits appropriate retrieval to study the long term effects of occupational exposures.

5.   Assuring the relevancy of the data being collected.

6.   An integrated program capable of responding to the need for the establishment of appropriate controls, both current and those resulting from technological advances and associated process changes.

The industrial hygienist at the corporate or equivalent level should have responsibility of reporting to top management. This involves appropriate input wherever product, technological, operational, process changes, or other considerations may have an influence on the nature and extent of associated health hazards so that adequate controls can be incorporated at the design stage.

5 HEALTH HAZARD RECOGNITION

An important aspect of a responsive industrial hygiene program is that it is capable of recognizing potential health hazards or, where new materials and operations are encountered, to exercise judicious judgment in maintaining an adequate surveillance program until the associated health hazards have been defined. This should not be a problem in cases involving operations, procedures, or materials where adequate knowledge is available, and it is primarily a matter of application of knowledge and techniques. In operations and procedures involving a new substance or material where relevant information is limited or unavailable, it may be necessary to extrapolate information from other kindred sources and use professional judgment in setting up a control program with a reasonable factor of safety and to incorporate an ongoing surveillance program to further define health hazards that may emerge. In some instances it may be necessary to undertake toxicological research prior to the production stage to better define parameters needed in setting up the control and surveillance program.

One of the basic concepts of industrial hygiene is that the environmental health stress of the workplace can be quantitatively measured and recorded in terms that relate to the degree of stress.

The recognition of potential health hazards is dependent on such relevant basic information as:

1. Detailed knowledge of the industrial process and any resultant emissions that may be harmful.
2. The toxicological, chemical, and physical properties of these emissions.
3. An awareness of the sites in the process that may involve worker exposure.
4. Job work patterns with energy requirements.
5. Other coexisting stresses that may be important.

This information may be expressed in a number of ways depending upon its ultimate use. A most effective form is the material-process flow chart that lists each step in the process along with the appropriate information just noted. This permits the pinpointing of areas of special concern. The effort in whatever form it may take, however, remains only a tool for the use of the industrial hygienist in the actual assessment of the stresses in the workplace. In the quantitation itself, many approaches may be taken depending on the information sought, its intended use, the required sensitivity of measurement, the level of effort and instrumentation available, and the practicality of the procedures.

Aside from the production workplace with the attended toxicological, physical, and other related health stresses, a new area of concern is rapidly gaining special attention where employees may be subjected to a high degree of stress from tension, physical and mental strain, fatigue, excessive concentration, and distraction such as may exist for operators of computers, word processors, video display terminals, and radar equipment. Off-the-job stresses, life-style factors, and the immediate room environment may become increasingly important for such operators. The recognition of associated health stresses and their evaluation require a special battery of psychological and physiological body reaction and response tests to define and measure factors of fatigue, tension, eyestrain, ability to concentrate, and the like.

6 EXPOSURE MEASUREMENTS

As shown in the Table 1.1, both direct and indirect methods may be used to measure worker exposure to stress agents.

6.1 Direct Measurement

To measure directly the quantity of the environmental stress actually received by the body, fluids, tissues, expired air, excreta, and so on must be analyzed

Table 1.1 Methods for Measuring Worker Exposure to Stress Agents

Direct	Indirect
Body dosage	Environment
Tissues	Ambient air
Fluids	Interface of body and
Blood	stress
Serum	Physiological response
Excreta	Sensory
Urine	Pulse rate and recovery
Feces	pattern
Sweat	Heart rate and recovery
Saliva[a]	pattern
Hair[a]	Body temperature and
Nails[a]	recovery pattern
Mother's milk[a]	Voice masking, etc.
Alveolar air	

[a] Not usually considered to be excreta; see, however, Volume 3 B, Chapter 3, Section 11.

to determine the agent per se or a biotransformation product. Such procedures may be quite involved, since the evaluation of the data at times depends on previous information gathered through epidemiological studies and animal research. Studies on animals, moreover, may have used indirect methods for measuring exposure to the stress agent, necessitating appropriate extrapolation in the use of such values. Examples are blood levels now in use for the evaluation of lead exposures and urinary fluoride levels for environmental fluoride exposures..

One decided advantage of biological monitoring is that a time-weighted factor is integrated that is difficult to estimate through ambient air sampling when the exposure is highly intermittent or involves peak exposures of varying duration. Conversely, it may fail to reflect adequately peak concentrations per se that may have special meaning. Urine analysis may also provide valuable data on body burden in addition to current exposures when the samples are collected at specific time intervals after exposure, such as at the end of the work shift and before returning to the job, to incorporate a suitable time lapse.

Sampling the alveolar air may be an appropriate procedure for monitoring exposures to organic vapors and gases. An acceleration of research in this area can be anticipated because of the ease with which the sample can be collected.

6.2 Indirect Measurement

The most widely used technique for the evaluation of occupational health hazards is indirect in that the measurement is made at the interface of the body and the stress agent, for example, the breathing zone or skin surface. In this approach the stress level actually measured may differ appreciably from the actual body dose. For example, all the particulates of an inhaled dust are not deposited in the lower respiratory tract. Some are exhaled and others entrapped in the mucous lining of the upper respiratory tract and eventually are expectorated or swallowed. The same is true of gases and vapors of low water solubility. Thus the target site for inhaled chemicals is scattered along the entire respiratory tract, depending on their chemical and physical properties. Another example is skin absorption of a toxic material. Many factors, such as the source and concentration of the contaminant, i.e., airborne or direct contact, and its characteristics, body skin location, and skin physiology, relate to the amount of the contaminant that reacts with or is absorbed through the skin.

The indirect method of health hazard assessment is, nevertheless, a valid one when the techniques used are the same or equivalent to those relied on in the studies that established the standards.

The sampling and analytical procedures must relate appropriately to the chemical and physical properties of the agents assessed, such as particle size, solubility, and limit of sensitivity for analytical procedures. Other factors of importance are weighted average values, peak exposures, and the job requirements with the energy demand which is directly related to respiratory volume and retention characteristics.

7 ENVIRONMENTAL EXPOSURE QUANTIFICATION

Procedures for measuring airborne exposure levels of a stress agent depend to a great degree upon the reasons for making the measurements. Some of these are (1) Obtaining worker exposure levels over a long period of time on which to base permissible exposure limits; (2) Compliance with standards; and (3) Performance of process equipment and controls. It is essential that the data be valid regardless of the purpose for which they were collected and that they be capable of duplication. This is a key factor in establishing exposure limits to be used in standards and in fact finding relating to compliance. Since judgment and action will in some way be passed on the data, validity is paramount if these are to be used as a bona fide basis for action.

7.1 Long-Term Exposure Studies

In an epidemiologic study in which the relationship between the stress agent and the body response is sought, the stress factor must be characterized in great detail. This may require massive volumes of data suitable for statistical analysis and a comprehensive data procuring procedure so that a complete exposure picture may be accurately constructed. The sampling procedures and strategy should be fully described, including number and length of time of samples, and their location, i.e., personal or area samples, and be adequate to cover the full work shift activities of the workers. Any departure from normal activities should be noted. These are important since the data may be used at a later date for a purpose not anticipated at the time of sample collection.

The collection of valid retrospective data may be extremely difficult. If available at all, the data may be scanty; the sample collection and analytical procedure may not have been well documented as to precise methodology and may have been less sensitive and efficient or may have measured different parameters in comparison to current procedures; sampling locations and types may not be well defined; and the job activities of the workers may have changed considerably even though the job designation may be the same. Other factors which need to be considered in securing retrospective data relate to contrasting past and current plant operations, including changes in technology and raw materials, effectiveness of control procedures and their surveillance, and housekeeping and maintenance practices. In many instances an attempt to accommodate these differences has been made through broad assumptions and extrapolations with an unknown degree of validity and without expressing the limitations of such derived data.

The effect of national emergencies may significantly change the nature and extent of worker exposure to associated stress agents. The experience during World War II is an example. The work hours per week were increased by four to eight hours in many industries. Control equipment was allocated to specified industries. Local exhaust ventilation systems at times became ineffective or completely inoperable due to lack of maintenance. Less attention was given to plant maintenance, housekeeping, and monitoring procedures. Substitute or lower quality raw material had to be used in many instances.

Although the major impact of World War II upon the levels of exposure to harmful agents occurred from around 1940 to the early 1950s, the effect of many of these exposures may not show up in the older work force and retired workers until the early 1980s.

Thus, expressing exposure levels in the past for more than a few years may be only extrapolated guesses unless factors such as the above can clearly be examined and data validity established.

7.2 Compliance with Standards

In contrast to the collection of data for epidemiological studies, data collected for the purpose of compliance with standards may require relatively few samples if the values are clearly above or below the designated value for that agent. If the values are borderline, the evaluation may call for a more comprehensive sampling strategy and may be a matter for legal interpretation. The nature and type of samples taken should meet the criteria upon which the standards were based. Scientifically, though, the data should be adequate to establish a clear pattern with no one single value being given undue weight and should meet data analysis requirements.

7.3 Spot Sampling

The exposure of a worker may arise from a number of sources, including the ambient levels of the agent in the general room air which in turn may be influenced by ambient levels of the agent in the community air, leaks from improperly maintained operating equipment such as from joints and flanges, the inadequate performance of control equipment, and the care which workers observe in performing job operations. Spot sampling can easily detect the effects of any one of these factors on the overall worker exposure level and point to the direction in which further control action should be taken.

8 DATA EVALUATION

The evaluation of airborne levels of a stress to determine compliance with a standard or to determine specific sources of the stress agent are generally uncomplicated and straightforward.

The evaluation of environmental exposure data that serve as a basis for determining whether a health hazard exists is more complicated and requires a denominator that characterizes a satisfactory workplace. Similarly, the use of environmental exposure data for establishing safe levels of exposure or a permissible exposure level, as in empidemiologic studies, requires their correlation with other parameters such as the health profile of the work force.

As pointed out earlier on, the early field studies of the 1920–1930s showed that when the very high exposures of workers were lowered, there was a corresponding lowering of the related disease incidence in the workers. These and other studies gave support to the dose–response rationale upon which the practice of industrial hygiene is primarily based, i.e., there is a dose–response relationship between the extent of exposure and severity of biological response to most stress agents and in which the response is negligible at some point above zero level.

There is great difficulty, however, in determining lower levels of exposure over a working lifetime to a specific agent and its effect on the health of the worker. Often, this is done through extrapolation of other data or trying to estimate past exposures. In the lower range of the dose–response region, the incidence of disease from exposure to an agent may be so low that it approaches or is within the existing reservoir for that disease in the community outside the industry under study. This results from exposure of the general population to stress agents such as from smoking, alcohol consumption, drug use, hobby activities, community and in-house noise, and the like, which may in many instances be similar to those on the job or may be additive to, accumulative, or synergistic with stresses from on-the-job exposures. Even the best available control studies are often not sensitive enough to give reliable data upon which to further extrapolate data for use in the lower dose–response region. Thus, at some point the effects on health from ever-present, extraneous, off-the-job stresses cannot be distinguished easily from low-level, on-the-job stresses.

It is known, however, that the body has protective mechanisms to guard against the entrance of many environmental agents and to handle those where invasion has occurred. For the vast number of agents encountered in the industrial environment, data on dose–response relationship support the industrial hygiene rationale that the level of the stress agent does not have to be zero over a lifetime of work to prevent injury to the worker's health. Some agents, however—such as those having genotoxic properties in being capable of directly damaging genetic material and those associated with hypersensitivity and hypersusceptibility—may have a biological response at such a low level of exposure that a minimal level of exposure should be observed.

Evaluation of data from exposure stresses relating to tension, fatigue, annoyances, irritation, ability to concentrate and discern, and the like are often subjective and may also involve the personal background, traits, habits, etc., of those being stressed for proper definition and control. This evaluation must be done by a different specialist brought on by a changing technology and society.

Thus, each exposure stress must be considered on an individual basis even though it may not be readily separated from the complex pattern of the 24-hour overall environment.

9 ENVIRONMENTAL CONTROL

The cornerstones of an acceptable industrial hygiene program can be described as follows:

1. Proper identification of on-the-job health hazards.
2. The exposure measurements of such hazards.
3. Data evaluation.
4. Environmental control.

In essence, the success of the entire program depends upon the degree and method of implementation, that is, the control strategy. The technical aspects of the program must encompass sound practices and must be related both to the worker and to the medical preventive program. This constitutes a challenge to the professionalism and to the ultimate contribution of the industrial hygienist. The effort and cost of the program must be commensurate with its effectiveness when examined in context with the other off-the-job health stresses.

The heart of the control program must rest with process and/or engineering controls properly designed to protect the workers' health. The most effective and economic control is that which has been incorporated at the stage of production planning and made an integral part of the process. With new processes this can be accomplished by bringing input at the bench design, pilot, and final stages of process development. It is neither good industrial hygiene practice nor sound economics to design minimal control into a process with the intention of adding supplemental control hardware piecemeal, as indicated by future production, or to comply with regulations. On the other hand, the need may exist for the judicious use of personal protective equipment under unique circumstances—for example, breakdowns, spills, accidental releases, maintenance, housekeeping, and certain repair jobs. The adequate control program must embrace a proper mix of process and/or engineering control hardware, personal protective equipment, and administrative control. No single design can be made to fit all circumstances. Rather, each program must be tailored to fit the individual situation without violating the basic tenets of industrial hygiene practice.

Engineering controls are being supported increasingly with automatic alarm systems to give an alert when the controls are malfunctioning and excessive air contamination is occurring.

The control of stresses associated with high technology in the operation of equipment such as computers, word processors, video display terminals, radar equipment, and the like requires a different engineering approach from that used in the control of toxic stresses. Providing optimal lighting contrast, preventing glare, adjusting the equipment to the operator's stature and capability, and maintaining an overall general room compatibility with the tasks are required where eyestrain, fatigue, tension, and concentration on details are encountered. Additive considerations including special rest periods and designated exercises may be indicated.

10 EDUCATIONAL INVOLVEMENT

In the late 1930s only a few universities in the United States offered courses leading to degrees in industrial hygiene per se at the undergraduate or graduate level. In contrast, in 1983 there were over 55 colleges and universities offering courses leading to undergraduate and graduate degrees in industrial hygiene. This attests to the enormous growth in the profession that has taken place over the past 40 years. The passage of the Occupational Safety and Health Act of 1970 had an impact on this growth.

The American Industrial Hygiene Association has increased in membership from 160 in 1940 to 5954 in 1982. In 1982 the Academy of Industrial Hygiene had 1865 diplomates, 800 industrial hygienists in training, and 126 industrial hygiene technologists.

Professional organizations such as the American Industrial Hygiene Association, the American Conference of Governmental Industrial Hygienists, and the American Academy of Industrial Hygiene offer an excellent opportunity for the interchange of professional knowledge and the continuing education of the industrial hygenist. These professional organizations invite participation through technical publications, lectures, committee activities, seminars, and referesher courses. As an example, the American Industrial Hygiene Conference of 1983 presented 485 technical papers covering a wide range of subjects. The same conference offered 43 professional development courses for the purpose of increasing knowledge and expanding skills in the practice of industrial hygiene. This participation by the experts in the many facets of the profession will enhance the overall performance of the profession and will permit members to keep abreast of newer industrial technology in the recognition, measurement and control of associated work stresses. These associations and the academy support the profession in its fullest concept.

The industrial hygienist is obligated to keep involved in the educational process by making professional information available to other groups having an interest in and a responsibility for the health of workers.

The industrial hygienist should have an active role in educating management concerning environmental stresses in the plant and the programs for their control. An alert management can bring pending situations to the industrial hygienist for study and follow-up and thus prevent the inadvertent occurrence of health problems.

The educational involvement of the worker is extremely important. The worker has a right to know the status of the job environment, the factors that may be deleterious to health if excessive exposures occur, and the control programs that have been instituted. Knowledgeable workers are in a position to enhance their own protection through the proper use of control equipment such as local exhaust ventilation and personal protective devices. A worker is often the first to observe that a control system is not functioning properly and can inform management of this situation. In cases of emergency spills and leaks or equipment breakdown, the worker who is informed of the hazardous nature of the materials involved can better follow prescribed procedures for such situations. The industrial hygienist is in an excellent position to participate in these types of educational programs.

11 SUMMARY

Gigantic strides have been made during the past four decades in characterizing and controlling environmental health hazards in the workplace. In many industries where full concept industrial hygiene programs are in effect, the off-the-job stresses such as smoking, alcohol consumption, drug use, and hobby activities often have a greater effect on worker health than that of on-the-job activities where the stresses associated with the job have been studied and controlled. These off-the-job stresses offer a tremendous opportunity for the development of relevant preventive programs in which the industrial hygienist can participate.

Industrial technology is rapidly changing the characteristics of many industries in that high technology is increasingly being applied with the emergence of different types of health stresses. The industrial hygienist must keep abreast of these changes along with the procedures for their recognition, evaluation, and control.

It is vital that the techniques used in measuring occupational health stresses cover all the relevant components of each stress and that these are incorporated into the control program strategy. The practice of industrial hygiene rests on proper judgment in evaluating valid data, combined with effective follow-through.

The high quality and relevancy of the industrial hygiene profession have been proven in meeting the challenge of maintaining a healthful working environment in the presence of an ever advancing and increasingly complex industrial technology.

The following chapters cover comprehensively the theoretical basis and rationale for the science of industrial hygiene.

CHAPTER TWO

Health Promotion in the Workplace

REBECCA S. PARKINSON, and
JONATHAN E. FIELDING, M.D.

1 INTRODUCTION AND HISTORICAL PERSPECTIVE

The presence of employee health promotion as an integral part of occupational medical services is testimony to a major change occurring within American industry. Traditionally occupational medicine had the role of controlling exposures, preventing accidents, and treating injuries. However, in the last five years, occupational medicine has become increasingly interested in what can be done to help employees improve their health habits. Starting with the provision of screening services such as periodic examinations, occupational medicine has begun to apply the principles of preventive medicine. Health risk appraisals and life-style education programs such as smoking cessation, high blood pressure control, exercise, and weight and cholesterol reduction are offered to prevent future disease among employees. These preventive measures, collectively called health promotion, focus on the relationship between life-style and disease. They offer traditional occupational medicine the opportunity to make a significant impact on the health status of workers.

During the twentieth century, medical and public health advances have conquered most of the serious communicable diseases and offered a medical technology that has enabled the United States to achieve an overall state of health enviable by most Third World and westernized countries. Opportunities for further improvement include preventing the occurrence of the remaining common diseases as well as providing better treatment when they develop. However, there is a growing consensus among the industrialized countries that further increases in medical expenditures will not bring commensurate changes in health status. In this sense society has reached a point of diminishing medical returns (1). New surgical procedures, transplants, and cardiac care units can extend the life of a patient after a disease event has manifested itself; but in reality, they provide only the opportunity for a person to adapt to the disease condition. They cannot prevent its first occurrence and in many cases do not improve the quality of life.

This sense of diminishing returns is being viewed by many as a signal to look elsewhere for methods to improve health status. Attention is being focused increasingly on the contribution of health habits to ill-health. One confirmation on the importance of life-style comes from the work of Belloc and Breslow. These researchers demonstrated increased longevity in individuals who followed the seven daily habits (2): eating three regular meals a day; eating breakfast; getting 7–8 hours of sleep; moderately exercising 2–3 times per week; keeping one's weight within normal bounds; drinking only in moderation or not at all; and not smoking.

Another confirmation is found in reviewing the causes for the decline in cardiovascular disease in the United States. Robert Levy, M.D., former head of the National Heart, Lung, and Blood Institute, while citing the contribution of medical technology to the decline of cardiovascular mortality, makes a strong case for the preeminent contribution of life-style changes such as reduced intake of saturated fats (leading to reduced blood cholesterol) and reduced smoking (3). Cardiovascular epidemiologists believe that, even if the increase in aerobic exercise mania in America is ignored and one just examines changes in cholesterol, smoking, and blood pressure control, these risk factor changes alone could explain the 125 percent decline in mortality due to cardiovascular disease during the past 10 years. They argue that improved medical treatment cannot be the primary, attributable cause of the reduction, since the decline in cardiovascular disease is not being seen in all industrialized countries. It is occurring only in countries such as the United States, Finland, Australia, and Canada, where risk factor modification and life-style change programs are actively encouraged (4).

There seems to be little doubt that well-constructed risk reduction efforts directed at specific groups of people can be successful. Farquhar and associates demonstrated that individuals in a community setting reduced smoking and improved their diet after receiving personal health education interventions (5). The results of this study are echoed in the life-style changes shown in North Karelia, Finland, where preventive health services were sponsored by government programs (6).

Common sense also argues that the prevention of disease through the identification of risk, self-care, and life-style change can yield greater rewards in terms of health status than the attempt to cure problems after they have been manifested (7). The top ten causes of death (illustrated in Tables 2.1 and 2.2) could each be reduced in frequency and/or severity if persons at risk would focus greater attention to self-imposed health risks. The available epidemiological data also support such a position. When one examines only hypertension

Table 2.1 Life-style Factors and Leading Causes of Death (12)

Cause of Death (1977)	Life-style Factors
Heart disease	Smoking, high blood pressure, elevated serum cholesterol, diabetes, obesity, lack of exercise, Type A behavior
Cancer	Smoking, alcohol, diet
Stroke	High blood pressure, elevated cholesterol, smoking
Accidents (other than motor vehicle)	Alcohol, smoking (fires), home hazards, handgun availability
Influenza/pneumonia	Vaccination status, smoking, alcohol
Motor vehicle accidents	Alcohol, no safety restraints, speed
Diabetes	Obesity (for adult/onset), diet
Cirrhosis of liver	Alcohol
Suicide	Handgun availability, alcohol or drug misuse, stress
Homicide	Handgun availability, alcohol, stress

and smoking (Table 2.2), a significant amount of chronic disease could be eliminated through such life-style change.

Why have industries chosen to integrate health promotion into their practice of occupational medicine? What is the scope of this health promotion venture? What type of programming is being offered to employees? What dollar benefits, if any, have been shown as the result of investments in these programs? And finally, what is the evidence that employee health promotion programs are successful at improving the health and productivity of employees? These are the questions this chapter will attempt to answer.

Table 2.2 Smoking and Hypertension Risk Factors (8–12)

Hypertension is associated with:
> $3\times$ risk of coronary heart disease in persons age 45–74 with blood pressure $\geq$ 160/95 mm Hg
> $4\times$ risk of congestive heart failure among persons age 45–74 with pressure $\geq$ 160/95 mm Hg
> $7\times$ risk of stroke among persons age 45–74 with pressure $\geq$ 160/95 mm Hg

Smoking is associated with:
> 2–$3\times$ risk of a coronary for males smoking 20+ cigarettes a day
> $5.7\times$ risk of subarachnoid hemorrhage in women
> $10\times$ risk of lung cancer
> >20% of all cancers
> 75% of all lung cancers
> 75% of all chronic bronchitis
> 80% of all emphysema

2 DEFINITIONS AND DISTINCTIONS

Health promotion is an emerging field. Definitions are rapidly coined to describe the branches of activity that grow from this basic disease prevention strategy. Three terms—*health promotion, wellness* and *holistic health*—are commonly used in this field as if they were interchangeable. While there are no standard definitions of any of these terms, their different connotations are important to note.

2.1 Wellness

Wellness is a term used to describe the positive attributes of health. It connotes a sense of well being, vitality, and energy, both physically and emotionally. Some of the more common characteristics of a "well" individual are (13):

1. Assuming self-responsiblity.
2. Improving the communication of emotions.
3. Enhancing creativity through talents and hobbies.
4. Relaxation and stress management.
5. Building self-esteem.
6. Setting life objectives.
7. Regular nutrition and exercise.

Wellness advocates personal-self actualization with the emphasis on well-being and on improving one's existing status of health.

2.2 Holistic Health

Holistic health is a term used to describe an approach to health and health care that integrates the individual's mind, body, and spirit. It was originally derived from the concept that it was more effective to focus on the whole person rather than a diseased organ. Thus, the term "(w)holistic health" or "holism" emerged. It uses both Eastern and Western philosophies and medicines to achieve its objectives. Although the term has many different meanings, holistic medicine or holistic health usually involves:

1. Meditation and spiritual growth.
2. The use of natural foods and medicines.
3. Healing by natural, mild methods.
4. Working toward a balance and harmony of the mind, body, and Spirit with the external environment (13).

2.3 Health Promotion

The term *health promotion* borrows concepts from wellness and holistic health. One broad definition sees it as the combination of activities designed to support behavior which enhances the health of individuals. As used here it incorporates both disease prevention activities (e.g., teaching breast self-examination, hypertension screening and control), as well as health improvement, or behaviorally oriented activities (e.g., stress management, weight reduction, nutrition education). Within the workplace, health promotion programs may include:

1. Health and health risk assessment.
2. Screening for the early detection of disease.
3. Health education and life-style change programs.
4. Environmental, organizational, and cultural support needed to enhance and reinforce healthy life-styles.
5. Evaluation strategies to measure changes in health risk, life-style, morale, and cost benefits.

A health and health risk assessment is usually carried out through a measurement of individual and group health and health risk by means of a written instrument. Risk is assessed based upon family history, life-style patterns, and sometimes biometric testing for height, weight, total cholesterol, high density lipoproteins (HDL), blood pressure, blood sugar, and fitness.

Screening programs for early detection of disease involve applying a specific technique (e.g., blood pressure determination or measurement of intraocular pressure) to an entire population or high-risk subsample, to identify remedial risks for important diseases. Referral to off-site medical resources or on-site treatment may be involved.

Health education programs provide employees with voluntary opportunities to reduce their risk of disease by learning new skills for quitting smoking, losing weight, adopting a regular exercise program, etc. Program content includes both information and behavioral change skills.

The process of *environmental, organizational, and cultural support* is often built into health promotion programs to promote, reinforce, and maintain healthy life-styles. Employee support groups and social networks are often established to assist employees during the life-style change process. Employee leadership committees serve to bring management and nonmanagement employees together to help establish health promotion goals for the company, facilitate new and existing program implementation, and create a sense of employee involvement in the programs. In addition to the social support employees themselves lend to this process, the endorsement of top management is critical to program success. Organizational commitment is also reflected in visible environmental support such as a smoking policy, showers and lockers for joggers, changes in cafeteria and vending machine foods, and accessible scales for weight measurement.

Evaluation strategies are used to assess whether health promotion programs have met their objectives, including the degree of health improvement which has occurred, and any discernible effect upon health related costs.

3 INDUSTRY'S COMMITMENT TO HEALTH PROMOTION

In an effort to accurately determine the number of companies with health promotion programs, several nationwide and state surveys have been conducted. A 1978 survey by the Washington Business Group Health sent to its 160 large employer members yielded 59 companies (36.9 percent) with health promotion and risk reduction programs. Forty-one percent of these companies offered stress management programs and 85 percent cardiopulmonary resuscitation (CPR) classes (14). In 1979, Fitness Systems sent a survey to 300 top industrial companies listed in the *Fortune* 500, and 50 of each of the life insurance, commercial banking, utilities, retailing, diversified financials, and transportation business sectors. Of the 22 percent that returned the questionnaire, approximately half of the companies offered diet/nutrition counseling and/or smoking cessation; a little more than one third offered stress management; two thirds had programs for alcohol and drug-related problems; and one quarter offered employees a physical fitness program (14).

A 1981 survey of California companies with over 100 employees was conducted by the University of California at Los Angeles. Of the 1,000 companies randomly selected, 511 met the criteria and 424 (83.0 percent) were interviewed by telephone. Of the sample, 47.9 percent were companies of 200 or fewer employees; about one third (30.4 percent) had 200–499 employees; one fifth (21.7 percent) had 500 or more employees; two sites had over 1,000 employees. In the total group, 78.3 percent (332) had one or more health promotion activities, the most frequent being: accident prevention (64.6 percent), CPR or choke saver (52.8 percent), alcohol and drug abuse (18.6 percent), and mental health counseling (18.4 percent). Programs such as hypertension control, smoking cessation, physical fitness, and stress management were available to employees in 8–12 percent of the companies (14).

Health promotion should be part of an industry's overall health care strategy (15). Such a strategy includes at a minimum the combination of appropriate medical and health services coupled with a health benefit package that encourages appropriate use of services. Part of the strategy also involves the identification of opportunities for both health improvement and cost containment. A basic framework for a corporate health care strategy usually includes:

1. Programs for education and monitoring of environmental health and safety.
2. Provision of medical care for illness and injury.
3. Monitoring of hospital and physician costs by employees and dependents.
4. Updating of employee benefit packages.
5. Offering enrollment in prepaid health care plans, collectively termed health maintenance organizations (HMOs).
6. Promoting the use of cost effective health care delivery (e.g., ambulatory surgery, second opinions, preadmission testing).

7.   Disability assistance programs to monitor appropriateness of care and hasten the employee's healthy return to work.

8.   Participation with other employers in health care coalitions to share experiences, data, and to formulate coordinated efforts to improve the availability and use of cost effective services.

9.   Establishment of employee health promotion programs.

10.   Establishment of programs to reduce the health and dollar consequences of mental health and drug abuse programs.

3.1   Industry's Rationale for Health Promotion

There is a variety of reasons industries have cited for their commitment to health promotion. One attractive feature of health promotion programs is their potential to reduce the rate of rising health care costs. But there are also expectations that these programs will impact positively on the other business concerns such as improving employee health, morale, and recruitment.

3.1.1   Health Care Costs

The United States is experiencing a geometric growth rate in its health care costs. From 1950 to 1982 national health expenditures increased from 4.6 to 10.5 percent of the gross national product. Total expenditures in 1982 were $322.4 billion dollars (1,16). Much of the disease burden contributing to these expenditures is now preventable through appropriate preventive strategies.

Up until the 1970s health care costs were generally considered by industry executives as too small to be of any major concern. Richman states in a 1983 interview of major employers by *Fortune* magazine that business, for a long time instead of intervening to ensure their monies were being spent effectively, chose not to get involved (17). Since that time, however, industries have earmarked health care costs as the largest, single, personnel related expenditure for which adequate control cannot be found, a complete reversal from the earlier position. Adequate control now means identifying workable cost containment strategies. Health promotion is one strategy that has been identified.

Industry pays the single largest share of the nation's health and health care bill. Industry pays the majority of the health care costs from its employees and dependents. In addition, industry pays the indirect costs of all sickness related absenteeism, turnover, premature retirement and death caused by acute and chronic illnesses (7).

A compound growth in health related benefit costs of 11–14 percent each year, without benefit plan additions, is now of major concern to employers (15). The employee bears only a small burden of the total health care cost. More and more these costs affect the cost of consumer products. One large auto manufacturer estimated that in 1970 health care costs per participating employee were $514, or an average of $55 per vehicle. In 1982 the costs had risen to $2,556 per employee and $392 per vehicle. Total company health care costs had risen from $144 million to $710 million during the same period (18).

Another company estimated it was spending $3,700 per employee or dependent for total health care expenses over a three-year period. This was for treatment, not prevention. They established a health promotion program that costs approximately $225 per employee over a three-year period. The company estimates that even if this program only reduces the health care costs by 10 percent each year, it will add about 8¢ to the earnings per share (19). This is a sample of the type of cost experience many employers are developing. Health promotion thus is being considered by many employers as a major opportunity to help reduce the health care cost problem.

3.1.2 Employee Health

Employees are a company's most valuable resource. A second common reason industries give for their investment in health promotion is to improve the health of their employees. Employee health impacts not only on the use of health care services (hospitals, physicians, etc.) but also on productivity and wages. Similar to other resources, such as equipment, human resources must be maintained. And similar to machines, the preventive maintenance of human resources is more cost effective than relying on later treatment or repair.

3.1.3 Employee Relations

Health promotion programs positively affect employee attitudes about the company. They demonstrate to employees that the company cares about them personally, and that the company will provide opportunities to improve the quality of their lives at work and at home.

While program effects on health status and medical utilization require a number of years to discern, positive effects on employee relations are quickly realized. Indeed, companies with health promotion programs have found a positive change in attitudes through unsolicited letters of praise, positive comments in interviews with the employees, or through more systematic evaluation efforts to measure such changes. Companies such as Johnson and Johnson and AT&T Communications have developed formal evaluation procedures as part of their health promotion programs to measure changes in job satisfaction, attitudes toward the company, and overall morale as a result of health promotion programming.

3.1.4 Employee Recruitment

An additional reason some companies use for initiating health promotion programs is recruitment. One of the first companies to recognize this was the Kimberly-Clark Corporation. Headquartered in Neenah, Wisconsin, the company's multimillion dollar fitness facility and health promotion program was recognized as one way to attract employees to work in their rural location (20). Many other companies consider health promotion programs, especially those which contain an active fitness component, a way to attract young, talented employees. Prospective employees are looking at exercise and other health promotion opportunities as carefully as other health benefits when offered as part of the recruitment package.

4 EMPLOYEE HEALTH PROMOTION PROGRAM MODELS

During the past five years a wide variety of employee health promotion programs has been established. These programs vary from periodic lectures on nutrition, or how to exercise, to extensive multicomponent programs with considerable staff and the construction of physical fitness facilities to accommodate the exercise and other health promoting activities (12). Despite the widespread variation, most programs can be classified into one of three models: comprehensive, physical fitness, or single-component programs. These models serve to identify only the major features of the health promotion program, not the many individual factors which make each company's program unique.

4.1 Comprehensive Model

The comprehensive health promotion program represents the most sophisticated of the efforts and usually requires the greatest resource and dollar investment. It also has the greatest potential for producing a significant effect on employee health and employee health care costs. The goal of the comprehensive program is to provide employees with multiple opportunities to change life-style behaviors voluntarily, while also creating a new organizational and cultural environment that supports the individual making such changes.

Comprehensive programs offer educational courses to employees which usually include, at a minimum, smoking cessation, weight control, nutrition, exercise, healthy back care, cancer prevention, stress reduction, and cardiopulmonary resuscitation (CPR). Hypertension control and employee assistance programs are included in the comprehensive model. Some employers are also offering courses such as parenting, safety, first aid, safe driving, "know your health care benefits," self-care, and assertiveness training. In general, instructional opportunities usually are provided at two levels: informational (e.g., lecture, film) and behavioral change programs designed to teach new life-style skills (e.g., group classes).

Courses on each topic are offered separately, usually over a period of weeks. Some courses intentionally overlap material in other courses. For example, both the smoking and weight control courses may include sections on the benefits and skills needed to start a regular exercise program. These skills help employees achieve their primary objective while also helping to motivate them to take the exercise course during the next round of program offerings. Employees not enrolled in a course are encouraged by the changes they see in other participants, thus stimulating future participation and creating a ripple effect.

In addition to promoting life-style change, the secondary objective of the comprehensive program is to change the corporate culture. To that end, a successful health promotion program tries to build a culture in which all employees believe, support, and actively work toward providing a social and physical environment that supports healthy life-styles. Some of the more common examples of corporate cultural support include instituting a non-smoking policy, providing showers for lunch-hour joggers, and putting healthy food selections in cafeterias and vending machines.

4.2 Physical Fitness Model

Fitness programs have often been the central focus of a company's investment in health promotion. In some cases, such as Kimberly-Clark, Xerox, and PepsiCo, multimillion dollar facilities were built for employees, families, and retirees. These facilities may include jogging tracks, pools, aerobic exercise rooms, Nautilus equipment, treadmills, bicycle ergometers (exercise bikes), weights, rowing machines, mats, and so forth. An exercise facility is frequently chosen as the cornerstone of the health promotion program because it visibly demonstrates a strong management commitment to employees.

Exercise is also a key factor from a motivational standpoint in health promotion. It provides many of the key elements needed to launch a person into a healthy life-style: noticeable, rapid, physical improvement, a practical stress reduction method, help in weight management, social support, camaraderie, and a sense of taking charge of one's life. Success in adopting a regular exercise program frequently motivates the individual to begin other life-style changes such as quitting smoking or losing weight. Exercise programs are a valuable asset to companies wanting to build immediate participation into their health promotion programs.

4.3 Single-Component Model

The third, and probably the most common approach to health promotion programming, is the single-component program. These programs are characterized by a company providing limited life-style activities. Often the primary emphasis is to provide information to employees through pamphlets, newspaper articles, lectures, and health screenings. Frequently, behavioral classes, such as smoking cessation, weight control, or hypertension programs are offered once a year or every other year. In other cases longer behavioral change programs are offered periodically to employees. The significant drawback of single-component programs is that their isolation tends to diminish their effectiveness. Because these programs do not carry the degree of organizational, social, and environmental support, as the more comprehensive programs do, they are less likely to affect corporate cultural norms regarding health. They usually have a smaller impact on overall health status than components that are integrated into an overall, coherent program with a strong identity and high visibility.

5   KEY ELEMENTS FOR SUCCESS

While each company's health promotion program is unique to the culture that inspires it, there are six basic elements which all successful programs share in common. These elements are:

1.   Support of top level management.
2.   Assessment methods for health and health risk—usually through computerized data collection systems.
3.   Multiple, life-style change and health improvement opportunities.
4.   Corporate culture efforts—environmental, social, and organizational.
5.   Evaluation methods to determine program success, cost effectiveness, and cost benefits.
6.   Organizational location, resources, and professional qualifications of program manager.

5.1   Top Level Support

Many of the employers with highly successful health promotion programs were significantly aided by the involvement of the industry's chief executive officer or board chairman. This support often was not only for the health promotion effort, but also for the development of an overall health care strategy of which health promotion was an integral part. In some companies the impetus came from the medical, benefits or personnel department and was presented to the company's management committee in the form of a business case as a cost containment strategy, an employee relations opportunity, or simply the right thing to do for employees. In other companies, the health promotion program has been developed by the medical or personnel department as part of their existing mandate. It may have had strong support from top management, or top management may have had very little involvement in its planning and implementation. In general, however, it appears that top level support is associated with more rapid development and greater program resources.

5.2   Health and Health Risk Assessment

Health risk assessment is the process of identifying an individual's or population's health status and risk of future health problems. Such assessment may include identifying an individual's family history of disease, clinical indicators (e.g., blood lipids, sugar, blood pressure, height, and weight), and certain life-style behaviors known to be associated with disease. These assessments are most commonly performed with the use of a computerized form called a health hazard or health risk appraisal.

Health risk appraisals were first developed by Lewis Robbins and Jack Hall in 1975 at the University of Indiana Medical School. This first appraisal was based on mortality data. Today the data collected in these appraisals varies, as does the set of epidemiological assumptions used to develop the algorithms by which the risk is estimated. While the risk of dying of certain diseases can be better estimated than the risk of becoming ill with the same disease, increasing attempts are being made to provide risk estimates for both mortality and morbidity.

More and more, assessment tools are attempting to provide individuals with feedback on more subjective aspects of their health, by asking how they feel about themselves, their level of energy and enthusiasm, if they are experiencing depression, the level and type of stress, etc. Health appraisals are also beginning to assess employee morale and attitudes. Several companies, including Johnson and Johnson and AT&T Communications, have undertaken elaborate assessment strategies to determine the "health" of their organization through attitude surveys. Such assessment provides management input about employee job satisfaction, organizational loyalty, perceived job stress, job commitment, and perception of the company's role in providing life-style programs. The positive impact of employee health promotion programs on these attitudes can be considered of great value to management. The rigorous measurement of such attitude changes serves to provide concrete evidence that health promotion is having a positive impact.

Health risk appraisals in the form of a questionnaire are usually completed by employees as the initial step of a health promotion program. Individual data from the employee completing the questionnaire provide a means of computing from age, sex, and race, specific data concerning the risk that person has of experiencing certain diseases or dying within a specified time period as well. The system for estimating risks is based on epidemiological studies which link risk indicators to morbidity and mortality predictions. For example, persons who smoke and have high blood pressure are told how much their risk of a heart attack can be reduced by quitting smoking and maintaining a controlled blood pressure. Risk appraisals cannot predict a person's future medical history since they only provide a probability, not a diagnosis. However, it is a prediction of risk in a statistical sense, based on comparisons with groups of persons in the past who shared the same risk indicator or group of risk indicators(21).

To the company, health risk appraisals offer an efficient data collection method that can be used in a variety of ways:

1. A referral point for individual counseling.
2. A method of providing an aggregate analysis of the health of the employee body, including those at highest risk based on their life-styles.
3. A base-line for quantifying preventable problems prior to and immediately after the initiation of a health promotion program.

4.   A strong case for providing life-style interventions or a method for measuring the health impact of merely doing nothing.

5.   A tool for planning health promotion programs.

6.   A basis for planning health care benefit packages.

If the risk of disease can be reduced, the risk of associated health care costs also has the potential to be reduced. Thus in theory there is potential for predicting cost benefits from life-style intervention programs based on potential reductions in aggregate risk. However, the data bases that permit estimates of direct and indirect health related savings based on reduced risk are not yet sufficiently developed to be widely used.

5.3   Life-Style Change Programs

Many of the twentieth century diseases are the result of individuals practicing what are considered to be unhealthy life-styles—smoking, lack of exercise, overeating, etc. To begin the process of changing to healthy life-styles most health promotion programs offer courses to employees which provide two levels of learning: information and behavioral change skills.

Informational programs are designed to create awareness of a healthful life-style. They provide only basic information. These programs often employ single lectures on a given topic supplemented by print and audiovisual media. For example, a lecture on weight control and nutrition might be offered to a company by a nutritionist or physician from a local hospital. The company newspaper might run a column on health featuring nutrition and weight control. Pamphlets and calorie counters may be made available in the medical office or cafeteria. Films and videotapes on these subjects may be offered during lunch hour or available on closed circuit TV in the company. These informational approaches are designed to impact primarily on the awareness level of the population and their motivation to make changes. Alone they very rarely lead to sustained changes in health behavior or affect overall health status.

Behavioral programs, on the other hand, provide individuals with new skills and habit patterns to be practiced or substituted for the previously unhealthy ones. Today many of these programs use behavior modification techniques. While behavior modification had its early roots in the work of Pavlov and Skinner, some of its principles have been adopted for use in life-style change programs where employees *voluntarily* learn to substitute new health behaviors for the ones that they wish to supplant.

Behavior modification is built upon a three-step process of providing: (1) *new cues or triggers* for (2) *new behaviors*, which are then reinforced with (3) *new positive consequences or rewards*. For example, if the trigger for a morning cigarette is the first cup of coffee, juice is substituted for the coffee, nonsmoking is substituted for the cigarette, and the reward is either an internal sense of control, or an external reward such as using the money saved to reinforce the change. Each unhealthy life-style is systematically broken down into the subsets of behaviors that compose it. Each sub-behavior is changed using this three-step process. In terms of human resources this becomes a very labor intensive process, as each participant must be helped to integrate new individualized behaviors into their life-style.

The same behavior modification principles used to help individuals quit smoking are used in teaching others how to control their eating and weight. To a lesser extent they can be used in the control of high blood pressure and in the initiation of and continued adherence to the exercise habit. With exercise and hypertension control, a new set of skills is being *added* rather than substituted. In exercise, the skills of taking one's pulse, warm-up, stretching, exercising, and cooling down slowly are learned. In blood pressure control, cutting back on sodium, exercising, practicing relaxation techniques, and integrating a regular medication taking routine are learned. The same principles apply: New information and skills are integrated and rewarded so that they continue on as daily habit patterns.

There is tremendous controversy as to what actually motivates individuals to adopt and maintain healthier life-styles. The research evidence clearly states that information is a necessary but not sufficient condition for achieving this objective (22). To date, the best methods for achieving long-term life-style change (one year or more) are through the use of behavioral change techniques including development of a system of internal and external support to facilitate adherence.

5.4 Program Administration

A variety of administrative decisions need to be made regarding the implementation of health education and life-style change programs, including eligibility, logistical arrangements, timetable, costs, and staffing. In most companies, all employees, management and non-management, are eligible for participation in health promotion programs. There is also a growing trend to include spouses and dependents since the employer pays most of their health care bills, and since social support within a family helps to influence and reinforce the behavioral change. In some companies, retirees and their spouses are able to avail themselves of the program offerings.

Most companies prefer not to interrupt the normal business operations with program activities. Many employees also find it difficult to leave during work to participate in most of these activities. For these reasons, many programs are offered before work, during lunch hour, or immediately after work. These times, however, compete with family commitments, car pools, other lunch-hour activities, and the full range of leisure time options. Unfortunately, there does not appear to be one perfect solution. Each company must work out this scheduling to be compatible with its corporate culture and policies and with the needs of its employees.

Health promotion programs at the worksite are financed in many ways. In some cases, the company totally finances the program. Other companies have their employees pay some of the expense of the program, believing this increases commitment. The employee often receives the convenience of the company location for programs and possibly some company time for participation. A partnership is thus formed between the employee and the company. Sometimes the tuition reimbursement approach is used. One company, IBM, as part of its program, pays for employee participation in community programs on a tuition refund basis. A maximum refund is allowed per course, giving the employee an option to pay part of the expense for choosing a more costly program. Another variation is seen in the case of the Kimberly-Clark Corporation, where employee spouses can use the fitness facility for an annual fee. The employee, however, must keep up a certain level of participation to make his/her spouse eligible for participation (20).

There are three methods for staffing health promotion programs: company health professionals, community organizations (profit and nonprofit), or a combination of the two. Each company, depending on such factors as its size, geographical location(s), presence of a medical department, and community relations policy, decides on one of these options. Obviously, the training and use of internal personnel provides a high degree of quality control. However, such a strategy is not easily applied to dispersed smaller units and lacks the flexibility derived from outside personnel who can be hired for as long as necessary. National and community organizations such as the American Cancer Society, Heart Association, Lung Association, and YMCA offer a variety of programs in most communities.

5.5 Corporate Culture

It is now well accepted that every organization has a culture shaped by its particular objectives, management style, and economic positioning. In general, an organization describes its culture as "the way we do things around here" (23). Researchers Deal and Kennedy point out the profound influence an organization's culture has upon its economic success. Their analysis of American business leads them to conclude that corporations with strong cultures remain marketplace leaders.

What is a strong culture? A business that creates an *environment* in which employees can be secure and thereby work to make the business a success (23). How does the culture influence the health of employees? Researcher Robert Allen has offered a practical method for organizations to assess the impact of their culture on the health of their employees—particularly the cultural norms that pertain to employee health. Allen's model asks organizations to assess the cultural norms that influence health behavior. His work forms the basis for the concept, now widely accepted by industry, that a successful health promotion program is dependent upon the support of the corporation's culture (24).

Practically, how does an industry support a culture that promotes healthy life-styles? In the case of health promotion, the new culture must first be endorsed by top management. Development of corporate policies on employee assistance programs for alcohol and drug abuse and smoking is another important step in demonstrating organizational support. Employee involvement through leadership committees that vertically integrate both management and nonmanagement personnel into the decision-making process for the health promotion program is another way of trying to integrate health promotion into the corporate environment. In some cases employees are trained to take on the responsibility for implementing the life-style change courses. Environmental support that visibly demonstrates the company's commitment to health promotion has been found to include changes in food offerings of the cafeteria and vending machines, showers for employees who jog during lunch hour, fitness trails, accessible weight scales to facilitate those losing weight, management and restriction of smoking areas, and so forth. These are a few examples of how companies are beginning to restructure their cultures to support health promotion activities. What is critical in this restructuring is that the changes be physically visible, recognized by all, and supported by top management.

5.6 Evaluation

Evaluation of a health promotion program provides an opportunity to measure success in achieving health and dollar objectives. However, in a world of limited resources evaluation does require the allocation of dollars which might be otherwise spent on programming. One notable corporation, IBM, in presenting their business case to management chose between providing the health promotion program, "A Plan for Life," to all employees, or providing the health promotion program to a few employees and completing a complex evaluation. They chose to spend the dollars on programming, limiting their evaluation mainly to the collection of participation data (25).

Companies tend to take one of two positions regarding evaluation dollars. Those companies that line up behind IBM assume that the program will yield improved employee relations and in the long run bear out the epidemiological predictions of health and cost savings. Other companies, such as AT&T, Johnson and Johnson, and Control Data Corporation, have concluded that many of the important questions are unanswered and have decided to invest in large scale studies, requiring sophisticated evaluation designs, including control groups. They have undertaken large data collection and analysis efforts covering health status attitudes, behaviors, risks, health insurance utilization, and costs.

Industries can consider collecting and analyzing at least three types of evaluation data:

1. Process measurements such as opinions of participants about the program, administrative quality control data, and participation figures.

2.   Measures of life-style, attitude change, and related physiological and biochemical changes.

3.   Information on program and dollar costs and benefits.

For process measurements participation rates represent the simplest and most common form of evaluation strategies. They are usually chosen by companies that want some form of program feedback but cannot allocate resources for more complex strategies. Participation figures provide information about the number of employees who enroll or register in health promotion program activities. They generally do not provide information about the outcomes of the participation, such as the number of people who quit smoking or lost weight over a year. Often the reporting of program participation figures is plagued by an inadequate definition of what constitutes real participation, such as whether it is measured by initial enrollment or adherence over a period of time. Participation data also provide information about the involvement of employees and departments in leadership committees, and program administration.

While most industries are ultimately interested in the cost impact of health promotion programs, they are usually satisfied with short term measurements of changes in life-style behaviors, physiological changes in health, and changes in employee attitudes until such outcomes can be obtained. The most accurate outcome measures are taken one year following the education program, reflecting not only the life-style change but also the maintenance of the change. Within an employee population these measures must take into account the effect of dropout, turnover rates, and the bias inherent in collecting data by self-report, without clinical verification. Some unreliability in the data must be expected, or what has been called the "lying factor," because of the participant's desire to please the instructor or evaluator. This is particularly true with the increase in social pressure to maintain a healthy life-style of nonsmoking, regular exercise, and controlled weight.

The measurement of employee attitudes is still very new to most health promotion programs. In its inception the measurement was primarily anecdotal, derived from letters, phone calls, and reports by employees to the company management about the program. More recently, because of the new emphasis on quality of worklife and corporate culture, measurement of these attitudes has become a vital part of health promotion evaluation. The value of this measurement is not just to provide a rationale for the continuance of the health promotion program. It also serves the corporation's need to maintain positive employee relations. Attitude measurement is most often completed by self administered questionnaires separately or in conjunction with the health risk appraisal. Results of the subsequent life-style risk and changes can then be compared with changes in morale to determine potential interrelationships.

The most significant impact a health promotion program can have is to ultimately affect the industry's bottom line by effecting a favorable benefit–

cost ratio. However, most industries are faced with practical barriers in such measurement. Disability absence data are frequently not collected at all, are not accurate in their diagnostic categorization, or their collection is not structured for computerization, making it difficult to link them with the individuals making the behavior changes. Health care utilization data, such as the use of hospitals, outpatient, physician, and ancillary services are often very difficult for a company to obtain even if the company is self-insured. Usually such data are collected for accounting purposes by type of service provided and by the family code, not by a diagnostic category or the specific individual using the service. Additionally, for both disability and health care utilization data, a considerable lag time occurs between the changes in life-style behaviors and discernible changes in utilization (26). Despite these barriers, several companies have made major investments in establishing data collection systems to provide them with needed answers about cost information.

5.7 Organizational Locale and Program Manager Qualifications

Both the correct organizational location and the assignment of program responsibilities to one appropriate individual are critical to the overall success of the program. Usually, the administration of these programs is placed within the human resource function of a company where responsibility for all other employee benefits is located.

At the functional level the day to day operations of health promotion programs are usually the responsibility of a human resource manager. In some companies this is a professionally trained health educator or an exercise physiologist. Depending on the size of the organization and the scope of the program, this manager may supervise both clerical and teaching staff.

While health promotion managers presently represent a great variety of backgrounds, they generally require skill in program development, management, budgeting, and evaluation. These jobs are facilitated by good communication skills since much time can be spent preparing and presenting the health promotion business case to company executives, managers, employees, and community groups interested in the program. A thorough understanding of risk reduction and behavioral science principles completes the major set of qualifications for these managers.

The primary exception to the location of the health promotion program in the human resource function occurs when the long-term objective for the company is to market a health promotion product. Here the program most often will be placed within the jurisdiction of the marketing personnel. In either the human resource or marketing location, the closer the health promotion function to top management, the higher the commitment, and the greater the impact on the resource allocation for the daily operation of the program.

Resources for health promotion always must be sufficient to achieve the program's objectives. Such resources include both the internal personnel needed and the financial base to purchase personnel and materials from the community. These resources must be available not only for the piloting of a health promotion program but also for its continuation and maintenance in future years.

6 EVIDENCE FROM LIFE-STYLE PROGRAMS THAT REDUCE HEALTH RISK

The epidemiology of health risk clearly points to the potential value of several life-style change programs, which offer immediate value to the individuals participating in them and companies paying their health care bills. High on this list are hypertension control and smoking cessation. The elimination of risk causally associated with these problems would reduce the prevalence and mortality of cardiovascular disease and many forms of cancer and respiratory disease (27–29). Accident prevention programs could also greatly decrease the frequency of accidental deaths and the high burden of disability from nonfatal accidents. However, several other health promotion activities are also of considerable potential value and are frequently more popular among employees and employers. For example, exercise programs are a visible activity with good potential for motivating and reinforcing other life-style changes. Stress reduction, weight control, and nutrition are among the most frequently requested and are commonly found as part of a broad based health promotion activity. The available information on the effectiveness of these programs in achieving the desired health and dollar outcomes anticipated by industry are reviewed below.

6.1 Hypertension

Elevated blood pressure, or hypertension, is associated with higher rates of illness and death, primarily from heart disease, stroke, and other circulatory problems. Individuals with high blood pressure develop three times as much coronary disease, six times as much congestive heart failure, and seven times as many strokes as those individuals with controlled blood or normal blood pressure (12, 29). In recent years, hypertension has been defined as a blood pressure $\geq$ 140/90 mm Hg, although in earlier literature it was defined as $\geq$ 160/95 mg Hg.

The prevalence of hypertension (controlled and uncontrolled) among the working population using the current definition is about 15–25 percent. However, both age and racial variables can considerably alter this percentage. For example, rates of hypertension in white male workers ages 55–64 are, on the average, about 37 percent, compared with 59 percent for same age black male workers (29). Hypertension is significantly associated with cardiovascular disease. About 26.7 million work days representing $1.3 billion in earnings are lost each year due to cardiovascular disease in which high blood pressure is a significant factor (8). There are 13.5 times as many deaths from hypertensive cardiovascular disease than deaths from industrial accidents (8).

Effective hypertension control programs in the worksite usually combine screening with an organized approach of referral and/or follow-up. In some cases the treatment is done on-site. In other programs, referral is made to private physicians or community organizations. The primary objective in such programs is sustained blood pressure control. Evidence from reported worksite blood pressure control programs indicates that control can be achieved, even at 80–90 percent, through programs that are well organized and administered (30–33).

Several significant studies have been conducted to evaluate worksite hypertension programs. The results are encouraging and practically applicable to businesses of varying size. The Massachusetts Mutual Life Insurance Company sponsored on-site screening, referral, and follow-up at their home office. After one year, they reported an increase of 36 to 82 percent in their hypertensives under control (34). Another study, based in New York City and involving three union Health and Security Plans, screened and treated hypertensive employees at a nearby site. Of those volunteering to have care delivered at a worksite clinic, 80 percent maintained blood pressure control for as long as two years. The same study also revealed a decrease in hospital days for those employees treated (32). A decrease in disability days for those employees receiving treatment on-site or by their physician was noted. A five-year study to measure the impact of educational interventions in two Baltimore clinics on hypertensive patients demonstrated a statistically significant increase in blood pressure control and a decrease in mortality for those individuals in the experimental groups in comparison to the usual care group (35).

Most recently a study by the University of Michigan with four manufacturing plants of the Ford Motor Company demonstrated that a much higher percentage of employees receiving screening, referral, and interventions achieved blood pressure control than those without these structured follow-up interventions. In one follow-up intervention a blood pressure counselor contacted employees semiannually to help with questions, encourage medication compliance, monitor blood pressure, and refer employees back to their physicians if needed. In the second intervention, a group counselor checked with the employee as needed, but at least semiannually, and also regularly checked with the employee's physician. A third group received treatment on-site and follow-up, and a fourth group served as the control. The largest blood pressure reduction occurred in the on-site treatment. However, all three intervention sites showed better blood pressure control than the group which received only screening and referral without follow-up (30).

Worksite hypertension programs are not without some controversy in their results. While at least one study showed that a worksite hypertension control program reduced absenteeism, a Canadian study among steel workers actually showed an increase in absenteeism after employees had been diagnosed as hypertensive. This was presumably because the employees adopted a "sick role" behavior pattern after diagnosis (36). The effect of worksite sponsored hypertension control programs on absenteeism therefore still remains unpredictable.

A number of studies are beginning to suggest that on-site treatment of hypertensives may be a more cost effective approach than community treatment (37, 38). However, the on-site treatment of Ford Motor Company employees that resulted in better control was also the most costly of the interventions (30).

6.2 Smoking

Smoking has been cited as the largest, single preventable cause of illness and premature death in the United States (11). It has been said that "no single measure would lengthen the life or improve the health of Americans more" than the cessation of smoking (9). The list of studies citing a relationship between smoking and premature disease and death is extensive. Smoking is associated with cardiovascular, respiratory, circulatory, and neoplastic disease in the smoker. It increases the risk of spontaneous abortion during pregnancy, low birth weight, and neonatal death (11, 39). (See Table 2.2). More respiratory illness and allergies are found in children living in households where one or more persons smoke (27). In addition, evidence is growing that the involuntary or passive smoking induced by side stream smoke may have adverse health consequences to allergic individuals and to those with compromised respiratory or cardiovascular function (40, 41).

The prevalence of smoking in the adult United States population is approximately 33 percent. Among employed persons it is 36 percent: 38.9 percent in males, and 32.9 percent in females. The percentages are somewhat higher for blue collar workers: 50 percent for males, and 39 percent for females (29, 42).

What is the impact of smoking on employers? More chronic disease, especially chronic bronchitis, is reported in employees who currently smoke. Acute conditions such as influenza, when adjusted for age, are 14 percent higher in smoking males and 21 percent higher in smoking females than in their nonsmoking counterparts. More lost workdays have been reported for smokers than for nonsmokers (33 percent more in males, and 45 percent more in females). Bed disability is higher in smokers (14 percent in males, and 17 percent in females) than in nonsmokers. The excesses in the disability figures are dose related (27). Table 2.3 summarizes some of the data.

In addition to the risks of smoking alone, smoking increases some of the

Table 2.3   Smoking Effects on the Workplace

	Percent Above Nonsmokers	
	Lost Workdays	Bed Disability
Males	33%	14%
Females	45%	17%

Source: Reference 27.

risks associated with occupational exposures. The National Institute of Occupational Safety and Health has identified six possible interactions of smoking with the work environment.

1. Toxic agents in tobacco and tobacco smoke may also occur in the workplace, thus increasing the exposure to these agents (e.g., hydrogen cyanide, carbon monoxide).
2. Workplace chemicals may be transformed into more harmful agents by smoking (e.g., polymer fume fever and the conversion of chlorinated hydrocarbons to phosgene).
3. Tobacco products may serve as vectors facilitating entry of toxic agents into the worker's body (e.g., lead, mercury, and polytetrafluoroethylene).
4. The toxic effects of smoking may combine additively with workplace toxic agents (e.g., cotton dust and coal dust).
5. The toxic effects of smoking may combine multiplicatively with workplace toxic agnets (e.g., asbestos, radon daughters, and rubber industry toxins).
6. Smoking may contribute to accidents in the workplace through loss of attention, eye irritation, risk of fire, and so forth (43).

Both the public and private sectors provide cessation programs available to employees. Cessation programs most often use one or more of the following techniques to help participants quit: information, behavior modification, aversive procedures, acupuncture, or hypnotherapy. The most common approaches are a series of behavior modification classes delivered to small groups over a one to eight-week period, often supplemented with several follow up sessions over a longer period of time. In the other companies one-on-one counseling, small groups, or self-quit booklets are used.

This array of smoking cessation methods has been transported into the work setting with a fair degree of success. However, controlled studies on worksite programs are still infrequent. In other settings which use intensive clinical and educational assistance 33–80 percent initial quit rates are achieved in comparison to the 15–27 percent quit rates using information, brief instructions, and encouragement (31). Even at these levels the sustained quit rates are usually down to 20–30 percent after one year (44).

Smoking contributes to higher employer health care benefit and training costs as well as higher lost productivity. Some studies report that smokers have a 50 percent higher rate of health care utilization, have an additional two to three days per year of absenteeism for smokers, and increased work accidents and disability reimbursements (27, 45, 46).

Estimated yearly costs to industry for its smokers range between 200 and 400 dollars. Luce and Schweitzer estimate $190 per year over an entire life span. This is based on estimating excess morbidity and mortality from neoplasms, circulatory, and respiratory diseases in smokers versus nonsmokers, using economic calculations of excess disease adjusted for inflation and dividing the economic costs by the number of smokers over 17 years of age (45). Other estimates of the overall costs to industry have taken into account such additional expense categories as fire and life insurance, energy for air filtration health effects, family suffering, and a higher turnover rates due to excess morbidity and mortality (47). One estimate of overall cost per year is $400 per smoker (1982 dollars) (31). Savings from absenteeism lost productivity, fewer accidents, involuntary smoking, and fire insurance can probably be found in the first several years after quitting. Savings in disability and health care costs will appear at different rates for different smoking related health problems over three to ten years.

6.3   Exercise

A belief in the significant benefits of regular exercise has existed since the time of Hippocrates (48). It is clear that more and more Americans are now adopting the habit of regular exercise. What is not clear is the number of persons exercising, the distribution of different levels of exercise, and the frequency of each.

A 1980 telephone survey of 1091 randomly selected adults in Massachusetts revealed the following data: 56 percent of the group reported that they exercised twice a week, and over one quarter said they got daily exercise. The most common forms of participation were: walking, 14 percent; jogging, 12.8 percent; and calisthenics, 8.7 percent (49). An earlier poll, taken by the Pacific Mutual Life Insurance Company, showed a total of 37 percent of the public exercising regularly, although regularly was not well defined. Business leaders showed a 75 percent exercise rate and labor union leaders a 51 percent rate (50). While no exact estimates can be found for the number of companies offering exercise programs to their employees, one estimate places the number at 500 companies which provide some type of a health facility. This ranges from outside showers to elaborate fitness centers. The estimate adds another 1500 companies which have made arrangements with local gyms and sports centers to encourage workers to exercise (51). In Canada, the Public Health Association undertook a survey of 800 companies with respect to physical fitness programs in 1981. Of the 26 percent that responded with information, 25.4 percent reported they had fitness programs. Of these companies, 86 percent had 500 or more employees (14).

What benefits can be achieved from regular exercise? Significant improvement in the level of fitness, as well as reductions in blood pressure, skinfold thickness, and percent body fat have been found (12). Health risk improvement is more difficult to assess. But there is growing evidence that the physically active employee has lower, age specific, heart attack and death rates than those who do not exercise, if all other heart health risks are held constant (52–55). In addition, the benefit appears to be dose related and is at least in part thought to be independent of the other established risk factors (56,57).

Reports of many studies from North America, Western Europe, and Russia support reductions in absenteeism related to exercise, but most have problems of self-selection and poor choice of controls, precluding definite conclusions (31). Two controlled studies by insurance companies provide more rigorous evidence. A Toronto insurance company found a 42 percent decline in average monthly absenteeism in those employees exercising at least twice a week compared with a 20 percent decline in the overall test group and a control insurance company (58). The Metropolitan Life Insurance Company studied two groups of 100 employees (unpublished data). Those in the exercise group reduced their absenteeism over a two year period from 6.3 days per year to 4.9. The absenteeism in the nonexercising group actually increased during the two year period from 5.6 to 7.0 days per year. Confounding the interpretation of these results is the fact that the control group initially had higher average risks for cholesterol and cigarette smoking while also having lower average, job groupings and absenteeism (31).

Improvement in morale, self-image, productivity, and attitudes toward the company and job have been reported by managers of exercise programs and participants (31). Participants in a National Aeronautics and Space Administration's 12-month exercise program reported better attitudes toward work, higher job performance, and improved physical stamina (56). Experience in other exercise programs has yielded similar results (59).

6.4 Weight Control

Traditional estimates of life insurance premiums have been based upon the relationship of height to weight. A 1971–74 Health and Nutrition Examination Survey revealed that 18 percent of men age 20–74 and 13 percent of women in same age range were 10–19 percent above their appropriate weight for height based on the standard tables. Fourteen percent of the men and 24 percent of the women were classified at 20 percent or more above their ideal weight (60). More recent estimates are that 20 percent of all Americans over age 30 are more than 20 percent over their desirable weight (61). In the worksite 13.6 percent men and 21.0 percent women are estimated to be 20 percent or more over their ideal weight (29). Excess body weight is significantly associated with an increase in blood pressure, cholesterol, and blood sugar (62, 63). It is also related to musculoskeletal conditions including lower-back and other health problems. Obesity is related to increased risk from surgery, anesthesia, accidents, pregnancy, and childbirth (60).

There are a number of available methods to determine appropriate weight for size or percent body fat. Some methods include simple calculation of height to weight ratio. Others involved direct skinfold measurement of thickness using calipers or formulas based on sex and height. Underwater weighing is the most accurately applied method to date. The most common approach to the management of overweight/obesity is the use of behavioral change techniques usually provided in a group. The objective of these techniques is to teach participants self-control over their eating behavior. Like other behavioral change programs, techniques for record keeping, including food diaries, are used.

Existing cues or triggers for eating are examined. New cues and eating behaviors are substituted. A new set of positive rewards is taught to reinforce the new eating behavior. Better management of the complete eating cycle is taught, including food purchasing, preparation and eating, both at home and out. Most programs devote considerable attention to good nutrition since improved nutrition is essential to sustained weight control. In addition, many weight control programs now include a strong exercise component, since regular exercise greatly facilitates weight management.

Weight control programs usually are conducted for 10–15 weeks in a group setting or more rarely on a one-to-one basis. At least one company has begun offering a weight maintenance program specifically for individuals whose objective is weight control and not weight loss (64).

Very few controlled studies have been conducted in the worksetting to evaluate the effectiveness of weight control programs. The first published worksite study, conducted in 1980, involved a controlled trial of 40 women assigned to four varying treatment groups. While the initial weight loss averaged approximately 8 lbs, for those staying with the program at the end of six months, average weight loss was significantly less (65). These results are consistant with those reported from clinical settings. A review of 26 groups in nine behavioral weight control studies in clinical settings showed that 14 groups had weight gains during the 12 months following treatment, and in the 12 groups which showed continued weight loss, the average additional loss was only 1.6 kg (3.5 lbs), hardly significant considering initial weight percentages of 20 to 78 over ideal body weight (65).

Weight control programs continue to be popular in health promotion programs because of the employees' high level of interest and their potential impact on other risk factors, such as high blood pressure and total cholesterol. More behavioral research is needed on how these programs can best help participants maintain new eating habits over a lifetime. Specifically more research is needed among worksite populations, including the value of increasing nutritional alternatives in worksite cafeterias and vending machines to encourage choice of lower calorie foods.

6.5 Nutrition

There is increasing evidence that nutritional habits relate to a large number of health problems, including heart disease, cancer, hypertension, stroke, diabetes, and other important causes of morbidity and mortality. For example, high saturated fat intake increases blood cholesterol which is a risk indicator for heart disease, stroke and other vascular problems related to atherosclerosis (67). Studies such as those conducted in Framingham, Massachusetts; Oslo, Norway; and North Karelia, Finland, have demonstrated how cholesterol reduction positively influences cardiovascular disease incidence and mortality. The 1983 results of one of the largest and most definitive study, the Lipid Research Clinics Primary Prevention Trial, provide even more definitive data on the preventive role of cholesterol reduction in heart disease (68). High fat intake is also believed to increase risk for several types of cancer, including breast, colon, and prostate (69).

Americans eat much more salt than is required, and this excess may contribute to a higher incidence of hypertension (70). However, the extent of sodium restrictions in the diet is still controversial. Given the present research findings, sodium restricted diets usually suggest limiting the amount of salt used in cooking, not adding table salt, and a careful reading of food and over the counter drugs for possible sodium content.

Dietary fiber, which is lacking in the diets in most Western industrialized countries as the result of food processing, can contribute to reduced gastrointestinal problems and may also facilitate better management of diabetes mellitus. Some researchers believe that a high intake of dietary fiber can reduce the frequency of colon cancer, although the evidence to date remains equivocal.

While the role of sugar in the causation of a number of diseases remains controversial, its role in the development of dental caries and also in fostering gum diseases are unassailable.

Based on current knowledge, it appears prudent to subscribe to the recommendations embodied in the Dietary Guidelines, jointly supported by the U.S. Department of Agriculture and Health and Human Services. These guidelines (71) are:

1. Eat a variety of foods.
2. Maintain ideal weight.
3. Avoid too much fat, saturated fat, and cholesterol.
4. Eat foods with adequate starch and fiber.
5. Avoid too much sugar.
6. Avoid too much sodium.
7. If you drink alcohol, do so in moderation.

Worksettings provide excellent opportunities for nutrition education for employees about the benefits of proper nutrition and how to improve nutritional choices. Informational programs on vitamins, fad diets, fiber, and the basics of nutrition are often offered. Special cholesterol reduction programs are also conducted for employees at high risk for heart disease. Labeled foods in the cafeterias and in the vending machines can make employees more aware of the types of choices they are making. Since about one third of daily calories may be eaten at the worksite or nearby, having an impact on the consumption of nutrients associated with the worksetting, could have a significant impact on overall nutrition. However, labeling food choices is only helpful if there are more nutritious alternatives available. Employers unsure of the nutritional value of the offerings should obtain assistance from a qualified nutritionist, preferably a registered dietitian.

Nutritional education courses can also be offered as part of the health promotion programs at the worksite. In general, such offerings are popular.

Often the weight management and nutrition offerings become synergistic, since improved nutrition involves reducing the intake of calorie dense fatty foods and snacks.

6.6   Employee Assistance Programs (EAP)

Employee assistance programs (EAPs) are worksite affiliated activities that provide assessment, referral, and sometimes short term counseling on a confidential basis to employees (and often to the dependents of employees) with a range of personal problems. These programs initially grew out of a strong interest in helping employers reduce the toll of alcoholism among their employees. Recovering alcoholics, trained in counseling and substance abuse, were used as counselors and referral agents for treatment.

6.6.1   Alcohol

Two thirds of all American adults consume alcoholic beverages (72). Alcohol acts directly upon the central nervous system causing a deterioration in coordination, memory, reflexes, and vision. This creates a special hazard for employees operating motor vehicles or equipment (66). The misuse of alcohol has been cited as a factor in over 50 percent of all traffic deaths. The toll for such death and injury is not limited to alcoholics and problem drinkers, but includes the occasional social drinker (11).

The adverse health effects of alcohol cover a broad spectrum. Inflammation and subsequent cirrhosis of the liver are among the serious results of alcohol misuse. Cancers of the liver, mouth, and esophagus are highly correlated with alcohol consumption. An even more dramatic increase in cancer of the esophagus is seen when alcohol consumption is associated with smoking (11). The effect of alcohol consumption upon the unborn fetus has been studied extensively in the past several years.. So obvious are the effects of alcohol misuse during pregnancy that the term "fetal alcohol syndrome" has been created to describe the collection of congenital birth defects. While such effects are seen in greater frequency among children of heavy drinkers, the children of moderate drinkers sometimes may also exhibit these effects.

Alcohol's effect upon the cardiovascular system is being carefully studied. Taylor and associates cite several research studies pointing to an increased death rate from cardiovascular and atherosclerotic conditions (66). While the data are not conclusive, they also cite two recent studies which describe moderate alcohol consumption having a positive effect on the level of high density lipoproteins (HDL), which are the lipid protein associated with lower age-specific rates of cardiovascular disease (73, 74).

Among the working population, recent figures place the prevalence of alcoholism at 6–10 percent (75). According to the National Institute of Alcohol Abuse and Alcoholism (NIAA) 25 percent of those employees with alcohol problems are white collar, 30 percent blue collar, and 25 percent professionals and managers (76). The annual cost of alcoholism to industry is estimated to be in the billions of dollars (75).

While the American Medical Association gave alcoholism the status of a disease in 1956, it has only been in the last 10–15 years that most employers have shed the fear of social stigma and begun to provide counseling and treatment services for employees. According to Trice, starting in the 1950s and 1960s, industries began to treat alcoholism as a disease. Disciplinary procedures for poor alcohol related job performance were suspended while the employee was obtaining assistance. Trice reports that in the early 1970s only about 600 firms reported using this type of a policy (77). However, in the 1981 NIAA Report to the United States Congress the number of industries with alcohol programs had grown in 1980 to 4,400, a considerable increase over 10 years (78).

The workplace can be an ideal place for identifying an alcoholism problem and intervening to control it. Early behavioral effects such as lateness and increased absenteeism may be easily monitored. Job performance, tied to the paycheck, and to the job itself, becomes a powerful motivating force to get the alcoholic into treatment (78). To date, a number of industry reports suggest that approximately 90 percent of the alcoholics to whom treatment is offered will participate (79). Of this, 60–80 percent recovered. Bethlehem Steel reports a 60 percent success, DuPont a 66 percent rehabilitation rate, and Minnesota Mining reports that 80 percent were either recovered or controlled so that marked improvement was shown (79).

6.6.2   Other Drug Abuse

Helping employees with other drug abuse problems in addition to alcoholism falls within the mission of most Employee Assistance Programs. It is difficult to obtain reliable estimates for the use of most drugs by employees. However, both anonymous employee surveys and experiences from EAPs point to drug problems that range from the abuse of legal prescription drugs such as tranquillizers and barbiturates to illicit drugs such as amphetamines, marijuana, cocaine, and heroin.

The health consequences of drug abuse are numerous. Of special concern to industry are the effects of drugs upon the central nervous system which include impaired motor and sensory functioning, both of which can affect the safe use of motor vehicles and equipment. The addictive qualities of some drugs make them also of special concern to employers due to their effects on health, health care costs, and productivity.

The abuse of drugs, both depressant and stimulant, can result in physical and psychological addition. Both psychological problems and premature death from overdose have been associated with cocaine. While many of the reported health effects of marijuana are disputed, one well established effect is its potential to reduce motivation and performance (11). Drug interactions resulting from the combined use of alcohol and prescription drugs and even with over the counter drugs can also yield potentially hazardous effects.

6.6.3 EAP Characteristics

The term "EAP" (Employee Assistance Program) now is used to describe a set of company policies and procedures used to identify and respond to employees with substance abuse problems, personal or emotional problems which interfere, directly or indirectly, with job performance. EAPs provided information and/or referrals to appropriate counseling, treatment and support services, financed in total or in part by the company (78). Some EAPs have broadened their scope to also include financial and legal counseling. EAP counselors are most frequently mental health professionals, usually with advanced degress in counseling psychology or social work.

In-house EAPs are most frequently placed in the medical or personnel department of a company. Some companies, especially those without medical clinics, prefer to contract for these services through local consultants and organizations. Employees enter EAP either by referral from their supervisor, through the medical department or by self-referral. Many companies offer training programs to assist supervisors in identifying and referring troubled employees.

More and more studies to determine the effectiveness of both alcohol and substance abuse programs are now providing industry managers with valuable information about these programs. Two studies, conducted in the Philadelphia fire and police forces, showed a decrease in the number of workdays lost to illness and injury following inpatient or outpatient care for alcoholism. A positive ratio of benefits to costs was reported in these studies, but evaluation design problems preclude definitive conclusions (80).

A study conducted at General Motors employees in Canada compared a group of 50 employees with alcohol problems, who were interviewed by a company doctor and referred to treatment, with 50 employees with similar problems, who also were referred, but did not undergo treatment. The treatment group's use of sickness and accident benefits decreased 48 percent as compared to an increase of 127 percent in the untreated group (80). Problems in comparablity of group selection, regarding the severity of the initial alcoholism problem and the use of benefits prior to the study, limit generalizing from the study.

The International Harvester Company, in cooperation with the United Auto Workers Union, initiated a program for alcoholic employees using an occupational health nurse for counseling and referral. Simultaneously, a revision in the benefit plan structure was made to include coverage for all inpatient and outpatient alcoholism treatment modalities. As a result of the program, disability days and disability income paid for employees with alcohol problems dropped 53 percent and 41 percent, respectively. Costs of medical benefits for these employees dropped 34 percent and employee absenteeism dropped 47 percent. Sobriety rates were calculated at one year to be 56 percent (39 percent for total sobriety and 17 percent drinking infrequently). Persons in the program who did not follow treatment recommendations showed an increase in disability days, benefit payments, and medical costs (81).

New York Telephone reports a rehabilitation rate of 85 percent of their alcoholism treatment program which enrolled 300 new cases annually for the past seven years. The employee assistance program is estimated to have saved the company $1,565,700 in the past seven years. Illinois Bell, reporting a nine year study of 752 cases in which absences were tracked five years before and after treatment, cites savings to the company of $1,272,240 (78).

Cost–benefit studies of comprehensive EAPs are also being reported. A recently published study by AT&T followed the progress of 100 EAP cases for a period of 22 months. Employees in the study presented a broad range of EAP problems: alcohol, drugs, emotional, family marital and work-related problems. Results of the AT&T study reveal that 86 percent of the sample could be considered "rehabilitated" or "improved" following treatment. Incidental absenteeism dropped from 421 to 92 days while occasions of absence declined from 172 to 35. Disability absence decreased from a total of 1,531 to 192 days. (These figures do not include disability absence associated with the initial treatment period.) Work related accidents prior to EAP intervention numbered 26 with 164 lost workdays. After intervention, the number of accidents was reduced to 11, with 49 associated workdays lost. Visits to the medical department also decreased 49 percent (82).

To measure the impact of changes resulting from their EAPs an analysis was conducted to determine the savings with respect to five variables: on the job accidents, incidental absence, disability absence, medical clinic visits, and anticipated job losses. In total, AT&T's program is estimated to have saved $448,000 which is considered a conservative figure because savings are only calculated on the five variables (82). Another company, Kennecott Copper, reports similar cost savings for its EAP. Their "Insight" psychotherapy program reports a 6 to 1 benefit–cost ratio (78).

Evaluation studies of EAP programs are faced with many of the same methodological problems as other health promotion programs. Specific to EAPs are problems in the definition of recovery, differing employee benefit packages affecting the choice of care, and the numerous cost variables, including the cost of rehabilitation, that must be entered into equations designed to compare benefits and costs. Yet despite all these barriers, the growing evidence is that EAP programs can produce significant cost savings to companies.

6.7 Stress Management

Much progress has been made in the past decade in defining the role of stress in the development of physical and mental disorders. As a result of this research and the increased public media coverage, stress management has become a popular health promotion program activity.

Researchers now suggest that stress not only plays a role in influencing the creation and progression of an illness, but also a person's susceptibility to a number of diseases (83). Specifically, stress has been associated with health conditions such as peptic ulcer, coronary heart disease, and ulcerative colitis.

Dr. Hans Selye, a pioneer in stress research, defines stress as the response of an organism to the demand for change (84). Much debate still exists as to how much stress is external, or internally caused, and how the personal coping skills of an individual affect his/her stress management. Important stressors exist in both the personal and work environments and their effects cannot always be separated. Stress is not always negative. Positive stressors, such as a job promotion, raise, marriage, or purchasing a house, can bring the same stress reaction as negative stressors such as death, divorce, job loss, etc.

Although the relationship between work related stress and illness is complicated and not well understood, the quality of worklife is recognized as having a major impact on employee health status. Much of the newer research points clearly to the multidimensional quality of stress, and thus the complexity of identifying and administering the appropriate interventions to reduce its toll. The worksite, being a potential cause of the stress, can also serve as a site for stress reduction programs to employees.

One review of occupational stress management programs reported that 12 interventions ranged from one to fifteen instructional sessions (81). Techniques utilized in worksite programs, including muscle relaxation, biofeedback, meditation, cognitive restructuring/behavioral skill training, and a counseling interview have shown the ability to reduce arousal level and psychological signs of stress (85–87). Most studies have demonstrated benefits on one or more of the self-reported stress measures: anxiety, stress, mood, coping, sleep (88), attitudes toward work (88), absenteeism (87), job satisfaction and job performance, directly measured electromyographic (EMG) readings, systolic and/or diastolic blood pressure (89), and urinary catecholamines.

Two other studies recently reported the effect of relaxation techniques on employee stress management. Peters and colleagues report a study of the effect of using daily relaxation breaks on self-reported measures of health, performance, well being, blood pressure, and heart rate. One experimental group practiced Benson's relaxation technique; another group was told to sit quietly and practice their own technique; a third group received no instruction. Those individuals in the first group achieved the greatest mean improvements in all of the measures on every subjective index, (symptoms, illness days, performance, and sociability-satisfaction) with several of the differences between the experimental and control groups achieving statistical significance. Blood pressure reductions were also directly correlated with increased intervention levels (90).

Another study conducted among 154 New York Telephone employees compared the relaxation techniques of Benson, Jacobson and Carrington. The three study groups and one control group all were similar in symptoms and personality profiles at study initiation. While the treatment groups reported clinical improvements in stress symptoms at five and a half months, only those using Carrington's or Benson's meditation techniques showed significantly greater reductions than the control group (91).

Manuso studied both clinical and economic data from a program utilizing biofeedback and other stress management techniques for employees experiencing headaches and general anxiety. Statistically significant decreases in symptoms, increases in job satisfaction and effectiveness were shown in all treatment groups. In addition, visits to the health center for anxiety and headaches decreased over six months (92).

Only a few systematic studies have been reported that bear directly on the impact occupational stress management has on the level of stress, other risk factors for illness, utilization of health benefits or other employer-sponsored health services. Assessment of the efficiency and effectiveness of occupational stress management programs is complicated by a lack of objective measures of stress, the number of factors which can affect perceptions of stress, both inside and outside the work environment, and the heterogeneity of individual, group and environment interventions which are considered "stress management." In one extensive review of the research literature, industrial psychologists Beehr and Newman repeatedly stress how these studies do not allow definitive conclusions about the effectiveness of these interventions due to methodological or other study problems (93).

Despite limitations in the general knowledge about stress, the use of stress management techniques is increasingly accepted within industry as a way of helping employees cope with stress. Unfortunately, as with other behavioral programs, what is not known is the ability to predict which individuals are at highest risk for adverse effects of various stressors and what combination of programs will best intervene for each person. The interrelationship between worksite and personal stress variables is still in need of study. Industry has a valuable opportunity to positively impact the quality of both the employee's work and personal life through such stress reduction programs and related research.

6.8 Accident Prevention

Lap and shoulder belts are about 60 percent effective in preventing fatalities from automobile accidents, but only 10–14 percent of drivers and a smaller percentage of passengers use them (95). Each accidental death causes an average loss of more than 20 years of working life, much longer than heart disease deaths (two years) and cancer deaths (slightly over five years) (95). In 1978 it is estimated that one-third of all work-related fatalities were caused by motor vehicle crashes, with an average employer cost for each death of $120,000 (94). While at least 19 states require state employees to use safety belts in vehicles used on the job (96), no state requires employees of private firms on or off the job to wear safety belts, despite an increasing number of states which require adequate restraints on infants and young children.

At least 20 private and public employers, have conducted incentive programs to increase safety belt usage of employees at one or more sites (97). While results vary considerably by site and program, in general, pre-program use based on direct observation was 10–25 percent, with notable exceptions in the 40–50 percent range in an Air Force base where use was required and a General Motors plant that had stressed safety belt usage prior to the program. Most incentive programs are able to double usage rates, at least during the period incentives are offered (97). Drivers using safety belts at entry to work have been given immediate rewards on a random basis or tickets good for drawings of large prizes. Prizes have included gift certificates, vacations, cars, lottery tickets, cash, and food coupons, among others (97). In general, blue collar workers have lower pre-program safety belt use but sometimes show a a greater percentage increase in their use than white collar employees. In those programs which have assessed compliance, post-program use has significantly declined, but almost always remained at a higher level than pre-program usage. Accident prevention, specifically that relates to motor vehicles, is an area of health promotion that requires coordination of the medical, benefits, and safety organizations in an industry. Its impact on life-style and cost savings is still in need of evaluation research.

6.9 Other Health Promotion/Disease Prevention Activities

It is not possible within the confines of this chapter to fully describe the total variety to health promotion activities offered to employees by American industries. However, a number of other types of program topics are frequently addressed including self-care, healthy back, CPR, first aid, cancer prevention, assertiveness training, and parenting. The evaluation of the effects and effectiveness of these program activities, like those previously described, while incomplete, is growing.

7 FUTURE DIRECTIONS

Many signs point to encouraging results for businesses investing in health promotion programs. The results of major epidemiological studies of worksite programs such as those being conducted at Johnson and Johnson and AT&T Communications should become available in the near future. Evidence is accumulating about the effect these programs are having upon employee morale, job commitment, and overall attitudes in the short and intermediate term. There will be considerably more data on the cost of life-style–related diseases in terms of premature illness, death, and dollar costs. Additional data on the cost effectiveness of differing health promotion program interventions in reducing risk and the frequency and severity of some diseases also should become available. If the results are positive, many of the industries currently "waiting for proof" are likely to initiate programs.

Health promotion programs have begun to spread beyond the small corporate family that first initiated them. Many managers of large corporate programs, academic centers, hospitals, and consulting firms are providing technical assistance and consultation to smaller companies or groups of companies on program content and cost benefit design. The dissemination of program information also has been the result of the information exchange between health care professional organizations and health care coalitions.

The National Center for Health Education, a private nonprofit organization devoted to promoting private sector initiatives in health education, is presently developing a nationwide resource network to provide resources and access services to small companies requesting assistance in health promotion program development. Voluntary organizations, such as the American Cancer Society, have been actively involved in workplace health promotion, most recently with data on the economic impact of cancer and cancer control on private industry and participation in the development of a model smoking policy. The Business Roundtable, a prestigious organization of companies, has devoted a considerable amount of time to cost containment activities. The Washington Business Group on Health, devoted to the health concerns of its corporate membership, has made education on health promotion a high priority for its membership.

Health promotion is here to stay. Life-style change has the potential to significantly alter the health status of the United States population. Industry can play an important role by encouraging, promoting, and supporting health improvement programs. Such support potentially could not only conserve corporate dollars, but also impact the quality of the American employee's life, and as a result, the health of society as a whole.

REFERENCES

1. T. Preston, *Biomed. Commun.*, 9–28 (1982); through *Pharos*, **39,** 64 (1976).

2. N. Belloc and L. Breslow, *Prev. Med.*, **1,** 409–421 (1972).

3. R. I. Levy, *Ann. Rev. Pub. Health*, **2,** 49–70 (1981).

4. J. T. Salonen, P. Puska, and H. Mustaniemi, "Changes in Morbidity and Mortality During Comprehensive Community Programs to Control Cardiovascular Diseases During 1972–77 in North Karelia," *Br. Med. J.* **2:**1178–1183, through R. I. Levy, *Ann. Rev. Pub. Health*.

5. J. Farquhar, N. Maccoby, and P. D. Wood, et al., *Lancet*, **1,** 1192–1195 (1977).

6. J. T. Salonen, P. Puska, and T. E. Kottle, et al., *Am. J. Epid.*, **114,** 81–94 (1981).

7. J. Fielding, *J. Occup. Med.*, **21,** 79–88 (1979).

8. National High Blood Pressure Education Program, "High Blood Pressure in the Workplace Fact Sheet," National Institutes of Health, National Heart, Lung and Blood Institute, September 1980.

9. Department Health, Education, and Welfare, *Health US 1980*, USDHHS Publication No. PHS-81-1232, U.S. Government Printing Office, Washington, D.C., 1980.

10. American Cancer Society, *Make Cancer Control Your Business*, New York, 1982.

11. Office of the Assistant Secretary for Health and Surgeon General, *Healthy People*, USDHHS Publication No. 79-55071, U.S. Government Printing Office, Washington, D.C., 1979.

12. J. Fielding, *Corporate Health Management*, Addison-Wesley, Menlo Park, Calif. 1984.

13. D. B. Ardell, *High Level Wellness, An Alternative to Doctors, Drugs and Disease*, Rodale Press, Emmaus, Pa., 1977.

14. J. E. Fielding and L. Breslow, *Am. J. Pub. Health*, **73**, 538–542 (1983).

15. R. N. Beck, "Health Promotion and Health Protection: The Role of Industry," paper presented at the Second Annual Lester Breslow Distinguished Lectureship, UCLA School of Public Health, Los Angeles, Calif., January 1982.

16. Department Health and Human Services, July 16, 1983; *National Journal*, July 30, 1983, Phone conversation Willis Goldbeck, President, Washington Business Group and Health, July 29, 1983.

17. L. K. Richman, *Fortune*, **107**, 95–110 (1983).

18. D. L. Block, "Executive Management and Medical Leadership Roles in Health Care Cost Containment: A Medical Director's Perspective," paper presented at the American Occupational Medical Association Annual Meeting, Washington, D.C., April 1983.

19. F. L. Provato, "Corporate Health Care Costs, A Medical Economic Problem at Southern New England Telephone" (unpublished) and A. Cantlon, Southern New England Telephone's Health Promotion Program, lecture presented at Bell System Nurses and Health Educators' Conference, Washington, D.C., April 25, 1983.

20. R. E. Dedmon and C. M. Smoczyk, "Kimberly Clark Health Management Program: Update 1983," paper presented at the American Occupational Medical Association, Washington, D.C., 1983.

21. A. Goetz, J. Duff, and J. E. Bernstein, *Publ. Health. Rep.*, **95**, 119–126 (1980).

22. D. L. Sackett, R. B. Haynes, E. S. Gibson, et al., *Lancet*, **1**, 1205–1207 (1975).

23. T. E. Deal and A. A. Kennedy, *Corporate Cultures*, Addison-Wesley, Menlo Park, Calif., 1982. (Underlining by authors.)

24. R. Allen, "A Culture-Based Approach to the Improvement of Health Practices," Proceedings of 13th Meeting of the Society of Prospective Medicine, September 29–October 2, 1977, San Diego, Calif.

25. O. B. Dickerson, and C. Mandelblit, *J. Occup. Med.*, **25**, 471–474 (1983).

26. J. Fielding, "Evaluation of Worksite Health Promotion Programs," paper presented at the Institute of Medicine Conference, Washington, D.C., June 1980.

27. Department Health, Education, and Welfare, *Smoking and Health, A Report of the Surgeon General*. USDHHS-PHS, Publication No. 79-50066, U.S. Government Printing Office, Washington, D.C.

28. "Five Year Findings of the Hypertension Detection and Follow-Up Program," *JAMA*, **242**, 2562–2571 (1979).

29. National Institute of Health, *Cardiovascular Primer For The Workplace*, DHHS-NIH, Publication No. (NIH) 81-2210, 1980.

30. A. Foote and J. C. Erfurt, *N. Engl. J. Med.*, **308**, 809–813 (1983).

31. J. E. Fielding, *J. Occup. Med.*, **24**, 907–916 (1982).

32. M. H. Alderman and T. K. Davis, *J. Occup. Med.*, **18**, 793–796 (1976).

33. M. H. Alderman and A. Melcher, *J. Occup. Med.*, **25**, 465–470 (1983).

34. National Heart, Lung, and Blood Institute, National High Blood Pressure Education Program at Massachusetts Mutual: Offsite Care and Good Monitoring Reduce Medical Costs, *Re: High Blood Pressure Control in the Worksetting*, Winter 1980, National Heart, Lung and Blood Institute, DHEW, through J. Fielding, *Corporate Health Management*, Addison-Wesley, Menlo Park, Calif., 1984,

35. D. E. Morisky and D. M. Levine, *Am. J. Pub. Health*, **73**, 153–162 (1983).

36. R. B. Haynes, D. L. Sackett, D. W. Taylor, et al., *N. Engl. J. Med.*, **299**, 741–744 (1978).

37. A. Logan, B. Milne, and C. Achber, et al., *Hypertension*, **3**, 211–219 (1981).

38. H. S. Ruchlin, and M. H. Alderman, *J. Occup. Med.*, **22**, 795–800. (1980).

39. Office of the Assistant Secretary for Health, *Promoting Health and Preventing Disease*, USDHHS Publication, Fall 1980.

40. J. R. White and H. F. Froeb, *N. Engl. J. Med.*, **302**, 720–723 (1980).

41. T. Hirayama, *Br. Med. J.*, **282**, 183–185 (1981).

42. Office on Smoking and Health, *Smoking, Tobacco and Health, A Fact Book*, USDHHS-PHS, Publication No. (PHS) 80-50150, 1981.

43. National Institute of Occupational Safety and Health, *Adverse Effects of Smoking in the Occupational Environment*, Current Intelligence Bulletin No. 31 NIOSH Publication No. 79-122, 1979.

44. W. A. Hunt and D. A. Bespalec, *J. Clin. Psych.*, **30**, 431–438 (1974).

45. B. R. Luce and S. O. Schweitzer, *N. Eng. J. Med.*, **298**, 569–571 (1978), through J. Fielding, *J. Occup. Med.*, **24**, 907–916 (1982).

46. M. M. Kristein, *Preven. Med.*, **6**, 252–264 (1977), through J. Fielding, *J. Occup. Med.*, **24**, 907–916 (1982).

47. M. M. Kristein, "How Much Can Business Expect to Earn From Smoking Cessation?" Presented at the National Interagency Council on Smoking and Health's Conference "Smoking and the Workplace," Chicago, January 9, 1980.

48. W. L. Haskell and S. N. Blair, "The Physical Activity Component of Health Promotion in Occupational Settings," in R. S. Parkinson and Associates, *Managing Health Promotion in the Workplace*, Mayfield Publishing, Palo Alto, Calif., 1982.

49. C. A. Lambert, D. R. Netherton, L. J. Fireson, J. N. Hyde, and S. J. Spaight, *N. Engl. J. Med.*, **306**, 1048–1051 (1982), through J. Fielding, *Corporate Health Management*, Addison-Wesley, Menlo Park, Calif., 1984.

50. The Pacific Mutual Life Insurance Company, *Health Maintenance*, 1978.

51. J. H. Andresky, *The Runner*, N. V., 40–44 (1982).

52. R. S. Paffenbarger, A. L. Wing, and R. T. Hyde, *Am. J. Epid.*, **108**, 161–175 (1978).

53. D. S. Siscovick, N. S. Weiss, A. P. Hallstrom, et al., *JAMA*, **248**, 3113–3117 (1982).

54. S. M. Fox, J. P. Naughton, and W. P. Haskell, *Annals of Clinical Research*, **3**, 404–432 (1973); through R. S. Parkinson and Associates, *Managing Health Promotion in the Workplace*, Mayfield Publishing, Palo Alto, Calif., 1982.

55. V. F. Froelicher, *The Effects of Chronic Exercise on the Heart and on Coronary Atherosclerotic Heart Disease, A Literature Review.* in: Cardiovascular Clinic, ed. by A. Brest, Philadelphia, F. H. Davis Co., through R. S. Parkinson and Associates, *Managing Health Promotion in the Workplace*, Mayfield Publishing, Palo Alto, Calif., 1982.

56. R. S. Paffenbarger, et al., *Am. J. Epid.*, **108**, 12–18 (1978), through R. S. Parkinson and Associates *Managing Health Promotion in the Workplace*, Mayfield Publishing, Palo Alto, Calif., 1982.

57. S. Shapiro et al., *Am. J. Pub. Health*, **59**, 1–101 (1969), through R. S. Parkinson and Associates, *Managing Health Promotion in the Workplace*, Mayfield Publishing, Palo Alto, Calif., 1982.

58. M. Cox, R. J. Shepard, and P. Corey, *Ergonomics*, **24**, 795–806 (1981), through J. Fielding, *J. Occup. Med.*, **24**, 907–916 (1982).

59. K. Cooper, *The Aerobics Program for Total Well Being*, M. Evans and Company, New York, 1982.

60. "Dietary Council Digest," **51**, (1980).

61. Gio Batta Gori, *Nutrition Today*, N.V., 18 (1981).

62. M. S. Sorlie, T. Gordon, and W. B. Kannel, *JAMA*, **243**, 1828–1831 (1981), through J. Fielding, *Corporate Health Management*, Addison-Wesley, Menlo Park, Calif., 1984.

63. *Build and Blood Pressure Study*, 1959, Society of Actuaries, **1**, 1–268 (1959), Chicago, through J. Fielding, *Corporate Health Management*, Addison-Wesley, Menlo Park, Calif., 1984.

64. Personal Communication, Susan Harris, R. D., M. S., Sandia Laboratories, Albuquerque, N.M., July 1983.

65. A. J. Stunkard and S. B. Penick, *Archives of General Psychiatry*, **36**, 801–806 (1979) through J. Fielding, *Corporate Health Management*, 1984.

66. K. B. Taylor, J. R. Ureda, and J. W. Denham, Eds., *Health Promotion Principles and Clinical Applications*, Appleton-Century-Crofts, Norwalk, Conn., 1982.

67. J. Stamler, *Circulation*, **58**, 3–19 (1978).

68. Lipid Research Clinics Program, *JAMA*, **251**, 351–365 (1984).

69. *Diet, Nutrition and Cancer*. Assembly of Life Sciences National Research Council, National Academy Press, Washington D.C., 1982.

70. "Statement on the Role of Dietary Management in Hypertension Control." Approved by the National High Blood Pressure Education Program Coordinating Committee, March 1979, *U.S. Department of Health and Human Services*.

71. U.S. Department of Agriculture, *Nutrition and Your Health, Dietary Guidelines for Americans*, U.S. Department of Health, Education and Welfare, Office of Governmental and Public Affairs, USDA, Washington, D.C.

72. E. P. Noble, Ed., Third Special Report to the U.S. Congress on Alcohol and Health, USDHEW, National Institute on Alcohol Abuse and Alcoholism, Rockville, Maryland, 1978, through K. B. Taylor, J. R. Ureda, and J. W. Denham, Eds., *Health Promotion Principles and Clinical Applications*, Appleton-Century-Crofts, Norwalk, Conn., 1982.

73. W. P. Castelli, T. Gordon, M. C. Hjortland, et al. *Lancet*, **2**, 153–155 (1977), through K. B. Taylor et al., *Health Promotion Principles and Clinical Applications*, Appleton-Century-Crofts, Norwalk, Conn., 1982.

74. C. H. Hennekens, W. Willet, B. Rosner, et al., *JAMA*, **242**, 1973–1974 (1979), through K. B. Taylor et al., *Health Promotion Principles and Clinical Applications*, Appleton-Century-Crofts, Norwalk, Conn., 1982.

75. W. N. Burton, P. R. Eggum, and P. J. Keller, *J. Occup. Med.*, **23**, 259–262 (1981).

76. P. Palisano, *Occupational Hazards*, N.V., 55–58 (1980).

77. H. M. Trice, Drug Use and Abuse in Industry, Washington, D.C., Office of Drug Abuse Policy, January 1979, through R. L. Dupont, M. M. Basen, "Control of Alcohol and Drug Abuse in Industry: A Literature Review," in: R. S. Parkinson and Associates, *Managing Health Promotion in the Workplace*, Mayfield Publishing, Palo Alto, Calif., 1982.

78. D. C. Walsh, Employee Assistance Programs, *Millbank Memorial Fund Quarterly, Health and Society*, **60**, 492–517 (1982).

79. "Alcoholism: New Victims, New Treatment," *Time*, April 22, 1974, through Solving Job Performance Problems, Dept. of Health and Rehabilitation Services, Florida Occupational Program Committee, N.D.

80. Office of Technology Assessment, "Health Technology Case Study 22, The Effectiveness and Costs of Alcoholism Treatment," March, 1983.

81. P. R. Eggum, P. J. Keller, and W. N. Burton, *J. Occup. Med.*, **22**, 545–548 (1980).

82. G. Gaeta, R. L. Lynn, and L. Grey, "AT&T Looks at Program Evaluation," *EAP Digest*, N.V. 22–31 (1982).

83. G. E. Schwartz, "Stress Management in Occupational Settings," in: R. S. Parkinson and Associates, *Managing Health Promotion in the Workplace*, Mayfield Publishing, Palo Alto, Calif., 1982.

84. P. J. Rosch and N. H. Hendler, "Stress Management," in: R. D. Taylor Ed., *Health Promotion: Principals and Clinical Applications*, Appleton-Century-Crofts, Norwalk, Conn., 1982.

85. L. R. Murphy, *Occupational Stress Management: A Review and Appraisal,* NIOSH, Centers for Disease Control, January 1983 (unpublished).

86. O. F. Pomerleu, J. P. Brady, *Behavioral Medicine, Theory and Practice*, Williams and Wilkins, Baltimore, 1979.

87. B. C. Semands, *J. Occup. Med.*, **24,** 393–397 (1982).

88. R. K. Peters, H. Benson, D. Porter, Daily Relaxation Response Breaks in a Working Population, I, Effects on Self-Reported Measures of Health, Performance and Well-Being, *AJPH*, **67,** 946–953 (1977).

89. R. K. Peters, H. Benson, J. M. Peters, *J. Occup. Med.* **22,** 221–231 (1980).

90. R. K. Peters, H. Benson, J. M. Peters, II, Effects on Blood Pressure, *AJPH*, **67,** 954–959 (1977) through, R. S. Parkinson and Associates, *Managing Health Promotion in the Work Place*, Mayfield Publishing, Palo Alto, Calif., 1982.

91. P. Carrington, et al., *J. Occup. Med.* **22,** 221–231 (1980).

92. J. Manuso "Corporate Mental Health Programs and Policies," in: *Strategies for Public Health*, L. Ng and D. Davies, Eds., New York, Van Nostrand Reinhold Co., 1980, through R. S. Parkinson, *Managing Health Promotion in the Workplace*, Mayfield Publishing, Palo Alto, Calif., 1982.

93. J. E. Newman and T. A. Beehr, *Personnel Psychol.* **32,** 2–43 (1979).

94. "Study Methods for Increasing Safety Belt Use," Study Report, National Highway Safety Administration, U.S. Department of Transportation, March 1980.

95. P. F. Waller, L. K. Li, B. J. Campbell, M. L. Herm, "Safety Belts: The Uncollected Dividends," Highway Safety Research Center, National Highway Traffic Safety Administration, May 1977.

96. "Protecting Your Assets: A Program for Government and Industry to Reduce Losses from Auto Accidents by Increasing the Use of Safety Belts," Michigan Office of Highway Safety Planning, Lansing, Mich., N.D.

97. E. S. Geller, "Corporate Incentives for Promoting Safety Belt Use. Rationale, Guidelines and Examples," Transportation Safety Institute, Oklahoma City, 1983.

CHAPTER THREE

Occupational Health Nursing: An Interdisciplinary Approach

CAROL SILBERSTEIN, R.N., Ph.D.

Concern for the health and safety of workers has been expressed since early times. In ancient times the slave, criminal, or prisoner was assigned to those tasks which were known to be hazardous or illness-producing. Hippocrates wrote of the need to ask "What do you do for a living?" as part of the initial physical assessment of individuals coming to the physician with a complaint or illness. Galen, a Greek physician who lived about A.D. 200, recognized the problems of acid mists in the workplace and the subsequent development of pulmonary edema. In the 1500s Agricola identified the hazards of mining and the resultant diseases of cancer and silicosis. Ramazzini in 1700 published a systematic study of trade diseases including the health hazards of corpse bearers, midwives, stone cutters, weavers, runners, and others. Interest in occupationally related diseases continued but had little impact on health professions as a whole until well into the industrial revolution. The first efforts toward solving some of the problems related to the work environment were legislative in nature. In 1836 Massachusetts passed the first child labor law, which forbade the hiring of children under the age of 10. Although this law was officially in the books, there was no effort made at enforcement. It was not until 1914, when the Office of Industrial Hygiene, a branch of the Public Health Service, was founded, that action to improve working conditions was taken at the federal level—in the form of education about occupational health, research into occupationally related problems, and consultation services. Other industrialized countries followed similar patterns in developing codes and practices to improve conditions in the workplace.

Along with the efforts of social reform in the workplace, nursing began to branch out into the specialty of occupational health nursing. Betty Moulder was the first nurse in the United States employed specifically to care for workers and their families. She was employed by a group of coal miners who felt the need for health care directed toward the hazards of coal mining. Other employers followed suit and companies such as the Vermont Marble Company, Wanamakers, and the Metropolitan Life Insurance Company began offering nursing care to workers and their families, both on the job and in the home. As the specialty of occupational health grew, it seemed appropriate that New England, the birthplace of American industry, should also be the birthplace of the first organized association of occupational health nurses in the United States. In 1915 the first Industrial Nurses club was formed in Boston. The purpose of this organization was to promote the specialty of occupational health nursing and to provide opportunities for nurses to develop expertise in the area of occupational health. Recognition came slowly; in 1917 Boston University College of Business Administration offered the first education course designed for industrial nurses. Meanwhile, the industrial nurses recognized the need to attain increased knowledge in their chosen specialty and began to organize local, state, and regional groups. These activities culminated in the formation of the American Association of Industrial Nurses. In 1977 the organization formally changed its name to the American Association of Occupational Health Nurses, Inc.

Looking at the specialty of occupational health nursing, one sees the initiation of a practice discipline specializing in a unique health care arena. Prior to the 1970s occupational health nursing had not been taught as an integral part of nursing education programs, be they diploma programs, associated degree programs, or baccalaureate programs. The nurses practicing the specialty of occupational health had to identify their own educational needs, roles and functions, develop job descriptions, and learn new skills essential to their practice area. Industrial hygienists and specialists in occupational safety were also experiencing to a lesser degree these same educational problems in their quest to seek recognition as health professionals within the occupational environment. The trend is changing. Undergraduate nursing programs around the country are integrating concepts of occupational health throughout the baccalaureate program. Educational Resource centers for Occupational Safety and Health provide graduate education for nurses as well as physicians, industrial hygienists, safety officers, toxicologists, and other disciplines essential to the planning, implementation, and evaluation of occupational health programs. The registered nurse who enters the specialty of occupational health nursing begins with the skills acquired in the basic educational program. These may include communication skills, physical assessment skills, counseling techniques, basic knowledge of pathophysiology, health education, health screening, rehabilitation, nursing process, and nursing diagnosis. These are a good beginning but for the nurse who wishes to be a clinical specialist in occupational health nursing she/he must keep abreast of knowledge and skills in physical assessment, epidemiology, biostatistics, environmental control, safety, toxicology, and occupational health. Additional courses in management, labor relations, budgeting, research, and consultation would be helpful. Nurses who wish to assume leadership positions in occupational health must have these skills and knowledge and therefore should be prepared at the graduate level. The accreditation document, *Standards, Interpretations and Audit Criteria for Performance of Occupational Health Programs*, recommends that those nurses holding supervisory positions, or nurses in one nurse units in industry, have earned a master's degree in occupational health nursing or in public health nursing.

In an effort to maintain a high quality of nursing care for individuals employed in a variety of occupational settings, a certification exam has been developed by the American Board for Occupational Health Nurses (ABOHN). The ABOHN certification program is designed for the registered nurse working in occupational health regardless of the type of basic educational program from which she graduated. In order to qualify for the exam, the nurse must give evidence that he/she has completed an accredited basic nursing program, as well as the required number of continuing education credits, and has worked in occupational health for at least five years. Upon verification of these factors the nurse may sit for the certification exam and if the exam is successfully completed, the nurse is then certified to practice in occupational health and may use the initials COHN (certified occupational health nurse). The completion of this exam identifies the nurse as having met specific criteria to be recognized as a qualified and competent occupational health nurse.

Occupational health is a unique discipline in that it requires an effective interdisciplinary team effort in order to have a well designed, well managed health promotion program. All peer disciplines: nursing, medicine, industrial hygiene, safety, labor and management must develop channels of communication which allow for careful program development to best serve the needs of the industry, the worker and the consumer.

Each member of the occupational health team brings unique skills and knowledge to the assessment, planning, implementation, and evaluation of the total occupational health program. An effective communications network is vital to the functioning of this team. The dissemination of information about new processes, the introduction of new chemicals or any change in the production process or job requirements between members of the occupational health team are essential. Similarly, providing team members with information about advances in a specialty area, the development of new technologies, or research findings will provide individual team members with the ability they need to recognize when they need the unique skills or knowledge of other team members. When new procedures, screening measures, or new processes are initiated, input should be sought from other members of the occupational health team to identify the impact of these changes on the worker. The occupational health nurse brings a special expertise to this team approach. The nurse is skilled in assessing many aspects of the worker and the working environment, and can assist in identifying the biologic and psychologic stressors which may be present. The occupational health nurse, through a systematic approach to problem-solving called the nursing process, is able to make a nursing diagnosis, develop a plan of nursing care, implement that plan, and evaluate the effectiveness of that interaction in terms of the health of the worker and the benefit to the industry. This problem-solving approach is used by the occupational health nurse in situations designed for primary health care, emergency planning, hazard control programs, protection programs, education programs, health promotion programs, and counseling sessions.

In order for the interdisciplinary team to provide a quality occupational health program, several factors are vital. The interdisciplinary team must agree on a common philosophy of the total occupational health program. All members of the team must reach consensus on the scope of the program, the direction of the program, and the objectives of the program. Such questions as: is the occupational health program to be holistic in nature?, will care for occupational illness be the only care provided?, will the program focus on health promotion or disease prevention?, need to be answered. Once the interdisciplinary team has made the decision concerning the direction and scope of the occupational health program, the program development can begin. The first decision to be made is whether to implement a holistic occupational health program, a secondary occupational health program, or a tertiary occupational health program. A holistic health program includes the secondary and tertiary programs.

A holistic health program is designed to anticipate physiologic and psychologic manifestations of disease and to take measures to decrease the threat to health. The program is designed to promote optimal health and thereby provide the worker with a safe and healthful work environment. Such a program is global in scope and initially costly but proves itself cost effective over time.

Holistic health programs focus on the workers' responsibilities of participating in decision-making about their own health care needs. Opportunities are provided for health screening programs, counseling sessions, employee assistance programs, protection and educational programs. A total health care program will also provide the workers with the opportunity to participate in health programs to identify risk factors or early stages of major diseases which may not be related to the work environment. These include programs such as weight control, hypertension screening, cancer screening, diabetes detections, smoking cessation, or stress reduction. The worker who participates in a holistic program must meet individually with the occupational health nurse to identify any health care deficits. Goals for improving the health care of the worker are then set and agreed upon by both the nurse and the worker. The nurse may facilitate the planning of the worker's health care program. Worksite programs, exercise programs for home or work, stress reduction, and screening programs run by community agencies such as the Red Cross, the local chapter of the American Heart Association, or the American Cancer Society, should be made available to the individual worker. The holistic occupational health program would include pre-employment examinations, follow-up physical examinations and other screening or monitoring programs. Workers then may be placed in a job compatible with their health status.

If the decision is to implement a secondary health care program, a different approach is necessary. A secondary program focuses on the prevention of illness, treatment, and rehabilitation of workers who have occupationally related illness or injury. This program should retain the pre-employment screening process since data obtained during this screening will provide both the employer and the worker with baseline information concerning the applicant's health status and ability to safely perform the designated work. Educational and protection programs are offered to workers in areas of specific hazard recognition rather than in total health care. Secondary health care programs, then, focus attention on workplace protection programs such as elimination of health hazards through engineering designs, the control of exposures through engineering approaches, personal respiratory and hearing protection, identification of dermatologic problems, hazards to the reproductive systems, and assessment of the impact of workplace design on the individual worker.

Finally, if the workplace is small and cannot afford a comprehensive occupational health program, a tertiary health care program may be appropriate. Specific services of the interdisciplinary team may be provided as needed through consultative sources and governmental and community agencies. Efforts toward health care are directed at immediate emergency care and disaster planning rather than disease prevention. The occupational health nurse in such instances may be responsible for the education of selected workers and management personnel in the skills of emergency care and disaster planning. This program would include the development of cooperative agreements with community based health care providers, including emergency physicians, Red Cross, Civil Defense and hospital emergency centers. It is essential to communicate to these agencies the particular needs of the industry, including information concerning the physical, chemical, and biologic hazards of the workplace. Thus informed, these community agencies can facilitate and coordinate the necessary health care agencies in providing emergency care to the workers should a disaster situation arise.

What then are the direct responsibilities of the occupational health nurse? Five areas are readily identifiable: administration function, theory development, independent nursing practice, worker education and research activities. The occupational health nurse is the direct care provider in most occupational health programs. The other members of the interdisciplinary health team are available for consultation but it is often the nurse who directs the care provision of such programs. Therefore, the nurse must have skills in administration functioning. Authority must be delegated through the table of organization and specific channels of communication identified. Job descriptions must be formulated to carefully delineate the responsibility of the occupational health professionals. The head of the designated program should have the authority for decision-making within that structure including budgetary decisions, program planning, and directing health care providers.

In the area of theory development the occupational health nurse must be familiar with the current nursing theorists and be able to identify how these theories relate to the practice of nursing in the occupational health setting.

In the practice setting the occupational health nurse is an independent practitioner functioning within the area of nursing practice defined by the Nurse Practice act of the state in which she/he is employed. The occupational health nurse provides direct care to ill and injured workers. The nurse also provides opportunities for workers to participate in health screening and health education programs either at the worksite or in the community. Counseling of workers regarding problems on the job or problems in the home situations which may impact the job is a major function of the occupational health nurse.

Employee Assistance Programs have been developed within industries to address some of the major health problems which arise among the working population. The occupational health nurse is a member of the health team providing services to these programs such as referring the worker to a local mental health clinic, a finance officer for budget problems, or a community group to help with problems of substance abuse. In order for these programs to function effectively it is essential that first line supervisors be trained to recognized changes in behaviors of workers and bring these issues to the attention of the staff supporting the employee assistance program.

Preparation for retirement is another area of nursing function. Older workers are usually well informed by the company of the benefits package which they have accumulated over the years of employment. Health issues related to the aging process and adjustment to retirement are usually not addressed. Educational programs planned and implemented by the occupational health nurse can provide valuable information to the retiring employee. These programs should cover such topics as the physiology of aging, sexuality and aging, and the psychological impact of retirement.

Assessment programs which center on identifying the impact of the workplace on women's health are essential. With close to half of the employed individuals in the United States today being women it is important to gather specific data on exposure levels and physiological changes in women. The current threshold limit values established by the American Conference of Governmental Industrial Hygienists based on a 40-hour week, as well as the permissible exposure limits established by OSHA, are values for both male and female workers. There are insufficient data available to determine if some of these exposure levels are safe for women. With the changing social system, many industries are working 10- or 12-hour shifts rather than the previously accepted 8-hour shift. The effects of unusual work schedules on workers are now receiving increased attention.

A vital component of the practice of occupational health nursing is the utilization of the nursing process by the occupational health nurse in the assessment of many aspects of the work environment. This problem-solving approach allows the nurse as a member of the interdisciplinary term to participate in assessing many aspects of the workplace, planning an occupational health program, implementing the program and evaluating the results in terms of worker health and cost effectiveness to management.

The occupational health nurse has the skills to recognize many aspects of hazards in the work environment and must participate in walk-through surveys of the workplace with other members of the interdisciplinary team. Observations should be made in such areas as condition of first aid equipment, housekeeping activities, use of protective equipment, availability of sinks, eyewashes, etc. If the nurse has any concerns about the workplace conditions, these should be brought to the attention of the industrial hygienist and/or safety officer or other responsible person. During a walk-through survey, the occupational health nurse also has the opportunity for some informal assessment and education of individual workers. Follow-up observations or compliance with a prescribed medical regime may then be brought to the attention of responsible persons.

Development and implementation of educational programs for workers is an important function of the occupational health nurse. If the worker is aware of the risks involved in the job and properly trained to function effectively, occupational illnesses and injuries will decrease. Well-defined educational programs then become extremely cost effective.

Decisions concerning the impact of occupational health exposures are based on clinical experience, case studies, epidemiologic studies, and animal studies. There is need to establish a sound data base concerning the physiologic responses of workers, male and female, to specific exposures. The occupational health nurse can participate in this process by assisting in the development of a sound data base which will provide needed data about the exposure, including time, concentration and duration, the health status of the worker, the job structure, and physiologic responses as measured by sound scientific methodology.

The occupational health nurse provides direct nursing care for a healthy working population. The focus of that care is primary prevention with the ultimate goal being the maintenance of the worker's level of wellness.

An interdisciplinary approach to the occupational health program is an essential commitment. Knowledge of the principles of safety, industrial hygiene, toxicology and epidemiology are needed for the optimal functioning of the occupational health nurse. The nurse as a member of the interdisciplinary team may function as an administrator, an independent practitioner, a consultant, an educator, and a researcher.

BIBLIOGRAPHY

M. L. Brown, *Occupational Health Nursing*, New York, Springer (1981).

J. Doull, Klaasen, C. D., and Amdur, M. O., *Casarett and Doull's Toxicology*, 2nd ed. New York, Macmillan Publishing Co. (1980).

A. J. Finkel, *Hamilton and Hardy's Industrial Toxicology*. Boston, John Wright Publishing (1983).

L. Jarvis, *Community Health Nursing*, New York, F. A. Davis Co. (1981).

J. Lee, The New Nurse in Industry, U.S. Department of Health, Education and Welfare, National Institute of Occupational Safety and Health (1978).

W. Rom, *Environmental and Occupational Medicine*, Boston, Little Brown and Co. (1983).

D. C. Walsh, and R. H. Egdahl, *Women, Work and Health: Challenges to Corporate Policy*. New York, Springer-Verlag (1980).

C. Zenz, *Occupational Medicine Principles and Practical Applications*, Chicago, Year Book Medical Publishers (1975).

CHAPTER FOUR

Detecting Disease Produced by Occupational Exposure

PHILIP E. ENTERLINE, Ph.D.

1 INTRODUCTION

It is the nature of modern man to speculate on the causes of disease. Often this speculation centers on factors associated with work and the recognition of disease producing agents in occupational settings have resulted in striking health improvements in some groups of workers. Such conditions as lead intoxication, mercury poisoning, and benzol poisoning have largely disappeared because they could be easily identified with a particular substance and industrial hygiene measures were relatively easy to apply (1). For some occupational disease, such as bladder cancer in the dye industry and phossy jaw in the match industry, the use of certain substances was simply stopped in order to protect the health of the worker.

As the more obvious industrial hazards have been identified and brought under control, the task of identifying disease produced by occupational exposure has become increasingly difficult. The skills of many disciplines are now being brought to bear on the problem of identifying hazards that yet exist. Epidemiologists, biostatisticians, biomathematicians, toxicologists, and industrial hygienists have joined occupational physicians in studies designed to detect or predict occupational hazards.

One way of detecting and evaluating disease clusters in working populations is by a well planned and conducted epidemiologic study. As early as 1928 the framework for epidemiologic investigations of the effects of work exposures was laid down by Bridge and Henry (2). In order to detect a causal relationship between exposure and occupational cancer they proposed "that the incidence rate in the occupations under review should exceed that in the general population and that in the occupation concerned, there should be sufficient association of a worker with a substance proved experimentally to have carcinogenic properties." For scientific workers of that period this proved to be an almost impossible set of conditions. For example, in 1935 a pathologist named Gloyne reported two cases where both asbestosis and lung cancer occurred in asbestos workers coming to autopsy (3). In this report he quoted the conditions laid down by Bridge and Henry and recognized that these had not been met. He had no idea as to the incidence of lung cancer in asbestosis cases or in asbestos workers and no work had been done which proved experimentally that asbestos was a carcinogen. His conclusion was: "It seems worthwhile to record these two cases, not in any attempt to make out an etiological association of these two diseases, but in order to emphasize certain histological points in which one disease appears to bear on the other." Actually it took nearly 30 years before Bridge and Henry's conditions for establishing a relationship between asbestos and lung cancer were met, and recognition of the cancer hazards of asbestos were greatly probably delayed because of this (4).

We now have many resources for conducting both experimental and epidemiologic investigations that were not available to Gloyne. We have also expanded on the rules originally proposed by Bridge and Henry for detecting a causal relationship between exposure and occupational disease. The following criteria are now generally recognized as important in concluding that exposure and disease are causally related (5):

1. The consistency of an association. The same association found in a variety of settings is strong evidence of a causal relationship. Of particular importance are studies which differ in their potential confounding factors but which agree on a particular association.

2. The specificity of an association. If a substance is associated with not one but many diseases evidence for a causal relationship is weakened. In occupational studies specific agents are often associated only with a single disease.

3. The strength of an association. A very large excess in a particular disease in the presence of a particular exposure is not likely to be due to the fact that the epidemiologic method does not usually permit control or adjustment for all confounders. Thus large excesses are likely to be due to a true association.

4. A dose–response relationship. This is probably the single most important criterion for concluding that disease and exposure are causally related.

5. The demonstration of a temporal relationship. Exposure must precede disease by a time period believed to be reasonable to be causally related to a particular disease.

6. Statistical significance. This relates to the size of the study and the magnitude of the excess. Tests of statistical significance are valuable in judging the meaning of a particular association.

7. The existence of experimental data. There are very few agents that produce disease in man that have not been shown experimentally to produce disease in animals. The existence of experimental data greatly reinforces epidemiologic observations.

Not only are the rules for carrying out epidemiologic studies now better understood than at the time of Bridge and Henry, but we have reasons for carrying out occupational epidemiologic studies which are perhaps more pressing than those that prevailed during that period. Progress in discovering cures for some diseases has been disappointingly slow and there is a well recognized need to identify hazardous substances in our environment as a preventive measure. Moreover there recently has been a special interest in epidemiologic studies on the part of industry since industry is now being held by the courts to "have the knowledge of an expert" about the health effects of their products (6). This means that industrial managers need to be among the first to be aware of any health effects associated with their products. One way to do this is to carry out epidemiologic investigations on workers exposed to their products or exposed to ingredients in their products.

2 WHAT IS OCCUPATIONAL EPIDEMIOLOGY

Epidemiology has been defined as the study of the distribution and determinants of disease in human populations (7). Occupational epidemiology simply limits the populations studied to persons employed and focuses on the occupational environment as a possible determinant of disease. Epidemiology focuses on risks of disease in groups of individuals rather than in any single individual and is observational as opposed to experimental.

In order to evaluate epidemiologic data it is important to understand the nature of the experimental method of study. In an experiment involving the role of a particular substance in disease causation, subjects would be randomly assigned to exposed and nonexposed groups and then observed for disease. In making this random assignment all factors related to disease can be assumed, within statistically defined limits, to be identical in both groups *except* for exposure to the substance of interest. In an epidemiologic investigation, on the other hand, allocation between exposed and nonexposed groups is not random and these groups are likely to differ on many factors related to disease. Some of these factors may be known, such as age, sex, race, and smoking habits, and matching or adjustment for these factors improves comparability between exposed and nonexposed groups. Others are unknown and cannot be adjusted for. As Hill (8) puts it, in constructing an epidemiologic study "one must have the experimental approach firmly in mind." That is to say, insofar as possible, one must consider whether the matching or adjustment adequately deals with all of the important variables. If an important variable has not been properly dealt with then the study is confounded by that variable and results may be impossible to interpret.

3 TYPES OF EPIDEMIOLOGICAL STUDIES

Epidemiological studies designed to detect disease produced by occupational exposure can be classified into five categories, each of which contributes a different level of inferential knowledge concerning disease etiology. These levels are directly related to the extent to which confounding factors can be taken into consideration:

1. *Ecological studies.* These are studies in which an evaluation is made of the spatial or temporal patterns of morbidity or mortality in human populations and where all classifications are made on the basis of aggregates of individuals as distinct from single individuals. In this type of study individuals are not classifiable according to the study parameters, and thus this type of study does not permit a direct measure of association. An example would be a comparison of cancer mortality in counties classified according to the presence or absence of selected industries and the conclusion that cancer differences are related to the presence of a particular industry.

2. *Demographic studies.* These are studies in which an evaluation is made of the risk of morbidity or mortality in human populations composed of individuals classifiable by demographic characteristics such as age, sex, occupation, income, or education. In this type of study the source of data often differs for the numerators and denominators used in calculating rates. An example is the occupational mortality publications for England and Wales (9). Here death rates are calculated for selected causes and occupation where occupational counts are as enumerated in a population census for the denominators and are as recorded on death certificates for the numerators.

3. *Cross-sectional studies.* These are studies in which an evaluation is made of the prevalence of disease at a specified time among two or more groups, the individuals of which are classified by exposure, or some index of exposure at that specified time. An example would be comparing the prevalence of X-ray abnormalities in a group of workers exposed to coal dust with a group of workers not so exposed.

4. *Cohort studies.* These are studies in which an evaluation is made of the incidence of disease among two or more groups of individuals classified by level of exposure to specific agent or agents, either inferred or actually measured, with each group and followed over some period of time. An example would be identifying all men employed in a steel plant as of 1953 and tracing these to see how their mortality experience relates to their jobs (10).

5. *Case-control studies.* These are studies in which a comparison is made of the past exposure to an agent comparing two groups of individuals defined according to the presence or absence of a specific disease. Controls are usually intended to be representative of the population from which the cases arose. An example would be a study which compares the smoking habits of a group of lung cancer cases entering a hospital with the smoking habits of all or a sample of other hospital admissions.

3.1 Ecological Studies

Geographic variations in mortality are the basis for most ecologic studies and have formed the basis for determining the etiology of disease for a very long time. In John Snow's classic study of the 1849 cholera epidemic in London, it was the variation in mortality from cholera in different districts of London that formed the basis for his observation that cholera was related to contamination of the water supply (11). Then as now, however, there was disagreement as to the cause of variations in mortality. William Farr, regarded by Raymond Pearl as "the greatest medical statistician who ever lived," believed that the geographic variation had something to do with elevation, since death rates were highest in low lying areas (12). Fortunately Snow prevailed, pointing out that most elevated towns in England had also suffered excessively from cholera. In London itself, he argued, one area was a full 56 feet above the "Trinity high water mark" and had a death rate for cholera of 55 per 1,000, while many other districts "of less than half the elevation did not suffer one third as much." Had Farr looked further into the situation, he would have noted that a third variable was at work—the source of the water supply—and that if adjustment were made for this, the geographic variation could be largely accounted for.

As an example of some uses of ecologic data for the United States, in 1957 a special tabulation of deaths and population was prepared for the years 1949–1951 which made possible the computation of death rates by cause and by county of residence, with appropriate adjustment for age, race, and sex. This formed the basis for data shown in Figure 4.1. This shows death rates for coronary heart disease for each of 116 economic subregions in the United States (13). An economic region is defined as a group of U.S. counties in which people make a living in about the same way, and this seems to be an interesting grid against which to display health related data. In Figure 4.1 metropolitan counties (urban areas) are shown in black so that all of the variation displayed is in non-urban areas. An ecological study would attempt to explain this kind of geographic variation by associating it with some enumerated factor or factors.

More recently the National Cancer Institute has published cancer death rates by county for the years 1950–1969, accompanied by an atlas of cancer mortality (14). A number of hypotheses have been generated from these data. An example is that the presence of chemical and petroleum industries may be related to cancer death rates (15).

A problem with ecological studies is that they deal with the characteristics of

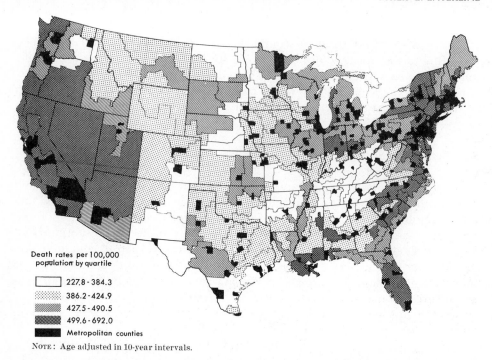

Death rates per 100,000
population by quartile

☐ 227.8 - 384.3
▨ 386.2 - 424.9
▨ 427.5 - 490.5
▨ 499.6 - 692.0
■ Metropolitan counties

NOTE: Age adjusted in 10-year intervals.

Figure 4.1. Coronary heart disease rates for 116 economic subregions, nonmetropolitan areas only, white males aged 45–64 (1949–51).

groups of people rather than with individuals. In a classic paper published in 1950 Robinson (16) pointed out that using states as the ecologic unit there is a strong positive correlation between the percentage of the population of each state that is foreign born and the income level in each state. From this, one might conclude that the reason is because foreign born persons have higher incomes than native born persons. Of course, just the opposite is true, and some other variables are obviously responsible for the positive association. The New England states have high incomes while the South has a relatively low income. The New England states also have a high percentage foreign born relative to the South. Clearly there are large differences between these two groups of states that need to be accounted for to make an ecologic study meaningful. The many problems with ecological studies have been extensively reviewed by Morgenstern (17).

Thus far, ecologic correlations have not provided much useful information with regard to occupational epidemiology (18). The problem is similar to the one faced by John Snow in 1849, and often we tend to work with about the knowledge of William Farr. Apparently we have been unable to adjust for the right variables. Until these can be identified and measured, progress in applying our knowledge of geographic variations in disease to the problem of disease etiology will be slow.

3.2 Demographic Studies

While William Farr missed the reason for the geographic variation in cholera observed by John Snow he is responsible for developing the basis for demographic studies of occupational mortality. In 1839 William Farr was appointed first head of the Office of the Registrar General for England and Wales. Using data available to him in that position he was able to combine counts of workers in various occupations derived from the federal census with counts from an item on occupation which appeared on death certificates. In his annual report for 1851 he made reference to mortality rates by occupation (19). Starting in 1861 mortality rates by occupation for England and Wales have been published every 10 years (except 1941). These publications are an excellent reference for providing clues to health hazards in various occupational groupings (9).

In the United States counts of deaths by occupation were published by the United States Bureau of the Census as early as 1890 and 1900, but no rates were computed. Complete tabulations were also made for the years 1910 and 1920 but were not released, probably because of the poor quality of information appearing on death certificates regarding the decedent's occupation. Just prior to 1930, State Vital Statistics offices took part in an intensive campaign to improve occupation information on death certificates in preparation for a study of mortality rates by occupation for the year 1930. Despite this emphasis, a large proportion of death certificates filed in 1930 contained unknown or nonspecific occupations. For this reason, mortality rates by occupation were published for only ten states where occupational data on death certificates appeared to be reasonably good (20). Death rates for selected occupations were derived by relating counts for items regarding occupation and industry appearing on death certificates filed during the year 1930 to counts on occupation and industry enumerated in the 1930 decennial census. Subsequent evaluation of the comparability of occupations, reported on death certificates and census returns for the same persons, showed a lack of correspondence, chiefly because of problems in classifying retired persons. For this reason plans for tabulations were dropped for 1940.

For 1950, State Vital Statistics offices were again encouraged to improve returns on the occupation and industry items appearing on the death certificate, and for deaths occurring in 1950, occupation and industry items were coded by the National Office of Vital Statistics of the United States Public Health Service for men 20 to 64 years of age (21). Earlier studies of the comparability of occupation and industry items on the death certificates, and on the census returns for the same individuals indicated that for males in this age group lack of comparability was not so serious. The only year for which mortality data are available by occupation and for the entire United States is 1950.

For 1960, an attempt was made to match a sample of death certificates filed during the months of May, June, July, and August with the April 1960 census returns, supplemented by additional information where matches could not be made (22). Classification of deaths by occupation and industry as enumerated in the 1960 census was thus possible, since the problem of noncomparability of occupation and industry items was largely eliminated. Unfortunately, the number of matches was disappointingly small, and numbers of deaths too small to provide very many useful details with regard to death rates by occupation.

Probably a better source of mortality data, from a technical standpoint at least, are statistics published by life insurance companies. These records have long been recognized as a valuable source of information on industrial hazards, particularly accidents. Two types of data have been published: data relating to experience with individual life insurance policies, and data relating to group life insurance policies. The earliest intercompany studies related to experience with individual policies. They were made possible by the pooling of records from several companies, coupled with the adoption in 1926 by insurance companies of a standardized method for classifying occupations. Studies of ordinary life policies provide mortality data by occupation, whereas studies of group policies provide mortality data by industry. Both types of data are published by the Society of Actuaries.

Data published by life insurance companies relate deaths to the occupation or industry, recorded at the time the policy was issued, and thus do not suffer from the defect of combining occupation and industry items from census returns and death certificates. On the other hand, they have the disadvantages that follow-up for death often relates only to the period an individual works and his life insurance policy is in effect. Three fairly comprehensive occupational studies have been published: the Joint Occupation Study of 1928 (23), the Occupational Study of 1937 (24), and the 1967 Occupation Study (25). In all three, major emphasis was placed on the experience of companies with ordinary policies considered substandard by reason of occupation but not otherwise substandard. That is they deal with individuals who pass the insurance physical examination but who must pay an extra insurance premium because of their occupation. The 1967 study also contains data on experience with group policies.

Demographic studies have offered some guidance in detecting disease produced by occupational agents. For example, in 1936 Kennaway and Kennaway identified several groups of coal carbonization and coal by-products workers with a high risk of lung cancer (26). These findings were later confirmed by cohort analysis (27). On the other hand, the 1936 study found not one single case of lung cancer in asbestos workers. This finding may have delayed the recognition of relationship between asbestos and cancer.

3.3 Cross-Sectional Studies

These are usually studies that relate to employee physical examinations and deal with disease prevalence at a particular point in time as opposed to incidence over some period of time. They are particularly appropriate for occupational diseases of long duration and usually involve relating physical examination findings to the duration and extent of exposure to various substances. Signs of exposure such as blood lead levels, urinary arsenic levels, and skin disorders are discoverable through cross-sectional studies. Signs of chronic respiratory disease including X-ray changes and pulmonary function abnormalities also lend themselves to study by cross-sectional methods. The identification of coal workers pneumoconiosis in American coal miners came about as the result of this kind of study (28).

Periodic physical examinations by employers are cross-sectional and usually deal with the prevalence of disease, although by linking successive examinations incidence data can often be obtained. The relationship between prevalence P and incidence I is

$$P = ID$$

where D is duration of disease. Whatever unit is selected for duration of disease it becomes the unit of time in which incidence is expressed. Duration is simply the period of time during which disease is detectable. It begins with the first detectable sign of disease and ends with death of the individual or the disappearance of the sign. Both ecologic and demographic methods can deal with either incidence or prevalence data.

Most diseases have an incidence, a prevalence, and a duration. Some statistics of interest, however, have no time dimension. Deaths and industrial accidents can be studied only as incidence data. A consequence of accidents, disability does, however, have a time dimension and can be presented as the prevalence of persons disabled due to accidents.

Cross-sectional studies have many advantages. One of these is the ability to plan for and collect a considerable amount of data in a fairly brief interval of time using uniform definitions for the characterization of disease and of exposure. In these studies it is also usually possible to enumerate and adjust for differences in the exposed and nonexposed segments of the population, so as to prevent confounding and facilitate statements about occupational factors causally related to disease.

One disadvantage in cross-sectional studies is that workers available for study at a particular point in time may have been selected in or out of the work force for health reasons. As a result, disease excesses related to work exposure can be muted or exaggerated.

3.4 Cohort Studies

Far better than ecological, demographic, or cross-sectional studies for most diseases, are cohort studies. These are investigations in which workers are individually identified and followed through time to determine their health experience. The term cohort refers to that part of a population born during a particular period and observed for its characteristics as it enters successive time and age intervals. In current usage, this notion has been broadened to describe any designated group of persons who are followed or traced over a period of time. Cohort analysis can be historical or concurrent. Historical cohort analysis involves identification of a cohort at some time in the past and observing its health experience up to the present. Concurrent cohort analysis involves identification of a cohort either historically or currently and observing its health experience on into the future.

Cohort analysis usually deals with the concept of life expectancy and can be directed to the question of whether there is any evidence that a particular environment has a life shortening or life enhancing effect. Secondarily it deals with the specific causes of death responsible for life shortening or enhancing effects and the environmental factors related to these causes. A life table method of analysis is used here (29). Where an internal comparison group is available, cohort analysis can also be viewed in the context of a case-control study. This kind of analysis involves developing four fold tables showing deaths and nondeaths for exposed and not exposed workers and summating these across stratifying categories such as age, time, and sex (30). Cohort studies may deal with morbidity as well as mortality (31). Some areas that need careful consideration in cohort analysis are discussed in the following sections.

3.4.1 Selection of a Cohort for Study

Usually only workers should be selected for whom a disease excess seems biologically pausable. If, for example, a possible effect is cancer, only populations which can be followed many years after some occupational experience should be studied since cancer is usually manifest only many years after exposure. Cancer occurring less than 5 years after initial employment is almost certainly not due to that employment. It would be pointless, therefore, to include recent hires in a situation where cancer is the health effect of interest—unless, of course, future follow-up of a cohort is anticipated and recent hires will be followed for a considerable time in the future. Populations selected for study should also:

1. Be a complete enumeration of all persons employed under a particular set of circumstances. Missing members of such a cohort may be those of most interest and result in bias of unknown magnitude and direction.

2. Have had a significant work period. Workers employed only a few days or weeks are usually not of interest—except perhaps for studying the effects of acute exposures. Short-term workers may have unusual personal or social characteristics which influence their health experience and make inferences about a particular exposure difficult.

3. Have available reliable demographic and work history data. In addition to job histories at the plant or location studied, job histories prior to that employment are often important. Where prior job histories are of possible importance but uncertain this might mean confining observations to workers hired at younger ages where extensive prior employment is unlikely. Demographic data should include sex, birthdate, social security number, and if possible race and country of birth.

4. Be traceable. Some working populations such as migratory farm laborers, World War II female industrial workers, foreign nationals and workers not covered by the U.S. Social Security System are very difficult to trace. Bias is likely in studies which include large numbers of such workers.

5. Have environmental exposures which distinguish the population studied from most other working populations. These exposures should be carefully evaluated and hypotheses developed in advance of the study. Such hypotheses might be based on animal investigations, case reports, or other epidemiologic studies. It is important here to separate the hypotheses testing from the hypotheses generated aspects of an epidemiologic investigation. Where epidemiologic studies are conducted because of individual case reports or disease clusters in the study population these cases cannot be used for hypotheses testing purposes (32).

6. Be large enough to yield meaningful results. This applies particularly where hypothesis testing is involved and requires calculation of the probability of detecting excesses or deficits in mortality of some specified size. These calculations are called power calculations and should be made not only for disease excesses or deficits, but also for slopes and trends to be used in the analysis. For example, the hypothesized slope of a dose–response relationship might form the basis for a power calculation. Power calculations are discussed in Chapter 3.

3.4.2 Abstracting Records

It is preferable that records of cohort members be selected and microfilmed by the person or organization assuming responsibility for the validity and scientific accuracy of a particular historical cohort epidemiologic investigation. This places the burden of proof as to the validity of observations on a single individual or organization. It also allows careful study of records in a research environment and reexamination of the data base as frequently as is necessary. It permits an external audit of data and facilitates case–control studies where these seem desirable. Abstracts should include job histories in sufficient detail to permit internal comparisons.

3.4.3 Verifying Cohort Completeness

As noted above, incomplete cohorts may provide misleading information— either missing excesses or deficits for certain causes of death or identifying excesses or deficits that do not in fact exist (33). Several methods for verifying cohort completeness exist (34):

1. Comparing names in company held records with names that appeared on Forms 941 (or SS-1) submitted to the federal government by industry for tax withholding purposes.

2. Comparing names in company held records with union seniority lists.

3. Searching for missing badge numbers where badge numbers are assigned consecutively, or in some other known order.

4. Comparing hire dates for the cohort with known hiring patterns.

5. Making internal comparisons—such as comparing names in personnel files with payroll lists or medical department files.

3.4.4   Entering Records for ADP

All records need to be put into machine readable form. This involves coding and data entry. Both coding and data entry should be verified. In addition, all data should be subject to a machine edit—looking for inconsistent items on demographic and other data. Where work histories are coded job or departmental codes need to be developed which are meaningful in the context of the study.

3.4.5   Follow-up

Cohort analysis usually involves establishing vital status for each cohort member as of some recent date. For cohort members, whose vital status is not known from company held records, the most productive resource for follow-up for deaths occurring prior to 1979 are records maintained by the U.S. Social Security Administration (SSA). Starting in 1979 the U.S. Department of Health, Education and Welfare has maintained a national death index, against which cases can also be cleared and which is more up-to-date than social security records (35). This may be used in lieu of or in addition to the Social Security files. Names, social security numbers, and birthdates submitted to SSA will be classified by them as (1) paying into the system as of some recent data and thus presumed to be alive; (2) death claim filed and presumed to be dead; (3) neither of the above. The third group must be traced by means other than the SSA. Some alternative sources include clearing with state drivers license bureaus, the Veterans Administration, or personal contact with the individual, his friends or relatives by mail or by phone. The goal is to account for every member of the cohort as of the study end date.

For known deaths, death certificates must be located and coded by a qualified nosologist. Death certificates not in company held records may be obtained from State Health Departments. Since place of death is not always certain a search is sometimes needed in several states.

Where deaths occur over a long period of time and where a comparison of the mortality experience in the cohort is to be made with some external population, say the entire United States, several revisions of the International Classification of Diseases may apply. This presents a problem since the rules for classifying diseases are changing and often there is no single set of rules which apply to the entire time period of the study. There are two views on how this can best be dealt with. One view is that somehow the lack of comparability in national death statistics among the several revisions can be overcome by combining diagnostic terms that seem to mean the thing across several revisions and then code all death certificates to the revisions in effect at the time the death occurred (36). The mortality experience of the cohort can then be expressed in terms of cause of death classifications that seem to mean the same thing across several revisions of the international lists. For many causes of death this seems to be a valid procedure.

The second view is that the existence of comparability ratios should be considered here. Comparability ratios are published for each revision and show for selected causes of death, the number of deaths that would be identified for each cause under the old and the new revision. These are calculated empirically by actually coding a sample of death certificates according to the old rules and according to the new rules, and then calculating the ratio of deaths coded to a particular cause under one revision to the number coded to the same cause under another. Given the existence of these comparability ratios, published vital statistics can be corrected to a single revision of the International Classification of Diseases and all deaths in a study, which extends over several revisions, can be coded to a single revision. For example, suppose deaths are observed over the period 1950–80. During these years there were four revisions of the International Classification of Diseases in effect. One took effect in the United States in 1949 (the sixth revision), one in 1958 (the seventh revision), one in 1968 (the eighth revision), and one in 1979 (the ninth revision). If all deaths occurring in the study population during the years 1950–80 were coded according to a single revision, a comparison with death rates published by the U.S. National Office of Vital Statistics for the entire period could be made by modifying these death rates, using the comparability ratios empirically derived for each new revision. If, for example, deaths in the study population were coded according to the seventh revision one would move forward in modifying U.S. rates for comparison purposes starting with deaths occurring in 1968 and backward starting with 1957.

Probably the second method is the most desirable since it is based on actual data rather than the judgments of an individual as to the meaning of terms under various revisions. Perhaps more important it requires coding cause of death only to a single revision of the international lists. It is difficult to find nosologists who can code to several revisions—particularly revisions dating back to 1948. Both methods are limiting, however, since neither method makes it possible to study every cause of death.

3.4.6 Analysis of Data

Data analysis should proceed in several ways. First it is important to clearly state the purpose of the study including hypotheses that are being tested, if any. A complete description of the history of the plant being studied, the processes used, and of the environment and exposures of interest is needed. The cohort must be carefully defined and some rationale for its selection given.

Assurance that the cohort is complete and follow-up adequate is needed. Demographic characteristics of the cohort should be given to the extent possible.

The mortality experience should be presented in relation to an internal control or to some standard population adjusted for confounding variables. Some of the important variables that could confound epidemiologic analysis in occupational studies include age, sex, race, time period, residence, smoking habits, and social class. The last three of these may be difficult to adjust for but can bias study results. Table 4.1 shows how cancer sites differ in their relationship to urban–rural residence (37). Table 4.2 shows how smoking habits relate to selected cancers (38). Table 4.3 shows how social class relates to cancer sites (37). Clearly, if the study population differs from the comparison population in these characteristics, differences in cancer will be observed which could be mistakenly attributed to some other characteristic, such as exposure.

Statistically significant excesses or deficits in deaths should be noted. Both internal and external comparison groups should be used if possible. For an exposure of interest, exposed and nonexposed members of the cohort should be compared if possible. In addition, comparisons with the mortality experience of total populations as published by health departments or the U.S. National Office of Vital Statistics are useful, since these are based on large numbers of persons and considerably increase the statistical power of the study. Local populations should be used for comparison wherever possible due to such geographic variations in mortality as shown in Figure 4.1. Where internal comparisons are possible these are often more useful than comparisons with external populations due to the selective nature of employment. However, cohorts may be too small to permit useful internal comparisons or the entire cohort may be considered exposed so that no internal comparison group is available.

Discussion should be directed to hypotheses or other reasons for the investigation. For causes of death which seem elevated, or which test some prior hypotheses, data should be displayed by duration of exposure and, if possible, type of exposure. Often job or department can be used as a surrogate

Table 4.1 Urban/Rural Ratios of Age-Adjusted Cancer Mortality Rates, Selected Sites, 1950–69

Site	Males	Females
All cancer	1.6	1.4
Esophagus	3.1	2.1
Rectum	2.7	2.1
Bladder	2.1	1.6
Lung	1.9	1.6
Breast	1.8	1.6
Stomach	1.4	1.4
Leukemia	1.1	1.2

Table 4.2 Smoker/Nonsmoker Ratios of Age-Adjusted Cancer Mortality Rates

Site	Ratio
Buccal cavity	8.9
Lung	6.8
Larynx	5.1
Esophagus	3.2
Pancreas	1.7
Bladder	1.0

for type of exposure while duration of work in that job or department can be used for duration of exposure. Sometimes the levels of exposure can be estimated. For cancer, time since first exposure or hire should be shown. A useful table is one that displays mortality ratios or excesses for a particular cause by time since first exposure and duration of exposure.

Special consideration in the analysis should be given to the possible consequence of worker selection (39). Some companies tend to hire unhealthy workers, or workers with habits that lead to poor health; while other tend to hire healthy workers with desirable personal health habits. In addition, workers tend to select themselves out of employment often on health or health related factors. Thus, working populations may tend to become healthy or unhealthy due to this second kind of selection, and this needs to be considered in the analysis (40). Wherever possible, information on health related personal or social factors should be presented and discussed.

When life table methods are used for the analysis, at least two computer programs are available and they should be strongly considered in preference to writing an entirely new and untested program (36, 41). Great care should

Table 4.3 Social Class Ratios of Age-Adjusted Cancer Mortality Rates, Selected Sites, 1950–69

Site	Ratio (high/low)	
	Male	Female
Rectum	2.1	1.7
Colon	1.7	1.5
Bladder	1.7	1.2
Kidney	1.5	1.1
Esophagus	1.5	1.0
Breast	1.2	1.5
Lung	1.1	1.3

be taken in setting appropriate starting points for person-year counts. In general, person-year accumulations should only take place while populations can be observed for death or other events of interest. If mathematical models are to be used for the analysis, care should be taken that the models are appropriate for the data set. In particular, consideration should be given as to whether response should be measured on a relative or an absolute scale.

Care should be taken to separate hypotheses tested from hypotheses generated. Regarding the latter, the multiple comparison problem needs to be considered. Simply stated, if enough comparisons are made something is sure to be statistically significant and findings need to be interpreted with this in mind. For a specific substance related hypothesis the general strategy should be to relate the mortality experience in the most highly exposed group to the least exposed group.

Deficits in mortality should not be overlooked. For nonhypothesized differences, and where the comparison population is appropriate, deficits should about balance excesses. When external comparisons are made the overall mortality rates in worker populations will usually be 70–90 percent of the mortality rates in the general population.

3.4.7 Credibility

It is important where there is an industrial sponsor whose products are being examined for health effects that study results not be subject to control by the sponsor. This includes stop/go rules where the sponsor is given the option of terminating the project depending on study outcomes. It is important that the study protocol be set in advance and that this be adequate to answer an agreed upon set of questions. The question addressed by the study should be considered carefully. For example, the question might be asked: Did the cohort have a death rate higher than the U.S. population living at the same ages and in the same time period? This may be superficial and is very different from the question: Is there any evidence that workers are harmed by a particular set of work exposures and conditions?

3.5 Case–Control Studies

Perhaps case–control studies are best described by examining the following table where N is the total number of individuals studied and the lowercase letters a, b, c, d refer to the numbers of individuals in a particular cell:

	Exposed	Not Exposed	
Diseased	a	b	$a + b$
Not diseased	c	d	$c + d$
	$a + c$	$b + d$	N

Usually case–control studies start, with workers known to have a particular disease and workers not known to have the disease, and compares their work exposures. These studies contrast with a cohort study which usually starts with workers believed to have a particular exposure and workers not believed to have had that exposure and where workers are followed and the incidence of disease in the two groups is observed.

In cohort analysis, disease incidence is expressed as:

$$\text{Exposed} \quad \frac{a}{a + c} \qquad \text{Not Exposed} \quad \frac{b}{b + d}$$

and where the ratio R of disease incidence in the exposed and nonexposed groups is calculated as

$$R = \frac{a/a + c}{a/b + d}$$

In a case–control study, the exposure of cases of disease is usually compared with the exposure of controls. Usually controls are selected so as to be representative of the population that produced the cases of disease. Unless the controls are an actual sample with a known sampling ratio, however, the actual incidence rate for disease cannot be calculated directly. Ordinarily what can be calculated is the ratio of the odds of exposures in the case group at the odds of exposure in the control groups O. This is equal to the ratio of the odds of disease in the exposed group to the odds of disease in the not exposed group:

$$O = \frac{a/b}{c/d} = \frac{ad}{cb} = \frac{a/c}{b/d}$$

This is often called an estimated relative risk R since $(a/c)/(b/d)$ approaches $(a/a + c)/(b/b + d)$ where controls are properly selected and the disease or condition is not highly prevalent. When controls are selected so as to be representative of the population that produced the cases, they are called population based controls. If the sampling ratios F were known, actual disease probabilities could be calculated along with the ratio of these probabilities as in cohort analysis:

$$R = \frac{a}{a + Fc} \bigg/ \frac{b}{d + Fd}$$

Usually, however, the sampling ratio F is unknown.

Where exposure is unique to a particular disease or cause of death, something less than a population based set of controls may be adequate. Many case–control studies, for example, compare the exposure of persons dying from a particular disease with the exposure of persons dying of other diseases, and in many situations these appear to be adequate.

Case–control studies have an advantage in that data collection efforts can be limited for the denominators that make up incidence or prevalence rates. For example, controls might be stratified random samples drawn from these denominators, with the total number of controls only a small multiple (2–5) of the cases and with stratification on variables which relate to disease. To illustrate, suppose age is related to disease in such a way that cases are older than the population that produced them. Selecting controls distributed by age like the cases would eliminate the age difference when cases and controls are compared.

Limiting the study size by drawing samples from the denominators and thus creating a case–control study that allows focus on a large number of variables, also introduces some efficiency in the application of analytic procedures. For example, logistic regression procedures or the Cox proportional hazards model are most efficiently applied to small number sets and deal inefficiently with a large data base (42).

A combination of a case–control with a cohort analysis can be very efficient. Such a study would involve collecting a limited amount of information for an entire cohort and much more detailed information on the diseases of interest and a sample of the entire cohort. The sample would then be used as a control group for the cases. Such a study is called a nested case-control study (43).

All of the detailed comments on cohort analysis apply to case–control studies and, in fact, when it is possible to select population based controls case–control studies should be thought of as efficient cohort studies.

3.6 Some Comments on Comparisons

Probably no aspect of occupational epidemiology is more important than comparing disease in populations at various dose levels using properly constructed historical dose estimates. This comment applies equally to cohort analysis and case-control studies. Obviously, cohorts can be assigned to exposure dose and type categories where adequate historical data are available. Perhaps not so obvious is the fact that cases and controls can also be placed in various dose type and level categories and an odds ratio calculated for each. Thus, an estimated relative risk would be available for each dose type and level.

Usually comparing with large external populations (say the entire U.S. population) greatly increases the probability of detecting diseases produced by occupational exposures, because of the magnitude of the external population and thus the stability of expected rates generated from such populations. Such comparisons are often inappropriate for a particular population, however, due to selective factors. In some industries, for example, these can be very powerful. Consider, for example, how a personnel manager might fill say 30 vacancies given 300 applicants. Many of the factors that predict good health such as a stable family situation, good job histories, no problem with alcohol and perhaps an associated problem with cigarette smoking, apparent good health, and enough drive and desire to be in the job market, all may enter into a decision to hire a particular individual. Rarely would it be possible to measure and control for, or even identify all of these variables. Under this kind of circumstance there is no reason to believe that the expected long term health experience of the employed individuals should be like that of the general population. Perhaps the only reasonable comparison group in this situation is an internal control group without exposure to the substance of interest.

4 SUMMARY

Occupational epidemiology presents an opportunity to identify substances which are primary in the causation of human disease. For some diseases, such as certain cancers, identification of such substances seems to offer the only hope for reducing the burden of these diseases in the next few decades. Some factors that need to be considered when carrying out occupational epidemiologic investigation have been presented.

REFERENCES

1. Donald Hunter, *The Diseases of Occupations*, sixth ed., Hodder and Stoughton, publisher, 1978.
2. J. C. Bridge, and S. A. Henry, "Industrial Cancers," in: *Report of the International Conference on Cancer, London, July 1928,* William Wood & Co., Baltimore, 1928.
3. S. R. Gloyne, "Two Cases of Squamous Carcinoma of the Lung Occurring in Asbestosis," *Tubercle,* **17,** 5 (1935).
4. P. E. Enterline, "Asbestos and Cancer: The International Lag" (Editorial), *Am. Rev. Resp. Dis.,* **118,** 975–978 (1978).
5. U.S. Department of Health, Education and Welfare, *Smoking and Health,* PHS Publication No. 1103. U.S. Government Printing Office, Washington, D.C., 1963.
6. *Borel* v. *Fibreboard,* 493 F 2d, 1976.
7. B. MacMahon and T. F. Pugh, *Epidemiology: Principles and Methods.* Little, Brown, Boston, 1980.
8. A. B. Hill, "Observation and Experiment," *New Engl. J. Med.,* **248,** 995–1001 (1953).
9. *Occupational Mortality.* The Registar General's decinnial supplement for England and Wales, 1970–72. Series DS no. 1. Her Majesty's Stationery Office, London, 1978.
10. J. W. Lloyd, "Long-Term Mortality Study of Steelworkers, V. Respiratory Cancer in Coke Plant Workers," *JOM,* **13**(2), 53–68 (1971).
11. J. Snow, *Snow on Cholera,* Hafner Publishing, New York, 1965.
12. R. Pearl, *Introduction to Medical Biometry and Statistics,* W. B. Saunders, Philadelphia, 1940.
13. P. E. Enterline, A. E. Rikli, H. I. Sauer, and M. Hyman, "Death Rates for Coronary Heart Diseases in Metropolitan and Other Areas," *Pub. Health Rep,* **75,** 759–766 (1960).
14. *Atlas of Cancer Mortality for U.S. Counties: 1950–1960,* U.S. Department of Health, Education and Welfare, DHEW Publication No. (NIH) 75–780.
15. W. J. Blot, L. A. Brinton, J. F. Fraumeni, and B. J. Stone, "Cancer Mortality in U.S. Counties with Petroleum Industries," *Science,* **198,** 51–53 (1977).

16. W. S. Robinson, "Ecological Correlations and the Behavior of Individuals," *Am. Sociol. Rev.,* **15,** 351–357 (1950).

17. H. Morgenstern, "Uses of Ecologic Analysis in Epidemiologic Research," *AJPH,* **72**(12), 1336–1344 (1982).

18. P. E. Enterline, "Some Observations on Geographic Variations in Mortality," Proceedings of the 13th Annual Conference on Trace Substances in Environmental Health, Columbia, Missouri, June 5, 1979, pp. 19–27.

19. *Registar General Fourteenth Annual Report of the Registar General of Births, Deaths, and Marriages in England,* His Majesty's Stationery Office, London, 1855.

20. L. Whitney, *Death Rates by Occupation Based on Data of the US Census Bureau,* National Tuberculosis Association (1934).

21. L. Guralnick, *Mortality by Occupation and Industry Among Men 20 to 64 Years of Age: United States, 1950,* U.S. Department of Health, Education and Welfare. Vital Statistics-Special Reports, Vol. 53, No. 2, September 1962.

22. U.S. Department of Health, Education, and Welfare, *The 1970 Census and Vital and Health Statistics—A Study Group Report of the Public Health Conference on Records and Statistics,* Public Health Service Publication No. 1000, Ser. 4, No. 10, 1969.

23. The Actuarial Society of America and the Association of Life Insurance Medical Directors, *Joint Occupation Study 1928.* The Actuarial Society of America and the Association of Life Insurance Medical Directors, New York, 1929.

24. The Actuarial Society of America and the Association of Life Insurance Medical Directors, *Occupational Study 1937,* Globe Printing Company, New York, 1938.

25. Society of Actuaries, *1967 Occupational Study,* Society of Actuaries, Chicago, 1967.

26. N. M. Kennaway and E. L. Kennaway, "The Incidence of Cancer of the Lung and Larynx," *J. Hyg. Camb.,* **36,** 236 (1936).

27. R. Doll, R. E. Fisher, E. J. Gammon, et al., "Mortality of Gasworkers with Special Reference to Cancers of the Lung and Bladder, Chronic Bronchitis, and Pneumoconiosis," *Br. J. Ind. Med.,* **22,** 1–12 (1965).

28. P. E. Enterline, "The Effects of Occupation on Chronic Respiratory Disease," *Arch. Environ. Health,* **14,** 189–200 (1967).

29. S. J. Cutler and F. Ederer, "Maximum Utilization of the Life Table Method in Analyzing Survival," *J. Chronic Disease,* **8,** 699–712 (1958).

30. N. Mantel and W. Haenszel, "Statistical Aspects of the Analysis of Data from Retrospective Studies of Disease," *J. Nat. Cancer Inst.,* **22,** 719 (1959).

31. W. C. Cooper, P. E. Enterline, and E. T. Worden, "Estimating Occupational Disease Hazards through Medical Care Plans," *Pub. Health Rep.,* **77**(12), 1065–1070 (1962).

32. P. E. Enterline, "Evaluating Disease Clusters," Ted Hatch Lecture, September 1983, in: *Proceedings of the Ted Hatch Symposium,* Princeton Scientific Publications, Inc. (in press).

33. P. E. Enterline and G. M. Marsh, "Missing Records in Occupational Disease Epidemiology," *JOM,* **24**(9), 677–680 (1982).

34. G. M. Marsh and P. E. Enterline, "A Method for Verifying the Completeness of Cohorts Used in Occupational Mortality Studies," *JOM,* **21**(10), 665–670 (1979).

35. *User's Manual, The National Death Index.* U.S. Department of Health and Human Services, DHHS Publication No. (PHS) 81-1148, 1981.

36. R. R. Monson, "Analysis of Relative Survival and Proportional Mortality," *Comput. Biomed. Res.,* **7,** 325–332 (1974).

37. R. Hoover, T. J. Mason, F. W. McKay, and J. F. Fraumeni, "Geographic Patterns of Cancer Mortality in the United States," in: *Persons at High Risk,* J. F. Fraumeni, Ed., Academic Press, New York, 1975.

38. E. C. Hammond, "Tobacco" in: *Persons at High Risk of Cancer, Etiology and Control,* J. F. Fraumeni, Ed., Academic Press, New York, 1975.

39. C. C. Seltzer and S. Jablon, "Effects of Selection on Mortality," *Am. J. Epidemiol.,* **100**(5), 367–372 (1974).

40. A. J. Fox and P. F. Collier, "Low Mortality Rates in Industrial Cohort Studies due to Selection for Work and Survival in Industry," *Br. J. Prev. Soc. Med.,* **30,** 225–230 (1976).

41. G. M. Marsh and M. E. Preininger, "OCMAP: A User Oriented Occupational Cohort Mortality Analysis Program," *Am Stat.,* **34,** 245 (1980).

42. N. E. Breslow and L. Pallan, *Case-Control Analysis of Cohort Studies in Energy and Health,* N. E. Breslow and A. S. Whittemore, Eds., SIAM, Philadelphia, 1979.

43. G. M. Marsh, "Mortality among Workers from a Plastics Producing Plant: A Matched Case Control Study Nested within a Cohort Study," *JOM,* **25,** 219 (1983).

BIBLIOGRAPHY

N. Breslow and N. E. Day, *Statistical Methods for Cancer Epidemiology,* Vol. 1, *The Analysis of Case–Control Studies.* IARC Scientific Publications No. 32, IARC, Lyon, 1980.

A. R. Feinstein, *Clinical Biostatistics,* C. V. Mosby, St. Louis, 1977.

D. G. Kleinbaum, L. L. Kupper, and H. Morgenstern, *Epidemiologic Research: Principles and Quantitative Methods,* Lifetime Learning Publications, Belmont, Calif., 1982.

J. M. Last, *A Dictionary of Epidemiology,* Oxford University Press, New York, 1983.

E. T. Lee, *Statistical Methods for Survival Data Analysis,* Lifetime Learning Publications, Belmont, Calif., 1980.

J. C. McDonald, *Recent Advances in Occupational Health,* Churchill-Livingstone, New York, 1981.

R. R. Monson, *Occupational Epidemiology,* CRC Press, Boca Raton, Fla., 1980.

P. Decoufle, *Occupation,* in: *Cancer Epidemiology and Prevention,* Philadelphia, D. Schottenfeld and J. F. Fraumeni, Eds., W. B. Saunders Co., 1982.

CHAPTER FIVE

Health Surveillance Programs in Industry

W. CLARK COOPER, M.D.

1 INTRODUCTION

Health surveillance programs in industry can be directed primarily at general health maintenance and the appropriate placement of workers, or the major emphasis can be on hazards in the workplace. In the latter case employees are periodically observed with the objective of prevention or early detection of harmful effects of such hazards in individual workers (1) or in groups of workers.

This chapter emphasizes hazard-oriented medical surveillance. Such activity, however, should be part of a broader program of health maintenance. Fragmentation of the medical management of employees should be avoided. The effective incorporation of specialized elements into a more general health examination program is the preferred approach. For this reason it is not feasible to discuss hazard-oriented examinations out of the context of examinations that detect preexisting conditions or intercurrent abnormalities that are unrelated to the work situation.

2 OBJECTIVES OF HEALTH SURVEILLANCE

2.1 General

An occupational health and safety program has many elements. Physical examinations of employees are only part of a comprehensive program. Examinations do not in themselves prevent illness or injury, though they may contribute directly to prevention. They should be designed and performed to secure maximum preventive benefits at minimal cost and inconvenience.

The following sections outline the types of examination that are commonly performed.

2.1.1 Preplacement Examinations

Preplacement examinations are those performed on all otherwise qualified applicants prior to initial employment, to aid in their proper job placement. However, the examinations often have been performed with the objective of obtaining the most physically fit work force that is available and to exclude individuals who have medical or psychological problems. This practice was defended by pointing out the cost of hiring and training individuals who could not perform available jobs, the risk of potential worker compensation costs, and the need to protect a medical care program from individuals with preexisting medical problems. Current policies for employing the handicapped have made such exclusions socially and legally unacceptable (2).

As will be pointed out in subsequent sections, the preplacement or preemployment examination serves an essential function in health surveillance, providing an historical record of previous exposures, state of health prior to joining a work force, and a baseline for comparison with later health observations. To a core preemployment history, physical examination, and laboratory appraisal can be added elements tailored to specific hazards of the plant or job under consideration.

2.1.2 Preassignment Examinations

When an individual is being transferred from an operation to another with a known hazard, or when a new process is to be begun, it may be necessary to carry out special inquiries, examinations, or tests. These have the objectives of sizing up individuals with respect to susceptibility and supplying baselines for later observations. They also offer an opportunity for worker education about potential health hazards associated with the new assignment or exposure.

2.1.3 Periodic Examinations

As related to individual workers, periodic examinations are administered to detect incipient disease, physiologic changes, biochemical deviations, or evidences of absorption of toxic agents, and to provide interim reappraisals of health. They provide opportunities for reinforcing education. Periodic surveillance of a group of workers having similar exposures may identify effects that are not of sufficient magnitude to be significant in an individual worker but which suggest or indicate an increased risk when they occur in several members of a group. The absence of positive findings is equally useful as part of the monitoring of preventive programs.

2.1.4   Termination Examinations

Examinations at the termination of employment are desirable to document health status at the end of exposure and provide evidence of any changes that have occurred during the employment period. There is, of course, no way to rule out the possibility that effects may show up at a later date. For example, asbestosis can progress years after the last exposure to asbestos, and cancer from a chemical carcinogen may not appear for decades.

2.1.5   Special Purpose Examinations

Other examinations, which may or may not coincide with and be supplementary to preemployment or periodic examinations, fill the requirements of special purposes, such as the following:

1.   Requirements of regulatory agencies, such as those of the Department of Transportation for vehicle operators in interstate commerce, or of the Federal Aviation Agency or other such body.
2.   Evaluation of the effects, if any, of accidental overexposure.
3.   Evaluation of recovery before return to work after an absence for illness or injury.
4.   Evaluation of the health status of an employee who has difficulty in performing work satisfactorily, in the absence of specific exposures to any known hazard.
5.   Determination of impairment of function or disability after complaint of such impairment or disability.
6.   Certification of fitness to wear a respirator.

3   SURVEILLANCE FOR GENERAL HEALTH MAINTENANCE

3.1   Content and Scope

3.1.1   Multiphasic Screening

The cost-effectiveness and the long-term benefits of periodic examinations of apparently healthy individual have been subjects for debate for many years (3–7). Modern technology and the efficient use of paramedical personnel make it possible to obtain a great deal of information in a short time at relatively low cost. The American Medical Association in a Council Report published in 1983 (6) provided general guidelines which included recommendations for medical evaluations at intervals of 5 years until age 40, with shorter intervals until age 65, when annual revaluations were suggested. Concern for the risks of unnecessary exposure to radiation has led to strong recommendations against routine chest radiography (7, 8). Most thoughtful students of the subject agree that multiphasic screening, if used judiciously, can provide the occupational physician with a valuable tool. When used periodically it may occasionally provide life-promoting information, such as the early detection of a curable cancer or asymptomatic hypertension. It is important that such screening be regarded as a tool, with regard for quality control, an understanding of the so-called normal or average values, a relaxed attitude toward minor deviations from the average range, and vigorous follow-up of the findings that are important.

3.1.2 Educational Value

As in all medical examinations, the opportunity for education of the patient or worker is one of the major benefits of periodic examinations. These opportunities are inseparable from those for hazard-oriented examinations.

4 HAZARD-ORIENTED MEDICAL EXAMINATIONS

It is necessary to consider hazard-oriented medical surveillance in terms of (1) what is legally required; (2) what has been recommended by official agencies such as the National Institute for Occupational Safety and Health (NIOSH); and (3) what has persuasive medical justification. The resulting programs are not necessarily the same. The objectives of medical surveillance are seldom clearly defined in regulations. One can question the wisdom of mandating detailed examinations of thousands of workers whose exposures are below a standard which is so low that hazard-derived abnormalities would be extremely rare. Nevertheless, if a mandatory medical examination is incorporated in a standard, the requirement must be met faithfully and with maximum attention to the quality of information and its relationship to the work exposures.

4.1 Legally Required Medical Surveillance

4.1.1 OSHA Requirements

All permanent standards of the Occupational Safety and Health Administration (OSHA) contain requirements for medical surveillance. As of mid-1983, there were 21 such permanent standards (9) as follows:

1. Asbestos
2. 4-Nitrobiphenyl
3. Alpha-naphthylamine
4. Methylchloromethyl ether
5. 3,3'-Dichlorobenzidine (and its salts)
6. bis-Chloromethyl ether

7. Beta-naphthylamine
8. Benzidine
9. 4-Aminodiphenyl
10. Ethyleneimine
11. Beta-propiolactone
12. 2-Acetylaminofluorene
13. 4-Dimethylaminoazobenzene
14. N-nitrosodimethylamine
15. Vinyl chloride
16. Inorganic arsenic
17. Lead
18. Coke oven emissions
19. Cotton dust
20. 1,2-Dibromo-3-chloropropane
21. Acrylonitrile

These stipulate essential components of the medical examination as well as requirements for access to records and record retention. They vary in degree to which the details of individual tests are defined. For a number of definite or suspect carcinogens there must be consideration of medical evidence for increased risk such as "reduced immunological competence, steroid or cytotoxic treatment, pregnancy and cigarette smoking." Others such as that for vinyl chloride specify the laboratory tests that must be carried out. The regulations for coke oven emissions and arsenic require periodic sputum cytologic examinations. A physician who has responsibility for workers with potential exposures to any of these substances at levels where examinations are required should consult the appropriate regulations for details. He must also be aware that in some states, as in California, the state occupational safety and health agency may have regulations which differ from and are more stringent than those of the federal agency.

4.1.2 Requirements of Other Regulatory Agencies

The Federal Coal Mine Health and Safety Act of 1969 (10) was landmark legislation in its requirement for medical examination of coal miners. It set schedules for chest roentgenograms, required the Secretary of Health, Education and Welfare to prescribe classification schemes for radiographic interpretation, and tied compensability to film interpretations. Regulations developed under the Act added provisions relating to the training and proficiency of physicians who interpreted films and specified methods of measuring pulmonary function, among other detailed requirements.

The legislative mandate (PL 95-164) for the Mine Safety and Health Administration in 1977 (11) included the provision that its standards shall "where appropriate, prescribe the type and frequency of medical examinations or other tests which shall be made available by the operator." In operations that fall under the jurisdiction of MSHA, therefore, the physician must be aware of its special requirements.

4.2 Medical Surveillance Recommended by NIOSH

The NIOSH criteria documents, which recommend standards for occupational exposures, always include recommendations to OSHA for medical surveillance. In nearly all cases these prescribe mandatory surveillance. The objectives (e.g., for epidemiologic studies, for early detection and treatment, for detection of group risk factors, or merely for good occupational medical practice) are rarely stated. These do not have the force of regulations, but constitute a powerful coercive influence, suggesting as they do a level of good practice. If used critically and with careful consideration of costs and benefits, they are useful in developing plans for surveillance.

In 1974 NIOSH commissioned the preparation of so-called mini-criteria documents, in which the salient toxicologic and occupational health control factors for nearly 400 chemicals in the occupational environment were summarized in Draft Technical Standards. Abbreviated recommendations for biologic monitoring and medical examinations were included.

Proctor and Hughes (12) used the foregoing to provide a valuable summary to aid the occupational physician in the preplacement, preemployment, or preassignment examination of workers, as well as for special tests and systems to be investigated in periodic surveillance. For 180 of 398 substances in their summary, the authors did not see a need for periodic physical examinations solely on the basis of exposures to a specific agent. They stressed that a brief interim history is usually sufficient for such agents.

4.3 Other Sources for Recommendations

The occupational physician will find that in the periodic revaluation of workers most frequently the systems of major concern are the respiratory tract, the liver, the blood-forming organs, and in recent years the reproductive function. In a general review it is impossible to expand upon currently recommended practices, but recommendations are available from many sources. The American Thoracic Society, for example, has developed detailed recommendations for surveillance for respiratory hazards (13).

4.4 Justification for Hazard-Oriented Medical Surveillance

There are several medical justifications for hazard-oriented surveillance, even though not all justify mandatory inclusion in standards. First, there are biologic indicators of absorption of toxic agents, based on analysis for the agent or a metabolic product in expired air, urine, or blood. Second, there can be the detection of minor physiologic changes, reductions of function, or early pathologic changes that may be early manifestations of toxicity in individual workers or groups of workers. Third, there may be indicators of hypersusceptibility, which can lead to special recommendations for certain individuals in the work force. Fourth, this history or findings may detect factors in the lifestyle of the worker, such as cigarette smoking, alcohol consumption, hobbies, or other activities, which should be discussed candidly with the individual as related to work exposures. Finally, every contact with the health establishment provides an opportunity for education of the workers.

4.4.1 Indicators of Absorption

Biologic monitoring is important in medical surveillance (14). Because the total exposure or absorption of a toxic chemical in an individual can result from a combination of inhalation, ingestion, and skin absorption on the job as well as absorption from nonoccupational sources, the measurements of blood or urinary concentrations can be very important for the individual. The best-established case is that for lead, but other metals such as mercury and cadmium are also examples. The relationship between blood or urine levels and exposure must be understood by the physician or industrial hygienist. The experience of a group is often more significant than that of the individual, but in either case abnormally high values show the need for more careful environmental evaluation.

Urine and blood are not the only biologic materials used for monitoring. Expired air can provide an index of absorption of a number of solvents. Hair has been used as an indicator of past exposures to heavy metals. This subject area is covered in more detail in Volume III B, Chapter 3.

4.4.2 Indicators of Early Effects

Indicators of early effects include reduction of red blood cell cholinesterase in individuals exposed to organic phosphates, reduced erythrocyte o-aminolevulinic acid dehydrase activity and increased nerve conduction time in lead workers, excessive amounts of low molecular weight proteins in the urine of cadmium workers, changes in the nasal mucosa of nickel workers, diminished sperm counts in those working with chemicals affecting fertility, changes in the gums of workers exposed to phosphorus, postexposure reduction in expiratory flowrates or forced expiratory volumes in cotton workers and those exposed to toluene-diisocyanate, anemia in lead workers and benzene workers, and genetic monitoring for chromosomal changes (15–17). The list can go on.

A major problem in interpreting early changes is that since normal variations may occur in most of the tests involved, serial tests are usually required. Another problem is the decision whether a suspected abnormality is sufficient to be reported as an occupational illness on OSHA log 200.

The detection of premonitory or early changes of cancer presents special problems, discussed in Section 4.4.4. For additional information on this subject see Volume III B, Chapter 3.

4.4.3   Indicators of Hypersusceptibility

All individuals exposed to a toxic agent do not respond alike, nor are all these differences the result of different levels of exposure. Some individuals are hypersusceptible. In some cases preexisting disease may be the cause of hypersusceptibility. In others, personal habits combined with occupational exposures may create synergistic effects, such as that between ethyl alcohol and chlorinated hydrocarbons, or between cigarette smoking and asbestos. There may also be inborn errors of metabolism that interfere with the detoxification of chemical toxins or augment their effects.

The search for inherited hypersusceptibility is commonly referred to as *genetic screening*, and should be distinguished from *genetic monitoring* referred to in Section 4.4.2.

Although the reality of such inherited factors is indisputable, it has been difficult to establish firmly their importance in occupational medicine. Cooper in 1973 (18) reviewed the status of the most promising indicators of hypersusceptibility and could not recommend their application at that time in routine screening. The tests considered were those for sickle cell trait, glucose-6-phosphodehydrogenase (G6PD), α-1-antitrypsin, and cholinesterase deviants. All appeared to be appropriate subjects for controlled research, but not for mandatory inclusion in regulations or as positive indicators for exclusion from jobs. In the intervening 10 years, there has been relatively little practical application of genetic screening (19). It has been the subject of a great deal of debate, often acrimonious, because of the possibility of genetic screening being used to discriminate unnecessarily against individuals or ethnic groups (2, 20, 21). See Volume III B, Chapter 3 for additional information.

4.4.4   Tests for Fertility

Monitoring for effects on the reproductive function has become necessary in occupational groups where potential risks are suspected. Levine et al. (22) have proposed screening with appropriate questionnaires as a prelude to more definitive testing.

4.4.5   Early Detection of Cancer

A major problem for the concerned occupational health physician, and a recurrent consideration of those mandating medical surveillance of workers exposed to suspected or proved carcinogens, has been the types of medical examination that might be useful for early detection of cancer. The reader is referred to guidelines developed for the American Cancer Society (23).

The current permanent asbestos standard (8) does not address itself to lung cancer or mesothelioma in its medical surveillance requirements, which are limited to annual chest films, measurements of pulmonary ventilatory function, and history.

The standard for vinyl chloride (8) contains provisions for a battery of liver function tests, presumably because vinyl chloride is a known hepatotoxin, and it has been suggested that such hepatoxic effects would precede angiosarcomas. There is no reason to be sure that this occurs. In any event, if exposures are kept below 1 ppm it is probable that thousands of exposed workers would be examined before any vinyl chloride-related disorder were discovered. In the meantime, many individuals with liver disorders resulting from other toxins, such as alcohol, would be detected and removed from exposure or advised to terminate employment.

The standards for coke oven emissions (8) and for arsenic (8) include provisions for sputum cytologic examination. The former also requires urine cytologic studies. Frequent sputum examinations in those who have worked in high risk areas for many years may detect an occasional operable lung cancer. This is probably an unproductive exercise in relation to individuals working under controlled conditions, even though it is required.

Urine cytology is also unlikely to be a useful diagnostic tool in coke oven workers, in view of the relatively low incidence of urinary tract cancers in the group. It is a useful test in those heavily exposed in the past to proved bladder carcinogens, such as β-naphthylamine and benzidine.

Specifying medical surveillance for cancer, particularly for chemicals whose effects in humans are speculative and based solely on findings in experimental animals, has produced difficulties for regulatory agencies. The published standards include a requirement for preplacement and annual physical exam-inations with a "personal history of the employee, family, and occupational background, including genetic and environmental factors. In all physical examinations the examining physician should consider whether there exist conditions of increased risk reducing immunologic competence, as undergoing treatments with steroids or cytotoxic agents, pregnancy, and cigarette smoking."

It is unclear how the examining physician would evaluate the information obtained or how it might affect employability.

Biochemical Markers of Cancer. The role of such biochemical markers of cancer as carcinoembryonic antigen (CEA) α-fetoprotein (AFP), and acid phosphatase for prostatic cancer, deserve mention. None is at a stage of proved validity, specificity, or sensitivity to warrant inclusion in regulations or routine testing. It is desirable that groups at high risk because of past exposure be included in studies for the evaluation of these so-called early warning signs (24). Additional information on this subject is covered in Volume III B, Chapter 2.

5 RELATIONSHIP OF MEDICAL SURVEILLANCE AND EXPOSURE DATA

Meeting the requirements of regulations and providing adequate medical surveillance of workers necessitates interlocking of qualitative and quantitative information on exposures with medical programs. This can be a complex process in a large plant with numerous and multiple exposures, but is becoming possible as computer programs are rapidly evolving. The reader is referred to several recent considerations of the problem (25–27).

6 RECORDKEEPING

6.1 Maintenance of Records

Because of the long latent periods between exposure and the appearance of chronic effects such as asbestosis or occupationally related cancers, there is a need for preservation of medical records longer than formerly was regarded as necessary. OSHA regulations and proposals, and NIOSH recommendations in criteria documents, contain a variety of record-retention periods, some as low as 5 years and many as long as 40 years from the termination of employment. The current trend is toward longer retention periods. It is recommended that systems be utilized that will permit all medical records to be kept for a minimum of 40 years. Duplicate storage in computer-accessible microfiche systems is currently the method that is most economical of space.

6.2 Confidentiality of Records

Records that contain personal information on workers must be protected from transmission to or perusal by those not responsible for medical services or care. Without assurance of confidentiality, it is impossible to maintain the professional relationship necessary to elicit needed information. This has not been easy, however, since management must know an individual's physical limitations as they relate to employment, and governmental agencies desire to know whether individual workers have been harmed in their employment. The reader is referred to some pertinent references (28, 29).

Full exploration of this issue is impossible in this section. In its simplest terms, the medical record should be retained by the physician, whether a member of a plant medical department, an external consultant, or part-time physician. He or she should provide the plant only with information that directly pertains to employment, and this should be done with the understanding of the employee. Any other information must be transmitted only with the express and written consent of the employee. Since in the occupational setting consent can be tantamount to a condition of employment, the physician must be zealous in protecting the worker's right to privacy.

6.3 Accessibility of Records

Provisions in regulations regarding accessibility and dissemination of medical records vary. Whereas the standard for asbestos requires that medical reports be sent to employers, later standards have been less explicit. It is clear that policy relating to this, and ultimately the law, is still evolving.

7 PROBLEM AREAS

7.1 Compulsory Versus Voluntary Examinations

All regulations so far promulgated provide that medical surveillance be made available by employers, but nowhere is it stated that employees shall be required to take the examinations. Individual employers may, however, make examinations a condition of employment. The issues here are obviously complex. If a regulatory agency believes that a given examination or test is so important to workers' health that every employer must be prepared to provide it, how can it not require employees to take it if they are to work with a specified chemical agent? Present regulations leave the question of such medical surveillance at the level of negotiation between the workers or their representatives and the industry, and necessitate assurances relating to protection of job rights. For that part of the work force which is not organized, such protection would have to be guaranteed by law.

7.2 The Problem of Small Industries

The multiplying problems of medical surveillance strike with particular force on small industries. Major corporations with medical departments gradually adjust to regulations and recommendations requiring medical surveillance. Many of them have managed similar and effective programs for decades. The small employer must meet these needs from a different baseline of operation. The regulations now in force and being proposed and promulgated leave no choice. The employees must be informed of any hazardous materials with which they are working and must be provided with appropriate safeguards. Also arrangements must be made to carry out medical surveillance. At present, the resources available to do physical examinations are unevenly distributed, and often necessary guidance is lacking.

7.3 Handling Abnormal Findings

A major problem for the occupational physician in the current climate of physical examinations is the proliferation of borderline abnormal findings. Regulations stipulate that a physician must certify that a worker has no condition that might be adversely affected by exposures on the job. It can be very difficult for a cautious physician to make such a certification if there are any abnormal findings. The presence of metaplastic cells in the sputum of an asbestos worker or a coke oven worker would make certification difficult, even though current exposures were low and controlled. Similarly, certification for wearing a respirator of an individual who years earlier had had a coronary occlusion would be difficult, even though the risk was very slight. Situations that can be handled easily in a doctor–patient relationship become matters with serious medicolegal implications because they have been the subject of certification under regulations.

7.4 Consequences of Overregulation

Inherent in all the foregoing discussion has been concern over the inclusion of detailed requirements for medical surveillance in regulations. Many practices that are highly desirable for physicians to follow in selected occupational groups, or for plants to carry out with full understanding of their implications, are not necessarily right for mass application mandated by law. When applied to individuals with very low exposures in adherence to environmental standards, they will result in an extremely low yield of abnormalities, a stultifying overuse of scarce physician time, and needless expense. Often it is former employees, no longer covered by regulations, who need the surveillance that is being carried out on new employees with relatively little exposure. Recourse to the judgment of industrial hygienists and occupational physicians would be desirable.

8 LEGAL CONSIDERATIONS

There are many legal burdens placed on the physician in carrying out medical surveillance of the worker. Some of these are discussed by Felton (30). No comprehensive or authoritative review of these is possible here, but some of the areas that can cause problems are discussed briefly.

8.1 Informing the Worker

There is general agreement at present that the worker must be given information about the hazards of the workplace. This is management's responsibility, but the physician must and will have a role. For many hazards the explanation is easily and readily understood by the worker. For carcinogens, particularly for suspect carcinogens, the message is difficult to impart. At present, the best that can be said is that the physician should be honest in giving an appraisal of the evidence, should create sufficient concern and anxiety to encourage observance of rules for containment and protection against known carcinogens, and should indicate the relative probability of effects from low exposures and for weak or suspect carcinogens.

8.2 Certification for Continued Employment

It is difficult to know the impact of regulations that require the examining physician to certify that workers will not be adversely affected by continued employment. There will be a tendency by some physicians to take no chances. There is relatively little information to say whether an individual with an elevated serum glutamic oxaloacetic transaminase (SGOT) or other liver enzyme would be harmed by exposure to 1 ppm of vinyl chloride. The probability is extremely high that it would make no difference. If, however, a worker should develop liver disease after certification, regardless of whether it was due to vinyl chloride, the certifying physician could conceivably be sued. A person certified to work with asbestos, even at low levels, after the finding of moderate metaplastic changes in his sputum, could present a similar problem. While the questions involved are being worked out, examining physicians can only use their best clinical judgments, attempt to learn something of the actual exposures their patients are experiencing, and realize that depriving a person of a job unnecessarily is a very serious matter.

8.3 Wearing of Respirators

In all current and proposed regulations and in NIOSH criteria documents, physicians are given the responsibility for certifying whether workers are physically able to wear nonpowered respirators. Objective criteria of ability to wear a respirator are limited, and the translation of pulmonary function tests results to respirator use is uncertain. Actual trial of individuals with respirators is the ultimate test. Differentiation must be made between situations when a respirator is worn briefly for emergency situations for a worker's own protection, when it is required for the worker to perform duties essential to the safety and health of others, and where the worker may be required to wear it over long periods of time. The recommendations of the American National Safety Institute's Z88 Committee (31) can be useful to the physician with responsibilities in this area.

9 SUMMARY

The health surveillance of workers requires preemployment evaluation and periodic examinations aimed both at general health maintenance and the prevention or early detection of effects from specific job hazards. Good practice calls for a comprehensive preemployment evaluation, with an occupational and medical history and review of all systems, baseline laboratory studies (including blood chemistry and urinalysis), study of visual and hearing acuity, simple tests of pulmonary ventilatory function, and a general physical examination. Periodic examinations should include a core examination directed toward general health maintenance and special hazard-oriented studies, such as biologic monitoring for determining levels of absorption of chemicals, early indicators of toxic or other biological effects, and audiometry when indicated by noise exposures. Proper scheduling and interpretation of hazard-oriented surveillance make it essential that environmental and medical data be closely interlocked.

There are no consistently effective methods for the early detection of most occupational cancers. Urinary cytologic screening can be useful in individuals who have been heavily exposed to bladder carcinogens. Sputum cytology, the only technique that is promising for the early detection of lung cancer, is recommended for those in high risk groups, but results to date have been discouraging.

An important part of a health surveillance program is the opportunity provided for periodic contact of the worker with a member of the health team. This should be utilized for education in the prevention of both occupational and nonoccupational disease.

REFERENCES

1. World Health Organization, "Early Detection of Health Impairment in Occupational Exposure to Health Hazards," Technical Report Series 571, WHO, Geneva, 1975.
2. C. R. Goerth, "Physical Standards: Discrimination Risk," *Occ. Health Saf.*, **52**(6)**,** 33–34 (June 1983).
3. N. J. Robert, "The Values and Limitations of Periodic Health Examinations," *J. Chronic Dis.*, **9,** 95–116 (February 1959).
4. G. S. Siegel, "Periodic Health Examinations. Abstracts from the Literature," Public Health Service Publication 1010, March 1963, U.S. Government Printing Office, Washington, D.C., 1963.
5. W. K. C. Morgan, "The Annual Fiasco (American Style)," *Med. J. Aust.*, **2,** 923–925. (November 1, 1969).
6. American Medical Association, Council on Scientific Affairs, "Medical Evaluations of Healthy Persons," *JAMA*, **249,** 1626–1633 (March 25, 1983).
7. N. J. Ashenburg, "Routine Chest X-ray Examinations in Occupational Medicine," *J. Occ. Med.*, **24,** 18–20 (January 1982).
8. Food and Drug Administration, "Chest X-ray Screening Statements." *FDA Drug Bull.*, **13,** 13–14 (August 1983).
9. U.S. Department of Labor, Occupational Safety and Health Administration, *OSHA Safety and Health Standards* (*29 CFR 1910*) *OSHA 2206*, revised March 11, 1983. Superintendent of Documents, U.S. Government Printing Office, Washington, D.C.
10. Federal Coal Mine Health and Safety Act of 1969, Public Law 91-173, December 30, 1969.
11. Federal Mine Safety and Health Amendments Act of 1977, Public Law 95-164, November 9, 1977.
12. N. H. Proctor and J. P. Hughes, *Chemical Hazards of the Workplace*, Lippincott, Philadelphia, 1978.
13. American Thoracic Society, Surveillance for Respiratory Hazards in the Occupational Setting. The Official ATS Statement Adopted by the ATS Board of Directors, June 1982. *Am. Rev. Respir. Dis.*, **126,** 952–956 (November 1982).
14. R. R. Lauwerys, *Industrial Chemical Exposure: Guidelines for Biological Monitoring*, Biochemical Publications, Davis, Calif. 1982.
15. J. D. Fabricant and M. S. Legator, "Etiology, Role and Detection of Chromosomal Aberrations in Man," *J. Occ. Med.*, **23,** 617–625 (September 1981).
16. B. J. Dabney, "The Role of Human Genetic Monitoring in the Workplace," *J. Occ. Med.*, **23,** 626–631 (September 1981).

17. P. A. Buffler and J. M. Aase, "Genetic Risks and Environmental Surveillance: Epidemiological Aspects of Monitoring Industrial Populations for Environmental Mutagens," *J. Occ. Med.,* **24,** 305–314 (April 1982).

18. W. C. Cooper, "Indicators of Susceptibility to Industrial Chemicals," *J. Occ. Med.,* **15,** 355–359 (April 1973).

19. G. S. Omenn, "Predictive Identification of Hypersusceptible Individuals," *J. Occ. Med.,* **24,** 369–374 (May 1982).

20. Office of Technology Assessment, Congress of the United States, *The Role of Genetic Testing in the Prevention of Occupational Disease,* OTA, Washington, D.C., 1983.

21. Marc Lappe, "Ethical Issues in Testing for Differential Sensitivity to Occupational Hazards." *J. Occ. Med.,* **25,** 797–808 (November 1983).

22. R. J. Levine, M. J. Symons, S. A. Balogh, D. M. Arndt, N. T. Kaswandik, and J. W. Gentile, "A Method for Monitoring the Fertility of Workers 1. Method and Pilot Studies." *J. Occ. Med.,* **22,** 781–791 (December 1980).

23. American Cancer Society, "Guidelines for the Cancer-Related Checkup. Recommendations and Rationale," *Ca—A Cancer Journal for Clinicians,* **30,** 193–240 (July/August 1980).

24. T. H. Maugh, II, "Biochemical Markers: Early Warning Signs of Cancer," *Science,* **197,** 543–545 (August 5, 1977).

25. M. G. Ott, "Linking Industrial Hygiene and Health Records," *J. Occup. Med.,* **19,** 388–390 (June 1977).

26. J. D. Forbes, J. P. Dunn, G. Hillman, L. L. Hipp, T. J. McDonagh, S. Pell, and G. F. Reichwein, "Utilization of Medical Information Systems in American Occupational Medicine, A Committee Report." *J. Occup. Med.,* **19,** 819–830 (December 1977).

27. R. D. Finucane and T. J. McDonagh (co-chairmen), "Medical Information Systems Roundtable," *J. Occup. Med.,* **24,** 781–866 (October 1982).

28. G. J. Annas, "Legal Aspects of Medical Confidentiality in the Occupational Setting," *J. Occup. Med.,* **18,** 537–540 (August 1976).

29. A. McLean, "Management of Occupational Health Records," *J. Occup. Med.,* **18,** 530–533 (August 1976).

30. J. S. Felton, "Legal Implications of Physical Examinations," *West. J. Med.,* **128,** 266–273 (March 1978).

31. American National Standards Institute, Z88 Committee for Respiratory Protection, *Z88.6 American National Standard Physical Qualification for Respirator Use,* (Draft available for final public comment) ANSI, 1983.

CHAPTER SIX

Occupational Exposure Limits, Pharmacokinetics, and Unusual Work Schedules

DENNIS J. PAUSTENBACH, Ph.D., CIH

1 INTRODUCTION

The concern about the adverse health effects of night shift work has existed for over 100 years. Around 1860, some scientists were worried that bakers might be at increased risk of physical and emotional illness because they always worked at night and, subsequently, some effort was made to try to regulate their work hours and their working conditions (1). Since then, it has been shown that some persons who have been very productive while working standard 8 hr/day, 40 hr/week daytime schedules can become fatigued, unhappy, less productive and perhaps even more susceptible to the effects of chemical agents and physical agents after they are placed on shift work (2–17). Even after much study, the degree to which shift work affects a worker's capabilities, longevity, mortality, morbidity and overall well-being is still not well understood.

So-called unusual work shifts and work schedules, which often involve a 10 or 12 hr workday, have been implemented in a number of industries in an attempt to eliminate or at least reduce some of the problems caused by normal shift work which requires three work shifts per day. These unusual or "other-than-normal" shifts have been termed odd, novel, extended, extraordinary, compressed, nonnormal, nonroutine, prolonged, exceptional, nonstandard, unusual, peculiar, weird and nontraditional (18, 19). In 1981, the American Industrial Hygiene Association (AIHA) established a permanent committee to address the potential occupational health aspects of shift work and the need to adjust exposure limits for persons who work schedules which were markedly different than the "normal" workweek which was defined as 5 consecutive 8-hr daylight workdays followed by 2 days off. At the committee's first meeting, it was agreed that the term "unusual work shift" should be used, for sake of uniformity, to describe these other-than-normal shifts. In general, most unusual work schedules will involve workdays markedly longer than 8 hours in duration, however, because many persons are regularly exposed to xenobiotics for very short periods during shifts, and because time-weighted average exposure limits were not necessarily intended for use during short exposures, schedules which contained numerous short periods of exposure to high concentrations were also classified as "unusual" by the committee. The assessment of the health aspects of unusual shifts can be complex, since some schedules require the worker to alternate between night and day work every few days (rapid rotation). These schedules have been called rapidly rotating, fast, or simply rapid-roto shifts (20–25).

In order for the health professional to protect the worker who is exposed to airborne chemicals during unusually long or unusually short periods, he or she must be familiar with the toxicology and pharmacokinetics of the chemical of interest as well as understand the rationale for its occupational exposure limit. Special consideration is suggested since most limits were developed with the intent of protecting only persons who work a normal 8 hr/day, 5 day/workweek.

This chapter reviews the history and rationale for the occupational exposure limits which have been established for normal work schedules, the history of unusual work schedules, the toxicologic and pharmacokinetic rationale for modifying limits for unusually long or short periods of exposure, and the various approaches which can be used for modifying existing limits so as to provide "equivalent" protection to exposed workers. The chapter should be most useful to professionals who are involved in setting exposure limits and/or applying them to settings other than those for which they were initially intended.

2 BACKGROUND ON SHIFT WORK

Traditionally, a fixed work schedule is established by the employer and consists of five 8-hr days each week starting at 8 a.m. and ending at 5 p.m. Initially, the need to operate certain manufacturing processes 24 hr/day served as the impetus for using three shifts of workers to cover the 24-hr workday. Later, shift work was implemented because the economics of having equipment idle 70 percent of the week was prohibitive. In addition, the use of three shifts per day allows uninterrupted production throughout the year so that a process never has to shut down. Many continuous-process operations, such as those found in oil refineries, chemical plants, steel and aluminum mills, pharmaceutical manufacturing, glass plants, and paper mills, *cannot* be shut down without causing serious production and financial losses; thus, they require round-the-clock staffing (18, 19, 21).

In many professions, adherence to the 40 hr/week schedule is unusual. For example, artisans and equipment repair persons frequently work beyond 8 hr/day (overtime) to accommodate routine equipment repair, fill-in for absent workers or deal with seasonal fluctuations in demand (21). For some persons, such as those in the railroad industry and the military, a workday of 16–24 hours is not uncommon. In certain industries, very complicated schedules which can involve both short and long periods of work have been implemented for a wide number of reasons. The advantages and disadvantages of alternative (unusual) work schedules, as well as the number of persons involved in these have been discussed (26–29).

Dozens of alternatives to traditional work schedules have been developed and are now used in a number of industries. As observed during the recession of the early 1980s, the length of time worked by a person was often altered by the use of part-time employees, and many workweeks were shortened or lengthened due to wide fluctuations in the demand for a given product. Overall, the high cost of manufacturing equipment, increased foreign competition, and the inability to halt certain chemical and physical processes after they have begun has forced industry to make shift work and modified work schedules a permanent part of manufacturing (21).

The shift worker employed in a 24 hr/day operation, usually on a rotating, 8-hr schedule, faces many disruptive events which arise due to his work schedule. Overall, the primary complaint of these workers surrounds the unsatisfactory impact shift work has on their social lives (21–23, 30–33). When they are home, most other persons are asleep, at work or at school. In general, these shift workers have only one full weekend per month during which they are not at work. As a result, shift workers and members of their families are frequently disappointed with the quantity and quality of time that they spend together (21, 23, 34–38).

Researchers who have studied the effects of shift work on human health have noted that shift workers may have trouble sleeping, often feel fatigued, frequently feel overly tired during days off work and are often chronically irritable (7–9, 36–37). Other undesirable effects of shift work which are more easily measured and are frequently reported include constipation, gastritis, gastroduodenal ulcers, peptic ulcers, high absenteeism and lessened productivity (11–17). Hoping to minimize these effects, numerous kinds of unusual work schedules have been developed and implemented by employers who use standard shift work. Throughout this chapter, standard shift work is always defined as a workweek of 8 hr/day and 5 days which occurs during daylight hours, usually between 8 a.m. to 5 p.m., and is followed by 2 days off work.

2.1 Unusual Work Schedules

One kind of schedule which is classified as unusual is the type involving work
periods longer than 8 hr and varying numbers of days worked per week (e.g.,
a 12 hr/day, 3 day/workweek). Another category of unusual work schedules are
those involving a series of brief exposures to a chemical or physical agent
during a given work schedule (e.g., a schedule where a person is exposed to a
chemical for 30 minutes, 5 times per day with 1 hour between exposures).
Another type of unusual schedule is that involving the "critical case" wherein
persons are continuously exposed to an air contaminant (e.g., in spacecraft,
submarines).

Compressed workweeks are a type of unusual work schedule that has been
used primarily in nonmanufacturing settings. They refer to full-time employment (virtually 40 hr/week) which is accomplished in less than 5 days/week.
Many compressed schedules are currently in use, but the most common are:
(1) 4-day workweeks with 10-hr days; (2) 3-day workweeks with 12 hr/days; (3)
$4\frac{1}{2}$ day workweeks with four 9-hr days and one 4-hr day (usually Friday); and
(4) the 5/4, 9 plan of alternating 5-day and 4-day workweeks of 9-hr day (27–
29).

Over the past two decades, unusual work shifts and schedules including
some form of the compressed workweek have been implemented in many
manufacturing facilities (18, 19). Examples of the types of schedules and one
type of industry which has used them include: four 10-hr workdays per week
(chemical); a 6-week cycle of three 12-hr workdays for 3 weeks followed by
four 12-hr workdays for 3 weeks (pharmaceutical); a 6-hr per day, 6-day
workweek (rubber); a 56/21 schedule involving 56 continuous days of work of
8 hr per day followed by 21 days off (petroleum); a 14/7 schedule involving 14
continuous days of work of 8 to 12 hr per day followed by 7 days off (petroleum);
a 3/4 schedule involving only three 12-hr workdays in 1 week followed by a
week of four 12-hr workdays (pharmaceutical); four 12-hr workdays followed
by three 10-hr workdays, followed by five 8-hr workdays then 4 days off; five
8-hr workdays, followed by two 12-hr days, then 5 more 8-hr workdays followed
by 5 days off (petrochemical); a 2/3 schedule involving 18 hr of work for 2
days then 3 days off (military); and numerous other variations of these (18, 19,
21, 39). Of all workers, those on unusual schedules represent only about 5
percent of the working population (29). Of this number, only about 50,000–
200,000 Americans who work unusual schedules are employed in industries
where there is routine exposure to significant levels of airborne chemicals. In
Canada, the percentage of chemical workers on unusual schedules is thought
to be greater.

The myriad problems associated with shift work (i.e., 3 shifts per day) have
been discussed in numerous books and journals and have filled hundreds of
pages. Many of these are cited in the reference section of this chapter and in
another chapter in Volume III B entitled *Biological Rhythms, Shift Work and
Occupational Health.* The purpose of this chapter is to review the rationale for

exposure limits, discuss those aspects of pharmacokinetics needed to understand the basis for adjusting limits for unusual schedules and to examine the various methods for adjusting them. The potential untoward effects of shift work on the physical or emotional well-being of workers is beyond the scope of this chapter and therefore it will not be discussed.

3 METHODS FOR ESTABLISHING OCCUPATIONAL EXPOSURE LIMITS

3.1 Introduction

Over the past 40 years, many organizations in numerous countries have proposed exposure limits for airborne contaminants. The limits or guidelines that have gradually become the most widely accepted both in this country and abroad are those issued annually by the American Conference of Governmental Hygienists (ACGIH) which are termed Threshold Limit Values, or (more commonly) TLVs.

The usefulness of establishing permissible limits of exposure for potentially harmful agents in the working environment has been demonstrated repeatedly ever since their inception. In short, whenever TLV's have been implemented in a particular industry, no worker has been shown to have sustained serious adverse effects on his health as a result of exposure to these concentrations of toxicant (40).

The ACGIH TLVs are limits which refer to airborne concentrations of substances and represent conditions under which it is believed that nearly all workers may be repeatedly exposed day-after-day without adverse effect (41). It is important to recognize that unlike some exposure limits set by other professional groups or regulatory agencies, exposure to the TLV will not necessarily prevent discomfort or injury for everyone who is exposed. The ACGIH recognized long ago that because of the wide range in individual susceptibility, a small percentage of workers may experience discomfort from some substances at concentrations at or below the threshold limit, and that a smaller percentage may be affected more seriously by aggravation of a preexisting condition or by development of an occupational illness (40–45). This limitation, although perhaps less than ideal, is a practical one since levels so low as to protect hypersusceptibles would be infeasible due to either engineering or economic limitations. This shortcoming in the TLVs has, in general, not been found to present a serious problem to either the employee or the employer.

Threshold Limit Values, like most other occupational exposure limits used in other countries, are based on the best available information from industrial experience, experimental human and animal studies and, when possible, from a combination of the three (45). The rationale for each of the established values differs from substance to substance; protection against impairment of health may be a guiding factor for some, whereas reasonable freedom from irritation, narcosis, nuisance or other forms of stress may form the basis for others. The age and completeness of the information available for establishing most permissible exposure limits also varies from substance to substance; consequently, the precision of each particular TLV is subject to variation and the most recent TLV and its documentation should always be consulted in order to evaluate the quality of the data upon which that value was set. The background information and rationale for a TLV (called the documentation) is published for each of the TLVs and some type of documentation is often available for limits set in other countries. It is especially important to review the rationale for a particular standard before interpreting or adjusting a limit since it will describe the goal of the limit as well as the exact data which were considered in establishing it (46).

Even though all of the publications which contain occupational limits emphasize that these are intended for use only in establishing safe levels of exposure for persons in the workplace, they have, unfortunately, been used at times in other situations. It is for this reason that all exposure limits should be interpreted and applied only by someone familiar with industrial hygiene and toxicology. The ACGIH TLV committee has warned users that TLV's are not intended for use, or in modification for use:

1.   As a relative index of hazard or toxicity;
2.   In the evaluation or control of community air pollution nuisances;
3.   Estimating the toxic potential of continuous uninterrupted exposures or other extended work periods;
4.   As proof or disproof of an existing disease or physical condition;
5.   For adoption by countries whose working conditions or substances and processes differ markedly from those in the United States (41).

3.2 Philosophy of Exposure Limits

An understanding of the philosophy used in setting exposure limits is critical to the professions of industrial hygiene and toxicology. Exposure limits for workplace air contaminants are based on the premise that, although all chemical substances are toxic at some concentration when experienced over a specific period of time, *a concentration (e.g., dose) does exist for all substances at which no injurious effect should result no matter how often the exposure is repeated.* A similar premise also applies for substances whose effects are limited to irritation, narcosis, nuisance or other forms of stress (40, 41).

This philosophy thus differs from that applied to physical agents such as ionizing radiation, and for some chemical carcinogens, since it is possible that there may be no threshold or no dose at which some risk would not be expected (40). Even though many would say that this position is too conservative given our poor understanding of the mechanism of the cancer, there are data on genotoxic chemicals which seem to support this premise (47). On the other hand, many respected scientists believe that a threshold does exist for those chemicals which have demonstrated carcinogenic activity in animals but act through a nongenotoxic (sometimes called epigenetic) mechanism (48–51). Still others maintain that a "practical" threshold exists for even genotoxic chemicals, although they agree that the threshold may occur at an extremely low dose (50–54). With this in mind, some exposure limits proposed by regulatory agencies in the early 1980s were established at levels which, although not "safe," presented risks so low as to be considered insignificant (55).

3.3 History of Exposure Limits

The role of occupational exposure limits in minimizing disease is now a widely accepted fact, but for many years such limits did not exist and even when they did, they were often not observed (40, 56–58). It was, of course, well understood that airborne dusts and chemicals could bring about illness and injury, but the concentrations and lengths of exposure at which this might be expected to occur were unclear due to the lack of documentation. A committee of the American Conference of Governmental Industrial Hygienists (ACGIH) met in early 1940 to set about the task of assembling all the data that they could locate which would relate the degree of exposure to a toxicant with the likelihood of producing an adverse effect (40, 58).

This task, as might be expected, was a formidable one. After four years of much painstaking research and labor intensive documentation, the first set of values were released in 1941 by this committee which was composed of Warren Cook, Manfred Boditch (reportedly America's first hygienist employed by industry), Bill Fredrick, Philip Drinker, Lawrence Fairhall and Alan Dooley (40). The overall impact of this effort to develop quantitative limits to protect humans from the adverse effects of workplace air contaminants and physical agents could not have been anticipated by the committee. To their credit, even though toxicology was then only a fledging science, their approach to setting limits has generally been shown to be correct even by today's standards. For this reason, many of the techniques for setting limits established by this committee are still in use today (56–60).

The benefits of setting limits are manifold; from the perspective of the hygienist, engineer and businessman. The establishment of limits, by their vary nature, implies that at some level, exposure to a toxicant can be expected to be safe and pose no concern to exposed persons. By incorporating this fundamental toxicological principle into the realm of business management, the practice of industrial hygiene has been able to make large strides. The key to the success of limits is not that they are established on solid scientific principles or because they represent the differences between safe and unsafe levels of exposure, rather, the setting of any goal gives a sense of purpose and direction to occupational medical programs which heretofore were too difficult to evaluate. The setting of goals, such as meeting a TLV, establishes an objective which can then be mutually pursued by the occupational health team, engineers and management. By introducing the concept of "safe level of exposure" and by establishing a kind of "management by objectives" (MBO), the focus of an occupational health program can be made clear (58). In fact, some persons believe that the use of limits remains relatively rare in the area of preventive medicine and, therefore, separates industrial hygiene and health physics from the areas of safety, fire prevention, nursing and epidemiology wherein such objectives are not as easily identified and, therefore, success is much more difficult to quantify.

3.4 Exposure Limits Set by Other Countries

The philosophical groundwork on which occupational exposure limits have been established varies between the various organizations and countries that developed them. For example, in the United States at least six groups recommend exposure limits for the workplace. These include the Threshold Limit Values (TLV) of the American Conference of Governmental Industrial Hygienists (ACGIH), the exposure limits recommended by the National Institute for Occupational Safety and Health (NIOSH) of the U.S. Department of Health and Human Services, the Workplace Environment Exposure Limits (WEEL) developed by the American Industrial Hygiene Association (AIHA), standards for workplace air contaminants suggested by the Z37 Committee of the American National Standards Institute (ANSI) and lastly, recommendations have been made by local, state or regional government. In addition to these recommendations or guidelines, permissible exposure limits (PEL), that must be met in the workplace because they are law, have been established in the United States by the Department of Labor (58).

Outside the United States, as many as fifty other countries or groups of countries have established workplace exposure limits (58–64). Many, if not most, of these limits are nearly or exactly the same as the ACGIH TLV's developed in the United States. In some cases, such as in the Soviet Union and other Soviet bloc countries, as well as Japan, the limits are dramatically different from those used in the United States. Differences among various limits recommended by other countries can be due to a number of factors:

1. Difference in the philosophical objective of the limit and the untoward effects they are meant to minimize or eliminate;
2. Difference in the predominant age and sex of the workers;
3. The duration of the average workweek;
4. The economic state of affairs in that country; or
5. A lack of enforcement therefore acting simply as a guide.

For example, limits established in the U.S.S.R. are often based on a premise that they will protect "everyone" rather than nearly everyone, from "any" rather than "most" toxic or undesirable effects of exposure (59–64).

The U.S.S.R. also establishes many of its limits with the goal of eliminating "any" possibility for even reversible effects, such as those involving subtle changes in behavioral response, irritation or discomfort. The philosophical differences between limits set in the U.S.S.R. and in the U.S. have been discussed by Letavet, a Russian toxicologist, who stated that:

The method of conditioned reflexes, provided it is used with due care and patience, is highly sensitive and therefore it is a highly valuable method for the determination of threshold concentrations of toxic substances.

At times, disagreement is voiced with Soviet MACs for toxic substances, and the argument is that these standards are founded on a method which is "excessively sensitive"—namely the method of conditioned reflexes. Unfortunately, science suffers not a surplus of excessively sensitive methods, but their lack. This is particularly true with regard to medicine and biology.

Although the methods of examination of the higher nervous activity are very sensitive, they cannot be considered to always be the most sensitive indicator of an adverse response and to enable us always to discover the harmful after-effects of being exposed to poison at the earliest time (61).

Such subclinical and fully reversible responses to workplace exposures have, thus far, been considered too restrictive to be useful in the United States and in most other countries. In fact, due to the economic and engineering difficulties in achieving such low levels of air contaminants in the workplace, there is little indication that these limits have actually been achieved in countries which have set them. Instead, the limits appear to serve more as idealized goals rather than limits which manufacturers are legally bound or morally committed to achieve (62, 63).

Clearly, occupational health professionals, who attempt to determine which one of the more than 50 different exposure limits for a chemical is appropriate, must understand the underlying rationale and documentation upon which the limit for any particular country was established. A number of fairly thorough discussions of the history and philosophical basis upon which occupational exposure limits have been set have been published and these should be reviewed (56–112). A comprehensive listing of the various occupational exposure limits used throughout the world can be found in two references which are frequently not well known amongst practitioners of industrial hygiene. They are: *Occupational Exposure Limits For Airborne Toxic Substances*, 2nd edition, published by International Labour Office of the World Health Organization, and *A Worldwide Compilation of Occupational Exposure Limits*, published by the American Industrial Hygiene Association (114, 115). Warren Cook, one of the "founding fathers" of the TLV concept, recently completed the latter book at the age of 86.

3.5   Basis for Current Exposure Limits

Occupational exposure limits established both in the United States and elsewhere are derived from a wide number of sources. As shown in Table 6.1, the 1968 TLVs (those adopted by OSHA as federal regulations) were based largely on human experience. This may come as a surprise to many hygienists who have

Table 6.1 Distribution of Procedures Used to Develop
ACGIH TLVs for 414 Substances Through 1968[a]

Procedure	Number	Percent Total
Industrial (human) experience	157	38
Human volunteer experiments	45	11
Animal, inhalation—chronic	83	20
Animal, inhalation—acute	8	2
Animal, oral—chronic	18	4.5
Animal, oral—acute	2	0.5
Analogy	101	24

From H. E. Stokinger, "Criteria and Procedures for Assessing the Toxic Responses to Industrial Chemicals." Permissible Levels of Toxic Substances in the Working Environment, *Occupational Safety and Health Series No. 20. International Labor Office, Geneva, Switzerland, 1970.*
[a] Exclusive of inert particulates and vapors.

recently entered the profession since it indicates that, in most cases, the setting of an exposure limit has been *after* it has been found to have toxic, irritational or otherwise undesirable effects on humans. As might be anticipated, many of the more current exposure limits for systemic toxins, especially those internal limits set by manufacturers, have been based primarily on toxicology tests conducted on animals prior to the chemical's manufacture, which is in contrast with waiting for observations of adverse effects in exposed workers. Of course, even as far back as 1940, animal tests were acknowledged by the TLV Committee to be very valuable and they do, in fact, constitute the second most common source of information upon which these guidelines have been established (59).

Many approaches for deriving occupational exposure limits from animal data have been proposed and put into use over the past 40 years. The approach used by the TLV Committee and others is not markedly different from that which has been used by the U.S. Food and Drug Administration (FDA) in establishing acceptable daily intakes for food additives and other toxic substances. An understanding of the FDA approach to setting exposure limits for food additives and contaminants is very useful to industrial hygienists who are involved in interpreting occupational exposure limits and should be reviewed (116, 117). Calabrase has discussed, in detail, various approaches for setting both environmental and occupational exposure limits from animal data (118). Discussions of methodological approaches which can be used to establish workplace exposure limits based exclusively on animal data have been presented (60, 108–110, 118–120). Overall, while these approaches have a high degree of uncertainty, they seem to be much better than directly extrapolating the results of animal tests to man.

The criteria used to develop the TLVs may be classified into four groups: morphologic, functional, biochemical, and miscellaneous (nuisance, cosmetic) (Table 6.2). As noted, approximately 50 percent of the TLVs have been derived from the human data, and approximately 30 percent are derived from animal data (Table 6.1). Of the human data, most of it is derived from the effects noted in workers who were occupationally exposed to the substance for many years. Consequently, most of the existing TLVs have been based on the results of workplace monitoring and qualitative and quantitative human responses (59, 121). In more recent times, TLVs for new compounds have been based primarily on the results of animal studies rather than human experience. The rationale by which most of the existing TLVs have been established are shown in Table 6.3. It is noteworthy that only about 50 percent of the TLVs are set so as to prevent systemic toxic effects. Roughly 40 percent are based on irritation and about 1–2 percent are intended to prevent cancer due to workplace exposure.

Table 6.2   Classification of Criteria for ACGIH TLVs Applicable to Man and Animals

Applied Criteria			
Morphologic	Functional	Biochemical	Miscellaneous
Systems of organs affected—lung, liver, kidney, blood, skin, eye, bone, CNS, endocrines, exocrines Carcinogenesis Roentgenographic changes	Changes in organ function—lung, liver, kidney, etc. Irritation Mucous membranes Epithelial linings Eye Skin Narcosis Odor	Changes in amounts biochemical constituents including hematologic Changes in enzyme activity Immunochemical allergic sensitization	Nuisance Visibility Cosmetic Comfort Esthetic (Analogy)
Potentially Useful Criteria			

Altered reproduction
Body–weight changes
Organ/body weight changes
Food consumption
	Behavioral changes
	  Higher nervous functions
	  Conditioned and uncondi-
	  tioned reflexes—learning
	  Audible and visual re-
	  sponses
	Endocrine glands
	Exocrine glands
		Changes in isoenzyme patterns
		Radiomimetic effects
		Teratogenesis
		Mutagenesis	

From H. E. Stokinger, "Criteria and Procedures for Assessing the Toxic Responses to Industrial Chemicals," Permissible Levels of Toxic Substances in the Working Environment, Occupational Safety and Health Series No. 20, International Labor Office, Geneva, Switzerland, 1970.

Table 6.3 Distribution of Criteria Used to Develop ACGIH TLVs for 414 Substances
Through 1968[a]

Criteria	Number	Percent	Criteria	Number	Percent[a]
Organ or organ system affected	201	49	Biochemical changes	8	2
Irritation	165	40	Fever	2	0.5
Narcosis	21	5	Visual changes (halo)	2	0.5
Odor	9	2	Visibility	2	0.5
Organ function changes	8	2	Taste	1	0.25
Allergic sensitivity	6	1.5	Roentgenographic changes	1	0.25
Cancer	6	1.5	Cosmetic effect	1	0.25

From H. E. Stokinger, "Criteria and Procedures for Assessing the Toxic Responses to Industrial Chemicals." Permissible Levels of Toxic Substances in the Working Environment, Occupational Safety and Health Series No. 20. International Labor Office, Geneva, Switzerland, 1970.
[a] Exclusive of inert particulates and vapors.
[b] Number of times a criterion was used of total number of substances examined × 100, rounded to nearest 0.25 percent. Total percentages exceed 100 because more than one criterion formed the basis of the TLV of some substances.

3.6 Limits for Chemical Carcinogens

The impetus to develop a classification for occupational carcinogens by the TLV Committee of the American Conference of Governmental Industrial Hygienists (ACGIH) arose in 1970, when it felt that the lists published by numerous agencies and different groups, claiming that a large number of substances were occupational carcinogens, was getting out of hand. Substances of purely laboratory curiosity, such as acetylaminofluorene and dimethylaminobenzene, which had been found to be tumorigenic in animals, were classed along with known human carcinogens of high potency, such as *bis*-chloromethyl ether. In short, no distinction was made between an animal tumorigen and a human carcinogen. Each came through to the union leader, lay worker, and public as equally worrisome—which was clearly not accurate or in anyone's best interest (52).

The committee felt that the finding of a substance to be tumorigenic, often in a half-dead mouse or rat which had been exposed to intolerable doses, as was the case for chloroform and trichloroethylene, is not *ipso facto* evidence that it will be carcinogenic in man under controlled working conditions. It was for this reason that the ACGIH Chemical Substance TLV Committee, in the 1972 booklet, made a clear distinction between animal and human carcinogens.

Human carcinogens were listed in two groups:

1. Those with an assigned TLV, four in number;
2. Those without an assigned TLV, eight in number.

By 1984, three categories had been established:

1. List Ala, human carcinogens with an assigned TLV, eight in number;
2. List Alb, human carcinogens without a TLV, four in number;
3. List A2, industrial substances suspected of carcinogenic potential for man, 38 in number.

By setting exposure limits, the ACGIH, as well as its sister organizations throughout the world, acknowledge that chemical carcinogens are likely to have a threshold, or at least a "practical threshold." Their belief is simply that at some level, exposure to that agent would not be expected to cause a significant incidence of tumors (perhaps one person in 1,000,000) in workers. In 1977, Herb Stokinger, chairman of the ACGIH TLV Committee, summarized the fundamental philosophy of the ACGIH with respect to carcinogen TLVs:

Experience and research findings still support the contention that TLVs make sense for carcinogens . . . First and foremost, the TLV Committee recognizes practical thresholds for chemical carcinogens in the workplace, and secondly, for those substances with a designated threshold, that the risk of cancer from a worker's occupation is negligible, provided exposure is below the stipulated limit. There is no evidence to date that cancer will develop from exposure during a working lifetime below the limit for any of those substances.

Where did the TLV Committee get the idea that thresholds exist for carcinogens? We have been asked "Where is the evidence?" . . . Well, the Committee thinks it has such evidence, and here it is.

It takes three forms:

1. Evidence from epidemiologic studies of industrial plant experience, and from well-designed carcinogenic studies in animals.
2. Indisputable biochemical, pharmacokinetic, and toxicologic evidence demonstrating inherent, built-in anticarcinogens and processes in our bodies.
3. Accumulated biochemical knowledge makes the threshold concept the only plausible concept (52).

Even though the TLV Committee, as well as many other groups that recommend exposure limits, may believe that there is likely to be a threshold for carcinogens at very low doses, another equally credible school of thought is that there is little or no evidence for the existence of thresholds for chemicals which are genotoxic (49–51). In an attempt to take into account the philosophical postulate that chemical carcinogens do not have a threshold even though a no observed effect level (NOEL) is often observed in an animal experiment, modeling approaches for estimating the carcinogenic response at low doses have been developed (52–55). The rationale for a modeling approach is that it is impossible to conduct toxicity studies at doses near those measured in the environment because the number of animals necessary to elicit a statistically significant response at these doses in a laboratory experiment would be too great. Consequently, results of animal studies conducted at high doses are extrapolated by mathematical models to those levels found in the workplace or the environment. By the early 1980s, modeling approaches for evaluating the risks of exposure to carcinogens became popular among various regulatory agencies even though the limits recommended by these models have rarely been the sole factor on which regulatory limits have been established. The reasons for this are two-fold: (a) The modelling approach is conservative since it does not account for biological repair of detoxification process and (b) consequently, the limits recommended by models are too low to be economically or practically feasible.

The most popular models for low-dose extrapolation are the one-hit, multistage, Weibull, multi-hit, Logit, and Probit. Since it is presumed for carcinogens that at any dose, no matter how small, a response could occur in a sufficiently large population, an arbitrary risk level is usually selected (i.e., 1 in 100,000 or 1 in 1,000,000) as presenting an insignificant or de-minimus level of risk. By identifying these de-minimus levels as virtually safe levels, regulatory agencies do not give the impression that there is an absolutely safe exposure or that there is a threshold below which no response would be expected. Having calculated an exposure level or virtually safe dose (VSD), regulatory agencies can begin to make judgments about the biologic and economic feasibility and reasonableness of establishing a standard based on mathematical modeling. Often the use of models to help assess risks of exposure to carcinogens has been erroneously called "risk assessment." In practice, however, modeling is only one part of the risk assessment process. A true risk assessment to determine safe levels of occupational exposure actually requires exhaustive analysis of all of the information obtained from studies of mutagenicity, acute toxicity, subchronic toxicity, chronic studies in animals, pharmacokinetics and metabolism data as well as human epidemiology before a limit is recommended (119).

At this time, the use of quantitative risk modeling can be useful in the overall process of setting occupational exposure limits, but because of the dozens of shortcomings associated with them, especially their inability to consider complex biological events which surely must occur at low dose levels, they should not be used as the sole basis for deriving these limits (119).

3.7 Shortcomings in Extrapolating Animal Data

Part of the difficulty in extrapolating data from animals to humans is the inability to predict species differences. Differences in the uptake, metabolism, and elimination, as well as in the sensitivity of the animal to chemicals, are well documented (122). In many cases, even when physiological differences are accounted for mathematically (123), the extrapolation of animal data to humans is not always predictable. Therefore, some arbitrary safety factor is usually used which allows for the possibility that humans will be more sensitive to the substance than the species tested.

There is another difficulty which is often not recognized in extrapolating toxicity data from animal tests in order to set occupational standards: Animal tests are conducted under different conditions than occupational exposures. For example, inhalation toxicity tests are often limited to a few weeks, while exposures in the workplace could last for as many as forty years. Another difference in testing is that animals are often exposed for only 4–6 hours a day. While these short exposure periods are convenient for laboratory personnel who must load and unload inhalation chambers or manually dose the animals, the results of these tests are used to set TLVs for humans exposed for as many as 6–12 hours per day. Another shortcoming is that the physical activity of animals during an inhalation toxicity test is minimal since test animals are usually confined to small cages that restrict movement. For example, rats, the most commonly used test animal, when exercised, have ventilation rates that are five times the basal level. The impact of exertion or strenuous work in the uptake and elimination of airborne toxicants in workers has been studied by a number of scientists in recent years (124–128). Lastly, by dosing during the sleep phase of the test animal, the potential impact of the circadian rhythm on toxicity are possible. The differences in acute toxic response between exposures which occur during the sleep phase and awake phase can be dramatic (129). Table 6.4 contains a list of some of the differences in the environment in which test animals are exposed to a chemical and the conditions in which persons work.

Another shortcoming of the toxicity testing is that the concentration of the test substance in the inhalation chamber is maintained at a constant level. In contrast, 94 percent of the TLVs are not constant exposure limits but are time-weighted average limits (59). This means that the workplace concentration of these substances could deviate above the TLV as long as the average concentration was at, or below, the TLV. In the past, a maximum excursion level was recommended for excessively high exposures for short periods of time and

Table 6.4   Parameters which Often Differ Between the Conditions under which Test Animals and Working Populations are Exposed to Airborne Chemicals[a]

Variables	Toxicity Test, Animals	Occupation, Humans
Length of exposure	Short (weeks to months)	Long (up to 40 yr)
Interval of exposure	often 4 to 6 hr/day	8 hr/day
Ventilation	Resting	Physical work
Concentration of test substance in the air	Constant	Usually a time-weighted average of a varying concentration

[a] A modification of that described by McGregor (1973).

these excursions were based on the magnitude of the TLV. The excursion factors recommended up until 1976 were: 3-fold for TLVs from 0–1 ppm; 2-fold for TLVs from 1–10 ppm; 1.5-fold for TLVs from 10–100 ppm; 1.25-fold for TLVs from 100–1000 ppm.

Because the general formula approach to adjusting TLVs for short periods had limitations, the ACGIH TLV Committee in 1976 began adopting short-term exposure limits (STELs) for each of the chemicals. Because the first set of STELs was established in a generic manner, the ACGIH TLV Committee does not claim that the present list (1983–84) has strong scientific merit. They are, however, currently considering pharmacokinetic approaches for setting new STELs (134). It is of interest that there is fairly recent evidence that the hazard posed by exposure to fluctuating air concentrations of chemical vapors might produce a greater degree of adverse effects than exposure to the same dose (ppm-hr) delivered through a constant exposure concentration (135). An exception to this handrule may be carcinogens, since the overall daily dose seems to be as important as the intermittent short-term high levels in predicting the likelihood of producing a tumor (47, 185).

There is clearly a need for more biological studies to evaluate the need to adjust exposure limits for unusual conditions of exposure. First, it has been well established that biological response to inhaled substances is a function not only of concentration and time, but of uptake characteristics. Some substances well tolerated under TLV conditions may become hazardous when exposure is continuous, while other substances may be well tolerated at TLV levels even for continuous exposure. Second, a significant number of nonnormal work schedules already exists or can be anticipated to become sufficiently large to justify special toxicity testing so that standard setting groups will have data on which to recommend appropriate exposure limits for these situations. Third, the flucuation of concentrations of substances in workroom air may be more hazardous than exposure to a constant level of an air contaminant, and it has been documented that variable concentrations appear to be the rule rather than the exception in most workplaces.

Because so many persons are exposed to airborne substances for unusually short or long periods of time, several organizations have begun setting limits for these situations. The state of Pennsylvania promulgated short-term exposure limits soon after the OSHA regulations took effect in 1971 (136). Brief and Scala (18) and OSHA (137, 138) have developed simple, mathematical models aimed at adapting exposure limits to unusual situations. The National Institute for Occupational Safety and Health (NIOSH) has applied its recommended standard for chloroform (140) and benzene (141) to both 8- and 10-hour work shifts. The National Research Council has described the problems of adapting exposure limits to appropriate limits for long-term continuous exposure, such as might be encountered in space exploration (142). ACGIH has incorporated short-term exposure limits into its TLV list since 1976. Numerous pharmacokineticists and other researchers (143–149) have proposed complex models for adjusting occupational limits and a few toxicologists have begun to investigate the differences in toxicologic response between administration of the same daily dose over particularly short and long periods of exposure. These will be discussed in detail later in the chapter.

4 WHY ADJUST EXPOSURE LIMITS FOR UNUSUAL WORK SCHEDULES

It has been speculated by a number of researchers that some workers may be at increased risk of injury during unusual work schedules. In particular, industrial hygienists and occupational physicians have been concerned that unusual work schedules and the potential effects on the circadian rhythm might eventually compromise worker's capacity to cope with exposure to airborne toxicants. In some cases, it is clear that adjustments to the TLV or any other limit are necessary to provide persons on long workdays the same degree of protection given those persons on normal 8-hr workdays. Establishing acceptable limits of exposure for unusual work schedules is a difficult task since, at best, each person's susceptibility to stressors is dependent on many factors which are unique to that individual (42, 43).

Two questions regarding the occupational health aspects of the 12-hr schedules surfaced shortly after its commencement. One of the initial concerns of occupational physicians and industrial hygienists was whether the recommended limit for exposure to noise in the workplace, 85 dBA for 8 hr, was sufficiently low to protect workers on the twenve-hour shift. There was speculation that regulatory agencies such as OSHA might arbitrarily lower noise standard by 3–5 dBA for workers on these shifts. Second, since most federal standards for occupational exposure to airborne toxicants are based on a work schedule of 8 hr/day, 5 day/week, there was concern that a marked decrease in the concentration of air contaminants in the workplace might be necessary to protect workers. The concern that the existing exposure limits might not protect those on unusual work schedules was legitimate since both the noise and air contaminant limits were based on the results of either animal testing, which was conducted for 6.0 or less hr/day, or the epidemiological experience of workers who were exposed during normal, 8 hr/day work shifts.

4.1 Limits for Exposure to Noise During Unusual Shifts

Currently, the American Conference of Governmental Industrial Hygienists (ACGIH) recommends that exposures to sound be limited to 85 dBA for periods of 8 hr/day and to no more than 80 dBA if exposed 16 hr per day (41). The setting of a special guideline for those persons working 12-hr shifts has not been specifically addressed by the ACGIH, but one can infer that some noise level between 80 and 85 dBA would be appropriate. The regulatory agency responsible for setting standards which protect worker health, the Occupational Safety and Health Administration (OSHA), proposed, in 1974, a legal limit for occupational exposure to noise of 85 dBA time-weighted average

(TWA). This proposed standard was to apply to all persons who work an 8-hr day or longer, but the total exposure could not exceed 50 hr/week.

At this time, it is the opinion of several experts in the field of noise-induced hearing loss that routine exposure to a 12 hr/day, 3 or 4 day/week work schedule probably does not require a special noise limit as long as the workers are exposed throughout the year an average of no more than 40 hr/week. This seems reasonable since there is evidence that injury due to exposure to certain physical agents such as noise or radiation is solely a function of the intensity and duration of exposure, therefore, the health risk should not be markedly affected by slight alterations in the work regimens. In part, this is not surprising since the potential for cumulative effects due to the biologic half-life of the stressor is not a property of most physical agents.

4.2 Exposure Limits for Air Contaminants During Unusual Schedules

In the preface of the ACGIH publication *TLVs Threshold Limit Values For Chemical Substances and Physical Agents in the Work Environment With Intended Changes for 1983–84*, it states that the TLVs "refer to airborne concentrations of substances" and "represent conditions under which it is believed that nearly all workers may be repeatedly exposed day after day without adverse effect." The values refer to "a time-weighted average concentration for a normal 8-hr workday and a 40-hr workweek" (41). Implicit in this evaluation is an assumption that a balance exists between the accumulation of a contaminant in the body while exposed at work, and the elimination of the contaminant while away from work (during which time there is presumably no exposure) (144). Because of these assumptions, it is appropriate to consider whether and to what extent the TLVs and other limits require modification for unusual work schedules.

It is readily apparent to a toxicologist or pharmacokineticist that some modification of the TLVs may be necessary if these limits are to provide an equivalent amount of protection to persons working unusual shifts since, for example, a 12-hr work shift involves a period of daily exposure which is 50 percent greater than that of the standard 8-hr workday, and because the period of recovery before reexposure is shortened from 16 to 12 hr (18). For certain systemic toxins having half-lives between 5 and 500 hr, it can be envisioned that shifts longer than 8 hr/day would present a correspondingly greater hazard than that incurred during normal workweeks if the exposure (i.e., dose) were not reduced by some albeit small amount.

The very basis of occupational exposure limits for inhaled toxicants is that there is a maximum concentration of the air contaminant in the workplace to which persons can safely be exposed and suffer no adverse effect. Consequently, it can be envisioned that there is a corresponding maximum body burden at which an adverse effect is not likely. TLVs represent conditions under which it is believed that nearly all workers may be exposed day after day without adverse effect, so exposure to a substance at its TLV should result in some maximum burden of the substance in a particular target oragn of the body (144). Therefore, it seems logical that TLVs for certain chemicals should be modified if persons are exposed more than 8 hr/day so as to prevent this maximum body burden from being exceeded (see Figure 6.1).

Since the rationales for the more than 600 ACGIH TLVs vary, modification to some, if not most, of the air contaminant guidelines for unusually long exposure periods is not a trivial task. Many factors must be considered. For example, the need to lower the average exposure limit for certain industrial chemicals would be especially important in those situations where the safety factor in the TLV is small, toxicity data is limited, the toxic effect is quite serious, accumulation of the chemical is possible following several days of repeated exposure or where there is the possibility for extreme variability in workers response to a toxicant.

The procedures or models for modifying occupational exposure limits which have been proposed are likely to appear cumbersome or complicated to many practicing industrial hygienists and occupational physicians. The goal of this chapter is to clarify the rationale for making adjustments and to untangle the complicated aspects of the adjustment formulas. Furthermore, because many occupational health professions are interested in how to make the fullest use of exposure limits and since most modification procedures are based on pharmacokinetic principles, a basic discussion of this topic follows.

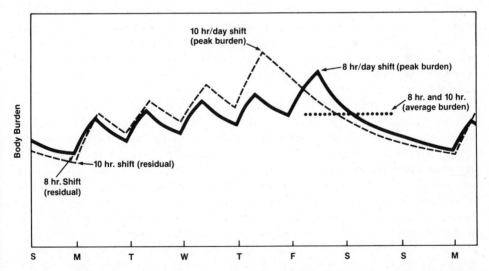

Figure 6.1 Comparison of the peak, average, and residual body burdens of an air contaminant following exposure during a standard (8 hr/day) and unusual (10 hr/day) work week. In this case, the weekly average body burdens are the same for both schedules since each involved 40 hr/week. However, the residual (Monday morning) body burden of the person who worked the 8-hr shift is greater than the 10-hr shift worker, and the peak body burden of the 10-hr shift worker is higher than the 8 hr worker. (Based on Hickey, 1977.)

5 CHEMICAL PHARMACOKINETICS

The principles of industrial pharmacokinetics (i.e., absorption, distribution, metabolism, and excretion) are reviewed with emphasis on those aspects which are necessary to understand the various models for adjusting occupational exposure limits for unusual periods of exposure.

5.1 Absorption of Chemicals

For workers, the two predominant routes of entry for industrial chemicals are absorption through the respiratory tract and the skin. Absorption which takes place through the lungs can be roughly assessed through analysis of the air which is breathed. The results can be interpreted by comparison with an exposure limit. Absorption via the skin, on the other hand, evades this method of measuring exposure. Uptake by the skin cannot be accounted for except by biological monitoring. It is acknowledged that methods for analyzing most chemicals in various excretory pathways or fluids have not been developed for most chemicals; therefore, when it is anticipated that a significant portion of the daily dose may be due to skin absorption, it can be assumed that a proportionally greater amount will be absorbed by persons who work on shifts longer than 8 hr/day.

Absorption of organic vapors may take place both in the upper and in the lower parts of the respiratory tract. Two basic processes may be observed here depending on the physical properties of the toxic substance. If the toxic substance is present in air in the form of an aerosol, absorption will often be preceded by deposition of the substance in the upper respiratory tract. When very small particles are present (less than 10 μm diameter), the deposition will be in the alveolar region (152).

The basic absorption mechanism for most industrial chemicals is gas diffusion and the factor which predicts the efficiency of this process is the partition coefficient between air and blood. Values for the coefficient between air and blood have been estimated for only a few substances, but the coefficients for partition between air and water are available for a far greater number of compounds. Usually, these have been obtained while working out sampling procedures involving absorption of a gaseous substance in water. By knowing the oil/water partition coefficient, one can often predict the relative efficiency of absorption through the skin and the lung. In addition, the partition coefficient and degree of solubility in water of a substance can give some insight as to the degree of distribution among the various tissues and the likely rate of elimination. For chemicals which have not been studied pharmacokinetically, these physical properties (especially the partition coefficient) can be very useful for predicting the chemical's behavior during unusually long exposures (152).

The respective values of the air/water partition coefficients for organic compounds may vary by several orders of magnitude: from about 10^0 for carbon disulfide, through 10^{-3} for acetone, acrylonitrile and nitrobenzene; to

10^{-5} for aniline and toluidine. The degree of pulmonary absorption (i.e., retention of vapors in the lungs) increases with a decreasing partition coefficient for air/blood (water); however, variation of the retention is much less than the variation of the respective partition coefficients. For example, lung retention of aniline vapor (partition coefficient 10^{-5}) is about 90 percent; nitrobenzene (partition coefficient 10^{-3}) up to 80 percent; benzene (partition coefficient 10^{-1}) up to about 50–75 percent, and finally, carbon disulfide (partition coefficient 10^{0}) up to about 40 percent (152).

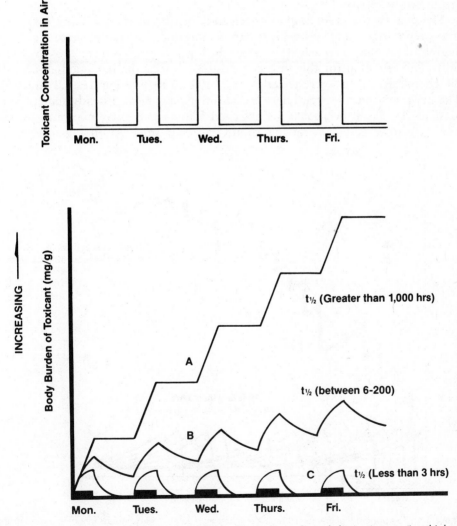

Figure 6.2 Body or tissue burden as a function of a chemical's uptake and elimination rate (i.e., biologic half).

5.2   Concept of Steady-State

The concept of a steady-state seems to be a difficult one for many persons to comprehend. It is easy to visualize how persons who are exposed to a chemical with a short half-life will rapidly eliminate the chemical so that the blood and tissue levels of the chemical return to zero before persons return to work the next day (e.g., carbon disulfide). The other extreme is represented by chemicals such as the polybrominated biphenyls, which have extremely long half-lives (whole body biologic half-life = 200 years). Here, the body burden never returns to zero and each successive dose adds to the burden. These phenomena are shown in Figure 6.2.

The goal of the modeling approach to adjust occupational limits, however, is to predict that daily or weekly accumulation does *not* take place to a greater degree during unusual schedules than during a normal 8 hr/day, 5 day/week shift. The potential for this to occur exists whenever the biologic half-life for the chemical in man is in the range of 3–400 hours (see Figures 6.2 and 6.3). The key point to understand here is that for any chemical, a steady-state plasma or tissue level will eventually be achieved following regular exposure to any work schedule and *even* during continuous exposure. At first glance, most

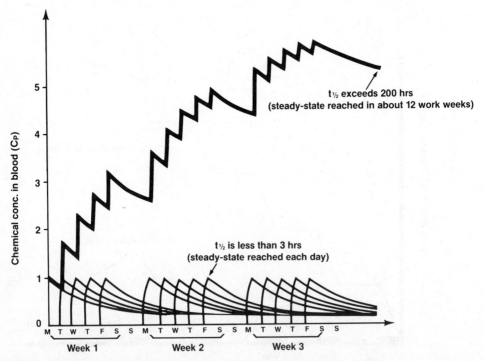

Figure 6.3   The principle of graphic summation assuming regular weekly periods free of exposure (from Piotrowski, 1971). Illustrates that biologic half-life determines the degree of day-to-day accumulation and the time to steady-state.

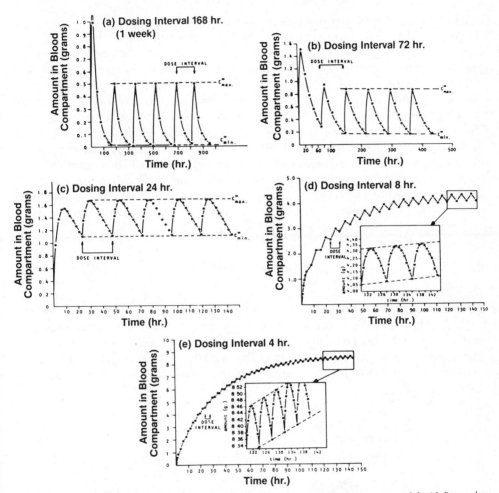

Figure 6.4 Multiple-dosing curves arising from one-compartment pharmacokinetic model with first-order absorption and elimination. These curves illustrate that the steady-state blood level and the time to reach steady-state are dependent on the period of time between exposures and the biologic half-life of the substance. (Reprinted with permission from Withey, 1979.)

persons might think that with continuous exposure, the body burdens will continuously increase as long as exposure is maintained. However, as shown in Figure 6.4, for any chemical, a steady-state blood or tissue level of the contaminant will be achieved when the rate of elimination of the drug equals the rate of absorption (153, 154).

During continuous inhalation exposure to volatile workplace gases and vapors, the concentration of the chemical in the blood increases toward an equilibrium between absorption, on the one hand, and metabolism and elimination, on the other. This is accompanied by a decreasing retention of the absorbed gases and vapors during each breath. This decrease of retention during the early periods of continuous inhalation exposure may be observed in practice for compounds whose air/blood (water) partition coefficient is of the order of 10^{-3} or greater. Decreasing retention is characteristic for carbon disulfide, tri- and tetra-chloroethylene, benzene, toluene, nitrobenzene and other chemicals (carbon disulfide is illustrated in Figure 6.5). If metabolism and distribution is rapid, this phenomenon is less pronounced and, for example, has not been observed for aniline (low partition coefficient) or for styrene (152). In the case of styrene, analogous to acrylonitrile, the explanation for lack of time-dependent decrease of retention seems to lie not so much in the magnitude of the physical partition coefficient as in the metabolism of the chemical.

The overall or average retention (R) of an organic vapor has been studied on human volunteers in chamber-type experiments, where it can be determined for a particular exposure period directly from the ratio of concentrations of the chemical in the inhaled and expired air:

$$R = \frac{C_i - C_e}{C_i} \tag{1}$$

C_i and C_e denote the concentrations in inhaled and exhaled air, respectively.

It is often useful for the industrial hygienist to know how to calculate an individual's anticipated daily uptake (dose) of a toxicant. One of the key determinants or elements in these calculations is the ventilation rate. In chamber-type experiments, in which volunteers are exposed in a sitting position and not

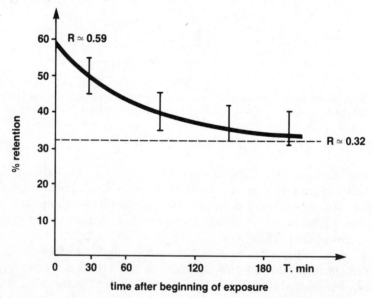

Figure 6.5 Retention of carbon disulfide vapors by the respiratory tract with continuous exposure. (Based on work of Jakubowski, 1966.)

subject to additional physical effort, the ventilation rate is usually of the order of 0.3–0.4 and 0.4–0.5 m^3/hr, in females and males, respectively (152). Differences are due to the differences in body weights and heart rates. It can be assumed that for people engaged in light work, corresponding to a slow walk, the rate is at least doubled. Specific ventilation rates have been determined for hundreds of tasks and should be used when applicable (155, 156).

In calculating the amount of a substance absorbed through the pulmonary tract during industrial work activities, various authors have used different average ventilation rates as typical for workers performing light work: from about 0.8 m^3/hr to about 1.25 m^3/hr. The latter figure (10 m^3 per 8-hr working shift) is most often used by industrial health professionals in the United States since it is almost always higher than actual levels and, therefore, conservative.

Taking the above factors into account, the total amount of a substance absorbed through the respiratory tract over a period of exposure would be the product of air concentration (C), duration of exposure (T), ventilation rate (V), and average retention rate (R) for the time of exposure (Figure 6.5):

$$C(\text{mg/m}^3) \times T(\text{hr}) \times V(\text{m}^3/\text{hr}) \times R(\%) = \text{Absorbed Dose (mg)} \qquad (2)$$

This simple formula is useful for estimating the acceptability of an air concentration of a toxicant in the workplace if the person is to be exposed for periods markedly longer than 8 hr/day or 40 hr/week. Even if the hygienist knows little else about the chemical's biologic or physical properties, he or she can limit the absorbed dose during the longer workday to that expected for a normal workday by calculating the air concentration for the nonnormal condition which would yield the same daily dose allowed for the normal schedule.

6   PHARMACOKINETIC MODELING OF INDUSTRIAL CHEMICALS

Pharmacokinetics is defined as the study of the rate processes of absorption, distribution, metabolism, and excretion of drugs and toxicants in intact animals. Many of the approaches or models which will be described for adjusting exposure limits for short, very long or continuous exposure periods are based on the pharmacokinetics of the inhaled toxicant, therefore, it is the objective of this section to review the principle concepts.

The rationale for making adjustments to ACGIH TLVs, OSHA PELs, European MACs, AIHA WEELs or any other conventional occupational exposure limit is the fundamental assumption that "exposure to a substance in air at its TLV (or equivalent limit) for five 8-hr days per week, results in some burden in the body which by experimental observation or experience has been inferred to be an acceptable exposure for most workers" (145). In order to help insure that persons who work on unusual shifts are not placed at greater risks of injury than workers on normal 8 hr/day shifts, pharmacokinetics have been incorporated into the models which have been developed to adjust (lower) the limit for persons on longer shifts or increase the limit for shorter shifts so that the "peak" body burden of the unusual shift worker is no greater than that of the normal shift worker.

Pharmacokinetic modeling is the science of describing, in mathematical terms, the time course of drug and metabolite concentrations in body fluids and tissues (157–159). Most often, the modeling of a chemical's behavior is based on the blood concentration of a contaminant versus the time course of its concentration in the blood following exposure. It is generally assumed that blood is in dynamic equilibrium with all tissues and fluids of the body, consequently, it is the biologic medium of choice for monitoring. In some cases, it may be necessary or useful to measure the output of the parent chemical or drug and the metabolites in the urine, breath and feces and then model this profile mathematically.

A pharmacokinetic model is regarded as being accurate if it correctly predicts the concentration of a substance or its metabolite in the blood, urine, feces, or expired air for as long as these substances can be measured analytically. The model may be tested by comparing its predictions with experimental data after exposure to the drug by intravenous, oral or inhalation administration. The effects of various dosing regimens can also be modeled. Usually, recovery of 95+ percent of the administered dose(s) as unchanged drug and metabolites in urine, breath and feces, and agreement between the recovered amounts and the pharmacokinetic model is considered confirming evidence that the model accounts for the fate of the absorbed chemical or drug (157).

6.1 Concept of Half-Life

Most biologic processes follow first-order kinetics. In other words, the rate of elimination or metabolism is known to be a function of the concentration of all reactive species and, where there is only one of these, the reaction is termed first-order and the rate is directly proportional to the concentration:

$$\text{rate} = dC/dt = kC \tag{3}$$

This is the differential form of a first-order reaction and is of little practical use since dC/dt, although it can be found from the tangent of the time–concentration curve, is very difficult to measure precisely.

The integrated form of Equation 3 is more useful:

$$C_t = C_0 e^{-kt} \tag{4}$$

C_0 is the initial concentration and is sometimes designated a; it is not unusual to find C_t, the concentration at time t, designated $a - x$, where x is the amount that has disappeared at the time t. Thus Equation 4 sometimes appears as

$$\ln \frac{a}{a - x} = kt \tag{5}$$

By definition, the half-life for any first-order kinetic process ($t_{1/2}$) is the time taken for one-half of the original amount or concentration to disappear. So from Equation 5, if $C_t = \frac{1}{2} C_0$ at time $t_{1/2}$ then

$$\ln \frac{C_0}{C_0/2} = kt_{1/2} \tag{6}$$

$$\ln 2 = kt_{1/2} \tag{7}$$

$$t_{1/2} = \frac{\ln 2}{k} = \frac{2.303 \times 0.3010}{k} \text{ (e.g., min, hr)} \tag{8}$$

Note that $t_{1/2}$ is independent of concentration for a first-order process and that the larger the half-life, the slower the elimination or reaction rate, i.e., the smaller the rate coefficient k. The fundamental concepts are illustrated in Figure 6.13.

When describing the elimination of a chemical, it is often convenient to use the term biologic half-life. The biologic half-life of a chemical is the time needed to eliminate 50 percent of the absorbed material; either as the parent compound or one of its metabolites. As will be shown, the biologic half-life is the most important criterion for assessing whether a TLV should be adjusted for either very short or very long durations of exposure. As will be shown later, the biologic half-life and the exact exposure regime (time exposed and recovery time) are the two factors which determine the degree of adjustment needed to provide equal protection.

7 PHARMACOKINETIC MODELS

In pharmacokinetics, the dynamic behavior of toxic substances can be conveniently described in terms of mathematical compartments in which a "compartment" represents all of the organs, tissues and cells for which the rates of uptake and subsequent clearance of a toxicant are sufficiently similar to preclude kinetic resolution. This simplification of the body introduces few errors in the final analysis and is much more realistic than attempting to describe the fate of a chemical in each major tissue (e.g., liver, heart, lung, etc.) (Figure 6.6). The pharmacokinetic or mathematical model is selected so as to yield the best fit of the data irrespective of whether the compartments have a physiologic identity.

In general, most models divide the body into from one to five compartments. Fiserova-Bergerova et al. (143–147) have suggested an approach to compartment categorization based on perfusion, ability to metabolize the inhaled substance, and the solubility of the substance in the tissue. Lung tissue, functional residual air, and arterial blood form the central compartment LG (lung group), in which pulmonary uptake and clearance take place. The partial pressure of inhaled vapor equilibrates with four peripheral compartments. Vessel-rich

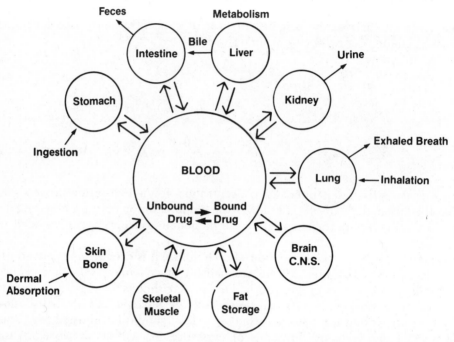

Figure 6.6 A diagram illustrating the potential distribution of a chemical following exposure. The difference in the rates of absorption or elimination between these tissues is the reason for using multicompartment pharmacokinetic models.

tissues form two peripheral compartments: BR-compartment (blood rich) includes brain, which lacks capability to metabolize most xenobiotics, and is treated as a separate compartment because of its biological importance and the toxic effect of many vapors and gases on the central nervous system. VRG-compartment (vessel rich group) includes tissues with sites of vapor metabolism such as liver, kidney, glands, heart, and tissues of the gastrointestinal tract. Muscles and skin form compartment MG (muscle group), and adipose tissue and white marrow form compartment FG (fat group). The FG-compartment is treated separately, since the dumping of lipid soluble vapors in this compartment has a smoothing effect on concentration variation in other tissues and these variations caused by changes in exposure concentrations, minute ventilation, and exposure duration would be more dramatic if not for the "buffering" effect of adipose tissue.

An example of a complex pharmacokinetic model which describes the possible fate of an inhaled substance is depicted in Figure 6.6. An illustration of the possible fate of a common solvent like tetrachloroethylene in the various "imaginary" compartments is shown in Figures 6.7 and 6.8. The key point illustrated here is that fat or lipid soluble chemicals will quickly reach eqilibrium in the highly perfused tissues and they will also be removed quickly following cessation of exposure. However, the less perfused tissues, which are usually high in lipid, *do not* reach saturation quickly; more importantly, they take a great deal of time to reach background levels. These graphs also illustrate that these differences in distribution can occur at levels near the TLV.

7.1 One-Compartment Model

The simplest pharmacokinetic model is the one-compartment open model illustrated in Figure 6.9. Here, the body is viewed as a single homogeneous box with a fixed volume. A drug or chemical entering the systemic circulation by any route is instantaneously distributed throughout the body. Although the absolute concentrations in all body tissues and fluids are not identical, it is assumed that they all rise and fall in parallel as drug is added and eliminated.

Overall elimination of drug from the body by urinary excretion and/or metabolism usually obeys first-order kinetics; that is, the rate of elimination is proportional to drug concentration in blood or plasma, Cp, as shown by the

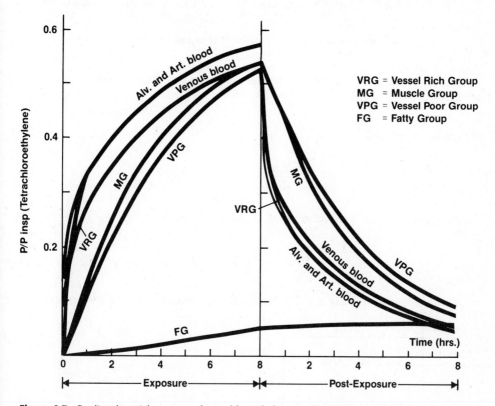

Figure 6.7 Predicted partial pressure of tetrachloroethylene in alveolar air, mixed venous blood, and tissue groups during and after 8 hrs of exposure to a constant air concentration of tetrachloroethylene. Partial pressure in alveoli, blood, and tissues (P) are expressed as a fraction of the constant partial pressure in ambient air during exposure (P_{insp}). (From Guberan and Fernandez, 1975.)

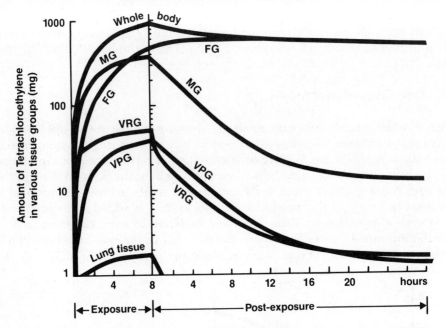

Figure 6.8 Predicted distribution of tetrachloroethylene to the tissue groups during and after 8-hr exposure to 100 ppm. In contrast with many other volatile chemicals, tetrachlorethylene demonstrates some persistence in fatty tissues. (From Guberan and Fernandez, 1974.)

following equation:

$$\frac{dCp}{dt} = -k_{el}Cp \tag{9}$$

Rearrangement and integration of Equation 9 gives

$$\log Cp = -\frac{k_{el}(t)}{2.3} + \log Cp^0 \tag{10}$$

which says that following an IV (intravenous) dose, a semilog plot of plasma

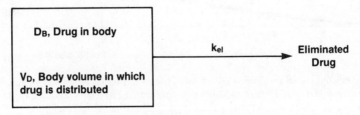

Figure 6.9 One-compartment open pharmacokinetic model.

concentrations versus time will be a straight line with a slope of $-k_{el}/2.3$ and an initial plasma concentration of Cp^0.

The best way to determine whether a one-compartment model is appropriate for a given drug is to plot the intravenous (IV) plasma concentrations against time on semilog paper and see if the plot is a straight line (Figure 6.10). This approach permits determination of the two key parameters of the model: k_{el}, the overall elimination rate constant, and V_D, the apparent volume of distribution of the drug in the body.

7.2 Two-Compartment Model

Often, it is clear from the blood plasma curves that there is a slower distribution to some tissues. In this case, the kinetic model that will allow the interpretation of the observed experimental data is a two-compartmental model, illustrated in Figure 6.11. In the two-compartment model, the body is essentially reduced to an accessible compartment, the blood, and a second less accessible and diffuse

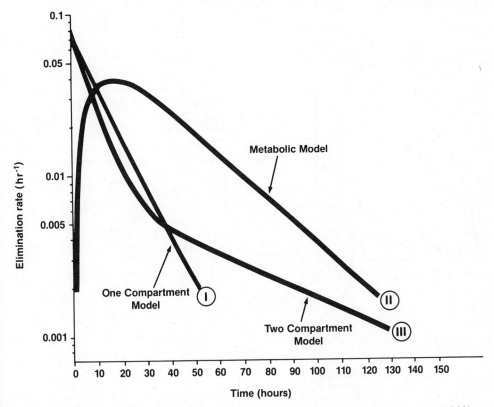

Figure 6.10 Three basic types of elimination curves following a single instantaneous exposure (I.V.) at times equals zero. (From Piotrowski, 1977.)

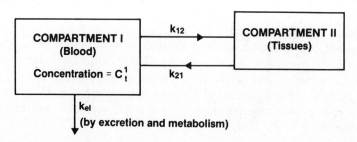

Figure 6.11 Schematic of the two-compartment pharmacokinetic model.

compartment, the tissues. In the physiological sense, as soon as the molecules in the administered dose mix with the blood, they will be rapidly carried to all parts of the body and brought into intimate contact (by perfusion) with organs, tissues, fat depots, and even bone. Some tissues, like the liver and the kidney, are very well perfused; others, such as muscle and fat, may be poorly perfused. Molecules that are carried to the perfused organs and tissues will then diffuse from the blood across cellular membranes and a dynamic equilibrium will usually be established (see Figure 6.12) (159).

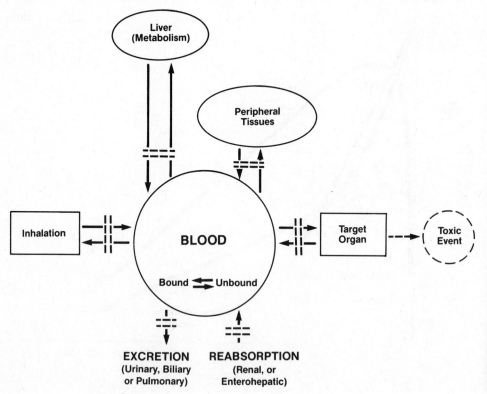

Figure 6.12 The physiologic model for toxin action proposed by Withey (1979) modified to emphasize the fate of inhaled substances. The dashed lines represent diffusion barriers.

The distribution of a chemical as described by a two-compartment model is readily apparent in a plot of blood versus time following exposure. A blood concentration–time curve following an intravenous bolus administration, which exhibits the characteristics of distribution, is shown in Figure 6.13. An inspection of this kind of plot reveals an initial rapid (steep slope) nonlinear portion followed after a time t (which depends on the nature and characteristics of the administered substance) by a slower linear portion. This linear portion, sometimes referred to as the terminal phase, has a definite slope from which a "rate coefficient" can be evaluated. If the rate coefficient for this linear terminal phase is defined as β, and the intercept of the line extrapolated back to the ordinate axis is B, then this part of the relationship can be described by

$$C_t^1 = Be^{-\beta t}. \tag{11}$$

From the initial curved portion of the elimination curve, the rate at which the central compartment releases the toxicant is the slope of the "feathered" line. This can be determined easily from the method of residuals (154). Having determined A, the intercept of the feathered line, and α, the slope, the whole experimentally observed curve can be described by the biexponential equation

$$C_t^1 = Ae^{-\alpha t} + Be^{-\beta t} \tag{12}$$

where C_t^1 = concentration of toxicant in blood at any time following exposure.

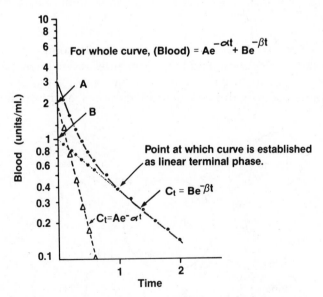

Figure 6.13 Semilogarithmic plot of blood concentration against time after intravenous administration illustrating biphasic elimination (two compartments) and the use of "curve stripping" to resolve the "A" intercept.

It is essential to note that in this case $Ae^{-\alpha t}$—that is, the initial phase of the curve, dominated by the rate coefficient α, is essentially over and completed when $e^{-\alpha t}$ approaches 0 and thereafter the curve is described by $C_t = Be^{-\beta t}$ (142).

The individual rate coefficients α and β, obtained empirically from the plotted data, and the intercepts A and B can then be used to determine the individual rate coefficients of the model (159):

$$k_{21} = \frac{A\alpha - B\beta}{\alpha - \beta} \tag{13}$$

$$k_e = \frac{\alpha\beta}{k_{21}} \tag{14}$$

$$k_{12} = \alpha + \beta - k_{21} - k_e \tag{15}$$

where k_{21} = rate transfer coefficient for the movement of molecules from compartment two to compartment one

$\quad\;\; k_{12}$ = rate transfer coefficient for the movement of two molecules from compartment one to compartment two

$\quad\;\; k_e$ = rate transfer coefficient for the elimination of the chemical from the system

Curve-fitting computer programs are available, which require only rough estimates of the four parameters, i.e., the slopes α and β and the intercepts A and B, to allow more precise calculation of the individual rate coefficients from the animal or human data.

To consider the body as one or two compartments involves considerable simplification of reality. To account for each of the 10 compartments shown in Figure 6.6 would involve considerable experimental and mathematical difficulty and could conceivably require nine different equations to resolve the chemical's behavior. A more complex three-compartment model, as shown in Figure 6.14,

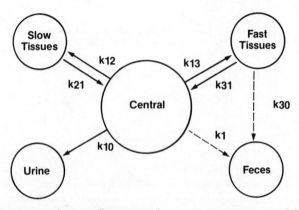

Figure 6.14   Schematic illustrating a three-compartment open model.

is usually the most rigorous model used for data analysis. Due to the minimal practical benefits of performing such rigorous mathematical analysis, combined with the comparatively less quantitative basis by which most occupational exposure limits are established, toxicologists should use Occam's razor in that the simplest model describing the available data is always the best one!

The one-compartment model (which is the one used in all pharmacokinetic models recommended for adjusting exposure limits) is consistent with the data in a great many cases (157), but in many others it provides a poor fit (165). For the purpose of adjusting exposure limits, when a particular chemical appears to exhibit the characteristics of a multicompartmental distribution, it is suggested that it be forcibly resolved into no more than two compartments. This can be accomplished through the use of several computer programs which are currently available (166, 191, 193).

7.3 Accumulation of Chemicals in the Body

Cumulation of a substance in the body is a process wherein the concentration of a particular substance increases following repeated or continuous exposure. When using biologic monitoring to evaluate occupational exposure, cumulation is likely when the concentration of a substance in the analyzed media (urine, feces, blood, or expired air) increase with each day or week of exposure. This phenomenon has been observed even during normal work schedules even when air contaminant concentrations were below the TLV (167). Therefore, for these chemicals exposure during unusually long work periods might be expected to increase the cumulation. In some cases, steady increases in the concentration of the inhaled contaminant in adipose tissue may be more indicative of any potential adverse effects following chronic exposure.

Cumulation, if it occurs, results from a slow turnover of the substance. Thus, it may take place under conditions of every kinetic model, provided the elimination rate constant is low (long biologic half-time). From theoretical considerations, it can be expected that the highest value of the rate constant (single compartment model) at which cumulation might occur would not exceed 0.1 hours^{-1} (143). The tendency of a chemical to cumulate in the body with repeated exposure can be due to several factors. For substances that are eliminated mainly unaltered in the breath, the cause of slow turnover may be the low air/blood partition coefficient or deposition in poorly perfused compartments such as fat; for substances excreted by the kidney, a low clearance may result from poor glomerular filtration, or intensive tubular reabsorption, or both. In practical situations, two other events may be of significance, namely; slow biotransformation (when excretion takes place in metabolized form) and deposition in adipose tissue. These factors can act in combination, as has been shown for nitrobenzene and DDT (152).

In short, the likelihood that a chemical will cumulate in the body or a key tissue is dependent on the overall rate at which a chemical is absorbed, metabolized and excreted. The time required for the concentration of the parent chemical or its metabolite to be reduced by one-half in a particular medium (e.g., blood) or a tissue (e.g., fat), is called the biologic half-life. It must be emphasized that the biologic half-life varies for each substance and is dependent on the animal species studied as well as the route of exposure. As a result, the estimation of the biologic half-life of a substance in humans based on animal data usually requires considerable effort. As shown in Figure 6.15, chemicals with moderate half-lives (greater than 4 hr and less than 400 hr) are likely to show some degree of day-to-day cumulation during the workweek even at exposures at or near the TLV (145, 150). Cumulation of a chemical during an exposure regimen (workweek or year) is not necessarily detrimental as long as the peak or steady-state tissue levels do not reach levels which are above the threshold concentration for such endpoints as cytotoxicity, behavioral toxicity, etc. This weekly increase in body burden in exposed workers has been demonstrated for some chemicals (167).

One of the primary concerns with unusual work schedules is the possibility that day-to-day increases in the toxicant concentration at the site of action might occur during unusually long work shifts to a much greater degree than that which occurs during standard 8 hr/day schedules. This potential problem is illustrated in Figure 6.16 where it is shown that a chemical with a 24-hr biologic half-life would clearly provide a greater peak body burden following 4 days of exposure on a 10 hr/day schedule than that observed in the workers exposed 8 hr/day for 5 days unless the chemical concentration in air was not lowered. The point here is that even when the number of hours of exposure per week are the same, for some chemicals, the peak body burden may be different when the work schedules vary. In Figure 6.16, C_T represents the concentration in the blood following a certain period of exposure and C_S is the concentration in the blood at the saturation level attained following continuous

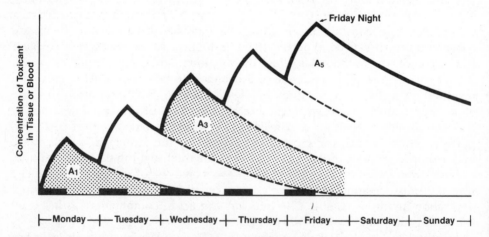

Figure 6.15 Day-to-day increase of toxicant concentration in tissue or blood following exposures of 8 hr/day, 5 day/week at TLV levels. For this type of cumulation to occur, the half-life of the air contaminant or its metabolite would be about 6–200, as shown in Figure 6.2.

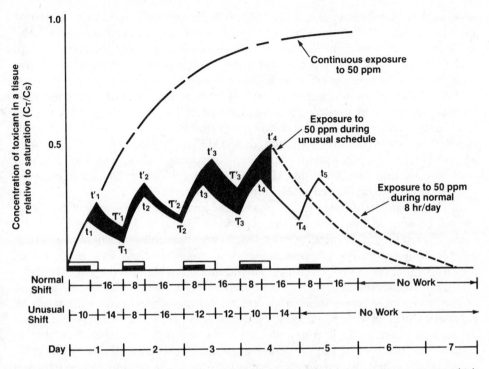

Figure 6.16 The accumulation of a substance during irregular periods of intermittent exposure which could occur during unusual work shifts and overtime. This plot illustrates that even when the total weekly dose (40 hr × 50 ppm) is unchanged, the peak body burden for two different shift schedules may not be equivalent. To help ensure that the peak tissue concentration for the unusual exposure schedule does not exceed the presumably "safe" level of the normal schedule, the air concentration of toxicant in the workplace may have to be reduced. K_{elim} has a value of 0.03 hr^{-1} in this illustration. (Example by Dr. J. Walter Mason.)

exposure to that same air concentration; t represents a schedule involving five 8-hr workdays per week, and t' represents a workweek consisting of four workdays: 10 hr/day, 8 hr/day, 12 hr/day, and another 10 hr/day followed by 3 days off work. Even though this unusual work schedule involves 40 hr of exposure to 50 ppm of a chemical vapor, exposed workers are likely to have greater peak body burdens than persons who are also exposed for 40 hr per week but only 8 hr per day.

7.4 Nonlinear Pharmacokinetic Models

All the models discussed so far have been based on the assumption that all pharmacokinetic pathways can be described by linear differential equations with constant coefficients and first-order rate constants. This need not always be true and it is important to recognize that in some cases nonlinear behavior may occur. This can have a great deal of effect on the estimation of risks associated with exposure (48, 159, 168, 169). In general, the concentration of an air contaminant at or near the occupational exposure limit will not be sufficient to saturate the blood or bring about nonlinear behavior (170). However, this could occur if exposures were continuous or at high concentrations for short periods, even though the acceptable 8-hr time-weighted average (TWA) concentration was not exceeded. This phenomenon must be understood by the hygienist and toxicologist when interpreting some oncogenicity studies wherein the only positive (carcinogenic) response was observed in those animals exposed to the maximum tolerated dose (MTD). In certain cases, it is only when saturation occurs that an oncogenetic response will be seen and, in these cases, the minimal likelihood that Michaelis-Menten kinetics will exist in exposed workers must be considered when setting exposure limits.

A dose-dependent change in pharmacokinetic behavior may arise from several causes. At "saturation," the capacity of some biological system in the overall process can become overloaded. Some examples of chemicals which exhibit dose-dependent pharmacokinetics at concentrations well in excess of the TLV include vinyl chloride, vinylidene chloride, methylene chloride, methanol, benzene and 2, 4, 5-T. Dose-dependent effects are of special interest to the toxicologist since some toxic effects are associated only with large or excessive doses of drugs. Alcohol pharmacokinetics is one example of dose-dependent kinetics. As is frequently the case, the dose-dependence is due to metabolic saturation of the particular enzyme; in this case, alcohol dehydrogenase (152). When this enzyme system is saturated, the person will be exposed to higher levels of unmetabolized alcohol via the blood than would normally occur at lower levels of alcohol ingestion.

In many of the cases cited in the literature, the observed dose-dependence has been attributed to changes in the metabolic reaction involved. The metabolism of the drug may be due to an enzymic reaction and thus would behave according to the classical Michaelis-Menten model of enzyme kinetics (Figure 6.17). Such a system involves reaction of drug as substrate with enzyme to produce a metabolite via an intermediate enzyme-drug complex. At low drug concentration, the reaction is overall first-order with a rate constant, $k_m = V_m/K_m$. At high drug concentration, the capacity of the limited amount of enzyme present for reaction is exceeded and the reaction rate becomes constant (V_m). Thus, the production of metabolite becomes a constant rate or zero-order process.

The transformation from first-order to zero-order reaction rate is not an abrupt jump but a graded transition. During this transition the kinetics may remain apparently first-order but the elimination constant changes. Thus, both types of dose-dependence effects mentioned earlier, may be explicable by a model of the Michaelis-Menten type. For those chemicals for which the rationale for the TLV or PEL is based on effects which occur only when the metabolic process is overloaded or saturated, such as vinyl chloride's carcinogenicity, it is suggested that no adjustment to the TLV is probably needed to protect workers on long shifts as long as the average weekly exposure throughout the month is

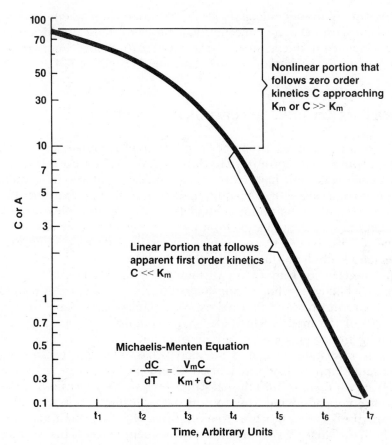

Figure 6.17 Simulated plasma concentration of a chemical (C) or the amount in the body (A) as a function of time for a chemical which displays dose-dependent or nonlinear pharmacokinetics described by the Michaelis-Menten equation. $\frac{dC}{dT}$ is the change in concentration with time. V_m is the maximum rate of the process, and K_m is the Michaelis constant. (From Gehring and Young, 1979.)

Table 6.5 Possible Causes of Dose-Dependent Pharmacokinetics

Metabolic saturation	a.	Self inhibition by excess drugs
	b.	Insufficient metabolic enzyme
	c.	Competition for co-enzyme or co-substrate
Excretory saturation	a.	Competition for tubular secretion or resorption mechanism
	b.	Changed renal function
Changed drug distribution	a.	Protein and tissue binding effect
	b.	Overflow into new volumes of distribution

From Withey (1979).

149

about 40 hours. This suggestion is based on the observation that the occupational exposure limits are set at air concentrations far below those at which saturation and dose-dependent kinetics are likely to occur (170). Some common causes of dose-dependent pharmacokinetics are shown in Table 6.5 (159).

8   MODELS FOR ADJUSTING OCCUPATIONAL EXPOSURE LIMITS

Several researchers have proposed mathematical formulae or models for adjusting occupational exposure limits (PELs, TLVs, etc.) for use during unusual work schedules and these have received a good deal of interest in the industrial and regulatory arenas (18, 143–149, 171–175). Although the Occupational Safety and Health Administration (OSHA) has not officially promulgated specific exposure limits applicable to unusual work shifts, they have published guidelines for use by OSHA compliance officers for adjusting exposure limits. These generally apply to shifts longer than 8 hr/day (138).

Under the General Duty Clause of the OSHA Act, Section 5A1, employers which have employees who, during long shifts are exposed to workplace air concentrations in excess of adjusted exposure limits, are citeable for noncompliance. All of the models which an OSHA compliance officer might use in determining the acceptability of an air contaminant level are presented and discussed in this chapter. Each of the models has limitations because a generalized formula approach (GFA) is used. These shortcomings in determining modified exposure limits for unusual schedules have been discussed (176, 177), but they are not considered as serious as not adjusting the limits at all (which is perhaps the current practice of most firms which use unusual shifts). It should be noted that OSHA has, thus far, not actively cited firms which have unusually long shifts and are not in compliance with these modified limits.

8.1   Brief and Scala Model

In the early 1970s, due to the increasingly large number of workers who had begun working unusual schedules, the Exxon Corporation began investigating approaches to modifying the occupational exposure limits for their employees on 12-hr shifts. In 1975, the first recommendations for modifying TLVs and Permissible Exposure Limits (PELs) were published by Brief and Scala (18), wherein they suggested that TLVs and PELs should be modified for individuals exposed to chemicals during novel or unusual work schedules.

They called attention to the fact that, for example, in a 12-hr workday the period of exposure to toxicants was 50 percent greater than in the 8-hr workday, and that the period of recovery between exposures was shortened by 25 percent, or from 16 to 12 hr. They noted that repeated exposure during longer workdays might, in some cases, stress the detoxication mechanisms to a point that accumulation of a toxicant might occur in target tissues, and that alternate pathways of metabolism might be initiated. If higher peak blood levels were to occur during long shifts, these could conceivably cause acute cellular damage that would not generally occur during standard work schedules.

Brief and Scala's approach was simple but important since it emphasized that unless worker exposure to systemic toxicants was lowered, the daily dose would be greater, and due to the lesser time for recovery between exposures, peak tissue levels might be higher during unusual shifts than during normal shifts. This concept is illustrated in Figure 6.1. Their formulae (16a and 16b) for adjusting limits are intended to insure that this will not occur during novel work shifts:

$$\text{TLV Reduction Factor } (RF) = \frac{8}{h} \times \frac{24\text{-}h}{16} \qquad (16a)$$

where h = hours worked per day.

For a 7-day workweek, they suggested that the formula be "driven by" the 40 hr exposure period; consequently they developed formula (16b), which accounts for both the period of exposure and period of recovery:

$$\text{TLV Reduction Factor } (RF) = \frac{40}{h} \times \frac{168\text{-}h}{128} \qquad (16b)$$

where h = hours exposed per week.

One advantage of this formula is that the biologic half-life of the chemical and the mechanism of action are not needed in order to calculate a modified TLV. Such a simplification has shortcomings since the reduction factor for a given work schedule is the same for all chemicals, even though the biologic half-life of the chemicals vary widely. Consequently, this formula over-estimates the degree to which the limit should be lowered.

The authors were cautious in describing the strength of their proposal and offered the following guidelines for its use. The following represent concepts which were offered by Brief and Scala (18) and should be considered when applying not only this model but also the others that follow:

1. Where the TLV is based on systemic effect (acute or chronic), the TLV reduction factor will be applied and the reduced TLV will be considered as a time-weighted average (TWA). Acute responses are viewed as falling into two categories: (a) rapid with immediate onset and (b) manifest with time during a single exposure. The former are guarded by the C notation and the latter are presumed time and concentration dependent and hence, are amenable to the modifications proposed. Number of days worked per week is not considered, except for a 7-day workweek discussed later.

2. Excursion factors for TWA limits (Appendix D of the 1974 TLV publication) will be reduced according to the following equation:

$$EF = (EF_8 - 1)RF + 1$$

where EF = desired excursion factor
EF_8 = value in Appendix D for 8-hour TWA
RF = TLV Reduction Factor

3. Special case of 7-day workweek. Determine the TLV Reduction Factor based on exposure hours per week and exposure-free hours per week.

4. When the novel work schedule involves 24-hour continuous exposure, such as in a submerged habitat or other totally enclosed environment designed for living and working, the TLV reduction technique cannot be used. In such cases, the 90-day continuous exposure limits of the National Academy of Science should be used, where applicable limits apply.

5. The techniques are not applicable to work schedules less than seven to eight hours per day or 35 hours per week.

Brief and Scala also correctly noted that:

The *RF* value should be applied a) to TLVs expressed as time-weighted average with respect to the mean and permissible excursion and b) to TLVs which have a *C* (ceiling) notation except where the *C* notation is based solely on sensory irritation. In this case the irritation response threshold is not likely to be altered downward by an increase in number of hours worked and modification of the TLV is not needed.

In short, the Brief and Scala formula is dependent solely on the number of hours worked per day and the period of time between exposures. For example, for any systemic toxin, this approach recommends that persons who are employed on a 12 hr/day, 3- or 4-day workweek, should not be exposed to air concentrations of a toxicant greater than one-half that of workers who work on an 8 hr/day, 5-day schedule.

In their publication, Brief and Scala acknowledged the importance of a chemical's biologic half-life when adjusting exposure limits, but because this information is rarely available, they felt comfortable with their proposal. These authors noted that a reduction in an occupational exposure limit is probably not necessary for chemicals whose primary untoward effect is irritation since the threshold for irritation is not likely to be altered downward by an increase in the number of hours worked each day (18). Although this appears to be a reasonable assumption, some researchers feel that it may not be entirely justified since duration of exposure could, in fact, be a factor in producing irritation in susceptible individuals who are not otherwise irritated during normal 8 hr/day exposure periods (178, 179).

ILLUSTRATIVE EXAMPLE 1 (Brief and Scala Model). Refinery operators often work a 6-week schedule of three 12-hr workdays for 3 weeks, followed by four 12-hr workdays for 3 weeks. What is the adjusted TLV for methanol (1983 TLV = 200 ppm) for these workers? Note that the weekly average exposure is only slightly greater than that of a normal work schedule.

Solution.

$$RF = \frac{8}{12} \times \frac{24 - 12}{16} = 0.5$$

New TLV = RF × TLV

$\qquad$ = 0.5 × 200 ppm

$\qquad$ = 100 ppm

Note: The TLV Reduction Factor of 0.5 applies to the 12-hr workday, whether exposure is for 3, 4, or 5 days per week.

ILLUSTRATIVE EXAMPLE 2 (Brief and Scala Models).   What is the modified TLV for tetrachloroethylene (1983 TLV = 50 ppm) for a 10 hr/day, 4 day/week work schedule if the biologic half-life in humans is 144 hr?

Solution.

$$RF = \frac{8}{10} \times \frac{24 - 10}{16} = 0.7$$

New TLV = 0.7 × 50 = 35 ppm

Note: This model and the one used by OSHA *do not consider* the pharmacokinetics (biologic half-life) of the chemical when deriving a modified TLV. Other models to be discussed later *do* take this information into account.

ILLUSTRATIVE EXAMPLE 3 (Brief and Scala Model).   In an 8-hr day, 7-day workweek situation, such as the 56/21 schedule, persons work 56 continuous days followed by 21 days off. What is the recommended TLV for H_2S (1983 TLV = 10 ppm) for this special case of a 7-day workweek? Assume that the biologic half-life in humans for H_2S is about 2 hr and the rationale for the limit is the prevention of irritational and systemic effects.

Solution.   Exposure hr per week = 8 × 7 = 56 hr
Exposure-free hr per week = (24 × 7) − 56 = 112 hr

$$RF = \frac{40}{56} \times \frac{112}{128} = 0.625$$

New TLV = RF × 10 ppm = 6 ppm.

ILLUSTRATIVE EXAMPLE 4 (Brief and Scala Model).   Ammonia has a TLV of 25 ppm and is an upper respiratory tract irritant. What is the modified TLV for this chemical for a work schedule of 14 hr/day for 3 day/week?

Solution.   Since the rationale for the limit for ammonia is the prevention of irritation, *no* adjustment (lowering) of the limit is needed.

8.2 OSHA Model

Many toxicologists believe that, in general, the intensity of a toxic reponse is a function of the concentration that reaches the site of action (164). This principle is simplistic and, while it *may not* apply to irritants, sensitizers and carcinogens, it is clearly true for most systemic toxics. This assumption is the basis for the OSHA model for modifying PELs for unusual shifts (138). The originators of the model assumed that for chemicals which cause an acute response, if the daily uptake (concentration × time) of a chemical during a *long workday* was limited to the amount which would be absorbed during a *standard workday,* then the same degree of protection would be given to workers on the longer shifts. For chemicals with cumulative effects (i.e., those with a long half-life), the adjustment model was based on the dose imparted through exposure during the *normal workweek* (40 hours) rather than the *normal workday* (8 hours).

OSHA recognized that the rationale for the occupational exposure limits for the various chemicals was based on different types of toxic effects. After OSHA adopted the some 500 TLVs of 1968 as the OSHA list (137) of Permissible Exposure Limits (29 CFR 1910.1000), they placed each of the chemicals into one of the six "Work Schedule" categories (Table 6.6) to assure that an appropriate adjustment model would be used by their hygienists whenever they evaluated employers who used unusual work schedules. As can be seen in illustrative Examples 5, 6, and 7, the degree to which an exposure limit is to be adjusted, if at all, is based to a large degree on the work schedule category in which a chemical is placed.

The parameters by which OSHA and their expert consultants categorized (Table 6.6) the various chemicals were the primary type of health effect to be prevented, biologic half-life (if known), and the rationale for the limit. The categories include (1) ceiling limit, (2) prevention of irritation, (3) technological feasibility limitations, (4) acute toxicity, (5) cumulative toxicity, as well as (6) acute and cumulative toxicity. A review of the degree of adjustment required for each type of chemical can be found in Table 6.6. Table 6.7 contains the names of the categories of the different health effects into which the 500 chemicals with PELs were placed. Table 6.8 identifies the adverse health effect of the chemicals placed into each of the categories. Table 6.9—the comprehensive table developed and published by OSHA in which the 500 most commonly used chemicals have been categorized according to the toxic effect to be avoided and the general mode of action (acute or systemic)—comprises the Appendix of this chapter. Using information from Tables 6.6 through 6.9, industrial hygienists and other health professionals can quickly determine whether an exposure limit requires adjustment. This should assist them in selecting the appropriate conceptual approach to modifying the limit for any unusual period of exposure.

Irrespective of the model which will be used to make the adjustments, including the pharmacokinetic models to be discussed, Table 6.9 should be consulted before the hygienist begins the task of modifying an exposure limit. This table, combined with some professional judgment, should prevent hygienists from requiring control measures when they are not necessary as well as minimize the risk of injury or discomfort from overexposure when persons work unusual schedules. For example, as noted by OSHA, substances in

Table 6.6 Work Schedule Categories

Category 1A. Ceiling Limit Standard. Substances in this category (e.g., butylamine) have ceiling limit standards which were intended never to be exceeded at any time, and so, are independent of the length of frequency of work shifts. The ceiling PELs for substances in this category should not be adjusted.

Category 1B. Standards Preventing Mild Irritation. Substances in this category have a PEL designed primarily to prevent acute irritation or discomfort (e.g., cyclopentadiene). There are essentially no known cumulative effects resulting from exposures for extended periods of time at concentration levels near the PEL. The PELs for substances in this category should not be adjusted.

Category 1C. Standards Limited by Technologic Feasibility. The PELs of substances assigned to this category have been set either by technologic feasibility (e.g., vinyl chloride) or good hygiene practices (e.g., methyl acetylene). These factors are independent of the length or frequency of work shifts. The PELs for substances in this category should not be adjusted.

Category 2. Acute Toxicity Standards

a. The substances in this category have PELs which prevent excessive accumulation of the substance in the body during 8 hr of exposure in any given day (e.g., carbon monoxide).

b. The following equation determines a level which ensures that employees exposed more than 8 hr/day will not receive a dosage (i.e., length of exposure × concentration) in excess of that intended by the standard.

$$\text{Equivalent PEL} = \text{8-hour PEL} \times \frac{8 \text{ hours}}{\text{hours of exposure in 1 day}} \qquad \text{Daily Adjustment}$$

c. The Industrial Hygienist should normally conduct sampling for the entire shift minus no more than 1 hr for equipment set-up and retrieval (e.g., at least 9 hr of a 10-hr shift). In situations where an employee works multiple shifts in a day (e.g., two 7-hr shifts), and the Industrial Hygienist can document sufficient cause to expect exposure concentrations to be similar during the other shifts, the sampling should be done during only one shift.

Category 3. Cumulative Toxicity Standards

a. Substances assigned to this category present cumulative hazards (e.g., lead, mercury, etc.). The PELs for these substances are designed to prevent excessive accumulation in the body resulting from many days or even years of exposure.

b. The following equation ensures that workers exposed more than 40 hours/week will not receive a dosage in excess of that intended by the standard.

$$\text{Equivalent PEL} = \text{8-hour PEL} \times \frac{40 \text{ hours}}{\text{hours of exposure in 1 week}} \qquad \text{Weekly Adjustment}$$

c. It is the responsibility of the Industrial Hygienist to conduct sufficient sampling to document exposure levels for the entire week when evaluating conditions on the basis of this equivalent PEL. For most operations, the Industrial Hygienist will be able to sample during one shift only and then document sufficient cause to predict exposure concentrations during the other shifts.

Table 6.6 *(Continued)*

Category 4. Acute and Cumulative Toxicity Standards. Substances in this category may present both an acute and a cumulative hazard. For this reason, the PELs of these substances should be adjusted by either equation; i.e., whichever provides the greatest protection.

Refined Adjustment Equations for Specific Standards. The adjustment equation presented for Categories 2 and 3 reflect an oversimplification of the actual accumulation and removal of a toxic agent from the body. Additional research, however, is needed in order to apply more complex equations to estimate resulting body burden and health risk due to prolonged exposure periods.

Source: OSHA, 1979.

Category 1A, 1B and 1C do not require adjustment during long shifts due to the rationale for those limits (Table 6.7).

As discussed briefly, OSHA has proposed two simple equations for adjusting occupational health limits. These equations are offered to their compliance officers as an alternative to the more complex models of Brief and Scala (18) or Hickey and Reist (145). The first equation which appears in Chapter 13 of the OSHA Field Manual (138) is recommended for use with chemicals posing a hazard for *acute injury* (Category 2, Table 6.6). In these cases, the objective is to modify the limit for the unusually long shift to a level which would produce a dose (mg) which would be no greater than that obtained during 8 hours of exposure at the PEL. Examples of chemicals with exclusively acute effects include carbon monoxide and phosphine. Equation 17 is recommended by OSHA for calculating an adjustment limit (Equivalent PEL) for substances with acute effects. This formula should *only* be used for those substances which have

Table 6.7 Prolonged Work Schedule Categories

Category[a]	Classification	Adjustment Criteria
1A	Ceiling standard	None
1B	Irritants	None
1C	Technologic limitations	None
2	Acute toxicants	Exposed—8 hr/day
3	Cumulative toxicants	Exposed—40 hr/week
4	Both acute & cumulative	Exposed—8 hr/day and/or Exposed—40 hr/week

Source: OSHA (1979).
Note: The health effects and classification of violation sections of this chapter have been reviewed by a panel of toxicologists and industrial hygienists from NIOSH using policy guidelines established by OSHA.
[a] This column indicates the code designation for prolonged work schedules which may require an adjustment to the PEL.

Table 6.8 OSHA Substance Toxicity Table (Rationale for Placing a Chemical into One of the Categories Listed in Table 6.9)

Health Code Number	Health Effect
1	Cancer—currently regulated by OSHA as carcinogens; chiefly work practice standards
2	Chronic (cumulative) toxicity—Suspect carcinogen or mutagen
3	Chronic (cumulative) toxicity—Long-term organ toxicity other than nervous, respiratory, hematologic or reproductive
4	Acute toxicity—Short-term high hazards effects
5	Reproductive hazards—Fertility impairment or teratogenesis
6	Nervous system disturbances—Cholinesterase inhibition
7	Nervous system disturbances—Nervous system effects other than narcosis
8	Nervous system disturbances—Narcosis
9	Respiratory effects other than irritation—Respiratory sensitization (asthma)
10	Respiratory effects other than irritation—Cumulative lung damage
11	Respiratory effects—Acute lung damage/edema
12	Hematologic (blood) disturbances—Anemias
13	Hematologic (blood) disturbances—Methemoglobinemia
14	Irritation—eye, nose, throat, skin—Marked
15	Irritation—eye, nose, throat, skin—Moderate
16	Irritation—eye, nose, throat, skin—Mild
17	Asphyxiants, anoxiants
18	Explosive, flammable, safety (No adverse effects encountered when good housekeeping practices are followed)
19	Generally low risk health effects—Nuisance particulates, vapors or gases
20	Generally low risk health effects—Odor

been classified as *acute toxins* in either Table 6.9 (Appendix) or in the *Documentation of Threshold Limits Values* (46) if the limit is not a PEL.

$$\text{Equivalent PEL} = \text{8-hour PEL} \times \frac{8 \text{ hours}}{\text{hours of exposure per day}} \qquad (17)$$

Table 6.9

Substance	Health Code No.	Health Effects	Work Category
Abate	6	Cholinesterase inhibition	3
Acetaldehyde	14	Marked irritation—eye, nose, throat, skin	1B
Acetic acid	14	Marked irritation—eye, nose, throat, skin	1B
Acetic anhydride	14	Marked irritation—eye, nose, throat, skin	1B
Acetone	16, 8	Mild irritation—eye, nose, throat/narcosis	1B
Acetonitrile	16, 4	Mild irritation—eye, nose, throat/acute toxicity (cyanosis)	4
2-Acetylaminofluorene	1	Cancer	1C
Acetylene	18, 17	Explosive/simple asphyxiation	1C
Acetylene dichloride	(See 1,2 Dichloroethylene)		
Acetylene tetrabromide	3, 10	Cumulative liver and lung damage	4
Acrolein	14	Marked irritation—eye, nose, throat, lungs, skin	1B
Acrylamide—skin	7, 3	Polyneuropathy, dermatitis/skin, eye irritation	4
Acrylonitrile—skin	2, 5	Suspect carcinogen	
Aldrin—skin	2, 3	Suspect carcinogen/cumulative liver damage	4
Allyl alcohol—skin	4, 14	Eye damage/marked irritation—eye, nose, throat, bronichi, skin	1B
Allyl chloride	3, 14	Liver damage/Marked irritation—eye, nose, throat	4
Allyl glycidyl ether (AGE)—Skin	14	Contact skin allergy/Marked irritation—eye, nose, throat, bronchi, skin	1B
Allyl propyl disulfide	14	Marked irritation—eye, nose, throat	1B
Aluminum oxide	18, 19	Nuisance particulate	1C
4-Aminodiphenyl—skin	1	Cancer	1C
2-Aminoethanol	(See Ethanolamine)		
2-Aminopyridine	4, 7	CNS stimulation/headache/increased blood pressure	4
Ammonia	11, 14	Marked irritation—eye, nose, throat, bronchi, lungs	1B
Ammonium chloride (fume)	16	Mild irritation—eye, nose, throat	1B

Table 6.9 (*Continued*)

Substance	Health Code No.	Health Effects	Work Category
Ammonium sulfamate (am-mate)	16	Mild irritation—eye	1B
n-Amyl acetate	15	Moderate irritation—eye, nose, throat	1B
sec-Amyl acetate	15	Moderate irritation—eye, nose, throat	1B
Aniline—skin	13, 4	Methemoglobinemia/acute toxic effects	4
Anisidine (o,p-isomers)—skin	13, 3	Methemoglobinemia/ Cumulative toxicity	4
Antimony & compounds (as Sb)	3, 2	Cumulative heart dam-age/suspect carcinogen	4
ANTU (alpha naphthyl thi-ourea)	3	Cumulative endocrine (thyroid and adrenal) damage	4
Argon	17	Simple asphyxiation	1A
Arsenic & compounds (as As)	2, 3	Suspect carcinogen/cumulative systemic poison	4
Arsine	4	Acute systemic toxicity	4
Asbestos (all forms)	1, 10	Cancer/asbestosis	3
Asphalt fumes (petroleum)	2	Suspect carcinogen/skin irritant	4
Azinphos-methyl—skin	6	Cholinesterase inhibition	3
Barium (soluble compounds)	3, 10	Cumulative heart, lung, and brain damage	3
Baygon (Propoxur)	6	Cholinesterase inhibition	3
Benzene—skin	2, 12	Suspect leukemogen/ Cumulative bone marrow damage	1C
Benzidine—skin	1	Cancer of bladder	1C
Benzidine derived dyes			
p-Benzoquinone	(See Quinone)		
Benzoyl peroxide	15	Moderate irritation—eye, nose, throat, skin	1B
Benzyl chloride	2, 4, 11	Suspect carcinogen/marked irri-tation—eye, nose, throat, skin/lung edema	4
Beryllium & compounds	2, 10	Suspect carcinogen/cumulative lung damage (berylliosis)	4
Biphenyl (diphenyl)	15	Moderate irritation—eye, nose, throat, bronchi, lungs	1B
Bismuth telluride	10	Accumulation in lungs	1C
Bismuth telluride (Se-doped)	10	Cumulative lung damage	3
Borates, tetra, sodium salt, anhydrous	15	Moderate irritation—eye, nose, throat, skin	1B

Table 6.9 *(Continued)*

Substance	Health Code No.	Health Effects	Work Category
Borates, tetra, sodium salt, decahydrate	15	Moderate irritation—eye, nose, throat, skin	1B
Borates, tetra, sodium salt, pentahydrate	15	Moderate irritation—nose, throat, skin	1B
Boron oxide	16	Mild irritation—eye, nose, throat, skin	1B
Boron tribromide	14	Marked irritation—eye, nose, throat, lungs	1B
Boron trifluoride	11, 14	Acute and chronic lung irritation (pneumonia)	1A
Bromine	14, 11	Marked irritation—eye, nose, throat, bronchi, lungs	1B
Bromine pentafluoride	14, 11	Marked irritation—eye, nose, throat, bronchi, lungs	1B
Bromoform—skin	14, 3	Marked irritation—eye, nose, throat/cumulative liver damage	4
Butadiene (1,3-butadiene)	16	Mild irritation—eye, nose, throat	1B
Butane	17, 8	Asphyxiant/narcosis	1C
Butanethiol	(See Butyl mercaptan)		
2-Butanone (MEK)	15, 8	Moderate irritation—eye, nose, throat/narcosis	1B
2-Butoxyethanol—skin (butyl cellosolve)	12, 16	Anemia/mild irritation—eye, nose, throat	4
n-Butyl acetate	15	Moderate irritation—eye, nose, throat	1B
sec-Butyl acetate	15	Moderate irritation—eye, nose, throat	1B
tert-Butyl acetate	15	Moderate irritation—eye, nose, throat	1B
Butyl alcohol—skin	15, 7	Moderate irritation—eye, nose, throat/hearing loss	2
sec-Butyl alcohol	16	Mild irritation—eye, nose, throat	2
tert-Butyl alcohol	15, 8	Moderate irritation—eye, nose, throat/narcosis	2
Butylamine—skin	2, 14	Suspect carcinogen/marked irritation—eye, nose, throat, lungs, skin	4
tert-Butyl chromate (As CrO₃)—skin	2, 14	Suspect carcinogen/marked irritation—eye, nose, throat, skin	1A
n-Butyl glycidyl ether (BGE)	2, 16	Suspect mutagen/mild irritation—eye, nose, throat, skin	4

Table 6.9 *(Continued)*

Substance	Health Code No.	Health Effects	Work Category
n-Butyl lactate	15	Moderate irritation—eye, nose, throat, bronchi, lungs	1B
Butyl mercaptan	15, 20	Moderate irritation—eye, nose, throat/odor	1B
p-tert-Butyltoluene	7, 3	CNS damage/cumulative liver, kidney damage	3
Cadmium dust (as Cd)	3, 2	Suspect carcinogen/cumulative kidney and lung damage	4
Cadmium fume (as Cd)	3, 2	Cumulative kidney and lung damage/suspect carcinogen	4
Calcium arsenate (as As)	2, 3	Suspect carcinogen/cumulative systemic poisoning	4
Calcium carbonate	19	Nuisance particulate	1C
Calcium cyanamide	15, 2	Moderate irritation—eye, nose, throat, skin/suspect carcinogen	4
Calcium hydroxide	14	Marked irritation—eye, nose, throat, skin	1B
Calcium oxide	14	Marked irritation—eye, nose, throat, skin	1B
Camphor (synthetic)	15, 4	Moderate irritation—eye, nose, throat/acute toxicity	1B
Caprolactam dust	15	Moderate irritation—eye, nose, throat, skin	1B
Caprolactam vapor	16	Mild irritation—eye, nose, throat, skin	1B
Captafol (difolatan[R])—skin	3, 9, 5	Dermatitis/respiratory sensitization (asrtm)/teratogen	4
Captan	2, 5	Suspect carcinogen, mutagen, and teratogen	4
Carbaryl (servin[R])	6, 5	Cholinesterase inhibition/teratogen	3
Carbofuran (furadan[R])	6	Cholinesterase inhibition	3
Carbon black	2, 3	Suspect carcinogen/cumulative heart damage	4
Carbon dioxide	17	Simple asphyxiation	1C
Carbon disulfide—skin	7, 5	Cumulative CNS damage/reproductive impairment	3
Carbon monoxide	17	Chemical anoxia, asphyxiation	2
Carbon tetrabromide	3, 14	Liver damage/potent lachrymator	4
Carbon tetrachloride—skin	3, 2, 5	Cumulative liver damage/suspect carcinogen/teratogen	4

Table 6.9 *(Continued)*

Substance	Health Code No.	Health Effects	Work Category
Catechol pyrocatechol	14, 3	Eye and skin irritation/kidney damage	4
Cellulose (paper fiber)	19	Nuisance particulate	1C
Cesium hydroxide	15	Moderate irritation—eye, nose, throat, skin	1B
1-chloro,2,3, epoxy-propane	(See Epichlorohydrin)		
Chlordane—skin	3, 2	Cumulative liver damage/suspect carcinogen	4
Chlorinated camphene (toxaphene)—skin	3	Cumulative liver damage	3
Chlorinated diphenyl oxide	3	Cumulative liver damage/dermatitis	3
Chlorine	11, 14	Lung injury/marked irritation—eye, nose, throat, bronchi	3
Chlorine dioxide	11, 14	lung injury/marked irritation—eye, nose, throat, bronchi	1B
Chlorine trifluoride	11, 14	marked irritation—eye, nose, throat, bronchi, lungs	1A
Chloroacetaldehyde	14	Marked irritation—eye, nose, throat, lungs, skin	1A
alpha-Chloroacetophenone (phenacyl chloride)	14	Marked irritation—eye, nose, throat, bronchi, lungs, skin	1B
Chlorobenzene (monochlorobenze)	3, 8	Cumulative systemic-toxicity/narcosis	4
o-Chlorobenzylidene malonitrile—skin	14	Marked irritation—eye, nose, throat, skn	1B
2-Chloro-1,3-butadiene	(See Chloroprene)		
Chlorobromomethane	3, 8	Cumulative liver damage/narcosis	3
Chlorodifluoromethane (F-22)	18	Good housekeeping practice	1C
Chlorodiphenyl (42% Cl)—skin	2, 3	Suspect carcinogen/ Chloracne/cumulative liver damage	4
Chlorodiphenyl (54% Cl)—skin	2, 3	Suspect carcinogen/ chloracne/cumulative liver damage	4
1,Chloro 2,3-epoxypropane	(See Epichlorohydrin)		
2-Chloroethanol	(See Ethylene chlorohydrin)		
Chloroethylene	(See Vinyl chloride)		
Chloroform (trichloromethane)	2, 3, 8	Suspect carcinogen/cumulative liver and kidney damage/narcosis	4

Table 6.9 (*Continued*)

Substance	Health Code No.	Health Effects	Work Category
bis-Chloromethyl ether	1	Cancer (lung)	1C
1-Chloro-1-nitropropane	15	Moderate irritation—eye, nose, throat, skin	1B
Chloropicrin	14, 11	Marked irritation—eye, nose, throat, bronchi, lungs, skin	1B
Chloroprene—(2-chloro-1,3 butadiene)—skin	5, 3, 2	Reproductive hazard/systemic toxicity/suspect mutagen	4
Chloropyrifos (Dursban^R)— skin	6	Cholinesterase inhibition	4
o-Chlorostyrene	3	Cumulative liver, kidney damage	3
o-Chlorotoluene—skin	2, 15	Mild irritant—eye, skin	1B
2-Chloro-6-trichloromethyl pyridine (N-Serve^R)	18	Good housekeeping practice	1C
Chromates, certain insoluble forms (as Cr)	2, 10, 3	Suspect carcinogen/cumulative lung damage/dermatitis	4
Chromic acid & chromates (as Cr)	2, 10, 3	Suspect carcinogen/cumulative lung damage/nasal preforation, ulceration	4
Chromium, soluble chromic, chromous salts (as Cr)	10, 3	Cumulative lung damage/dermatitis	3
Clopidol (coyden^R)	18	Good housekeeping practice	1C
Coal dust	10	Pneumoconiosis	3
Coal tar pitch volatiles	2, 10	Suspect carcinogen/cumulative lung changes	4
Cobalt, metal, fume & dust (as Co)	9, 10, 3	Asthma/cumulative lung changes/dermatitis	3
Coke oven emissions	1, 3	Cancer—lungs, bladder, kidney/skin sensitization	1C
Copper dusts & mists (as Cu)	16	Mild irritation—eye, nose, throat, skin	4
Copper fume (as Cu)	15, 11	Moderate irritation—eye, nose, throat, lung	4
Corundum (Al_2O_3)	19	Nuisance particulate	1C
Cotton dust (raw)	9, 10	Asthma/cumulative lung damage (bysinosis)	4
Crag^R herbicide	3	Cumulative liver damage	3
Cresol (all isomers)—skin	14, 4, 3	Marked irritation—eye, skin/acute toxicity (CNS), liver and kidney damage	1B
Cristobalite	10	Pneumoconiosis	3
Crotonalydehyde	14	Marked irritation—eye, nose, throat, lungs	1B
Crufomate^R	6	Cholinesterase inhibition	3

Table 6.9 (*Continued*)

Substance	Health Code No.	Health Effects	Work Category
Cumene—skin	8, 15	Narcosis/moderate irritation—eye, skin	3
Cyanamide	14, 4	Marked irritation—eye, nose, throat, skin/acute toxicity	2
Cyanides (as CN)—skin	15	Marked irritation—skin, eye, nose, throat	1B
Cyanogen	15, 4	Moderate irritation—eye, nose, throat/acute toxicity (cyanosis)	2
Cyclohexane	15	Moderate irritation—eye, nose, throat	1B
Cyclohexanol	16, 3	Mild irritation—nose, throat/cumulative liver and kidney damage	1B
Cyclohexanone	15, 3	Moderate irritation—eye, nose, throat/cumulative liver and kidney damage	4
Cyclohexene	15, 3	Moderate irritation—eye, nose, throat/cumulative systemic toxicity	4
Cyclohexylamine—skin	14, 2	Marked irritation—eye, nose, throat, skin/suspect mutagen	4
Cyclopentadiene	15	Moderate irritation—eye, nose, throat	1B
DBCP	(See 1,2, Dibromo-3-chloroproane)		
2,4-D (2,4 dichlorophenoxy-acetic acid)	5	Suspect teratogen	3
DDT—skin	7, 2, 3	Cumulative toxicity CNS suspect/carcinogen and mutagen	4
DDVP—skin	6	Cholinesterase inhibition	3
Decaborane—skin	4, 7	Acute and chronic CNS toxicity	4
Demeton^R (systox)—skin	6, 5	Cholinesterase inhibition/suspect teratogen	3
Diacetone alcohol (4-hydroxy-4 methyl-2 pentanone)	15, 3	Moderate irritation—eye, nose, throat/cumulative kidney damage	1B
1,2 Diaminoethane	(See Ethylenediamine)		
2,4-Diaminoanisole	(See NIOSH CIB No. 19)		
4,4 Diaminodiphenyl-methane (DDM)	(See NIOSH CIB No. 8)		
Diazinon—skin	6, 5	Cholinesterase inhibition/suspect teratogen	3

Table 6.9 *(Continued)*

Substance	Health Code No.	Health Effects	Work Category
Diazomethane	2, 11, 14	Suspect carcinogen/acute lung damage/marked irritation—eye, nose, throat	4
Diborane	11, 7, 14	Acute respiratory damage, irritation/nervous system damage	4
Dibrom^R	6	Cholinesterase inhibition	3
1,2-Dibromo-3-chloro-propane (DBCP)	5, 2	Male sterility/suspect carcinogen	1C
1,2-Dibromoethane (ethylene dibromide)—skin	2, 3, 5	Suspect carcinogen, mutagen, teratogen/cumulative kidney damage/reproductive hazard	4
2-n-Dibutylamino-ethanol—skin	15, 3	Moderate irritation—eye, nose, throat, lungs, skin/cumulative liver	4
Dibutyl phosphate	16	Mild irritation—eye, nose, throat, lungs	1B
Dibutylphthalate	19, 5	Apparent low toxicity/suspect teratogen	1C
Dichloroacetylene	4, 11, 7	Acute toxicity—nausea, headache, lung edema/cumulative CNS effects	4
o-Dichlorobenzene	14, 4	Marked irritation—eye, nose, throat/liver damage	1A
p-Dichlorobenzene	3, 7	Cumulative systemic toxicity/cataracts	3
3,3′-Dichlorobenzidine—skin	1	Cancer—bladder	1C
Dichlorodifluoromethane (F-12)	18	Good housekeeping practice	1C
1,3-Dichloro-5, 5-dimethyl-hydantion	14	Mild irritation—eye, nose, throat, lungs	1B
1,1-Dichloroethane	3	Cumulative liver damage	3
1,2-Dichloroethylene	3, 7, 2	Cumulative liver, kidney damage/CNS effects/suspect carcinogen	4
1,2-Dichloroethylene	8, 7	Narcosis/CNS effects	2
Dichloroethyl ether—skin	14, 2, 11	Marked irritation—eye, nose, throat, lungs/suspect carcinogen/lung edema	1A
Dichloromethane	(See Methylene chloride)		
Dichoromonofluoromethane (F-21)	3	Cumulative liver damage	1C
1,1-Dichloro-1-nitroethane	4, 11	Acute systemic toxicity—lungs, heart, liver, kidneys	1A

Table 6.9 (*Continued*)

Substance	Health Code No.	Health Effects	Work Category
1,2 Dichloropropane	(See Propylene dichloride)		
Dichlorotetrafluoroethane	18	Good housekeeping practice	1C
Dicrotophos (bidrin^R)	6	Cholinesterase inhibition	3
Dicyclopentadiene	3, 16	Cumulative kidney, liver damage/mild irritation—eye, nose, throat	3
Dichylopentadienyl iron	18	Good housekeeping practice	1C
Dieldrin—skin	2, 3	Suspect carcinogen/cumulative liver damage	4
Diethylamine	14, 3	Marked irritation—eye, nose, throat, lungs, skin/myocardial degeneration	1B
Diethylaminoethanol—skin	14	Marked irritation—eye, nose, throat	1B
Diethylcarbamoyl chloride (DECC)	(See NIOSH CIB No. 12)		
Diethylene triamine	14, 9	Marked irritation—eye, nose, throat, lung, skin	1B
Diethyl ether	(See Ethyl ether)		
Diethyl phthalate	16	Mild irritation—nose, throat	1B
Difluorodibromomethane (F-12B2)	16, 3	Respiratory irritation/ cumulative liver and CNS damage	1B
Diglycidyl ether (DGE)	14, 3, 2	Marked irritation—eye, nose, throat, lungs, skin/cumulative systemic toxicity/suspect mutagen	1A
Dihydroxybenzene	(See Hydroquinone)		
Diisobutyl ketone	16, 8	Mild irritation—eye, nose, throat/narcosis	1B
Diisopropylamine—skin	7, 15	CNS Effects/moderate irritation—eye, nose, throat, lungs	1B
Dimethoxymethane	(See Methylal)		
Dimethyl acetamide—skin	3, 5	Cumulative liver damage/ suspect teratogen	3
Dimethylamine	14, 3	Marked irritation—eye, nose, throat, bronchi, lung, skin/cumulative liver, testicular damage	4
Dimethylaminobenzene	(See Xylidene)		
4-Dimethylaminoazobenzene	1	Cancer—liver	
Dimethylaminopropionitril	(See ESN)		
Dimethylaniline—skin	13, 7	Methemoglobinemia CNS effects	4

Table 6.9 *(Continued)*

Substance	Health Code No.	Health Effects	Work Category
Dimethylbenzene	(See Xylene)		
Dimethyl carbamoyl chloride (DMCC)	(See NIOSH CIB No. 11)		
Dimethylformamide—skin	3, 7	Cumulative liver damage/CNS effects	3
2,6 Dimethylheptanone	(See Diisobutyl ketone)		
1,1-Dimethylhydrazine—skin	2, 7, 12	Suspect carcinogen/CNS effects/anemia	4
Dimethylphthalate	5, 16	Suspect teratogen/mild irritation—nose, throat	4
Dimethyl sulfate—skin	2, 4	Suspect carcinogen/acute eye and lung effects	4
Dinitrobenzene (all isomers)—skin	13, 12, 3	Blood disturbances/liver, kidney damage	4
Dinitro-o-cresol—skin	3	Cumulative systemic (metabolic) toxin	3
3,5-Dinitro-o-toluamide (zoalene[R])	3	Cumulative liver damage	3
Dinitrotoluene—skin	13, 12, 3	Methemoglobinemia/anemia/liver damage	4
Dioxane (diethylene dioxide) technical grade—skin	2, 3	Suspect carcinogen/cumulative liver, kidney damage	4
Dioxathion (delnav[R])	6	Cholinesterase inhibition	3
Diphenyl	(See Biphenyl)		
Diphenylamine	3, 5	Cumulative liver, kidney bladder damage/suspect teratogen	3
Diphenylmethane diisocyanate	(See Methylene biphenyl isocyanate)		
Dipropylene glycol methyl ether—skin	15, 3	Moderate irritation—eye	4
Diquat	3, 5	Cumulative effects (cataracts)/suspect teratogen	4
Di-sec, octyl phthalate (Di-2 ethylhexylphthalate)	16	Mild irritation—eye, nose, throat	1C
Direct Black 38, Direct Blue 6, and Direct Brown 95 (Benzidine derived dyes)—See NIOSH CIB No. 24			
Disulfiram (tetraethylthiuram disulfide)	4, 2	Acute toxicity—antabuse effects with alcohol/suspect carcinogen	4
Disyston—skin	6	Cholinesterase inhibition	3
2,6-Di-tert-butyl-p-cresol	18	Good housekeeping practice	1C
Diuron	19	Apparent low toxicity	1C
Dyfonate	6	Cholinesterase inhibition	3

Table 6.9 (*Continued*)

Substance	Health Code No.	Health Effects	Work Category
Emery	19	Nuisance particulate	1C
Endosulfan (thiodan^R)—skin	4, 3, 2	Acute CNS toxin/cumulative kidney damage/suspect carcinogen	4
Endrin—skin	4	Acute toxicity	4
Epichlorohydrin (1-chloro,2,3-epoxypropane)—skin	14, 2, 3	Marked skin irritation, sensitization/suspect carcinogen and mutagen/kidney and liver damage	4
EPN—skin	6	Cholinesterase inhibition	3
1,2-Epoxypropane	(See Propylene oxide)		
2,3-Epoxy-1-propanol	(See Glycidol)		
ESN^R	(See NIOSH CIB No. 26)		
Ethane	18, 17	Explosive/simple asphyxiation	1C
Ethanethiol	(See Ethyl mercaptan)		
Ethanolamine	14, 3	Marked irritation—skin/cumulative liver, lung and kidney damage	3
Ethion (nialate^R)—skin	6	Cholinesterase inhibition	3
2-Ethoxyethanol (cellosolve)—skin	15, 12	Moderate irritation—eye, nose/cumulative blood disturbances	3
2-Ethoxyethyl acetate (cellosolve acetate)—skin	3, 16	Cumulative liver, kidney damage/mild irritant—eye, nose, throat	3
Ethyl acetate	16, 20	Mild irritation—eye, nose, throat, lungs/odor	1B
Ethyl acrylate—skin	14, 11	Marked irritation—eye, nose, throat, lungs/lung edema	1B
Ethyl alcohol (ethanol)	14, 8	Mild irritation—eye, nose, throat/narcosis	1B
Ethylamine	14, 3	Marked irritation—eye, nose, throat, lungs/corneal injury	4
Ethyl sec-amyl ketone (4-methyl-3 heptanone)	15	Moderate irritation—eye, nose, throat	1B
Ethyl benzene	15	Moderate irritation—eye, nose, throat	1B
Ethyl bromide	8, 3	Narcosis/cumulative liver, kidney and heart damage	4
Ethyl butyl ketone (3-heptanone)	16, 8	Mild irritation—eye, nose, throat/narcosis	2
Ethyl chloride	8	Narcosis	2

Table 6.9 (*Continued*)

Substance	Health Code No.	Health Effects	Work Category
Ethyl ether	8, 16	Narcosis/mild irritation—eye, nose, throat	1B
Ethyl formate	16, 8	Mild irritation—eye, nose, throat/narcosis	1B
Ethyl mercaptan	20, 4	Odor/acute systemic toxicity	1A
Ethyl silicate	3, 16	Cumulative kidney damage/mild irritation—eye, nose, throat	4
Ethylene	17, 18	Explosive/simple asphyxiation	1C
Ethylene chlorohydrin—skin	4	Acute toxicity (local and systemic)	4
Ethylenediamine	15, 3, 9	Moderate irritation—eye, nose, throat, skin/contact dermatitis, asthma	2
Ethylene dibromide	(See 1,2,-Dibromoethane)		
Ethylene dichloride	(See 1,2-Dichloroethane)		
Ethylene glycol dinitrate—skin	3	Cumulative blood pressure lowering/headache	1A
Ethylene glycol, particulate	15	Moderate irritation—eye, nose, throat	1B
Ethylene glycol, vapor	15	Moderate irritation—eye, nose, throat	1B
Ethylene glycol monomethyl ether acetate	(See Methyl cellosolve acetate)		
Ethyleneimine—skin	1	Cancer	1C
Ethylene oxide	15, 3, 2	Moderate irritation—eye, nose, throat/cumulative lung, liver and kidney damage/suspect mutagen	4
Ethylene thiourea	(See NIOSH CIB No. 22)		
Ethylidene chloride	(See 1,1-Dichlorethane)		
Ethylidene norbornene	15, 3, 5	Moderate irritation—eye, nose, throat/cumulative liver and testicular damage	3
N-Ethyl—morpholine—skin	4, 15	Acute CNS effects/moderate irritation—eye, nose, throat	1B
Fensulfothion (dasanit[R])	6	Cholinesterase inhibition	3
Ferbam	16, 2	Mild irritation—eye, nose, upper respiratory tract/suspect carcinogen	4
Ferrovanadium dust	16	Mild irritation—upper respiratory tract	1B
Fluoride (as F)	14, 3	Marked irritation—eye, nose, throat/cumulative bone damage	4

Table 6.9 (*Continued*)

Substance	Health Code No.	Health Effects	Work Category
Fluorine	11, 13	Lung edema/kidney damage	1B
Fluorotrichloromethane (F-11)	7	Acute CNS effects	2
Formaldehyde	14, 2	Marked irritation—eye, lungs, skin/suspect carcinogen	4
Formamide	3	Cumulative systemic toxicity	3
Formic acid	14	Marked irritation—eye, nose, throat, lungs	1B
Furfural—skin	15	Moderate irritation—eye, nose, throat	1B
Furfuryl alcohol—skin	15, 8	Moderate irritation—eye, lungs/narcosis	2
Gasoline	16, 7, 18	Mild irritation—eye, nose, throat/CNS effects/flammable	4
Germanium tetrahydride	4	Acute systemic toxicity	2
Glass, fibrous or dust	15	Moderate irritation—nose, throat, skin	1B
Glycerin mist	19	Nuisance particulate	1C
Glycidol (2,3-Epoxy-1-propanol)	15, 7	Moderate irritation—eye, nose, throat, skin, CNS effects	2
Glycol monoethyl ether	(See 2-Ethoxyethanol)		
Guthion	(See Azinphos methyl)		
Glycidyl ethers	(See NIOSH CIB No. 29)		
Graphite (natural)	10	Cumulative lung damage (pneumoconiosis)	3
Graphite (synthetic)	19	Nuisance particulate (accumulation in lungs)	1C
Gypsum	19	Nuisance particulate (accumulation in lungs	1C
Hafnium	3	Cumulative liver damage	3
Helium	17	Simple asphyxiation	1C
Heptachlor—skin	2, 3	Suspect carcinogen	4
Heptane	15, 7, 8	Moderate irritation—eye, nose, lungs/CNS effects/narcosis	4
Hexachlorocyclopentadiene	14, 11, 3	Marked irritation—eye, throat, lungs/lung edema/cumulative organ damage	4
Hexachloroethane—skin	3, 7	Cumulative organ damage/CNS effects	3
Hexachloronaphthalene—skin	3	Cumulative liver damage/chloracne	3
Hexafluoroacetone	3, 5	Multiple cumulative organ damage	3

Table 6.9 (*Continued*)

Substance	Health Code No.	Health Effects	Work Category
Hexamethylphosphoric triamide (HMPA)	(See NIOSH CIB No. 6)		
n-Hexane	7, 8	Polyneuropathy/narcosis	4
2-Hexanone (MBK)—skin	7, 15	Polyneuropathy/moderate irritation—eye, nose, throat	4
Hexone (MBK)—skin	15	Moderate irritation—eye, nose, throat	1B
neo-Hexyl acetate	16	Mild irritation—eye, nose, throat	1B
Hexylene glycol	16	Mild irritation—eye, nose, throat, skin/narcosis	1B
Hydrazine—skin	14, 3, 2	Marked irritation—respiratory tract/cumulative organ damage/suspect carcinogen	4
Hydrogen	17, 18	Explosive/simple asphyxiation	1C
Hydrogenated terphenyls	3, 10	Cumulative liver, kidney, lung damage	3
Hydrogen bromide	14, 11	Marked irritation—nose, throat/acute lung damage	1B
Hydrogen chloride	14, 11	Marked irritation—eye, nose, throat/lung edema	1A
Hydrogen cyanide—skin	4, 3	Acute and cumulative systemic toxicity (cyanosis)	4
Hydrogen fluoride	14, 11, 3	Marked irritation—eye, nose, throat/acute lung damage/cumulative bone damage	4
Hydrogen peroxide, (90%)	14, 11, 18	Marked irritation—eye, nose, throat, skin/acute lung damage/explosive	1B
Hydrogen selenide	11, 7, 3	Acute lung damage/CNS effects/liver damage	4
Hydrogen sulfide	4, 15, 7	Acute systemic toxicity/moderate irritation—eye (conjunctivitis), lungs/CNS effects	2
Hydroquinone	3, 7	Cumulative corneal damage/CNS effects	3
Indene	15, 3	Moderate irritation—eye, nose, throat/cumulative liver and kidney damage	4
Indium & compounds (as In)	10, 3, 5	Cumulative lung, other organ damage/suspect teratogen ($InNO_3$)	3

Table 6.9 (*Continued*)

Substance	Health Code No.	Health Effects	Work Category
Iodine	14, 11	Marked irritation—eye, nose, throat/lung edema	1A
Indoform	15, 4	Moderate irritation—eye, nose, throat, lungs/acute CNS effects	2
Iron oxide fume	10	Lung changes (siderosis)	3
Iron pentacarbonyl	4, 11	Acute toxicity CNS and lungs	2
Iron salts, soluble (as Fe)	15	Moderate irritation—upper respiratory tract, skin	1B
Isoamyl acetate	15	Moderate irritation—upper respiratory tract	1B
Isoamyl alcohol	16, 8, 2	Mild irritation—eye, nose, throat/narcosis/suspect carcinogen	4
Isobutyl acetate	15	Moderate irritation—eye, nose, throat	1B
Isobutyl alcohol	15, 2	Moderate irritation—eye, nose, throat/suspect carcinogen	4
Isophorone	14, 7	Marked irritation—eye, nose, throat/chronic CNS effects	4
Isophorone diisocyanate—Skin	9, 14	Respiratory sensitization/marked irritation—eye, nose, throat, lungs, skin	4
Isopropyl acetate	16	Mild irritation—eye, nose, throat	1B
Isopropyl alcohol—skin	16, 8	Mild irritation—eye, nose, throat/narcosis	4
Isopropylamine	14	Marked irritation—eye, nose, throat, lung	1B
Isopropyl ether	16	Mild irritation—eye, nose, throat	1B
Isopropyl glydidyl ether (IGE)	15, 3	Moderate irritation—eye, nose, throat, skin/skin sensitization	4
Kaolin	19	Nuisance particulate/accumulation in lungs	1C
Ketene	11	Marked irritation, edema—lungs	2
Lead arsenate (as Pb)	3, 2	Cumulative organ toxicity/suspect carcinogen	4
Lead, inorganic fumes & dusts (as Pb)	12, 7, 5	Cumulative blood and neurologic effects/reproductive hazard	3
Limestone	19	Nuisance particulate/accumulation in lungs	1C

Table 6.9 (*Continued*)

Substance	Health Code No.	Health Effects	Work Category
Lindane—skin	7, 3, 2	Cumulative CNS and liver damage/suspect carcinogen	4
Lithium hydride	14, 11, 7	Marked irritation—eye, nose, throat, skin/lung damage/CNS effects	1B
LPG (liquified petroleum gas)	18, 17, 8	Explosive/asphyxiant/narcosis	2
Magnesite	19	Nuisance particulate/accumulation in lungs	1C
Magnesium oxide fume	11	Lung effects (fume fever)	2
Malathion—skin	6	Cholinesterase inhibition	3
Maleic anhydride	14, 9, 2	Marked irritation—eye, nose, throat, lungs (edema), skin/asthma	2
Manganese & compounds (as Mn)	7, 10	Cumulative CNS damage/lung damage	1A
Manganese cyclopentadienyl tricarbonyl (as Mn)—skin	4, 7, 3	Acute CNS and blood effects/cumulative kidney damage	4
Marble	19	Nuisance particulate/accumulation in lungs	1C
Mercury, (organo) alkyl compounds, (as Hg)—skin	7, 3, 14	Acute and cumulative CNS damage/marked skin irritation	4
Mercury, inorganic (as Hg)—skin	7, 3, 2	Acute and cumulative CNS damage/gastrointestinal effects/gingivitis/suspect carcinogen	4
Mesityl oxide	16	Mild irritation—eye, nose, throat	1B
Methane	18, 17	Explosive/simple	1C
Methanethiol	(See Methyl mercaptan)		
Methomyl (lannate^R)—skin	6	Cholinesterase inhibition	3
Methoxychlor	3	Cumulative kidney damage	3
2 Methoxyethanol	(See Methyl cellusolve)		
4-Methoxy-m-phenyleneclia-mine	(See 2,4-Diamioanisole)		
Methyl acetate	16, 8, 7	Mild irritation—nose, throat, lungs/narcosis/CNS effects	4
Methyl acetylene (propyne)	18, 8	Explosive/narcosis	1C
Methyl acetylene—propadiene mix (MAPP)	18	Flammable	1C
Methyl acrylate—skin	14, 4, 3	Marked irritation—eye, nose, throat, skin/acute lung dam-	1B

173

Table 6.9 (*Continued*)

Substance	Health Code No.	Health Effects	Work Category
		age/cumulative lung, liver and kidney damage	
Methyl acrylonitrile—skin	7, 16	Cumulative CNS effects/mild irritation—eye, skin	3
Methylal (dimethoxymethane)	3	Cumulative systemic toxicity	3
Methyl alcohol—skin	7, 8, 16	Narcosis/cumulative CNS effects/mild irritation—eye, nose, throat	4
Methylamine	14	Marked irritation—eye, nose, throat, skin	1B
Methyl amyl alcohol	(See Methyl isobutyl carbinol)		
Methyl n-amyl ketone (2-heptanone)	15, 8	Moderate irritation—eye, nose, throat/narcosis	1B
Methyl bromide—skin	4, 11, 13	Acute lung damage/cumulative CNS and organ damage	1A
Methyl butyl ketone	(See 2-Hexanone)		
Methyl cellosolve (2-methoxyethanol)—skin	12, 7	Blood disorders/CNS effects	3
Methyl cellosolve acetate (ethylene glycol monomethyl ether acetate)—skin	12, 7, 3	Blood disorders/CNS effects/kidney damage	3
Methyl chloride	4, 7, 3	Acute and chronic CNS effects/liver and kidney damage	4
Methyl chloroform	16, 8	Mild irritation—eye, nose, throat/narcosis	1B
Methyl chloromethyl ether	1	Cancer—lung	1C
Methyl 2-cyanoacrylate	15	Moderate irritation—eye, nose, throat	1B
Methylcylohexane	8	Narcosis	4
Methylcyclohexanol	16, 8, 3	Mild irritation—eye, respiratory tract/narcosis/cumulative liver and kidney damage	4
o-Methycyclohexanone—skin	16, 8	Mild irritation—eye, nose, throat/narcosis	2
Methylcylopentadienyl manganese tricarbonyl (as Mn)—skin	4, 3, 15	Acute CNS effects/cumulative liver, kidney damage/moderate eye irritation	4
Methyl demeton—skin	6	Cholinesterase inhibition	3
4,4'-Methylene bis (2-Chloroaniline) (MOCA)—skin	1	Cancer	1C

Table 6.9 (*Continued*)

Substance	Health Code No.	Health Effects	Work Category
Methylene bis (4-cyclohexyl-isocyanate)	14, 3, 9	Marked irritation—skin/skin sensitization/asthma	4
Methylene bisphenyl iso-cyanate (MDI)	9, 14	Asthma/marked irritation—eye, nose, throat, skin	1A
Methylene chloride (dichlo-romethane)	17, 3, 8	Chemical anoxia (metabolic conversion to CO)/chronic liver damage/CNS effects/narcosis	4
Methyl ethyl ketone (MEK)	(See 2-Butanone)		
Methyl ethyl ketone perox-ide	14, 3, 2	Marked irritation—eye, nose, throat, lungs/cumulative liver and kidney damage/suspect carcinogen	4
Methyl formate	8, 15	Narcosis/moderate irritation—eye, nose, throat, lungs	1B
Methyl iodide—skin	4, 3, 2	Acute and cumulative CNS effects/suspect carcinogen	4
Methyl isoamyl ketone	15, 8	Moderate irritation—eye	1B
Methyl isobutyl carbinol—skin	15, 8	Moderate irritation—eye, nose, throat/narcosis	1B
Methyl isoamyl ketone	(See Hexone)		
Methyl isocyanate—skin	9, 14, 11	Asthma/marked irritation—eye, nose, throat, skin/lung edema	1B
Methyl mercaptan	20, 15	Odor/moderate irritation—eye, nose, throat	1A
Methyl methacrylate	16, 2	Mild irritation—eye, nose, throat/suspect carcinogen	4
Methyl propyl ketone	(See Pentanone)		
alpha-Methyl styrene	15, 7, 8	Moderate irritation—eye, nose, throat/CNS effects/narcosis	2
Methyl parathion—skin	6, 5	Cholinesterase inhibi-tion/suspect teratogen	3
Methyl silicate	4, 14, 3	Severe eye damage/marked irri-tation—eye, nose, throat, lungs/kidney damage	4
Mica (less than 1% quartz)	10	Accumulation in lungs (pneu-moconiosis)	3
Mineral wool fiber	15	Moderate irritation—nose, throat, skin	1B
Molybdenum (as Mo) (insol-ubles)	3, 16	Cumulative liver and kidney damage/blood disorders/mild irritation—eye, nose, throat, lung	3

Table 6.9 *(Continued)*

Substance	Health Code No.	Health Effects	Work Category
Molybdenum (as Mo) (solubles)	3, 16	Cumulative liver and kidney damage/blood disorders/mild irritation—eye, nose, throat, lung	3
Monocrotophos (axodrin^R)	6	Cholinesterase inhibition	3
Monomethyl aniline—skin	13, 12	Methemoglobinemia/anemia	3
Monomethyl hydrazine—skin	4, 2, 5	Acute lung/CNS and blood damage/suspect carcinogen and teratogen	1A
Morpholine—skin	15, 3	Moderate irritation—eye, nose, throat/cumulative liver and kidney damage	1B
Naphtha (coal tar)	15, 8	Moderate irritation—eye, throat/narcosis	2
Naphthalene	14, 3, 2	Marked irritation—eye, nose, throat/ocular damage/anemia/CNS damage/suspect carcinogen	4
alpha-Napthylamine	1	Cancer—bladder (suspect)	1C
beta-Napthylamine	1	Cancer—bladder	1C
Neon	17	Simple asphyxiation	1C
Nickel (soluble compounds)	2, 10, 3	Suspect carcinogen/cumulative lung damage/dermatitis	4
Nickel carbonyl	2, 4	Suspect carcinogen/acute	4
Nickel, metal & insoluble compounds	2, 10, 3	Suspect carcinogen/cumulative lung damage/dermatitis	4
Nicotine—skin	4, 7, 5	Acute systemic toxicity/CNS damage/suspect teratogen	3
Nitric acid	4, 14	Acute lung damage/marked irritation—eye, nose, throat, skin	2
Nitric oxide	13, 7	Methemoglobinemia/CNS effects	4
p-Nitroaniline—skin	13, 3	Methemoglobinemia/cumulative liver damage	3
Nitrobenzene—skin	13, 12, 7	Methemoglobinemia/anemia/CNS effects	3
4-Nitrobiphenyl	1	Cancer—bladder	1C
p-Nitrochlorobenzene—skin	13, 12	Methemoglobinemia/anemia	3
Nitroethane	15, 8	Moderate irritation—respiratory tract/narcosis	2
Nitrogen	17	Simple asphyxiation	1C
Nitrogen dioxide	10, 11	Cumulative lung damage (bronchitic, emphysema)/lung edema	3

176

Table 6.9 (*Continued*)

Substance	Health Code No.	Health Effects	Work Category
Nitrogen trifluoride	13, 3	Methemoglobinemia/ cumulative liver and kidney damage	3
Nitroglycerin—skin	3	Cumulative effect on blood pressure (lowering)/ headache	1A
Nitromethane	16, 8	Mild irritation—eye, nose, throat/narcos2-Nitropropane	2
1-Nitropropane	15, 3	Moderate irritation—eye, nose, throat/cumulative liver damage	4
2-Nitropropane	3, 2, 15	Cumulative liver damage/ suspect carcinogen/ moderate irritation—eye, nose, throat	4
n-Nitrosodimethylamine— skin	1	Cancer	1C
Nitrotrichloromethane	(See Chloropicrin)		
Nitrotoluene—skin	13	Methemoglobinemia	3
Nitrous oxide	5, 7	Reproductive hazard (male and female)/CNS effect	3
Nonane	16, 8	Mild irritation—eye, nose, throat/narcosis	1B
Octachloronaphthalene— skin	3	Cumulative liver damage/ chlorance	3
Octane	16, 8	Mild irritation	1B
Oil mist (mineral)	18, 10	Good housekeeping practice/accumulation in lungs (pneumonitis)	1C
Osmium tetroxide (as Os)	14, 11	Marked irritation—eye, nose, throat, bronchi, lungs/lung edema	1B
Oxalic acid	14	Marked irritation—eye, nose, throat, skin	1B
Oxygen difluoride	14, 11, 3	Marked irritation—respiratory tract/marked edema— lungs/cumulative kidney damage	4
Ozone	14, 11	Marked irritation—respiratory tract/lung edema	4
Paraffin wax fume	16	Mild irritation—eye, nose, throat	1B
Paraquat—skin	3, 16, 5	Cumulative systemic lung damage/mild irritation— eye, nose, throat/suspect teratogen	3

177

Table 6.9 (*Continued*)

Substance	Health Code No.	Health Effects	Work Category
Parathion—skin	6, 5	Cholinesterase inhibition/suspect teratogen	3
Pentaborane	4, 7	Acute and cumulative CNS damage	4
Pentachloronaphthalene—skin	3, 14, 2	Acute systemic toxicity—vascular and CNS injury/marked irritation—eye and nose/suspect carcinogen	4
Pentachlorophenol—skin	4, 3, 7	Acute systemic toxicity/vascular and nervous system injury/chloracne	4
Pentaerythritol	19	Nuisance particulate	1C
Pentane	18, 8	Flammable/narcosis	1B
2-Pentanone	15, 8	Moderate irritation—eye, nose, throat/narcosis	2
Perchloroethylene (tetra-chloroethylene)—skin	3, 8, 2	Cumulative liver and CNS damage/narcosis/suspect carcinogen	4
Perchloromethyl mercaptan	14, 2	Marked irritation—eye, nose, throat/suspect carcinogen	4
Perchloroyl fluoride	13, 12, 15	Methemoglobine-mia/anemia/moderate irritation—eye, nose, throat	4
Perlite (less than 1% quartz)	19%	Nuisance particulate—accumulation in lungs	1C
Petroleum distillates (naphtha)	15, 8	Moderate irritation/narcosis	1B
Phenacylchloride	(See alpha-Chloroacetophenone)		
Phenol—skin	14, 4, 2	Marked irritation—eye, nose, throat, lungs/acute and chronic systemic toxicity/suspect carcinogen	4
Phenothiazine—skin	15, 3	Moderate irritation—skin/photosensitization—skin	2
Phenyl ether (vapor)	7, 16, 3	Nausea/mild irritation—eye, skin/cumulative liver and kidney damage	4
Phenyl ether-biphenyl mix (vapor)	7, 16, 3	Nausea/mild irritation—eye, skin/cumulative liver and kidney damage	4
p-Phenylene diamine—skin	9, 3	Respiratory sensitization (asthma)/contract skin irritant sensitizer	2
Phenylgycidyl ether (PGE)	15, 3, 8	Moderate irritation—eye, nose, throat, skin/skin sensitization/narcosis	2

Table 6.9 *(Continued)*

Substance	Health Code No.	Health Effects	Work Category
Phenylhydrazine—skin	12, 3	Hemolytic anemia/skin irritation and sensitization	4
Phenylphosphine	12, 7, 5	Hemolytic anemia/CNS effects/ testicular damage	3
Phorate (Thimet[R])—skin	6	Cholinesterase inhibition	3
Phosdrin (Mevinphos[R])	6	Cholinesterase inhibition	3
Phosgene (carbonyl chloride)	11, 10	Marked edema-lungs/chronic lung disease	4
Phosphine	4, 7, 11	Acute and chronic systemic toxicity (CNS effects, lung edema, anemia)	4
Phosphoric acid	14	Marked irritation—eye, nose, throat	1B
Phosphorus (yellow)	3	Cumulative bone and liver damage	3
Phosphorus pentachloride	14, 10, 11	Marked irritation—eye, nose, throat, bronchitis/lung edema	2
Phosphorus pentasulfide	14, 4	Marked irritation—respiratory tract/H_2S hazard	1B
Phosphorus trichloride	14, 10, 11	Marked irritation—eye, nose, throat, bronchi, lungs/bronchial pneumonia	1B
Phthalic anhydride	14, 9, 3	Marked irritation—eye, nose, throat, lungs/asthma/contact skin irritant and sensitizer	2
m-Phthalodinitrile	19	Particulate (apparent low toxicity)	1C
Picloram (Tordon[R])	19	Particulate (apparent low toxicity)	1C
Picric acid—skin	3	Skin irritant and sensitizer/ cumulative liver, kidney and red blood cell damage	3
Pival[R] (2 Pivalyl-1 3-indandione)	3	Cumulative anticoagulant effect (Warfarin analogy)	3
Plaster of Paris	19	Nuisance particulate	1C
Platinum (soluble salts as Pt)	9, 3	Respiratory sensitization (asthma/dermatitis)	1B
Polychlorinated biphenyls (PCB)	(See Chlorodiephenyl)		
Polytetrafluoroethylene decomposition products	4	Acute toxic effects (polymer fume fever)	2
	(See Fluorocarbon polymers)		
Portland cement (Less than 1% quartz	19, 16	Nuisance particulate/mild irritation—eye and nose	2
Potassium hydroxide	14	Marked irritation—eye, nose, throat, lungs, skin	1B

Table 6.9 (*Continued*)

Substance	Health Code No.	Health Effects	Work Category
Propane	18, 7	Explosive/CNS effects	1C
Propargyl alcohol—skin	14	Marked irritation—eye, nose, throat, skin	1B
beta-Propiolactone	1	Marked irritation—eye, nose, throat, skin	1B
n-Propyl acetate	16, 8	Mild irritation—eye, nose, throat/narcosis	1B
Propyl alcohol—skin	16, 8, 2	Mild irritation—eye, nose, throat/narcosis/suspect carcinogen	1B
n-Propyl nitrate	3, 13	Cumulative systemic effects (methemoglobinemia)	3
Propylene dichloride	3	Cumulative liver damage	3
Propylene glycol monomethyl ether	15, 8	Moderate irritation—eye, nose, throat/narcosis	2
Propylene imine—skin	15, 4,	Moderate irritation—eyes, nose, throat/acute kidney and lung damage/suspect carcinogen	4
Propylene oxide	15, 3, 2	Moderate irritation—eye, nose, throat, lungs, skin/cumulative CNS/kidney and liver damage/suspect carcinogen	4
Propyne	(See Acetylene)		
Pryethrum	3, 16	Contact and allergic dermatitis/mild irritation—lungs	3
Pyridine	3, 7	Cumulative liver, kidney and bone marrow damage/CNS effects	3
Quinone (p-benzoquinone)	4, 3, 2	Acute and cumulative—eye (corneal) damage/suspect carcinogen	4
Radon daughters	(See NIOSH CIB No. 10)		
RDX (cyclotrimethylene trinitramine)—skin	7	Chronic CNS effects (nausea, convulsions)	3
Resorcinol	15, 3	Moderate irritation—eye, nose, throat, skin/cumulative systemic toxicity	4
Rhodium, metal fume	9	Respiratory sensitization	1B
Rhodium, soluble salts (Rh)	9, 2	Respiratory sensitization (asthma)/suspect carcinogen	4
Ronnel	6	Cholinesterase inhibition	3
Rosin core solder pyrolysis products (as formaldehyde)	14	Marked irritation—eye, nose, throat	1B

Table 6.9 (*Continued*)

Substance	Health Code No.	Health Effects	Work Category
Rotenone (commercial)	3, 16, 2	Cumulative systemic toxicity/mild irritation—nose, throat/suspect carcinogen	4
Rouge	19	Nuisance particulate (accumulation in lungs)	1C
Rubber solvent	7, 8	Cumulative central and peripheral nervous system damage/narcosis	3
Selenium compounds (as Se)	15, 3, 2	Moderate irritation—eye	4
Selenium hexafluoride (as Se)	11	Lung edema	4
Silica (amorphous)	19, 10	Good housekeeping practice/possible pneumoconiosis	3
Silica (fused)	10	Pneumoconiosis	3
Silica (quartz), respirable	10	Pneumoconiosis (silicosis)	3
Silicon	19	Nuisance particulate (accumulation in lungs)	1C
Silicon carbide	19	Nuisance particulate (accumulation in lungs)	1C
Silicon tetrahydride (silane)	4	Acute systemic toxicity by (analogy with other metal hydrides)	2
Silver, metal & soluble compound (as Ag)	3	Cumulative skin pigmentation and organ accumulation	3
Soapstone	10	Pneumoconiosis	3
Sodium azide	4, 15	Acute CNS and blood pressure effects/mild irritant—eye	2
Sodium fluoroacetate (1080)—skin	4	Acute systemic toxicity (metabolic poison)	2
Sodium hydroxide	14	Marked irritation—eye, nose, throat, lungs, skin	1B
Starch	19	Nuisance particulate	1C
Stibine	4	Acute systemic toxicity	4
Stoddard solvent	16, 8	Mild irritation—eye, nose, throat/narcosis	4
Strychnine	4	Acute systemic toxicity, CNS (convulsions and paralysis)	4
Styrene monomer (phenylethylene)	15, 7, 8	Moderate irritation—eye, nose, throat/CNS effects/narcosis	2
Subtilisins (proteolytic enzymes)	9, 10, 16	Respiratory allergy (asthma and lung damage/)/mild skin irritant	2
Sucrose	19	Nuisance particulate	1C

Table 6.9 *(Continued)*

Substance	Health Code No.	Health Effects	Work Category
Sulfur dioxide	14, 4	Marked irritation—eye, nose, throat, lungs/broncho-constriction	1B
Sulfur hexafluoride	19	Apparent low toxicity	1C
Sulfuric acid	14, 10, 3	Marked irritation—eye, nose, throat, skin, bronchi/dental erosion	1B
Sulfur monochloride	14	Marked irritation—eye, nose, throat, lung	1B
Sulfur pentafluoride	11	Marked irritation—lung (edema)	2
Sulfur tetrafluoride	11	Marked irritation—lung (edema)	2
Sulfuryl fluoride	3, 10, 4	Cumulative kidney and lung damage/acute CNS effects	3
Systox	(See Demetonr)		
2,4,5-T (2,4,5-Trichloro-phenoxyacetic acid)	5, 2	Suspect teratogen and carcinogen	4
Talc (total)	10	Pneumoconiosis (talcosis)	3
Talc, fibrous tremolite	1	Cancer (lung)	3
Talc, fibrous non-tremolite	1	Cancer (lung)	3
Tantalum	19	Apparent low toxicity	1C
TEDP—skin	6	Cholinesterase inhibition	3
Teflon^R decomposition products	4	Acute systemic toxicity (polymer fume fever)	2
Tellurium	4, 3	Acute CNS effects/cumulative organ damage	4
Tellurium hexafluoride (as Te)	11	Lung edema	4
TEPP—skin	6	Cholinesterase inhibition	3
Terphenyls	15	Moderate irritation—eye, nose, throat, lungs	1A
1,1,1,2-Tetrachloro-2,2-difluoroethane	11, 4	Lung edema/respiratory failure	4
1,1,2,2-Tetrachloro-1,2-difluoroethane	3, 12, 11	Cumulative liver damage/decreased white blood cell count/lung edema	3
1,1,2,2-Tetrachloro-ethane—skin	3	Cumulative liver and other organ damage	3
Tetrachloroethylene	(See Perchloroethylene)		
Tetrachloromethane	(See Carbon tetrachloride)		
Tetrachloronaphthalene—skin	3	Cumulative liver damage/chloracne	3
Tetraethyl lead (as Pb)—skin	3, 7, 4	Cumulative liver, CNS and kidney damage/acute CNS effects	3

Table 6.9 *(Continued)*

Substance	Health Code No.	Health Effects	Work Category
Tetrahydrofuran	15, 8	Moderate irritation—eye, nose, throat/narcosis	2
Tetramethyl lead (as Pb)—skin	3, 7, 4	Cumulative liver, CNS and kidney damage/acute CNS effects	3
Tetramethyl succinonitrile—skin	4	Acute systemic toxicity (CNS)—headache, nausea, convulsions	4
Tetranitromethane	14, 4, 3	Marked irritation—eye, nose, throat/acute CNS and lung effects (edema)/cumulative systemic damage	4
Tetryl (2,4,6-Trinitrophenyl Methylnitramine)—skin	3	Contact dermatitis, skin sensitization/cumulative systemic toxicity	3
Thallium (soluble compounds)—skin (as Tl)	3	Cumulative systemic toxicity	3
4,4'-Thiobis (6-tert-butyl-m-cresol)	19	Apparent low toxicity	1C
Thiram^R	4, 5	Acute systemic toxicity (antabuselike effects)/suspect teratogen	4
Tin (Inorganic compounds, except oxide) (as Sn)	4, 3	Acute and chronic systemic toxicity	4
Tin (Organic compounds) (as Sn)	14, 3	Marked irritation—skin/cumulative systemic toxicity	3
Tin oxide	10	Pneumoconiosis (stannosis)	3
Titanium dioxide	19	Nuisance particulate (accumulation in lungs)	1C
Toluene—skin	15, 8	Moderate irritation—eye, nose, throat/narcosis	2
Toluene-2,4-diisocyanate (TDI)	9, 14, 3	Asthma/marked irritation—eye, nose, throat, bronchi, lungs/dermatitis	1A
o-Toluidine—skin	13, 4, 2	Methemoglobinemia/acute systemic effects/suspect carcinogen	4
Toxaphene	(See Chlorinated camphene)		
Tributyl phosphate	15, 7	Moderate irritation—nose, throat, lungs/headache	4
1,1,1-Trichloroethane	(See Methyl chloroform)		
1,1,2-Trichloroethane—skin	3, 8	Cumulative liver damage/narcosis	4
Trichloroethylene	8, 3, 2	Narcosis/cumulative systemic toxic effects/suspect carcinogen	4

Table 6.9 (*Continued*)

Substance	Health Code No.	Health Effects	Work Category
Trichloromethane	(See Chloroform)		
Trichloronaphthalene—skin	3	Cumulative liver damage/ Chloracne	3
1,2,3-Trichloropropane	15, 3	Moderate irritation—eye, nose, throat/cumulative liver damage	4
1,1,2-Trichloro-1,2,2 trifluoroethane (F-113)	19	Apparent low toxicity	1C
Tricyclohexyltin hydroxide (Plictran^R)	19	Apparent low toxicity	1C
Triethylamine	14, 11, 3	Marked irritation—eye, nose, throat, lungs/lung edema/corneal damage	4
Trifluoromonobromo-methane	19	Apparent low toxicity	1C
2,4,6-Trinitrophenol	(See Picric acid)		
2,4,6-Trinitrophenylmethyl-nitramine	(See Tetryl)		
Trimellitic anhydride (TMA)	(See NIOSH CIB No. 21)		
Trimethylbenzene	14, 7	Marked irritation—lungs, skin/cumulative CNS effects	4
Trinitrotoluene—skin (TNT)	13, 12, 3	Methemoglobinemia/A plastic anemia/cumulative eye (cataracts) and liver damage	3
Triorthocresyl phosphate	7	Polyneuropathy	3
Triphenyl phosphate	6	Cholinesterase inhibition	3
Tungsten & compounds (Insoluble) (as W)	19	Lung accumulation/apparent low toxicity	1C
Tungsten compounds (soluble) (as W)	4	Acute CNS effects (metabolic poison)	2
Turpentine	15, 3	Moderate irritation—eye, nose, throat, bronchi, lungs, skin/cumulative kidney damage	1B
Uranium (insoluble compounds)	3	Cumulative kidney damage/ lung accumulation	3
Uranium (soluble compounds)	3	Cumulative kidney damage	3
Vanadium (V_2O_5) dust (as V)	14, 11, 10	Marked irritation—eye, nose, throat, bronchi, lungs, skin/acute and chronic bronchial and lung damage	1A
Vanadium (V_2O_5) fume (as V)	14, 11, 10	Marked irritation—eye, nose, throat, bronchi, lungs, skin/acute and chronic bronchial damage	1A

Table 6.9 (*Continued*)

Substance	Health Code No.	Health Effects	Work Category
Vinyl acetate	16	Mild irritation—eye, nose, throat	1B
Vinyl benzene	(See Styrene)		
Vinyl bromide	3, 2	Cumulative bromide intoxication (CNS effects)/suspect carcinogen	4
Vinyl chloride	1	Cancer—liver	1C
Vinyl cyanide	(See Acrylonitrile)		
Vinyl cyclohexene dioxide	14, 2	Marked irritation—skin/suspect carcinogen	4
Vinylidene chloride	3, 2	Cumulative liver and kidney damage/suspect carcinogen	4
Vinyl toluene	15	Moderate irritation—eye, nose, throat	1B
VM & P naphtha	16, 8, 3	Mild irritation—eye, nose	1B
Warfarin	3	Cumulative anticoagulant effect	3
Welding fumes (total particulate)	15, 11, 3	Moderate irritation—nose, throat, bronchi, lungs/acute and chronic toxicity from metal oxides	4
Wood dust, hardwood (non-allergenic)	10, 3, 2	Lung damage/dermatitis/ suspect carcinogen	4
Wood dust, softwood	19	Nuisance particulate (accumulation in lungs)	1C
Xylene (o-, m-. and p-isomers)—skin	15, 8	Moderate irritation—eye, nose, throat/narcosis	2
m-Xylene, alpha, alpha'-Diamine	3, 15	Contact skin sensitizer/ moderate irritation—skin	2
Xylidene—skin	13, 4	Methemoglobinema/acute systemic toxicity	4
Yttrium	10	Pneumoconiosis (diffuse fibrosis)	3
Zinc chloride fume	14, 11, 2	Marked irritation—eye, nose, throat, lungs/acute lung damage/suspect carcinogen	4
Zinc chromate (as Cr)	2	Suspect carcinogen	4
Zinc oxide fume	4	Acute systemic toxicity (metal fume fever)	2
Zinc stearate	19	Nuisance particulate (accumulation in lungs)	1C
Zirconium compounds (as Zr)	10, 3	Pneumoconiosis/lung and skin granulomas	3

The other formula recommended by OSHA applies to chemicals for which the PEL is intended to prevent the cumulative effects of repeated exposure. For example, PCBs, PBBs, mercury, lead and DDT are considered cumulative toxins because repeated exposure is usually required to cause an adverse effect and the overall biologic half-life is clearly in excess of 12 hours. The goal of PELs in this category is to prevent excessive cumulation in the body following many days or even years of exposure. Chemicals whose rationale is based on *cumulative toxic effects* are placed in Category 3 in Table 6.6. Accordingly, Equation 18 is offered to OSHA compliance officers as a viable approach for calculating a modified limit for chemicals whose half-life would suggest that not all of the chemical will be eliminated before returning to work the following day. Its intent is to ensure that workers exposed more than 40 hr/week will not eventually develop a body burden of that substance in excess of persons who work normal 8 hr/day, 40 hr/week schedules.

$$\text{Equivalent PEL} = \text{8-hour PEL} \times \frac{40 \text{ hours}}{\text{hours of exposure in one week}} \tag{18}$$

The specific approach to be used by an OSHA compliance officer for evaluating a workplace using unusual work shifts is described in detail in the OSHA Field Operations Manual (138). The OSHA models, although less rigorous than the pharmacokinetic models which will be discussed, have certain advantages since they do account for the kind of toxic effect to be avoided, require no pharmacokinetic data, and tend to be more conservative than the pharmacokinetic models.

It is interesting that in the first OSHA occupational health regulation which discusses the long workday (unusual shifts), OSHA prohibited the use of its own adjustment scheme to establish acceptable levels of exposure (174). Because regulatory agencies must make decisions based on political, social and scientific information, and, especially because they must survive legal scrutiny, it is important that the industrial hygienist, toxicologist or physician not expect regulatory actions to have considered each and every exposure condition in the rulemaking. Consequently, these professionals should take the time to become familiar with the rationales for the TLV, PEL, AIHA WEEL, European Maximum Allowable Concentration (MAC) and any other occupational exposure limit before adjusting it for an unusually short or long period of exposure. In this respect, Tables 6.6 through 6.9, as well as the various books which document the rationales for the various limits, are very important.

ILLUSTRATIVE EXAMPLE 5 (OSHA Model).   An occupational exposure limit of 1 microgram per cubic meter has been suggested by NIOSH for polychlorinated biphenyls (PCBs). In animal tests, it has been found that the biologic half-life of PCBs could be as long as several years. What adjustment to the occupational exposure limit might be suggested by NIOSH for workers on the standard 12-

hr work shift involving 4 days of work per week if they adopted the simple OSHA formulae?

Solution.   Recommended Limit $= 8$ hr PEL $\times \dfrac{8 \text{ hr}}{12 \text{ hr}}$

Recommended Limit $= 1\mu g/m^3 \times 0.667 = 0.667 \ \mu g/m^3$

Note: Since PCBs (chlorodiphenyl) are listed as both cumulative and acute toxicants (Category 4) in Table 6.9, Equation 17 rather than 18 should be used since it yields the more conservative results.

ILLUSTRATIVE EXAMPLE 6 (OSHA Model).   Many industries such as boat manufacturing are seasonal in their workload. During the months of January, February, March and April, the builders of boats work 5 days per week, 14 hrs per day and could be exposed to concentrations of Toluene Diisocyanate (TDI) at the TLV of 0.005 ppm. What occupational exposure limit is recommended for TDI for a person who works 14 hr per day for 5 days per week but only works 8 weeks per year?

Solution.   No adjustment is made.

Note: TDI is categorized in Table 6.9 as a Category 1A chemical (i.e., one that has a ceiling limit). Substances in this category have limits which should never be exceeded and consequently the limits are independent of the length or frequency of exposure. Exposure limits for chemical irritants such as these are currently thought not to require adjustment. Until more is known about man's response to irritants during unusually long durations of exposure, the physician, nurse, and hygienist should make note of the employee tolerance to the presence of irritants at levels at or near the TLV. Eventually, human experience will tell us whether irritation is a time dependent phenomena.

ILLUSTRATIVE EXAMPLE 7 (OSHA Model).   The permissable exposure limit for elemental mercury is 50 ug/m³. It has a half-life in humans in excess of several days. What adjustment to the limit would be recommended by OSHA for workers on a shift involving 4 days at 12 hr/day followed by 3 days of vacation, then 3 days of 12 hr/day followed by 4 days off?

Solution.   Equivalent PEL $= 8$ hr PEL $\times \dfrac{40 \text{ hr}}{48 \text{ hr}}$

$= 50 \ \mu g/m^3 \times 0.833 = 40 \ \mu g/m^3$

Note: Since elemental mercury is classified as a cumulative toxin (Category 3) according to Table 6.9, Equation 18 should be used. The 48-hr workweek was used in this example since it yields a more conservative adjustment than the 36-hour workweek. It could be argued that a lesser adjustment factor based on the average number of hours worked every two weeks more accurately reflects the exposure [i.e., $(48 + 36)/2 = 42$] since the chronic effects of mercury are due to many weeks or years of excess exposure. A more detailed discussion of how to deal with toxins which tend to accumulate can be found in Hickey (174).

One major drawback to the simple formulae suggested by Brief and Scala as well as OSHA is that they are generally conservative (i.e., they suggest a modified TLV or PEL that is lower than that predicted by presumably more accurate pharmacokinetic models). This occurs because they *do not* take into account (quantitatively) the toxicant's overall biologic half-life (i.e., metabolism and elimination). As will be shown later in this chapter, pharmacokinetic models usually recommend a lesser degree of reduction in the air contaminant limit and, therefore, the achievement of the adjusted limit should be less costly yet still provide adequate protection.

8.3 Iuliucci Model (1982)

Robert Iuliucci of Sun Chemical proposed a formula (182) for adjusting limits for long workdays which is similar to Brief and Scala's, except that it accounts for the number of days worked each week as well as the number of hours worked each day. It is mentioned only for sake of completeness since it poses no particular advantages over Brief and Scala's approach and because it is limited only to the schedule Iuliucci described (12 hr/day, 4 day/week schedule). For that work schedule, Iuliucci recommended the following equation for modifying the TLV:

$$TLV_x = TLV_s \times \frac{8 \text{ hr worked}}{12 \text{ hr worked}} \times \frac{12 \text{ hr recovered}}{36 \text{ hr recovered}} \times \frac{4 \text{ day workweek}}{5 \text{ day workweek}} \quad (19)$$

In addition to this formula, he recommended that exposure to carcinogens during 12-hr shifts always be reduced by 50 percent. Although this may be prudent, it would seem to be unnecessarily strict based on pharmacokinetic considerations. Equally important, adjustments to limits should be made as carefully as possible using all available scientific information because recommendations to reduce the air contaminant concentrations in some industries are often economically stressful.

8.4 Pharmacokinetic Models

Pharmacokinetic models for adjusting occupational limits have been proposed by several researchers (143–149, 171–175). These models acknowledge that the maximum body burden arising from a particular work schedule is a function of the biological half-life of the substance. Pharmacokinetic models, like the other models, generate a correction factor, however, it is based on the pharmacokinetic behavior of the substance as well as the number of hours worked each day and week. This factor is applied to the standard limit in order to determine the modified limit. Unlike the OSHA, Brief and Scala, and Iuliucci models, by accounting for a chemical's behavior, the kinetic models can identify those exposure schedules where a reduction in the limit is *not* necessary. As noted previously, all of the approaches protect the person who is exposed during an atypical work schedule by generating a modified limit which will prevent the maximum body burden of the toxicant from rising above the level reached following exposures of 8 hr/day, 5 days/week.

The rationale for a pharmacokinetic approach to modifying limits is that during exposure to the TLV for a normal workweek, the body burden rises and falls by amounts governed by the biologic half-time of the substance (Figures 6.2 and 6.3). A general formula provides a modified limit for exposure during unusual work shifts so that the peak body burden accumulated during the unusual schedule is no greater than the body burden accumulated during the normal schedule. All of the pharmacokinetic models which have, thus far, been developed have the same goal (145).

It is worthwhile to note that the maximum body burden arising from continuous uniform exposure under the standard 8 hr/day work schedule *always* occurs at the end of the last work shift before the 2-day weekend. On the other hand, the maximum body burden under an extraordinary work schedule *may not* occur at the end of the last shift of that schedule (see Example 14). This is especially true when the duration and spacing of work shifts which precede the last shift differ markedly from the standard week. No generalization regarding the time of peak body burden can be offered, since unusual work schedules can be based on a 2-week, 3-week, 4-week, or even 11-week cycle and the work shift may be 10, 12, 16, or even 24 hours in duration. The time of peak tissue burden for an unusual schedule must therefore be calculated for each specific schedule.

8.4.1   Mason and Dershin Model (1976)

Mason and Dershin (143) were the first to propose a pharmacokinetic model for adjusting exposure limits for unusual exposure schedules. Apparently, due to the manner in which the model was presented, their publication did not receive the attention and use which later researchers enjoyed. In spite of this, their approach is entirely accurate and is still useful. Like the other pharmacokinetic models to follow, it accounted for the biologic half-life of the chemical and the number of hours of exposure per day and per week. Their mathematical approach, although perhaps more cumbersome than others, is quite general and yields exactly the same results as the other pharmacokinetic models.

The pharmacokinetic model which they developed accounted for a number of factors known to influence the rate of accumulation of a chemical. These factors included the toxicant concentration to which the individual is exposed, the physiochemical form of the material, the rate of metabolism and excretion, as well as the distribution of the material in the body following absorption. The major drawback of the model, and of all those which have followed it, is that it assumes the body to act as a single compartment. The one-compartment approach (discussed previously) assumes the chemical to be uniformly distributed throughout blood and aqueous body fluids without significant storage in specific tissues except where such tissues may constitute the rate limiting step. In Mason and Dershin's model, the overall respiratory exchange, metabolism, and renal excretion were accounted for by using a single effective clearance constant, k.

In their manuscript, Mason and Dershin noted that for a simple, single-compartment model, Ruzic (181) had shown that the overall rate of change in accumulation can be expressed as

$$\frac{d[A]}{dt} = k_i^*[M] - k_c[A] \tag{20}$$

where $[M]$ = the concentration of the contaminant in the environment (alveolar spaces) mg/1,
$[A]$ = the concentration in the compartment, mg/1,
k_i^* = the effective rate constant for uptake, h^{-1},
$k_c[A]$ = an overall clearance constant, h^{-1}.

Note: The effective rate constants may also include other factors: e.g., changes in vital capacity, minute volume, membrane permeability, or absorption from other sources as the cutaneous absorption and loss of carbon disulfide.

In a cyclic pattern of exposure and recovery, Equation 20 has a general solution following the final period of recovery of

$$[A]_{(tm)} = \frac{k_i^*}{k_c}[M][e^{-k_c(t_m - t_{m-1})} - e^{-k_c(t_m - t_{m-2})}$$

$$+ e^{-k_c(t_m - t_{m-3})} \ldots - e^{-k_c(t_m - t_{m-i})}] \tag{21}$$

for $t_{(m-i)} \geq 0$,

where k^*, k, and M are held constant,

$t_{(m)}$ = the time elapsed from the onset of the initial exposure, hours, including recovery following the last exposure,
$t_{(m-1)}$ = the time elapsed from the onset of exposure through the completion of the ith phase of uptake or recovery (loss),

and the initial body burden is assumed to be negligible.

Similarly, Equation 21 may be solved for the body burden obtained at the close of the last exposure by substituting $t_{(m-1)}$ for $t_{(m)}$ so

$$[A]_{(t_{m-1})} = \frac{k_i^*}{k_c}]M] [1 - e^{-k_c(t_{m-1}-t_{m-2})}$$

$$+ e^{-k_c(t_{m-1}-t_{m-3})} \ldots - e^{-k_c(t_{m-1}-t_{m-i})}] \qquad (22)$$

As noted by the authors, in intermittent exposure and recovery, an upper limit to accumulation should be achieved for even many of the typical, lipid soluble solvents, within five days, provided that the periods of recovery and rate of excretion are sufficiently large. The fraction of the saturation value (from continuous exposure) attained at the close of a series of cycles of exposure and recovery varies with the clearance coefficient k_c and the pattern and duration of exposure according to the function

$$[e^{-k_c(t_m-t_{m-1})} - e^{-k_c(t_m-t_{m-2})} + e^{-k_c(t_m-t_{m-3})} \ldots - e^{-k_c(t_m-t_{m-i})}] \qquad (23)$$

which is dependent on both the concentration of exposure and final equilibrium position. Using Henderson and Haggard's definition (143, 184) of the equilibrium distribution coefficient,

$$D = \frac{C}{C_1} \qquad (24)$$

where D = the distribution coefficient, dimensionless,
C = the concentration of contaminant in the fluid phase (mg/1),
C_1 = the concentration of contaminant in the vapor phase of the alveolar air (mg/1),

and solving Equation 24 at equilibrium,

$$\frac{[A]}{[M]} \approx \frac{k_i^*}{k_c} \approx D \qquad (25)$$

This completes the data requirements for calculation of an expected body burden due to a series of intermittent exposures.

The authors acknowledged the limitations inherent in the assumption regarding the use of a one-compartment model but noted that in most cases this limitation had little practical significance with respect to adjusting exposure limits. Other kineticists, although not all of them, would probably agree with this observation (157). In addition, Mason and Dershin supported the prior recommendations of Brief and Scala wherein a modeling approach should not be used to adjust limits whose goal is to minimize the likelihood of irritation, sensitization, or a carcinogenic response. Illustrative Examples 8, 9, and 10 demonstrate the use and general applicability of the model. Mason and Hughes have also proposed approaches for using this model in other situations (185, 186).

ILLUSTRATIVE EXAMPLE 8 (Mason and Dershin Model).   Several workers exposed to methanol in a printing operation complained that their exposure left them dizzy and with optic neuritis at the end of work on Wednesday, 56 hr after reporting to work Monday morning. If the distribution coefficient for methanol at saturation is 1700 to 1 for body water over the concentration in alveolar air; and the concentration in workroom air is 350 parts per million (TLV 200 ppm), what concentration would a worker have obtained in blood at: 1) the end of the last exposure and 2) the time at which they would report to work on Thursday morning?

A schematic diagram illustrating the behavior of methanol for this situation is shown in Figure 6.18.

Part I.   What is the saturation fraction at the close of the last shift, 56 hrs after initial exposure? Shifts are 8 hr in length and are separated by 16 hr without exposure (recovery).

A. $C_T/C_S = [1 - e^{-k(T)} + e^{-k(T-t_1)} - e^{-k(T-\tau_1)} + e^{-k(T-t_n)} - e^{-k(T-\tau_n)}]$

where: $t_1 = 8h$, $\tau_1 = 24h$, $t_2 = 32h$, $\tau_2 = 48h$, $t3 = 56h = T$ (which is quitting time on Wednesday), k for methanol (MeOH) $= 0.03h^{-1}$,

$$C_T/C_S = [1 - e^{-k(56)} + e^{-k(56-8)} - e^{-k(56-24)} + e^{-k(56-32)} - e^{-k(56-48)}]$$

$$C_{56}/C_S = [1 - e^{-.03(56)} + e^{-.03(48)} - e^{-.03(8)}]$$

$$C_{56}/C_S = 0.3678$$

B.   To calculate the tissue concentration (whole body) at saturation:

$$C_S = D \cdot C_{MeOH} = 1700 \cdot 0.44 \text{ mg/l} = 747 \text{ mg/l (body water) @ } 37°C,$$

where D is the distribution coefficient for MeOH (body fluids/alveolar air) and C_{MeOH} the ambient methanol concentration

$$C_{56} = 747 \text{ mg/l} \times 0.3678 = 275 \text{ mg/l.}$$

Note: If this was the only exposure to MeOH, exposure to $C_{8,5}/C_{Uns} \times$ TLV$_{8,5}$ would not allow an increase in the tissue concentration over that experienced under the 8-hr day, 5-day week. This provides the general solution for adjusting standards to unusual shifts as:

$$\text{TLV}_{unusual} = C_{5,8}/C_{Uns} \times \text{TLV}_{5,8}$$

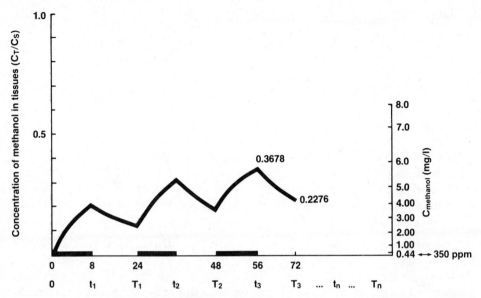

Figure 6.18 Graphical illustration of the behavior of methanol following the exposure schedule described in Example 8. Day-to-day increases in body burden were predicted by pharmacokinetic models. Exposure begins at $t = 0$, the first period ends at T_1, followed by first period of recovery (beginning at t_1) which ends at T_1; and repeats through t_n if the last point (T) is during exposure, and t_n if in recovery. (Courtesy of Dr. Walter Mason.)

where

$$C_{5,8}/C_{\text{Uns}} = F$$

Then,

$$\text{TLV}_{\text{unusual}} = F \cdot \text{TLV}_{5,8}$$

Part II. What is the saturation fraction at the end of the final period of recovery 72 hr after the onset of the first exposure? Shifts are 8 hr as before.

A. $C_T/C_S = [-e^{-k(T)} + e^{-k(T-t_1)} - e^{-k(T-\tau_1)}$

$$+ e^{-k(T-t_2)} - e^{-k(T-\tau_2)} + e^{-k(T-t_n)}]$$

where: $T = 72$, $t_1 = 8$, $\tau_1 = 24$, $t_2 = 32$, $\tau_2 = 48$, $t_3 = 56$, $\tau_3 = 72 = T$ (which is the starting time for work on Thursday), k for MeOH $= 0.03\text{h}^{-1}$,

$$C_T/C_S = [-e^{-k(72)} + e^{-k(72-8)} - e^{-k(72-24)}$$

$$+ e^{-k(72-32)} - e^{-k(72-48)} + e^{-k(72-56)}]$$

$$C_{72}/C_S = [0 - e^{-.03(72)} + e^{-.03(64)} \ldots + e^{-.03(16)}]$$

$$C_{72}/C_S = 0.2276$$

B.   The tissue concentration is then found by:

$$C_S = D \cdot C_{MeOH} = 747 \text{ mg/l (body water) @ 37°C,}$$

$$C_{72} = 747 \text{ mg/l} \times 0.2276 = 170 \text{ mg/l}$$

ILLUSTRATIVE EXAMPLE 9 (Mason and Dershin Model).

Part I.   Workers are exposed to methanol for 8 hr/day for 5 days/week. What is the body burden of those workers at the end of the 5th day knowing that the distribution coefficient (blood/air) is 1700, the effective clearance constant, k_c, is 0.03 h^{-1} ($t_{1/2} = 24$ hr) and the concentration to which they are exposed (1983 TLV) is 0.26 mg/l (200 ppm)?

Solution.   Setting, $1700 = D$, $k_c = 0.03$h^{-1}, and the periods of exposure and recovery at eight hours and sixteen hours respectively; for which $t_{(m)} = 5(24) = 120$, $t_{(m-1)} = 120 - 16 = 104$, $t_{(m-2)} = 104 - 8 = 96$ etc.; the body burden remaining at the close of recovery on the fifth day may be calculated as a function of the concentration of exposure.

$$[A]_{(t_m)} = 1700[M] \, [e^{-.03(16)} - e^{-.03(24)} + e^{-.03(40)} - e^{-.03(48)}$$

$$+ \, e^{-.03(64)} - e^{-.03(72)} \ldots]$$

$$[A]_{(t_m)} = (1700 \; M)(0.249) = 420M$$

Similarly, if the concentration at the end of the last exposure is of interest $t_{(m)}$ is set equal to $t_{(m-1)}$ and:

$$[A]_{(t_{m-1})} = 1700[M][1 - e^{-.03(104-96)} + e^{-.03(104-80)} \ldots - e^{-.03(104-72)}]$$

$$A_{(t_{m-1})} = 1700M(0.401) = 680M \text{ (Peak Burden during 8 hr/day schedule)}$$

Part II.   Having calculated the body burden (peak) at the end of 5 days of exposure during a normal 8 hr workday, what modified TLV would be recommended for a 14 hr workday and 4 day workweek?

Solution.   If the modified exposure limit is chosen so that the tissue concentration attained at the close of the last work phase under standard conditions is equal to the accumulation allowed under the novel shift arrangement:

$$[A]_{(t_{m-1})n} = [A]_{(104)5}$$

where $[A]_{(t_{m-1})n}$ = the concentration that would be obtained at the close of the final exposure period in a novel shift arrangement

$[A]_{(104)5}$ = the concentration obtained after a five day work week under standard conditions.

The accumulation at the close of the last exposure may be obtained directly with Equation 25 or from tables constructed with the exponential term of the same equation. The final form is then reduced to:

$$[M]_n = \frac{[A]_{(104)5}}{D} \times \frac{1}{\text{saturation fraction for } (t_{m-1})_n}$$

where $[M]_n$ = the alveolar concentration of the contaminant resulting in a body burden equal to that attained by exposure under standard conditions,

and

TLV$_n$ = $[M]_n$ adjusted to ambient conditions,

therefore

$[A]_{96-56}$ = body burden of toxicant following 4 days of exposure during 14 hr/day schedule (i.e., 10 hr/day or recovery).

By substitution into Equation:

$$[A]_{96} = 1700M[1 - e^{-0.03(96-82)} + e^{-0.03(96-72)} - e^{-(0.03(96-58)} + e^{-0.03(96-48)}....]$$

$$[A]_{96} = (1700M)[1 - 0.657 + 0.487 - 0.320 + 0.237 - 0.156 + 0.115 - 0.075]$$

$$[A]_{96} = (1700M)\,0.630 = 1070M \text{ (peak burden during 14 hr/day schedule)}$$

To calculate the modified TLV for the 4-day, 14 hr/day schedule:

$$F \cdot [A]_{(t_{m-1})n} = [A]_{(104)5}$$

$$F = \frac{A_{(120)}}{A_{(96)}} = \frac{680M}{1070M} = 0.636$$

Adjusted TLV = 0.636 (8 hr TLV) = 0.636 (0.26 mg/l) = 0.165 mg/l.

Conclusion. The air concentration for the 14 hr/day schedule should be reduced from 0.260 mg/1 to 0.165 mg/1 (37% lower) in order to have the same peak body burden as that noted during the standard workweek.
Figure 6.19 shows how a series of curves can be generated for a particular chemical which would permit the rapid identification of an adjustment factor for a number of schedules. The factor suggested by Mason and Dershin's formula is compared to that recommended by Brief and Scala (dotted line) in this figure.

ILLUSTRATIVE EXAMPLE 10 (Mason and Dershin). In the oil producing regions of Canada and in the North Sea, work schedules can become very complex. Calculate a modified exposure limit for a 2-week work schedule where persons

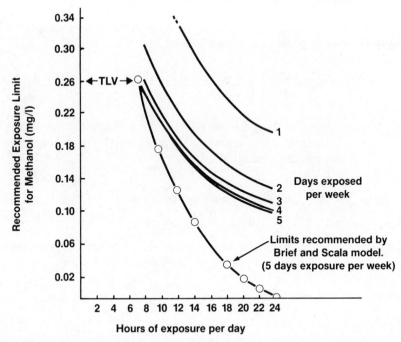

Figure 6.19 Modified exposure limits for methanol for various work schedules as determined by Mason and Dershin's formula. The dotted line illustrates the limits recommended by the Brief and Scala models. (From Mason and Dershin, 1976.)

will be exposed to cyclohexane (assume biologic half-life in humans of 23 hr) for four 10-hr workdays, followed by 4 days off, then four 12-hr workdays followed by 2 days off, to complete 2 calendar weeks, which involved 88 hr of exposure.

Equation 1.

$$[A]_{(t_m)} = \frac{k_i^*}{k_c} [M] [e^{-k_c(t_m - t_{m-1})} - e^{-k_c(t_m - t_{m-2})} + e^{-k_c(t_m - t_{m-3})} \ldots - e^{-k_c(t_m - t_{m-i})}]$$

Equation 2.

$$[A]_{(t_{m-1})} = \frac{k^*}{k_c} [M] [1 - e^{-k_c(t_{m-1} - t_{m-2})} + e^{-k_c(t_{m-1} - t_{m-3})} - e^{-k_c(t_{m-1} - t_{m-i})}]$$

$$\text{for } t_{(m-i)} \geq 0,$$

where k_i^*, k_c, and $[M]$ are held constant

t_m = time elapsed since initial onset of exposure ($t = 0$) through the last phase of recovery = 336 hr

t_{m-1} = time elapsed to the point at which work ends = 276 hr.

This schedule and the day-to-day increase in blood plasma concentration are shown in Figure 6.20. The solid line was calculated by iteration (as a check against Equation 1). But, the point at 252 hr and 336 hr were obtained *via* Equations 2 and 1, respectively. Data points for the iteration are:

$A(t)/A_{sat}$ [a]	t, hr	$A(t)/A_{sat}$ [a]	t, hr	$A(t)/A_{sat}$ [a]	t, hr
0	0	0.3132	96	0.3178	240
0.2592	10	0.1525	120	0.5240	252
0.1703	24	0.0742	144	0.3656	264
0.3854	34	0.0361	168	0.5574	276
0.2532	48	0.0176	192	0.3889	288
0.4468	58	0.3146	204	0.1893	312
0.2935	72	0.2195	216	0.0921	336
0.4766	82	0.4555	228		

[a] This is the "saturation fraction."

The construction of a table is the easiest way to solve Equations 1 and 2. For $A(t_m)$ use Equation 1 $[e^{-k\Delta t} - e^{-k\Delta t} \ldots]$. For $A(t_{m-1})$ use Equation 2 $\ldots [1 - e^{-k\Delta t} \ldots]$.

t_m	t_{m-n}	Δt	$-k$	$-k\Delta t$	$e^{-k\Delta t}$
336 − 276_1	= 60	−.03		−1.80	+0.1653
336 − 264_2	= 72	"		−2.16	−0.1153
336 − 252_3	= 84	−.03		−2.52	+0.0805
336 − 240	= 96	"		−2.88	−0.0561
336 − 228	= 108	−.03		−3.24	+.039200
336 − 216	= 120	"		−3.60	−.027300
336 − 204	= 132	−.03		−3.96	+.019100
336 − 192	= 144	"		−4.32	−.013300
336 − 82	= 254	−.03		−7.62	+.0005
336 − 72	= 264	"		−7.92	−.0004
336 − 58	= 278	−.03		−8.34	+.0024
336 − 48	= 288	"		−8.64	−.000177
336 − 34	= 302	−.03		−9.06	+.000116
336 − 24	= 312	"		−9.36	−.000086
336 − 10_{15}	= 326	−.03		−9.78	+.000057
336 − 0_{16}	= 336	"		−10.08	−.000042
				$\Sigma e^{-k\Delta t}$	= 0.0915

A diagram comparing the behavior of the body (blood) levels for the work schedule described in this example with that of a standard schedule is shown in Fig. 6.20. Each day's exposure or "daily additions" are represented by the dashed lines at the bottom of the plot. The solid line which leads up to point t represents the sum of the dashed line values for the same point. Piotrowski

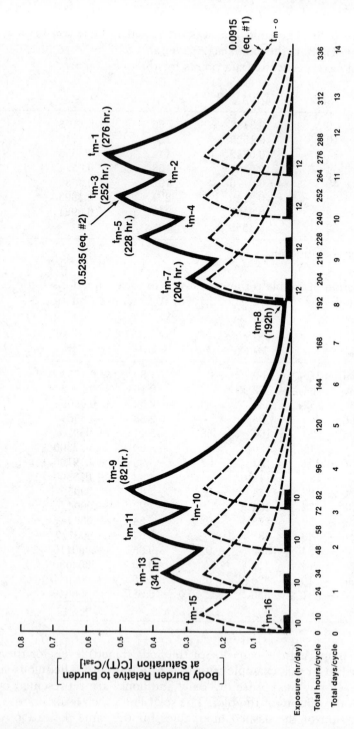

Figure 6.20 The likely body burden of cyclohexane in persons who work an unusual schedule involving four 10-hr days followed by four 12-hr days described in Example 10. The dashed line represents the behavior of the chemical during each day of exposure. And the dark lines represent the overall behavior of the chemical due to repeated exposure.

and others have used this approach to illustrate the principle of summation (167).

Examples 8 through 10 illustrate one of the advantages of the Mason and Dershin model in that some persons feel that it is more flexible than the other models for calculating a modified exposure limit for complex work schedules. Specifically, it is useful whenever there is no fixed number of hours worked each day or a fixed number of days worked per week. In short, the best aspect of this approach is that the periods in the cycles need not be of equal duration or number. Dr. Mason and Dr. Hughes (187) have recently developed a computer program which calculates the modified limit for not only any work schedule, but also for any starting and stopping time (e.g., 8:00 a.m. or 5:00 p.m.)

8.4.2 Hickey and Reist Model (1977)

In 1977, Hickey and Reist (145) published a paper describing a general formula approach to modifying exposure limits which was equivalent to that of Mason and Dershin. The benefits of their work were manifold. They confirmed the soundness of the previous model but, equally importantly, they also validated it to some extent by comparing the results with published biological data. In addition, they proposed broader uses of the pharmacokinetic approach to modifying limits and presented a number of graphs which could be used to adjust exposure limits for a wide number of exposure schedules. The graphs were based on (1), the biologic half-life of the material, (2) hours worked each day, and (3) hours worked per week. Over the following 3 years they wrote publications which illustrated how their model could be used to set limits for persons on overtime (171) and for seasonal workers (172). Hickey's treatment of the topic of adjusting exposure limits is quite thorough and his publications are primarily responsible for most of the interest and research activity in this area.

As discussed, it is clear that for any schedule, the degree of toxicant accumulation in tissue is a function of the biologic half-life of the substance. Figure 6.21 illustrates how a toxicant might behave in a biologic system or a tissue following repeated exposure to a given average air concentration during a typical work schedule. Note that the *peak body burden*, rather than the average or residual body burden, is the parameter of interest. The biologic half-life not only dictates the level to which a chemical accumulates with repeated exposure, it dictates the time at which steady-state will be reached for any given exposure regimen (normal, unusual or continuous). For example, for moderately volatile substances (e.g., solvents) which have half-lives in the range of 12–60 hours, and for most work schedules, the steady-state tissue burden will be reached after approximately 2–6 weeks of repeated exposure (187, 188). For most volatile chemicals (low-molecular-weight solvents) with shorter half-lives, the steady-state blood levels will be reached after about 1 to 4 workdays. Under conditions of continuous uniform exposure, most chemicals will be within 10 percent of the steady-state levels following about 4 times the biologic half-life of the chemical, and after 7 half-lives it will be within 1 percent of the plateau (steady-state) levels (154).

Several indices of body or tissue burden could have been chosen as the basis for "predicting equal protection" for any two different exposure regimens. These indices are the peak, residual, and average body or tissue burden of a substance. Figure 6.21 illustrates these three potential criteria from which one must choose in order to build a mathematical model. As in Mason and Dershin's model, Hickey and Reist selected the peak body burden as the criterion since it is more likely to predict the occurence of a toxic effect than either the average or residual tissue concentrations (180). A thorough discussion of the rationale for selecting the peak burden for building the models rather than the residual or average can be found in Hickey's dissertation (144).

In short, other choices for modeling are problematic. For most chemicals, the residual body burden goes to virtual zero for most chemicals after a weekend away from exposure. Consequently, modeling to control this criterion would not prevent excessive peak burdens. The use of the average burden reduces the model to Haber's Law. This, or coures, would allow high tissue burdens to occur for long periods even though the time-weighted average burden might be acceptable. Peak burden, therefore, is the best criterion; however, it may not be appropriate when the goal of an exposure limit is to avoid a carcinogenic hazard. In these cases, control of the average weekly or daily exposure (at the TLV or PEL) should prevent any excess risk (185).

The Hickey and Reist model can, like Mason and Dershin's approach, be

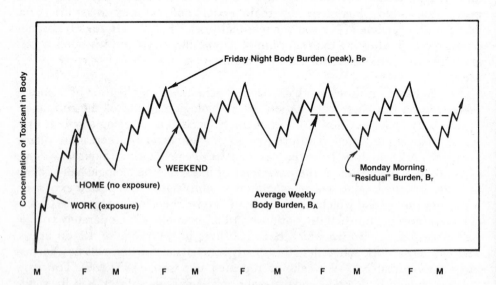

Figure 6.21 Illustration of the weekly fluctuation of body burden resulting from occupational exposure to an inhaled substance. Peak, (B_P), residual, (B_r) and average body burdens (B_A) are shown. (From Hickey and Reist, 1977.)

used to determine a special exposure limit for workers on extraordinary schedules, which will prevent peak tissue or body burdens from being greater than that observed during standard shifts. This special limit is expressed as a decimal adjustment factor which, when multiplied by the appropriate exposure limit, would yield the "modified" limit. It is worthwhile to note that all of the researchers have been careful to note that they *did not assert that currently prescribed or recommended occupational health limits are safe, but only that the special limit which can be predicted from their models should yield "equal protection" during a special exposure situation!* Example problems 11 through 15 illustrate the use of the Hickey and Reist model.

One limitation of the models of Hickey and Reist (144, 145), Mason and Dershin (143), and Roach (148) is that they assume that the body acts as one compartment. Although this simplification may not pose many shortcomings for the task of adjusting occupational exposure limits, it is well known that many, if not most, chemicals do not exhibit one-compartment behavior (165, 188). This is not surprising since after a substance (particulate, gaseous, or vapor) is inhaled in air, it is taken up by the body, distributed, perhaps metabolized, and excreted by complex processes. Even though these processes can now be modeled quite well through the use of complex mathematical models, the one-compartment model has been used by these scientists since, in most cases, even this simple approach will yield results similar to those which incorporate more complex approaches.

As discussed by Hickey, the one-compartment model is the simplest one and it assumes that the body is a homogeneous mass, comparable to a room or compartment containing a clean fluid such as air. More of the air, bearing a contaminant, enters and flows continuously through the compartment, mixing en route with the air therein in a process analogous to inhalation. If contaminated air continues to enter, the contaminant concentration in the compartment increases until it reaches equilibrium with that of the incoming air; that is, as much contaminant is leaving as entering (see Figure 6.22). When the contaminated air supply is replaced with clean air, the process is reversed, and the contaminant concentration in the compartment decreases exponentially (145). An anology can be shown using fluid in tanks as shown in Figure 6.23. The mass transfer phenomena are described by the following equations:

$$\text{For uptake: } B_t = \text{CWK}(1 - e^{-kt}) + B_0(e^{-kt}) \tag{26}$$

$$\text{For excretion: } B_r = B_t(e^{-kt_r}) \tag{27}$$

in which B_t = body burden of substance at time t (mass),

$\qquad B_0$ = initial body burden of substance at time zero (mass),

$\qquad B_r$ = residual body burden of substance at time t (mass),

$\qquad C$ = substance concentration in air (mass/volume),

$\qquad K$ = ratio of the substance's equilibrium solubility in the body to that in air, or "partition coefficient" (dimensionless),

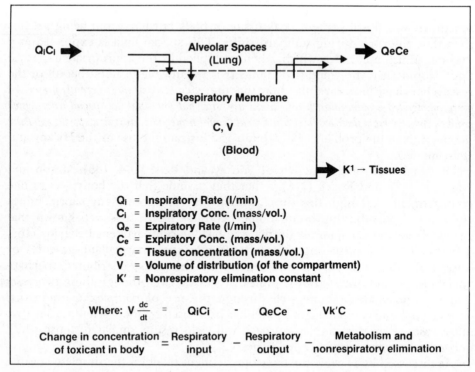

Figure 6.22 Diagram illustrating the driving forces for the pharmacokinetic behavior of an inhaled air contaminant. This simple description is the basis of the models developed by Mason and Dershin (1976) as well as Hickey and Reist (19770.

W = volume of body (volume),
k = uptake and excretion rate of substance in the body, equal to L/WK, in which L is the flow rate of air to the body time (time, h^{-1}),
t = time of exposure to substance in air (hrs).
t_r = time since cessation of exposure to substance in air (hrs).

It should be noted that k may also be expressed in terms of half-life or half-time of the substance in the body, $T_{1/2}$, where $k = (\ln 2)/T_{1/2}$.

Hickey and Reist have noted that while the predictive capability of the pharmacokinetic models is limited by the shortcomings caused by simplification to a one-compartment system, there are practical circumstances which minimize these drawbacks (144). First, many of the body tissues which are important targets for inhaled substances ("critical" tissues) are highly perfused (150–154, 170, 181), and the concentration of the contaminant in these tissues may follow that of the arterial blood closely, thus in effect, becoming part of the lung-arterial blood compartment. The opposite case occurs when the buildup of contaminant in the critical or target tissue is extremely slow compared to the buildup in the rest of the body. In such a case, the remainder of the body, or more specifically the arterial blood, may be assumed to reach saturation relatively quickly and remain at a virtually constant concentration. In effect, the body (except for the critical tissue) becomes part of the ambient environment, and the critical tissue becomes the one-compartment body (145).

Figure 6.21 illustrates how the body takes up and excretes an inhaled air contaminant as described by according to Equations 26 and 27, where exposure to contaminated air occurs during working hours and clean air is inhaled during nonworking hours. The body takes up the contaminant according to rate k during periods of exposure, and during nonworking hours the body excretes the contaminant according to negative rate $-k$ (145). The rate constant k is assumed to remain unchanged for each chemical regardless of the duration of exposure or whether there are repeated exposures. Small changes in k have been reported following repeated exposure to unusual shifts, however these are not usually large enough to justify mathematical correction (19). It should be remembered that for a given exposure schedule and any k, the body *will* eventually reach some equilibrium level with the contaminated air after continuous or repeated exposures (Figure 6.4). This is true even for substances which have very long half-lives.

The variation in body burden upon exposure to an air contaminant for five workdays per week for a period long enough to reach equilibrium (steady-state) is also illustrated in Figure 6.21. Equilibrium implies that the "Monday morning" body burden (B_r) remains the same from week to week for a given exposure schedule. Each schedule also has a characteristic "Friday afternoon" peak body burden (B_p), and average body burden (B_a). This is illustrated in Figure 6.1 where two different exposure schedules and the resulting body burdens are described for a chemical with a moderately long half-life.

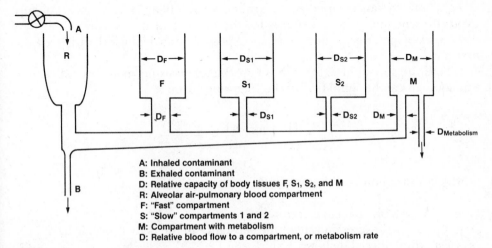

A: Inhaled contaminant
B: Exhaled contaminant
D: Relative capacity of body tissues F, S_1, S_2, and M
R: Alveolar air-pulmonary blood compartment
F: "Fast" compartment
S: "Slow" compartments 1 and 2
M: Compartment with metabolism
D: Relative blood flow to a compartment, or metabolism rate

Figure 6.23   Simulation of body uptake of an air contaminant using fluid in tanks as analogy. (Reprinted from Hickey, 1977, with permission.)

In spite of the fact that nearly all volatile chemicals will demonstrate some degree of two and three compartment behavior, the one-comparment assumption is probably, in general, satisfactory. Dittert (157) has noted that in many, if not most situations, simplification to one-compartment behavior poses a minimal source for error when calculating most pharmacokinetic parameters.

In their publications, Hickey and Reist (145) described the derivation and use of the following equation for adjusting limits:

$$
F_p = \frac{(1 - e^{-kT_s}) \left[1 - \exp\left(-kt_n\right) + \exp\left(-k \sum\limits_{i=n-1}^{n} t_i\right) - \ldots + \ldots - \exp\left[\left(-k \sum\limits_{i=1}^{n} t_i\right)\right] n \right]}{(1 - e^{-kl_n}) \left[1 - \exp\left(-kt_s\right) + \exp\left(-k \sum\limits_{j=s-1}^{s} t_j\right) - \ldots + \ldots - \exp\left(-k \sum\limits_{j=1}^{s} t_j\right) \right] s}
$$

(28)

in which t values represent duration of sequential work and rest periods in cycle T for normal n and special s exposure schedules. The authors noted that in their model, the use of ratios causes many of the imponderable and unknown terms to cancel, leaving only the special work schedule, which will be known, and the substance half-life (or uptake/excretion rate), which may or may not be known.

Equation 28 can be used to determine a modified TLV or PEL for any exposure schedule since it accounts for the number of hours worked per day, days worked per week, time between exposures and biologic half-life of the toxicant.

Where the special or extraordinary work cycle uses normal days and weeks, Equation 28 can be simplified to the following form:

$$
F_p = \frac{(1 - e^{-8k})(1 - e^{-120k})}{(1 - e^{-hk})(1 - e^{-24dhk})}
$$

(29)

in which, using hours as the time unit,

F_p = TLV or PEL reduction factor,
k = uptake and excretion rate of the substance in the body (biologic half-life),
h = length of special daily work shift,
d = number of workdays per "workweek" in the special schedule.

The general Equation 28 for regular repetitive schedules simplifies to

$$F_p = \frac{[1 - e^{-kt_{1n}}]\,[1 - e^{-k(t_{1n}+t_{2n})n}]\,[1 - e^{-kT_s}]\,[1 - e^{-k(t_{1s}+t_{2s})}]}{[1 - e^{-kt_{1s}}]\,[1 - e^{-k(t_{1s}+t_{2s})m}]\,[1 - e^{-kT_n}]\,[1 - e^{-k)t_{1n}+t_{2n})}]} \tag{30}$$

in which, using hours as the time unit,

t_{1n} = length of normal daily work shift (8 hours),
t_{2n} = length of normal daily nonexposure periods (16 hours),
$t_{1n} + t_{2n}$ = length of normal day (24 hours),
T_n = length of normal week (168 hours),
n = number of workdays per normal week (5),
t_{1s} = length of special "daily" work shift, hours,
t_{2s} = length of special nonexposure periods between shifts, hours,
$t_{1s} + t_{2s}$ = length of basic work cycle, analogous to the "day," hours,
T_s = length of periodic work cycle, analogous to the "day," hours,
m = number of work "days" per work "week" in the special schedule.

The model may be used to predict the permissible level and duration of exposure necessary to avoid exceeding the normal peak body burden during intrashift, short, high-level exposures. The model does this by establishing excursion limits which will provide equal protection for these situations. Equation 31 is used to do this:

$$F_p = \frac{1 - e^{-kt_n}}{l - e^{-kt_e}} \tag{31}$$

where $k = \ln 2 / T_{1/2}$
t_e = exposure time (hrs)
t_n = normal shift length (8 hrs).

Hickey and Reist have noted that for substances with short biologic half-lives (less than 3 hr), no adjustment needs to be applied for workers on most extraordinary work shifts since there is no opportunity for accumulation. In Figure 6.24, the normal workweek is compared to workweeks of from one to seven 8-hr days. It can be seen that exposure limits may not be increased, even if exposure is for only one day per week, unless the substance half-life is greater than 6 hr. Similarly, limits need not be decreased for 6- or 7-day workweeks involving exposures of 8 hr/day unless the substance half-life is greater than about 16 hr. For substances with very long half-lives, those in excess of 400 hr, F_p is simply proportional to the number of hours worked per week, as compared to 40 hours.

In an effort to simplify the process of adjusting limits for unusual shifts, Hickey and Reist developed a number of graphs which are shown in Figures 24–26 and 28–30. Many health professionals have found these to be very useful

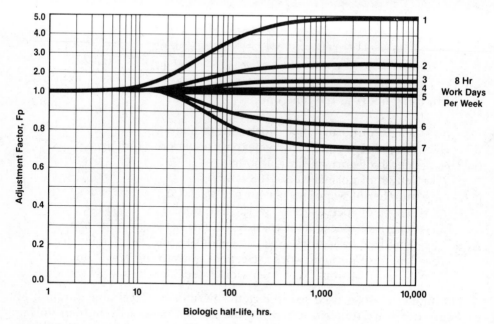

Figure 6.24 Adjustment factor (F_p) as a function of substance half-life ($T_{1/2}$) for various work weeks. (From Hickey and Reist, 1977.)

when estimating safe levels of exposure for chemicals for which they have little or no pharmacokinetic data. In these graphs, the adjustment factor, F_p is usually plotted as a funtion of substance half-life, $T_{1/2}$, for a particular work schedule(s). For example, Figure 6.25 shows the difference in the occupational exposure limit between a normal workweek and a workweek of four 10-hr days, a workweek of three 12-hr days, and a single 40-hr shift per week.

It is clear from Figure 6.25 that for substances with very short half-lives (less than one hour), the peak body burden is reached very quickly and is the same for a normal workweek as for any special schedule with longer shifts. Therefore, if B_p is chosen as the predictor of equal protection, no reduction in OSHA limits is necessary for longer-than-normal work shifts as long as the weekly exposure is less than 40 hr.

For substances with very long half-lives in the body, the adjustment factor is proportional merely to the number of hours exposed, not the daily or weekly exposure schedule. Thus, all 40-hr weeks have a special exposure limit for such substances equal to the normal limit, or an F_p of unity. Since three 12-hr days total only 36 hr per week, F_p for that schedule is 40/36, or 1.1, for a substance with a very long half-life.

Hickey offered sound advice when he noted that one need not resort to the conservative approaches of Brief and Scala, OSHA or Iuliucci when the biologic half-life of the substance is not known. By assuming that the chemical has a half-life which would cause the greatest degree of day-to-day accumulation for that particular work schedule, the *worst-case* F_p can be calculated for any exposure schedule. Some of these worst case values of F_p for selected schedules are shown in Figure 6.25. For example, since F_p varies as a function of the half-life, the worst case condition is 0.84 for four 10-hr days, 0.75 for three 12-hr days, and 0.54 for the single 40-hr shift. Whenever the half-life is not known, the worst case F_p can be used. Consequently, the pharmacokinetic models can accurately protect workers on any schedule even when the pharmacokinetic behavior of the specific chemical is not known.

Other curves may be generated from Equations 28 or 30 to compare any two schedules. In Figure 6.26, the normal workweek is compared to continuous exposures for several periods of time from 1 to 1,024 days, followed by rest periods equal to three times the exposure periods. Again, for substances with very short half-lives, no adjustment to exposure limits is necessary when B_p is the criterion. For substances with very long half-lives, F_p approaches proportionality with number of hours exposed (145).

ILLUSTRATIVE EXAMPLE 11 (Hickey and Reist Model).   In cities where commuting distances are a burden to the worker, one of the more frequent work schedules is the four day, 10 hr/day "compressed workweek." Assuming that this workweek is used in the textile industry and that persons are routinely exposed to aniline at the PEL of 5 ppm, what adjusted occupational exposure limit would be recommended? Assume that aniline has an overall (beta phase) half-life of about 2 hr in humans.

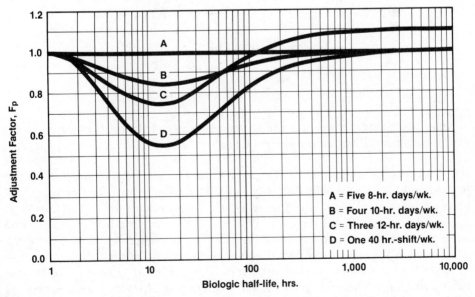

Figure 6.25   Adjustment factor (F_p) as a function of substance half-life ($T_{1/2}$) for various exposure regimens (shift schedules). (From Hickey and Reist, 1977.)

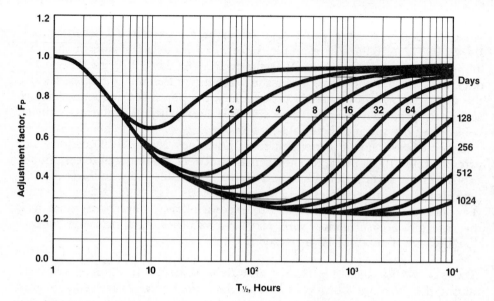

Figure 6.26 Adjustment factor (F_p) as a function of substance half-life ($T_{1/2}$) for continuous exposure schedules (days). (From Hickey and Reist, 1977.)

Solution. Since the workweeks are equal and both schedules have all workdays consecutive, The simplified form of the general Equation 29 can be used.

$$F = \frac{C_s}{C_n} = \frac{(1 - e^{-8k})(1 - e^{-120k})}{(1 - e^{-10k})(1 - e^{-96k})} \text{ for } T_{1/2} = 2 \text{ hr,}$$

$$k = \ln2/2 \text{ hr} = 0.347 \text{ hr}^{-1}$$

$$F = 0.9677 \equiv 0.97$$

$$C_s = C_n{\cdot}F = 5 \times 0.97 = 4.85 \approx 5.0$$

Note: No change is needed for chemicals with a half-life this short unless exposure is for 24 hr/day for several days.

ILLUSTRATIVE EXAMPLES 12 (Hickey and Reist Model). Some persons in the petrochemical industry will routinely work 12-hr shifts for 5 consecutive days before having a 4-day period of no work; then they return for three 12-hr days again to be followed by 4 days off. This 5/4, 3/4, 12-hr schedule requires some adjustment to the normal exposure limit for certain systemic toxins if the peak body burden for this unusual shift is not to exceed the normal shift.

Based on Piotrowski's work (152), the overall half-life (B phase) in the human for trichloroethylene and its metabolites is about 9 hr. What modifications of the 1983 TLV of 50 ppm for trichloroethylene would be recommended for the workers on this new shift schedule during their first week of 5 consecutive 12-hr days of work?

Solution. Begin by determining the body burden for a normal week (Bp_n)

$$\text{For } T_{1/2} = 9 \text{ hr}, k = 0.077$$

$$Bp_n = C_nWK \frac{(1 - e^{-8k})(1 - e^{-120k})}{(1 - e^{-168k})(1 - e^{-24k})}$$

$$Bp_n = 0.546 \, C_nWK$$

We do not want the first-week special body burden, Bp_s (after switching schedules) to exceed Bp_n (normal burden), so

$$Bp_s \,(\text{Week 1}) = C_sWK(1 - e^{-12k})(\;1\;\; +\;\; e^{-124k} +\;\; e^{-48k} +\;\; e^{-72k} +\;\; e^{-96k})$$

$$\begin{array}{ccccc}
\uparrow & \uparrow & \uparrow & \uparrow & \uparrow \\
\text{Day} & \text{Day} & \text{Day} & \text{Day} & \text{Day} \\
5 & 4 & 3 & 2 & 1
\end{array}$$

$$+ \quad Bp_n \,(e^{-178k})$$

$$\uparrow$$
Residual left
from last week
of old schedule.

We set $Bp_s = Bp_n$ because we want the body burden from the "special" schedule (Bp_s) to be equal to the body burden for the normal schedule (Bp_n).

$$Bp_n - Bp_n(e^{-178k}) = Bp_n(1 - e^{-178k}) = 0.546 \, C_nWK(1 - e^{-178k})$$

$$0.546 \, C_nWK(1 - e^{-178k}) = C_sWK(1 - e^{-12k})(1 + e^{-96k} + e^{-72k} + e^{-48k} + e^{-24k})$$

$$\frac{C_s}{C_n} = \frac{\text{TLV special}}{\text{TLV normal}}$$

$$= \frac{0.546 \,(1 - e^{-178k})}{(1 - e^{-12k})(1 + e^{-96k} + e^{-72k} + e^{-48k} + e^{-24k}},$$

$$= \frac{0.546}{0.716} = 0.76$$

$$C_s = C_n \times 0.76$$

$$C_n = 50 \, \text{ppm Trichloroethylene.}$$

Recommended TLV for Trichloroethylene for a 12 hr/day, 5-day schedule = 38 ppm.

Note: Due to the short half-life and 40 hr/week schedule, a short cut approach would yield same result:

$$F = \frac{1 - e^{-8k}}{1 - e^{-12k}} = 0.76.$$

ILLUSTRATIVE EXAMPLE 13 (Hickey and Reist Model).   The NIOSH recommended occupational exposure limit for PCB is 1 microgram per cubic meter. In tests with animals it has been found that the biologic half-life of PCBs is roughly 12.6 years. What adjustments to the occupational exposure limit would you recommend for workers on the standard 12-hr shift involving 4 days of work followed by 3 days of vacation, then 3 days of work followed by 4 days off, etc?

Solution.

$$F_P = \frac{\dfrac{(1 - e^{-8k})(1 - e^{-120k})}{(1 - e^{-168k})(1 - e^{-24k})}}{(1 - e^{-12k} + e^{-24k} - e^{-36k} + e^{-48k} \ldots e^{-228k})/(1 - e^{-336k})}$$

$$F_p = \frac{0.2381}{0.2501} = 0.9524.$$

As pointed out earlier, as $t_{1/2}$ gets very large,

$$F = \frac{40}{\text{hr/wk special schedule}}$$

$$F = \frac{40}{84/2} = 0.9525$$

Adjusted Limit = TLV$_S$ = 0.95 × 1 µg/m^3 (essentially no change is needed).

ILLUSTRATIVE EXAMPLE 14 (Hickey and Reist Model).   In Canada, unusual work shifts have become more commonplace than in the United States. In one industry, the unions and management decided that a combination of the 8 hr/ day and 12 hr/day work schedule best fit their needs. Assuming that exposure to benzene is the primary hazard in this industry, what modified occupational exposure limit would be called for if the limit imposed in the plant were the most rigorous one to which they must adhere during the month (i.e., 12 days of repeated exposure)?

The exact schedule used in this industry involved 5 days of exposure for 8 hr/day followed by 2 days of 12 hr/day then 5 days of exposure for 8 hr/day shift followed by 6 days off work. The schedule then repeats itself so that workers average only 40 hours per week each 4-week cycle. Figure 6.27 illustrates the qualitative behavior of benzene in the body during this exposure schedule.

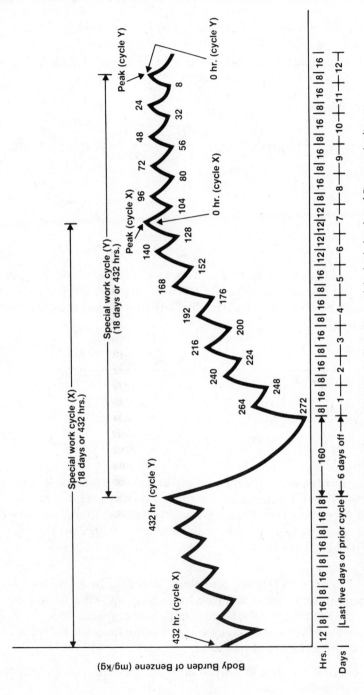

Figure 6.27 Graphical illustration of the likely fluctuations of the body burden of Benzene (or its metabolites) following repeated exposure to the complex 8 hr/day and 12 hr/day work shedule described in Example 16.

Solution.   Two cycles must be examined to determine which gives the lower F_p:

a.   Cycle of 12 on and 6 off.
b.   Cycle of 5 on, 6 off and 7 on, with 6th and 7th days having 12 hr shifts.
c.   Benzene $T_{1/2} = 10$ hr, $k = \ln 2/10$

$$F_p = \frac{\dfrac{(1 - e^{-8k})(1 - e^{-120k})}{(1 - e^{-168k})(1 - e^{-24k})}}{(1 - e^{-8k} + e^{-24k} - e^{-32k} + e^{-48k} \ldots - e^{-272k})/(1 - e^{-432k})}$$

$$F_p = \frac{0.52502}{0.52524} = 1.0$$

Note: The 12 on, 6 off cycle, gives an F_p of unity.

Why? The residual levels from the 2 extra exposures from the previous Saturday and Sunday add only 0.0002 CWK to the 0.52502 CWK from the normal exposure burdens. To confirm, check the burden at the end of the 12 hr Sunday shift.

$$F_p = \frac{0.52502 \leftarrow \text{normal week}}{(1 - e^{-12k} + e^{-24k} - e^{-36k} + e^{-52k} \ldots - e^{-420k})/(1 - e^{-432k})}$$

$$F_p = \frac{0.525}{0.686} = 0.765$$

$\text{TLV}_S = 0.765 \times 10 \text{ ppm} = 7.65 \text{ ppm}$

The peak body burden occurs at the end of the second 12 hr shift. The moral is that the time of peak burden must be chosen correctly. Otherwise, as could have occurred here, the incorrect factor would be applied to the exposure limit. If it is not obvious, more than one peak time must be tested. (Problem was developed by Dr. John Hickey.)

ILLUSTRATIVE EXAMPLE 15 (Hickey and Reist Model).   In an 8-hr day, 7-day workweek situation, such as the 56/21 or 14/7 schedules, what should the TLV for H_2S be (1983 TLV = 100 ppm)? This is the special case of a 7-day workweek. Biologic half-life in humans is 2 hr, but the rationale for the standard is based on systemic effects and irritation.

Solution.   There is no need to reduce limits to prevent excess irritation, but for system effects it should be.

For 14 days on and 7 days off:

$$F_p = \frac{\left[\dfrac{(1 - e^{-8k})(1 - e^{-120k})}{(1 - e^{-168k})(1 - e^{-24k})}\right]}{\left[\dfrac{(1 - e^{-8k})(1 - e^{-(14 \times 24)k})}{(1 - e^{-(3 \times 168)k})(1 - e^{-24k})}\right]}$$

where Total week = 21 days or 168 × 3 hrs
       Workweek = 14 days or 14 × 24 hrs

$$F_p = \frac{\left[\dfrac{1 - e^{-120k}}{1 - e^{-168k}}\right]}{\left[\dfrac{1 - e^{-336k}}{1 - e^{-504k}}\right]} = 1$$

With a $T_{1/2}$ of 2 hr, virtually all of the chemical is lost during the 16 hr of recovery each day, so there is no need to lower the TLV. Also, even though the average hours worked per week is 37.3, the TLV may not be raised by 40/37.3 or by 1.07X.

For a 56/21 schedule,

$$F_p = \frac{(1 - e^{-120k})/(1 - e^{-168k})}{(1 - e^{-(56 \times 24)k})/(1 - e^{-(7 \times 168)k})} = 1$$

Likewise, with 56/21 there is no need to reduce TLV, even though average hr/week is 40.7.

8.4.3   Roach Model (1978)

Roach (148) also proposed a mathematical model for use during extraordinary workshifts. His model, although developed independently, was virtually identical to that proposed by Mason and Dershin, as well as Hickey and Reist. His general equation is shown below:

$$R = \frac{(1 - e^{-8a})(1 - e^{-120a})(1 - e^{-la})}{(1 - e^{-24a})(1 - e^{-168a})(1 - e^{-ma})\Sigma e^{-na}} \tag{32}$$

In this formula the shifts included are those in one complete work cycle prior to the shift end in question and

l = total number of hours for a complete work cycle,
m = number of hours duration of the work shift,

n = number of hours from a prior work shift end to the shift end in question,

e = the exponent of natural logarithms, 2.718,

$a = \dfrac{\log 2}{T} = \dfrac{0.693}{T}$,

T = biologic half-time in hours.

The minimum value of this ratio, R_{min}, is the value of R obtained for the particular workshift in the cycle in which the maximum body burden occurs.

Persons who wish to use this model are referred to the original article (148). Since it is functionally the same as the prior models, no examples of its use are provided. Table 6.10, however, was developed by Roach and can serve as a useful guide for quickly approximating the modified exposure limit for a number of types of unusual shifts where the biologic half-life, T, is known. Roach has shown that for any given work schedule, no matter how complex, a generalized graphical solution which yields the adjustment factor for any chemical can be developed. To illustrate this point, Roach developed Figure 6.28 for the particular complex work schedule shown.

Like previous writers, Roach noted that the limits for substances which have a very short biologic half-time, such as irritants and carcinogens, may require no alteration when the standard work schedule is altered. Roach also suggested that the TLV for substances which have a very long biological half-life, such as mineral dusts, should be modified in proportion to the average hours worked per week, which has also been the recommendation of OSHA, Mason and Dershin (143), and Hickey and Reist (145). With such substances, the duration of any practical work cycle is short in comparison with their biologic half-time, therefore it is appropriate that the limit would be unaltered so long as persons only work an average of 40 hr/week. Roach suggested that as an additional precaution, appropriate medical surveillance to detect any adverse effects would

Table 6.10 Examples of Exposure Limit Adjustment Factors for a Variety of Unusual Work Shifts Based on the Chemical's Biologic Half-Life in Humans. These Were Calculated by Roach (1978) Using a Pharmacokinetic Model.

Work Shifts/Week	Hr/Work Shift	R_{min} when $t_{1/2}$ is:			
		1 hr	10 hr	100 hr	1000 hr
4	10	1.00	0.85	0.94	0.99
5	9	1.00	0.92	0.89	0.89
5	10	1.00	0.85	0.81	0.80
5	12	1.00	0.75	0.68	0.67
6	6	1.01	1.26	1.18	1.12
6	8	1.00	1.00	0.89	0.84
6	10	1.00	0.85	0.72	0.67
7	6	1.01	1.26	1.09	0.97
7	8	1.00	1.00	0.82	0.73

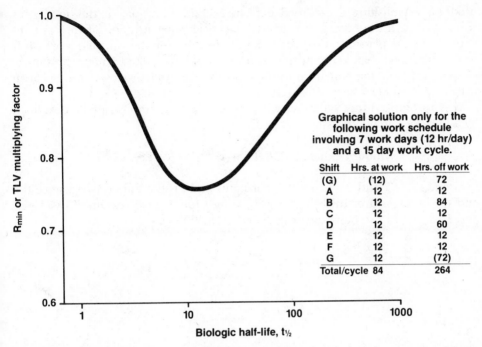

Figure 6.28 Graph showing the adjustment factor for any chemical to which workers are exposed during this specific work schedule. This approach is useful for hygienists who must set limits for dozens of chemicals for a given shift schedule. (From Roach, 1978.)

be advisable if the work schedule is such that R_{min} is less than 0.5 or greater than 2.0.

8.4.4 Veng-Pedersen Model (1984)

As mentioned previously, one of the shortcomings of the existing models for adjusting exposure limits is that a one-compartment model is used to determine the adjustment factor. Recently, a model by Veng-Pedersen (149) has been proposed which takes into account any number of compartments exhibited by a chemical. The equations for adjusting the TLV are based on a linear systems approach. The merits of a linear systems approach have been discussed by several authors (190–192). At this time, the usefulness of this proposal is limited since a true "working" formula for adjusting exposure limits has not been fully developed. The model, however, does have the capability of predicting, based on animal data, whether or not a chemical will accumulate with repeated exposures if the standard kinetic parameters are known (149). The approach has on one occasion been used to evaluate the results of animal toxicologic studies of 12 hr/day exposures (189).

The following equations describe how both alpha and beta half-lives are used for establishing a modified exposure limit. The goal of this model, as contrasted with the others, is to insure that peak concentration of the contaminant in blood plasma rather than "body burden" for the unusual work shift does not exceed that of the normal work shift. These will, of course, be similar or identical. The mathematical foundation for the model follows and has been fully discussed elsewhere (149).

It is recognized that pulmonary excretion data can empirically be described by a biexponential expression of the form

$$\text{Amount exhaled} = K_\alpha(1 - e^{-\alpha t}) + K_\beta(1 - e^{-\beta t}) \tag{33}$$

where K_α, α, K_β and β are positive constants and t is the time elapsed since the end of the exposure. In this approach, values for two expressions $K_1 K_2 A_T(T)$ and $K_1 K_2 B_T(T)$ can be readily calculated according to

$$K_1 K_2 A_T(T) = \frac{\alpha}{\phi_n(L\alpha)} K_\alpha \tag{34}$$

$$K_1 K_2 B_T(T) = \frac{\beta}{\phi_n(L\beta)} K_\beta \tag{35}$$

The predicted maximum plasma level, $\max_{n,m}$, which will be achieved during a given exposure period can then be calculated through Equation 36:

$$\begin{aligned}
\max c_{n,m} = {} & \phi_N(L\alpha)[1 + \phi_{m-1}(7L\alpha)e^{-(M+n)L\alpha}]A_T(T) \\
& + \alpha_N(L\beta)[1 + \alpha_{m-1}(7L\beta)e^{-(M+n)L\beta}]B_T(T)
\end{aligned} \tag{36}$$

where $c(t)$ = plasma level of compound during or following a single exposure,
$c_n(t_r)$ = plasma level of compound during or following nth dosing cycle,
$M(t)$ = cumulative amount of pulmonary excretion,
$M_n(t)$ = cumulative amount of pulmonary excretion following the nth exposure.

If there is no more than one workweek exposure, $m - 1$, the expression in Equation 36 reduces to the regular dosing cycle case:

$$\max c_{n,1} = \phi_N(L\alpha)A_T(T) + \phi_N(L\beta)B_T(T) \tag{37}$$

The final equation needed for determining the modified limit is simply the ratio of the peak concentration of contaminant in plasma calculated for the standard 5-day, 8 hr/day shift versus that calculated for the unusual shift:

$$P = \frac{\max c_{n,m}^{\text{I}}}{\max c_{n,m}^{\text{II}}}. \tag{38}$$

As in previous models, the modified TLV or PEL is simply

$$\text{TLV}_{\text{(modified)}} = (P)\text{TLV}_{\text{(normal)}} \tag{39}$$

The original purpose of this model was to analyze pharmacokinetic parameters determined from the breath of animals which had been exposed for 8 hr/day and 12 hr/day to an air contaminant. The objective was to evaluate what degree, if any, of adjustment to the limit would be needed in order to insure that during repeated exposure, the contaminant concentration in plasma of the 12 hr/day group would not exceed that of the 8 hr/day group. Although the above equations which were used to calculate the adjustment factor may appear complex and demanding in kinetic terms, P can be determined directly from two sets of pulmonary excretion data: one set from a single or multiple exposure with an exposure period T^{I} and another set with an exposure period T^{II}. By fitting a simple two exponential expression to the two data sets, values for α, β, $K_1K_2A_T(T)^{\text{I}}$, $K_1K_2A_T(T)^{\text{II}}$, $K_1K_2B_T(T)^{\text{I}}$ and $K_1K_2B_T(T)^{\text{II}}$ can then be determined as previously discussed. In calculating P, it is important to realize that it is not necessary to know $A_T(T)$ and $B_T(T)$ as such. It is sufficient to know the products $K_1K_2A_T(T)$ and $K_1K_2B_T(T)$ since in forming the ratio in Equation 38 it makes no difference if $A_T(T)$ and $B_T(T)$ are replaced with the products.

At the present time, this approach is only useful for calculating the modified exposure limit when some description of the chemical's behavior following exposure during the unusual exposure period is known. It is useful in that it can predict the peak and residual concentrations of a chemical in plasma following repeated exposure when only the exhalation concentration time profile is known. A method for calculating the likelihood of accumulation during very long workdays when only the pharmacokinetic data following 6 hr/day animal exposures are known.

Basic kinetic parameters:

   K_1—Mass transfer constant for transport of the inhaled compound into blood plasma.

   K_2—Partition coefficient between air and blood plasma for inhaled compound. The term $K_2c(t)$ represents the partial pressure of the compound in the plasma.

A,B,α,β—Parameters defining the unit impulse response—that is, the parameters describing the plasma concentration profile $Ae^{-\alpha t} + Be^{-\beta t}$ resulting from a unit amount introduced momentarily in the plasma.

Parameters defining the exposure:

C_g—Concentration of gas in the inhaled air.

T—Duration of exposure(s).

H—Duration without exposure in an exposure cycle.

L—$T + H$ Length of exposure cycle.

n—The number of the current exposure cycle.

t—Time.

t_r—Time elapsed since the start of the most recent exposure.

Parameters used in calculating TLV *adjustment factor:*

C_g^I—Average concentration of toxicant in air to which workers may be safely exposed for 8 hr per day (TLV or PEL).

C_g^{II}—Average concentration of toxicant in air which will not produce plasma levels in workers on unusual shifts which are no higher than that of 8-hr workers.

$\phi_i(x)$—Function introduced to simplify the mathematical notation.

8.5 Determining a Chemical's Biologic Half-Life

Effective half-lives of substances are very difficult to determine precisely because of the complex manner in which many substances behave in the body. As discussed previously, the body can be envisioned to be made up of many compartments. Therefore, a major complicating factor is that different parts of the body take up and excrete substances at different rates and consequently have different half-lives for each substance. The difficulties in determining the half-life of a contaminant in a tissue is the reason why overall or apparent half-lives which represent clearance of contaminants from the blood are used. Figure 6.29 illustrates how the degree of perfusion of various tissues and the exposure schedule will influence the peak concentration of toxicant which will be achieved as well as the biologic half-life of the chemical in that tissue.

Effective half-lives for human uptake and excretion have been determined for many substances. Roach (147) and Hickey (144) have listed the half-lives of several industrial chemicals which they compiled from various sources. Table 6.11 is a slightly more comprehensive list of substances for which half-lives

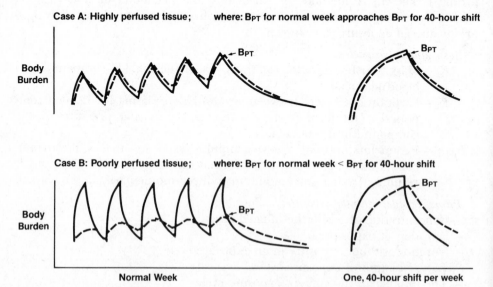

Figure 6.29 Variation in tissue uptake as a function of exposure schedule and blood flow (B_{PT} is peak tissue concentration of toxicant). (From Hickey, 1977.)

Table 6.11 Estimated Half-lives of Various Chemicals or Their Metabolites in Humans

Substance	Compartment (Media Collected)	Half-Life	Reference
Acetone	Overall (blood)	3 hr	Piotrowski (152)
Aniline	Overall (urine)	2.9	Piotrowski (152)
Benzene	Overall (blood)	3.0 hr	Piotrowski (152)
			Nomiyama (209)
			Hunter (198)
Benzidine	Overall (urine)	5.3 hr	Piotrowski (152)
Carbon monoxide	Overall (breath)	1.5 hr	Peterson (212)
Carbon disulfide	Overall (breath)	0.9 hr	Piotrowski (152)
Carbon tetrachloride	Fast (breath)	20 min	Stewart et al. (214)
	Slow (breath)	3.0 hr	
Dichlorodifluoromethane	Overall (blood)	9.4 min	Adir et al. (195)
Dimethyl formamide	Overall (urine)	3.0 hr	Krivanek et al. (203)
Ethyl acetate	Overall (breath)	2.0 hr	Nomiyama and Nomiyama (209)
Ethyl alcohol	Overall (breath)	1.5 hr	Nomiyama and Nomiyama (209)
Ethyl benzene	Overall (urine)	5.0 hr	Piotrowski (152)
Hexane	Overall (breath)	3.0 hr	Nomiyama and Nomiyama (209)
Methanol	Overall (urine)	7.0	Piotrowski (152)
Methylene chloride	Overall (blood)	2.4 hr	DiVincenzo (197)
Nitrobenzene	Overall (urine)	86.0 hr	Piotrowski (152)
Phenol	Overall (urine)	3.4 hr	Piotrowski (213)
p-Nitrophenol	Overall (urine)	1.0 hr	Piotrowski (152)
Styrene	Overall (urine)	8.0 hr	Ikeda et al. (200)
			Ramsey and Young (202)
			Stewart et al. (217)
Tetrachloroethylene	Overall (breath)	70 hr	Stewart et al. (215, 219)
1,1,1-Trichloroethane	Overall (urine)	8.7 hr	Monster et al. (205)
Trichloroethylene	Fast (breath)	About 30 min	Stewart et al. (218)
	Slow (breath)	24 hr	Kimmerle (204)
Trichlorofluoroethane	Overall (blood)	16 min	Adir et al. (195)
Toluene	Fast (urine)	4 hr	Ogata et al. (210)
	Slow (urine)	12 hr	Carlson (196)
Xylene	Overall (urine)	3.8 hr	Ogata et al. (210)

have been determined in humans. Where available, half-lives from human studies should be applied to the models, and where the half-life is not known, the worst case approach described by Hickey and Reist (139) can be used. Whenever both the α and β phase half-lives are known, the beta phase half-life should be used in the proposed models since the terminal or beta phase half-life best represents the behavior of the chemical in the body (for purposes of this discussion). There are numerous pitfalls in accurately determining biologic half-life, and these have been reviewed by Gibaldi and Weintraub (194). Mason and Hughes have recently developed a model wherein information about both the α and β phases can be utilized (193).

8.6 Comparing the Various Models

In the way of a review, the models which modify exposure limits for varied exposure regimens without consideration of pharmacokinetic behavior are those of OSHA (138) and Brief and Scala (18). The Brief and Scala model equations are given below:

$$\text{R.F.} = \frac{8}{h} \times \frac{24 - h}{16} \qquad \text{on a daily basis}$$

$$\text{R.F.} = \frac{40}{H} \times \frac{168 - H}{128} \qquad \text{on a weekly basis}$$

$$\text{E.F.} = [(\text{E.F.}_8 - 1) \times \text{R.F.} + 1] \qquad \text{for excursions}$$

where R.F. = reduction factor to be applied to the TLV or OSHA limit,
  h = hours worked per day
  H = hours worked per week
  E.F. = adjusted excursion factor
  E.F._8 = normal excursion factor

Note: ACGIH abandoned the use of excursion factors in 1976. OSHA never adopted them.

As noted by Hickey (144), in applying the reduction factor to determine allowable exposure limits for nonnormal schedules, Brief and Scala give separate weightings to increased exposure time and decreased recovery time between exposures. Their model incorporates the concept of the "excursion factor" and reduces it by a factor proportional to the decrease in the allowable exposure limit. The model is not intended to be applied to shorter-than-normal exposures, only to longer-than-normal daily or weekly exposures. It cannot be applied to continuous exposures, as it devolves to zero at this point. The model does not take into account substances with very short half-lives, but it is evident that rapidity of toxic response was considered in development of the model.

By comparison, the OSHA model (Equations 17, 18) if applied to other than 8-hr shifts, could be recast in its simplest form using the Brief and Scala symbols:

$$\text{R.F.} = \frac{8}{h} \tag{40}$$

This represents the OSHA model as it would be used to determine limits for a substance (with no peak or ceiling limits) if a single uniform exposure occurred for h hours. For example, if a person were exposed only for 4 hr of an 8-hr shift, the adjustment factor would be 2; that is, a concentration of up to twice the TWA limit would meet OSHA regulations. Assuming application of OSHA regulations to 10-hr shifts, the adjustment factor for a 10-hr exposure would be 0.8; that is, the OSHA 8-hr limit would be reduced to 0.8 of its value to meet regulations.

Neither of these models accounts for uptake rate, so the reduction (or adjustment) factors derived from these models can be compared only to the worst-case adjustment factor for the model under development. It must also be assumed that OSHA regulations would apply to exposures longer than 8 hr. With these qualifications, several comparisons are made in Table 6.12.

It can be seen that in every case, the pharmacokinetic models call for *less* reduction in TWA limits for longer-than-normal exposure periods and for *more*

Table 6.12 Comparison of Adjustment Factors Derived from Different Models

Condition	Worst Case Pharmacokinetic Models, F_p	Brief & Scala R. F.	OSHA Current Practice	OSHA Adjustment Factor for Longer Shifts
Five 8-hr days per week	1	1	1	1
Four 10-hr days per week	0.84	0.7	1	0.8
Three 12-hr days per week	0.75	0.5	1	0.67
Five 16-hr days biweekly	0.56	0.25	1	0.5
Alternating weeks of three and four 12-hr days	0.72	0.5	1	0.67
Four hrs of exposure daily[a]	1	[b]	2	[b]
Two hrs of exposure daily[a]	1	[b]	4	[b]
One-half hr exposure daily[a]	1	[b]	16	[b]

Note: This chart was developed by Hickey (1977).

[a] For substances with only TWA limits.

[b] Method not applicable.

reduction in short-term exposure limits than those required by the OSHA model. As noted by Hickey (144), this is a direct reflection of the pharmacokinetic model's use of peak body burden rather than average burden as a criterion to predict equal protection, and of the fact that it takes into account the uptake and excretion rate of substances.

Even considering the limitations of the pharmacokinetic models as discussed in this chapter, the adjustment factors derived by the model are considered to reflect more realistically the needed protection from exposure than does the OSHA model, and to be a further extension of the concept of the Brief and Scala model (18, 144).

The various mathematical models which have been proposed each have advantages and disadvantages. For chemicals with acute or chronic toxicity, the OSHA model restricts the daily dose or the weekly dose, respectively, during an unusual work shift to the same amount as that obtained during a standard 8-hr shift. It does not acknowledge the lesser recovery period or biologic half-life of the compound. The Brief and Scala model does not permit the daily dose of the toxicant under a novel work shift to be greater than that for a standard shift and it also accounts for the lessened time for elimination; however, it does not consider biologic half-life. Consequently, the Brief and Scala model is the most conservative of all of the models. The Mason and Dershin, Hickey and Reist, and Roach models account for biologic half-life, increased daily dose, lessened recovery between exposures as well as the weekly dose. As a result, these pharmacokinetic models yield less conservative results and they are presumed to be more accurate. The Veng-Pedersen model yields essentially the same results as the other pharmacokinetic models but it is slightly more complex than the one-compartment models and, as such, should be more accurate. Table 6.13 contains some general guidelines for adjusting occupational exposure limits for persons who work unusually long work shifts. Example 16 shows the potential differences in the results of the various models.

ILLUSTRATIVE EXAMPLE 16 (Comparing the Popular Models).   Assuming that 1,1,2-trichloroethane has a biologic half-life of 16 hr in man, what modified TLV or PEL would be appropriate for persons who wished to work a 3 day, 12 hr/day workweek? Note that the dose for the unusual workweek (360 ppm-hr) would be less than for the normal 8 hr/day, 5-day workweek (400 ppm). The present PEL and TLV for 1,1,2-trichloroethane is 10 ppm.

OSHA Model: Modified PEL $= 10.0 \text{ ppm} \times \dfrac{8 \text{ hr}}{12 \text{ hr worked/day}}$

Modified PEL $= 6.66$ ppm.

Brief and Scala Model: Modified TLV $= \dfrac{8 \text{ hr}}{12 \text{ hr}} \times \dfrac{24 - 12}{16} \times 10.0 \text{ ppm}$

Modified TLV $= 5.0$ ppm.

Table 6.13   Handrules for Adjusting Occupational Exposure Limits for Persons Working Unusual Shifts

(1)   Where the goal of the occupational exposure limit is to minimize the likelihood of a systemic effect, the concentration of toxicant to which persons can be exposed should be less than the TLV if they work more than 8 hr/day or more than 40 hr/week and the chemical has a half-life between 4 and 400 hrs.

(2)   Exposure limits whose goals are to avoid excessive irritation or odor will, in general, not require modification to protect persons working unusual work shifts.

(3)   Adjustments to TLVs or PELs are not generally necessary for unusual work shifts if the biological half-life of the toxicant is less than 3 hr or greater than 400 hr.

(4)   The biologic half-life of a chemical in humans can often be estimated by extrapolation from animal data.

(5)   The four most widely accepted approaches to modifying exposure limits will recommend adjustment factors which will vary. In order of conservatism, the Brief and Scala model will recommend the lowest limit and the kinetic models will recommend the highest.

$$\text{Brief and Scala} > \text{OSHA} > \text{ACGIH} > \text{Pharmacokinetic}$$

(6)   Whenever the biologic half-life is unknown, a "safe" level can be estimated by assuming that the chemical has a biologic half-life of about 20 hours. (Note: This will generally yield the most conservative adjustment factor for typical 8-, 10-, 12-, and 14-hr workdays.)

Hickey and Reist Model:

$$\text{Modified TLV} = 10.0 \text{ ppm} \times \frac{(1 - e^{-8k})(1 - e^{-120k})}{(1 - e^{-t_1 k})(1 - e^{-t_2 k})}$$

$$\text{Modified TLV} = 10.0 \text{ ppm} \times \frac{(1 - e^{-8(0.04)})(1 - e^{-120(0.04)})}{(1 - e^{12(0.04)})(1 - e^{72(0.04)})}$$

$$\text{Modified TLV} = 7.5 \text{ ppm}.$$

Note: $k = \ln 2/t_{1/2} = \dfrac{0.693}{16} = 0.04,$

t_1 = hrs worked per day on unusual schedule,

t_2 = 24 × days worked per week on unusual schedule.

It is apparent from Example 16 that the various models can recommend markedly different limits of exposure. In all cases, the pharmacokinetic approach recommends a less strict time-weighted average limit than that generated by models that do not consider the biologic half-life of the chemical.

8.7   A Generalized Approach to the Use of Pharmacokinetic Models

As discussed, four different researchers have proposed pharmacokinetic models for adjusting limits and since they are based on the same assumptions, the results will essentially be the same. Roach (148) and Hickey (145) have presented charts which can be used to quickly determine the adjustment factor for many common shifts. However, for all other situations, the hygienist must begin with the basic equations in order to calculate a modified limit.

Because most persons have found Hickey and Reist's approach and their publications to be most easily understood, the following generalized scheme for determining the adjustment factor for any schedule will be based on their equations. The limits derived from their model will be virtually identical to those obtained by use of the other models. Where the nonnormal work exposure schedule does not fit a curve derived by Hickey and Reist, one of several equations may be used:

1. For any regular weekly schedule, Equation 30.
2. For a sporadic schedule, Equation 28.
3. For an excursion in a normal shift, Equation 31.
4. Where continuously rising exposure is expected, Equation 41, Figure 6.30.

$$F_p = \frac{(1 - e^{kt_e})4k}{kt_e - (1 - e^{-kt_e})} \tag{41}$$

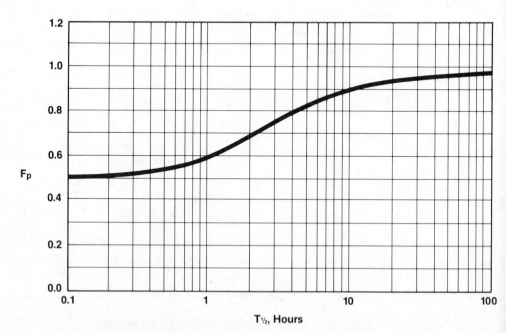

Figure 6.30 Adjustment factor (F_p) for continuously rising contaminant levels during work shift. (From Hickey, 1977.)

5.   For discrete variations in exposure levels, Equation 42.

$$F_p = \frac{(1 - e^{-kt})}{f_c(1 - e^{-kt_c}) + f_b(1 - e^{-kt_b})(e^{-kt_c}) + f_a(1 - e^{-kt_a})(e^{-kt_b})(e^{-kt_c})} \quad (42)$$

where t = total exposure time period,
   f_i = air concentration of substance as fraction of special exposure level
       (concentrations f_a, f_b, f_c),
   t_i = time of exposure at f_i (periods t_a, t_b, t_c)
   k = clearance factor, ln $2/t_{1/2}$

As noted by Hickey (144), Equation 42 has a drawback in that F_p is determined on the *assumption that peak body burden occurs at the end of a shift*. If the actual peak occurs within a shift, F_p must be determined on that basis. Thus, to use this equation correctly, it must be known or calculated in advance at which exposure level the peak body burden will occur. If exposure levels *do not decrease* during a shift, the peak burden may be predicted to occur at shift's end. Where levels of contaminant decrease during the shift, F_p may be determined for the end of each discrete period and the lowest one applied.

It should be noted that different exposure regimens can affect how the body accumulates a chemical during a particular work shift. To illustrate how the pharmacokinetic models can predict variations in body burden due to unusual exposure periods, Hickey (144) has offered the following Examples 17 and 18.

ILLUSTRATIVE EXAMPLE 17 (Evaluating Short-Term Exposures).   Assume that three work schedules involve exposure to trichloroethylene (TCE). In Situation 1, the worker is exposed to the TWA limit for 8 hr per day (normal). In Situation 2, the TCE concentration rises linearly from 0 to 200 ppm during the shift. In Situation 3, the workweek is the same, but the worker is exposed to a "worst-case" situation: a discretely rising TCE concentration with peaks of 300 ppm for 5 min every 2 hr. These situations are depicted in Figure 6.31.

What can be said about compliance to the various OSHA limits? Would one expect the peak body burdens to vary with the different schedules even though the absorbed dose (ppm-hr) is the same for all three? Is the peak body burden for any of these short-term exposure schedules likely to exceed the peak body burden observed during continuous 8-hr exposure to 100 ppm?

Note: TCE has an OSHA PEL of 100 ppm (TWA), a ceiling limit of 200 ppm, and an OSHA peak limit of 300 ppm of 5 min every 2 hr. TCE has a fast compartment biologic half-life of 15 min and a slow compartment biologic half-life of 7.6 hr. (This example was developed by Dr. John Hickey.)

Solution.
Part I.   In each case, exposures are within OSHA limits, since all have TWA averages of 100 ppm and none of the short-term exposures exceed an OSHA limit for short exposure periods.

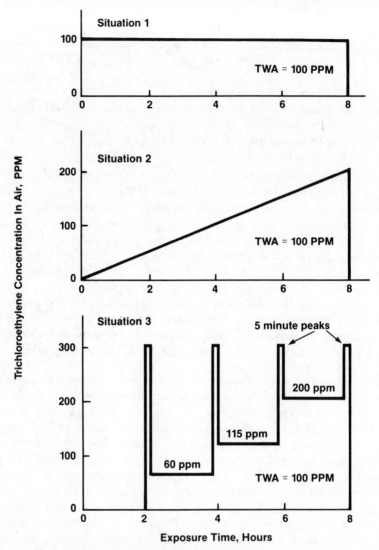

Figure 6.31 Comparison of three different trichloroethylene exposure situations, wherein the 8 hr TWA concentration is always 100 ppm, but the resulting peak body burden could vary between the exposure schedules. Based on Example 17. (From Hickey, 1977.)

Part II. Using Equation 41 for Situation 2 and Equation 42 for Situation 3, the F_p values for these two exposure regimens can be calculated. F_p is the adjustment applied to the TWA limit in a special exposure situation which will result in a predicted peak body burden equal to that resulting from a normal exposure at the TWA limit. When the adjustment is not made, and C_s is left equal to C_n, then R_p can be determined. R_p is the predicted ratio of special and normal body burdens (Bp_s/Bp_n), and is the reciprocal of the predicted F_p.

In the situations at hand, both F_p and R_p values are predicted for Situations 2 and 3 relative to normal Situation 1. This is shown in Table 6.14 for both the fast compartment of TCE ($t_{1/2}$ = about 15 min) and its slow compartment ($t_{1/2}$ = 7.6 hr). Results are interpreted as follows:

If the slow compartment is the critical tissue, the predicted peak tissue burdens will not exceed the peak burden for normal exposure if the OSHA TWA limit (C_n) is reduced to $0.89 \cdot C_n$ (or 89 ppm) in Situation 2, and to $0.86 \cdot C_n$ (or 86 ppm) in Situation 3. Stated in terms of R_p, if the concentration is not reduced but is left at C_n (100 ppm), the predicted peak body burden will be 1.12 times greater than desired in Situation 2 and 1.16 times greater in Situation 3. It is clear that departures from normal exposure have only a small effect on the accumulated burden in a compartment with a low uptake rate (long half-life) for TCE. Also, the additional short peak exposures in Situation 3 have little further effect on peak burden over that resulting from Situation 2.

Part III.   If the fast compartment is the critical tissue group, the OSHA TWA limit must be reduced from 100 to 56 ppm for Situation 2 and to 48 ppm for Situation 3, if the predicted peak compartment burden is not to exceed that resulting from normal exposure. Failure to reduce the concentration will result in a peak burden of 1.8 times greater than normal in Situation 2, and 2.1 times greater than normal in Situation 3. Note that in the fast compartment, the brief peaks in Situation 3 add considerably (about 16%) to the burden accumulated in Situation 2. Part of this increase is due to the longer exposure at the "ceiling" limit of 200 ppm in Situation 3.

Table 6.14   Adjustment Factors (F_p) and the Predicted Ratio of Body Burdens (R_p) for the Three Exposure Schedules Described in Example 17.[a]

Predictive Index	Compartment[b]	Exposure Situation[c]		
		Situation 1	Situation 2	Situation 3
F_p	Slow	1	0.89	0.86
	Fast	1	0.56	0.48
R_p	Slow	1	1.12	1.16
	Fast	1	1.8	2.1

Note: Proposed OSHA standard of 10/20/75 (40 FR 49032) would set a TWA limit of 100 ppm, a ceiling of 150 ppm, and allow no peak.

[a] This example readily illustrates how the manner in which one is exposed to a particular airborne toxicant can affect the peak body burden even though the total amount of toxicant absorbed each workday remains the same for all three situations.

[b] Fast compartment $T_{1/2}$ = 15 minutes; slow compartment $T_{1/2}$ = 7.6 hours.

[c] Situations 2 and 3 are compared to Situation 1 (normal).

This example illustrates how various exposure regimens can result in widely different body accumulations of a substance, and how the model predicts adjustment factors.

ILLUSTRATIVE EXAMPLE 18.   To illustrate how pharmacokinetic models can predict equal peak body burdens for different shift lengths, let us assume two work situations with exposure to TCE. In Situation 1, workers are exposed for five 8-hr days per week at the OSHA TWA limit of 100 ppm, and in Situation 2, workers are exposed to the same level for four 10-hr days per week (illustrated in Figure 6.1).

What adjustment factor (F_p) must be applied to the normal limit (100 ppm) so that the exposure in Situation 2 will result in a body burden no greater than that resulting from exposure in Situation 1?

Solution.

Part I.   Using Equation 29 or Figure 6.25, it can be determined that for the fast compartment $(T_{1/2} = 15$ min), $F_p = 1$, and for the slow compartment $(T_{1/2} = 7.6$ hr), F_p 0.87. Thus, if the fast compartment is critical, no reduction in the OSHA limit is required in Situation 2 to avoid an increase in fast-compartment accumulation. Since the peak burden is reached in the fast compartment in a few hours, the peak tissue concentration should be no greater with 10 hours exposure than with 8 hr.

Part II.   If the slow compartment is critical, the OSHA limit must be reduced to 0.87 of the normal limit, or 87 ppm, to avoid a higher predicted peak in Situation 2 than in Situation 1. The prudent course would be to make the reduction where there is doubt as to which compartment is critical. If the $t_{1/2}$ values are unknown or in doubt, a "worst-case" F_p of 0.84, or 84 ppm, can be determined from Figure 6.25.

As shown in the previous examples, the use of the pharmacokinetic approach has a good deal of flexibility in that it can predict modified exposure limits for both short and long periods of exposure.

8.8   Special Application to STELs

Hickey (144) has discussed the use of the pharmacokinetic approach to setting acceptable limits for very short periods of exposure. Figure 6.34 illustrates the model's predicted F_p values for exposure to only a single excursion each day, as a function of substance half-life and the short-term exposure time. Hickey (144) has suggested an approach to determining the effect on F_p when there are multiple excursions during a shift. For sake of simplicity, ACGIH has published limits applicable to short-term exposures (STELs) with the following restrictions:

". . . workers can be exposed for a period of up to 15 minutes continuously—provided that no more than four excursions per day are permitted, with at least 60 minutes between exposure periods, and provided that the daily TLV-TWA also is not exceeded."

Example 19 illustrates how F_p can be calculated for short exposures which are spaced at intervals other than 60 minutes. From this example, it is readily seen (Figure 6.32) that depending on the chemical's half-life, the spacing of the exposures can influence the recommended short term limit.

ILLUSTRATIVE EXAMPLE 19.   To illustrate how the time between high exposures (above limit) can influence the degree of adjustment, solve F_p for the following two situations.

Case A.   Assume that a person works in a foundry and is exposed to carbon monoxide for only 5 minutes, 4 times per day when he opens an oven. There is a 1-hr interval between the times he opens the oven (shown in Figure 6.32, Case A). What adjustment factor (F_p) would be suggested according to the Hickey and Reist model?

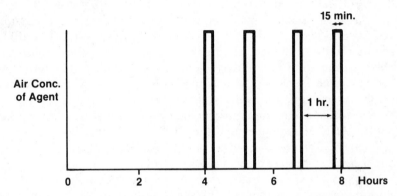

A: Four 15-min. peak exposures; One hour between exposures

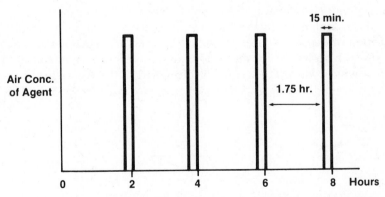

B: Four 15-min. peak exposures; 1.75 hrs. between exposures

Figure 6.32 Comparison of two short-term exposure schedules wherein the peak and TWA concentration(s) are the same but the peak body burden could in some cases be different. Consequently, a lower short-term exposure limit might be needed. Based on Example 19. (From Hickey, 1977.)

Case B.   Assume that the worker who opens the ovens can space the times between exposures at 1.75 hr rather than 1.0 hr (Figure 6.32). What F_p is needed to show the level of protection as when the exposures last one hr?

Answer.   As shown in Figure 6.33, F_p for Case A and Case B will vary with the time between short periods of exposure even though the total dose (ppm-hr) remains the same each day. In the case of carbon monoxide, which has a biologic half-life in man of 3.5 hr, the F_p for the 4 excursions with a one-hr interval is 5.7. When the rest interval is 1.75 hr, the F_p is 6.7.

Case A of Figure 6.32 illustrates air concentrations of a substance over an 8-hr shift, using the ACGIH statement as the basis for the short-term exposure schedule, as compared to normal exposure. Similarly, Case B of Figure 6.32 depicts four 15-min excursions, but spaced equally over the entire shift, with 1.75 hr rest periods between exposures.

Hickey (144) has noted that generally, as shown in Figure 6.33, it makes little difference in resultant peak body burden whether the excursions are separated by one-hr or by 1.75-hr intervals of nonexposure. The difference increases markedly, however, as the rest intervals between excursions are diminished, culminating in a single one-hr exposure (four 15-min excursions with zero rest time between). This situation is shown in the dotted curve in Figure 6.33 for contrast, and is the same as the 1-hr excursion curve in Figure 6.34. Note also that the model's predictions reach a limiting STEL of 8.0, thus satisfying the ACGIH statement that the TLV-TWA may not be exceeded. *The curves in Figure 6.33 indicate once again that for long half-life substances, the exposure schedule is of little consequence when establishing STEL's.*

Hickey has pointed out that the pharmacokinetic models predict equal protection at somewhat higher STELs than many of those recommended by the ACGIH. This is because many of the ACGIH STEL values are based generally on the previous excursion limits of ACGIH. By contrast, the pharmacokinetic models calculate STELs solely on first-order uptake rates.

The pharmacokinetic model may be used to predict the permissible level and duration of exposure necessary to avoid exceeding the normal peak burden during short, high-level exposures. The model does this by establishing excursion limits which will predict equal protection for these situations. Assuming exposure is limited to a *single daily excursion* of duration t_e, Equations 28 and 31 reduce to

$$F_p = \frac{1 - e^{-k(8)}}{1 - e^{-kt_e}} . \tag{43}$$

Adjustment factors derived from Equation 43 have been plotted in Figure 6.34 as a function of substance half-life and excursion time. To illustrate, if a worker is exposed for one 30-min period during an 8-hr shift to a substance with a half-life in the body of 4 hrs ($k = 0.17$), the adjustment factor would be

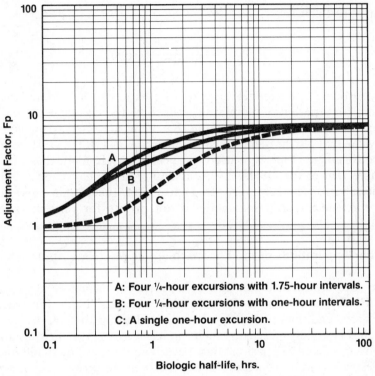

Figure 6.33 Adjustment factor (F_p) as a function of substance half-life $(T_{1/2})$ for three excursion schedules. Based on exposure schedule shown in Figure 6.32. (From Hickey, 1977.)

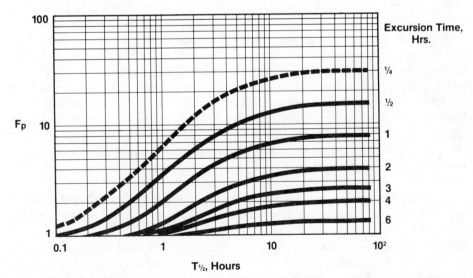

Figure 6.34 Adjustment factor (F_p) as a function of substance half-life $(T_{1/2})$ and excursion time (hrs). Dotted line is used to express the high level of uncertainty involved in the prediction. (From Hickey, 1977.)

determined as follows:

$$F_p = \frac{1 - e^{-(0.17)8}}{1 - e^{-(0.17)1/2}} = 9.0.$$

For this substance, the model suggests that exposure nine times the OSHA TWA limit for a period of 30 minutes is likely to be acceptable as long as such a concentration were not irritating and the chemical caused systemic toxicity.

This value may also be determined from Figure 6.34. If the OSHA cumulative exposure formulae (138) were used to calculate a safe limit, that value would be 16 times that of the OSHA TWA, assuming no ceiling or peak limits were recommended for the substance. In Figure 6.34, the 15-min curve is dotted because the one-compartment model is not precise for very short exposure times.

8.9   Other Applications of the Pharmacokinetic Modeling Approach

Thus far, the use of modeling has been restricted to only inhaled gases and vapors. Hickey and Reist (145) have suggested that limits for particulates, reactive gases and vapors can also be adjusted using their approach. Mason and Hughes (185) have discussed their models for carcinogenic substances. For these substances and other situations, peak body burden, B_p is still deemed to be the best criterion on which to develop a scheme to provide equal protection for the unusual workshift. The following sections describe these other situations and are based exclusively on the work of Hickey (144) and Mason and Hughes (185). The usefulness of kinetic modeling for very short durations, for seasonal workers, for mixtures, off-the-job exposure and carcinogens will be discussed.

For many chemicals, only the results of animal studies will be available when calculating modified exposure limits. These *data can* be quite useful for estimating the biologic half-life of specific chemicals in humans. It is, however, *always* inappropriate to use the half-life of a chemical obtained in a mouse, rat, hamster, dog, rabbit, or even a monkey to describe its likely elimination in a human. Unfortunately, animal data have, in the past, been directly used to adjust exposure limits for unusual work schedules. The major problem is that, the biologic half-life of an industrial chemical in an animal will often be much less than in a human and it will, therefore, *underestimate* the hazard of longer shifts.

Quantitative approaches for extrapolating animal metabolism and excretion data to humans are being developed and could be used in the existing approaches (123). As shown in Table 6.15, it is readily apparent why the biologic half-life of a chemical in the smaller animals can be much shorter than in humans. Exceptions to this handrule occur when a human cannot easily metabolize the parent compound—but the animal can—and therefore the human metabolite has a long biologic half-life. Fortuitously, this will rarely occur. In general, the rate of metabolism or elimination by an animal is often

Table 6.15 Interspecies Scaling of Hexane Clearance Based on Alveolar Ventilation Rates

Species	Body Weight (g)	V_{alv}[a] (l/hr/kg)	Expected CL_{hexane}[b,c] (l/hr/kg)	Ratio (Species/Man)
Mouse	30	35.9	11.8	6.9
Rat	250	21.1	7.4	4.1
Rabbit	3000	11.4	3.8	2.2
Man	70000	5.2	1.7	1.0

From M. E. Andersen, 1981 (123).

[a] $V_{alv} = [0.084 \ l/hr \ (b.wt.)^{0.75}]/(b.wt.)$.

[b] For these clearance calculations, it is asumed that S_b for n-hexane is about equal from species to species and that E_t at low inhaled concentrations is 0.25 in all species.

[c] Rate of metabolism is CL times C_{inh}. Relative ratio is given by the ratio of CL in animals/Cl in man.

a function of the alveolar ventilation rate and the cardiac rate. Table 6.15 illustrates the difference in the cardiac rate and the expected rate of clearance of hexane from four species, including humans (123).

The use of physiological pharmacokinetic models (Figure 6.35) is the preferred method for determining the likely biologic half-life in man from animal data. Physiological modeling can be used to describe the animal in terms of particular organs with the associated blood flows, volumes and partition coefficients. This approach is mathematically complex, but the procedure can frequently be used to extrapolate data obtained in one species to another by accounting for differences in physiology and biochemistry between the tested and untested species. The approach models the movement of a chemical throughout the body by sets of mass-balance differential equations that account for the disposition of chemical entering the various components of the model. These models represent the mammalian system in terms of specific organs (or groups of tissues lumped together based on a common characteristic), all of which are defined. It is the coherent relationship of the anatomical and physiological characteristics between different species that provides the basis for extrapolation of pharmacokinetic data from laboratory animals to man. The advantages of physiological pharmacokinetic modeling have been demonstrated in a recent paper on styrene (223). This approach to extrapolating animal data to humans, sometimes called "animal scale-up," will in all likelihood soon become a routine part of the interpretation and extrapolation process in toxicology.

Lastly, there are three additional caveats which need to be expressed regarding the determination of a biologic half-life for a substance. First, the half-life in the urine, breath, or feces *is not* necessarily the same as that for blood. Secondly, unless only one of these routes of elimination are predominant, none may correlate to the blood. Therefore, when available, the blood plasma

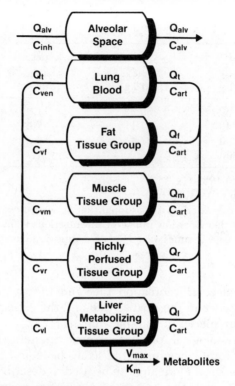

Where:

Q_{alv} Alveolar ventilation rate (liters air/hr)
C_{inh} Concentration in inhaled air (mg/liter air)
C_{alv} Concentration in alveolar air (mg/liter air)
C_{exh} Concentration in exhaled air (mg/liter air)
N Blood air partition coefficient (liters air/liter blood)
Q_t Cardiac output (liters blood/hr)
C_{art} Concentration in arterial blood (mg/liter blood)
C_{ven} Concentration in mixed venous blood (mg/liter blood)
V_{max} Maximum enzymatic reaction rate (mg/hr)
K_m Michaelis constant for enzymatic reaction (mg/liter blood)
Q_i Blood flow rate to tissue group (liters blood/hr)
V_i Volume of tissue group (liters i)
C_i Concentration in tissue group (mg/liter i)
A_i Amount in tissue group (mg)
C_{vi} Concentration in venous blood leaving tissue group (mg/liter blood)
P_i Tissue: blood partition coefficient (liters blood/liter i)

Where subscripts (i) for tissue groups or compartments:
l Liver (metabolizing tissue group)
f Fat tissue group
m Muscle (lean) tissue group
r Richly perfused tissue group

Figure 6.35 Illustration of how the body is described in physiologic pharmacokinetic modeling. (From Ramsey and Andersen, 1984.)

half-life is the one which should be used. Third, the biologic half-life of a particular chemical in animals and humans can vary with repeated exposure. That repeated exposure can vary the biologic half-life was noted in the work of Paustenbach et al. (220–221) which addressed the effects of carbon tetrachloride on rats which were exposed for periods of 11.5 hr/day and in the work of O'Flaherty et al. (222) who noted progressive changes in the half-life of lead in exposed workers with increasing years of exposure. The many potential pitfalls in determining the biologic half-life of a chemical have been discussed and these are worthy of careful review by those who are trying to calculate elimination rates from animal data (123, 223).

8.9.1   Particulates

1.   Chemical.   This entire section on particulates and reactive gases was published in Hickey and Reist (145). In their manuscript, they noted that the modeling approach for particulates is likely to be similar to vapors. Some researchers have suggested that more data need to be gathered to support the use of their model for particulates. Because it seemed reasonable and appropriate, given our state of knowledge, Hickey and Reist asserted that the one-compartment model as applied to particulates is likely to be analogous to inert gases and vapors with one exception; deposition of particulates is presumed to occur in the body at some linear rate proportional to air concentration. In their derivation, Hickey (144) assumed that the clearance of particulates is presumed to conform to a first-order exponential.

$$\text{For uptake,} \qquad B_t = (CLf/k)(1 - e^{-kt}) + B_0(e^{-kt}), \qquad (44)$$

$$\text{For clearance,} \qquad B_r = B_t(e^{-kt_r}), \qquad (45)$$

where L = flow rate of air to the body (volume/time),
  f = fraction of particulates deposited,
  k = clearance rate of deposited particulates (k = ln $2/t_{1/2}$).

  Note: The other symbols are as defined in Equations 26 and 27.

As is the case for inert gases, when F_p is determined, L, f, and the k in the denominator cancel, leaving F_p for particulates identical to that for inert gases. The mechanisms are different, but the model, being a ratio, requires only that L, f, and k be the same for normal as for special schedules.

The retention of inhaled particulates has traditionally been considered to vary directly with their concentration in air. Models for such retention which have been developed or reviewed by Hatch and Gross, Harris, and the ICRP Task Group on Lung Dynamics are rate-independent of the air concentration of particulates, whether deposition is by inertia, gravity or diffusion (145, 224).

Retention varies significantly with other factors, however. The simple expression "Lf" in Equation 44 masks a complex combination of variables including:

inhalation rate, which in turn affects air velocity in respiratory passages; particulate size, size distribution, density, and shape; respiratory frequency; and breathing habits, such as depth of breathing and mouth or nose breathing. The model does not consider any of these factors in predicting equal peak burdens, as they are canceled out by the assumption that they do not change with exposure time (shift length). In spite of these shortcomings, the assumption of a linear deposition rate appears valid.

As noted by Hickey (144), the assumption of first-order clearance is tenuous. However, the lung clearance rate apparently varies depending on the magnitude of lung burden, although there is no general agreement on this point. It is thus not definite that clearance follows a single exponential, although half-lives for particulate clearance have been published. Because of the uncertainty of the half-life or half-lives for body clearance of any particulate substance, Hickey and Reist have suggested that the use of the worst case F_p would seem to be prudent in using the model for adjusting particulate limits to predict equal protection.

When the model is applied to short exposures for particulates at high concentrations, there is the implicit assumption that the predicted allowable higher concentration limit is not so high as to overwhelm the deposition or clearance mechanisms of the body. This can occur at very high concentrations, and application of the model to particulates is limited to this extent.

2. Microbial Aerosols. Particulate aerosols may contain viable microorganisms and the use of models for this hazard have been discussed by Hickey (144). These aerosols behave physically as any other airborne particulate until deposition in the host. There they may exhibit the unique characteristic of being able to multiply, either at the deposition site or some secondary site in the host. The resultant adverse effect may be an infection.

The number of viable organisms which must reach the host in order to initiate an infection depends on many factors, including host susceptibility, deposition site, and organism virulence. However, if one presumes that such a number exists, and that this number is analogous to peak body burden, the model may be appplied to viable particulates.

There are no TLVs or OSHA limits or any other occupational exposure limits for specific microorganisms or viable particulates in air. There are however, recommended limits applicable to particular locations, such as hospital areas. Also, these have not been correlated with, nor are they claimed to be based on, an infectious dose. As Hickey noted, in the absence of such limits, the determination of F_p (or F_a or F_r) for microorganisms in air becomes academic, as there is no limit to adjust. However, the potential for application of the model exists, and awaits the development of relevant limits.

8.9.2 Reactive Gases and Vapors

1. Systemic Poisons. Two types of reactive substances are discussed in relation to the model; those which are both metabolized and excreted through respiration, and those which are only metabolized. The equations for substances which are both metabolized and expired have been derived as follows:

For uptake, $B_t = \text{CWK}[1 - e^{-(k_1 + k_2)t}][k_1/(k_1 + k_2)] + B_0[e^{-(k_1 + k_2)t}]$ (46)

For excretion, $B_r = B_t[e^{-(k_1 + k_2)t_r}]$,   (47)

where k_1 = the uptake and excretion rate by respiration,
    k_2 = the rate by metabolism.

The other symbols are as described in Equations 43 and 44.

Again, when F_p is calculated, the additional factor, $k_1/(k_1 + k_2)$, cancels, and F_p is identical so that for inert gases and vapors, except that the effective half-life is described by $k_1 + k_2 = \ln 2/t_{1/2}$. Use of the prepared curves, such as Figure 6.25, would require knowledge of the combined effective half-life of a substance or use of the worst case, as before. For substabces which are only metabolized and not exhaled through the lungs, Equations 44 and 45 apply.

2. Local Irritants and Allergens.   As noted by Brief and Scala, Mason and Dershin, Hickey and Reist, and Roach, the models *do not* appear amenable to deriving adjustment factors for exposure to primary irritants or allergens. The actions of these substances appear to be based on such a small local compartment, as contrasted to the entire body, that the predicted equal protection would be inapplicable. It is also likely that B_p is unsuitable as a criterion for predicting the likelihood or severity of an adverse response. By the same token, however, a prolonged exposure, beyond 8 hr, might not require any reduction in exposure limits. Fiserova-Bergerova (161) has suggested that a predictive model could be derived for irritants.

8.9.3   Radioactive Material

The mechanism of uptake and excretion of radioactive substances has been studied thoroughly by many researchers and its discussion is outside the scope of this chapter. However, the fate of inhaled radioactive gases and particulates has been modeled thoroughly and maximum allowable concentrations in air for them have been published, based in a large part on effective half-life. One marked difference here is that for many radioactive substances, the actual dose to which one is exposed is much easier to calculate and measure biologically, thus making the modeling and validation much more straightforward.

8.9.4   Mixtures

Hickey (144) has discussed mixtures and he has noted that OSHA regulations state that exposure limits for mixtures of substances in air (except for some dusts) shall be such that

$$C_1/L_1 + C_2/L_2 + \cdots + C_n/L_n \leq 1$$ (48)

in which C_i is the concentration of a substance in air and L_i is its OSHA limit. The equation presumes strictly an additive effect of inhaled substances.

The model does not take into account potentiation or the possible synergistic effects of mixtures. However, if it is assumed that additive effects exist in proportion to peak body burdens, the model may be modified to accommodate mixtures. Instead of equating peak body burden for a special schedule (B_{p_s}) to that for a normal one (B_{p_n}), as in Equation 28, the model would set

$$B_{p_s} = B_{p_n}(C_i/L_i) \tag{49}$$

and the adjustment factor for any substance in a mixture, $F_{p/m}$, would be its F_p acting alone reduced by whatever factor is needed to meet the limits of Equation 48, or

$$F_{p/m} = F_p(C_i/L_i). \tag{50}$$

In practice, F_p would merely be calculated from Equations 28 through 30 or read from graphs, but applied as an adjustment to the reduced limit, C_i, as determined for the OSHA formula (Equation 18), rather than to the normal OSHA limit. Example 20 is provided to illustrate Hickey's conceptual approach to adjusting limits for exposure to mixtures.

ILLUSTRATIVE EXAMPLE 20. EXPOSURE TO MIXTURES.   The pharmaceutical industry was the first to pioneer the use of the 12 hr work shift. In many firms the typical workweek involves 4 days, 12 hr/day then 3 days off followed by 3 days of 12 hr/day followed by 4 days off. Every two weeks, everyone will have worked about 40 hr/week.

Acknowledging that persons are often exposed to more than one chemical at a time, what modification would be recommended for exposure to isoamyl alochol (TLV = 100 ppm) and carbon tetrachloride (TLV = 5 ppm) for this type of shift?

It is assumed that the biologic half-life in humans for isoamyl alcohol is 12 hr and for carbon tetrachloride it is about 5 hr. Recognize that each is a systemic toxin and that the intent of the TLV for isoamyl alcohol is to prevent CNS depression and liver toxicity, while the intent of the carbon tetrachloride standard is primarily to prevent liver toxicity.

Solution.

A.   For both isoamyl alcohol and carbon tetrachloride, the following approach could be used:

$$F = \frac{\dfrac{(1 - e^{-8k})(1 - e^{-120k})}{(1 - e^{-168k})(1 - e^{-24k})}}{[1 - e^{-12k} + e^{-24k} - e^{-36k} + e^{-48k} - e^{-60k} + e^{-72k} - e^{-84k} + e^{-192k} \cdot}{- e^{-204k} + e^{-216k} - e^{-228k} + e^{-240k} - e^{-252k}/(1 - e^{-336k})]}$$

B.   For CCl_4, $t_{1/2} = 5$ hr; so $k = 0.139$ hr^{-1},

By substitution, $F = \dfrac{0.695}{0.841} = 0.83$.

Therefore, $TLV_S = TLV_N(.83) = 5 \times .83 = 4.2$ ppm for CCl_4.

C.   Using the same rationale, the modified TLV for isoamyl acetate is found using $k = \ln 2/12$ hr, or $k = 0.058$ hr^{-1}.

By substitution, $F = \dfrac{0.493}{0.664} = 0.74$.

Therefore, special TLV = 74 ppm for isoamyl alcohol.

D.   This does not consider the potential additive or synergistic effect of simultaneous exposures to both chemicals. If it is desired to apply the ACGIH approach for additive effects of exposure to both chemicals then one could apply their approach for assessing mixtures. The modified TLVs for this situation would be:

$$\frac{C_1}{4.2} + \frac{C_2}{74} \leq 1.0$$

and the concentration of one or both chemicals should be reduced until the equation is less than, or equal to, unity.

9   ADJUSTING LIMITS FOR CARCINOGENS

Unlike other toxic effects caused by xenobiotic substances, many scientists believe that there is no safe level of exposure (no threshold) to chemical carcinogens which appear to be genotoxic agents (255). Although such a phenomenon has minimal "practical" significance for most industrial chemicals, since exposure to very low doses of the less potent carcinogens would require an exposed human population of enormous size before a response would be observed, this theory must be considered in setting and modifying exposure limits. The adjustment of exposure limits for carcinogenic materials has been addressed by Mason and Hughes (185) and their work is the basis for the following discussion.

9.1   Rationale

In general, toxic substances have been shown to exert their effect in proportion to the concentration in the body, or within specific tissues. With most of these substances, it is common to find an "all or none" toxic response above and below a critical or threshold concentration. In other words, at higher concentrations there will be a correlation between an increase in the response and increasing concentrations in tissue, and below this threshold a given effect will not be observed. For toxicants which act in this fashion, the peak concentration in tissue is generally thought to be the most important parameter in assessing toxicity (180). This group includes systemically acting substances such as chemical asphyxiants, narcotics and anesthetics, hemolytic agents, and probably some carcinogens and co-carcinogens. It *does not* include irritants which act without absorption, *nor does* it include allergens, for which the severity of the response, once triggered, frequently appears to be independent of the magnitude of exposure.

The threshold principle *may not* apply, however, to a small group of chemicals (e.g., carcinogens and mutagens) in which the biological response appears to result from chance molecular interactions which are independent of a tissue threshold. Response to these substances is in some ways similar to the response observed at low levels of ionizing radiation (226, 227). For the most part, this latter group of substances act by forming covalent bonds with the genetic material of cells, causing an alteration of the genetic code, which in the case of somatic tissue may lead to cancer (228). With many of these substances, the production of a diseased state appears to take place in separate phases. These begin with the initial chemical reaction or "hits" at a sensitive target (initiation) but end in a complicated series of interactions involving cell transformation, survival and replication. Once the process of proliferation is initiated, progression to a diseased state may be independent of the concentration of the initiating substances, but may also be affected by the presence of other (promoting) substances. The promotor may affect genetic repair, cell regulation or the immune system and subsequently the survival of a transformed cell line. As a result, the overall pattern of dose–response, especially at very low doses, is unclear.

Some chemicals can cause cancer in animals yet have no apparent genotoxicity or ability to initiate cell transformation. These chemicals, which can be promotors, are often called epigenetic or nongenotoxic carcinogens. Although the exact mechanism of action is unclear for these substances, they apparently act in a manner much different than that of initiators (48–51, 169, 229–233). A diagram contrasting, in a very simplistic fashion, the differences in the process is shown in Figure 6.36. Since many experts believe that a threshold exists for nongenotoxic carcinogens, it has been suggested that any approach to adjusting the TLV for these substances should be similar to that used for systemic toxins (i.e., the pharmacokinetic approach) (49).

9.2 Method for Adjusting Exposure Limits for Carcinogens

The approach suggested by Mason and Hughes assumes that the most sensitive or critical step in the carcinogenic process is that of initiation, since initiation

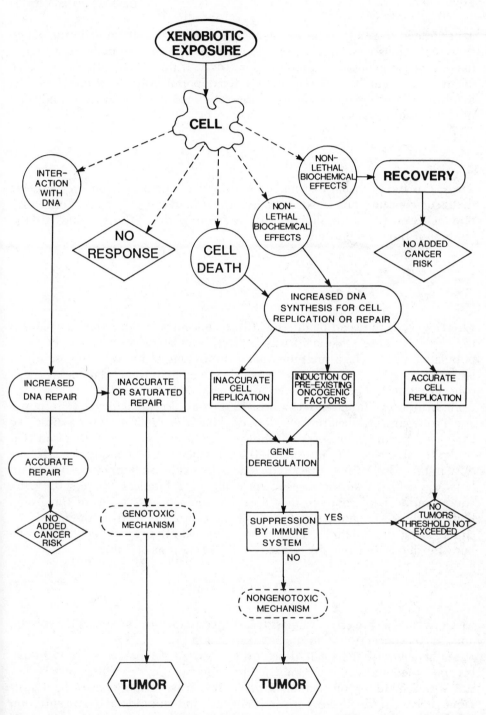

Figure 6.36 Simple illustration of the distinctly different biologic mechanisms which are believed to exist when tumor induction is brought about by exposure to a genotoxic and nongenotoxic chemical.

is predicated to result from the chance interaction of a single molecule of the substance, or its metabolic derivatives, with an appropriate molecule within a tissue which is generally presumed to be DNA. It then follows that the chance of such an event occurring among a fixed population of receptors will be determined by the concentration of the substance which is available, and the time which it is available:

$$P = f(d, t) \tag{51}$$

where P is the probability of initiation occurring, and d and t are respectively the tissue concentration (dose) of the substance and duration of the concentration in tissue (234). In slightly different terms, the "effective dose" is the integral of the body burden:

$$D_{eff} = \int_{t_1}^{t_2} C \, dt \tag{52}$$

where: t_1-t_2 is the span of interest, C the concentration in tissue, and D_{eff} is the "effective dose" associated with a given level of response in the exposed population (226). The assumption that it is the "long-term" average dose which is more important in assessing risk to carcinogens than day-to-day peak concentrations has been supported in a recent study by Bolt (47).

According to Mason and Hughes (185), if an acceptable level of response and a concomitant, albeit probably low, dose has been established for a substance to which exposure occurs over a 40-hr (5-day, 8 hr/day) workweek, that standard could be extrapolated to unusual shift work schedules by limiting the effective dose in the novel shift to that predicted at the exposure limit in the 5-day, 40-hr workweek. Unlike the case of nonstochastic substances wherein the peak concentration is of concern, the "dose" attributable to a series of integrated body burdens will be the simple sum of the contributions from the individual shifts. Using the indefinite integral, the effective life time contribution to the body burden (CT) from any one exposure lasting time t_0 is thus

$$(CT) = (k_i^*(M)/k_0)t_0 \tag{53}$$

where: k^* and k_0 are effective mass transfer constants for the substance in man, and (M) its concentration in the environment (143).

Mathematically, the total "dose" over a series of exposure periods simply becomes the sum of the individual shift contributions, regardless of whether the work schedule is unusual or normal. Obviously, this influences the adjustment process, since all exposure cycles of the same duration and concentration yield identical doses. Consequently, the authors have suggested that the TLV under unusual shift conditions should be

$$TLV_{(unusual)} = TLV_{(std)} \times \frac{\text{Total exposure time (std shift sequence)}}{\text{Total exposure time (unusual shift sequence)}}. \quad (54)$$

Graphical depiction of the solution to this equation is shown in Figure 6.37.

The mechanics of shift arrangement then become a factor only if the unusual work schedule results in a longer (or shorter) workweek, workmonth, workyear or working lifetime. This is quite different from the case of nonstochastic agents for which the novel TLV is determined by iteration of the exponential function $(1 - e^{-k_o t_o})$ to obtain the peak body burdens, which would result in the respective series (143). Consequently, biologic kinetics become important only in making comparisons between substances, and not in the extrapolation process.

In spite of the possible shortcomings and our lack of biologic understanding of the carcinogenic processes, the use of this approach to modifying limits

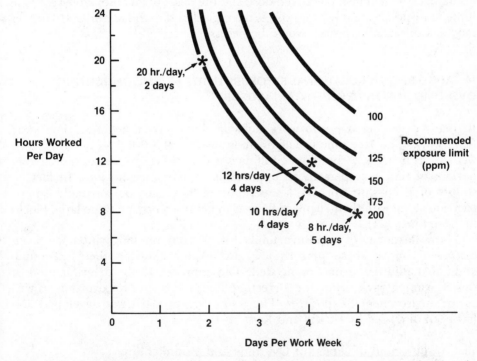

Figure 6.37 Graphical approach to adjusting exposure limits for genotoxic carcinogens based on the method of Mason and Hughes (1984). Curves for adjusting limits for these substances are based on limiting the time integral of the tissue concentration for the work schedule rather than limiting the peak tissue concentration. Consequently, the three 40-hr workweeks illustrated have the same allowable concentrations. Exposure for longer periods, e.g., in four 12-hr shifts, requires lowering the degree of exposure. (Courtesy of Dr. J. Walter Mason.)

seems appropriate for any chemical which has been shown to be positive in animal tests and also has demonstrated genotoxic potential in tests for muta- genicity.

9.3 Potential Shortcoming of the Proposed Approach for Adjusting TLVs for Carcinogens

As discussed by Mason and Hughes, the adjustment of TLVs for substances which produce a response in proportion to the time integral of the body burden is simply accomplished by arithmetic calculations. However, the extent to which integrated body burdens adequately describe biologic response for chemical carcinogens in man is at present unknown (235–237). For some electrophilic substances, e.g., ethylene oxide, this approach appears to be reasonable (238), while for others, such as benz(a)pyrene, it may not be appropriate because of the apparent role that enzyme induction, caused by repeated exposure, may play in modifying the biologic response. Also, for some chemicals, there may be an inverse relationship between dose and the onset of an observable response (time to response), and because this is not considered in the Mason and Hughes model, a different approach would be necessary.

10 ADJUSTING OCCUPATIONAL EXPOSURE LIMITS FOR MOONLIGHTING, OVERTIME, AND ENVIRONMENTAL EXPOSURES

In many cases, persons who work unusual shifts have much free time away from work. This free time usually includes as few as 3 full days off work each week or as many as 6 continuous full days off work every 2 weeks. These long periods of time give persons an opportunity to work second jobs. In fact, in studies of 12-hr-shift workers, it was reported that many persons hold a part- time job along with their regular job so as to gainfully occupy these large blocks of "free time" (21).

The adjustment of exposure limits for persons on long shifts has been discussed, but if these persons are also exposed during their off-hours, additional adjustment may be needed. The approach to the determination of this adjustment is the same for correcting limits to account for overtime as well as for environmental exposures. The following approach was developed and has been discussed by Hickey and Reist (171).

10.1 Adjustment of Limits for Overtime and Moonlighting

As before, the one-compartment biological model is used in this approach. The one-compartment model predicts that no adjustment is necessary to exposure limits for substances with very long (over 1,000 hr) or very short (less than 1 hr) half-lives, but that adjustment is necessary for substances with intermediate half-lives (usually 6–100 hr).

For regular moonlighting or overtime on a 5-day/week basis, Equation 30 devolves to

$$F_p = \frac{1 - e^{-8k}}{1 - e^{-t_s k}} \qquad (55)$$

in which t_s is the daily exposure in hours. Adjustment factors derived from Equation 55 for this situation are shown in Figure 6.38 which was developed by Hickey and Reist (171). Similar equations and curves can be developed from Equation 30 for any schedule. For regular weekend moonlighting or overtime, for example, working six or seven 8-hr days, Equation 55 becomes

$$F_p = \frac{1 - e^{-120k}}{1 - e^{-124km}} \qquad (56)$$

where m is the number of workdays per week. F_p values from Equation 56 are shown in Figure 6.39.

Hickey has noted that the model is more complex when dealing with irregular or unplanned overtime or moonlighting added to an otherwise normal schedule. Suppose an employer decides Friday afternoon that workers must work overtime

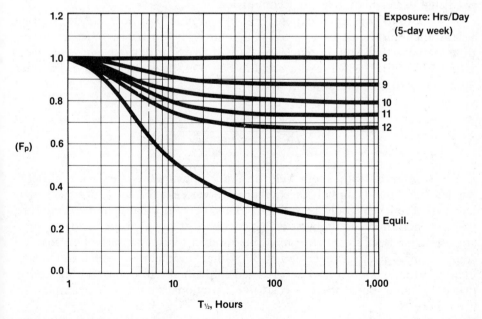

Figure 6.38 Adjustment factor (F_p) as function of substance half-life ($T_{1/2}$) for various daily exposures. The bottom line (equil.) represents the recommended adjustment factors for continuous exposure (24 hr/day). (From Hickey and Reist, 1979.)

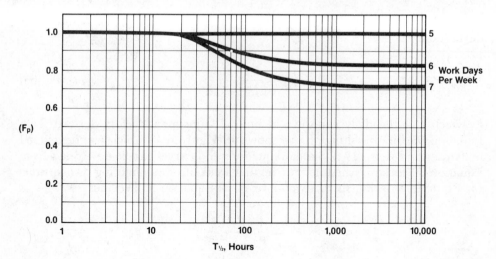

Figure 6.39 Adjustment factor (F_p) as a function of substance half-life ($T_{1/2}$) for work weeks of 5, 6, and 7 days. (From Hickey and Reist, 1979.)

that same day or on Saturday. Recalling that in this model, workers are normally presumed to have accumulated their allowable peak body burden of a substance by Friday afternoon, how does one adjust the exposure limit to prevent a predicted excess body burden accumulation?

If work (and presumed exposure) is to continue past normal quitting time Friday, the limit should be adjusted so that the body burden becomes no greater than the peak that would occur as a result of five 8-hr day exposures per week at the TLV (PEL), which is the usual Friday p.m. peak. In this case, Equation 30 devolves to an equilibrium situation:

$$F_p = \frac{(1 - e^{-8k})(1 - e^{-120k})}{(1 - e^{-168k})(1 - e^{-24k})}. \tag{57}$$

For substances with short half-lives, no adjustment to exposure level is needed (the body is already at equilibrium), whereas for substances with long half-lives, the level should be reduced to 40/168 of normal. The problem does not end there because the worker has lost part of his weekend recovery time, and the next week's exposure must be lowered to compensate. This involves complex manipulation of Equation 30. Other irregular overtime and moonlighting situations must also be modeled individually to determine an appropriate F_p value. Since there are infinite variations, only a few examples are given here.

The point to be emphasized is that these exposures do add to the body burden and should, whenever possible, be taken into account when adjusting exposure limits. An illustrative example of the mathematical approach is shown in Example 21. Figure 6.40 and others have been developed by Hickey (171) for use during second shifts and are useful.

ILLUSTRATIVE EXAMPLE 21 (Calculating an Adjustment Factor with Consideration Given to Overtime and Moonlighting).

Many persons work two jobs. If an employee were self-employed as a furniture stripper in his off-hours, what modification to the normal daily TWA exposure limit would be necessary for methylene chloride given the following information?

A person works 10 hr/day, 4 days/week at a paint plant and is exposed to methylene chloride. He usually strips furniture as a second job for only 2 hr on the days he also works at the factory, but he usually strips furniture 8 hr/day during 2 of the 3 days off work each week. He is not exposed to methylene chloride on Sunday. The biologic half-life for methylene chloride's in man is 6 hours (1983 TLV = 100 ppm).

Solution.   The model considers only exposure time, not total work period, so F_p is based on four 2-hr daily exposures followed by two 8-hr daily exposures per week.

$$F_p = \frac{\dfrac{(1 - e^{-8k})(1 - e^{-120k})}{(1 - e^{-24k})}}{\begin{aligned}1 &- e^{-8k} + e^{-24k} - e^{-32k} + e^{-54k} - e^{-56k} + e^{-78k}\\ &- e^{-80k} + e^{-102k} - e^{-104k} + e^{-126k} - e^{-148k}\end{aligned}}$$

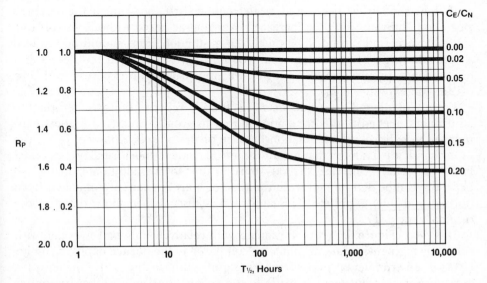

Figure 6.40   Predicted effect of off-the-job exposures to contaminants on adjustment factor (F_p) and peak body burden (R_p). (From Hickey and Reist, 1979.)

where $k = \ln 2/6$ hr $= 0.1155$,

$$F_p = \frac{0.6434}{0.6413} = 1.0, \qquad \text{TLV}_{\text{special}} = 1 \times 100 = 100 \text{ ppm}.$$

The shorter exposure week (32 hr versus 40 hr) would allow an increase in exposure limit except that the $t_{1/2}$ is short (6 hr), the daily weekend exposure is the governing factor. Since weekend exposure is 8 hr/day, as in the normal schedule, no increase in TLV is predicted. This is illustrated in Figure 6.24.

Note that in the absence of weekend exposure, the model predicts an allowable increase in regular job exposure:

$$F_p = \frac{(1 - e^{-8k})(1 - e^{-120k})}{(1 - e^{-2k})(1 - e^{-96K})} = 2.92$$

The modified limit for 2 hr/day, 4 days/week exposures is 2.92×100 or 292 ppm.

10.2   Simultaneous Occupational and Environmental Exposures

Regulatory limits and recommended guidelines often assume zero "off-the-job" exposure to a substance. This is not always the case. The models which have been discussed derive an expression for F_p, assuming a situation wherein a person is exposed to an exposure limit of a substance during normal working hours and to its environmental limit for the remainder of the time. The Hickey and Reist model has been used to determine, first, how much the worker's peak body burden would be increased over that acquired without any off-the-job exposure, and second, how much the on-the-job limits should be reduced so that the peak body burden would be no higher with both on-the-job and off-the-job exposure than it would be with normal on-job exposure and zero off-the-job exposure (171).

From Equation 30 a value of F_p is found, representing, as before, the adjustment needed to the occupational limit to avoid a predicted higher-than-normal body burden accumulation because of the off-job exposure. If, on the other hand, it is assumed that no adjustment is made to the normal occupational limit, the relative increase in body burden due to the additional off-the-job exposure can be determined from the ratio of body burdens accumulated from normal and dual exposure. This approach is discussed in more detail in the original article (171).

It is clear (Figure 6.40) that for agents with relatively long half-lives and with relatively high environmental limits (as compared to the normal exposure limits, C_n), the environmental exposure adds significantly to the body burden of a substance if no adjustment is made to the occupational limit. Likewise, under these conditions, significant reductions in occupational limits are necessary to avoid any predicted increase in body burden as a result of the environmental exposure.

As noted by Hickey, the Environmental Protection Agency environmental limits for SO_2 and NO_2 are less than 1 percent of the OSHA occupational exposure limits, so dual exposure makes little difference to body burden accumulation. However, the EPA limit for carbon monoxide of 10 mg/m³ (9 ppm), an 8-hr limit but in effect a ceiling limit, is 18 percent of the OSHA limit. This has some effect on F_p and R_p. For example, as shown in Figure 6.40, if a worker is exposed to 50 ppm CO on the job and 9 ppm the remainder of the time, his predicted peak body burden ($T_{1/2}$ in man for carbon monoxide = 3–4 hrs) would be 1.06 times his burden, with off-the-job exposure. If this environmental exposure occurs, the on-the-job exposure should be reduced to 0.94 of normal to avoid the predicted excessive peak body burden.

ILLUSTRATIVE EXAMPLE 22 (Hickey and Reist Approach for Combined Environmental and Occupational Exposure).   In certain regions of the country, the ambient concentration of carbon monoxide averages 9.0 ppm which is the current EPA limit for environmental exposure. Many persons are occupationally exposed to carbon monoxide at concentrations at or near the TLV (50 ppm in 1983). What modified occupational exposure limit would be suggested if a person worked 12 hr/day for 4 days each week if one also wanted to take into consideration the background concentration to which the person would be exposed when away from his job? Carbon monoxide has a biologic half-life of about 4 hr in humans.

Solution.   Our objective is to find the allowable CO concentration at work (C_s) which will not result in a *Bps* greater than *Bpn*. In determining C_s, you account for both the work-related CO and the background CO. Of course, you can reduce only the work-related contribution of CO since the environmental levels are fixed. Consequently, the additional CO concentration during work is $C_s - C_e$.

If there were no additional environmental exposure, the TLV adjustment required by the special schedule would be

$$B_{p(\text{normal})} = C_N WK\ (f\colon t_n, k)$$

where

$$f\colon t_{n,k} = \frac{(1 - e^{-8k})(1 - e^{-120k})}{(1 - e^{-168k})(1 - e^{-24k})},$$

$$B_{p(\text{special})} = (C_s - C_e)WK\ (f\colon t_s, k) + C_e WK\ (f\colon t_e, k)$$

and

$$F_p = \frac{C_{\text{special}}}{C_{\text{normal}}}$$

The additional reduction from environmental exposure is

$$F_p = \frac{(f: t_n, k)}{(f: t_s, k)} - \frac{C_e}{C_n}\left[\frac{1}{f: t_p, k} - 1.0\right]$$

where $t_{1/2} = 4$ hr

$$F_p = \left[\frac{\dfrac{(1 - e^{-8k})(1 - e^{-120k})}{(1 - e^{-168k})(1 - e^{-24k})}}{\dfrac{(1 - e^{-120k})(1 - e^{-96k})}{(1 - e^{-168k})(1 - e^{-24k})}}\right] - \frac{9}{50}\left[\frac{1}{\dfrac{(1 - e^{-12k})(1 - e^{-96k})}{(1 - e^{-168k})(1 - e^{-24k})}}\right].$$

Combined $F_p = 0.8571 - 0.0225 = 0.835$.

New TLV for the combined exposures: TLV $= 0.83 \times 50$ ppm $= 41.7$ ppm.

Since 9 ppm of carbon monoxide is already present in the ambient air, the work environment should contain no more than 32.7 ppm in order to maintain a peak body burden for this special situation (B_{ps}) at the same level as a person occupationally exposed to 50 ppm (B_{pn}).

11   ADJUSTMENT OF OCCUPATIONAL EXPOSURE LIMITS FOR SEASONAL OCCUPATIONS

Seasonal occupations are particularly prevalent in agriculture and related activities, including fertilizer and pesticide manufacture and use, cotton ginning, and food canning, as well as construction. Occupational Safety and Health Administration (OSHA) regulatory exposure limits have not taken seasonal exposure patterns directly into account. In response to the lack of information in this area, Hickey (172) developed an approach to adjusting TLVs and PELs to accommodate particular seasonal exposures. The following discussion is based on this work.

A one-compartment model was used to determine a "special" exposure limit which would predict protection for a worker in some special exposure situation equivalent to that provided by the TLVs in a "normal" exposure situation. As before, this special limit is expressed as an adjustment factor (F_p). That is, the normal exposure limits times F_p equals the special exposure limit, or

$$\text{(TLV normal)}(F_p) = \text{(TLV special)}. \tag{58}$$

F_p is a function of the work schedule and the biological half-life ($t_{1/2}$) of a substance in the body. The general equation for F_p as modified to apply to *seasonal exposure* is

$$F_p = \frac{(1 - e^{-8k})(1 - e^{-120k})(1 - e^{-8400k})}{(1 - e^{-kH})(1 - e^{-24kD})(1 - e^{-168kW})} \tag{59}$$

where H = length of daily work shift in the special schedule (hours/day),
$\quad D$ = length of the special workweek (days/week),
$\quad W$ = number of weeks in the special work season (weeks/year),
$\quad k$ = the excretion rate of a particular substance in the body (k = ln $2/t_{1/2}$),
$\quad t_{1/2}$ = the biological half-life of the substance in the body.

In this general equation, the special work schedule factors in the denominator are balanced against the "normal" work schedule factors in the numerator. These are a workday of 8 hr, 5 workdays per week (120 hr), and fifty 168-hr weeks per year (8,400 hr). This schedule assumes a 2-week vacation annually.

11.1 Application of the Hickey and Reist Model for Seasonal Shift Work

Hickey has offered the following example to illustrate the use of his model for seasonal shift work.

ILLUSTRATIVE EXAMPLE 23. (Adjusting Limits for Seasonal Occupations). Take, for example, a seasonal job in which workers are exposed to a substance six 16-hour days/week for a 17-week season. Values of H = 16, D = 6, and W = 17 may be substituted in Equation 59 and F_p found for any k value.

In this example, F_p goes from a value of 1.0 (no adjustment to TLVs) for exposure to substances with very short half-lives (high k values) to a low point of 0.43 for substances with a 400-hr half-life, and then rises to 1.2 for substances with very long half-lives (low k values) (172).

This phenomenon is explained as follows. If $T_{1/2}$ = 1 hr, the peak body burden is reached in 5 or 6 hr of exposure and gets no higher even if exposure continues for 16 hr/day. As the $T_{1/2}$ gets longer, the abnormally long workday and workweek require a reduction in the TLV to avoid a predicted higher-than-normal body burden. For a substance with a $T_{1/2}$ of 400 hr, F_p is 0.43, or nearly as low as the ratio of the normal 40-hr to the special 96-hr workweek (40/96 = 0.42).

To this point, there has been no effect on F_p from the short season. For substances with a half-life longer than 400-hr, as shown in Figure 6.41, F_p increases (Curve B) to 1.2 at $T_{1/2}$ = 11 years. This value approaches the ratio of the normal (2,000 hr) to the special (1,632 hr) work year (2,000/1,632 = 1.22). F_p can exceed unity (i.e., TLVs can be adjusted upward) for substances with very long half-lives, because with such substances, the total intake/year is more important than whether the intake takes place over a 17-week or a 50-week period. Since the seasonal work year has fewer work hours than a normal

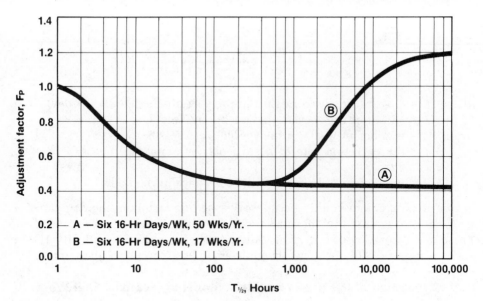

Figure 6.41   Adjustment factor (F_p) as a function of substance half-life ($T_{1/2}$) for the two seasonal exposure schedules indicated. (From Hickey, 1980.)

work year, the total intake and thus the predicted peak body burden will be no more than for a normal year even if the exposed level is increased by 20 or 22 percent.

Let us now consider the adjustment factor if the workday and workweek are normal and only the work year varies. In this case, Equation 59 reduces to

$$F_p = \frac{(1 - e^{-8400k})}{(1 - e^{-168kW})}.$$ (60)

F_p values from Equation 60 are plotted in Figure 6.42 against $T_{1/2}$ for work seasons of from 10 to 50 weeks. It can be seen that no "credit" (i.e., increase in TLV) can be taken for a short season unless that $T_{1/2}$ of the substance involved exceeds 30 hrs (many particulate substances do have much longer half-lives). For substances with very long half-lives, the F_p again becomes the ratio of hours worked per normal year to hours worked per special season.

To avoid calculating each schedule separately, a quick approximation may be used to determine whether F_p is significantly different from 1.0 for a particular agent and work season. One may arbitrarily state that the effect of the shorter season on F_p is not worth considering if it permits raising the TLV by no more than 5 percent, or

$$1/(1 - e^{-168kW}) < 1.05.$$ (61)

Solving this equation for $t_{1/2}$ ($k = \ln2/t_{1/2}$):

$$t_{1/2} \text{ in hours} < 38 \; W. \tag{62}$$

That is, if the substance half-life in hours is less than 38 times the number of weeks (W) in the work season, the short season does not materially change the predicted peak body burden from that expected in a normal work year. Stated another way, the TLV may not be raised by virtue of the shorter work year unless the substance half-life in hours is greater than 38W. Example 24 illustrates the use of this approach.

ILLUSTRATIVE EXAMPLE 24 (Approach to Adjusting Limits for Seasonal Occupations).   Seasonal jobs are common in many industries. Assuming that persons are exposed to styrene 16 hr/day for 6 days/week for an 18-week season, what modified TLV would be suggested if the biologic half-life in humans were 6 hr? (1983 TLV = 50 ppm.)

Solution.
a.   Since $t_{1/2}$ in hours $<38 \times$ (18 wk), or <684 hr, the exposure limit may not be raised by virtue of a shorter work *year*.
b.   Must it be lowered because of the longer workday?

$$F = \cfrac{\cfrac{(1 - e^{-8k})(1 - e^{-120k})}{(1 - e^{-168k})(1 - e^{-24k})}}{\cfrac{(1 - e^{-16k})(1 - e^{-144k})}{(1 - e^{-168k})(1 - e^{-24k})}} = \frac{(1 - e^{-8k})(1 - e^{-120k})}{(1 - e^{-16k})(1 - e^{-144k})}$$

$$F = \frac{(0.6031)\,(1)}{(0.8425)\,(1)} = 0.72$$

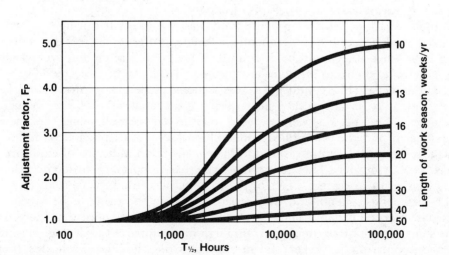

Figure 6.42 Adjustment factor (F_p) as a function of substance half-life ($T_{1/2}$) for various work seasons. (From Hickey, 1980.)

This calculation suggests that only the length of the work *day* is important in determining TLV adjustment in this particular case.

Note that this adjustment is the same as that predicted by Figure 6.41 for a 16 hr/day and a chemical with a 6-hr biologic half-life.

12   BIOLOGIC STUDIES OF UNUSUAL EXPOSURE SCHEDULES

Very few experiments have investigated the potential effects of unusual exposure regimens on the severity of toxic response. It appears that the toxicological effects that might occur if an exposure limit were not lowered during an unusual work shift would, in all likelihood, be too subtle to be measured quantitatively in most animal studies. There is, however, some evidence that, in general, exposures which are intermittent or unusually long are likely to potentiate the response (135, 240). There is evidence that measurable but perhaps clinically insignificant changes in the rates and routes of elimination of some inhaled substances can be expected during long work shifts (157, 165, 170, 221–223). Only a few biologic monitoring studies (241) have thus far been conducted to demonstrate whether a dramatic difference in effects between exposure during two different exposure regimens is likely to occur for any chemical at levels near its TLV. *It will, in all likelihood, be many years before clinical studies in humans, or toxicologic studies in animals, will clearly show whether significant problems can occur if exposure limits are not lowered for unusual periods of exposure.*

The rationale for such adjustments is therefore, in part, philosophical in nature and involves a judgment about risk. Our knowledge of the pharmacokinetics of chemicals clearly tells us that when concentrations in air are not lowered during extra long periods of exposure, the peak levels of the toxicant in tissue *will be higher* during that period than would occur during "normal" 8 hr/day work schedules. Consequently, if health professionals feel that workers on both schedules should be at equivalent levels of risk, models are available to determine the concentrations at which equal risk can be expected.

Lehnert et al. (167) conducted a cross-sectional epidemiology study of workers on normal 8 hr/day shifts which involved the analysis of 6,126 biological samples. These persons were exposed to either trichloroethylene, benzene, or toluene during their workday, and trichloroacetic acid, phenol, or hippuric acid, respectively, were measured in their urine as an indicator of exposure as well as body burden. They found that even during normal 8 hr/day schedules, exposures at or near the TLVs of these chemicals caused the concentration of the metabolites phenol and trichloroacetic acid in the urine to be higher on Friday afternoon than Monday afternoon. Their data suggest that there may be some degree of accumulation, although of no apparent toxicological significance, occurring in workers during normal work schedules. From this, it can be expected that for those chemicals where day-to-day accumulation normally occurs, accumulation might be exaggerated when the daily exposure period is longer and the recovery period shorter.

A recently reported biological monitoring study was specifically conducted in an effort to determine whether persons exposed to Dimethylformamide (DMF) for 12 hr/day accumulated the substance to a greater degree than persons who were exposed during a normal workweek (241). The conditions of the study were ideal in that a baseline set of urinary excretion data were collected on 80 employees who had been working the 8-hr shift for at least several months, and this was compared to the urinary data of the same group after they were placed on the 12 hr/day shift (only day time workers). Through use of the paired-T test, the differences in the elimination for each person as well as the group were determined. Their results indicated that *no difference* was observed in the concentration or the quantity of the urinary metabolites after workers were placed on the 12 hr/day, 4 days/week shift. It should be noted that the implications of this study are not general in nature because exposures to DMF for these persons averaged about 2 ppm, which is markedly below the TLV of 10 ppm. Secondly, the biologic half-life of DMF in humans is only about 4 hrs; therefore, it would not be expected to accumulate from day-to-day.

Few pharmacokinetic and toxicological studies have been conducted in animals exposed for 12 hr/day with the intent of determining any differences due to the longer exposure period. It appears that MacGregor conducted one of the first toxicity studies comparing the response due to 12 hr and 8 hr of exposure (242). She determined uptake factors for hexafluoroacetone using rats and predicted uptakes for periods of 6, 8 and 12 hr. MacGregor then developed a first-order model for determining uptake of inert gases in which she determined uptake constants for carbon monoxide using human volunteers. These predictions were compared to the data of Peterson and Stewart (206) and close agreement was reported between the observed and predicted values of CO uptake. The Hickey and Reist model predictions of peak CO burdens have been compared to the predictions of the MacGregor model (144). In Hickey's validation procedure, exposure to 50 ppm CO was assumed for five 8-hr days/week, three 12-hr days/week, and 24 hr/day, 7 days/week. The comparisons were made in terms of F_p. The predictions of both models agreed quite closely. Hickey has noted that MacGregor also developed a predictive model for reactive vapors but that it has not been validated (144).

Paustenbach et al. (220, 221) conducted an experiment to compare subtle differences in toxicity, distribution and pharmacokinetics due to exposures of 8 hr/day and 11.5 hr/day. One group of rats was exposed for 8 hr/day for 10 of 12 consecutive days (simulating two weeks on standard work schedule) and another for 11.5 hr/day for 7 of 12 consecutive days (simulating the 12 hr/day schedule). Thus, each group received essentially the same dose (ppm-hr) of toxicant (carbon tetrachloride) during the two-week test period. Overall, the 11.5 hr/day exposure schedule produced no significant changes in the distribution and concentration of CCl_4 in various tissues as compared to rats exposed 8 hr/day. There was no significant difference in hepatotoxicity between the groups following each week of exposure as measured by histopathology. However, exposure to the 11.5 hr/day dosing regimen consistently produced significantly higher levels of serum sorbitol dehydrogenase (SDH), an enzyme which indicates liver damage, than exposure to the 8 hr/day schedule (Table 6.16).

It is noteworthy that the rates and routes of elimination were measurably different for the two groups of animals (Figures 6.43 and 6.44). Following 2 weeks of exposure to the 8 hr/day schedule, ^{14}C activity in the breath and feces of the rats comprised 52 percent and 41 percent of the total ^{14}C excreted. Following 2 weeks of exposure to the simulated 12 hr/day work schedule, the values were 32 percent and 62 percent, indicating that the longer work shift altered both the rate and route of elimination of CCl_4. It was found that 97–98 percent of the ^{14}C activity in the expired air was $^{14}CCl_4$. The elimination of $^{14}CCl_4$ and $^{14}CO_2$ in the breath followed a two-compartment, first-order pharmacokinetic model ($r^2 = 0.98$). For rats exposed 8 hr/day, the average half-life for elimination of $^{14}CCl_4$ in the breath for the fast (α) and slow (β) phases for the 2-week schedule averaged 85 min and 435 min, respectively. For rats exposed 11.5 hr/day, the average half-lives for the α and β phases over the 2 weeks averaged about 95 min and 590 min, respectively. Differences in the rates of elimination of $^{14}CO_2$ and ^{14}C activity in the urine and feces were also observed (Figure 6.40).

The results of Paustenbach et al. suggest that even if the weekly dose (ppm-

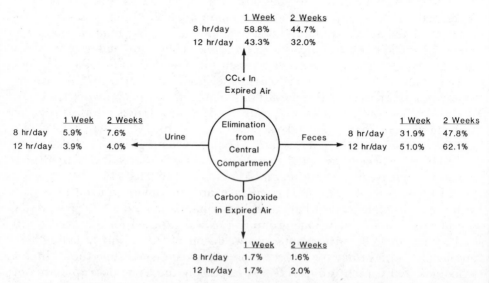

Figure 6.43 Differences in excretion of ^{14}C activity following one and two weeks of exposure to 100 ppm of carbon tetrachloride during an 8 hr/day and 11.5 hr/day exposure schedule. (From Paustenbach et al., 1985.)

Table 6.16 Statistical Comparison of Serum Sorbitol Dehydrogenase (SDH) Activity in Selected Groups of Rats Following Exposure to 100 ppm of Carbon Tetrachloride for Either 8 Hr/Day or 11.5 Hr/Day Under a Number of Different Dosage Regimens[a]

Treatment Groups Compared	Dosage Regimen	SDH Activity Mean ± SE (IU/ml)
1	1 day, 8 hr	7.0 ± 1.5[b]
2	1 day, 11.5 hr	14.8 ± 3.7
3	2 days, 8 hr/day	11.5 ± 2.2
4	2 days, 11.5 hr/day	18.3 ± 4.0
5	3 days, 8 hr/day	21.0 ± 3.2
6	3 days, 11.5 hr/day	29.0 ± 6.2
7	5 days, 8 hr/day	22.5 ± 2.7[b]
8	4 days, 11.5 hr/day	68.3 ± 9.3
9	5 days, 8 hr/day[c]	20.3 ± 4.4[b]
10	4 days, 11.5 hr/day[d]	12.3 ± 0.6
11	5 + 5 days, 8 hr/day	39.0 ± 8.8[b]
12	4 + 3 days, 11.5 hr/day	65.0 ± 16.2
13	5 + 5 days, 8 hr/day[e]	4.9 ± 1.1[b]
14	4 + 3 days, 11.5 hr/day[f]	14.3 ± 1.3
7	5 days, 8 hr/day	22.5 ± 2.7[b]
11	5 + 5 days, 8 hr/day	39.0 ± 8.8
8	4 days, 11.5 hr/day	68.3 ± 9.3
12	4 + 3 days, 11.5 hr/day	65.0 ± 16.2
9	5 days, 8 hr/day[c]	20.3 ± 4.4[b]
13	5 + 5 days, 8 hr/day[e]	4.9 ± 1.1
10	4 days, 11.5 hr/day[d]	12.3 ± 0.6[b]
14	4 + 3 days, 11.5 hr/day[f]	14.3 ± 1.3
12	4 + 3 days, 11.5 hr/day	65.0 ± 16.2[b]
15	4 + 4 days, 11.5 hr/day	110.0 ± 26.7
12	4 + 3 days, 11.5 hr/day	65.0 ± 16.2[b]
16	4 + 5 days, 11.5 hr/day	102.3 ± 14.3

From Paustenbach (1982).
[a] SDH activity was determined immediately after exposure except where indicated. Values shown are the mean of four rats per group.
[b] Significant difference in SDH activity between groups (p < 0.05).
[c] SDH activity was determined 64 hrs after exposure.
[d] SDH activity was determined 84 hrs after exposure.
[e] SDH activity was determined 64 hrs after exposure.
[f] SDH activity was determined 108 hrs after exposure.

hr) is held constant, subtle changes in dosage regimen, like those involving unusual (12 hr/day) work schedules, can have an effect on toxicant distribution (Figure 6.45), the degree of toxic response (Table 6.16), the pharmacokinetics of elimination (Figure 6.43) and, perhaps, the metabolism. The markedly longer elimination half-lives of $^{14}CO_2$ and $^{14}CCl_4$ observed in the groups exposed to the simulated 12 hr/day work schedule compared to the groups exposed 8 hr/day indicates that the four additional hours of daily exposure places a greater percentage of the absorbed dose to poorly perfused lipophilic depots such as the fat (Figure 6.45). The likely behavior of lipophilic substances in fat following repeated exposure is shown in Figure 6.46. The results of this study cannot be generalized since only a lipophilic substance was investigated. Consequently, it is still unclear whether it is clinically or toxicologically necessary to adjust exposure limits.

A comparative toxicity study involving 8 hr/day and 12 hr/day exposures has also been conducted by Yim and Carlson (243) for dichloromethane (methylene chloride). In this study, carboxyhemoglobin (COHb) formation and elimination

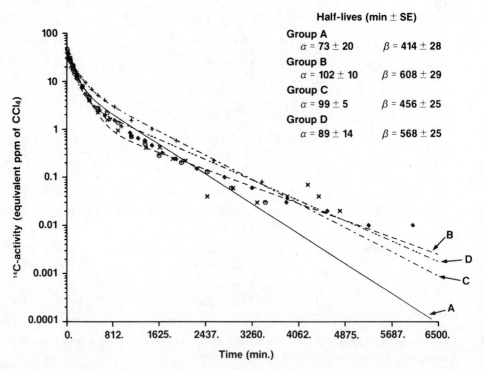

Figure 6.44 Elimination of ^{14}C activity (98% CCl_4) in the expired air of four groups of rats (4 per group) exposed to 100 ppm of carbon tetrachloride. Two groups were exposed for 8 hr/day for either 5 of 7 days (A) or 10 of 14 days (C). The others were exposed for 11.5 hr/day for either 4 of 7 days (B) or 7 of 10 days (D) and have markedly longer half-lives than those exposed 8 hr/day. (From Paustenbach et al., 1985.)

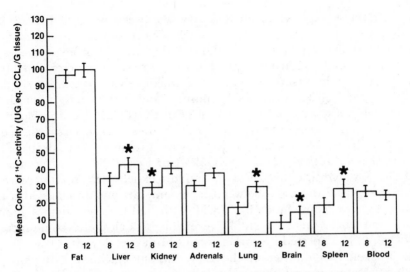

Figure 6.45 Average concentration ± SE of ^{14}C-activity (expressed as CCl$_4$) in various tissues of the rat following one week of exposure which consisted of 8 hr/day for 5 days and 11.5 hr/day for 4 days. Four rats per treatment group. (From Paustenbach et al., 1985.)

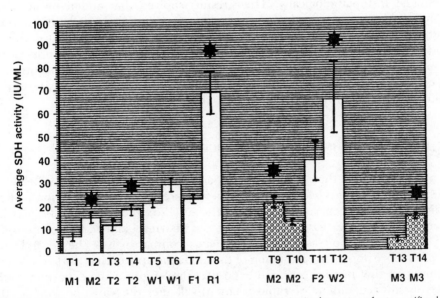

Figure 6.46 Average level ± SE of sorbitol dehydrogenase activity in the serum of rats sacrificed immediately after exposure to either an 8 hr/day or 12 hr/day dosing schedule. Dotted parts indicate that SDH activity was determined on Monday morning after a weekend without exposure. Four rats per treatment group. Description of treatment groups (T1–T14) is shown in Table 6.16. (From Paustenbach et al., 1985.)

in rats and mice exposed to an 8 hr/day workweek or a 12 hr/day, 4-day simulated workweek at dichloromethane (DCM) concentrations of 200, 500 or 1000 ppm were compared. They showed that the 12-hr exposure period produced no greater hazard than the 8 hr/day exposures as determined by the COHb level after the first day's exposure, immediately prior to the second days exposure, after the last workday's exposure, and 2 or 3 days after the last exposure. They also measured the half-lives of COHb and DCM in blood. The relatively short half-lives of COHb and DCM in these two species indicated that neither COHb nor DCM would be present for prolonged periods after DCM exposure ceased. Treatment with SKF-525A or CoCl2 did not affect the half-life of DCM, suggesting that DCM was rapidly exhaled. Even after correcting for physiological differences between mice and rats and humans, this study indicated that for compounds like DCM, with half-lives less than 4 hours in humans, and where there are readily reversible biological effects, no increased toxicity would be expected in workers on 12 hr/day work shifts.

The Haskell Laboratories of the E. I. duPont Chemical Company have recently evaluated several chemicals to determine the influence of exposure duration on toxicological response (244). In their study of aniline vapor, they exposed adult male rats to either 10, 30, or 90 ppm for either 3, 6, or 12 hr per day for 2 weeks. Daily indices of toxicological response included body weight and methemoglobin measurements. After the final exposure, red blood cell counts as well as spleen and liver weights were measured and these were examined histopathologically. Their results showed that aniline-induced hemolysis and consequent splenic enlargement and deposition of hemosiderin is slightly related to exposure duration, and strongly related to exposure concentration. Aniline-induced methemoglobin formation, however, was not related to exposure duration, but was linearly correlated with exposure concentration. Their results suggest that concentration rather than duration of exposure to aniline predominantly influence toxicological response.

13 UNCERTAINTIES IN PREDICTING TOXICOLOGICAL RESPONSE

It has been noted that any model for adjusting exposure limits will have a number of limitations because models, per se, are based on assumptions. These limitations have been reviewed by Calabrese (176) and Hickey (144). As has been noted, one key assumption is that the pharmacokinetic models consider the body to function as one homogeneous compartment. The second, and possibly more important, limitation of the modeling approach is the toxicological assumption that neither repeated exposure, the length of the exposure, or the type of shift schedule (e.g., rapid rotation) will alter the way a person absorbs, metabolizes, and eliminates the substance. In many cases, this will probably not be true. The shortcomings of the existing models are discussed in the following paragraphs and should be understood since in some cases sufficient information may, for certain chemicals, soon be available thus permitting further refinements.

The first shortcoming in the modeling approach is that it is assumed that a good deal is known about all of the toxicological effects of the substance. This is all too often not the case. Ideally, a battery of tests of reasonable size, cost and duration have been performed which permit the investigator to "predict" the toxic action of a drug under a number of circumstances. Often toxicologists attempt to predict the acute and chronic effects of a compound based on the results of only the following tests: 1-day oral LD_{50}, 4 hr LC_{50}, patch test for guinea pig sensitization, ocular test, dermal LD_{50}, 30-day subacute test and/or a battery of tests for mutagenicity. Some have said the application of complex pharmacokinetic models to exposure limits is unnecessary and even worthless since most TLVs are based on such limited data that they are little better than guesswork. In part, it may be true that there may be a high margin of safety in many of the limits, however, the objective of adjusting limits for unusual exposure settings (of both short and long duration) is not necessarily to protect them from known injury but, rather, to provide an equivalent degree of protection to these persons as that of normal shift workers.

Another potential shortcoming of the modeling approach is the potential error involved in assuming that lengthening the exposure does not change the way the body handles the chemical. In an effort to predict the effects of changes in toxicity due exclusively to changes in dosage regimen (such as that involved in 12-hr shifts), toxicologists have frequently used Haber's Law as a handrule. Haber's Law claims that it is the dose, which is of central interest, rather than the time over which that dose is administered. A thorough study of the limitation of Haber's Law was published by David et al. (246), and they showed that the severity of liver toxicity was markedly influenced by the time period over which a dose (ppm-hr) was administered. Although the conclusions are similar to those obtained by other researchers (247) who have studied the limitations of extrapolating data from short-term inhalation exposures, this study is unique in that four different dosing regimens were used. Although the heavy emphasis on the total dose, rather than dose per unit time, has not been relied upon for nearly 40 years, industrial hygienists must all too often evaluate risks based on the amount of toxicant taken up per day rather than consider the pharmacokinetics. For example, this principle is used when hygienists extrapolate the results of standard 4–6 hr inhalation tests to estimate the adverse effects of exposure to 12-hr work schedules.

The existing modeling approaches assume that the distribution of a toxicant amongst key tissues does not vary between normal and unusual work shifts. Although numerous studies have examined the effects on distribution with respect to changes in dose, and other studies have addressed the effect of route of administration on toxicant distribution, little attention has been aimed at defining the effect of duration of exposure on distribution amongst tissues.

The effect of repeated exposure and long periods of exposure on a chemical's pharmacokinetic behavior are not accounted for in the models. This phenomena could have some effect on the toxicity of the material if the shift schedule were markedly different than an 8 hr/day, 5-day week. MacGregor (242) has conducted research which indicated that the rate of metabolism and elimination during 12 hr/day exposure periods is measurably different than that observed during 8 hr/day exposure periods. Paustenbach et al (221, 222) also showed that measurable differences in the rates and routes of elimination in the feces, breath, and urine between two different exposure schedules can occur.

There are several reasons why alterations in excretion rates and routes with repeated exposure have not been frequently reported. Colburn and Matthews (248) have noted that, unless everyone of the inhalation exposures involve radiolabeled material, rather than using labeled material only during the first and last weeks, the "last in, first out" phenomena may take place. When this occurs, potential effects on distribution, metabolism and elimination due to repeated exposure may not be detected because nonlabeled chemical may not be uniformly distributed or excreted with the labeled material. In short, the first dose may be equilibrated more deeply in the fat than the second or subsequent doses.

The observation that repeated exposure to unusual work schedules may affect the distribution and excretion of some chemicals will not be limited to carbon tetrachloride, carbon monoxide and 2-HFA. For example, similar effects would be expected for such industrial solvents such as cyclopropane, cyclohexane, 1,1,1-trichloroethane, and perchloroethylene which are similar to classic anesthetics (such as halothane) in that they are quite lipid soluble and not appreciably metabolized. For example, Fiserovea-Bergerova and Holaday (170) have reported that the half times of uptake for halothane for the Vessel Rich Group (VRG), Muscle Group (MG), and Fatty Groups (FG) are in the magnitude of 2 min, 30 min, and 20 hr, respectively. They noted that most clinical anesthesia lasts 1–4 hr and that by the end of anesthesia the partial pressures of anesthetic agents in tissues of the FG compartment are far from equilibrium and that a steady state has not been reached in that compartment. The result is a redistribution of vapor in the body after the offset of anesthesia. While clearance of the VRG and MG compartments starts instantly, the FG compartment continues uptake until the partial pressure in arterial blood declines to the partial pressure in the FG compartment. Consequently, desaturation curves are very much affected by the duration of exposure (Figures 6.7 and 6.8). This is shown schematically in Figure 6.47 where the concentration of a lipophase substance reaches a different peak level in fat following 12 hours of exposure than that reached after 8 hours. The different quantities stored in the fat for the two exposure periods subsequently affects the elimination (desaturation curves) as shown in Figure 6.44.

Another potential shortcoming of the modeling approach is the assumption that exposure to the toxicant does not inhibit or induce the microsomal enzyme system responsible for its metabolism (245). If this occurs, subsequent doses of the substance will alter the rate of toxication or detoxification. Along the same

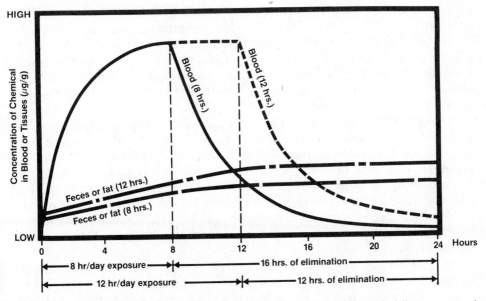

Figure 6.47   Predicted distribution of a lipid-soluble chemical between blood and fat following repeated inhalation exposure. (From Paustenbach et al., 1985.)

lines, models assume that one or more metabolic pathways will not be saturated. Although saturation is *not* very likely for exposure at or near the TLV, the likelihood that the metabolism of a compound will remain constant following repeated dosing is less likely. The induction or inhibition of its own metabolism by previous exposure has been demonstrated for nitrobenzene, acetone and many other xenobiotics (245).

All of the aforementioned factors, as well as those involving the likely effects of the circadian rhythm on toxic response (chronopharmacology and chronokinetics) make the modeling approach to the setting of modified exposure limits a crude approximation of the likely biologic processes which probably take place during unusually long periods of exposure. However, even though the current models may not account for the dozens of biologic phenomena that may be occurring, their use is still appropriate since they are probably a close approximation of the actual behavior of the chemical. Lastly, even though the existing biological data seem to suggest that at levels near the TLV the actual increased risk of injury may not be appreciable, the many unpredictable events that may occur during these exposure schedules, which could potentiate the effects, make the small degree of recommended adjustment worthwhile.

14   SUMMARY

It is clear that unusual shift work schedules have grown in their importance over the last 15 years. In 1984, at least 250,000 persons in the United States were regularly working more than 8 hr/day in jobs that involved exposure to airborne toxicants. Interestingly, all of the countries which have established occupational exposure limits assume that exposures last about 8 hr/workday and are limited to about 40 hr/workweek. Consequently, the current limits were not established with consideration for protecting people who work on the many kinds of shift schedules which are longer than 8 hr/day or 40 hr/week (i.e., unusual).

It has been demonstrated mathematically that for certain chemicals, exposure at levels which would normally be considered acceptable for periods of 8 hrs per day for 5 days/week, may not be acceptable if the length of the daily exposure were to be markedly longer yet limited to only 40 hrs/week (Figures 6.1 and 6.16). In other words, exposures to airborne chemicals during long shifts at the TLV can be expected to produce blood plasma levels or tissue burdens in excess of that which would be noted during standard work shifts if the biologic half-life of the chemical is between 4 and 200 hr. Most industrial chemicals fall into this region.

In an effort to limit the degree of risk for persons who work unusual schedules to a level that is no greater than persons who work normal shifts, a number of models have been proposed for adjusting occupational exposure limits. At this time, there are insufficient data to suggest that modification of limits based on irritation, sensitization, esthetic or discomfort is required. It has also been suggested that exposure to carcinogens should be limited only to the degree necessary to prevent the weekly or monthly uptake (dose) from exceeding that permitted under normal exposure limits. It is recognized that there is a school of thought which suggests that exposure to irritants should be kept to a minimum since these may potentially exacerbate certain toxic effects including carcinogenicity however the evidence is not yet sufficient to require further work.

At least five different models have been proposed for adjusting occupational exposure limits for both short periods of time, as little as 15 minutes, and for periods of continuous (24 hr) exposure. The most technically accurate and rigorous approaches require an understanding of a chemical's pharmacokinetic behavior in a mammalian system, preferably the human. With this information, pharmacokinetic models can be used to mathematically determine the degree to which a particular exposure limit should be raised or lowered and to predict those situations in which no adjustment will be needed (e.g., chemicals with very short and very long half-lives). The OSHA and Brief and Scala models tend to be more conservative than the more complex models because they do not incorporate pharmacokinetic behavior. However, because of our lack of understanding of a certain chemical's toxicity or pharmacokinetic behavior, these conservative models are a fast and easy method for estimating an appropriate limit. Due to the dearth of information on most chemicals, such conservatism could be considered prudent. However, at least from a pharmacokinetic standpoint, this reliance on prudence is not necessary since an adjustment factor *can be calculated* by assuming a "worst-case" half-life for any schedule, therefore eliminating the "scientific grounds" for using less rigorous, more conservative approaches.

It cannot be stressed too much that efforts to modify occupational exposure limits for situations other than 8 hr/day, 40-hr workweeks require a clear understanding of the rationale for a particular limit. *Blind use of any of the modeling approaches to modifying limits for either very short or very long periods of exposure can lead to either a lack of protection for those workers on unusual shifts or, as is more likely, a good deal of overprotection.* Overprotection, although perhaps prudent, is not an optimal use of the limited resources allotted for minimizing occupational disease and, in some cases, could bring about undue economic hardships for both the employer and the employee. Consequently, professionals who are faced with the challenge of modifying exposure limits for atypical periods of exposure must discipline themselves to be thoroughly familiar with the toxicological aspects of the chemical, the rationale for a particular exposure limit and, to whatever degree possible, the employee's work schedule and appropriate air sampling results.

The existing models offer occupational health professionals an approach to protecting workers from the potential adverse health effects of exposure during unusually long and short periods. They represent a significant improvement over the current state of affairs which typically involves no adjustment in the occupational exposure limits for workers on these schedules. Their objective, as has been noted, is to provide an equal level of protection to persons who work nonnormal schedules as those who work standard work shifts.

ACKNOWLEDGMENTS

The permission of Dr. John Hickey (University of North Carolina) to use many unpublished portions of his dissertation as well as reproduce graphs from the publications he coauthored with Dr. Parker Reist was greatly appreciated. Dr. Hickey's critical review of the manuscript was quite useful. The encouragement of Dr. John Christian, Hovde Distinguished Professor of Pharmacy (Purdue University), Dr. Walter Mason (University of Alabama), and Dr. Jane Gallagher (National Institute of Environmental Health Sciences) during the past three years has been greatly appreciated. I have also appreciated the encouragement of Dr. Ralph Smith (University of Michigan), Dr. Noel Moore (Rose-Hulman Institute of Technology), Dr. Ken Sanning and Dr. Byron Landis (Eli Lilly and Company), Dr. Herb Northrop (Stauffer Chemical Company), Dr. Mike Smolensky (University of Texas), Dr. Robert Scala and Dr. Richard Brief (Exxon) and Dr. Herb Stokinger (retired-Chairman, ACGIH TLV Committee). The superb technical illustrations were provided by Ms. Karen Larocque. My sincere thanks to Ms. Debra Berescik for her errorless typing and proofreading as well as her continued perseverance.

REFERENCES

1. P. G. Rentos and R. D. Shepard, Eds., *Shift Work and Health—A Symposium,* U.S. HEW PHS, National Institute Occupational Safety and Health, Washington, D.C., 1976.
2. D. L. Tasto, M. J. Culligan, E. W. Skjei, and S. J. Polly, *Health Consequences of Shift Work,* U.S. DHEW PHS, National Institute of Safety and Health, Washington, D.C., 1978.
3. C. A. Czeisler, M. C. Moore-Ede, and R. M. Coleman, "Rotating Shift Work Schedules that Disrupt Sleep Are Improved by Applying Circadian Principles," *Science,* **217,** 460–463 (1982).
4. E. A. Higgins, W. D. Chiles, J. M. McKenzie, P. F. Iampietro, et al., *The Effects of a 12 Hour Shift in the Wake-Sleep Cycle on Physiological and Biochemical Responses and on Multiple Task Performance.* Report No. FAA-AM-75-10, National Technical Info. Services, Springfield, Va., 1975.
5. H. Allenspach, *Flexible Working Hours,* Inter Labor Office, WHO, Geneva, 1975.
6. G. Felton and M. G. Patterson, "Shift Rotation Is Against Nature," *Am. J. Nursing,* **71,** 4–8 (1971).
7. M. H. Smolensky, "Human Biological Rhythms and Their Pertinence to Shift Work and Occupational Health," *Chronobiologia,* **7,** 378–390 (1980).
8. M. H. Smolensky, J. A. Jovonovich, G. M. Kyle, and B. Hsi, "Chronotoxicity in Rodents Challenged with Propranolol HCL (Inderal®)," *Chronopharmacology,* Proceedings of the Satellite Symposium of 7th Intercongress of Pharmacology, Paris, July 21–24, 1978.
9. M. H. Smolensky, *The Chronoepidemiology of Occupational Health and Shift Work,* Advances in the Biosciences, Vol. 30, Night and Shift Work, Biological and Social Aspects, 1980.
10. A. W. Gardner and B. D. Dagnall, "The Effect of 12-Hour Shift Working on Absence Attributed to Sickness," *Br. J. Ind. Med.,* **34,** 148–160 (1977).
11. B. Bjerner, A. Holm, and Swensson, "Diurnal Variation in Mental Performance: A Study of 3-Shift Workers," *Br. J. Ind. Med.,* **12,** 103–110 (1955).
12. B. Bjerner, A. Holm, and A. Swensson, "Studies on Night and Shiftwork," in: *Shiftwork and Health,* A. Aanonsen, Ed., Scandinavian University Books, Oslo, 1964.
13. W. P. Colquhoun, M. J. F. Blake, and R. S. Edwards, "Experimental Studies of Shiftwork. I: A Comparison of 'Rotating and Stabilized' 4-Hour Shift System," *Ergonomics,* **11,** 437–447 (1968a).
14. W. P. Colquhoun, M. J. F. Blake, and R. S. Edwards, "Experimental Studies of Shiftwork. II: Stabilized 8-Hour Shift Systems," *Ergonomics,* **11,** 527–537 (1968b).
15. W. P. Colquhoun, M. J. F. Blake, and R. S. Edwards, "Experimental Studies of Shiftwork. III: Stabilized 12-Hour Shift Systems," *Ergonomics,* **12,** 865–875 (1969).
16. W. P. Colquhoun, "Circadian Variations in Mental Efficiency," in: *Biological Rhythms and Human Performance,* W. P. Colquhoun, Ed., Academic Press, London and New York, 1971.
17. L. C. Johnson, D. I. Tepas, W. P. Colquhoun, and M. J. Colligan (Eds.), *The Twenty Four Hour Workday: Proceedings of a Symposium on Variations in Work Sleep Schedules.* U.S. Dept. of Health and Human Services, Cincinnati, Ohio, 1981.
18. R. S. Brief and R. A. Scala, "Occupational Exposure Limits for Novel Work Schedules," *Am. Ind. Hyg. Assoc. J.,* **36,** 467–471 (1975).
19. D. J. Paustenbach, "The Effect of the Twelve-Hour Workshift on the Toxicology, Disposition and Pharmacokinetics of Carbon Tetrachloride in the Rat," Doctoral dissertation, Purdue University, West Lafayette, Ind., 1982.
20. A. Brandt, "On the Influence of Various Shift Systems on the Health of the Workers," *XVI Int. Congr. Occup. Health,* Tokyo, 106 (1969).
21. J. T. Wilson and K. M. Rose, *The Twelve-Hour Shift in the Petroleum and Chemical Industries of*

the United States and Canada: A Study of Current Experience. Wharton Business School, University of Pennsylvania, Philadelphia, 1978.

22. T. A. Yoder and G. D. Botzum, *The Long-Day Short-Week in Shift Work—A Human Factors Study,* 16th Annual Meeting Proceedings, 1971.

23. G. D. Botzum and R. L. Lucas, "Slide Shift Evaluation—A Practical Look at Rapid Rotation Theory," *Proceedings of Human Factors Society,* 207–211 (1981).

24. Z. Vokac, P. Magnus, E. Jebens, and N. Gundersen, "Apparent Phase-Shifts of Circadian Rhythms (Masking Effects) During Rapid Shift Rotation," *Int. Arch. Occup. Environ. Health,* **49,** 53–65 (1981).

25. P. Knauth, B. Eichhorn, I. Lowenthal, K. H. Gartner, and J. Rutenfranz, "Reduction of Nightwork by Re-designing of Shift-Rotas.," *Int. Arch. Occup. Environ. Health,* **51,** 371–379 (1983).

26. U. S. Bureau of Labor Statistics, *News, Number of Days in the Workweek* (Press release) (March 1979).

27. S. D. Nollen and V. H. Martin, *Alternative Work Schedules. Part 3: The Compressed Workweek,* AMACOM, New York, 1978.

28. S. D. Nollen, "The Compressed Workweek: Is It Worth the Effort?" *Ind. Eng.* **101,** 58–63 (1981).

29. S. D. Nollen, "Work Schedules," In: *Handbook of Industrial Engineering,* G. Salvendy, Ed., Wiley-Interscience, New York, 1982.

30. E. Thiis-Evensen, "Shift Work and Health," *Ind. Med. Surg.,* **27,** 493–513 (1958).

31. M. D. Fottler, "Employee Acceptance of Four Day Workweek," *Acad. Management J.,* **20** (December 1977).

32. J. N. Hedges, "How Many Days Make a Workweek?" *M. Labor Rev.,* **98** (1975).

33. R. B. Dunham and D. I. Hawk, "The Four-Day/Forty-Hour Week: Who Wants It?" *Acad. Management J.,* **20** (1977).

34. A. Henschel, "NIOSH Point of View on Shift Work," in: *Shift Work and Health,* ed. P. G. Rentos and R. D. Shepard, NIOSH Publication 76–203, Cincinnati, Ohio.

35. K. Wheeler, R. Gurman, and D. Tarnowieski, *The Four-Day Week,* AMACOM, New York, 1972.

36. P. J. Taylor, "The Problems of Shift Work." Proceedings of an International Symposium on Night and Shiftwork, Oslo, Sweden, 1969.

37. R. F. Landry, "Off-Beat Rhythms and Biological Variables," *Occup. Health Saf.,* **50,** 40–43 (1981).

38. M. Kenny, "Public Employee Attitudes Toward the Four-Day Workweek," *Publ. Personnel Management,* **3** (1974).

39. W. J. Astleford, "Unusual Work Schedules in the Marine Chemical Transport Industry," Paper #44, presented at the 1982 American Industrial Hygiene Conference, Cincinatti, Ohio, June 6–11, 1982.

40. H. E. Stokinger, "Threshold Limit Values: Part I," in: *Dangerous Properties of Industrial Materials Report,* p. 8–13 (May–June 1981).

41. *Threshold Limit Value for Chemical Substances and Physical Agents in the Workroom Environment with Intended Changes for 1984.* American Conference of Government Industrial Hygienists, P.O. Box 1937, Cincinnati, Ohio, 45201, 1983.

42. W. Clark Cooper, "Indicators of Susceptibility to Industrial Chemicals," *J. Occup. Med.,* **15**(4), 355–359 (1973).

43. G. S. Omenn, "Predictive Identification of Hypersusceptible Individuals," *J. Occup. Med.,* **24,** 369–374 (1982).

44. R. L. Zielhuis, "Permissible Limits for Occupational Exposure to Toxic Agents: A Discussion on Differences in Approach between US and USSR," *Int. Arch. Arbeitsmed.*, **33** (1974).

45. R. G. Smith and J. B. Olishifski, "Industrial Toxicology," in: *Fundamentals of Industrial Hygiene*, J. Olishifski, Ed., National Safety Council, Chicago, 1979.

46. *Documentation of Threshold Limit Values.* American Conference of Governmental Industrial Hygienists, P.O. Box 1937, Cincinnati, Ohio, 1981.

47. H. M. Bolt, J. G. Filser, and A. Buchter, "Inhalation Pharmacokinetics Based on Gas Uptake Studies, III: A Pharmacokinetic Assessment In Man of Peak Concentrations of Vinyl Chloride," *Arch. Toxicol.*, **48**, 213–228 (1981).

48. P. G. Watanabe, J. D. Young, and P. J. Gehring, "The Importance of Non-Linear (Dose-Dependent) Pharmacokinetics in Hazard Assessment," *J. Environ. Path. Tox.*, **1**, 147–159 (1977).

49. P. G. Watanabe, R. H. Reitz, A. M. Schumann, M. J. McKenna, and P. J. Gehring, "Implications of the Mechanisms of Tumorigenicity for Risk Assessment," in: *The Scientific Basis of Toxicity Assessment*, Elsevier/North-Holland Press, M. Witschi, Ed., 1980, p. 69–88.

50. R. H. Reitz, A. M. Schumann, P. G. Watanabe, T. F. Quast, and P. J. Gehring, "Experimental Approach for Evaluating Genetic and Epigenetic Contributions to Chemical Carcinogenesis," *Proc. Am. Assoc. Cancer Res.*, **20**, 266–281 (1979).

51. W. T. Stott, R. H. Reitz, A. M. Schumann, and P. G. Watanabe, "Genetic and Nongenetic Events in Neoplasia," *Food. Cosmet. Toxicol.*, **19**, 567–576 (1981).

52. H. E. Stokinger, "The Case for Carcinogen TLV's Continues Strong," *Occup. Health Safety*, **46**, 54–58 (March/April 1977).

53. D. J. Schaeffer, K. G. Janardan, and H. W. Kerster, "Threshold Estimation from the Linear Dose-Response Model Method and Radiation Data," *Environ. Mgt.*, **5**, 515–520 (1981).

54. J. P. Seiler, "Apparent and Real Thresholds: A Study of Two Mutagens," in: *Progress in Genetic Toxicology*, D. Scott, B. A. Bridges, and F. H. Sobels, eds., Elsevier Biomedical Press, New York, 1977.

55. B. Crouch and H. W. Wilson, "Calculating and Comparing Various Acceptable Levels of Risk," *Risk Anal.*, **1**, 42–51 (1981).

56. W. A. Cook, "Maximum Allowable Concentrations of Industrial Contaminants," *Ind. Med.*, **14**, 11, 936–946 (1945).

57. H. F. Smyth, "Improved Communication; Hygienic Standard for Daily Inhalation," *Am. Ind. Hyg. Assoc. Q.*, **17**, 129–185 (1956).

58. M. E. LaNier, *Threshold Limit Values: Discussion and 35 Year Index with Recommendations (TLVs, 1946–81).* American Conference of Governmental Industrial Hygienists, Cincinatti, Ohio, 1984.

59. H. E. Stokinger, "Criteria and Procedures for Assessing the Toxic Responses to Industrial Chemicals," in: *Permissible Levels of Toxic Substances in the Working Environment*, ILO, Geneva, 1970.

60. World Health Organization, *Methods Used in Establishing Permissible Levels in Occupational Exposure to Harmful Agents.* Technical Report 601, International Labor Office, WHO, Geneva, 1977.

61. A. A. Letavet, "Scientific Principles for the Establishment of the Maximum Allowable Concentrations of Toxic Substances in the USSR," in: *Proceedings of the 13th Annual Congress on Occupational Health July 25–29, 1960*, Book Craftsmen Assoc., New York, 1961.

62. H. E. Stokinger, "International Threshold Limits Values," *Am. Ind. Hyg. Assoc. J.*, **24**, 469 (1963).

63. H. B. Elkins, "Maximum Acceptable Concentrations, A Comparison in Russia and the United States," *AMA Arch. Environ. Health*, **2**, 45 (1961).

64. E. I. Lyublina, "Some Methods Used in Establishing the Maximum Allowable Concentrations," in: *MAC of Toxic Substances in Industry, IUPAC,* 1962, p. 109–112.

65. L. Hardy and B. Elkins, "Medical Aspects of Maximum Allowable Concentrations: Benzene," *J. Ind. Hyg. Toxicol.,* **30,** 196–200 (1948).

66. American Conference of Governmental Industrial Hygienists, Threshold Limit Values for 1950. *Arch. Ind. Hyg. Occup. Med.,* **2,** 1, 98–100 (1950).

67. H. E. Stokinger, "Standards for Safeguarding the Health of the Industrial Worker," *Pub. Health Rep.,* **70,** 1 (1955).

68. J. H. Sterna, "Methods of Establishing Limits," *Am. Ind. Hyg. Assoc. J.,* **17,** 280–284 (1956).

69. Warren A. Cook, "Symposium on Threshold Limits: Present Trends in MAC's," *Am. Ind. Hyg. Assoc. J.,* **17,** 273–274 (1956).

70. M. Sachs, "The Need for Threshold Limits," *Am. Ind. Hyg. Assoc. J.,* **17,** 274–278 (1956).

71. I. Pacseri, M. Ladiscaus, and I. Batskor, "Threshold and Toxic Limits of Some Amino and Nitro Compounds," *AMA Arch. Ind. Health,* **18,** 1–8 (1958).

72. H. E. Stokinger, "Toxicologic Methods for Establishing Drinking Water Standards," *J. Am. Water Works Assn.,* 515–529 (April 1958).

73. H. F. Smyth, "The Toxicological Basis of Threshold Limit Values: 1. Experience with Threshold Limit Values Based on Animal Data," *Am. Ind. Hyg. J.,* **20,** 341–345 (1959).

74. W. K. Rowe, M. A. Wolf, C. S. Weil, and H. F. Smith, Jr., "The Toxicological Basis of Threshold Limit Values: 2. Pathological and Biochemical Criteria," *Am. Ind. Hyg. Assoc. J.,* **20,** 346–349 (1959).

75. John A. Zapp, Jr., Ph.D., "The Toxicological Basis of Threshold Limit Values: 3. Physiological Criteria," *Am. Ind. Hyg. Assoc. J.,* **20,** 350–356 (1959).

76. W. L. Ball, "The Toxicological Basis of Threshold Limit Values: 4. Theoretical Approach to Prediction of Toxicity of Mixtures," *Am. Ind. Hyg. Assoc. J.,* **20,** 357–363 (1959).

77. U. C. Pozzani, C. S. Weil, and C. P. Carpenter, "The Toxicological Basis of Threshold Limit Values: 5. The Experimental Inhalation of Vapor Mixtures by Rats, with Notes upon the Relationship between Single Dose Inhalation and Single Dose Oral Data," *Am. Ind. Hyg. Assoc. J.,* **20,** 364–369 (1959).

78. W. L. Ball, "The Toxicological Basis of Threshold Limit and Values: 6. Report of Prague Symposium on International Threshold Limit Values," *Am. Ind. Hyg. Assoc. J.,* **20,** 370–373 (1959).

79. C. R. Williams, "Scientific Basis for the Establishment of Tolerable Limits Adopted in the United States for the Principal Industrial Toxins," in: *Proceedings of the 13th Annual Congress on Occupational Research,* July 25–29, 1960. Book Craftsmen Assoc., New York, 1961.

80. H. E. Stokinger, "Threshold Limits and Maximal Acceptable Concentrations. Their Definition and Interpretation, 1961," *Am. Ind. Hyg. Assoc. J.,* **23,** 45–47 (1962).

81. W. F. Von Oettingen, "Methods Used by the Committee Z-37 of the American Standards Association in Developing Maximal Acceptable Concentrations of Toxic Dusts and Gases," in: *IUPAC Publication on Exposure Limits,* p. 115, 116, 1962.

82. H. E. Stokinger, "Threshold Limits and Maximal Acceptable Concentrations," *AMA Arch. Environ. Health,* **4,** 115 (1962).

83. J. M. Barnes, "Experimental Methods Used as a Basis for Determining Maximum Allowable Concentrations," *MAC. of Toxic Substances in Industry, IUPAC,* 97–107, 1962.

84. Toxicology Committee, American Industrial Hygiene Association, "Emergency Exposure Limits," *Am. Ind. Hyg. Assoc. J.,* **25,** 578 (1964).

85. H. E. Stokinger, "Modus Operandi of Threshold Limits Committee of ACGIH," *Am. Ind. Hyg. Assoc. J.,* **25,** 589–594 (1964).

86. H. L. Magnuson, "Industrial Toxicology in the Soviet Union Theoretical and Applied," *Am. Ind. Hyg. Assoc. J.*, **25**, 185 (1964).

87. H. E. Stokinger, "Industrial Contribution to Threshold Limits Values," *Arch. Environ. Health*, **10**, 609–613 (1965).

88. H. B. Elkins, "Excretory and Biological Threshold Limits," *Am. Ind. Hyg. Assoc. J.*, **28**, 305 (1967).

89. H. E. Stokinger, "Current Problems of Setting Occupational Exposure Standards," *Arch. Environ. Health*, **19**, 277–284 (1969).

90. R. L. Zielhuis, "Tentative Emergency Limits for Sulphur Dioxide, Sulphuric Acid, Chlorine and Phosgene," *Ann. Occup. Hyg.*, **13**, 171 (1970).

91. R. L. Zielhuis, and M. M. Verberk, "Validity of Biological Tests on Epidemiological Toxicology," *Int. Arch. Occup. Health.* **32**, 167 (1974).

92. R. L. Zielhuis, "Threshold Limit Values and Total Work Load," *J. Occup. Med.*, **13**, 30 (1971).

93. T. F. Hatch, "Permissible Levels of Exposure to Hazardous Agents in Industry," *J. Occup. Med.*, **14**, 134–137 (1972).

94. L. Ulrich and L. Rosival, *The Concept of Maximum Permissible Concentrations of Chemical Substances in the Working Environment,* Bratislavske, Lekarski City, **58**(3), 366–373 (1972).

95. I. Sanotzki, "Concept of Maximum Allowable Concentrations and Criteria of Harmful Effects of Environmental Chemicals," in: *Adverse Effects of Environmental Chemicals and Psychotropic Drugs,* M. Horvath, Ed., Elsevier, Amsterdam, 1973.

96. R. L. Zielhuis, "Health and Environmental Quality Standards," in: *Proceedings III International Clean Air Congress,* p. A74 (1973).

97. K. W. Nelson, "The Place of Biological Measurements in Standard Setting Concepts," *J. Occup. Med.*, **15**, 439 (1973).

98. T. F. Hatch, "Criteria for Hazardous Exposure Limits," *Arch. Environ. Health*, **27**, 231 (1973).

99. H. E. Stokinger, "Industrial Air Standards, Theory and Practice," *J. Occup. Med.*, **15**, 429 (1973).

100. R. L. Zielhuis and M. M. Verberk, "Validity of Biological Tests on Epidemiological Toxicology," *Int. Arch. Occup. Health*, **32**, 167 (1974).

101. A. L. Linch, *Biological Monitoring for Industrial Chemical Exposure Control,* CRC Press, Cleveland, 1974.

102. D. Turner, "The Development of Hygiene Standards," *Ann. Occup. Hyg.*, **19**, 147–152 (1976).

103. R. L. Zielhuis, "Permissible Limits for Chemical Exposures," in: *Occupational Medicine,* C. Zenz, Ed., Year Book Medical Publishers, Chicago, 1975, pp. 579–88.

104. Richard A. Carpenter, "Scientific Information, Expert Judgment and Political Decision Making," *J. Occup. Med.*, **18**, 292–296 (1976).

105. *Biologic Standards for the Industrial Worker by Breath Analysis: Trichloroethylene.* A Study by the Department of Environmental Health, Medical College of Wisconsin (Directed by R. D. Stewart, M. D.), NIOSH Pub. 74–133, **3**, 278–402 (1977).

106. R. L. Zielhuis, "Standards Setting for Work Conditions as Risky Behaviour," in: *Standards Setting,* P. Grandjean, Ed., Arbejdsmiliofondet, Copenhagen, 1977a.

107. B. Holmberg and M. Winell, "Occupational Health Standards, An International Comparison," *Scand. J. Work Environ. Health*, **3**, 1–15 (1977).

108. E. J. Calabrese, *Methodological Approaches to Deriving Environmental and Occupational Health Standards,* Wiley, New York, 1978.

109. R. L. Zielhuis and F. van der Kreek, "Calculations of a Safety Factor in Setting Health Based Permissible Levels for Occupational Exposure. A Proposal I," *Int. Arch. Occup. Environ. Health*, **42**, 191–201 (1979).

110. R. L. Zielhuis and F. W. van der Kreek, "Calculations of a Safety Factor in Setting Health

Based Permissible Levels for Occupational Exposure. A Proposal II, Comparison of Extrapolated and Published Permissible Levels," *Int. Arch. Occup. Environ. Health* **42**, 203–215 (1979).

111. H. Glubrecht, "Dose Comparisons in the Effects of Radiation and of Chemical Pollutants," *Atomkernenergie Kerntechnik Bd.,* **33, 126**–129 (1979).

112. M. Lippman and R. B. Schlesinger, *Chemical Contamination in the Human Environment,* Oxford University Press, New York, 1979.

113. L. M. Lowe and D. B. Chambers, "TLV's for Non-Standard Work Schedules," *Pollution Eng.,* 36–37 (November 1983).

114. World Health Organization, *Occupational Exposure Limits for Airborne Toxic Substances,* 2nd ed., Occ. Safety and Health Series, No. 37, International Labor Office, WHO, Geneva, 1980.

115. W. A. Cook, *A Worldwide Compilation of Occupational Exposure Limits,* American Industrial Hygiene Association, Akron, Ohio, 1984.

116. J. Rodericks and M. R. Taylor, "Application of Risk Assessment to Food Safety Decision Making," *Regul. Toxicol. Pharm.,* **3,** 275–307 (1983).

117. M. L. Dourson and J. F. Stara, "Regulatory History and Experimental Support of Uncertainty (Safety Factors)," *Regulatory Toxicol. Pharm.,* **3,** 224–238 (1983).

118. E. J. Calabrese, *Principles of Animal Extrapolation,* Wiley, New York, 1983.

119. R. Reichsman, and E. Calabrese, "Animal Extrapolation in Environmental Health; Its Theoretical Basis and Practical Applications," *Rev. Environ. Health,* **3,** 59–78 (1979).

120. C. S. Weil, "Statistics Versus Safety Factors and Scientific Judgment in the Evaluation of Safety for Man," *Toxicol. Appl. Pharm.,* **21,** 454–463 (1972).

121. C. N. Park and R. D. Snee, "Quantitative Risk Assessment State-of-the-Art for Carcinogenesis," *Fund. Appl. Toxicol.* **3,** 320–333 (1983).

122. R. T. Williams, "Inter-species Variations in the Metabolism of Xenobiotics," *Biochem. Soc. Trans.,* **2:** 359–377 (1974).

123. M. E. Andersen, "Pharmacokinetics of Inhaled Gases and Vapors," *Neurobehavioral Toxicol. Tertol.,* **3,** 383–389 (1981).

124. I. Astrand and F. Gamberale, "Effects on Humans of Solvents in the Inspiratory Air: A Method of Estimation of Uptake," *Environ. Res.,* **15,** 1–4 (1978).

125. I. Astrand, "Uptake of Solvents in the Blood and Tissues of Man–A Review," *Scand. J. Work Environ. Health,* **1,** 199–218 (1975).

126. P. O. Droz and J. G. Fernandez, "Effect of Physical Workload on Retention and Metabolism of Inhaled Organic Solvents—A Comparative Theoretical Approach and its Applications with Regards to Exposure Monitoring," *Int. Arch. Occup. Environ. Health,* **38,** 231–240 (1977).

127. C. Zenz and B. A. Berg, "Influence of Submaximal Work on Solvent Uptake," *J. Occup. Med.,* **12,** 367–369 (1970).

128. I. Astrand, H. Ehrner-Samuel, A. Kilbom, and P. Ovrum, "Toluene Exposure. I. Concentration in Alveolar Air and Blood at Rest and During Exercise," *Work. Environ. Health,* **9,** 119–130 (1972).

129. B. F. Craft, "The Effects of Phase Shifting on the Chronotoxicity of Carbon Tetrachloride in the Rat." Doctoral dissertation, University of Michigan, Ann Arbor, 1970.

130. R. T. W. L. Conroy and J. N. Mills, "Circadian Rhythms and Shiftworking," *Proc. Int. Symposium on Night and Shiftwork,* Oslo, **42** (1969).

131. R. T. W. L. Conroy, A. L. Elliott, and J. N. Mills, "Circadian Excretory Rhythms in Night Workers," *Br. J. Ind. Med.,* **27,** 356–369 (1970).

132. R. T. W. L. Conroy and J. N. Mills, *Human Circadian Rhythms,* J. and A. Churchill, London, 1970.

133. A. Reinberg and M. H. Smolensky, "Circadian Changes of Drug Disposition in Man," *Clinical Pharmacokinetics,* **7,** 401–420 (1982).

134. Personal Communication with Dr. Mel Andersen, member, 1984 ACGIH TLV Committee.

135. E. W. Van Stee, G. A. Boorman, M. P. Moorman, and R. A. Sloane, "Time-Varying Concentration Profile as a Determinant of the Inhalation Toxicity of Carbon Tetrachloride," *J. Toxicol. Env. Hlth.*, **10**, 785–795 (1982).

136. Commonwealth of Pennsylvania, Threshold Limit Values and Short-term Limits. Title 25, Part 1, Subpart D, Article IV, Chapter 201, Subchapter A, Threshold Limits, *Rules and Regulations, 1 Pa. B. 1985* (Nov. 1, 1971).

137. Occupational Safety and Health Standards, Occupational Safety and Health Administration, *Title 29, Code of Federal Regulations, Part 1910,* 1976.

138. Occupational Safety and Health Administration, *Compliance Officers: Field Manual,* Department of Labor, Washington, D.C., 1979.

139. Threshold Limit Values and Short-Term Limits: Title 25, Part I, Subpart D, Article IV, Chapter 201, Subchapter A, Threshold Limits, *Rules and Regulations, 1 Pa. B. 1985,* Commonwealth of Pennsylvania (Nov. 1, 1971).

140. Criteria for a Recommended Standard-Occupational Exposure to Chloroform, HEW Publication (NIOSH) 75–114, National Institute for Occupational Safety and Health, Rockville, Maryland (1975).

141. Criteria for a Recommended Standard-Occupational Exposure to Benzene, HEW Publication (NIOSH) 74–137, National Institute for Occupational Safety and Health, Rockville, Maryland (1974).

142. *Atmospheric Contaminants in Spacecraft,* Panel for Air Standards for Manned Space Flight. Space Science Board, National Academy of Sciences, Washington, D.C. 1968.

143. J. W. Mason and H. Dershin, "Limits to Occupational Exposure in Chemical Environments Under Novel Work Schedules," *J. Occup. Med.*, **18**, 603–607 (1976).

144. J. L. S. Hickey, "Application of Occupational Exposure Limits to Unusual Work Schedules and Excursions." Doctoral dissertation, University of North Carolina at Chapel Hill, 1977.

145. J. L. S. Hickey and P. C. Reist, "Application of Occupational Exposure Limits to Unusual Work Schedules," *Am. Ind. Hyg. Assoc. J.*, **38**, 613–621 (1977).

146. S. A. Roach, "A More Rational Basis for Air Sampling Programs," *Am. Ind. Hyg. Assoc. J.*, **27**, 1–19 (1966).

147. S. A. Roach, "A Most Rational Basis for Air Sampling Programs," *Ann. Occup. Hyg.*, **20**, 65–84 (1977).

148. S. A. Roach, "Threshold Limit Values for Extraordinary Work Schedules," *Am. Ind. Hyg. Assoc. J.*, **39**, 345–364 (1978).

149. P. Veng-Pedersen, "Pulmonary Absorption and Excretion of Compounds in the Gas Phase. A Theoretical Pharmacokinetic and Toxicokinetic Analysis," *J. Pharm. Sci.*, **73**, 1136–1141 (Aug. 1984).

150. E. Guberan and J. Fernandez, "Control of Industrial Exposure to Tetrachloroethylene by Measuring Alveolar Concentrations: Theoretical Approach Using a Mathematical Model," *Br. J. Ind. Med.*, **31**, 159–167 (1974).

151. R. Handy and A. Schindler, *Estimation of Permissible Concentrations of Pollutants for Continuous Exposure.* U.S. Environmental Protection Agency, Pub. No. EPA-600/2-76-155 (1976).

152. J. K. Piotrowski, *Exposure Tests for Organic Compounds in Industrial Toxicology,* Department of Health, Education and Welfare, NIOSH 77–144 (1977).

153. R. E. Notari, *Biopharmaceutics and Pharmacokinetics, 2nd ed.*, Marcel Dekker, New York, 1975.

154. M. Gibaldi and D. Perrier, *Pharmacokinetics,* Marcel Dekker, New York, 1975.

155. J. V. G. Durnin and R. Passmore, *Energy, Work and Leisure,* Heinemann Educational Books, London, 1967.

156. Ergonomics Guide to Assessment of Metabolism and Cardiac Costs of Physical Work, *Am. Ind. Hyg. Assn. J.*, **32**, 560–564 (1971).

157. L. W. Dittert, "Pharmacokinetic Prediction of Tissue Residues," *J. Tox. and Environ. Health,* **2,** 735–756 (1977).

158. W. J. O'Reilly, "Pharmacokinetics in Drug Metabolism and Toxicology," *Can. J. Pharm. Sci.,* **7,** 66–77 (1972).

159. Withey, J. R., "Pharmacokinetic Principles," in: *First International Congress of Toxicology.* G. Plaa., Ed., Academic Press, New York, 1979.

160. V. Fiserova-Bergerova and J. Teisinger, "Pulmonary Styrene Vapor Retention," *Ind. Med. Surg.,* **34,** 620–622 (1965).

161. V. Fiserova-Bergerova, *Simulation of Uptake, Distribution, Metabolism, and Excretion of Lipid Soluble Solvents in Man.* Aerospace Medical Research Laboratory Report No. AMRL-TR-72-130 (Paper No. 4), Wright-Patterson Air Force Base, Ohio (1972).

162. V. Fiserova-Bergerova, J. Vlach, and K. Singhal, "Simulation and Prediction of Uptake, Distribution and Exhalation of Organic Solvents," *Br. J. Ind. Med.,* **31,** 45–52 (1974).

163. V. Fiserova-Bergerova, "Mathematical Modeling of Inhalation Exposure," *J. Combust. Toxicol.,* **3,** 201–209 (1976).

164. V. Fiserova-Bergerova, J. Vlach, and J. C. Cassady, "Predictable Individual Differences in Uptake and Excretion of Gases and Lipid Soluble Vapours Simulation Study," *Br. J. Ind. Med.,* **37,** 42–49 (1980).

165. S. Riegelman, J. C. Loo, and M. Rowland, "Shortcomings in Pharmacokinetic Analysis by Conceiving the Body to Exhibit Properties of a Single Compartment," *J. Pharm. Sci.,* **57,** 117–125 (1968).

166. A. J. Sedman and J. G. Wagner, "CSTRIP, A Fortran IV Computer Program for Obtaining Initial Poly Exponential Parameter Estimates," *J. Pharm. Sci.,* **65,** 1006–1020 (1976).

167. G. Lehnert, R. D. Ladendorf, and D. Szadkowski, "The Relevance of the Accumulation of Organic Solvents for Organization of Screening Tests in Occupational Medicine. Results of Toxicological Analyses of More Than 6000 Samples," *Int. Arch. Occup. Environ. Health,* **49,** 95–102 (1978).

168. J. C. Ramsey and P. J. Gehring, "Application of Pharmacokinetic Principles in Practice," *Fed. Proceedings,* **39,** 60–65 (1980).

169. P. J. Gehring and G. E. Blau, "Mechanisms of Carcinogenesis: Dose Response," *J. Env. Path. Tox.,* **1,** 163–179 (1977).

170. V. Fiserova-Bergerova and D. A. Holaday, "Uptake and Clearance of Inhalation Anesthetics in Man," *Drug Metab. Reviews,* **9**(1), 43–60 (1979).

171. J. L. S. Hickey and P. C. Reist, "Adjusting Occupational Exposure Limits for Moonlighting, Overtime, and Environmental Exposures," *Am. Ind. Hyg. Assoc. J.,* **40,** 727–734 (1979).

172. J. L. S. Hickey, "Adjustment of Occupational Exposure Limits for Seasonal Occupations," *Am. Ind. Hyg. Assoc. J.,* **41,** 261–263 (1980).

173. J. W. Mason and J. M. Hughes, "The Application of First Order Kinetics to Standard Adjustment in Overtime and Rotating Shift Work. (Submitted to *Scand. J. Work Env. Hth.*)

174. J. L. S. Hickey, "The 'TWAP' in the Lead Standard," *Am. Ind. Hyg. Assoc. J.,* **44**(4), 310–311 (1983).

175. P. Veng-Pederson, D. J. Paustenbach, G. P. Carlson, and L. Suarez, "A Linear Systems Approach to Analyzing the Pharmacokinetics of Carbon Tetrachloride in the Rat Following Exposure to an 8 hr/day and 12 hr/day Simulated Workweek." Submitted to *Toxicol. Appl. Pharm.* (1985).

176. E. J. Calabrese, "Further Comments on Novel Schedule TLVs," *Am. Ind. Hyg. Assoc. J.,* **38,** 443–446 (1977).

177. E. J. Calabrese, A Critical Assessment of High Risk Workers Who Work Unusual Shifts, Paper #35, presented at the 1982 American Industrial Hygiene Conference, Cincinatti, Ohio, June 6–11, 1982.

178. H. E. Stokinger, Personal communication to D. Paustenbach (1979).

179. Y. Alarie et al., "Estimating Safe Levels of Exposure Based on Sensory Response," *Toxicol. Appl. Pharm.*, **42,** 100–105 (1978).

180. M. O. Amdur, *Industrial Toxicology in the Industrial Environment—Its Evaluation and Control*, National Institute for Occupational Safety and Health, Rockville, Maryland (1973).

181. A. Ruzic, "Pharmacokinetic Modeling of Various Theoretical Systems," *J. Pharm. Sci.*, **11,** 110–150 (1970).

182. R. L. Iuliucci, "12 Hour TLV's," *Pollution Eng.*, 25–27 (November 1982).

183. J. Piotrowski, *The Application of Metabolic and Excretion Kinetics to the Problems of Industrial Toxicology*, National Library of Medicine, U.S. Government Printing Office, Washington, D.C., 1971.

184. Y. Henderson and H. H. Haggard, *Noxious Gases and the Principles of Respiration Influencing Their Action*, Reinhold, New York, 1943.

185. J. W. Mason and J. Hughes, "The Use of First Order Kinetics to Extend Occupational Health Standards to Unusual Work Schedules Involving Carcinogens," submitted to *Am. Ind. Hyg. Assoc. J.* (1985).

186. J. W. Mason and J. M. Hughes, The Application of First Order Kinetics to Standard Adjustment in Overtime & Rotating Shift Work. In preparation (1985).

187. J. W. Mason and J. M. Hughes, A Computer Program for Adjusting Occupational Exposure Limits for Any Exposure Period or Work Schedule. Paper #51, presented at the American Industrial Hygiene Association Conference, Detroit, 1984.

188. J. R. Withey and B. T. Collins, "Chlorinated Aliphatic Hydrocarbons Used in the Foods Industry: The Comparative Pharmacokinetics of Methylene Chloride, 1,2 Dichloroethane, Chloroform, and Trichloroethylene After I.V. Administration in the Rat," *J. Environ. Pathol. Toxicol.*, **3,** 313–332 (1980).

189. P. Veng-Pederson, D. J. Paustenbach, G. P. Carlson, and L. Suarez, "A Linear Systems Approach to Analyzing the Pharmacokinetics of Carbon Tetrachloride in the Rat Following Exposure to an 8 hr/day and 12 hr/day Simulated Workweek." Submitted to *Toxicol. Appl. Pharm.* (1985).

190. C. D. Thran, "Linearity and Superposition in Pharmacokinetics," *Pharmacol. Rev.*, **26,** 3–31 (1974).

191. P. Veng-Pederson, "Curve Fitting and Modeling in Pharmacokinetics and Some Practical Experiences with NONLIN and a New Program FUNFIT," *J. Pharmacokin. Biopharm.*, **5,** 513–531 (1977).

192. D. J. Cutler, "Linear System Analysis in Pharmacokinetics," *J. Pharmacokin. Biopharm.*, **6,** 265–282 (1978).

193. J. W. Mason and J. M. Hughes, "A Two-Compartment Modelling Approach to Adjusting Exposure Limits for Unusual Work Schedules." Submitted to *Am. J. Ind. Hyg.* (1985).

194. M. Gibaldi and H. Weintraub, "Some Considerations as to the Determination and Significance of Biologic Half-Life," *J. Pharm. Sci.*, **60,** 624–626 (1971).

195. J. Adir et al., "Pharmacokinetics of Fluorocarbon 11 and 12 in Dogs and Humans," *J. Clin. Pharmacol.*, **15,** 760–770 (1975).

196. A. Carlsson, "Exposure to Toluene: Uptake, Distribution, and Elimination in Man," *Scand. J. Work Environ. Health*, **8,** 43–56 (1982).

197. G. D. DiVincenzo, F. J. Yanno, and B. D. Astill, "Human and Canine Exposure to Methylene Chloride Vapor," *Am. Ind. Hyg. Assoc. J.*, **33,** 125–135 (1972).

198. C. G. Hunter and D. Blair, "Benzene: Pharmacokinetic Studies in Man," *Arch. Occup. Hyg.*, **15,** 193–199 (1972).

199. M. Ikeda and T. Immamura, "Biological Half-Life of Trichloroethylene and Tetrachloroethylene in Human Subjects," *Int. Arch. Arbeitsmed.*, **21,** 209–224 (1973).

200. M. Ikeda, T. Immamura, M. Hayashi, T. Tabuchi, and I. Hara, "Biological Half-Life of Styrene in Human Subjects," *Int. Arch. Arbeitsmed.,* **32,** 93–100 (1974).

201. M. Ikeda, H. Ohtsuji, T. Immamura, and Y. Komosike, "Urinary Excretion of Total Trichloro-Compounds, Tetrichloroethanol and TCAS as a Measure of Exposure to TRI and Tetra-chloroethylene," *Br. J. Industr. Med.,* **29,** 328–333 (1972).

202. J. C. Ramsey and J. D. Young, "Pharmacokinetics of Inhaled Styrene in Rats and Humans," *Scand. J. Work. Environ. Health,* **4,** suppl. 2, 84–91 (1978).

203. N. Krivanek et al., "Monomethylformamide Levels in Human Urine after Repetitive Exposure to Dimethylformamide Vapor," *J. Occup. Med.,* **20,** 179–187 (1978).

204. G. Kimmerle and A. Eben, "Metabolism, Excretion and Toxicology of Trichloroethylene after Inhalation, II. Experimental Human Exposure," *Arch. Toxicol.,* **30,** 127–138 (1973).

205. A. C. Monster, "Difference in Uptake, Elimination and Metabolism in Exposure to trichloro-ethylene, 1,1,1-trichloroethane and tetrachloroethylene," *Int. Arch. Occup. Environ. Health,* **42,** 311–317 (1979).

206. A. Morgan, A. Black, and D. R. Belcher, "The Excretion in Breath of Some Aliphatic Hydrocarbons Following Administration by Inhalation," *Ann. Occup. Hyg.,* **13,** 219–233 (1970).

207. A. Morgan, A. Black, and D. R. Belcher, "Studies on the Absorption of Halogenated Hydrocarbons and Their Excretion in Breath Using [38]Cl Tracer Techniques," *Ann. Occup. Hyg.,* **15,** 273–282 (1972).

208. K. Nomiyama and H. Nomiyama, "Metabolism of Trichloroethylene in Humans: Sex Difference in Urinary Excretion of Trichloroacetic Acid and Trichloroethanol," *Int. Arch. Arbeitsmed.,* **28,** 37–48 (1971).

209. K. Nomiyama and H. Nomiyama, "Respiratory Retention, Uptake and Excretion of Organic Solvents in Man: Benzene, Toluene n-hexane, Ethyl Acetate and Ethyl Alcohol," *Int. Arch. Arbeitsmed.,* **32,** 75–83 (1974).

210. M. Ogata, K. Tomokuni, and Y. Takatsuka, "Urinary Excretion of Hippuric Acid and *m*- or *p*-methylhippuric Acid in the Urine of Persons Exposed to Vapours of Toluene and *m*- or *p*-xylene as a Test of Exposure," *Br. J. Ind. Med.,* **27,** 43–50 (1970).

211. M. Ogata, Y. Takatsuka, and K. Tomokuni, "Excretion of Organic Chlorine Comoounds in the Urine of Persons Exposed to Vapours of Trichloroethylene and Tetrachloroethylene," *Br. J. Ind. Med.,* **28,** 386–391 (1971).

212. J. E. Peterson, "Absorption and Elimination of Carbon Monoxide by Inactive Young Men," *Arch. Environ. Health,* **21,** 165–171 (1970).

213. J. K. Piotrowski, "Evaluation of Exposure to Phenol: Absorption of Phenol Vapors in the Lungs and Through the Skin and Excretion in Urine," *Br. J. Ind. Med.,* **28,** 172–178 (1971).

214. R. D. Stewart, H. H. Gay, D. S. Erley, C. L. Hake, and J. E. Peterson, "Human Exposure to Carbon Tetrachloride Vapor-Relationship of Expired Air Concentration to Exposure and Toxicity," *J. Occup. Med.,* **3,** 586–590 (1961).

215. R. D. Stewart, A. Arbor, H. H. Gay, D. S. Erley, C. L. Hake, and A . W. Schaffer, "Human Exposure to Tetrachloroethylene Vapor," *Arch. Environ. Health,* **2,** 516–522 (1961).

216. R. D. Stewart and H. C. Dodd, "Absorption of Carbon Tetrachloride, Trichloroethylene, Tetrachloroethylene, Methylene Chloride and 1,1,1-trichloroethane Through Human Skin," *Am. Ind. Hyg. Assoc. J.,* **25,** 438–443 (1964).

217. R. D. Stewart, H. C. Dodd, E. D. Baretta, and A. W. Schaffer, "Human Exposure to Styrene Vapor," *Arch. Environ. Health,* **16,** 656–662 (1968).

218. R. D. Stewart, H. C. Dodd, H. H. Gay, and D. S. Erley, "Experimental Human Exposure to Trichloroethylene," *Arch. Environ. Health,* **20,** 64–71 (1970).

219. R. D. Stewart, E. D. Baretta, H. C. Dodd, and T. R. Torkelson, "Experimental Human Exposure to Tetrachloroethylene," *Arch. Environ. Health,* **20,** 224–229 (1970).

220. D. J. Paustenbach, G. P. Carlson, J. E. Christian, and G. S. Born, "Pharmacokinetics of

Carbon Tetrachloride in the Rat Following Repeated Inhalation Exposure for 8 hr/day and 11.5 hr/day." Accepted by *Fund. Appl. Toxicol.* (1985).

221. D. J. Paustenbach, G. P. Carlson, J. E. Christian, and G. S. Born, "The Effect of the Simulated Twelve Hour Work Schedule on the Distribution and Toxicity of Inhaled Carbon Tetrachloride in the Rat." Accepted by *Fund. Appl. Toxicol.* (1985).

222. E. J. O'Flaherty, P. B. Hammond, and S. I. Lerner, "Dependence of Apparent Blood Lead Half-Life on the Length of Previous Lead Exposure in Humans," *Fund. Appl. Toxicol.*, **2**, 49–54 (1982).

223. J. C. Ramsey and M. E. Andersen, "A Physiologically Based Description of the Inhalation Pharmacokinetics of Styrene in Rats and Humans." *Toxicol. Appl. Pharm.*, **73**, 159–175 (1984).

224. NCRP, Maximum Permissible Body Burdens and Maximum Permissible Concentrations of Radionuclides in Air and in Water for Occupational Exposure. *NBS Handbook No. 69*, U.S. Government Printing Office, Washington, D.C., 1959.

225. Richard Peto, Carcinogenic Effects of Chronic Exposure to Very Low Levels of Toxic Substances, *Env. Health Perspectives*, Vol. 22, p. 155–159 (1978).

226. G. C. Butler, "Estimation of Doses and Integrated Doses," in: *Principles of Ecotoxicology*, G. C. Butler, Ed., Scientific Committee on Problems of the Environment (SCOPE), Wiley, New York, 1979.

227. U. Saffioti, "Identification and Definition of Chemical Carcinogens: Review of Criteria and Research Needs," *J. Toxical. Environ. Health*, **6**(5), 1029–1058 (1980).

228. R. G. Harvey, "Polycyclic Hydrocarbons and Cancer," *Am. Sci.*, **70**, 386–393 (1982).

229. A. M. Schumann, J. F. Quast, and P. G., Watanabe, "The Pharmacokinetics and Macromolecular Interactions of Perchloroethylene in Mice and Rats as Related to Oncogenicity," *Toxicol. Appl. Pharm.*, **55**, 207–219 (1980).

230. G. M. Williams, "Classification of Genotoxic and Epigenetic Heptatocarcinogens Using Liver Culture Assays," *Ann. N.Y. Acad. Sci.*, **349**, 273–282 (1980).

231. G. M. Williams, Epigenetic Mechanisms of Action of Carcinogenic Organochlorine Pesticides, ACS, Symp. Series, *ISS Pest. Chem. Mod. Toxicol.*, **160**, 45–56 (1981).

232. G. M. Williams, J. H. Weisburger, and D. Brusick, The Role of Genetic Toxicology in a Scheme of Systematic Carcinogen Testing, *The Pesticide Chemist and Modern Toxicology*, ACS, p. 57–87 (March 2, 1981).

233. R. A. Squire, "Ranking Animal Carcinogens: A Proposed Regulatory Approach," *Science*, **214**, 877–880 (1981).

234. W. Jacobi, "Basic Concepts of Radiation Protection," *J. Ecotox. Environ. Safety*, **4**(4), 434–443 (1980).

235. B. D. Dinman, "Non-concept of No-Threshold Chemicals in the Environment," *Science*, **175**, 495–497 (1972).

236. E. M. Rupp, D. C. Parzyck, R. S. Booth, R. J. Ravidon, and B. L. Whitfield, Composite Hazard Index for Assessing Limiting Exposures to Environmental Pollutants: Application through a Case Study, *Environ. Sci. Technol.*, **12**(7): 802–807 (1978).

237. S. Osteman-Golkar and L. Ehrenberg, "Dosimetry of Electrophilic Compounds by Means of Hemoglobin Alkylation," *Ann. Rev. Pub. Health*, **4**, 317–402 (1983).

238. L. Ehrenberg, K. D. Hieschke, S. Osterman-Golkar, and I. Wennberg, "Evaluation of Genetic Risks of Alkylating Agents: Tissue Doses in the Mouse from Air Contaminated with Ethylene Oxide," *Mutation Research*, **24**, 83–103 (1974).

239. N. Ya. Yansheva, Yu. G. Antomonov, R. E. Albert, B. Altshuler, and L. Friedman, "Approaches to the Formulation of Standards for Carcinogenic Substances in the Environment," *Environ. Health Perspect.*, **30**, 81–85 (1979).

240. D. L. Coffin, D. E. Gardner, G. I. Sidorenko, and M. A. Pinigin, "Role of Time as a Factor

in the Toxicity of Chemical Compounds in Intermittent and Continuous Exposures. Part II. Effects of intermittent exposure," *J. Toxicol. Environ. Health*, **3,** 821–828 (1977).

241. S. W. Dixon, G. J. Graepel, D. L. Leser, and L. F. Percival, "Effect of a Change From an 8-Hr to a 12-Hr Shift on the Levels of DMF Metabolites in the Urine." Submitted to *J. Occup. Med.* (1985).

242. J. A. MacGregor, "Application of Pharmacokinetics to Occupational Health Problems." Doctoral dissertation, University of California, San Francisco, 1973.

243. Y. Kim and G. P. Carlson, "Comparison of a Standard and Novel Workshift on the Formation of Carboxyhemoglobin and Elimination of Dichloromethane in Rats and Mice Exposed by Inhalation." Paper presented at the Annual Meeting of the Society of Toxicology. Abstract published in *The Toxicologist,* **4** (March 1984).

244. T. P. Pastoor and B. A. Burgess, "Effect of concentration and duration of exposure on the inhalation toxicity of aniline for periods of 3, 6, 9 and 12 hrs." Presented at the 1983 Joint Conference on Occupational Health.

245. J. M. Wisniewska-Knypl, J. K. Jablonska, and J. K. Piotrowski, "Effect of Repeated Aniline, Nitrobenzene, and Benzene on Liver Microsomal Metabolism in the Rat," *Br. J. Ind. Med.*, **32,** 42–48 (1975).

246. A. David, E Frantik, R. Holvsa, and O. Novakova, "Role of Time and Concentration on Carbon Tetrachloride Toxicity in Rats," *Int. Arch. Occup. Environ. Health*, **48,** 49–60 (1981).

247. N. P. Kazmina, "Study of the Adaptation Processes of the Liver to Monotonous and Intermittent Exposures to Carbon Tetrachloride (in Russian)," *Gig. Tr. Prof. Zabol.*, **3,** 39–45 (1976).

248. W. A. Colburn and H. B. Matthews, "Pharmacokinetics in the Interpretation of Chronic Toxicity Tests: The Last-In, First-Out Phenomena," *Toxicol. Appl. Pharm.*, **48,** 387–395 (1979).

CHAPTER SEVEN

Data Automation

ROBERT L. FISCHOFF, and
FRED G. FREIBERGER

1 THE NEED FOR DATA AUTOMATION

Over the years the emphasis of industrial hygiene has gradually changed from discovering job-related causes of ill health to monitoring and controlling potentially harmful work environment situations before they result in injury to workers or the public. Associated with this modification in concept has been a significant change in industrial hygiene methodology, namely, an increasing requirement for data collection, recordkeeping, statistical analysis, and reporting. These activities are essential in determining what controls should be implemented to ensure employee health. Added to this professional responsibility for data management are the requirements of the Occupational Safety and Health Administration (OSHA) for recordkeeping and reporting on a growing list of harmful or suspected substances used in modern industrial processes.

The purpose of OSHA is to ensure, as far as possible, safe and healthful working conditions for every industrial employee in the nation by providing mandatory occupational safety and health standards. Basically the Secretary of Labor is responsible for issuing regulations requiring the employer to protect the employee. These standards also oblige the employer to maintain records of employee exposures, to give employees access to the records, to allow employees the opportunity to observe monitoring or measuring being conducted, and to notify the employees of excessive exposures, as well as to inform them of corrective action being taken. The government is also allowed access to all of the foregoing records. In some cases record retention requirements may be 20 or more years.

In addition to recording and reporting requirements, OSHA standards prescribe, as necessary, the training of the employee, suitable protective equipment, control procedures, type and frequency of medical exams, and use of warnings to ensure employee awareness of hazards, symptoms, emergency treatment, and safe use conditions. At the time of writing there are 17 specific OSHA standards, and the 1976 Registry of Toxic Effects lists 218 chemicals as recommended standards from National Institute for Occupational Safety and Health (NIOSH) criteria documents. This publication lists 100,000 chemicals, 4,000 of which are carcinogens, mutagens, or teratogens. Thus the responsibility of industrial hygiene management can become exceedingly complex. In protecting the health of employees it must recognize potential health hazards, have them evaluated, assure that controls are in place, initiate exposure monitoring procedures, enter and delete employees from the recordkeeping system, and comply with changing government requirements. Industrial hygiene management is not the only area that is in need of the information just outlined. Other persons or functional areas that are affected by the OSHA regulations and need hazard exposure evaluations and reports are:

1. Line management.
2. Employees.
3. Safety.
4. Transportation.
5. Medical.
6. Development engineering.
7. Manufacturing engineering.
8. Facilities engineering.
9. Chemical control and disposal.
10. Purchasing.
11. Shipping and receiving.
12. Personnel.
13. Laboratory.

Because of the rapidly increasing workload of recordkeeping, data analysis, and reporting, many industrial hygiene managers are turning to electronic data processing (EDP). An EDP system can greatly increase the productivity and quality of most information collecting, storage, and retrieval operations, while providing more timely and economical data analyses and reports. However, automating a large industrial hygiene program is a complex task requiring careful planning and evaluation if costly mistakes are to be avoided.

This chapter presents the basic concepts of EDP systems and describes the steps necessary to plan, design, and implement an effective computer application. Two specific examples of industrial health computer applications are given; one is a relatively simple, single data base system to control plant ventilation, and the other is a more complex data base involving an environmental health information and control system. Concepts of other industrial health applications are also discussed and references are provided to assist those interested in pursuing additional aspects of data automation. A data processing glossary is included for reference.

2 ELECTRONIC DATA PROCESSING CONCEPTS

2.1 Computer Equipment (Hardware)

EDP is the handling of data by an electronic computer and associated (periph-eral) devices in a planned sequence of operations to produce a desired result. The many types of EDP system range in size from desk-top units to systems that fill several large rooms with interconnected devices. But regardless of the information to be processed or the complexity of equipment used, all EDP involves four basic functions:

1. Entering the source data into the system (input).
2. Storing the data in addressable locations (storage).
3. Processing the data in an orderly manner within the system (processing).
4. Providing the resulting information in a usable form (output).

These functions are performed by an input device, a storage unit, a central processing unit (CPU), and an output device (Figure 7.1).

2.1.1 Input Devices

Input devices read or sense coded data that are recorded on a prescribed medium and make this information available to the computer. Data for input

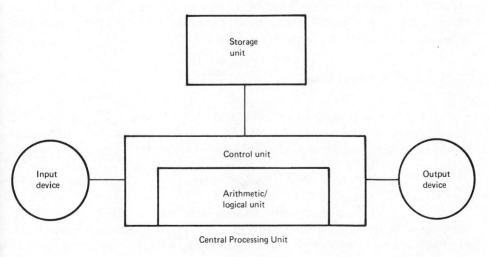

Figure 7.1 Central organization of an EDP system.

are recorded on cards and paper tape as punched holes, on magnetic tape, disks, or drums as magnetized spots, or on paper documents as characters or line drawings. Section 2.3 discusses the method of recording data for machine use and the characteristics of each type of recording medium.

2.1.2 Storage

Storage is somewhat like an electronic filing cabinet, completely indexed and instantaneously accessible to the computer. All data must be placed in storage by an input device before they can be processed by the computer. Each position of storage has a specific location, called an address, so that the stored data can be located by the computer as needed.

The computer can rearrange data in storage by sorting or combining different types of information received from a number of input devices. The computer also can take the original data from storage, calculate new information, and place the result back in storage.

The size or capacity of storage determines the amount of information that can be held in the system at any one time. In some computers, storage capacity is measured in millions of digits or characters (bytes) that provide space to retain entire files of information. In other systems storage is smaller, and data are held only while being processed. Consequently the capacity and design of storage affect the method in which data are handled by the system.

Storage capacity built into a computer is called main storage. This usually consists of main data storage, for programs or other data; control storage, which often contains special built-in "microprograms" to assist the computer in carrying out its own operations; local storage, consisting of high speed working areas (registers) for performing calculations and other processing; and large capacity storage. In addition, much more storage can be provided by auxiliary magnetic tape drives or disk storage drives connected to the computer. Examples of auxilliary storage units appear in Figure 7.2.

2.1.3 Central Processing Unit

The CPU is the controlling center of the entire EDP system. It is usually divided into two parts as shown in Figure 7.1: the control section, and the arithmetic/logical unit. The control section directs and coordinates all computer system functions. It is like a traffic cop that schedules and initiates the operation of input and output devices, arithmetic/logical unit tasks, and the movement of data from and to storage.

The arithmetic/logical unit performs such operations as addition, subtraction, multiplication, division, shifting, moving, comparing, and storing. It also has a capability to test various conditions encountered during processing and to take action accordingly.

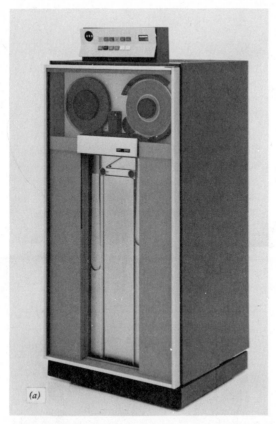

Figure 7.2 Auxiliary storage units. (a) IBM 3420 magnetic tape unit. (b) IBM 3350 direct access storage.

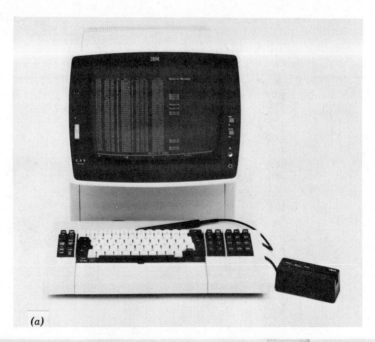

Figure 7.3 Example of I/O devices: (a) IBM 3278 display station; (b) IBM 3505 card reader; (c) IBM 3211 printer.

(c)

Figure 7.3 *(Continued)*

2.1.4 Output Devices

Output devices record or write information from the computer onto cards, paper tape, magnetic tape, disks, or drums. They may print information on paper, generate signals from transmissions over teleprocessing networks, produce graphic displays or microfilm images, or take other specialized forms.

Frequently the same physical device, such as a card reader/punch or a tape drive, is used for both input and output operations. Thus input and output (I/O) functions are generally treated together. The number and type of I/O devices that may be connected directly to a CPU depends on the design of the system and its application. Note that the functions of I/O devices and auxiliary storage units may overlap—thus a tape drive or disk file may be used both for I/O operations and for data storage. Figure 7.3 presents some examples of I/O devices.

The console (Figure 7.4) is an I/O device that provides external control of a data processing system. Keys or switches allow the computer operator to turn power on or off, start or stop operation, and control various devices in the system. There are usually lights, permitting data in the system to be displayed visually.

Figure 7.4 CPU and operator console.

On some systems a console printer and keyboard provide limited output or input capability. The I/O device may print messages, signaling the end of processing or an error condition. It may also print totals or other information that enables the operator to monitor and supervise operation, or it may give instructions to the operator. On the other hand, the console keyboard may be used to key in meaningful information (such as altering instructions) to a data processing system that is programmed to respond to such messages.

A remote console may offer increased efficiency and flexibility by providing duplicate operator controls at a station removed from the CPU.

2.2   Computer Programs (Software)

2.2.1   Application Programs

Each EDP system is designed to perform a specific number and type of operations. It is directed to perform each operation by an instruction. The instruction defines a basic operation to be performed and identifies the data, device, or mechanism needed to carry out the operation. The entire series of instructions required to complete a given procedure is known as an application (or problem) program.

For example, the computer may have the operation of multiplication built into its circuits in much the same way that the ability to add is built into a simple desk-top adding machine. But there must be some means of directing the computer to perform multiplication, just as the adding machine is directed by depressing keys. There must also be a way to instruct the computer where in storage it can find the factors to multiply.

Furthermore, the comparatively simple operation of multiplication implies other activity that must preceed and follow the calculation. Assume that the multiplicand and the multiplier are read into storage by an input device. Once the calculation has been performed, the product must be returned to storage at a specified location, from which it may be written out by an output device.

Any calculation, therefore, involves reading, locating the factors in storage, performing the required computation, returning the result to storage, and writing out the completed result. Even the simplest portion of a procedure involves a number of planned steps that must be spelled out to the computer if the procedure is to be accomplished.

An entire application program is composed of these individual steps grouped in a sequence that directs the computer to produce a desired result. Thus a complex problem must be reduced to a series of basic machine operations before it can be solved. Each of these operations is coded as one instruction or as a series of instructions, in a form that can be interpreted by a computer, and is placed in the main storage unit as a portion of a stored program.

The possible variations of a stored program afford the EDP system almost unlimited flexibility. A computer can be applied to a great number of different procedures simply by reading in, or loading, the proper program into storage. Any of the standard input devices can be used for this purpose, because instructions can be coded into machine language just as data can.

The stored program is accessible to the computer, giving it the ability to alter the program in response to conditions encountered during an operation. Consequently the program selects alternatives within the framework of the anticipated conditions.

2.2.2   Control Programs

To make possible the teleprocessing networks and the orderly operation of many types of I/O devices that may be on-line with a computer, control programs have been developed. Control programs, also known as monitor programs or supervisory programs, act as traffic directors for all the application programs (which solve a problem or carry out a particular operation or process on a set of data), then relinquish control of the computer to the control program. The control program may be constructed to allow the computer to handle random inquiries from remote terminals, to switch from one problem program within the computer to another, to control external equipment, or to do whatever the application requests.

The concept of maintaining optimum computer usage by interleaving and interspersing application programs under the direction of control programs gives rise to the use of two terms—time sharing and multiprogramming.

Briefly, time sharing may be thought of as the cooperative use of a central computer by more than one user (company, division, branch of a company, institution, or government agency). Each user receives a share of the time available, with the result that many jobs are being performed within a congruent time (either simultaneously or seemingly simultaneously). This service may be achieved by interspersing programs rapidly on one computer system, by multiprogramming, or by using two computers that are connected. Multiprogramming is usually thought of as a system of control programs and computer equipment that permits many application programs to go on concurrently. This is accomplished by interleaving the programs with each other in their use of the CPU, storage, and I/O devices.

2.3 Data Representation

2.3.1 Recording Media

Symbols convey information; the symbol itself is not the information but merely represents it. Presenting data to a computer is similar in many ways to communicating with another person by letter. The intelligence to be conveyed must be reduced to a set of symbols. In the English language, these are the familiar letters of the alphabet, numbers, and punctuation marks.

Similarly, communication with a computer system requires that data be reduced to a set of symbols that can be read and interpreted by data processing machines. The symbols differ from those commonly used by people because the information to be represented must conform to the design and operation of the machine. The choice of these symbols (and their meaning) is a matter of convention on the part of the designers. Just as there are rules of grammar that dictate the proper use of a language for clear communication among people, so there are rules of syntax that prescribe how a set of symbols must be used for communication between people and machines. Use of the assigned set of symbols in accordance with the prescribed rules constitutes a programming language.

When a computer program is written, it must be recorded in a medium that can be read by a machine. It may be put in the form of punched cards or paper tape, magnetic tape, direct access storage devices (DASD) such as magnetic disks or drums, magnetic ink characters, optically recognizable characters, microfilm, display screen images, communication network signals, and so on. Figure 7.5 illustrates some of these data recording forms.

Data are represented on the punched card by small rectangular holes in specific locations of the card. In a similar manner, small circular holes along a paper tape represent data. On magnetic tape, or DASD, the symbols are small magnetized areas, called spots or bits, arranged in specific patterns. Magnetic ink characters are printed on paper. The shape of the characters and the magnetic properties of the ink permit the printed data to be read by both man and machine. Each medium requires a code or specific arrangement of symbols to represent data.

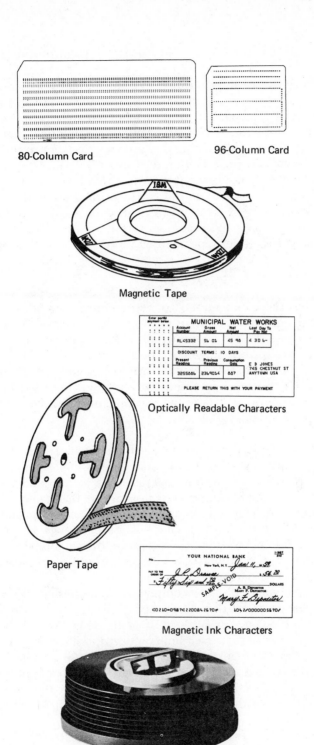

80-Column Card   96-Column Card

Magnetic Tape

Optically Readable Characters

Paper Tape

Magnetic Ink Characters

Direct Access Storage Device

Figure 7.5 Data recording media.

An input device of the computer system is a machine designed to sense or read information from one of the recording media. In the reading process, recorded data are converted to, or symbolized in, electronic form; then the data can be used by the machine for data processing operations.

An output device is a machine that receives electronic information from the computer system and records it on the designated output medium.

All I/O devices cannot be used directly with all computer systems. However, data recorded on one medium can be transcribed to another medium for use with a different system. For example, data on cards or paper tape can be transcribed onto magnetic tape. Conversely, data on magnetic tape can be converted to cards, paper tape, printed reports, or plotted graphs.

2.3.2 Machine Data

Not only must there be a method of representing data on physical media such as cards or magnetic tape, there must also be a method of representing data within a machine. In the computer, data are represented by many electronic components: transistors, magnetic cores, wires, and so on. The storage and flow of data through these devices are represented as electronic signals or indications. The presence or absence of these signals in specific circuitry is the method of representing data in the machine, much as the presence of holes in a card represents data.

Binary States. Digital computers function in binary states; this means that the computer components can indicate only two possible states or conditions. For example, the ordinary light bulb operates in a binary mode; it is either on or off. Likewise, within the computer, transistors are maintained either conducting or nonconducting; magnetic materials are magnetized in one direction or in the opposite direction; and specific voltage potentials are present or absent (Figure 7.6). The binary states of operation of the components are signals to the computer, as the presence or absence of light from an electric light bulb can be a signal to a person.

Representing data within the computer is accomplished by assigning a specific value to a binary indication or group of binary indications. For example, a device to represent decimal values could be designed with four electric light bulbs and switches to turn each bulb on or off as illustrated in Figure 7.7. The bulbs are assigned decimal values of 1, 2, 4, and 8. When a light is on, it represents the decimal value associated with it. When a light is off, the decimal value is not considered. With such an arrangement, the single decimal value represented by the four bulbs will be the numeric sum indicated by the lighted bulbs.

Decimal values 0 through 15 can be represented. The numeric value 0 is represented by all lights off; the value 15, by all lights on; 9, by having the 8 and 1 lights on and the 4 and 2 lights off; 5, by the 1 and 4 lights on and the 8 and 2 lights off; and so on.

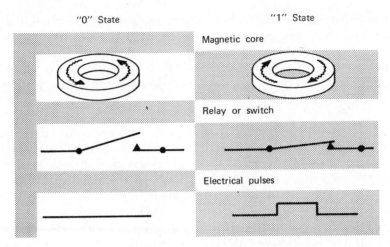

Figure 7.6 Binary components.

The value assigned to each bulb in the example could have differed from the respective values used. This change would involve assigning new values and determining a new scheme of operation. In a computer the values assigned to a specific number of binary indications become the code or language for representing data.

Because binary indications represent data within a computer, a binary method of notation is used to illustrate these indications. The binary system of notation uses only two symbols, zero (0) and one (1), to represent specific values. In any one position of binary notation, the 0 represents the absence of a related or assigned value, and the 1 represents the presence of a related or assigned

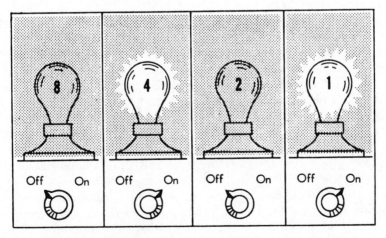

Figure 7.7 Representing decimal data with binary components.

value. For example, to illustrate the indications of the light bulbs in Figure 7.7, the following binary notation would be used: 0101.

The binary notations 0 and 1 are commonly called bits. For computers using the binary system of data representation (typified by the IBM System/370), the basic unit of information is contained in eight consecutive bit positions, called a byte. Four bytes (32 consecutive bits) constitute a word.

Computer Codes.   The method used to represent (symbolize) data is called a code. In a computer the code relates data to a fixed number of binary indications (symbols). For example, a code used to represent alphabetic and numeric (alphameric) characters may use eight positions of binary indication. By the proper arrangement of the binary indications (0 bit, 1 bit), all numbers and characters can be represented with these eight binary positions.

Some computer codes in use are six-bit alphameric code, eight-bit alphameric code, two-out-of-five-count code, and six-bit (packed) numeric code. Most computer codes are self-checking; that is, they have a built-in method of checking the validity of the coded information. Code checking occurs automatically within the machine as the data processing operations are carried out. The method of validity checking is part of the design of the code.

3   STEPS IN EDP APPLICATION DEVELOPMENT

In designing and implementing any new computer application there are logical steps of analysis, planning, development, testing, and installation that must be carried out. Usually these steps are accomplished in the sequence described below.

3.1   Defining and Sizing

When a new computer application is desired, the first task is to define the problem. One should be able to state clearly and concisely just what is to be performed by the proposed system, and the scope (size) of the effort. A short written report should be prepared to describe the problem in terms of subject, scope, objectives, and recommendations. This report will establish the basis for communicating needs to the system analyst and to management.

3.2   Analysis

The analysis phase of a design effort is not really an isolated step. The system analyst, who views a new computer application in terms of its scope and objectives, must work closely with the users to determine their needs, the information in use in the present system, and the information needed in the new system. The analyst must also consider what equipment would be the most cost-effective for the necessary functions of the new system. The basic specification for the new system may then be modified or expanded as necessary during the development (programming) phase. Often, as the result of the preliminary analysis and discussions with the users and programmer(s), the original problem statement is redefined.

3.3 Charting of Tasks

System design is a creative process in which the analyst must identify each activity or procedure required to arrive at the desired result. A flow chart is used to help thinking through the entire process and keeping track of each step, (Figure 7.8). Once the new system has been fully designed, the analyst may use a PERT (program evaluation review technique) chart to define and schedule the key system development tasks. A PERT chart is a graphic representation of the interrelationships and chronological dependencies of all activities required to complete a project.

3.4 Data Base

A data base is a collection of related, nonredundant data files that can be accessed by more than one user. It is usually shared by several application areas such as personnel, marketing, and medical, but it may be dedicated to a single application. When the data base is shared by various users, it is necessary to restrict access of each type of user to the part of the data needed for the respective jobs, so that personal or company confidential information can be controlled.

The process of structuring a data base to accommodate the needs of different users can become quite complex. The system analyst usually defines the data base requirements in terms of the input data sources available and the type of output (reports) needed by the users.

3.5 Programming

Traditionally, the system analyst will prepare a system, or program, specification document, which is a detailed blueprint for the development programmer to follow in writing the computer program needed. The specification explains the overall function of the system and includes detailed descriptions of the input data sources, data base structure, processing requirements, security considerations, and output requirements (reports and new or updated data files). The programmer then prepares the detailed computer instructions to accomplish the specified tasks. Often, however, experienced programmers do both jobs—analysis and programming; this is especially likely to happen on smaller applications.

The programmer develops the logical sequence of computer instructions and functions using a flow chart; or a form of shorthand called "pidgin" may be used with the newly developed technique of structured programming. Then

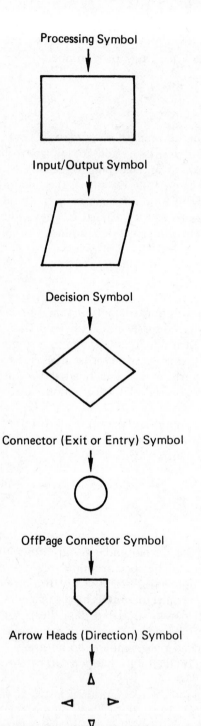

Processing Symbol

Input/Output Symbol

Decision Symbol

Connector (Exit or Entry) Symbol

OffPage Connector Symbol

Arrow Heads (Direction) Symbol

Figure 7.8 Flow chart symbols.

the detailed computer instructions are written in program segments and each section is tested as it is completed.

3.6 Testing and Debugging

The programmer works closely with the system analyst, or with the ultimate users of the new system, to be sure that the new programs are processing the unit data correctly and producing the required data on reports or other output desired. The user's help may be requested in developing adequate test data to cover every conceivable condition or error in the input data. Finding syntax or logic errors in a new program and eliminating them is called "debugging." Each program segment is tested individually as it is developed; then all the pieces are put together for final testing.

When all required programs have been written and tested and the necessary hardware has been installed, all components of the new system are put together for an integration test. The hardware, software, operating procedures, and security controls are tested for validity and reliability using test and actual or "live" data. It is very unusual for all aspects of a new system to run smoothly the first time.

The system must be debugged, and procedures and controls may need to be modified. Backup procedures (to enable the system to recover in case of a system failure or accidental loss of data files) must also be tested and any potential problems identified and corrected. Audit procedures and security controls should be checked out by persons other than those who designed or developed the system, to avoid the possibility of intentional fraud.

3.7 Pilot Test

The pilot test is the last phase of implementing a new system. This is also called conversion, because the old system or procedure is converted to the new system. There are two commonly used types of pilot tests: parallel and phase-in.

In the parallel method the old system or procedure is continued while the new system is installed and checked out. This method may be more expensive because two systems are being run, but it is usually considered the best approach. The old system is a backup to guarantee that the required reports and processes will be carried on even if the new system does not run smoothly, and the output of the two systems can be compared for completeness and accuracy. When the reliability of the new system has been demonstrated, all files and procedures are converted to the new system and the old system is eliminated.

In the phase-in method of conversion the new system is installed in segments. As each new segment or function is put in place, the corresponding function of the old system or procedure is dropped, until, eventually, the new system is in full operation and the old system has been eliminated. This method is well suited to the installation of a new computerized system to replace a manual system.

3.8   Modify/Extend System

Once the new system has been installed, users may see ways to improve or expand it. Also, computer operations personnel, programmers, or system analysts may have ideas for cutting costs or improving performance that were not obvious during the development of the system. Any new system should be formally evaluated after installation to determine whether it is performing to expectations and within anticipated costs. Any modifications or additional functions should be thoroughly analyzed to determine the potential benefits and impact on the system, the users, or related systems. If approved, the changes should be implemented in the same process of planning, development, testing, and installation used for the introduction of the original system.

Actually, any operating system should be constantly monitored to determine whether it is performing properly. New ideas for restructuring the data base, eliminating unnecessary files or reports, or adding additional capabilities should be encouraged, to maintain the system in an efficient manner and to ensure that it stays current with the needs of the users.

4   APPLICATION CONSIDERATIONS

As discussed in Section 3, the computer system analyst is usually the first person contacted when one wants to utilize a computer. This person determines whether the program is feasible and ascertains whether computer time is justifiable and available. Thus in many organizations it is necessary to justify the use of the computer with both direct management and computer operations management, since resources and time may be limited. However, certain basic questions should first be answered about the potential application to determine whether data computerization is practical and economically feasible. The "user" questions are:

1.   What do I want to accomplish?
2.   How should it be accomplished?   •
3.   How do I obtain the information?
4.   How will the data be used?

Once these questions have been answered satisfactorily, the proposed application must be examined by management, and these additional questions answered:

1.   What other methods are available to accomplish the program?
2.   Is the computer the best method?
3.   What are the advantages and disadvantages of computerization?
4.   And, most important, is there a cost savings by using the computer?

One must fully understand the purpose and functions to be accomplished before they can be explained to a system analyst who is not familiar with the industrial hygiene field. Often the scope of a new computer application is unnecessarily limited because insufficient thought was given to the functions that could be designed into the system. Some typical industrial hygiene data system responsibilities are:

1. Making available and maintaining safety data.
2. Maintaining medical/industrial hygiene data on employees and their exposures.
3. Providing selective data to medical systems and the personnel department.
4. Establishing medical/industrial hygiene requirements.
5. Maintaining company records with legally required documentation, under OSHA and Toxic Substance Regulation, at both federal and state levels.

Early system planning should also be concerned with future growth and flexibility of the system. Some considerations for future development are:

1. The system should have the ability to process the anticipated work load effectively and economically within current limitations imposed by technology.
2. It should possess the flexibility to accommodate changing requirements to meet business needs.
3. It should allow for capacity growth at little increase in cost.
4. Long-range expansion should be accommodated readily.

Appendix 7A lists typical industrial hygiene system requirements.
Answers to the following questions about the application should be helpful.

1. What do I really want to accomplish (scope and purpose)?
2. How is the concept defined, including size and complexity?
3. What information is required?
4. Who should obtain information?
5. What is the best method to input data to the computer?
6. Who is responsible for information?
7. What methods of checking validity are feasible?
8. Who needs the data?
9. How will data be used?
10. How should data be presented?

Although such information is necessary for the thought process, the system analyst will be concerned with data base design, how the data will be acquired, how managed, and finally, how presented. These aspects of a potential application are discussed below.

4.1 Data Base Design

The essence of any information system is the data base. But one of the most difficult exercises faced by the industrial hygienist and management is determining what data are to be included in the data base. Some considerations are:

1. How will data be used?
2. What format is needed?
3. How will access be obtained and protected?
4. When are data needed (is instant retrieval a requirement, or will there be lead time)?

Design aids in the form of design standards and programs can be used to improve data base quality. Data base design aids (DBDA) is a productivity tool that allows interim data design information to be catalogued and stored in a dictionarylike system for quick recall during the design phase. Such an automated analysis can reduce quite significantly the time required for generating a working, structural model. Better design quality will result, since the DBDA program performs a more thorough analysis of the data requirements than is normally possible with manual methods. This increases the likelihood of attaining a consistent and effective design.

Standardization of the new data base is achieved by using requirements from various information source applications in the same company or classification. If the personnel department records employees at work in days, and industrial hygiene documents exposures in hours, needless complexity in cross-referencing exists. DBDA greatly reduces this possibility. Careful data design is critical to future effective use. The design aid program is a software package created for use with a particular system. This package can be purchased commercially or prepared by a programmer. Use of the design aid will provide a sound data base by detecting omissions, inconsistencies, and redundancies in the data requirements.

In addition, the design aid helps to create new data bases, redesign and integrate existing data, add new applications to existing data, and add new elements or associations to existing information. During this process a structural model is created to identify human decision points. The process of designing a data base can be generally divided into the following six steps:

1. Gathering requirements.
2. Generating a structural model (organization).
3. Constructing a physical model (hardware).
4. Design evaluation.
5. Physical implementation.
6. Performance evaluation.

In summary, generating a structural model combines analysis of the data requirements of the applications and synthesis of these needs into a single network. Such a network will help the designer address the questions that arise during construction of a physical or working model.

Data base design activities normally are performed in the following order:

1. Gathering the data requirements for the data base.
2. Identifying and correcting inconsistencies, omissions, and duplications of data elements.
3. Selecting the elements and associations that must be included in the data base and those that can be derived from them.
4. Grouping elements and keys into segments.
5. Determining physical and logical relationships.
6. Arranging segments into hierarchical structures.
7. Selecting the access method.
8. Providing data elements for retrieval, insertion, update, and deletion.
9. Updating, reorganizing, and backup requirements.

To illustrate data base design considerations, the outline of a typical application could be as follows:

1. *Goal.* System will provide and assemble chemical health and safety information for all company locations. Chem-Safe.
2. *Basis.* Corporate chemical material data bank.
3. *Responsibility.* Corporate data center and procurement office.
4. *Concerns.*
   a. Standarized material identification.
   b. Hazardous ingredients.
   c. Labeling.
   d. Medical examination requirements.
   e. Workplace measurement rules.
   f. Compatibility.
   g. Ventilation.
   h. Handling and storage.
   i. Emergency control.
   j. Protective equipment.
   k. Overexposure effects.
   l. Disposal.
   m. Toxicology.
5. *Access by.*
   a. All locations—management/medical.
   b. Corporate technical review center.
6. *Data sources.*
   a. Personnel records.
   b. Medical files.
   c. Purchasing.

    d.   Manufacture engineering.
    e.   Facility engineering.
    f.   EPA and OSHA.
7. *System configuration.*
    a.   Magnetic storage.
    b.   Keyboard input.
    c.   Printer/cassette output (1) punch card option; (2) phone patch.

Starting with development of a master chemical numbering system, from uniform labels to format design, an information system of this magnitude requires a lot of basic data gathering before any electronic data processing is inolved. These steps must be in place before the system analyst and other EDP professionals are approached.

    Note. Business justification for an industrial hygiene data system can be based on the premise that many pieces of data already exist in computer format. Personnel records include job description and assigned workplace—add materials each employee worked with and air sampling data and an essential part of an industrial hygiene record system is present.

4.2 Data Acquisition Methods

For the purpose of this discussion data acquisition methods have separated into three categories: manual, automated, and sensors. Actually, sensors would fall into the automated category but because of their value to the industrial health field, a separate section is devoted to these devices.

4.2.1 Manual Methods

The cost of data entry has increased primarily because manual methods of entry involved labor. Producing faster computers and printers could not completely offset this cost. Manual transcriptive data entry involves discrete steps, such as source document handling, recording and transcribing, and transporting, which are slow. If manual methods must be used, graphic display units are in most cases the best method, because they offer the most direct entry into the computing system by use of a keyboard. This aspect is especially relevant for entry of scientific data. Because these display units can be connected remotely to the computer by means of telephone lines, the data can be entered at the point of origin, which is of value to the scientist. Another advantage is that data can be entered directly from the source document, by the person who generated the data, if necessary.

    If errors are detected in the data entry, the operator can correct them dynamically. Certain display consoles allow entry also by selector pen that enables the operator to enter several commands with one action. Once the transaction has been processed, the data in the working field are updated immediately. The computer is always current. Thus if manual methods must be used, greater system efficiency results from video input because of higher input rates, lower error rates, ease of update, and data handling.

4.2.2 Automated Methods

The automated methods have the advantage of putting the data directly into the computer. As such they are considered "real time" in computer terminology, with obvious advantages as compared to manual input. For example, a chemical operator presses a button to start a motor. The computer searches for safety conditions that must be met if the motor is to be turned on. If the conditions are met, it turns the motor on; otherwise it informs the operator by signal or printout why the motor cannot be turned on, and corrective action can be taken. A more responsive computer system might be able to correct the situation before the motor is turned on, bypassing the chemical operator.

In industrial hygiene time studies are important, especially where there is a need to calculate a time-weighted exposure. This could be done by the individual recording the time spent in an exposure area. However, this is not a dependable method because the person may not keep an accurate record. It would also be expensive to have someone assigned to this recording task. One method that has been very helpful with collection of data of this type is the badge reader. Into the door of the location is inserted a card that includes the workers identifying characteristics. The door does not open unless a card is inserted into the system. The time of entry is then recorded. When the worker leaves the area the same process is repeated and the computer calculates the time in the area. This method is very useful for keeping track of exposure times to maintenance personnel who can be found working in any part of the plant. Once the times in a specific area are known, the computer can calculate a time-weighted average (TWA) for the person based on the air sampling data that have also been programmed.

The same type of system can also be used to exclude persons from an area. That is, the door will open only when an approved badge is placed in it. Such a system is also of value in work with carcinogens, where the law requires a logging of persons entering the area. It can also be used to exclude from an area persons not having proper chemical training. Examples of sensors-based applications appear in Figure 7.9.

Another automated form of data collection makes use of machine-readable forms. It seems redundant to have to transcribe data from one form to another, or to a card to be placed in the computer. The government, as well as the medical and legal aspects of industrial health, requires significant recordkeeping. Machine-readable systems have been of value with air sampling data, analytical results, medical histories, and questionnaires of all types. The machine-readable form takes information from the source directly to the computer and eliminates much of the intermediate handling.

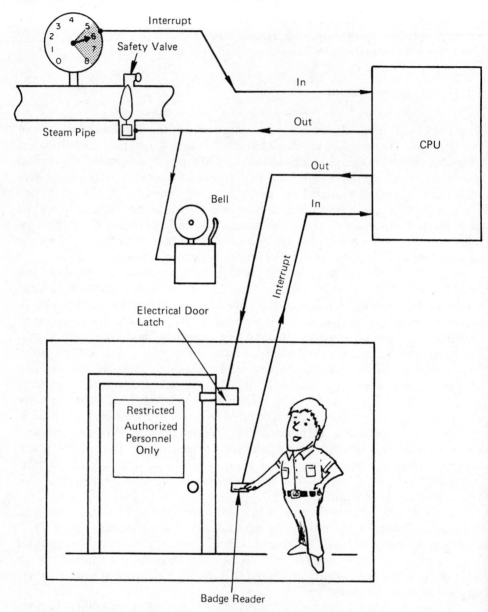

Figure 7.9 Examples of sensor-based applications.

4.2.3 Sensors

Sensors receive data from the real world. Since the sensor-based computer receives input from these sensors, it has distinct advantages over the non-sensor-based system, which receives input prepared manually for entry into the system, and the output must be interpreted manually before action can be taken. The sensor-based computer on the other hand can also send output directly through a sensing device to control a physical action. The most obvious advantage to a sensor-based system is time, since the interval between inputting the data and using the output is reduced; another advantage is accuracy. With the sensor-based computer, human intervention is eliminated at several points, thus decreasing the possibility of error. Figure 7.10 indicates the differences between sensor-based and non-sensor-based computers.

What Is A Sensor-Based Computer

One That Can Interact Directly With The Real World

The Difference

Sensor-Based	Non-Sensor-Based
● Receives Input Directly From Sensing Device	● Input Prepared Manually For Entry To The System
● Sends Output Directly Through Sensing Device To Affect A Physical Process	● Output Interpreted Manually Before It Can Be Used

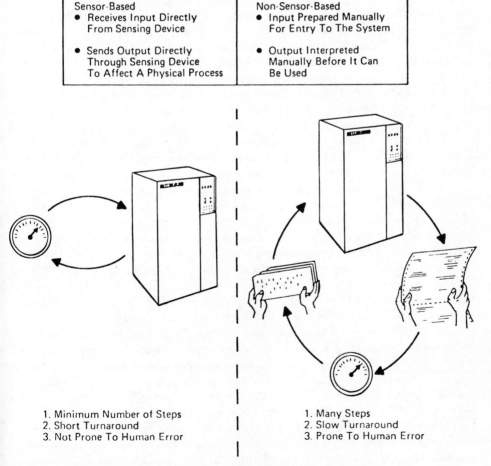

1. Minimum Number of Steps
2. Short Turnaround
3. Not Prone To Human Error

1. Many Steps
2. Slow Turnaround
3. Prone To Human Error

Figure 7.10 Differences between sensor-based and non-sensor-based computers.

Sensors in use today are able to measure temperature, pressure, force, flow, acceleration, velocity, and sound. A sensor is often called a transducer because it converts one form of energy to another. For computer usage we need to convert mechanical or thermal energy to digital electrical signals.

There are three types of sensor signal: contact, voltage, and process interrupt. For contact sense, if a switch is activated the contact closure would be sensed and interpreted by a digital input device as being a bit or no-bit condition (1 or 0). This is useful information for the computer, as described for input/output devices in Section 2.3. The same is true for voltage sense (a voltage is presented to the digital input circuits) and for process interrupt (a specific features is being measured). From a programming standpoint, one must decide when and how the digital input device should read the data.

When transferring data in the opposite direction, that is, from the CPU, the sensor-related hardware is a digital output device that can convert digital data into a voltage capable of operating various pieces of equipment. These outut devices fall into two general categories: contact output and voltage power. The contact output device can perform functions by activating a circuit to close or open a switch. Thus it provides the switching power to turn a given piece of equipment "on" or "off." Voltage power, as the word implies, has the ability to supply voltage to a given piece of equipment (e.g., to turn on a light or operate a motor). Depending on the equipment to be operated, voltage is supplied or removed by the digital output device to operate or stop the equipment.

On the other hand, analogue signals from sensors must be converted to digital input for computer usage. This is accomplished by an analogue-to-digital converter (ADC). If the output signal needs to be analogue, then a digital-to-analogue converter (DAC) is needed for the equipment to function. In the latter process the digital data are converted to a voltage, which could operate a motor or other piece of equipment as previously described.

Thermocouples, resistance thermometers, thermistors, strain gauges, and flow meters are commonly used analogue input/output sensors. The first three measure temperature and the last two measure respectively, force and flow. All these devices have been used accurately, successfully, and reliably with the computer. The thermocouple, for example, is a sensing device capable of measuring temperatures as high as 4000°F. Since it is also able to detect temperature changes, it could be coupled with other logic and parameters being measured to raise or lower the temperature as required.

Each type of sensor has advantages and disadvantages that make each one more suited for certain applications. For example, resistance thermometers are more accurate than thermocouples, but the measuring temperature range is less. Thermistors are very small, relatively inexpensive, and can be made into various shapes and sizes. However they are not suitable for all operations because they are less stable than the other devices.

There are many applications in industrial health for sensor-based computers. Section 5 has information on both sensor-based and non-sensor-based applications. Acquisition of data from instruments, processes, physiological functions, and other sources is of importance in industrial health because with the sensor-based computer, the data acquisition is direct and produces real-time information. The response to the information can also be direct if output sensors are used. The sensor-based applications can be separated in three general categories: data acquisition, control, and automation.

A sensor-based data acquisition type of system, as the name implies, is primarily concerned with the collection of information from the various sensors attached to the system. The sensor-based computer can apply some degree of control over the data acquisition system, but this control function would be secondary to the application of data acquisition. Quite often in data acquisition systems data are only collected, or only partially processed, with complete processing done at a later date. This may be necessary because most sensor-based computers are small. In the case of the data collected from an electrocardiogram, for example, the data could be interpreted by a physician, interpreted by another computer, or stored for comparison with another electrocardiogram taken from the patient at a later date.

Another example applying to a sensor-based computer data acquisition system could be water and air pollution monitoring (see also Section 5.2.4.). A computer could be attached to a wind speed guage or an overfall weir. When the sensor measured a certain pressure or flow, the water or air pollution sampling equipment would be turned on to collect data. This simple case could be controlled manually. However, the computer could monitor from 1 to 1000 sensors, predict impending pollution alerts, or operate a large number of sampling devices at very remote locations. Thus preprogrammed sampling procedures, with predetermined limits, are placed in the sensor-based computer. When the limits are met, the data acquisition process starts.

If the sensor-based computer application is for control of an operation or process, data acquisition is still important. The control application involves the acquisition of data, either analogue or digital, from a given operation. The computer processing of the data could make control correction calculations to ensure the proper functioning of the process or operation. Depending on the application for the computer and the parameters of the computer program, the computer would produce output control signals, either analogue or digital, for control of the process. In this case the task of the sensor-based computer would be to receive input from the attached sensors, analyze the data, and effect a change in the process or operation by activating one of its sensor-based output devices. The sensor-based computer has advantages over other types of systems that can accomplish the same control because it can collect data, make calculations, and initiate control in real time (i.e., during the actual time that the operation or process is transpiring).

A sensor-based control system could be employed in many industrial applications. One application receiving much attention today is the use of a sensor-based computer control system for energy management. The computer monitors sensors attached to chillers, standby boilers, air-handling systems, and cooling louvers. The computer regulates air conditioning and heat, and controls humidity and temperature by zones, using data from inside and outside weather stations. The computer can also automatically turn off lights at specific times. Although the computer has the logic for the application desired, its use should be carefully monitored where industrial health aspects are evident. For example, one would not want to use an energy management system that automatically turned off the lights in an industrial building if safety of health hazards could occur, if the lights went off while operations were still being conducted. Likewise one would not want to automatically turn fans off in laboratory hoods, or cycle fans in areas that require makeup air.

Another use for a sensor-based control system could be in controlling air and water pollution from a process. In this case direct reading water and air pollution equipment would be continuously operated. Data from the sensors in the equipment would be directed to the computer for analysis. If concentrations reached a certain predetermined level, the computer would take control over the process to ensure that concentrations did not exceed legal limits. With this type of control the process could be operated closer to the allowable pollution level, obtaining, in most cases, maximum production and also providing the capability to shut the process down during a malfunction or accidental release.

Computer-based sensor control systems are often termed "closed loop" or "open loop" systems. The air and water pollution control example just cited is a closed loop system because the computer obtains all the data, makes calculations, and produces control of the operation. It is faster than the open loop, which requires manual intervention. If delays of several minutes can be tolerated by the desired application, or if direct sensor-based control of the operation is not practical, the open loop method would have to be used. In this case the computer acquires the data, performs complex calculations, and gives the operator an audible or visual indication that something has happened. It is then up to the operator to make adjustments to the process.

Given rising costs, it is important to bear in mind that use of the computer should result in a manpower savings, since one computer could be used to control any number of operations throughout a plant. From the industrial hygiene standpoint, a chemical operator in a computer control center will not receive a significant exposure as compared to the number of persons that would have to work near the operation, with the resultant high exposure in a manually controlled situation. Thus sensor-based computer applications result in more effective use of personnel, quicker diagnosis of operation changes, and concise presentation of process status, and exposure of personnel to hazardous materials is reduced, as well.

The final type of sensor-based computer application, automation, is a combination of data acquisition and process control. However, in automation the need for data acquisition perhaps exceeds the need for process or operational control. Automation is concerned with monitoring individual items; determining how they operate and what is happening as they move from one step to another, testing each step to ensure proper operation in an expeditious manner, and collecting the resulting data. Not only is the information obtained of value to the operator of the equipment in regard to timely corrective measures, periodic exception reports, and summary reports, but the information supplied to management on performance of personnel and equipment allows management to make corrective decisions.

For example, an automated system would be of value in an industrial health application involving the attachment of sensors to laboratory equipment. Some of the incentives for the laboratory to use a sensor-based automation computer system are precise timing and control of equipment, buffering of large quantities of data, validity checking, simulation, decreased response time, and better utilization of personnel. More detailed aspects of laboratory automation, as well as expanding the computer system for information reporting, are discussed in Sections 5.2.3 and 5.2.4.

4.3 Data Management

This section instructs potential computer users in the concepts of data management as related to information processing. Data management is related to the control, retrieval, and storage of information to be processed by a computer.

4.3.1 Control

To make the process of "data management control" understandable, the relationship between user and information system must be defined in computer terminology. A computer information system provides information to a user, on request, by processing the data through the computer. Therefore a user of an information can obtain data from the computer. Control is the authorization and supervision of the data management process. Authorization is the validation of a user's right to access (read) the data. A higher level of user authorization may permit the user to alter (update) the data. User rights are established by the person responsible for the program. A user is never allowed to alter the system without permission of the person responsible for the program. The data manager for the program is responsible for monitoring the location of the information ensuring against data loss (data integrity), and ensuring that the information in the system is current. On the other hand, the person responsible for the program must ensure that the data placed into the computer are valid.

4.3.2 Data Retrieval

Before we can discuss data retrieval, or for that matter data storage as related to management, it is important to examine information and how it is represented and used in an information processing system. Information is ideas and facts about things, people, places, chemicals, operations, machines, equipment, and so on. Each of these ideas or facts is referred to as an entity.

The user of an information system wishes to know about certain entities.

Some facts may be off-limits to certain users. This is very important in the field of industrial health, since some facts, such as personal medical data, should be retrieved only by authorized persons. To retrieve the facts needed, an authorized user must define the entities of interest and supply for each entity a list of information attributes.

For example, an industrial physician is concerned about an embryotoxin being used in the plant; the entities of interest are female employees exposed to the chemical "DMAC." The information attributes would be defined as:

1. Name.
2. Sex.
3. Department.
4. Employee Number.
5. Chemical.
6. Exposure level.

The list of information attributes for each entity is a logical record. A logical record may have meaning only to the person using the system to obtain information. For the information to be obtained by the user, it must be programmed so that the entities, as well as the information attributes, can be retrieved, and only that information. In this case, for example, the physician might not be authorized to obtain salary data for the female employees.

Thus it is the function of data management to build meaningful information by bringing together the proper context, data, and data representation as well as user control. The retrieval function is the process of locating, structuring, and ordering the information in the system for the user. Locating information involves determining what data are needed and where they may be found. If the information is not in a form suitable to the user, it must be structured to meet the user's needs and perhaps ordered in a different sequence.

4.3.3 Storage

The last function of data management is storage, which is the technique for representing the information both logically and physically on a storage device such as a disk, tape, or punched card. It also includes the order in which the information is stored. Consideration must also be given to the way it may be physically accessed or addressed as well as the physical method of representing the data. Depending on usage, certain data may be placed in the computer main storage for quick recall rather than being stored outside the computer. Decisions concerning control, retrieval, and storage of data are made by the computer programmer and data manager and may be modified, as the need arises, to provide the best possible data management for the program.

4.3.4 Security/Privacy

Protection of data in the computer is as important as the reliability of the data placed in the computer. The security aspect of this is the necessary protection of unwanted use of the data, whereas personal data protection is concerned primarily with the protection of the individuals privacy related to the data placed in the computer. Since 1981 data protection laws have been passed in Austria, Canada, Denmark, France, Germany, Luxemburg, Norway, Sweden, and the United States. The number of countries as well as the requirements are expected to increase. Regardless of the legal requirements, it is advisable that your security/privacy program be extended not only to your employees but also to customers and suppliers of data to you. Such information in the computer can also be considered a company "asset", since it may contain proprietary information and confidential manufacturing information that should be protected.

To protect the information, certain steps must be taken. In this regard you should establish for your computer programs a *security management program, personal data protection program,* and *records management program.* As a responsible data manager of computer programs you have the requirement to protect *personnel and company "assets",* protect *sensitive business information,* and *comply with legislation.* The security portion refers to the necessity to put programs and procedures in place to provide methods for the protection of these assets and resources. Security/privacy related to record management means that you should provide standards and operating requirements for defining and classifying the data and its subsequent usage.

To be effective, in this very important phase of computer data management, it is necessary to ensure that all users, owners, and suppliers of data are aware of their responsibility to the data and the issues that require the protection of the data. The primary "owners" responsibility for the data processing information is to classify the data related to the value, kind, and type of data as well as its importance. Once this is established the owner of the data must ensure that the users and suppliers of the data understand and comply with the preestablished criteria for that type of data. The owner must also be responsible for authorizing access to any requesting users. The users and suppliers of data are primarily responsible for knowing, administering, and complying with the owner's established policy for the data. No computer data management method in the security/privacy area should be considered a foolproof system. It is therefore very important that each case be considered separately and all possible means be used to prevent unauthorized access, disclosure, modification or loss of data that could be detrimental to the individual's privacy or company's "assets."

4.4 Data Presentation Methods

Any computerized system produces a significant amount of data. For example, the monitoring of radioactivity may involve the collection of hundreds of samples weekly, which could be checked for the alpha, beta, and gamma activity by a scintillation counter. The results, such as sample number and total counts, are placed into the computer, which sorts and edits the data, calculates counts per minutes, arranges the data accordingly, and sends the data to the laboratory for further editing. It is essential that the computer do as much work as possible rather than furnish an unorganized mass of data that must be worked manually to obtain desirable results.

Data screening, reduction, and analysis by the computer are factors that must be considered in the presentation of information. An unorganized array of data, which is possible because of the mass of information that can be retrieved from the computer, may mean little or nothing unless it can be put into a compact and reviewable form. Methods of screening the data must be employed so that useful information can be obtained to convey some notion of the nature and dimensions of the entire aggregate of data. For example, controls in industrial hygiene are very expensive. Management responsible for approving the expenditures for controls does not want to review a mass of unorganized data. Management is usually interested in the "bottom line" or net results. Presenting the data properly is important and can augment the point to be made. This can be done effectively with the computer, thus saving time in analysis of the data.

The most frequently used method of screening is to arrange the data into a statistical distribution. A person desiring to set up a computer system should be thoroughly versed in statistics because such methods of analysis can be programmed into the computer. In a well-defined data screening operation the computer can not only compute mean, median, standard deviation, and so on, but also can point out items with missing data. Statistical analysis can also be performed on data entered in the computer in a completely random design (randomized block design) or in a Latin square design.

Screening is very important when collecting physiological data. Even a preliminary screening often can reduce the number of computing problems attributable to poor or missing data.

The averaging technique is another method of screening data. For example, a large number of bioelectric responses from a repetitive stimulus can be averaged together to reduce or eliminate random errors. A filtering technique can be used also to eliminate frequencies outside a specified bandwidth. By applying these techniques to data files before sophisticated analysis is undertaken, time and resources and be saved and meaningful information produced.

4.5   Responsibilities of an Industrial Health Data System

What should an industrial hygiene data system be capable of doing? Let us first review some typical responsibilities:

1.   Making available and maintaining material safety data.
2.   Maintaining medical/industrial hygiene data on employees and their exposures.
3.   Providing selective data to medical systems and personnel.
4.   Establishing medical/industrial hygiene requirements.
5.   Keeping corporate records with required legal documentation, under OSHA and Toxic Substance Regulation at both federal and state levels.

Requirements for future development include the following:

   1.   The system should have the ability to process the known and to anticipate current work load effectively and economically within limitations imposed by technology.
   2.   It should possess the flexibility to accommodate changing requirements to meet business needs.
   3.   Small increases in cost should suffice to secure necessary capacity growth.
   4.   Long-range expansion should be accommodated readily.

These responsibilities are user oriented and quite obvious to the experienced system analyst. The subtle needs of a system are more difficult to ensure.

System security is a great concern. Aside from the conflicts that arise when personal data are divulged to an unauthorized source, think for a moment what would happen if a competitor were to have access to supplier contracts. Data security is complex and should be part of the basic programming effort. Only those with a valid "need to know" should be able to access the system.

Both the programmer and user have a severe responsibility to keep the confidence of those whose data are recorded in the system by ensuring that only authorized persons can retrieve such data. Anything less would be a breach of the trust and confidence placed in the occupational health specialists and the supporting data processing team. For example, OSHA records for annual lost time, injuries, and sickness do not need to include details of employee name (as long as management can, if necessary, obtain the information).

Accuracy and timeliness are probably the most important systems functions. Why get data fast by automation if they are not up-to-date and accurate? Simplicity should also be a goal of an effective system. The more complex an information system becomes, the more restrictive it is to its users.

Documented programming is essential to the longevity of an information data base. People come and go, but the effectiveness of the system should not be dependent on the skills of specific individuals. Standard note taking and the programming instruction should be understandable by any successive series of programmers. A test of this is the ability to duplicate the program by those who have never seen the original package. An additional safeguard is the locked-in aspect of critical data. A power failure, user mistake, or programming error should not be able to cause a major loss to the system. Depending on the sophistication of the data, dedicated phone lines, codes, and ciphers may be necessary to assure data integrity.

5   DATA BASE APPLICATIONS AND CONCEPTS

5.1   Example of a Single Data Base System

Ventilation is a primary method in industrial hygiene for controlling exposures to chemicals. The following example of a computer application involves a single program for the testing of ventilation systems. The same principles and adaptations, however, could be used for other control systems to prevent excessive exposures to employees.

5.1.1   Establishing a Ventilation Testing Program

The example data base system, designed and installed at an IBM Corporation manufacturing facility in Rochester, Minnesota, has been found to be very effective in conducting an exhaust ventilation testing program. Testing and inspection of the ventilation systems are performed manually with ventilation measuring instrumentation. The data are recorded in the computer for recall as required. This simple method of recordkeeping is used to document, in compliance with government regulations, that the system was tested quarterly and within 5 working days of any process or ventilation change that might affect workplace employee exposure. Data placards are placed on the ventilation system to indicate that the system was tested and meets specifications. The ventilation readings and the next inspection date are also recorded. This gives the user of the system and line management a visual record of the information being placed in the computer system.

Although this relatively simple recordkeeping function of the computer could have been performed manually, the use of the computer has advantages. It can produce, in a short interval of time, a clear and concise report of all the exhaust systems at the location. It can also produce historical facts about any single system. Most important, the manager can be assured that a system will receive its required testing. The computer notifies Maintenance when testing and inspection are required, and it prints delinquency notifications concerning any system not inspected until the work is completed.

A more advantageous type of testing would be a system that continually monitors the operation and function of the environmental controls. This can best be accomplished by use of the sensor-based computer. Gauges can be placed in the ventilation system and coupled to an on-line computer for sensor-based monitoring or control. If the computer cannot correct a fault condition by an output sensor response, it notifies Maintenance that abnormal conditions or functions exist, to permit corrective action to be taken. As with the less complex system just described, the computer records data for legal purposes and information regarding system efficiency.

In automating the recordkeeping and reporting functions for the manual measurement and control system described above, Ronald Long and William Karoly (at IBM's Rochester, Minnesota location) had to devote much time and energy to achieve the results they wanted. Liaison with the system analyst, the computer programmer, the engineering and maintenance departments, and line management was necessary to ensure that each understood its responsibilities. Then a procedure was written. Appendix 7B gives portions of this procedure to demonstrate the complexity of putting together the computer and noncomputer procedures for a simple ventilation testing program.

5.1.2 Workings of the System

In this program the computer is utilized to notify Maintenance when a particular ventilation unit requires inspection and measurements. The measurements are made manually. For the system to work, Environmental Safety and Health, Facility Engineering, Facility Maintenance, and the using line management at the location must know and complete their responsibilities as required in the program. The computer is programmed to carry out its responsibilities and functions. For example, Facility Maintenance must be trained in proper air measurement technique and in the evaluation of each system. If this department fails to measure or send the data to the computer for recording, the computer report received by Environmental Safety and Health will so indicate, and corrective action can be initiated.

The action sequence for the entire program appears in Figure 7.11, that is, the action to be taken and what is to be done when the system does not meet requirements. These actions are not computer-required responses. The computer printout of the ventilation data work completed is sent to the industrial hygienist on a monthly basis for review. The computer printout (Figure 7.12) is interpreted in Appendix 7B. The industrial hygienist also receives a weekly computer printout indicating the ventilation systems that did not meet requirements, the reasons for the discrepancies, and the action that is being taken. With this system the industrial hygienist has an up-to-date record of all plant ventilation systems and can do whatever is deemed necessary to control conditions.

5.2 Other Industrial Health Applications and Concepts

5.2.1 Information Search and Recovery

A very important facet in conducting any type of health-related work is reliance on information. Information and the communication of the information to required areas is the backbone of any industrial health program. For the program to be effective, the information must be obtained in an expedient manner.

As industrial hygienists with responsibility for multiple locations, we receive calls daily from industrial hygienist colleagues, physicians, engineers, and others, asking such questions as: What is the compatibility of acetone and formaldehyde? Is the chemical "MDS" a carcinogen? Should pregnant women be exposed to lead? We have switched from using the chemical "TDI" to "MDI," the toxicology is similar—should the medical surveillance be the same? What are the analysis and sampling procedures for chlorobenzene?

In some cases the answer is available in a standard text and the necessary facts can be recovered quickly. With more complex questions textbooks are of little help and may be outdated. Perhaps the best method for information search and recovery for "routines" is a computer, to provide a viable response

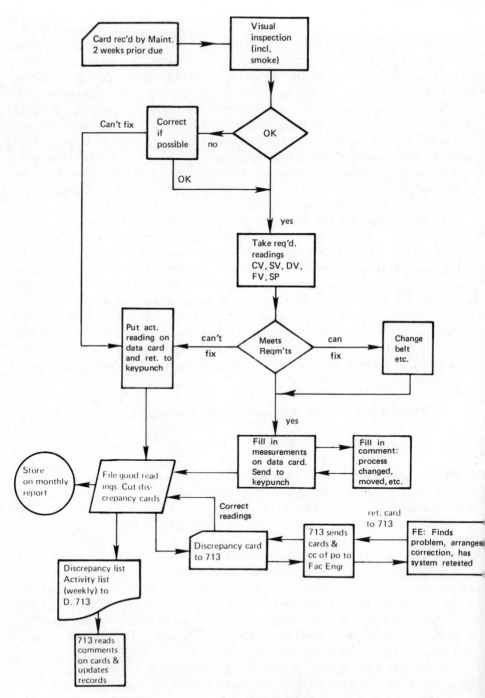

Figure 7.11 Ventilation testing action sequence.

VENT HOOD	STK	FAN	OPERATION	MATERIALS AND CONDITIONS	OSHA CLASS HAZ/LEVEL	TYPE	L1	L2	W1	W2	FACE/SLOT D	TANK DUCT A1	A2	A3	MEASUREMENTS	MIN VEL	AIR AVE VEL	FLOW VOL	ST PS	FQ/ DATE
106036	26	000	HOT WATER RINSE	H2O 130F		SV	30	27	3	24	0000	63	0000	.000	REQUIRED	900	1000	630	0	3
												.00	.00	.000	MEASUREMENTS	1500	1660	1045	0	780309
															MEASUREMENTS	1150	1200	756	0	771212
106037	26	000	CUPOSIT CU 9F	928502514 RT	CORR	SV	31	27	3	24	0000	65	454	0000	REQUIRED	400	500	325	0	3
															MEASUREMENTS	1800	1900	1235	0	780309
															MEASUREMENTS	1150	1200	780	0	771212
106038	26	000	IMMERSION TIN	915101204	C3 CORR	SV	30	27	3	24	0000	63	454	0000	REQUIRED	400	500	315	0	3
															MEASUREMENTS	1600	1750	1102	0	780309
															MEASUREMENTS	1150	1200	756	0	771212
106039	26	000	CUPOSIT 19	93440044 RT	C4 CORR	SV	29	27	3	24	0000	57	454	0000	REQUIRED	400	500	285	0	3
															MEASUREMENTS	1700	1800	1026	0	780309
															MEASUREMENTS	1150	1200	684	0	771212
106040	26	000	HCL TANK	922012555 DRT	C4 CORR	SV	32	27	3	24	0000	67	454	0000	REQUIRED	900	1000	670	0	3
															MEASUREMENTS	1800	2000	1340	0	780309
															MEASUREMENTS	1150	1200	804	0	771212
106041	26	000	ELESS CU	910010013 RT	TOXC	SV	60	0000	3	0000	0000	126	0000	0000	REQUIRED	900	1000	1260	0	3
															MEASUREMENTS	1500	1625	2047	0	780309
															MEASUREMENTS	1150	1200	1512	0	771212
106042	26	000	NEUTRA CLEAN 7	928118105	D4 CORR	SV	30	0000	3	0000	0000	63	0000	0000	REQUIRED	400	500	315	0	3
															MEASUREMENTS	2000	2300	1449	0	780309
															MEASUREMENTS	1150	1200	756	0	771212
106043	26	000	SULFURIC ACID	922017734 DRT	B4 CORR	SV	30	0000	3	0000	0000	63	0000	0000	REQUIRED	400	500	315	0	3
															MEASUREMENTS	1700	1850	1165	0	780309
															MEASUREMENTS	1150	1200	756	0	771212
106044	26	000	CU PYROPHOSPHAT	924296806 120F	C3 CORR	FV	72	0000	36	0000	0000	1800	0000	0000	REQUIRED	250	300	5400	0	3
															MEASUREMENTS	50	78	1404	0	780309
															MEASUREMENTS	250	300	6042	0	771212

Figure 7.12 Ventilation system inspection computer printout.

time. For example, why not have a computer list of all chemicals that are carcinogenic, mutagenic, or embryotoxic? Likewise, why not have the medical surveillance information for all chemicals computerized? The sampling and analysis procedures that may be needed by chemists or industrial hygienists also can be reduced to program form. The use of computer information system also assists in maintaining uniformity between locations.

The computer files just discussed are rather simple and could be undertaken by the responsible person programming the data on a portable computer. Updates to the system could also be made by the coordinator in addition to sending the hard copy to the requestor of information. If the requests are too frequent, this information system would occupy a significant portion of the person's time, and it becomes advisable to have the responsible persons separated from the request for information. This could be accomplished by providing the ancillary locations with cathode ray terminals for on-line retrieval of information.

An advantage of computerized information is that the information is always current. With the computer, because the obsolete data are removed, an update is available to everyone immediately.

It is also obvious that when such electronic files of information become very large, the number of users or frequency of use must increase for the computer to be competitive with other methods. In industrial health practice also, significant data are stored for medical/legal reasons. This reduces storage cost when compared with bulk storage of paper. The retrieval of computerized information is of course more efficient. The factors that make electronic files of information worthwhile are implicit in the following questions: Is the real-time response needed? Are labor costs with other methods prohibitive? Are other users available to reduce the cost of the computer usage?

An example of a computerized information data base is the computer-derived exposure list of common contact dermatitis antigens (1). By use of this system, information can be obtained concerning product ingredients. Although its original concept was to supply information to dermatologists, the system would be of value to industrial physicians and hygienists.

When programming information systems, ease of user interaction for the data must be addressed. Therefore it is important to select the "best" key words and synonyms for retrieval purpose. Information systems that involve chemicals can become quite complex. Programming systems of single ingredients are not too complex provided all generic and trade names are programmed for the user. Mixtures with chemicals in various proportions increase the programming complexity as well as the ability to retrieve the information the user wants. Another factor to be considered is the loss or inadequacy of data. This results because many manufacturers will not give complete chemical ingredients to protect proprietary information. New legislation may resolve some of these problems.

Group usage definitely brings the cost of computerization to an acceptable level in relationship to the benefit derived. Bibliographic research is an area where computerization is of extreme value. Expanding technology produces myriads of information for publication. Related to any field of interest are many journals and government, industry, and academic publications, as well as proceedings from technical symposia. It is physically impossible to keep current manually without a large budget, time, and manpower. Even after the retrieval of the articles, time must be spent to read as well as classify information. Many are of little use and may be outdated. Thus it is advantageous to have access to a system that has computerized bibliographies of interest.

The computer can deal effectively with the foregoing type of information and data handling if it has been adequately formalized for this special usage. The advantages of computerization are better collection, storage, retrieval, and tabulation. Once the data have been programmed, the computer is able to search through abstracts containing a key work—for example, "epoxies." One computer may have more information than most large libraries and can retrieve information faster than the most efficient noncomputerized library. The computer relieves much of the administrative and clerical burden and also reduces errors.

Because of the need for quick and efficient bibliographical research, many organizations are programming such information on computers, and technical information retrieval centers are becoming quite commonplace in very large organizations. For example, chemical information from the widely used chemical abstracts of the American Chemical Society are now on a computer data base. The National Library of Medicine has medical and toxicology information in the computer data base called "Toxicon." IBM and many other companies have computer files and abstracted bibliographies for their own internal technical activity. In this way technical and professional personnel can keep up with the state of the art more quickly and economically. To recover the cost of programming, some companies that have established large electronic libraries are offering access to these facilities on a fee basis.

Computerized technical information retrieval centers usually operate under the following format. Descriptions of documents are keyed into machine-readable form. The summaries include title, author, and other bibliographic information. In most cases the abstract text is the same as written by the author. A word text searching technique is usually used for retrieval and dissemination of information. The computer searches every word of input from title through bibliographic data, including index terms in addition to the complex text of the abstract. As a result, searching strategy is extremely flexible and accurate. The "answer" to an information search may be a complete bibliography on the effects of lead poisoning, or the effects on the central nervous system alone. Likewise, one can search for only a single reference.

One of the principal advantages of electronic libraries is that a standing profile can be left in the system by a user. Each time new data on a subject (e.g., benzene) are placed in the data files, the subscriber automatically receives the reports. Figure 7.13 is an example of a computer printout from the IBM Technical Information Retrieval Center. The subscriber's key indicators, listed

DOCUMENT NUMBER: HHH76H002292

TITLE   PB-245 851/1SL NIOSH ANALYTICAL METHODS FOR SET B STANDARDS
COMPLETION PROGRAM. OCT 75.

SEQNO   76H 02292

AUTHOR   STANFORD RESEARCH INST., MENLO PARK, CALIF.*NATIONAL INST. FOR
OCCUPATIONAL SAFETY AND HEALTH, CINCINNATI, OHIO.

SOURCE   PB-245 851/1SL
NIOSH-SCP-B

ABSTRACT   53P. INDUSTRIAL HYGIENE SAMPLING AND ANALYTICAL MONITORING
METHODS VALIDATED UNDER THE JOINT NIOSH/OSHA STANDARDS COMPLETION
PROGRAM FOR SET B ARE CONTAINED HEREIN. MONITORING METHODS FOR THE
FOLLOWING COMPOUNDS ARE INCLUDED: CAMPHOR; MESITYL OXIDE;
5-METHYL-3-HEPTANONE; ETHYL BUTYL KETONE; METHYL (N-AMYL) KETONE;
AND OZONE.

SUBJECT   CHEMICAL ANALYSIS   GAS ANALYSIS   GAS SAMPLING
INDUSTRIAL HYGIENE
OCCUPATIONAL SAFETY AND HEALTH

COSATI   FIELD 07D, 06J, 99A, 57U, 68G

Figure 7.13 IBM Technical Information Retrieval Center abstract.

at the bottom of the abstract, are chemical analysis, gas analysis, gas sampling, industrial hygiene, and occupational safety and health. Any new abstracts of articles containing those key words are forwarded to the subscriber monthly.

Table 7.1 lists data that were collected from two computerized technical libraries. The main purpose of the data gathering was to discover industrial health applications using the computer. The key word in the information search was "computers." Since this would bring information from many sources, modifier words had to be projected for retrieval purposes, to reduce the number of publications that the computer would print and the user would have to review. The computer was instructed by the library specialist to only print abstracts that also contained the words "health and safety," "air pollution," "chemicals," and so on.

The computer "run" produced more than 2000 abstracts. Since Table 7.1 contains approximately 200 references, it is seen that only 10 percent of the abstracts were of value. Because of programming, the computer was not able to break the data into the categories listed in the table. This was accomplished manually by reading each abstract and in many cases going to the library to obtain the article for further information. If the interest in computer applications involved only industrial hygiene sampling, the references as listed in the table would provide such information. The same would be true for any of the other categories.

If the computer had been programmed specifically to retrieve the word headings in the table, the task of producing the table would have been less time-consuming. This points out the need for proper programming and search methods. For example, in the case of audiometry it would have been necessary to review 2000 abstracts to find the five references. Because some of the reference searching is complex, many of the computer libraries have cathode ray terminals for direct user interaction with the information. With this method the user can directly modify the key search words to obtain the information desired.

Table 7.2 lists data bases available from external sources. The data bases described represent those commercially available that are most vital for occupational health purposes and have the largest potential user audience within this profession. However, they constitute just the "tips of the iceberg." For example, there is an "on-line data base" called EVENTS. This data base is "current" and contains information on technical conferences, meetings, courses, and other events significant to the technical professionals. Thus the creation of unique on-line data bases in many disciplines and for informational requirements is a fast-growing activity. There are now literally hundreds of them for your use, available from a growing number of on-line vendors. The table is just an important sampling of those available and required by your profession.

Table 7.1   A Bibliography of Computer Applications Arranged by Various Subjects

General Category	Subject	Reference
Industrial hygiene	Information systems	1, 4, 5, 13–20, 208, 210–213
	Sampling	2–12, 88, 208, 209
	Noise	21–28
	Ventilation/energy	29–36
	Heat	36–40, 215, 235
Medical	Information systems	1, 21, 61–69, 110, 111, 189, 208, 210–213, 216–219, 230–232
	Diagnosis	41–60, 70, 75, 86
	Radiology	50, 53, 55, 70–76, 103
	Heart	72, 77–86, 104
	Lung	47, 51–53, 55, 56, 69, 79, 87–108, 193, 194, 220–222
	Electroencephalogram	42, 109
	Audiometry	118–122
	Multiphasic health screening	63, 64, 66, 112–117
Analytical	Spectrograph	123–131, 134, 135, 137
	Microscopy	9, 10, 132, 154, 158
	Gas chromatograms	11, 69, 133–145, 223, 224
	Other methods	9, 67, 73, 95, 126, 146–156, 229
Miscellaneous	Safety	159–166, 214, 225–228
	Environmental	3, 37, 128, 155, 167–176
	Statistics	48, 68, 176, 177
	Toxicology/research	1, 38, 42, 87, 95, 178–186
	Epidemiology	18, 65, 68, 187–194
	Bioengineering	38, 47, 49, 202–204
	Training	51, 174, 205–207, 257
	Health physics	20, 55, 73, 88, 91 92, 175, 195–201, 235

Table 7.2 Data Bases Available from External Sources

Data Base	Subject Coverage	Producer
Biosis	Worldwide coverage of research in the life sciences	BioSciences Information Services
CA Search	Chemistry; bibliographic information appearing in Chemical Abstracts	Chemical Abstracts Service
Chemline	Dictionary File; 500,000 chemical records	National Library of Medicine
Computer and Control Abstracts	Worldwide coverage of computer and control engineering	Institution of Electrical Engineers, London
Energyline	Scientific, technical, socioeconomic, government policy, and current affairs aspects of energy	Environment Information Center, Inc.
Enviroline	Interdisciplinary approach to environmental information; science, technology, politics, sociology, commerce, law	Environment Information Center, Inc.
Environmental Impact Statements	Environmental issues including air transportation, energy, urban and social programs, wastes and water	Information Resources Press
Physics Abstracts	Worldwide coverage of physics literature	Institution of Electrical Engineers, London
Product Specifications	Indexes to vendor product and component specifications (as needed by purchasing departments)	Information Handling Services
Psychological Abstracts	Citations and abstracts to literature in psychology and behavioral sciences	American Psychological Association
RTECS—"Registry of Toxic Effects of Chemicals"	50,000 citations of potentially toxic chemicals compiled by NIOSH	National Library of Medicine
SCISEARCH	Multidisciplinary data base that indexes literature of science and technology	Institute for Scientific Information

Table 7.2 (*Continued*)

Data Base	Subject Coverage	Producer
SPIN	Most current coverage of a selected set of the world's significant literature on physics	American Institute of Physics
TBD—"Toxicology Data Bank"	Database composed of approximately 4,000 peer reviewed, chemical records	National Library of Medicine
Toxiline	Bibliographic database covering the pharmacological, biochemical, physiological, and toxicological effects of drugs and chemicals. 1,500,000 citations, almost all with abstracts	National Library of Medicine
U.S. Statistical Abstracts	Abstracts of published forecasts for the U.S. from journals, reports, special studies	Predicasts, Inc.
U.S. patents	All patents registered through U.S. Patent Office	Pergamon International Information Corp.

Requests for searches of these external data bases can be handled by some of the major libraries. The library personnel should be able to recommend which of the many available data bases are appropriate to search for your particular questions. Where frequent use of a specific data base is required for your informational needs, it may be advantageous as well as a cost and time savings necessity to purchase direct or indirect access to the data base.

5.2.2 Physiological Monitoring

Conducting intensive physiological monitoring for patient care or research can be tedious and time-consuming. The physiological monitoring also requires highly trained personnel, which also results in significant cost. Many of the physiological data being collected are redundant or irrelevant. On the other hand it may be difficult to tell which variables are significant without measuring and studying all the information. This results in the collection of a significant amount of data to find useful information. Likewise, to produce effective research or to make valid patient diagnosis, all the short and long-term trends of the data collected must be studied and compared. In most cases this requires complex analysis and calculations involving several related or secondary factors.

Physiological monitoring, therefore, is amenable to computerization, especially, if real-time information or analysis is needed.

Since physiological systems usually emit bioelectric impulses of an analogue or continuous nature that can be measured and quantified into digital or discrete form for computer processing, the computer, which offers the advantages of accuracy, unlimited storage, mathematical computation, and logic capabilities, is suitable for physiological monitoring. The data can be stored for long periods so that trends can be studied. The data can also be compressed, ensuring that only relevant information is stored. When data acquisition is on-line, by positioned sensors on the subject, real-time or extended-period measurements of physiological parameters can be made. In addition, alarms or notification can be based on the immediate situation (e.g., ventricular fibrillation, increased blood pressure, decreased respiration). Another use is for the analysis of special tests such as electroencephalogram or electrocardiograms.

The rapidly advancing state of sensor development now makes available a wide selection of accurate and reliable devices. These can be transducers, thermistors, or an array of electrodes and other components, which can present the variable being measured in the form of an electrical signal that is proportioned to the value of the variable. The choice of the sensor is important. As with any detection system, the device should not adversely affect the signal or variable being measured. In addition, it should introduce little or no distortion, be relatively insensitive to movement, and offer minimum restriction to activity and comfort.

The nervous, cardiovascular, respiratory, and digestive systems are areas in which data acquisition by computers can be accomplished. Table 7.3 breaks down by system the typical physiological measurements that can be made. For example, in cardiovascular research blood pressure can be measured by placing a needle in a subject's artery and recording the data on analogue tape. Selected portions of the tape can be placed in an analogue computer to convert the data to digital form for transmittal to a digital computer. Upon receipt, calculations are made, as well as comparisons with stored mathematical models. Once normal conditions have been projected, experimental conditions can be undertaken. An electroencephalogram or other physiological system could be tied into the same program if these parameters are related and must also be measured and analyzed.

Complex mathematical models of physiologic systems have also been developed using the computer. The general procedure is to develop a set of different equations that describe the proposed model. By using the computer these equations can be solved, and least-square techniques (to determine the unknown variables and constraints in the equations) allow testing of the validity of the model. For example, a model of lung clearance could be developed and tested for a dust or radioisotope. The area in which the most progress has been made with computers is radionuclide scanning of organs. Scanning programs using isotopes and computers are available for almost all organs.

**Table 7.3 Typical Physiological Measurements for
Computerization**

Cardiovascular system
  Electrocardiogram
  Phonocardiogram
  Arterial pulse
  Venous pulse
  Blood counts
  Peripheral blood flow
Digestive system
  ph
  Motility
  Electrogastrogram
Nervous system
  Electroencephalogram
  Electromyogram
  Sensory signals
Respiratory system
  Rate and/or depth of breathing
  Oxygen, carbon dioxide, and pH levels in the
    blood
  Lung volumes

5.2.3 Laboratory Analysis, Control, and Information Systems

Automated analytical instruments, virtually nonexistent a few years ago, are
becoming commonplace in the laboratory. Since the output of this type of
equipment is an electrical signal, it can be accepted in either analogue and
digital form by a sensor-based computer. Sensors are discussed in Section 4.

The continuous flow and multiple channel types of laboratory analyzer are
very suitable for computer analysis. For reasons already given, the computer
has demonstrated its capability to automatically acquire data from a variety of
instruments, to perform analysis, and to make calculations without error. In
addition, since the computer can present data in myriad ways, it lends itself to
a total laboratory information system. Data presentation methods, which include
screening, reduction, and analysis, are given in Section 4.4. Laboratory instrument analysis data that are not available for direct computer input by way of
sensors can be processed by batch methods of data handling. Specifically, many
instruments, such as colorimeters, chromatographs, and spectrophotometers
can be computerized by the on-line sensor method. A bibliography of application
examples and uses can be found in Table 7.1. The gas chromatograph and its
subsequent advantages for computerization to the analyst are briefly discussed
here. The reader can make the same analogies to other laboratory instrument
and equipment types. For example, since the introduction of gas chromato-

graphs to the laboratory more than 20 years ago, these instruments have continually undergone redesign and improvements (new and more sensitive detectors, columns, materials, and liquid phases, etc.).

Perhaps one of the more important reasons for computerization is that analysis of chromatographic data, because of the new technology, can become increasingly complex and time-consuming. Consider the drifting of the baseline or the occurrence of several peaks. Although the new developments have resulted in better information, laboratory managers find that output has decreased. Thus in the face of increased cost per sample, computerization is the best solution, especially in a laboratory with a high volume of sample analysis.

The major benefits of computerizing gas chromatographic analysis are: faster results, increased reproducibility, greater accuracy, increased production, increased utilization of equipment, and better utilization of analyst's time as well as complete control over all functions. Some of the computer equipment available today has the ability to handle up to 20 chromatographs simultaneously. As previously stated most of the advantages above can also be related to other laboratory equipment, for example, clinical laboratory instruments.

Once the laboratory equipment has been computerized, the output of data from the laboratory will increase. Even in the most advanced noncomputerized laboratory, the analyst is still bogged down with paperwork for a significant portion of time. The most efficient utilization of both equipment and laboratory personnel is an on-line data acquisition and analysis system with information reporting. Where complete automation cannot be accomplished, data from other instruments can be introduced into the computer system by way of batch processing (Section 2). This enables all laboratory instrumentation to take advantage of the information reporting portion of the computer program. Input into the information system for non-on-line instruments can be accomplished semidirectly by way of keyboard input or other devices, also mentioned in Section 2.

For programming purposes and to obtain a better view of the entire process, complete laboratory computerization can be separated into the following steps or tasks:

1. Instruments to be completely automated.
2. Instruments requiring batch processing.
3. Entry of analytical request.
4. Development of master log.
5. Generation of work list.
6. Automatic instrument reading and data logging.
7. Entry of test results—automatic and manual.
8. Mathematical computation and analysis.
9. Verification of instrument performance and quality control.
10. Verification of test results.
11. Reporting of test results.

12.   Production of statistical information used for laboratory management control.

13.   Storage of data for future recall.

The overall computer system of analysis and control as well as information can be viewed diagrammatively (Figure 7.14); its overall advantages are increased analyst production, improved control, and improved laboratory responsiveness.

5.2.4   Environmental Protection/Sampling and Energy Management Programs

Environmental protection is of primary consideration in the design and operation of industrial facilities, especially those using chemicals in their processes. Environmental protection includes air and water sampling and analysis and operational control procedures. Today there are stringent laws regarding air and water effluents from the plant and hygiene standards for in-plant air quality.

Another management problem is the continual downward revision of "acceptable" levels for effluent and in-plant air quality. Thus managements of industrial facilities have the dilemma of complying with increasing stringent control standards with as little as possible impact on production. Therefore many industrial facilities are utilizing the computer for environmental protection, sampling, and energy management programs. This provides real-time control over pollutants with less impact on the day-to-day operation of the plant. As the laws change, the computer program can be altered to meet the new parameters.

Important elements in an integrated environmental protection strategy include information obtained from the monitoring of stacks, outfalls, and pollution abatement facilities, as well as the actual production processes. By use of the computer, the facility is able to collect and analyze continuously plant emissions and effluent data. A computer system can also be used to calibrate automatically the pollution instruments, to maintain overall integrity of the data.

The computer relates the foregoing information to the environmental protection control parameter established for the plant. The computer can also initiate action to ensure that pollution does not occur. In Figure 7.15, for example, shutdown of operation is one possible result of activation of the pressure sensor. A computer that could act in this manner would have to be sensor based: a non-sensor-based computer can do no more than give a signal for manual interruption. For the compliance reporting portion, the computer automatically generates required detail and summary reports and maintains a complete historical record of the environmental data.

In setting up environmental protection programs the control of effluents is usually more easily defined because the sources are easily identified. In particular, needed sampling locations of water effluents can be identified without difficulty because of the "static" type of source. Air emission control

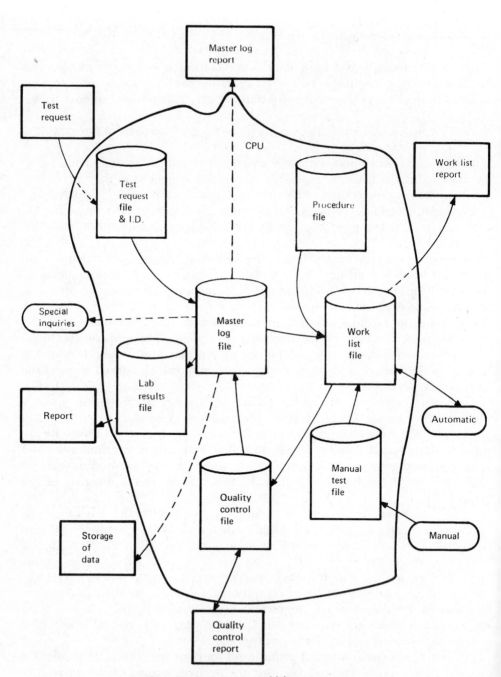

Figure 7.14 Computerized laboratory system.

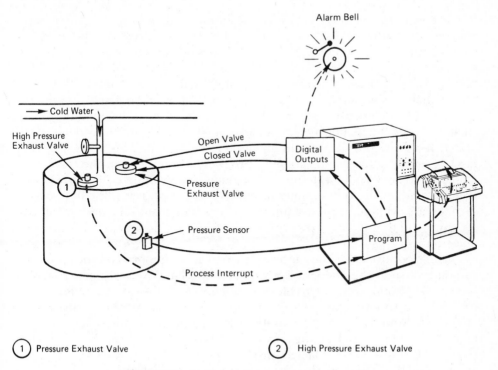

Figure 7.15   Sensor-based computer process control.

represents a slightly more difficult problem, except for stack monitors, which can also be considered to be static. For other air emission control information, multiple monitoring stations must be positioned over a wide area. In the simplest case these locations are at the boundary of the plant. More sophisticated placement can be undertaken by use of the computer. Utilizing air pollution theory and formulas, the location of the maximum downstream concentration from a single or multiple source can be predicted under average climatic conditions. The computer has also been used successfully for in-plant monitoring and control. A recent article gave useful information regarding the use of the computer for automatic monitoring of vinyl chloride in working atmospheres (209). The use of the computer was justified on the need to comply with the OSHA "Emergency Temporary Standard for Exposure to Vinyl Chloride."

The sampling and analysis portion of the vinyl chloride monitoring system consisted of Union Carbide Model 1200 chromatographs, with flame ionization detectors located throughout the plant in 19 areas that would be representative of employee exposures. The locations were also selected on the basis of greatest potential for process leaks into the worker environment. The sampling and analysis regime was under the control of a Union Carbide Model 2800 Microcomputer used in conjunction with a General Electric TermiNet 300 Printer.

For result viewing at the monitoring location for employee or management, a strip-chart recorder was used. The microcomputer accepted the same raw data from the analyzer and performed validations to ensure the reliability of the data. A computer program was established to calculate probable employee exposures as well as to provide work shift, daily, and monthly reports of information. In addition, the computer was also used to signal undesirable conditions for corrective action to reduce exposures. The operational and analytical instrumentation personnel associated with industrial hygiene collaborated to define the data requirements for the system. This is an important facet of any computer and sampling program. However even though considerable effort was made to define all situations, one parameter was overlooked. In most cases once programming has been established, it is difficult to correct such errors without reprogramming. This often results in additional cost. Sections 3 and 4 indicate procedures in data base design to minimize such omissions.

On the positive side, the authors felt that most important function of the computer system they designed was preventing employees from exposure to high concentrations of vinyl chloride monomer. If the concentration of the chemical was above either of the two target conditions, an alarm notified employees. Another important contribution of the computer system was that by locating sources and timing occurrences of leaks, the company was able to reduce emissions. This enabled production engineers to redesign problem areas. This also resulted in cost savings, since valuable product loss to the atmosphere was reduced. In addition, the system provided a permanent record of the estimated exposure for each job for industrial hygiene and medical evaluations and for compliance purposes.

In most industrial plant facilities engineering is responsible for environmental protection as well as energy conservation. If further justification is needed for the purchase of a computer or usage, energy conservation management should be considered. Facility computer-controlled power management systems have been found to save energy consumed by preprogrammed criteria. Such energy management programs have resulted in cost savings. These programs could also be placed in the same computer, thus reducing its overall cost to use.

In some cases ready-to-use programs can be obtained from computer software vendors. The computer is programmed to turn on or shut down operations at prescribed times. It can also interrupt for prescheduled short periods selected energy devices that do not have to operate continuously. Also by continuously monitoring the rate of electrical consumption in the facility and comparing the rate to variable and/or preestablished targets, it can inform management of problem areas. Management can then take action regarding personnel or equipment requiring correction action.

5.2.5 Medical Examination and Surveillance

Periodic health examination to aid in early detection of disease, as well as medical surveillance techniques such as bioassay, are being administered with increasing frequency. Many industries have inaugurated programs in which employees are periodically given medical exams because of exposures to hazardous materials or processes, or to comply with legal requirements. In the latter case positive tracking is needed to ensure that appropriate exams are conducted. This poses increased administrative and clerical demand on the medical department.

To provide the best possible preventive medicine for the employees and to ensure that legal and management requirements are met, the programming of medical examinations and surveillance has been found to be very helpful. It offers an easy method of tracking to ensure that physicals are conducted as required. Manual methods of tracking are very prone to error, especially in a busy medical department. When programmed accordingly, the computer can also act as an audit to ensure that the physicals are indeed conducted after notification. The computer can be programmed to print delinquency reports until the physical is entered into the computer as being completed. Computerization also aids in scheduling and enables the medical department to be more efficient. The medical department administrator can see at a glance what type of exams are scheduled for next period and can allot proper time for their completion. The manufacturing managers are extremely happy with this type of efficiency because it reduces unproductive work time caused when employees are waiting needlessly for appointments.

Figure 7.16 is a printout of a very basic type of computerized scheduling arrangements. It is used at a small facility and does not require all the additional programming as indicated earlier. It is described to give an idea of a computerized medical tracking program. Such a system could be easily modified by reprogramming to meet all the special needs previously mentioned. This system is effective because the medical department manager knows who is scheduled for what exams and where each person works, so that the proper manager can be notified to have the employee make an appointment with the medical department. The printout across the top line gives all that information in addition to the birthdate of the employee. The birthdate appears because certain additional health measurements such as electrocardiograms are administered on the basis of the age of the employee.

The computer printout is rather simple to interpret, except for the last column across the top line (PHY/SCH). For example, the first employee listed is James O. (last name was deleted to declassify the information). His employee serial number is 047564 and he was born on January 11, 1943. He also works in Department 092, where he is notified to come to the medical department for his exam. The last number in the "department" column indicates which shift the individual works. In this case "1" corresponds to the first shift and "2" is the second shift. [The printout indicates that three people to be scheduled for exams work on the second shift. In most medical departments special plans must be made because exams cannot be conducted on the second shift. Simple items like this often become important in computerization of data. If shift designation were needed but not supplied, there would be two alternatives: to go back into the data base and revise all the information, or to get the

EMPLOYEE PHYSICALSBY DOB/M -

SERIAL	EMPLOYEE	NAME		DEPT-S	BIRTHDATE	PHY/SCH
047564		JAMES	O	092-1	01/11/43	AHY
478152		HAROLD		706-1	01/21/31 [1]	ART
301754		ROBERT	E	716-1	01/11/41 [1]	ATH
373433		DALLAS	L	641-1	01/21/25	ATY
679602		ROBERT	W	707-2	01/04/35	ATY
682363		CLAYTON	M	707-2	01/08/25 [1]	ATY
235337		WILLIAM	J	406-1	01/14/53	AZY
753162		EDWARD	E	716-1	01/08/37	AZY
173494		GORDON	H	920-1	01/15/24	AZY
783552		DONALD	P	41D-1	01/11/33 [3]	AZY
403709		JAMES	E	528-1	01/05/51 [4]	EBE
902641		THOMAS	J	49F-1	01/12/31 [1]	EBO
841771		MICHAEL	J	42A-1	01/17/53	EBO
086541		FREDERICK	J	49E-1	01/22/39	EBO
987770		RAYMOND	G	948-1	01/06/23	EBO
788862		LAWRENCE	C	945-1	01/08/50	EBO
403704		EDWARD	L	421-1	01/12/35 [5]	ECO
152512		LAVERN	F	446-1	01/08/22	ECO
379645		CHARLES	D	424-1	01/24/45	ECO
895866		WILLIAM	S	429-1	01/19/45	ECO
341825		JAMES M		959-1	01/01/42 [4]	ENY
446832		MICHAEL	L	946-1	01/13/48	ENY
956663		JOHN	D	948-1	01/15/35	ENY
723932		JANICE	M	867-1	01/18/42 [3]	FZY
359284		SHARON	K	867-1	01/11/41	FZY
268522		CAROLE	E	867-1	01/25/39	FZY
088415		DONNA	G	867-1	01/08/29	FZY
103413		GAIL	H	707-2	01/03/38 [4]	HRT
216961		THOMAS	J	572-1	01/18/27 [1]	LID

Figure 7.16 Computer printout for scheduling of medical exams.

information from each chart. To save embarrassment regarding reprogramming, it is wise to ensure that all the information needed can be obtained from the items included prior to programming the data.]

The PHY/SCH column indicates to the person scheduling the exams the type of physical James is to have. To reduce "space," which may be needed for future programming, the physical tape and the scheduled frequency has been letter coded, (Table 7.4). Thus "AHY" for James means an audiogram and a "heat treat" physical yearly. When the physical is completed an input card is forwarded to the computer center requesting a recall next year for a repeat physical and indicating that the physical this year was completed as required.

Some employees who work in environmental chambers or with lasers may require different types of physical depending on the degree of exposure. This is covered in the classification block (2nd) of the PHY/SCH column. For example, Lavern F. (152512) is required to enter an environmental chamber. She should receive the C class physical, and she is scheduled for reexamination

Table 7.4   Computer Coding Information for Medical Exams

Type of physical (1st, 2nd or 3rd block of exam code)
  A = Audio
  F = Cafeteria
  M = Chemical materials
  E = Environmental chambers
  H = Heat treat
  X = Radiation (CBC—no fasting)
  S = Security
  T = Trucker
  R = Respirators
  Z = Filler
  L = Laser
  W = Three wheel scooter (regular trucker nurse assessment) sit down
        scooter (regular trucker nurse assessment)
Classification (2nd block)
  N = A chamber
  B = B chamber
  C = C chamber
  I = A laser
  J = B laser
  K = C laser
Schedule dates (3rd block)
  Y = Yearly
  E = Even years
  O = Odd years
  U = Three year recheck
  D = No recheck
Instructions
  One periodic physical   Fill in 1st block with appropriate symbol
                                 Fill in 2nd block with Z
                                 Fill in 3rd block with Y, E, or O
  Two periodic physicals   Fill in 1st and 2nd blocks with symbol
                                 Fill in 3rd block with Y, E, or O
  Three periodic physicals   Fill all 3 blocks with symbols
  Chamber physical   Fill in 1st block with physical symbol
                                 Fill in 2nd block with classification
                                 Fill in 3rd block with Y, E, or O
  Laser exams   Fill in 1st block with physical symbol
                                 Fill in 2nd block with classification
                                 Fill in 3rd block with Y, U, or D

every other year. For persons requiring annual physicals, the third block can also be used to indicate the type of physical. One example of this coding is Gail H. (103413), who works in heat treating, is a trucker, and may be required to wear respirators. She therefore must receive all three physical examinations annually. Thus a simple computer program and system can be expanded to indicate when bioassay tests for chemicals (lead, mercury, etc.) should be conducted.

Although the actual physical exams in the example above were conducted manually, many of the portions of the examination processes, including the history and summary report, can be computerized. For example, the person requiring the examination is given prepunched computer cards with the employee number indicating the test required. The history can be taken by presenting questions to the employee on a cathode ray tube or an optical image unit. Using a special electrical probe, the subject can respond in "yes" or "no" to the questions. Depending on the response, the unit is advanced to another question. For example, if a positive response is given to the question "Have you ever been exposed to chemicals?" the next question will elicit the duration of exposure. A negative response to the first question will cause the program to skip to the next category. With this type of branching logic, an almost complete medical history could be taken on-line with the computer.

Other on-line testing with the computer can also be conducted as part of the medical exam. For example, the technician prepares the patient for an electrocardiogram. The test card with the employee information is entered into the terminal, notifying the computer that the subject is ready for the test. The computer then activates the device, collects and analyzes the data, and records the results in the employee record. Other physical examination tests that are amenable to on-line testing are audiometry, blood pressure, clinical laboratory chemistry, hematology (red blood count, white blood count, hemoglobin, hematocrit), and spirometry.

Off-line tests involving the input response to the computer of a physician or paraprofessional can also be obtained. Such tests include chest X-rays, differential blood cell counts, measurements of height and weight, vision checks, and urinalysis. The results are entered into the test booklet on a mark-sense card, which is placed in a reader connected to the computer. The computer reads the data and records the results of the employee examination.

When all the tests have been completed, the physician is given a patient summary report, which is standardized by the computer. The physician can then conduct any additional measurements deemed necessary or ask for the additional information based on the test results or the history. This information can also be computerized if necessary or simply placed in the employee file along with the computerized data. Figure 7.17 diagrams the entire process.

When computerized, the medical examination will afford greater accuracy and efficiency in history taking, laboratory testing, and the physical measurement portion of the exam. Advantages include better utilization of professional and paraprofessional personnel as compared to traditional examination meth-

Figure 7.17 Sequence of a computer-assisted medical examination.

ods. Since the clerical functions are drastically reduced, personnel can devote more time to the actual medical aspects of the examination.

5.2.6 Portable Computers/Personal Computers/Software

To persons in the industrial health field, perhaps one of the most important computer developments is the availability of the small scientific and engineering computers that are truly portable. These computers, which can be hand carried to a location, have all the abilities of the large computers: significantly less storage is their only limitation. Yet when the equipment is used in a remote-type location for a problem-solving need, great storage of data or voluminous program libraries are usually not required. The most valuable attribute of the portable computer is that it can be taken to a work site for data entry or problem solving with little prior planning or facility installation cost. Thus the small computer with the ability of a larger computer can be placed at the location on a surface no larger than a desk top; it does not require any special communication lines or environmental control or electrical facilities, as are needed with the larger computer.

The benefits of using the small portable computer are somewhat analogous to the advantages that are evident in industrial hygiene by having direct reading instrumentation for field use. By the collection of data at the site either directly or indirectly through the computer, on-the-spot evaluation of information can be made. A personal computer, readily available for use with the full-function capabilities of high level computers for problem-solving, offers many uses. The ability to collect accurate data at the source for immediate answers is an advantage to any problem-solving professional. The main limitation in using the computer in an industrial environment is that there are "hostile" environments for computers, just as exposure to certain substances is hazardous to humans. In some locations, depending on the environmental conditions, it may be impossible to install some of the portable computers if environmental controls cannot be provided. Where this is a problem, remote sensor-based relays or other data acquisition methods, as discussed in Section 4, should be used. Another solution is to use the slightly larger sensor-based computer equipment, which can operate in a more hostile environment without damage to the electronic components.

The small portable computer illustrated in Figure 7.18 is about the size and weight of a large typewriter and has main storage capacity of up to 64,000 storage positions. This computer comes equipped with a familiar typewriterlike keyboard, which is easy to use for data entry or programming. A small visual display is usually provided to show the data being placed in the program or the information being retrieved. When a permanent copy of the data is needed, a printer can be attached to produce such "hard" copy. The data, information, or program can also be placed on a magnetic tape cartridge for storage and retrieval later. To reduce weight, the visual display provided with the system is small and really usable only by the operator and perhaps one or two other persons. Where there is a need for demonstrations or computer-assisted instruction for larger groups, the unit should be attached to a larger television monitor. The usefulness of demonstrations and the advantages of computer-assisted instruction in the industrial health field are discussed in Section 5.2.7.

The removable magnetic tape cartridge enhances the small computer by providing a source of increased storage of data and programs beyond the main capacity. As very sophisticated programs are developed, future use without time-consuming reprogramming becomes possible. Having data or programs on magnetic tape cartridges also is advantageous insofar as the tape can easily be transported to a remote location where a compatible computer is available for use. Also available from the vendors of portable computers are problem-solving programs already on magnetic tape. Thus programmed routines of mathematical or statistical equations such as complex differential equations, analyses of variance or regression, and design analyses may be readily available

Figure 7.18 Small portable computer.

for use. Cartridges are also available to show how to use the computer as well as its programs. Presentation techniques used for plotting line graphs, bar charts, histograms, and curve fits, have also been preprogrammed. Here it is necessary to attach the computer to a matrix printer. Data presentation methods are discussed in Section 4.4.

Enhancing the uniqueness and advantages of the small computer is the capability of connecting it to a teleprocessing facility, offering communications to larger data bases and voluminous program libraries in other computers. Thus information and answers are available from the portable computer or other machines where the need is greatest. Directly, this can be accomplished by attaching the computer to teleprocessing connections or to an industrial process or location by a sensor, or with a research device coupler, allowing it to be tied into laboratory or a variety of instrumentation devices. With direct data entry, immediate solutions at the source of concern are available with unmatched speed and precision. Real-time control, not as fast, can also be obtained by manual input of data directly at the site to provide most scientific and engineering problem-solving information or to indicate needed control methods. Of extreme value in signaling when real-time controls are needed is the ability to use the computer to tie into remote areas that are interrelated to the solution. This can be accomplished by using two or more computers tied into each other by connecting transmission lines or multiple sensors from the location to a single computer.

The "portable computer" has been found to be a very useful tool for Safety/ Health data applications. At the IBM International Environmental Health and Safety Technical Seminar in October 1979, D. E. Swanson presented the "Safety and Health Applications for the IBM 5100 Portable Computer." The programming was done on the computer by Swanson. Important factors for using a portable computer were: user owned, stand-alone capability, direct user control, improved information quality. The entire rationale concerning why the portable computer was used and several program printouts are given in Appendix 7-D. As you get more involved in data automation, our experience has shown that data automation programs are usually placed in larger computer systems which offer more capability and memory. However, as a starting base, even the small personal computer has very good applications and versatility for automation of occupational health data.

As indicated above, many professionals have found that the smaller personal computer offers a good starting base for automation of data. Because manufactures of the hardware are numerous and dealer competition prevails, the personal computer has reached a desirable scale related to size, usage, and cost. A "complete" personal computer can be purchased for several thousand dollars. This includes the system unit, keyboard, display, printer, disk drive, hardware option adapters, plus several software programs. Thus, because of the present cost and productivity realized from usage, even the smallest safety and health or medical department of a plant of reasonable size can justify the item in its budget.

Because of the millions of personal computers in use today, a significant number of useful and unique commercially available software programs have been developed. The amount of software available will increase because of the number of potential buyers of the computer hardware. Many of the software programs can be utilized to automate your occupational health records in their present or specific format. Similarly, these readily available software programs can be adopted for your specialized use or you can develop your own program on the personal computer. Although it is easy to program your data by reading the computer manual or software instructional manual, time and frustration can be eliminated by having someone experienced in the personal computer to assit you in getting started. Once programmed, "loading" the computer with the data can also be time consuming as well as frustrating. However, once data has been programmed, being able to retrieve the data in a specific format for a presentation or report or to do calculations, trend analysis or predictions will bring back rewards related to the time-consuming computer tasks or the time it would take to accomplish this manually. Likewise, the "simple" filing of the record disks represent another savings as well as space needed for this task. M. Devine and P. Guastella, both IBM Industrial Hygienists, using commercially available "P.C. Files" and "Lotus" software were able to automate their Environmental Health and Safety Department's data. Examples of some of their programmed printouts are shown in Appendix 7-E.

With suitable software and some personal experience with computers almost any informational data system can be programmed. Software is also available for the graphic presentation of information. By utilizing a Color/Graphics unit, very interesting charts of all types can be produced from the programmed data. Thus on-the-spot presentation material for conferences or reports of current data can be obtained in a matter of minutes without the time-consuming delay of going through the location reproduction department. Several examples of graphics are given in Appendix 7.F. Another interesting development in this area of graphics is the computer-aided design system. This allows the user to create an apparent 3-D object on the screen and to perform various tests and analysis as if the object were real. The object, in conjunction with others, can then be used to put complex assemblies together before a prototype is ever constructed. Potential areas of usage of this type of computer system include ventilation design and control as well as noise problems and control and ergonomics.

5.2.7   Computer-Based Training

Computer-based training (CBT) is an attempt to resolve two major industrial training problems, outlined as follows.

1. Rapidly rising training requirements to:
    a. New products and technological changes requiring a significant amount of employee retraining.
    b. New systems or procedures.
    c. Lack of experienced, qualified new employees.
    d. Changing, more demanding regulations.
2. Continually increasing training costs.

With this background in mind, it is understandable that in every technical endeavor, the search continues for more effective approaches that will enable instructors to increase their productivity. New technologies are being employed, and interactive-self-study methods are being adapted for training.

Industrial health and safety training requirements require a significant amount of time. CBT, when properly conceived and utilized, can provide the protective knowledge and documentation required by federal regulations and sound business sense. Tables 7.5 through 7.7 describe of chemical safety, electrical safety, and laser safety courses for which IBM has utilized CBT.

Two types of CBT can be structured.

1.   Computer assisted instruction (CAI), in which the majority of instructions are presented through the terminal, possibly with some off-terminal assignments.

Table 7.5   Outline of a CBT Course in Chemical Safety

Course title	Chemical Safety
Course number	6006 (CAI)
	5007 (Hazardous Material—Department 713)
Length of course	11 hr (CAI)
	1 hr (Hazardous Material—Department 713)
	As required for on-the-job training
Method of instruction	CAI (CHEMSAFE)
	Classroom
	On-the-job training
Courses appropriate for	Recommend CAI course for department technicians and personnel with prolonged contact with chemicals. Recommend the "Hazardous Materials" presentation or on-the-job training for production people and those with limited exposure to chemicals.
Abstract	CAI course includes instruction on the terminal, slide/tape presentation regarding chemicals in the eye, and a film on lab safety. The content of CAI and classroom courses covers protective equipment, chemical hazards, safety practices, and laboratory safety.
Responsible for implementation	Department manager monitors need and initiates request for training.
Instruction given by	Education Department (CAI)
	Department manager and/or Department 713 (Hazardous Materials)
Documentation	On formal classes held, forward the completed rosters to the Education Department for processing into the PSD system. Department manager maintains on-the-job training records.
Refresher training	Annually

2.   Computer-managed instruction (CMI), in which the majority of instruction is off-terminal through audiovisual (A-V) materials (films, audio or video tapes, slides, etc.), reading assignments, lectures, or perhaps practical laboratory assignments. The student may be tested through the terminal and directed to appropriate off-terminal material. Special, individualized instruction can be provided by the terminal to fit each student's needs, based on ability to master the subject and required repetition of course information.

There are no rigid distinctions between these two approaches. Both can be used in the same course. However, the course developer usually is in a better position to select the technique and materials than the training group or department.

One of the most significant benefits of CBT is its potential contribution to cost-effective training programs. It can help increase productivity by reducing such costs as instructor time, travel-time and expense, and non-productive waiting for training. In the business environment, CBT can be integrated with existing computer hardware on terminal-based systems. CBT can be effectively piggybacked into the main data processing operation of a firm.

In multilocation businesses the employee "to be trained" can take CBT courses at work location. To end a training session, a simple course sign-off returns the same terminal to production applications for the same or another employee. While one terminal (or several) is engaged in training, others in the system are independent to provide business support. Terminals may be as close

Table 7.6   Outline for CBT Course in Electrical Safety

Course title	Electrical Safety
Course number	6010 (CAI)
	4021 (708-Electrical Safety)
Length of course	3 hr (CAI)
	1 hr (Department 708 Safety Presentation)
	As required for on-the-job training
Method of instruction	CAI Course (ELSAFE)
	Classroom or department meetings
	On-the-job training
Course appropriate for	Recommend CAI course for test equipment, builders, electrical engineers, manufacturing engineers, and Facility engineers, Recommend 1 hr presentation or on-the-job training for production people and those with limited exposure to electrical hazards.
Abstract	CAI contains instruction on terminal, audio tape/slide presentation. Describes basic safety precautions, effects of electricity on the body, and action to be taken in case of an electrical accident. Also includes instruction for high potential operations and insulating testing. Classroom discussion of UL-approved equipment, grounding, and potential electrical hazards and corrective action.
Responsible for implementation	Department manager monitors need and initiates requests for training
Instruction given by	Education Department (CAI)
	Department 708 (1 hr presentation)
	Department manager (on-the-job)
Documentation	On formal classes held, forward the completed rosters to the Education Department for processing into the PDS system. Department manager maintains on-the-job training records.
Refresher training	Every 18 months

Table 7.7 Outline for CBT Course in Laser Safety

Course title	Laser Safety
Course number	6018 (CAI)
	4019 (classroom)
Length of course	3 hr (CAI)
	1 hr (classroom)
Method of instruction	CAI (LASAFE)
	Classroom
	On-the-job training
Course appropriate for	Recommend the CAI course for Development personnel, Test Equipment Building personnel, and those who are closely associated with an in prolonged contact with lasers.
	For production personnel and those with minimal exposure, the one hour classroom program will be adequate.
Abstract	CAI course includes two motion pictures, an audio tape/slide presentation, plus questions and instructions from the terminal. This training will provide the ability to:
	1. Identify laser terminology.
	2. Identify laser hazards.
	3. Identify control measures to minimize hazards.
	The classroom program will simplify and condense the major points and highlights of the CAI course for those with minimum exposure.
Responsible for implementation	Department manager monitors needs and initiates requests for training.
	Ref. Safety Manual Index 36-8-02
Instruction given by	Education Department (CAI)
	Department 713 for classroom program
	Department manager for on-the-job training
Documentation	On formal classes held, forward the completed rosters to the Education Department for processing into the PDS system. Department manager maintains on-the-job training records.
Refresher training	18 months

or remote as business needs dictate. Normally telephone lines provide the necessary "remote" link to the base operating unit.

Depending on the technology for which the CBT is based, the course context can be formulated and implemented through the combined efforts of the course developer, data processing staff, and the training function. Though several CBT programs can coexist, a central coordinating responsibility should be assigned. This helps to reduce duplication and inconsistencies, and contributes to efficient use of the system facilities.

For the sake of our discussion, the CBT control function is addressed as the "training group." Figure 7.19 shows the communications and functional responsibilities required for a CBT system. CBT is usually part of training, along with classroom instruction, audiovisual aids, and other functions.

Applications for CBT.   The applications for computers in occupational training are almost unlimited. For example, CBT can be used to:

1.   Orient and test new employees to identify each one's training needs. (The computer could be used to input unique occupational health hazards when keyed by each employee's department number at the time of signing on.)

2.   Provide a common background to a group of students before they enter a class (or undertake a new process).

3.   Teach operators (almost any process) how to track their productivity, equipment, or material use through a terminal.

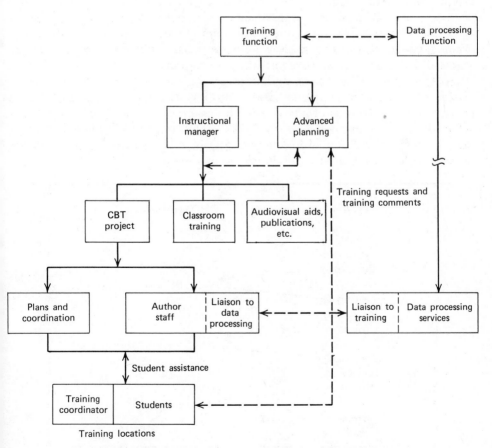

Figure 7.19 Functional responsibilities for a CBT system.

    4.   Provide basic skill training to new (or reassigned) employees as necessary before technical training (e.g., in the use of protective equipment prior to learning required skills).

In addition to the instructional uses, the same computer data processing system can have some other applications, such as:

    1.   Maintaining training records of each employee (an OSHA requirement).
    2.   Providing performance certification testing.
    3.   Recording employee survey or questionnaire responses.
    4.   Recording employee hazard awareness.
    5.   Testing to identify the general training needs of a given group, to help trainers decide what new training must be developed.
    6.   Comparing legal training requirements with employee hazard recognition.

Getting Started with CBT.   The adoption of CBT does not mean that existing training materials must be discarded or must undergo major revision. Instead CBT should be viewed as just one more training medium to be combined with other appropriate techniques. For example; CMI could be used to optimize material for a given subject, consisting of programmed instruction (PI) test, audio tape, laboratory, slides, and lectures. Figure 7.20 illustrates the incorporation of several instructional modes into a CMI training program. In each given subject, the results of a terminal-administered test are used to determine what additional instruction is needed. The pretest (Figure 7.20) is a useful first step in sizing the additional instruction. Figure 7.21 reflects the course development process as it applies to several training media.

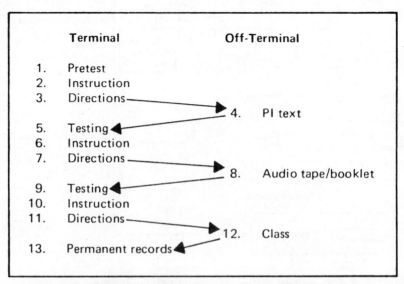

Figure 7.20 Sample CMI course sequence.

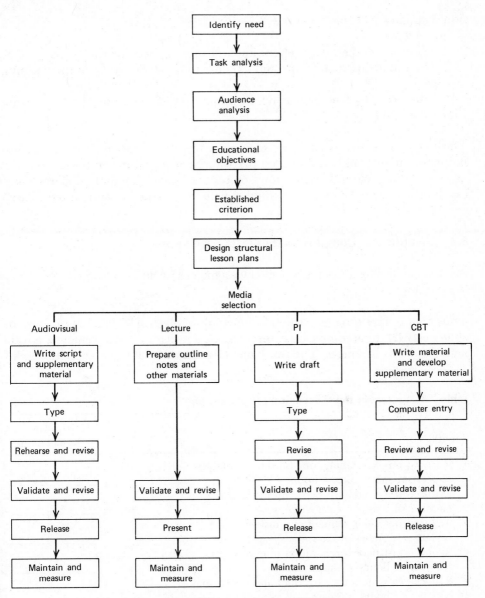

Figure 7.21 CBT course development process.

Benefits of CBT.   A benefit must be measurable if it is to be assessed accurately and balanced against a cost. To do this, the benefit must be translated into a tangible cost saving. For example, a reduction in training time, can be measured in terms of the student and instructor salaries saved.

Compared to non-self-study media, CBT should reduce (perhaps eliminate) instructor involvement once the course has been developed (except for occasional updates). This is because CBT can:

1. Eliminate repetitive preparation and presentation for each class.
2. Eliminate instructor time and expense to travel to and from the teaching location.
3. Reduce class time by placing introductory and common basics of existing courses on CBT.

The test or measure is not only savings in travel and instrutor time but also the new courses (CBT and others) that can be achieved in the time gained by use of CBT. Table 7.8 shows how some of the major tangible dollar savings can be calculated. Many items apply equally to instructor- and student-based cost reduction.

5.3   Example of a Complex/Shared Data Base System

5.3.1   Establishing a Computerized Environmental Health and Control Information System

Thus far we have discussed the fundamentals of electronic data processing, the ability of EDP to accomplish required tasks, and its use, requirements, advantages, and disadvantages. The fundamentals of industrial hygiene are the ability

Table 7.8   Factors in the Calculation of CBT Benefits

Benefit	Dollar Calculation
• Reduced on-job-training, one-on-one training (supervisor) Supervisor salary rate × hours =	_____
• Reduced travel time (student and instructor) Number of hours × salary rate =	_____
• Reduced per diem (student and instructor) Dollars × number of days =	_____
• Reduced instructor preparation time per class Number of hours × salary rate =	_____
• Reduced classroom time (student and instructor) Number of hours × salary rate =	_____
• Reduced time waiting for training (student) Waiting time × salary rate × fraction (ineffectiveness) =	_____
• Reduced publication (printing) costs Printing costs × number of copies =	_____
• Reduced publication (distribution) Costs × unit wrapping and postage cost =	_____
• Reduced cost of changes (additions/deletions) Cost of changes × printing and distribution =	_____

to recognize, evaluate, and provide controls, as needed, to prevent health hazards. This is the basis of solving any single industrial health problem, but an overall industrial health program is more complex. A complete industrial health program consists of problem solving, inspections, control procedures, routine exposure monitoring, and other functions. These programs produce the basic goal of industrial health, which is to ensure that employees, the general public, and the environment are not adversely affected by company operations.

The same analogy can be made to the sections of this chapter. Up to this point, specific applications and possible uses for the computer have been given. Other uses for the computer as related to specific expertise, operations, and needs are evident. To couple the best possible utilization of the computer and benefits with the most effective control, a complete industrial health computer program for these applications should be combined as much as possible, or at least coordinated under one management system. This section puts the parts together to indicate the concept of a complete package, termed a computerized environmental health information and control system. First, to develop the necessary computer systems, we discuss the noncomputer programs. Although some of the parts of the concept for the overall system may not be needed for a particular operation or problem, our main goal is to provide background information with the concept, to start the reader toward the design and development of a computerized environmental health information and control system to answer individual requirements.

Noncomputer Program Aspects. Before discussing the computerization aspects of an environmental health information and control system it is necessary to list all the individual applications or programs in the total package that could be computerized. This section treats the programs from the industrial health standpoint. The next section reviews them from the computer processing point of establishment. Both sections reflect the general philosophy that individual computer applications are necessary and advantageous, but it is the interrelationship or tie-in coordination between applications that provides the superior total program and the best operational results.

In viewing any program, total or individual, each of the steps or requirements must be indicated and reviewed from creation to realization. Table 7.9 lists the individual industrial health programs required for an environmental health information and control system aimed at protecting the employees, the general public, and the environment. Once operational it should be able to provide information to the company areas affected by the program as listed in Section 1 as well as complying with government requirements and maintaining adequate recordkeeping.

Some of the other programs named in Table 7.9 were discussed earlier. For example, Section 5.2.5 (Medical Examinations and Surveillance) is related to programs 7 and 15, and Section 5.2.7 (Computer-Based Training) is especially related to programs 5 and 8 as well as most of the other programs. In the same

Table 7.9   Noncomputer Industrial Health Programs Required

1. Complete chemical data information.
2. Advance evaluation before approval for use.
3. Control mechanisms to limit exposure.
4. Regulated area (carcinogens, etc.).
5. Protective equipment authorization and control.
6. Emergency plans and programs.
7. Physician certification and selective job placement.
8. Management and employee training programs.
9. Procurement and inventory control, and labeling and storage.
10. Shipping controls.
11. Authorization, distribution, and quantity control.
12. Monitoring and maintenance of control mechanisms.
13. Exposure measurements and determination.
14. Laboratory methods and analysis.
15. Medical surveillance.
16. Employee exposure data bank.
17. Employee notifications.
18. Exception reporting system.
19. Epidemiology studies.
20. Emergency incidents and releases.
21. Waste disposal programs.
22. Compliance audits.
23. Complete recordkeeping.

manner it may not be necessary or possible to computerize all the programs listed in Table 7.9.

However, for the best operational and computerization results it is necessary to review and separate each of the individual programs into three elemental parts. Once these parts of each program have been defined and established, the programs can be blended into the overall computer system because the analyst has a better view of their interrelationships. The first element is the administrative aspects of each program. In this portion of the review we must define what is necessary and assign to management and support functions their responsibilities and instructions for implementing and maintaining each program.

The second element is to provide specific operating procedures for each program. For example, programs 9, 10, and 11 would call for well-defined procedures for approval, ordering, handling, storage, and use of chemicals, as well as hazardous materials. There would also be an interrelationship between these procedures and other procedures (e.g., programs 2, 3, and 8). In this context ensure that there is not conflict in procedures.

To be an effective program monitoring and auditing must be provided; this is the last element as well as the most important part of each program. To obtain compliance with corporate as well as governmental regulations, a mechanism to ensure the effectiveness of the administration and operating procedures is needed. This could be called the check and balance of the individual as well as overall program. It provides us an indication of where corrections, additions, deletions, and revisions should be made. Thus all the programs given in Table 7.9 are related to programs 22 and 23. Putting together the three elements above is by no means easy. The foregoing elements were utilized to establish the computer program application indicated in Section 5.1.1 for establishing a ventilation testing program. This can also be used as a guide for establishing each of the three elements for an individual program. But once each of the elements has been established and where possible computerized, a very effective program should be in place.

Computerization Concepts and Requirements. Now we turn to concepts and requirements for a computerized environmental health information and control system. Although we discuss the computer aspects of the system in this section, our objectives remain the same as before. Simply restated, these objectives are:

1.   To protect the employees from health hazards.
2.   To protect the general public from health hazards.
3.   To protect the corporation and management from undue citations and litigation.

Likewise the programs remain the same as indicated in Table 7.9, but for computerization the purpose changes. Now the purpose becomes the utilization of the programs in Table 7.9 to develop an environmental health program with management system to be responsive to:

1.   The individual operational needs.
2.   Management and government requirements.
3.   Cost effectiveness, where possible.

To accomplish this purpose the 23 programs of Table 7.9 would be reduced seven program or systems (Table 7.10). The information that was developed for the 23 programs would be utilized to establish the seven required systems. All the aspects of each system must be developed utilizing the 23 "noncomputer" programs established to put them in a "computerized form." Each task for every system must be determined, reviewed, and established before computerization can begin. As was evident with the 23 programs, interrelationships exist and coordination is necessary among the seven systems. For example, all the first five systems are related to systems 6 and 7.

The task involved in system 1 for chemicals is to relate the person to the chemicals to the job and department. If the personnel computer system has an employee numbering and job coding system, part of the work is already accomplished. If there are no computer descriptive terms for operation and code as shown in Table 7.11, these must be developed. The next task is putting the data into the computer. A form similar to Figure 7.22, which is machine

Table 7.10 Computer Programs

1. System to identify chemical, physical, and biological hazards; their location, use, and personnel exposed.
2. System that establishes baseline data measurements and monitors to assure that they are evaluated periodically.
3. System that assures that hazard information is generated and kept current for management, support operations, and employees.
4. System that assures that necessary controls are monitored and alerted to process changes.
5. System that assures that necessary and appropriate medical surveillance and training are conducted.
6. System that assures that review of results of items 1–5 and where indicated proper and expeditious action will take place.
7. System of recordkeeping in compliance with federal regulations that is functional and allows ready retrieval of data.

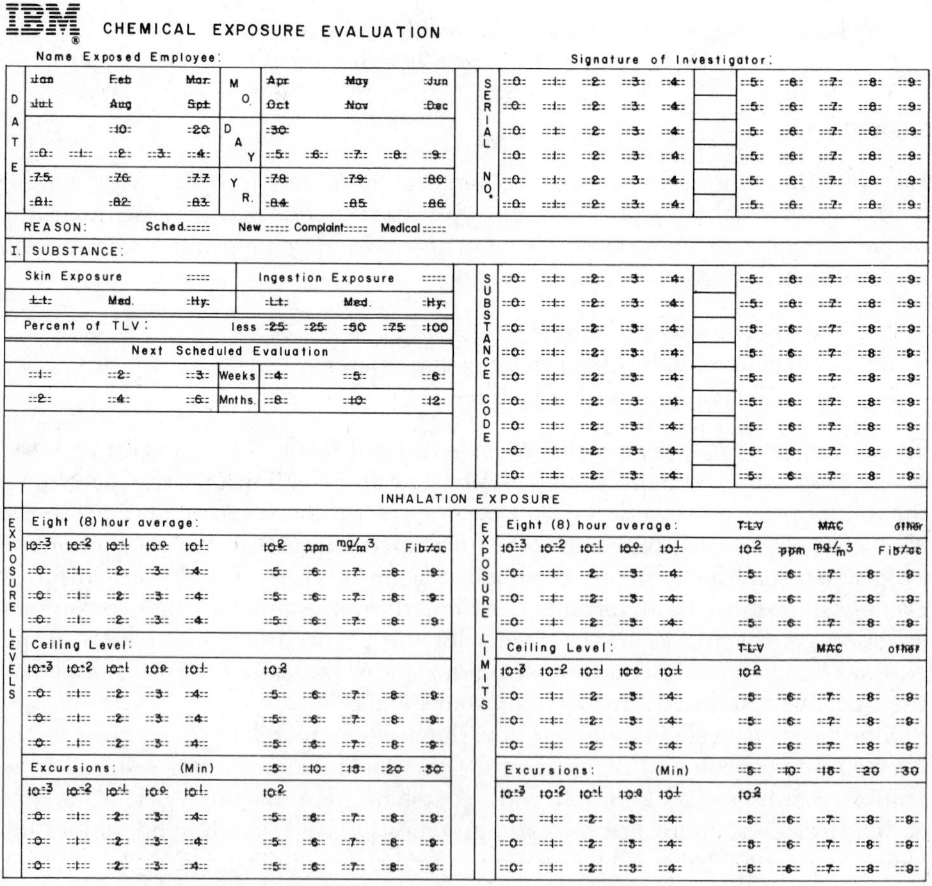

IBM CHEMICAL EXPOSURE EVALUATION

Name Exposed Employee: Signature of Investigator:

REASON: Sched.   New   Complaint   Medical

I. SUBSTANCE:

Skin Exposure   Ingestion Exposure

Percent of TLV: less :25: :25: :50: :75: :100

Next Scheduled Evaluation

INHALATION EXPOSURE

Eight (8) hour average: Eight (8) hour average: TLV   MAC   other

Ceiling Level: Ceiling Level: TLV   MAC   other

Excursions: (Min)   Excursions: (Min)

Figure 7.22 Computer input form for employee exposure recordkeeping.

348

Table 7.11 Descriptive Terms for Operations

Code		Code	
101	Abrading	124	Handling
102	Assembling	125	Heating
103	Blasting	126	Impregnating
104	Burning	127	Irradiating
105	Casting	128	Laminating
106	Cementing	129	Machining
107	Cleaning	130	Melting
108	Coating	131	Milling
109	Condensing	132	Mixing
110	Crystal growing	133	Molding
111	Crystal slicing	134	Packaging
112	Cutting	135	Paint removing
113	Deburring	136	Painting
114	Degreasing	137	Photolithographing
115	Developing	138	Plating
116	Diffusing	139	Polishing
117	Dipping	140	Pouring
118	Doping	141	Soldering
119	Encapsulating	142	Spraying
120	Etching	143	Stripping
121	Gassing	144	Trucking
122	Glass blowing	145	Welding
123	Grinding	146	Wiping

readable, must be developed for line management use. Special cases may also occur—for example, an OSHA-regulated area, necessitating indication by the computer system of persons authorized to enter the area. The computer system, by use of a sensor badge reader, could permit access to the area and at the same time maintain the daily company roster as required by the regulation (Table 7.5). Thus in this special case system 1 is also related to system 4.

In the same manner as in the Table 7.9 programs, the noncomputer tasks of how, when, and where to obtain exposure measurements are developed which are related to system 2 in Table 7.10. Now some of the computerization task for system 2 are:

1. To relate persons to their exposure data.
2. To calculate time-weighted averages for multiple exposure conditions.
3. To record all exposures, calculate the mean, and indicate the maximum for the year.
4. To print the entire work exposure history for a person.
5. To indicate persons exceeding legal requirements and whether they were notified of exposure results within 5 working days.

The last two tasks are of course related also to systems 6 and 7. See also

Appendix 7A for additional tasks. Likewise the data obtained from system 1 and system 2 must be meshed with the noncomputer information obtained from system 5. Thus flow charting similar to that in Figure 7.23 must be developed for particular needs. Additional tasks include the development of medical forms similar to those in Figures 7.24 and 7.25.

For establishing the computer program for system 3, the information in Section 5.2.7 on CBT will be helpful regarding training of management and

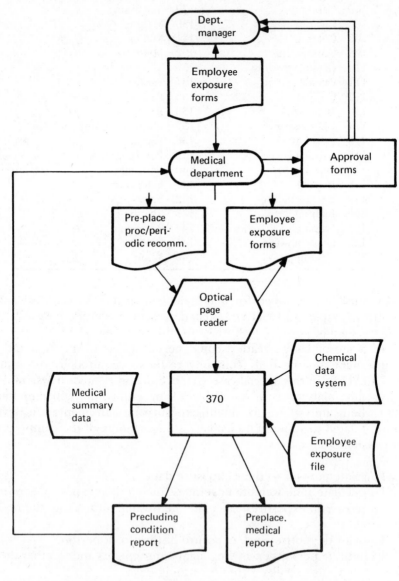

Figure 7.23 Flow chart of medical surveillance program.

Name Internal Zip City & State

To:

IBM

Date:

Name & Tie/Ext.:

Title/Dept. Name: Medical Department

Zip/City, State: 400/9M3/044

U.S. mail address: FSD Manassas

Subject: Periodic Physical

Reference: Name: Serial:

Your employee has been scheduled for a ______________________ physical.

In order to meet the requirements for this examination, he/she must have a preliminary work-up by the nurse and a review and examination by Dr. Nichols.

Please notify your employee that he/she is scheduled for their preliminary work-up on ____________________ at ______________.

The appointment for the review and examination by Dr. Nichols is on ____________________________ at _____________________________.

Since it is a job requirement, it is your responsibility to see that your employee keeps this appointment.

Special Instructions: ( ) None

( ) 12 hour fast - nothing to eat or drink except sips of water.

Figure 7.24 Medical appointment card.

IBM

MEDICAL DEPARTMENT EXAMINATION REPORT

INITIALS NAME DEPT. DATE

YOU RECENTLY PARTICIPATED IN A MEDICAL DEPARTMENT EXAMINATION CONCERNING ______________________

RESULTS INDICATE THAT:

1. ( ) THE REPORT REVEALED NO SIGNIFICANT ABNORMALITY.

2. ( ) PLEASE CONTACT ______________________ IN THE MEDICAL DEPARTMENT.

M04-0011-0 IBM C67743

Figure 7.25 Medical department examination report.

SUPPOSE A PHYSICIAN WANTS TO KNOW WHAT CHEMICALS MAY BE CANCER
CAUSING.

INPUT	EXPLANATION
SELECT	
CODES: 9.	
SELECTION, VALUES, OR FUNCTION ? S	
PROPERTIES: DESC CANCER.	DISPLAY CHEMICAL NAME.
EXPRESSION: CANCER HAS 'YES'	CANCER IS PROPERTY NAME CON- TAINING 'YES/NO' CARCINOGEN -- INDICATOR.

Figure 7.26 Input to computer necessary to obtain information.

employees. As well as training, a computer program to supply health and safety information to other areas of the company (chemical control, shipping, purchasing, etc.) needs to be established. For example, suppose a physician wants to know what chemicals being used at a given location may be carcinogenic. Input of the data as in Figure 7.26 into the computer terminal is all that is

```
CODES:    9.
SELECTION, VALUES, OR FUNCTION? S
PROPERTIES:    DESC CANCER.
EXPRESSION:
      CANCER HAS 'YES'

IBM CODE.    : 924336332
DESCRIPTION.     : ARSENIC TRIOXIDE
                 : ARSENOLITE, WHITE ARSENIC, CRUDE ARSENIC
                 : CLAUDETITE, ARSENIOUS ACID, ARSENOUS ANHYDRIDE
CARCINOGEN.    : YES

IBM CODE.    : 943010610
DESCRIPTION.     : BETA NAPHTHALAMINE
CARCINOGEN.    : YES

IBM CODE.    : 951835296
DESCRIPTION.     : DICHLORO BENZIDINE
CARCINOGEN.    : YES

IBM CODE.    : 970160103
DESCRIPTION.     : BETA PROPRIOLACTONE
CARCINOGEN.    : YES

EXPRESSION:
```

Figure 7.27 Output from computer answering question.

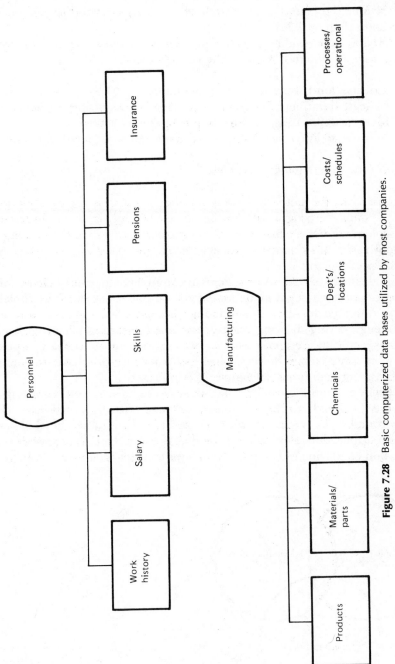

Figure 7.28 Basic computerized data bases utilized by most companies.

353

necessary to obtain the information immediately. Once the data have been imputted into the computer, a search of the files will produce the answer to the question (Figure 7.27).

As previously mentioned, all the programs above are related to system 6. For example, in system 6 some of the related tasks are:

1. Indicate control program documentation.
2. Indicate situations (emergencies or incidents) requiring reports to OSHA.
3. Identify exposure situations requiring special action.
4. Indicate delinquency reports of work not completed for any of the systems.
5. Conduct epidemiology studies.

In the last task, the method of programming the data is extremely important. Often the data are programmed to be retrievable only in a certain format. For epidemiology studies the hierarchy of programmed data should be accomplished to permit retrieval of any segment (chemical, job, bioassay, exposures greater than certain level, etc.).

When the tasks for the seven systems are completed, one must decide whether to have a completely separate or combined computer system. Traditionally data files are designed to serve individual applications, such as inventory control, payroll, engineering drawing release, and manufacturing planning. Each data file was specifically designed with its own storage space in the computer, on tape, or in direct access devices. Most companies already have personnel and manufacturing data bases, as Figure 7.28 indicates.

It is evident from Figure 7.28 that some of the information regarding health and safety is already in the personnel and manufacturing data bases. Thus if an environmental computer system is established the same data will reside in different application files. What is needed is a data base that provides for the integration of sharing of common data. This concept is depicted in Figure 7.29.

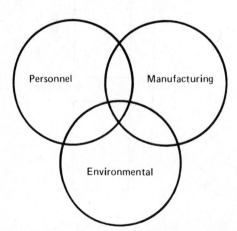

Figure 7.29 Integration of company data base systems.

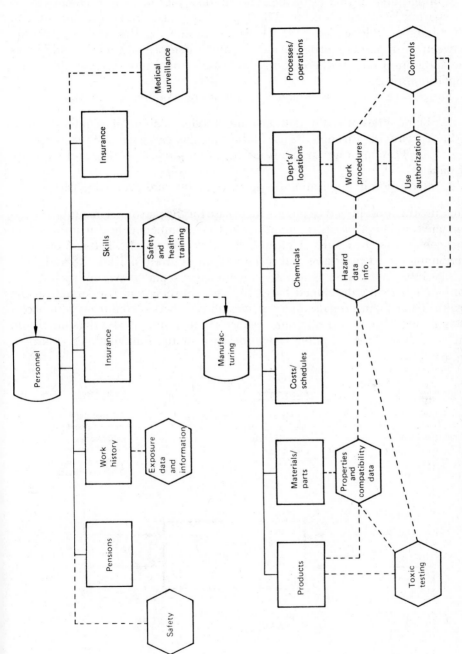

Figure 7.30 Computerized environmental health information control system.

355

The undesirable attributes of the separate data files have been eliminated by the sharing or overlap concept. This type of data base offers flexibility of data organization. Independence is achieved by removing the direct association between the application program and the physical storage of data. This facilitates the addition of data to the existing data base without modification of existing application programs.

The advantages of this common data base concept are:

1.   Elimination of redundant data and implied redundant maintenance.
2.   Consistency through the use of the same data by all parts of the company.
3.   Application program independence from physical storage and sequence of data.
4.   Reduction in application costs, storage costs, and processing costs.

If we use the concept of the common data base for the personnel, manufacturing, and environmental areas, a computerized environmental health information and control system results (Figure 7.30). This concept also had the effect of combining the 23 programs and seven systems as given in Table 7.8 and 7.9.

The other alternative is to have a completely separate system, such as the computer network developed by (210). The entire system is operated in the health department domain and provides all members of the team with access to vital data. Figure 7.31 diagrams Amoco's basic system health data input into the computer. The computer system designs for the four modules appear in Figure 7.32.

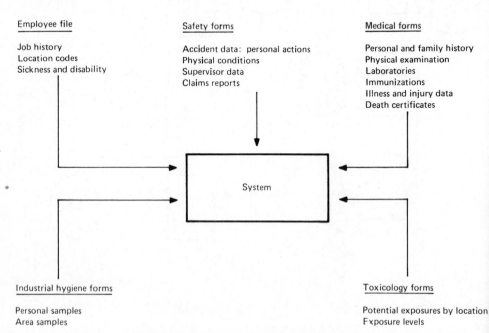

Figure 7.31   Completely separate data base system.

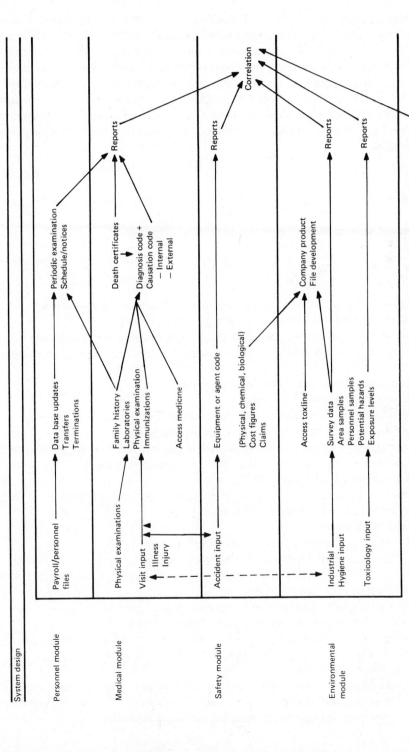

Figure 7.32 Modules designed for a separate data base system.

The literature contains few descriptions of computerized environmental health information and control systems. Westinghouse (211), DOW (212), and IBM (213) have published information concerning their computer systems. Other companies, such as DuPont, Western Electric, and CIBA-Geigy are in the process of developing computerized recordkeeping and information systems. As we stated in Section 1 and throughout the chapter, there is a definite need for data automation industrial health management systems. Computerization does have its advantages as related to "manual" recordkeeping and control systems.

APPENDIX 7A

7A.1 Typical System Requirements: Industrial Hygiene

Typical systems requirements are as follows:

1. Relate person to chemical and exposure and job/or department.
2. Calculate time-weighted averages for multiple exposure.
3. Record all exposures, calculate mean, and indicate maximum for year.
4. Print entire work history while working for this firm as related to items 1, 2, and 3.
5. Relate person to any bioassay performed.
6. Repeat 3 and 4 for bioassay.
7. Indicate quarterly bioassay or exposure evaluations not conducted.
8. Indicate routes of exposure (e.g., skin, lungs, etc.).
9. Hierarchy of programmed data should permit retrieval of any segment of the above for epidemiology studies (e.g., a certain chemical, job exposures greater than certain level, action level, etc.).
10. Indicate persons exceeding legal requirements and whether they were notified of exposure results within 5 working days.
11. Regulated areas:
    a. Indicate persons authorized to enter.
    b. Indicate daily roster—tie badge reader into system.
12. Indicate for each person the physician's work certification for chemicals and physical exposures (laser, etc.).
13. Indicate person received training annually.
14. Indicate quarterly deliquency regarding training.
15. System should be amenable to including physical and biological exposures immediately or at future date as indicated in items above.
16. Compliance program:
    a. Indicate control program documentation (engineering controls investigated) for exposures exceeding limits—OSHA requirement for documentation of wearing personal protection.

      b. Indicate situations (emergencies or incidents) requiring reports to OSHA.
      c. Identify exposure situations requiring special action or correction.
      d. Indicate quarterly delinquency regarding above.
17. Emergency program:
      a. Indicate that above have been accomplished for areas requiring OSHA and company requirements.
      b. Indicate delinquency quarterly.
18. Determine persons wearing respirators.
19. Indicate if medical surveillance performed.
20. Indicate if training conducted.
21. Indicate quarterly delinquencies regarding items 19 and 20.
22. Record and track ventilation measurements quarterly.
23. Indicate that measurements meet control values and/or the corrections being made.
24. Measurement made within 5 days of changes to meet new control values and OSHA ventilation requirements.
25. Indicate delinquencies to ventilation measurements quarterly (items 22–24).

APPENDIX 7B

7B.1 Ventilation Testing Program

7B.1.1 Responsibilities

The responsibilities are as follows:

1. Environmental safety and health (D/713-D/708).
      a. Determine the need for installation of local exhaust ventilation.
      b. Determine required velocity readings for exhaust systems in accordance with OSHA and/or recommended practice (joint effort between D/705 and D/713).
      c. Review installation of new exhausts, alteration and/or relocation of existing exhausts (normally included in project folder sign-off).
      d. Assign testing frequency schedule to all systems based on regulations and/or recommended practice.
      e. Develop and maintain a computerized ventilation program for the periodic maintenance of exhaust systems.
      f. Assign identification numbers to new systems and enter on computer.
      g. Apply for and obtain permits.
2. Facility engineering.
      a. Design, write specifications for, and classify all new or altered exhaust systems for inclusion in project folders.

    b.   Balance exhaust systems in accordance with OSHA and/or recommended practice (joint effort between D/705 and D/713).

    c.   Supervise comprehensive initial testing for each system. Refer to recommended sequence of testing for initial test.

    d.   Create field data form and file in central location accessible to both Facilities and Environmental Safety and Health.

    e.   Notify Environmental Safety and Health of changes initiated by Facility Engineering that would not appear on a project.

3.   Facility maintenance.

    a.   Perform initial and periodic testing on all systems. Refer to measurement sequence flow chart.

    b.   Determine the need for maintenance and make repairs.

    c.   Notify Environmental Safety and Health of failures or under specified conditions in existing exhaust systems.

4.   Using department line management.

    a.   Contact Environmental Safety and Health if a need for local exhaust appears.

    b.   Advise Maintenance Department 706 of any deterioration of exhaust systems in their area or systems that have not been tested when due.

    c.   Advise Environmental Safety and Health of any process changes where materials may be substituted, temperatures changed, or any adjustment to exhaust requirements might be needed.

7B.1.2   Measurements

All exhaust systems should be thoroughly tested upon installation and periodically tested throughout the life of the system. A valid comparison between the design basis and optimum system performance can be made only at the time of the initial survey of the system. Experience has shown that periodic surveys are required to assure that the system is performing adequately.

Basis for Evaluation of Ventilation Systems.

1.   To assure adequacy and performance.
2.   To assure that system performance is maintained.
3.   To determine the feasibility for expanding the system.
4.   To establish improved design parameters for new systems.
5.   To assure compliance with federal, state, and/or local regulations.

Initial Tests.

1.   Review system specifications and drawings.
2.   Inspect systems (fan rotation, belts, dampers, etc.).
3.   Select and identify test locations; make drawing on back of data sheet.

4.   Measure air volume, fan static pressure, motor rpm, and pressure drops across all components.

5.   Record test data and design specifications on test data sheet.

6.   Compare test data with design specifications, make alterations to meet specifications, codes, and so on.

7.   Retest system after adjustments have been made, record final test data; note on sketch the changes made.

8.   Retain test data for life of system.

9.   Use data sheet as base for inputting system on computer and updating master map.

Periodic Test for all Systems.

1.   Computer will issue maintenance cards based on frequency required. Refer to master map for location.

2.   Visual inspection.
   a.   Physical damage (corroded ducts, etc.).
   b.   Proper operation of components.
   c.   Smoke test.

3.   Take measurements (method specified on data card).
   a.   Same locations as initial tests or
   b.   Static pressure measurement (only when system is operating adequately).*

4.   If system is performing as required (per data card):
   a.   Fill in blanks on card.
   b.   Comment if process has changed at all.
   c.   Send card to Keypunch.

5.   If system is not performing adequately:
   a.   See if corrections can be made quickly (putting on new fan belt, removing hood blockage, etc.).
   b.   If corrected, reread, fill out card, and send to Keypunch.
   c.   If engineering help is required to find and arrange correction of problem, put actual readings on the card along with any comments and return to Keypunch.
   d.   If hood performance significantly changed (≥ 50 percent), place "do not operate" tag on hood face and alert using department manager and Environmental Safety and Health.

6.   Facility Engineering will receive discrepancy cards indicating systems that are not performing adequately.
   a.   System should be corrected as soon as possible.
   b.   Once corrected, remeasurement is essential.
   c.   The card is then filled in and sent to Department 713.

* Whenever alterations are made to a system, new initial testing is necessary as outlined in initial tests.

Computer Functions.

1. Information available from computer:
    a. Vent/hood.
    b. Stack.
    c. Fan.
    d. Operation.
    e. Materials and conditions (temperature, etc.).
    f. OSHA class (open surface tank).
    g. Principal hazard.
    h. Toxicity level (TLV or measured).
    i. Dimensions face/slot, tank, duct.
    j. Frequency of measurements required.
    k. Data of next measurement.
    l. Required readings: average velocity.
    m. Latest readings: minimum velocity.
    n. Previous readings: volume, static pressure, type (SP, DV, FV, SV, CV).
2. Measurement designations (see Appendix 4C).
    a. Capture or control velocity (CV).
    b. Slot velocity (SV).
    c. Duct velocity (DV).
    d. Face velocity (FV).
    e. Static pressure (SP).
    f. Velocity pressure (VP).
3. Maintenance cards sent to Maintenance every week identifying systems that need measuring in 2 weeks.
4. Weekly list of systems measured and discrepancies found sent to Department 713.
5. Discrepancies sent to Facility Engineering with master list weekly.
6. Monthly update listing sent to Department 713.

APPENDIX 7C

7C.1 Data Processing Glossary

The following definitions will assist the person interested in data processing in obtaining insight into the specialized language of data processing.

ADDRESS: An identification for a register, location in storage, or other data source or destination; the identification may be a name, a label, or a number.

ALGOL: Algorithmic language. A data processing language utilizing algebraic symbols to express problem-solving formulas for machine solution.

ALGORITHM: A prescribed set of well-defined rules or processes for the solution of a problem in a finite number of steps.

ANALOGUE DATA: Data represented in a continuous form, as contrasted with digital data in which are represented in a discrete (discontinuous) form. Analogue data are usually represented by means of physical variables, such as voltage, resistance, and rotation.

APL: A programming language. A problem-solving language designed for use at remote terminals; it offers special capabilities for handling arrays and for performing mathematical functions.

APPLICATION PROGRAM: A program written for or by a user that applies to the individual's own work.

AUXILIARY (PERIPHERAL) EQUIPMENT: Equipment not actively involved during the processing of data, such as input/output equipment and auxiliary storage utilizing punched cards, magnetic tapes, disks, or drums.

BASIC: An algebralike language used for problem solving by engineers, scientists, and others who may not be professional programmers.

BATCH PROCESSING: A system approach to processing: similar input items are grouped for processing during the same machine run.

BINARY: (1) The number representation system with a base 2. (2) A characteristic or property involving a selection, choice, or condition in which there are only two possibilities.

BINARY CODED DECIMAL (BDC):   A type of notation in which each decimal digit is identified by a group of binary ones and zeros.

BIT: A binary digit.

BYTE: A continguous set of binary digits operated on as a unit.

CENTRAL PROCESSING UNIT (CPU):   The unit of a computing system that contains the circuits that calculate and perform logic decisions based on a manmade program of operating instructions.

CHARACTER: One of a set of elementary symbols acceptable to a data processing system for reading, writing, or storing.

CHARACTER RECOGNITION: The technique of reading, identifying, and encoding a printed character by optical means.

CLEAR: To put a storage or memory device into a state denoting zero or blank.

COBOL: Common business-oriented language. A data processing language that resembles business English.

COMPILE: To convert a source-language program such as COBOL to a machine-language program.

COMPUTER PROGRAM: A series of instructions or statements, in a form acceptable to a computer, prepared to achieve a certain result.

CONSOLE: The unit of equipment used for communication between the operator or service engineer and the computer.

CONTROL PROGRAM: A program that is designed to schedule and supervise the performance of data processing work by a computing system.

CORE STORAGE: A form of magnetic storage that permits high speed access to information within the computer. See *Magnetic core*.

CPU: Central processing unit.

CRT: A display device on which images are produced: cathode ray tube.

DATA PROCESSING SYSTEM: A network of machine components, capable of accepting information, processing it according to man-made instructions, and producing the computed results.

DEBUG: To detect, locate, and remove errors from a programming routine or malfunctions from a computer.

DIGITAL DATA: Information expressed in discrete symbols.

DIRECT ACCESS: See *Random access*.

DISK STORAGE: A method of storing information in code, magnetically, in quickly accessible segments on flat rotating disks.

DOWNTIME: The elapsed time when a computer is not operating correctly because of machine or program malfunction.

DRUM STORAGE: A method of storing information in code, magnetically, on the surface of a rotating cylinder.

EDP: Electronic data processing.

EXECUTIVE: To perform a data processing routine or program, based on machine-language instructions.

FILE: A collection of related records—for example, in inventory control, one line of an invoice forms an item, a complete invoice forms a record, and the complete set of such records forms a file.

FILE MAINTENANCE: The processing of information in a file to keep it up to date.

FLOW CHART: A graphic representation for the definition, analysis, or solution of problem in which operations, data flow, and equipment are depicted symbolically.

FORTRAN: Formula translating system. A data processing language that closely resembles algebraic notation.

HARD COPY: A printed copy of machine output (printed reports, listings, documents, etc.).

HARDWARE: The mechanical, magnetic, electrical, and electronic devices of a computer.

INPUT/OUTPUT: Commonly called I/O. A general term for the equipment used to communicate with a computer; the data involved in such communication.

INQUIRY: A request for information from storage (e.g., a request for the number of available airline seats on a given flight).

INSTRUCTION: A statement that calls for a specific computer operation.

INTEGRITY: Preservation of data or programs for their intended purpose.

LANGUAGE: A defined set of symbols that can be converted to machine language. (ALGOL, FORTRAN, COBOL, etc.).

LIBRARY ROUTINE: A special-purpose program that can be maintained in storage for use when needed.

MACHINE LANGUAGE: A code used directly to operate a computer.

MACHINE READABLE: A medium that can convey data to a given sensing device.

MAGNETIC CORE (MAIN CORE): A configuration of tiny doughnut-shaped magnetic elements in which information can be stored for use at extremely high speed by the CPU.

MAGNETIC INK CHARACTER RECOGNITION (MICR): A method of storing information in characters printed with ink containing particles of magnetic material. The information can be detected or read at high speed by automatic devices.

MAGNETIC TAPE: A plastic tape with a magnetic surface on which data can be stored in a code of magnetized spots.

MARK-SENSE: To mark a position on a card or paper form with a pencil. The marks are interpreted electrically for machine processing.

MATHEMATICAL MODEL: A set of mathematical expressions that describes symbolically the operation of a process, device, or concept.

OBJECT PROGRAM: The machine-language program that is converted to electrical pulses that actually guide operation of a computer.

ON-LINE SYSTEM: In teleprocessing, a system in which the input data enter the computer directly from point of origin or in which output data are transmitted directly to where used.

OPERATING SYSTEM: An integrated collection of computer instructions that handle secretion, movement, and processing of programs and data needed to solve problems.

OPTICAL READER: A device used for machine recognition of characters by identification of their shapes.

OUTPUT: (1) The final results after data have been processed in a computer.(2) The device or set of devices used for taking data out of a computer system and presenting them to the user in the form desired.

PAPER-TAPE READER: A device that senses and translates the holes in a roll of perforated paper tape into machine-processable form.

PL/1: A high-level programming language, designed for use in a wide range of commercial and scientific computer applications.

PRINTER: A device that prints results from a computer on paper.

PROGRAM: (1)(n) The plan and operating instructions needed to produce results from a computer. (2)(v) To plan the method of attack for a defined problem.

PUNCHED CARD: A card punched with a pattern of holes to represent data.

RANDOM ACCESS: A technique for storing and retrieving data: it does not require a strict sequential storage of the data nor a sequential search of an entire file to find a specific record. A record can be addressed and accessed directly at its location in the file.

RAW DATA: Data that have not been processed or reduced.

REAL TIME: Computing that occurs while a process takes place so that results can be used to guide operation of the process.

ROUTINE: A sequence of machine instructions that carry out a specific processing function.

SIMULATE: To represent the functioning of a system or process by a symbolic (usually mathematical) analogous representation of it.

SOFTWARE: (1) The collection of man-written solutions and specific instructions needed to solve problems with a computer. (2) All documents needed to guide the operation of a computer (manuals, programs, flow charts, etc.).

SOLID STATE COMPONENT: A component whose operation depends on electrical activity in solids (e.g., performance of a transistor or cystal diode).

SORT: To arrange data in an ordered sequence.

SOURCE LANGUAGE: A language nearest to the user's usual business or professional language, which enables the user to instruct a computer more easily. FORTRAN, COBOL, ALGOL, BASIC, PL/I are a few examples.

STORAGE: Pertaining to a device in which data can be entered and stored and from which they can be retrieved at a later time.

STORED-PROGRAM COMPUTER: A digital computer that stores instructions in main core and can be programmed to alter its own instructions as though they were and can subsequently execute these altered instructions.

TELEPROCESSING: The use of telecommunications equipment to transmit data between two computers in different locations, or between input/output devices and a centralized computer when I/O is at a location remote from the computer.

TERMINAL UNIT: A device, such as a key-driven or visual display terminal, which can be connected to a computer over a communications circuit and can be used for either input or output from a location either near or far removed from the computer.

APPENDIX 7D

7D.1 Portable Computer Applications

7D.1.1 Safety and Health 5100 Computer Programs

This is a brief overview of the 5100 application at GSD Rochester.

I'm sure some of you would like to be able to see a program and if you like it, pick it up, take it back to your location, and run it. But it never is quite that way. There are a lot of subtle things that must be determined to find out whether or not the program will work at your location (i.e., computer availability, programming support, space, etc.).

I will be showing some reasons why Safety and Health at GSD Rochester is using computers to assist in the day-to-day Safety Engineering tasks.

As you know, the business environment in the past few years has been one of rapid change, with increasing competition and shortened product cycle time. This, together with new technologies requiring special facilities and working environments, has made information accessibility vital to Safety and Health!

While use of interactive language in the early 1970s was very productive, there was still a need to get closer to the user. In early 1976, IBM's first portable computer known as the 5100 was announced.

The lack of I/S support forced us to look at other ways to solve the need for better information—we chose the 5100 to fill that need.

Some of the features of the 5100 are user availability—owned by user, stand-alone capability, direct user control as a communication terminal with other systems make it an ideal system for our use.

The Safety and Health strategy for using a computer is to improve productivity, to improve information quality, develop inhouse use, have programming capability in Environmental Safety and Health area, and develop systems-oriented skills by using a computer.

The 5100 portable computer is self-contained and weighs 24 kg (approximately 50 pounds). It operates on ordinary 115 volt household current, and except for extreme cases, is insensitive to environmental conditions. No air conditioning is required. The computer includes the following components:

1. A processing unit.
2. A main storage unit.
3. A magnet tape unit.
4. A keyboard.
5. A display screen.
6. Various dials, lights, and switches.

Safety/Environmental Engineering has generated a number of programs for use on the 5100. A brief explanation follows:

1. Fire extinguisher tracking program which generates printouts of all extinguishers on site. The program can select by building, by hydrostatic test date, by extinguisher type and then generate a summary report of extinguishers. This has proven to be a real time-saver for the site fire marshal.

2. Internal power truck licensing program is used by the person that handles truck licensing. The program maintains a list of licensed operators and will generate a monthly expiration date listing for determining retraining/relicensing requirements.

3. Occupational injury/illness programs are a series of programs that compile statistics on injury and illness cases. We compile our accident experience monthly with additional summaries quarterly and annually.

The programs compute accurate accident rates quickly and also compare rates and types of cases from the previous year to determine accident trends.

4.   Safety administration programs consist of audit summary by site and department, safety glasses issued and costs, and material safety data sheet attainment report.

5.   Industrial hygiene programs that we are using are the respiratory authorization and training, and environmental sampling.

6.   Environmental Engineering programs are the monthly discharge report which is the results of chemical analysis on 25 constituents of wastewater compiled into a meaningful report to meet the state, federal, and corporate requirements, and the Facility Engineering licensing report which consists of the various licenses and permits needed in our plant.

What are conclusions?

The 5100 provides the flexibility needed in our everchanging safety and health field. It has stimulated technical vitality and pride of ownership in the individuals that use the system. Since the IBM 5100 is designed to be an easy-to-use portable computer that can be used by persons with varying levels of computer knowledge, there are many intangible benefits for the user.

Printout examples follow.

RESPIRATOR AUTHORIZATION AND TRAINING

EMPLOYEE NAME	SERIAL	DEPT/S	DATE OF TRAINING	RESP. ISSUED	DATE AUTHORIZED	AUTHORIZED FOR		
						SELF	AIR	CART
RJ	675801	712-0	0479	Y	0779	X	Y	X
GJ	676313	712-0	0379	Y	0778			X
A	377120	314-1	0479		0578	X		
H	478152	704-1	0379		0478	X	X	
VE	678834	712-1	0479	Y	0779	X	X	X
R	679602	707-2	0178		0478	X	X	X
KC	479711	706-1	0379		0478	X		
T	981301	090-3	0479	Y	0279			X
LS	881614	713-1	0379	Y	0578	X	X	X
F	282642	108-1	0679		0679			X
CM	682363	707-2	0679		0578	X	X	X
JG	782833	707-3	0178		0378	X	X	X
CS	082986	263-1	0779	Y	0679			
MN	483447	706-1	0379		0478	X	X	X
JF	783633	712-0	0479	Y	0778	X		X
LK	486059	454-1	0379		0379	X		
M	986193	454-1	0479		0479	X		
AL	486652	712-0	0479	Y	0778			X
RE	086754	703-1	0479		0578	X	X	
C	187444	706-1	0379		0678	X	X	X
EE	892042	707-3	0178		0278	X	X	X
TK	397121	675-1	0679		0679	X		
RW	599811	675-1	0479		0178	Y		Y

NUMBER OF PEOPLE TRAINED FOR RESPIRATOR USAGE : 73

MONTHLY INDUSTRIAL INJURY ILLNESS REPORT

	CURRENT YEAR		PREVIOUS YEAR		PERCENT
	CASES	RATE	CASES	RATE	CHANGE/RATE

OSHA LOST WORKDAY CASES

    AWAY FROM WORK CASES

    RESTRICTED WORK CASES

MEDICAL TREATMENT CASES(OSHA)

FIRST AID CASES

TOTAL NUMBER OF DAYS 'AWAY FROM WORK' : 1979 =   1978 =

CURRENT HOURS =   ,PREVIOUS HOURS =

COMPARISION OF OCCUPATIONAL INJURIES AND ILLNESSES FOR AUGUST 1979

INJURY TYPES	TOTALS		FIRSTAID		OSHA AFW		OSHA RWA		OSHA MT	
	79	78	79	78	79	78	79	78	79	78
BACK STRAIN										
BRUISE										
BURN										
DERMATITIS										
EYE INJURY										
FOREIGN BODY										
FRACTURE										
LACERATION										
INFECTION										
PUNCTURE										
STRAINS										
OTHER										

GRAND TOTAL

```
        SUMMARY REPORT OF EXTINGUISHERS


THE  NUMBER  OF  ABC  2.5  LB  EXTINGUISHERS:..............   1

THE  NUMBER  OF  ABC  10  LB  EXTINGUISHERS:..............  15

THE  NUMBER  OF  CO2  10  LBS  EXTINGUISHERS:..............  24

THE  NUMBER  OF  CO2  15  LBS  EXTINGUISHERS:.............. 186

THE  NUMBER  OF  CO2  18  LBS  EXTINGUISHERS:..............  80

THE  NUMBER  OF  DRY  CHEMICAL  10  LBS  EXTINGUISHERS:.....   5

THE  NUMBER  OF  DRY  CHEMICAL  20  LBS  EXTINGUISHERS:......  31

THE  NUMBER  OF  DRY  CHEMICAL  75  LBS  EXTINGUISHERS:......   4

THE  NUMBER  OF  HALON  5  LB  EXTINGUISHERS:. ...........  10

THE  NUMBER  OF  H2O  EXTINGUISHERS:.................... 314

THE  NUMBER  OF  METAL  X  EXTINGUISHERS:.................   1

THE  AMOUNT  OF  OPEN  NUMBERS  (NOT  ISSUED):.............   4

THE  TOTAL  NUMBER  OF  EXTINGUISHERS:....................,675


            ................SELECT BY BUILDING....................


NUMBER TYPE   SIZE   BLDG    LOCATION   INS/DATE  HYDRO/DATE
------ ----   ----   ----               ----       ----
 054   H2O           0202   DOCK         0179       0878
 056   H2O           0202   K 12         0179       0978
 110   H2O           0202   E12          0179       0173
 118   H2O           0202   B17          0179       0173
 246   H2O           0202   C19          0179       0376
 276   H2O           0202   S AISLE      0179       0579
 301   CO2    15     0202   J15 MACHRM   0179       1178
 303   CO2    15     0202   H12          0179       1178
 325   CO2    15     0202   G16 COMPRM   0179       0173
 392   CO2    15     0202   K19          0179       0173
 397   CO2    15     0202   F 19         0179       0978
 427   CO2    18     0202   F16 COMPRM   0179       0173
 429   CO2    15     0202   C12          0179       0173
 506   CO2    15     0202   D21 DOCK     0179       0273
 508   CO2    18     0202   L16          0179       1178
 512   HAL    5      0202   TAPE LIBRY   0179       0976
 523   CO2    15     0202   H17 MAC RM   0179       0673
 606   CO2    15     0202   G14 COMPRM   0179       0579
 608   CO2    15     0202   F14 COMPRM   0179       0579
 622   CO2    15     0202   W AISLE N    0179       0778
 624   CO2    15     0202   W AISLE S    0179       0778
 625   CO2    15     0202   S AISLE W    0179       0778


TOTAL EXTINGUISHERS ON THIS FLOOR......22
```

ALPHABETICAL LISTING OF LICENSED OPERATORS OF POWER TRUCKS

NAME	INT	SER #	DEPT	EXPIR	C/W	STRD	S/UP	WALK	SD/S	P/L	D/C
	K	14223	924-1	09/79	*	*	*	*	*		
	C	14664	716-1	12/79	*		*	*	*		
	B	14833	920-1	10/79	*	*					
	M	15624	710-3	03/81					*		
	S	16392	837-1	02/81						*	
	J	17091	920-1	04/80	*	*					
	D	17993	094-2	12/79		*	*				
	D	17964	674-1	01/80	*	*	*	*			
	K	19152	671-1	05/80		*	*	*			
	J	19261	715-1	10/79	*	*	*	*	*		
	R	19941	704-1	11/80					*		
	J	20296	641-1	11/79	*						
	J	20672	706-1	05/81	*	*	*	*	*		
	D	21880	716-1	05/80	*	*	*	*	*		
	D	21831	710-1	02/81					*		
	G	23901	674-1	08/79	*	*		*			
	J	23812	920-1	12/79	*	*					
	T	23412	857-1	01/81					*		
	R	23382	715-1	10/79	*	*	*	*	*		
	D	24434	706-1	09/80	*	*	*	*	*		
	D	24512	924-1	03/81	*			*			
	H	24921	671-1	02/80	*	*	*	*			
	H	24923	715-1	06/80	*	*	*	*	*		
	M	25235	675-1	05/80	*	*	*	*	*		
	K	26185	672-1	03/81	*	*		*			
	N	26735	674-1	02/80				*			
	R	27254	137-2	04/80	*		*				

SAFETY/HEALTH ENVIRONMENTAL SURVEY REPORT FOR MAY 1979

SAMPLE #	DATE	TYPE	TIME MINS	RATE LPM	DPT/S	BLD/F	OPERATION	CHEMICAL	RESULTS
00510	05/09/79	SOURCE	354	1.6	090-1	106-1	TANKS 22+23	ZNCL	
00520	05/09/79	SOURCE	143	1.1	090-1	106-1	TANK 16	CHROMIC ACID	
00530	05/09/79	SOURCE	128	1.9	090-1	106-1	TANK 608	CHROMIC ACID	
00540	05/09/79	SOURCE	180	1.9	090-1	106-1	TANK 608	CHROMIC ACID	
00550	05/09/79	SOURCE	170	2.09	090-1	106-1	TANK 614	NAOH	
00560	05/09/79	SOURCE	337	2.1	090-1	106-1	TANK 607	NAOH	
00570	05/09/79	SOURCE	28	.76	090-1	106-	TANK 13	NITROGEN DIOXIDE	
00580	05/09/79	SOURCE	32	.76	090-1	106-1	TANK 16	NITROGEN DIOXIDE	
00590	05/09/79	SOURCE	123	1.1	090-1	106-1	TANK 27+28	CHROMIC ACID	
00600	05/09/79	SOURCE	128	2.09	090-1	106-1	TANK 619	NICKEL	
00610	05/14/79	SOURCE	22	.98	920-1	104-1	FOAM IN PLACE	MDI	
00620	05/14/79	SOURCE	22	1.09	920-1	104-1	FOAM IN PLACE	MDI	
00630	05/16/79	SOURCE	109	1.5	090-1	106-1	TANK 608	SULFURIC ACID	
00640	05/16/79	SOURCE	15	.97	090-1	106-1	TANK 609	HCL	
00650	05/16/79	SOURCE	67	1	090-1	106-1	TANK 16	NITROGEN DIOXIDE	
00660	05/16/79	SOURCE	67	.98	090-1	106-1	TANK 16	HF	
00670	05/16/79	SOURCE	15	.97	090-1	106-1	TANK 26	HCL	
00680	05/16/79	SOURCE	97	.97	090-1	106-1	TANK 5	NITROGEN DIOXIDE	
00690	05/16/79	SOURCE	97	1.49	090-1	106-1	TANK 5	SULFURIC ACID	

NAME	INT	SER #	DEPT	EXPIR	C/W	STRD	S/UP	WALK	SD/S	P/L	D/C
	A	05022	713-1	11/79				*			
	R	06758	094-2	11/79		*	*				
	J	20296	641-1	11/79	*						
	A	28773	716-1	11/79	*	*	*	*	*		
	B	31281	920-1	11/79	*	*					
	E	42721	707-3	11/79	*	*	*	*	*		
	M	48739	867-1	11/79					*		
	G	53233	707-2	11/79	*	*	*	*	*		
	D	56749	674-1	11/79	*						
	D	66625	672-1	11/79				*			
	H	67064	715-1	11/79	*	*	*	*	*		
	P	74833	672-1	11/79				*			
	C	74672	670-1	11/79	*	*					
	D	75735	715-1	11/79	*	*	*	*	*		
	D	84003	675-1	11/79	*	*	*	*	*		
	B	90401	921-1	11/79	*	*					
	R	92925	672-1	11/79	*	*	*	*			
	R	93669	670-1	11/79	*	*	*	*			

THE NUMBER OF TRUCKERS WITH LICENSES EXPIRING DURING 1179 = 18

PRESCRIPTION SAFETY GLASSES ISSUED

DATE	SERIAL	NAME	DEPT	COST	TYPE
05/24/79	30175	G A	313	15.71	BIFOCAL
05/03/79	60352	W H	084	15.71	SINGLE
05/01/79	40370		439	19.26	SINGLE
05/24/79	30384	L J	707	16.56	BIFOCAL
05/10/79	80418	T	048	15.71	SINGLE
05/23/79	10451	P	031	15.71	SINGLE
05/24/79	20504	D	290	15.71	SINGLE
05/14/79	80532	J	261	16.81	SINGLE
05/15/79	30612	J T	117	15.71	BIFOCAL
05/01/79	70606	J	528	6.75	SINGLE
05/22/79	20743	D	131	16.56	BIFOCAL
05/17/79	90771	W	45L	16.56	BIFOCAL
05/08/79	31045	J	49B	19.26	SINGLE
05/17/79	11072	D L	284	22.51	SINGLE
05/03/79	21170	D	712	15.71	BIFOCAL
05/29/79	01435	R	116	24.26	SINGLE
05/17/79	61765	D J	117	15.21	SINGLE
05/10/79	31796	D J	674	16.56	SINGLE
05/22/79	72021	J	312	16.56	SINGLE

FACILITY ENGINEERING LICENSING REPORT

CTRL NO.	LICENSE TYPE	DESCRIPTION	BLDG NO.	PERMIT NO.	AGENCY	RQMT FREQ	DATE OBTAINED	RENEWAL DATE	DEPT RESP.	PERSON RESP.
001	PERMIT	COLAG WATER SCRUBBER	020-1		MPCA	RENEWAL	04/14/78	04/14/83	713	BAKER
002	PERMIT	WATER FALL SCRUBBER	104-0		MPCA	RENEWAL	04/14/78	04/14/83	713	BAKER
003	PERMIT	POWER HOUSE BOILERS	315		MPCA	RENEWAL	06/06/74	06/06/79	713	HARMON
004	PERMIT	FLAME SPRAY SCRUBBER 1,2	106-0		MPCA	RENEWAL	09/03/75	09/03/80	713	BAKER
005	PERMIT	WATER FALL SCRUBBER	104-0		MPCA	RENEWAL	03/18/76	03/18/81	713	BAKER
006	PERMIT	MAIN PLANT EMISSION FACILITY	ALL		MPCA	RENEWAL	09/03/75	09/03/80	713	BAKER
007	EXEMPTION		ALL		MPCA	RENEWAL	04/30/78	11/30/82	713	BAKER
008	REPORT		ALL		MPCA	ANNUAL		01/31/80	713	BAKER
009	PERMIT	LIQUID STORAGE	SITE		MPCA	ONE TIME	09/05/73		705	JORDON
010	PERMIT	LIQUID STORAGE	SITE		MPCA	ONE TIME	11/01/73		705	JORDON
011	PERMIT	FOAM SPRAY PACKAGING	104		MPCA	ONE TIME	01/07/75		713	HARMON
012	PERMIT	FOAM SPRAY PACKAGING	104		MPCA	RENEWAL	03/12/75	03/12/80	713	HARMON
013	PERMIT	MAGNETIC INK COATER	030		MPCA	ONE TIME	03/27/75		713	HARMON
014	PERMIT	WELL	SITE		MCD	ONE TIME	06/02/75		713	HARMON
015	PERMIT	EXHAUST VENT	020		MPCA	ONE TIME	06/18/75		713	BAKER
016	PERMIT	PAINT BOOTH	104		MPCA	ONE TIME	06/30/75		713	BAKER
017	PERMIT	IWT	204		MPCA	ONE TIME	10/27/75		713	BAKER
018	PERMIT	CHROME ETCH	020		MPCA	ONE TIME	03/18/76		713	BAKER
019	PERMIT	LAB TEST CHAMBER	004		MPCA	ONE TIME	04/23/76		713	BAKER
020	PERMIT	WIRE SOLDER	103		MPCA	ONE TIME	06/04/76		713	BAKER
021	PERMIT	PLASMA ETCH	020		MPCA	ONE TIME	06/15/76		713	BAKER

APPENDIX 7E

7E.1   Personal Computer Applications

7E.1.1   Printouts Using P.C. Files

```
Which drive (ABCD) for the Data:_

FILE:REPROHAZ



                                    (F1)   ADD a record
                                    (F2)   MODify  a record
                                    (F3)   DELete  a record
                                    (F4)   DISplay a record
                                    (F5)   FINd a record
                                    (F6)   LISt or clone
                                    (F7)   SORt the index
                                    (F8)   see the record LAYout
                                    (F9)   alter a Field NAMe
                                    (F10)  END or change database
```

```
                    RADIATION SOURCES ON SITE
      01-01-1980 AT 00:38
      TYPE SOURCE       MANAGER            BLD DEPT ROOM  ACTIVITY
      ================= ================== === ==== ===== ==========
      I
      J
      TH 232            AUSTEN             110 PN6  LL10  .1 UCI
      PO 210            AUSTEN             110 PN6  LL10  .1 UCI
      PO 210            SCHNUR             110 T42        20 MCI


              TOTALS:
      ------------------------------
```

RADIATION EQUIPMENT ON SITE

```
01-01-1980 AT 00:40
MFG                  MAXVOL MODEL        TYPE                    BT            BLD DEPT
================     ====== ===========  ====================    ============  === ====
                                                                 10
                                                                 2
                                                                 3
                                                                 4
                                                                 5
                                                                 6
                                                                 7
                                                                 8
                                                                 9
VARIAN/EXTRON        200    200CF4       ION IMP                 2950200       110 T20
VARIAN/EXTRON        200    DF3000       ION IMP                 306177A       110 T20
HEWLETT PACKARD      130    43804N       RADIOGRAPHIC            1849207-AA    110 C5H
HEWLETT PACKARD      100    43805N       RADIOGRAPHIC            2950288       110 T91
HEWLETT PACKARD      110    43804N       RADIOGRAPHIC            9369584-AB    120 C8R
CAMBRIDGE            25     100          SEM                     11            110 T90
CAMBRIDGE            30     S-4          SEM                     1698210       110 PN9
AMRAY               30     1200B        SEM                     1849374AA     110 C5H
AMRAY               30     1400         SEM                     2991380       110 PN9
CAMBRIDGE            40     250          SEM                     4136989       110 T90
HOLGER ANDREASEN    250    A2501        X-RAY                   1694257       120 EQ2
```
EMPLOYEES ON RADIATION BADGE PROGRAM
```
01-01-1980 AT 00:42                              Page 1
NAME              SS NO     SEX    BADNO1  MREM8 DATE8
================  ========  =====  ======= ===== =======
```

```
        TOTALS:
-----------------------------
Printed 0 of the 1 records.
```

GENOTOXINS

```
01-01-1980 AT 00:22
CHEMICAL                          MUTAGEN CARCINOGEN TERATOGEN EMBR/FETOTOX M REPRO HAZ TLV
================================  ======= ========== ========= ============ =========== ==========
NICKELOUS CHLORIDE                   X                                                  .1 MG/M3
NICKELOUS NITRATE                    X                                                  .1 MG/M3
NITROUS OXIDE                                                               X
ORGANIC MERCURY COMPOUNDS                                                   X           .01 MG/M3
PHENOL                               X                                                  5 PPM
PHENOL                               X
PHOSPHOROUS                                                                 X           .1 MG/M3
POLYCHLORINATED BIPHENYLS            X         X          X                 X           .5 MG/M3
PYROCATECHOL VIOLET                  X
SILANE A 186                         X                                                  1 PPM
SODIUM ARSENITE                      X         X
SULFANILAMIDE                        X
TETRACHLORO ETHLYENE         X       X         X                            X           50 PPM
THIOACETAMIDE                        X
TOLUENE                                                                     X           375 MG/M3
TRICHLOROETHYLENE                                                           X           50 PPM
TRICHLOROTRIFLOUROETHANE/FREON       X                                                  1000 PPM
XYLENE                                                                      X           100 PPM
```

```
        TOTALS:
--------------------------------------
Printed 72 of the 72 records.
```

VENTILATION PROGRAM

01-01-1980 AT 00:14

BT NO	LOCATION	DEPT	DEPT NAME	DATE	AVG FAC VEL	TARG VEL	MIN ACPT	AIR VEL	CHEMICALS	RECHEK	RECK FAC NAIR
VEL REMARKS											
1388247	110-D69	C5J	CHEM LAB	021583	122.5 FPM	100 FPM	90 FPM				
2164253	110-D69	C5J	CHEM LAB	021583	109.38 FPM	100 FPM	90 FPM				
21642533											
2180809	110-D69	C5J	CHEM LAB	021583	131 FPM	100 FPM	90 FPM				
2414679	110-Y29	TE2	VLSI	130	178.3 FPM	100 FPM	90 FPM		PHOSPHORIC ACID		
2464816	110-T31	TE2	VLSI	130	72.5 FPM	100 FPM	90 FPM				
2466804	110-T31	TE2	VLSI	130	132.5 FPM	150 FPM	135 FPM				
2950208	110-Y2	TE2	VLSI	021683	105 FPM	100 FPM	90 FPM		ACETONE		
RINSE AND DRY											
2950243	110-10Y	TE2	VLSI	013083	153.8	100 FPM	90 FPM		IPA, NMP		
2950244	110-	TE2	VLSI	013083	147.5 FPM	100 FPM	90 FPM				
2950245	110-Y16	TE2	VLSI	013083	156.25 FPM	100 FPM	90 FPM				
RINSE AND DRY											

CALIBRATION SCHEDULE

01-01-1980 AT 00:19

Page 1

BT NO	EQUIP TYPE	MFG	FREQ	ACTION DATE	DUE DATE 1
2949276	THERMOANEMOMETER	ALNOR	1/YR	3/84	2/4/82

SAMPLING RESULTS

01-01-1980 AT 00:00

SAMPNO	DIV	DEPT	BLDG/RM	CONTAMINANT	DATE	SAMPLE LOCAT	TLV	RESULTS	REMARKS
0032	FSD	068	105	TOLUENE	06/02/82		375 MG/M3		BEFORE VENT FIX
0033	FSD	068	105	TOLUENE	06/02/82		375 MG/M3		
0070	FSD	068	105	TOLUENE	08/17/82		375 MG/M3		
0071	FSD	068	105	TOLUENE	08/17/82		375 MG/M3		
0072	FSD	068	105	TOLUENE	08/17/82		375 MG/M3		
0073	FSD	068	105	TOLUENE	08/17/82		375 MG/M3		
0074	FSD	068	105	TOLUENE	08/17/82		375 MG/M3		
0130	GTD	C8W	110	TOLUENE	02/17/83		375 MG/M3		BEFOR VENT INCR
0131	GTD	C8W	110	TOLUENE	02/17/83		375 MG/M3		BEFOR VENT INCR
0132	GTD	C8W	110	TOLUENE	02/17/83		375 MG/M3		BEFOR VENT INCR
0133	GTD	C8W	110	TOLUENE	02/17/83		375 MG/M3		

EMPLOYEES ON RESPIRATOR PROGRAM

01-01-1980 AT 00:11

NAME	SER NO	DEPT	MANAGER	MEDAPPR	RETRAIN	TRAIN1	TRAIN2	TYPE	X
,PA	464710	0682	,GARY	07/08/83	08/84	08/19/83		MSA	S
,LINDA	934202	068	,BONNIE	06/16/83	06/84	06/23/83		MSA	M
,SUSAN	153510	068	,BONNIE	02/02/83	NONE	03/30/83		DEVILB	M
,GEORGE	189778	068	,BONNIE	03/05/81	07/84	03/31/81	7/29/83		

378

7E.1.2   Printouts Using LOTUS

```
Lotus Access System  V.1A  (C)1983 Lotus Development Corp.          MENU
----------------------------------------------------------------------------
1-2-3  File-Manager  Disk-Manager  PrintGraph  Translate  Exit
Enter 1-2-3 -- Lotus Spreadsheet/Graphics/Database program
============================================================================
```

```
                        Tue  01-Jan-80
                          0:00:42am
```

```
       Use the arrow keys to highlight command choice and press [Enter]
   Press [Esc] to cancel a choice; Press [F1] for information on command choices
```

```
                    INJURY/ILLNESS SUMMARY
==================================================================================================
DIV:                           |               TYPE OF OSHA CASE
MANAGER:                        |#################################################################
DEPT: C9E                       |             RESTRICTED
PERIOD COVERED: JAN. TO OCT., 1983 |  AWAY FROM     WORK                        FIRST
SOURCE OF INJURY/ILLNESS         |     WORK      ACTIVITY       MEDICAL          AID         TOTAL
==================================================================================================
TOTAL                                   0           0              0             1             1
SLIPS,SAME LEVEL           A1            0           0              0             0             0
SLIPS,DIFFERENT LEVEL      A2            0           0              0             0             0
STRUCK BY OBJECT           B1            0           0              0             0             0
STRUCK AGAINST OBJECT      B2            0           0              0             0             0
CAUGHT IN/BETWEEN OBJECT   B3            0           0              0             0             0
ABRASIONS,PUNCTURES        B4            0           0              0             0             0
LIFTING                    C             0           0              0             0             0
PUSH,PULL,OVEREXERT        D             0           0              0             1             1
ELECTRICAL CONTACT         E             0           0              0             0             0
INHALE,INGEST,ABSORB       F             0           0              0             0             0
HAND/POWER TOOLS,MACHINERY G             0           0              0             0             0
CHEMICAL EXPOSURE          H             0           0              0             0             0
AUTO ACCIDENT              I             0           0              0             0             0
RADIATION EXPOSURE         J1            0           0              0             0             0
CONTACT TEMPERATURE        J2            0           0              0             0             0
NOT INDUSTRIALLY RELATED   J3            0           0              0             0             0
OTHER                      J4            0           0              0             0             0
==================================================================================================
                      TOTAL             0           0              0             1             1
==================================================================================================
```

```
=====================================================================================================
DIV:                         │                            UNSAFE CONDITIONS
MANAGER:                     │ ▌▌▌▌▌▌▌▌▌▌▌▌▌▌▌▌▌▌▌▌▌▌▌▌▌▌▌▌▌▌▌▌▌▌▌▌▌▌▌▌▌▌▌▌▌▌▌▌▌▌▌▌▌▌▌▌▌▌▌▌▌▌▌▌▌▌
DEPT: C9E                    │
PERIOD COVERED: JAN. TO      │ INEFFECTIVE      NO    HAZARDOUS                        IMPROPER
                OCT., 1983   │   SAFETY       SAFETY    HOUSE   EQU/TOOLS  IMPROPER     ILLUM.
SOURCE OF                    │   DEVICE       DEVICE   KEEPING  DEFECTIVE  APPAREL      VENT.      TOTAL
INJURY/ILLNESS               │
=====================================================================================================
TOTAL                              0            0        0         0         0          0          0
SLIPS,SAME LEVEL        A1         0            0        0         0         0          0          0
SLIPS,DIFF LEVEL        A2         0            0        0         0         0          0          0
STRUCK BY OBJECT        B1         0            0        0         0         0          0          0
STRUCK AGAINST          B2         0            0        0         0         0          0          0
CAUGHT IN/BETWEEN       B3         0            0        0         0         0          0          0
ABRASIONS,PUNCTURES     B4         0            0        0         0         0          0          0
LIFTING                 C          0            0        0         0         0          0          0
PUSH,PULL,OVEREXERT     D          0            0        0         0         0          0          0
ELECTRICAL CONTACT      E          0            0        0         0         0          0          0
INHALE,INGEST,ABSORB    F          0            0        0         0         0          0          0
HAND/POWER TOOL         G          0            0        0         0         0          0          0
CHEMICAL EXPOSURE       H          0            0        0         0         0          0          0
AUTO ACCIDENT           I          0            0        0         0         0          0          0
RADIATION EXPOSURE      J1         0            0        0         0         0          0          0
CONTACT TEMPERATURE     J2         0            0        0         0         0          0          0
NOT IND. RELATED        J3         0            0        0         0         0          0          0
OTHER                   J4         0            0        0         0         0          0          0
=====================================================================================================
                TOTAL              0            0        0         0         0          0          0
=====================================================================================================
```

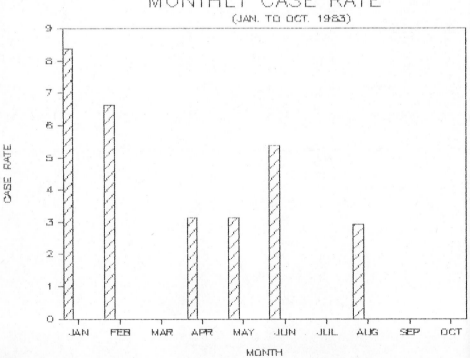

MONTHLY CASE RATE
(JAN. TO OCT. 1983)

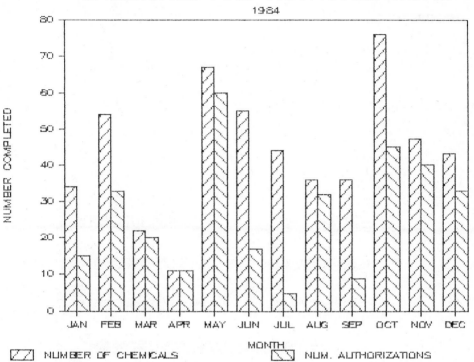

CHEMICALS AND AUTHORIZATIONS
1984

NUMBER COMPLETED

MONTH

 NUMBER OF CHEMICALS    NUM. AUTHORIZATIONS

INDUSTRIAL HYGIENE AIR SAMPLING
1984

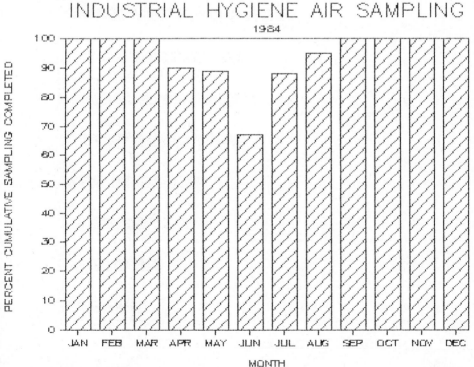

PERCENT CUMULATIVE SAMPLING COMPLETED

MONTH

381

APPENDIX 7F

7F.1   Computer Graphics Applications

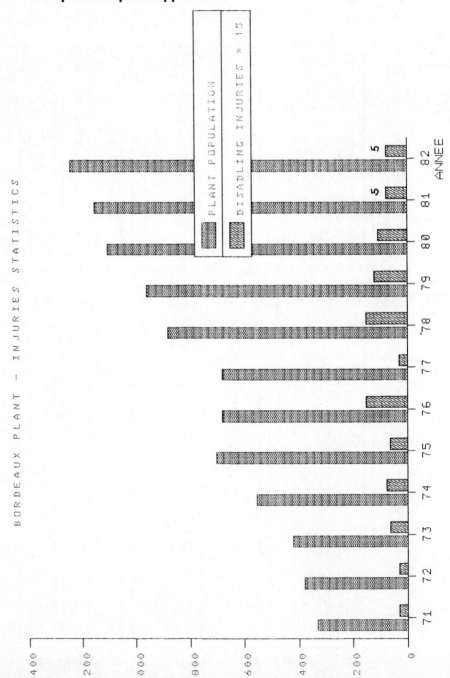

PREPARED ON DECEMBER 21, 1983 AT 4:59 PM

ENERGY MANAGEMENT: ENERGY SUMMARY
- -

ENERGY UNIT : US
CURRENCY UNIT: US DOLLARS

CHILLED WATER: GENERATED
STEAM   : GENERATED

ELECTRICITY COST AND CONSUMPTION PROFILES
- -

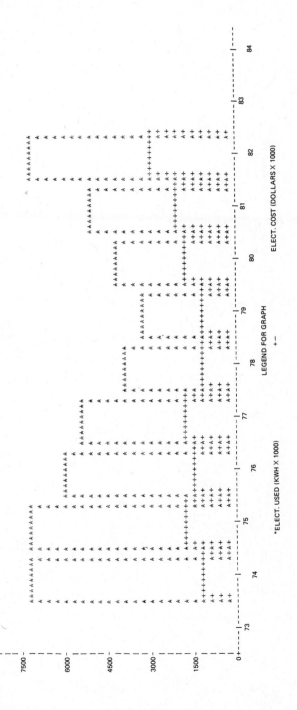

LEGEND FOR GRAPH

*ELECT. USED (KWH X 1000)

+ —

ELECT. COST (DOLLARS X 1000)

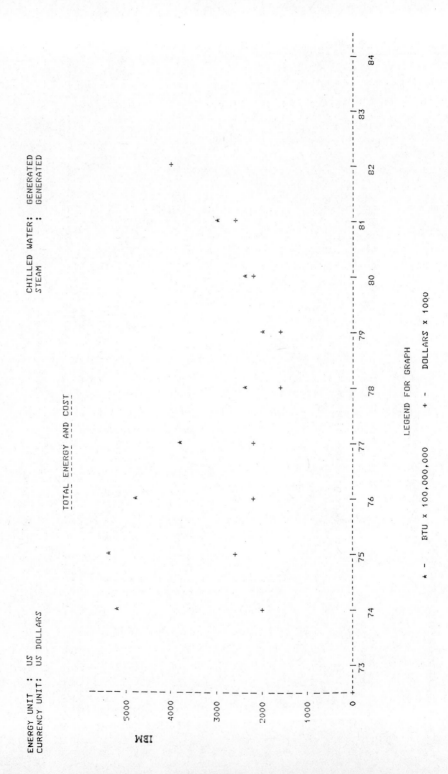

ENERGY UNIT : US
CURRENCY UNIT: US DOLLARS

CHILLED WATER: GENERATED
STEAM : GENERATED

TOTAL ENERGY AND COST

LEGEND FOR GRAPH

* - BTU x 100,000,000   + - DOLLARS x 1000

384

REFERENCES

1. J. H. MacEachran, W. E. Clendenning, and R. E. Gosselin, "Computer-Derived Exposure Tests for Common Contact Dermatitis Antigens," *Contact Dermatitis*, **2,** 239 (1976).

2. R. O. Moss and H. J. Ettinger, "Respirable Dust Characteristics of Polydisperse Aerosols," *Arch. Env. Health*, **31,** 546 (1970).

3. J. R. Goldsmith, J. Terzaghi, and J. D. Hackney, "Evaluation of Fluctuating Carbon Monoxide Exposures," *Arch. Environ. Health*, **7,**33–49, 647–663, (1963).

4. J. E. Peterson, H. R. Hoyle, and E. J. Schneider, "The Application of Computer Science to Industrial Hygiene," *Am. Ind. Hyg. Assoc. J.*, **27,** 180 (1966).

5. "Computers Turn on for Industrial Hygiene," *Occup. Haz.*, **30,** 37 (1968).

6. R. G. Edwards, Jr., C. H. Powell, and M. A. Kendrick, "Dust Counting Variability," *Am. Ind. Hyg. Assoc. J.*, **27,** 546 (1966).

7. D. C. Stevens, W. L. Churchill, D. Fox, and N. R. Large, "A Data Processing System for Radioactivity Measurements on Air Samples," *Ann. Occup. Hyg.*, **13,** 177 (1970).

8. F. J. Haughey and R. M. Maganelli, "An Experimental System for Aerosol Research," *Am. Ind. Hyg. Assoc. J.*, **29,** 268 (1968).

9. M. E. Jacobson, E. W. White, P. B. Denee, K. M. Morse, et al., "Dust Measurement and Control, Coal Workers' Pneumoconiosis," *Ann. N. Y. Acad. Sci.*, **200,** 661 (1972).

10. E. J. Jones, "Practical Aspects of Counting Asbestos on the Millipore, PIMC," *Microscope*, **23,** 93 (1975).

11. J. F. Remark, "Computerized Three-Dimensional Illustrations of Gas Equations," *J. Chem. Educ.*, **52,** 61 (1975).

12. J. E. Evans, and J. T. Arnold, "Monitoring Organic Vapors," *Environ. Sci. Technol.*, **9,** 1134 (1975).

13. D. P. Schlik and R. G. Peluso, "Respirable Dust Sampling Requirements under the Federal Coalmine Health and Safety Act of 1969," U. S. Bureau of Mines, Information Circular 8484, Publication Distribution Branch, Pittsburgh, Information Circular, July 1970.

14. R. Barry, "Computer Control," *Safety*, **40,** 13 (1968).

15. M. Robert, "The International Occupational Safety and Health Center," *Ann. Occup. Hyg.*, **16,** 267 (1973).

16. A. S. Reid, "Classification and Sources in Environmental Safety and Health," *Ann. Occup. Hyg.*, **16,** 257 (1973).

17. E. B. Duncan, B. McGovern, and S. A. Hall, "BOHS Symposium on Cooperation in Information Handling and Retrieval in Environmental Hygiene," *Ann. Occup. Hyg.*, **16,** 305 (1973).

18. B. Herman, "Computer Analysis in the Evaluation of Occupational Medical Records," *Ann. Occup. Hyg.*, **16,** (1973).

19. R. F. Cumberland and M. D. Hebden, "A Scheme for Recognizing Chemicals and Their Hazards in an Emergency," *J. Hazardous Mat.*, **1,** 35 (1975).

20. W. C. McArthur and B. G. Kneazewycz, "A Nuclear Power Plant Radiation Monitoring System," *Health Phys.*, **29,** 427 (1975).

21. E. N. Corlett, V. J. Morcombe, and B. Chanda, "Shielding Factory Noise by Work-in-Progress Storage," *Appl. Ergonomics*, **1,** 73 (1970).

22. R. C. Miller and A. Sagar, "Analysis of Machining Noise and Its Potential Application to Adaptive Control," *Mach. Prod. Eng.*, **121,** 840 (1972.

23. F. C. Hagg, "Using Computer Programs as Noise Control Tools," *Sound Vib.* **5,** 22 (1971).

24. J. S. Bendat, "Modern Methods for Random Data Analysis," *Sound Vib.*, **3,** 14 (1969).

25. D. W. Merritt and R. R. James, "Isograms Show Sound-Level Distribution in Industrial Noise Studies," *Sound Vib.*, **7,** 12 (1973).

26. K. Drechsler, "Studies on Vibration Transmission to the Driver's Cab of Automotive Agricultural Machinery," *Agratechnik,* **24,** 553 (1974).

27. J. Donovan, "Community Noise Criteria," *Am. Ind. Hyg. Assoc. J.,* **36,** 849 (1975).

28. J. F. Bell, A. C. Johnson, and J. C. K. Sharp, "Pulse-Echo Method of Investigating the Properties of Mechanical Resonators," *J. Acoust. Soc. Am.,* **57,** 1085 (1975).

29. B. Stanton, "The Computer in Development Testing and Evaluation," *ASHRAE J.,* October 1973, p. 35.

30. G. W. Dunn, "Central Plant Design for Utilization of Computer Controls," *ASHRAE J.,* July 1973, p. 55.

31. E. S. Rubin, "A New Application of Building Programs," *ASHRAE J.,* February 1973, p. 46.

32. L. W. Nelson, "Reducing Fuel Consumption with Night Setback," *ASHRAE J.,* August 1973, p. 41.

33. R. S. Bycraft, "Ventilation Energy System Analysis by Computer," *ASHRAE J.,* June 1973, p. 46.

34. K. N. Feinberg, "Use of Computer Programs to Evaluate Energy Consumption of Large Office Buildings," *ASHRAE J.,* January 1974, p. 73.

35. H. Rentsch, "The Use of Computers for Calculating the Soundproofing Requirements of Ventilation Systems," *Luft-Und Kalletechnic,* **7,** 11 (1971).

36. B. I. Medvedev and V. A. Pavlovskii, "Computer Simulation of Emergency Ventilation Conditions in Miners with Intense Heat Transfer," *Izv. Vyssh. Uchebn. Zavode. Forn. Zh.,* **15,** 81 (1972).

37. K E. Hicks, "Computer Calculation and Analysis of the P4SR Heat Stress Index," *Environ. Res.,* **4,** 253 (1971).

38. P. E. Smith, and E. W. James II, "Human Responses to Heat Stress," *Occup. Med.,* **9,** 332 (1964).

39. R. L. Harris, Jr., "Computer Simulation and Radiant Heat Load and Control Alternatives," *Am. Ind. Hyg. Assoc. J.,* **35,** 75 (1974).

40. H. M. Berlin, L. Stroschein, and R. F. Goldman, "A Computer Program to Predict Energy Cost, Rectal Temperature, and Heart Rate Response to Work, Clothing, and Environment," Edgewood Arsenal Special Publication ED-SP-74011, Aberdeen Proving Ground, Aberdeen, Md.

41. "A Study of Hypertension During Work and Computerized Analysis of Data Obtained," *Cah. Med. Interprof.,* **43,** 25 (1971).

42. W. S. Neisel and D. C. Colins, "Structural Languages and Biomedical Signal Analysis Using Interactive Graphics," Air Force Office of Scientific Research, Report AFOSR-TR-72-0616, Arlington, Va., March 1972.

43. C. D. Jenkins, R. H. Rosenman, and S. J. Zyzanski, "Prediction of Clinical Coronary Heart Disease by a Test for Coronary-Prone Behavior Pattern," *New Eng. J. Med.,* **290,** 1271 (1974).

44. R. E. Birk et al., "Approach to a Reliable Program for Computer-Aided Medical Diagnosis," *Aerosp. Med.,* **45,** 659 (1974).

45. M. Helberman et al., "Complex Man-Machine System for Delivery of Outpatient Medical Care," *Aerosp. Med.,* **45,** 975 (1974).

46. G. N. Bycroft and L. Seaman, "Mathmatical Models of Head Injuries," Stanford Research Institute, SRI Project 1633, Menlo Park, Calif., July 1973.

47. D. P. Discher, F. J. Massey, and W. Y. Halleti, "Quality Evaluation and Control Methods in Computer-Assisted Screening," *Arch. Environ. Health,* **19,** 323 (1969).

48. H. R. Newman and M. L. Schulman, "Renal Cortical Tumors: A 40-Year Statistical Study," *Urol. Surv.,* **15,** 2 (1969).

49. E. Edwards and F. P. Lees, "The Influence of the Process Characteristics on the Role of the Human Operator in Process Control," *App. Ergonomics,* **5,** 21 (1974).

50. C. H. Suh, "The Fundamentals of Computer Aided X-Ray Analysis of the Spine," *J. Biomech.,* **7,** 161 (1974).

51. J. A. Crocco et al., "A Computer-Assisted Instruction Course in the Diagnosis and Treatment of Respiratory Diseases," *Am. Rev. Respir. Dis.,* **111,** 299 (1975).

52. W. R. Ayers et al., "Description of a Computer Program for Analysis of the Forced Expiratory Soirogram. II, Validation," *Comput. Biochem. Res.,* **2,** 220 (1969).

53. M. V. Merrick, "The Role of Radioisotopes in the Diagnosis of Bronchial Cancer, *Scand. J Respir. Dis.,* **85,** 106 (1974).

54. A. L. Rector and E. Ackerman, "Rules for Sequential Diagnosis," *Comput. Biomed. Res.,* **8,** 143 (1975).

55. N. Konietzko, W. E. Adam, and H. Matthys, "Use of Radioisotopes in Modern Diagnosis of Pulmonary Function Impairment," *Munch. Med. Wochensch.,* **116,** 159 (1974).

56. F. Wiener, "Computer Simulation of the Diagnostic Process in Medicine," *Comput. Biomed. Res.,* **8,** 129 (1975).

57. B. N. Feinberg and J. D. Schoeffler, "Computer Optimization Methods Applied to Medical Diagnosis," *Comput. Biol. Med.,* **5,** 3 (1975).

58. R. G. Mancellas and A. Ward, "Machine Recognition in Pathology," *Comput. Biol. Med.,* **5,** 39 (1975).

59. A. W. Sills, V. Honrubia, and W. E. Kumley, "Algorithm for the Multi-Parameter Analysis of Nystagmus Using a Digital Computer," *Aviat., Space Environ. Med.,* **46,** 934 (1975).

60. E. Anzaldi and E. Mira, "An Interactive Program for the Analysis of ENG Tracings," *Acta Otolaryngol.,* **80,** 120 (1975).

61. J. L. Craig and C. M. Derryberry, "Applied Concepts of Automation in an Occupational Medical Program," *Ind. Med. Surg.,* **40,** 9 (1971).

62. J. Ortega, "The Application of Data Processing Techniques to Occupational Health Records," *Cah. Notes Doc. Secur. Hyg. Trav.,* **62,** 61 (1971).

63. M. R. Scott and W. S. Frederik, "Electronic Data Processing and Multiphasic Health Screening," *J. Occup. Med.,* **14,** 457 (1972).

64. W. S. Frederik and M. R. Scott, "Medical Statistics, System Monitoring and Provisional Normals," *J. Occup. Med.,* **14,** 466 (1972).

65. S. Pell, "Epidemiological Studies in a Large Company Based on Health and Personnel Records," *Public Health Rep.,* **83,** 399 (1968).

66. J. Planques, "Report of the Commission for Medical Records in Occupational Medicine," *Arch. Mal. Prof.,* **34,** 25 (1973).

67. I. Cavill et al., "A System for Data Processing in Haematology," *J. Clin. Pathol.,* **27,** 330 (1974).

68. J. Planques et al., "Statistical Evaluation on Medical Records for Occupational Health Purposes," *Arch. Mal. Prof.,* **32,** 129 (1971).

69. J. P. Horwitz et al., "Adjunct Hospital Emergency Toxicological Service," *JAMA,* **235,** 1708 (1976).

70. H. LaRocca and I. Macnab, "Value of Pre-Employment Radiographic Assessment of the Lumbar Spine," *Ind. Med. Surg.,* **39,** 31–36, 253–258 (1970).

71. A. Beck and J. Killus, "Normal Posture and Spine Determined by Mathematical and Statistical Methods," *Aerosp. Med.,* **44,** 1277 (1973).

72. R. S. Sherman, C. A. Bertrand, and J. C. Duffy, "Roentgenographic Detection of Cardiomegaly in Employees with Normal Electrocardiograms," *Am. J. Roentgenol. Radium Ther. Nuclear Med.,* **119,** 493 (1973).

73. F. Bockat and S. N. Wiener, "An Electrostatic Printer Display for Computerized Scintiscans," *Am. J. Roentgenol.,* **117,** 146 (1973).

74. R. S. Sherman, "An Automated System for Recording Reports of Chest Roentgenograms," *Am. J. Roentgenol. Radium Ther. Nuclear Med.,* **117,** 848 (1973).

75. J. R. Jagoe and K. A. Paton, "Reading Chest Radiographs for Pneumoconiosis by Computer," *Br. J. Ind. Med.,* **32,** 267 (1975).

76. Digital Film Library NIOSH 54638, National Institute for Occupational Safety and Health, Rockville, Md., August 1975.

77. F. Yanowitz et al., "Quantitative Exercise Electrocardiography in the Evaluation of Patients with Early Coronary Artery Disease," *Aerosp. Med.,* **45,** 443 (1974).

78. W. H. Walter et al., "Dynamic Electrocardiography and Computer Analysis," *Aerosp. Med.,* **44,** 414 (1973).

79. R. A. Taha et al., "The Electrocardiogram in Chronic Obstructive Pulmonary Disease," *Am. Rev. Respir. Dis.,* **107,** 1067 (1973).

80. C. A. Bertrand et al., "How the Computer Helps the Cardiologist Interpret ECG's," *Ind. Med. Sur.,* August 1973, pp. 14–19.

81. S. Talbot et al., "Normal Measurements of Modified Frank Corrected Orthogonal Electrocardiograms and Their Importance in an On-Line Computer-Aided Electrocardiographic System," *Br. Heart J.,* **36,** 475 (1974).

82. M. L. Simoons et al., "On-Line Processing of Orthogonal Exercise Electrocardiograms," *Comput. Biomed. Res.,* **8,** 105 (1975).

83. R. W. Morris, "The Value of the Effort Electrocardiogram," *South Afr. Med. J.,* **49,** 1553 (1975).

84. H. Karlsson, B. Lindberg, and D. Linnarsson, "Time Courses of Pulmonary Gas Exchange and Heart Rate Changes in Supine Exercise," *Acta Physiolog. Scand.,* **95,** 329 (1975).

85. P. Franket et al., "A Computerized System for ECG Monitoring," *Comput. Biomed. Res.,* **8,** 560 (1975).

86. A. A. Sarkady, R. R. Clark, and R. Williams, "Computer Analysis Techniques for Phonocardiograms Diagnosis," *Comput. Biomed. Res.,* **9,** 349 (1976).

87. K. Horsfield et al., "Models of the Human Bronchial Tree," *J. Appl. Physiol.,* **31,** 207 (1971).

88. P. G. Voilleque, "Computer Calculation of Bone Doses Following Acute Exposure to Strontium-90 Aerosols," Health Services Laboratory, U. S. Atomic Energy Commission, National Reactor Testing Station, Idaho Falls, Idaho, 1970.

89. J. M Beeckmans, "The Deposition of Aerosols in the Respiratory Tract—Mathematical Analysis and Comparison with Experimental Data," *Can. J. Physiol. Pharmacol.,* **43,** 157 (1965).

90. D. P. Discher and A. H. Palmer, "Development of a New Motivational Spirometer—Rationale for Hardware and Software," *J. Occup. Med.,* **14,** 679 (1972).

91. J. D. Brain and P. A. Valberg, "Models of Lung Retention Based on ICRP Task Group Report," *Arch. Environ. Health,* **28,** 1 (1974).

92. D. B. Yeates et al., "Regional Clearance fo Ions from the Airways of the Lung," *Am. Rev. Respir. Dis.,* **107,** 602 (1973).

93. A. W. Brodey et al., "The Residual Volume—A Graphic Solution to the Functional Residual Capacity Equation," *Am. Rev. Respir. Dis.,* **109,** 87 (1974).

94. G. J. Trezek, "Predictions of the Dynamic Response of the Lung," *Aerosp. Med.,* **44,** 8 (1973).

95. M. K. Loken et al., "Dual Camera Studies of Pulmonary Function with Computer Processing of Data," *Am. J. Roentgenol. Radium Ther. Nuclear Med.,* **121,** 761 (1974).

96. W. R. Beaver, "On-Line Computer Analysis and Breath-by-Breath Graphical Display of Exercise Function Tests," *J. Appl. Physiol.,* **34,** 128 (1973).

97. R. S. Sherman, "An Automated System for Recording Reports of Chest Roentgenograms," *Am. J. Roentgenol. Radium Ther. Nuclear Med.,* **117,** 848 (1973).

98. W. R. Ayers et al., "Description of a Computer Program for Analysis of the Forced Expiratory Spirogram. I. Instrumentation and Programming," *Comput. Biochem. Res.,* **2,** 207 (1969).

99. L. Jansson and B. Jonson, "A Method for Studies of Airway Closure in Relation to Lung

Volume and Transpulmonary Pressure at a Regulated Flow Rate," *Scand. J. Respir. Dis.,* **85,** 228 (1974).

100. M. Paiva, "Gaseous Diffusion in an Alveolar Duct Simulated by a Digital Computer," *Comput. Biomed. Res.,* **7,** 533 (1974).

101. A. A. Smith and E. A. Gaensler, "Timing of Forced Expiratory Volume in One Second, "*Am. Rev. Respir. Dis.,* **112,** 882 (1975).

102. J. H. Ellis et al., "A Computer Program for Calculation and Interpretation of Pulmonary Function Studies," *Chest,* **68,** 209 (1975).

103. W. A. Barrett, "Computerized Roentgenographic Determination of Total Lung Capacity," *Am. Rev. Respir. Dis.,* **113,** 239 (1976).

104. H. Pessenhofer and T. Kennes, "Method for the Continuous Measurements of the Phase Relation between Heart and Respiration," *Pfluegers Arch.,* **335,** 77 (1975).

105. C. Simecek, "Formulae for Calculation of Membrane Diffusion Component and Pulmonary Capillary Blood Volume," *Bull. Physiol-Pathol. Respir.,* **11,** 349 (1975).

106. J. N. Davis and D. Stagg, "Interrelationship of the Volume and the Time Components of Individual Breaths in Resting Man," *J. Physiol.,* **245,** 481 (1975).

107. A. Crockett and R. L. Smith, "Use of On-Line Computer Facilities in a Respiratory Function Laboratory," *Med. J. Aust.,* **2,** 486 (1975).

108. H. Guy et al., Computerized, Noninvasive Tests of Lung Function," *Am. Rev. Respir. Dis.,* **113,** 737 (1976).

109. C. Xintaras et al., "Brain Potentials Studied by Computer Analysis," *Arch. Environ. Health,* **13,** 223 (1966).

110. J. C. G. Pearson and D. Radwanski, "Principles of Design of Occupational Health Records," *J. Soc. Occup. Med.,* **24,** 17 (1974).

111. L. Nottbohn, "Collection and Electronic Processing of Data from Routine Medical Examinations," *Arbeitsmedizin-Solialmedizin-Arbeitshygiene,* **5,** 127 (1970).

112. G. H. Collings et al., "Follow-up of MHS," *J. Occup. Med.,* **14,** 462 (1972).

113. A. Yedidia, "California Cannery Workers Program, Multiphasic Testing as an Introduction to Orderly Health Care," *Arch. Environ. Health,* **27,** 259 (1973).

114. "Multiphasic Health Screening . . . A Tool for Industrial Hygiene?" *Nat. Saf. News,* **102,** 72, (1970).

115. H. A. Haessler, "Industrial Experience with Automated Multiphasic Health Examinations," *Ind. Med. Surg.,* **39,** 24–26, 335–337 (1970).

116. S. Bangs, "Multiphasic Screening: Wave of the Future or Droplet?" *Occup. Hazards,* **31,** 51 (1969).

117. D. F. Davies, "Progress Toward the Assessment of Health Status," *Prev. Med.,* **4,** 282 (1975).

118. N. Righthand et al., "A Computer-Oriented Hearing Conservation Program," *J. Occup. Med.,* **16,** 654 (1974).

119. R. A. Campbell, "Computer Audiometry," *J. Speech Hear. Res.,* **17,** 134 (1974).

120. H. M. Sussman, P. F. MacNeilage, and J. Lumbley, "Sensorimotor Dominance and the Right Ear Advantage in Mandibular—Auditory Tracking," *J. Acoust. Soc. Am.,* **56,** 214 (1974).

121. I. Klockhoff et al., "A Emthod for Computerized Classification of Pure Tone Screening Audiometry Results in Noise-Exposed Groups, *Acta Otolaryngol.,* **75,** 339 (1973).

122. V. Mellert, K. F. Siebrasse, and S. Mehrgardt, "Determination of the Transfer Function of the External Ear by an Impulse Response Measurment," *Anal. Spectrogr.,* **134,** 135, 137 (1974).

123. W. Niedermeier, J. H. Griggs, and R. S. Johnson, "Emission Spectrometric Determination of Trace Elements in Biological Fluids," *Appl. Spectrosc.,* **25,** 53 (1971).

124. D. W. Lander et al., Spectrographic Determination of Elements in Airborne Dirt," *Appl. Spectrosc.,* **25,** 270 (1971).

125. A. G. Schoning, "A Computer-Based Storage and Retrieval System for Electronic Absorption Spectra," *Anal. Chem. Acta,* **71,** 17 (1974).

126. H. G. Langer, T. P. Brady, and P. R. Briggs, "Formation of Dibenzodoxins and Other Condensation Products from Chlorinated Phenols and Derivatives," *Environ. Health Perspect.,* **5,** 3 (1973).

127. R. J. Gelirke and R. C. Davies, "Spectrum Fitting Technique for Energy Dispersive X-ray Analysis of Oxides and Silicates with Electron Microbeam Excitation," *Anal. Chem.,* **47,** 1537 (1975).

128. B. M. Golden and E. S. Yeung, "Analytical Lines for Long-Path Infrared Absorption, Spectrometry of Air Pollutants," *Anal. Chem.,* **47,** 2132 (1975).

129. D. Schuetzle, "Analysis of Complex Mixtures by Computer Controlled High Resolution Mass Spectrometry. I. Application to Atmospheric Aerosol Composition," *Biomed. Mass Spectrom.,* **2,** 288 (1975).

130. H. W. Dickson et al., "Environmental Gamma Ray Measurements Using *in Situ* and Core Sampling Techniques," *Health Phys.,* **30,** 221 (1976).

131. B. Versino et al., "Organic Micropollutants in Air and Water," *J. Chromatogr.,* **122,** 373 (1976).

132. I. Harness, "Airborne Asbestos Dust Evaluation," *Ann. Occup. Hyg.,* **16,** 397 (1973).

133. M. V. Sussman, K. N. Astill, and R. N. S. Rathore, "Continuous Gas Chromatography," *J. Chromatogr. Sci.,* **12,** 91 (1974).

134. M. Axelson, G. Schumacher, and J. Sjovall, "Analysis of Tissue Steroids by Liquid-Gel Chromatography and Computerized Gas Chromatography–Mass Spectrometry," *J. Chromatogr. Sci.,* **12,** 535 (1974).

135. P. J. Arpino, B. G. Dawkins, and F. W. McLafferty, "A Liquid Chromatrgraphy/Mass Spectrometry System Providing Continuous Monitoring with Nanogram Sensitivity," *J. Chromatogr. Sci.,* **12,** 574 (1974).

136. B. S. Finkle, R. L. Foltz, and D. M. Taylor, "A Comprehensive GC-MS Reference Data System for Toxicological and Biomedical Purposes," *J. Chromatogr. Sci.,* **12,** 304 (1974).

137. E. Jellum, O. Stokke, and L. Eldjarn, "Application of Gas Chromatography, Mass Spectrometry, and Computer Methods in Clinical Biochemistry," *Anal. Chem.,* **45,** 1099 (1973).

138. H. Nau and K. Biemann, "Computer-Assisted Assignment of Retention Indices in Gas Chromatography–Mass Spectrometry and Its Application to Mixtures of Biological Origin," *Anal. Chem.,* **46,** 426 (1974).

139. A. Bye and G. Land, "Determination of 3-(5-Tetrazolyl) Thioxanthone 1010-Dioxide in Human Plasma, Urine, and Feces," *J. Chromatogr.,* **115,** 93 (1975).

140. R. C. Lao, R. S. Thomas, and J. L. Monkman, "Computerized Gas Chromatographic–Mass Spectrometric Analysis of Polycyclic Aromatic Hydrocarbons in Environmental Samples," *J. Chromatogr.,* **112,** 681 (1975).

141. N. Buchan, "Computer Analysis of Amino Acid Chromatograms," *J. Chromatogr.,* **103,** 33 (1975).

142. J. Einhorn et al., "Computerized Analytical System for the Analysis of the Thermal Decomposition Products of Flexible Urethane Foam," Flammability Research Center, University of Utah, Salt Lake City, December 1973.

143. G. F. Gostecnik and A. Zlatkis, "Computer Evaluation of Gas Chromatographic Profiles for the Correlation of Quality Differences in Cold Pressed Orange Oils," *J. Chromatogr.,* **106,** 73 (1975).

144. W. Cautreels and K. Van Cauwenberghe, "Determination of Organic Compounds in Airborne Particulate Matter by Gas Chromatography–Mass Spectrometry," *Atmos. Environ.,* **10,** 447 (1976).

145. S. R. Heller, J. M. McGuire, and W. L. Budde, "Trace Organics by GC/MS," *Environ. Sci. Technol.,* **9,** 210 (1975).

146. T. Mamuro et al., "Activation Analysis of Polluted River Water," *Radioisotopes,* **20,** 111 (1971).

147. A. H. Qazi et al., "Identification of Carcinogenic and Noncarcinogenic Polycyclic Aromatic Hydrocarbons Through Computer Programming," *Am. Ind. Hyg. Assoc. J.,* **34,** 554 (1973).

148. K. D. Hapner and K. R. Hamilton, "Basic Computer Program for Amino Acid Analysis Data," *J. Chromatogr.,* **93,** 99 (1974).

149. M. A. Fox, "Progress for Use with Automatic Amino Acid Analyser to Identify, Compute, and Correlate Amino Acid Concentrations in Biological Samples," *J. Chromatogr.,* **89,** 61 (1974).

150. I. Cavill and C. Ricketts, "Automated Quality Control for Haematology Laboratory," *J. Clin. Pathol.,* **27,** 757 (1974).

151. D. M. Linekin, "Multielement Instrumental Neutron Activation Analysis of Biological Tissue Using A Single Comparator Standard and Data Processing by Computer," *Int. J. Appl. Radiat. Isot.,* **24,** 343 (1973).

152. J. S. Ploem et al., "A Microspectrofluorometer with Epi-Illumination Operated Under Computer Control," *J. Histochem. Cytochem.,* **22,** 668 (1974).

153. B. Sarkar and T. P. A. Kruck, "Theoretical Considerations and Equilibrium Conditions in Analytical Potentiometry," *Can. J. Chem.,* **51,** 3541–3548 (1973).

154. B. T. Dew, T. King, and D. Mighdoll, "An Automatic Microscope System for Differential Leukocyte Counting," *J. Histochem. Cytochem.,* **22,** 685 (1974).

155. A. L. Linch et al., "Nondestructive Neutron Activation Analysis of Air Pollution Particulates," *Health Lab. Sci.,* **10,** 251 (1973).

156. L. Kryger, D. Jagner, and H. J. Skov., "Computerized Electroanalysis. Part I. Instrumentation and Programming," *Anal. Chem. Acta.,* **78,** 241 (1975).

157. C. Overby, et al, "Some Human Factors and Related Socio-Technical Issues in Bringing Jobs to Disable Persons via Computer-Telecommunications Technology," *Human Factors,* **20,** 349 (1978).

158. E. W. White and P. B. Denee, "Characterization of Coal Mine Dust by Computer Processing of Scanning Electron Microscope Information," *Ann. N.Y. Acad. Sco.,* **200,** 666 (1972).

159. E. A. Curth, "Causes and Prevenfion of Transportation Accidents in Bituminous Coal Mines," Bureau of Mines, Information Circular 8506, Pittsburgh, 1971.

160. W. R. Miller, "System Analysis Approach to Safety," *Nat. Saf. Congr. Trans.,* **11,** 19 (1972).

161. "How Safety/Security Directors Use EDP," *Occup. Hazards,* **30,** 41 (1968).

162. S. E. Hall, "Procedure for Army Safety Sampling (PASS)," National Technical Information Service, Springfield, Va., December 1970.

163. V. Steinecke, "Electronic Data Processing for Safety," *Sicherheitsingenieur,* **4,** 364 (1973).

164. "How a Computer System Cuts Accident Costs," *Occup. Hazards,* **35,** 45 (1973).

165. G. W. Radl, "Can Visibility Be Improved on a Fork-Lift Truck?" *Mod. Unfallverhüt.,* Vulkan-Verlag Haus, Der Techniq. Essen, Germany, **17,** 41 (1973).

166. J. Wanat, "A System for Computer Analysis of Work Accidents Using a Modern Form of Information Bank Utilization in the Area of Industrial Safety," *Ochr. Pr.,* **26,** 4 (1972).

167. R. V. O'Neil and O. W. Burke, "A Simple Systems Model for DDT and DDE Movement in the Human Food Chain," Ecological Sciences Division, Publication 415, Oak Ridge National Laboratory, Oak Ridge, Tenn., October 1971.

168. W. J. Stanley and D. D. Cranshaw, "The Use of a Computer-Based Total Management Information System to Support an Air Resource Management Program," *J. Air Pollut. Control Assoc.,* **18,** 158 (1968).

169. E. B. Cook and J. M. Singer, "Predicted Air Entrainment by Subsonic Free Round Jets," *J. Spacecr. Rockets,* **6,** 1066 (1969).

170. H. A. James and H. Currie, "Punched Card Information Retrieval System for Air Pollution Control Data," *J. Air Pollut. Control Assoc.,* **14,** 118 (1964).

171. N. M. Rochkind, "Infrared Analysis of Gases: A New Method," *Environ. Sci. Technol.,* **1,** 434 (1967).

172. B. J. Huebert, "Computer Modelling of Photochemical Smog Formation," *J. Chem. Educ.,* **51,** 644 (1974).

173. M. I. Hoffert et al., "Laboratory Simulation of Photochemically Reacting Atmospheric Boundary Layers: A Feasibility Study," *Atmos. Environ.,* **9,** 33 (1975).

174. F. C. Hamburg and F. L. Cross, Jr., "A Training Exercise on Cost-Effectiveness Evaluation of Air Pollution Control Strategies," *J. Air Pollut. Control Assoc.,* **21,** 66 (1971).

175. H. A. Hawthorne et al., "Cesium-137 Cycling in a Utah Dairy Farm," *Health Phys.,* **30,** 447 (1976).

176. E. W. Crampton, "Husbandry Versus Fluroide Ingestion as Factor in Unsatisfactory Dairy Cow Performance," *J. Air Pollut. Control Assoc.,* **18,** 229 (1968).

177. R. G. Edwards, Jr., C. H. Powell, and M. A. Kendrick," Dust Counting Variability," *Am. Ind. Hyg. Assoc. J.,* **27,** 546 (1966).

178. J. L. Spratt, "Computer Program for Prohibit Analyses," *Toxicol. Appl. Pharmacol.,* **8,** 110 (1966).

179. C. Xintaras et al., "Brain Potentials Studied by Computer Analysis," *Arch. Environ. Health,* **13,** 223 (1966).

180. J. K. Raines, M. Y. Jaffrin, and A. H. Shapiro, "A Computer Simulation of Arterial Dynamics in the Human Leg," *J. Biomech.,* **7,** 77 (1974).

181. D. A. Hobson and L. E. Torfason, "Optimization of Four-Bar Knee Mechanisms—A Computerized Approach," *J. Biomech.,* **7,** 371 (1974).

182. R. Penn, R. Walser, and L. Ackerman, "Cerebral Blood Volume in Man," *JAMA,* **234,** 1154 (1975).

183. P. J. Lewi and R. P. H. M. Marshoom, "Automated Weighing Procedure for Toxicological Studies on Small Animals Using a Minicomputer," *Lab. Anim. Sci.,* **25,** 487 (1975).

184. M. Salzer, "Model for Describing Tremor," *Eur. J. Appl. Physiol. Occup. Physiol.,* **34,** 19 (1975).

185. S. Schottenfeld and S. Rothenberg, "An Automated Laboratory Control System: Collection and Analysis of Behavior and Electro-Physiological Data," *Comput. Programs, Biomed.,* **5,** 296 (1976).

186. N. H. Sabah, "A Presettable Multichannel Digital Timer," *J. Appl. Physiol.,* **38,** 757 (1975).

187. A. L. Henschel et al., "An Analysis of Heat Deaths in St. Louis During July 1966," *Am. J. Pub. Health,* **59,** 2232 (1969).

188. F. Burbank, "A Sequential Space-Time Cluster Analysis of Cancer Mortality in the United States: Etiologic Implications," *Am. J. Epidemiol.,* **95,** 393 (1972).

189. S. Pell, "Epidemiological Studies in a Large Company Based on Health and Personnel Records," *Pub. Health Rep.,* **83,** 399 (1968).

190. W. E. McConnell, "The Wave File," *Aerosp. Med.,* **44,** 210 (1973).

191. S. Milham, Jr., *Occupational Mortality in Washington State 1950–1971,* Vol. 3, National Institute for Occupational Safety and Health, Department of Health Education and Welfare Publication (NIOSH) 76-175-C, Cincinnati, Ohio, April 1976.

192. D. J. Kilian, D. J. Picciano, and C. B. Jacobson, "Industrial Monitoring: A Cytogenetic Approach," *Ann. N.Y. Acad. Sci.,* **269,** 4 (1975).

193. Q. T. Pham et al., "Methodology of an Epidemiological Survey in the Iron Ore Mines of Lorraine—Research into the Long-term Effects of Potentially Irritant Gases on the Pulmonary System," *Ann. Occup. Hyg.,* **19,** 33 (1976).

194. C. A. Mitchell, R. S. F. Schilling, and A. Bouhuys, "Community Studies of Lung Disease in Connecticut," *Am. J. Epidemiol.,* **103,** 212 (1976).

195. P. G. Voilleque, "Computer Calculation of Bone Dose Following Acute Exposure to Strontium-90 Aerosols," U. S. Atomic Energy Commissions, Idaho Falls, Idaho, 1970.

196. J. R. Mallard and T. A. Whittingham, "Dielectric Absorption of Microwaves in Human Tissues," *Nature*, **218**, 366 (1968).

197. S. A. Beach, "A Digital Computer Program for the Estimation of Body Content of Plutonium from Urine Data," *Health Phys.*, **24**, 9 (1973).

198. R. E. Ellis et al., "A System for Estimation of Mean Active Bone Marrow Dose," Food and Drug Administration, Rockville, Md., August 1975.

199. A. I. Burhanov, "Combined Effects of the Basic Constituents of Mixed Metal Dusts," *Gigi. Tr. Prof. Zabol.*, **3**, 30 (1975).

200. R. E. Goans, "Two Approaches to Determining Pu-239 and Am-241 Levels in Phoswich Spectra," *Health Phys.*, **29**, 421 (1975).

201. W. R. Wood, Jr., and W. E. Sheehan, "Evaluation of the Pugfua Method of Calculating Systemic Burdens," *Am. Ind. Hyg. Assoc. J.*, **32**, 58 (1971).

202. B. Bergstrom et al., "Use of a Digital Computer for Studying Velocity Judgments of Radar Targets," *Ergonomics*, **16**, 417 (1973).

203. D. M. S. Peace and R. S. Easterby, "The Evaluation of User Interaction with Computer-Based Management Information Systems," *Hum. Factors*, **15**, 163 (1973).

204. A. B. Trump-Thorton and R. Daher, "The Prediction of Reaction Forces from Gait Data," *J. Biomech.*, **8**, 173 (1975).

205. H. R. Warner, F. R. Woolley, and R. L. Kane, "Computer Assisted Instructions for Teaching Clinical Decision-Making," *Comput. Biomed. Res.*, **7**, 564 (1974).

206. D. N. Ostrow, N. Craven, and R. M. Cherniack, "Learning Pulmonary Function Interpretation: Deductive Versus Inductive Methods," *Am. Rev. Respir. Dis.*, **112**, 89 (1975).

207. A. U. Valish and N. J. Boyd, "The Role of Computer Assisted Instruction in Continuing Education of Registered Nurses: An Experimental Study," *J. Contin. Educ. Nurs.*, **6**, 13 (1975).

208. D. P. Schlick and K. R. Werner, "Computerized Programming of Respirable Dust Sampling Data," U. S. Department of the Interior, Bureau of Mines, Report 8504, March 1971.

209. G. L. Baker and R. E. Reiter, "Automatic Systems for Monitoring Vinyl Chloride in Working Atmospheres," *Am. Ind. Hyg. Assoc. J.*, **38**, 24 (1977).

210. P. S. Kerr, "Standard Oil's Computer Network Makes Health Management Possible," *Occup. Health Saf.*, November–December 1977, pp. 44–46.

211. H. R. Jennings and K. L. Rohrer, "A Computerized Industrial Hygiene Program," *Plant Eng.*, October 14, 1976, pp. 149–151.

212. M. G. Ott et al., "Linking Industrial Hygiene and Health Records," *Am. Ind. Hyg. Assoc. J.*, **10**, 760 (1975).

213. N. J. Gilson, W. I. Bitter, and H. G. Barrett, "Automated Monitoring System for Exposure to Toxic Materials," IBM Research Center, Yorktown Heights, N.Y., February 8, 1972.

214. M. Ayoub and K. Kushner, "A Computerized Safety Management Information System for State OSHA Inspection and Enforcement Programs," *J. Saf. Res.*, **12**, 21 (1980).

215. J. H. Hagopian, "A Computer Program for Recirculation System Design," Division of Physical Sciences and Engineering, NIOSH, 210-77-0154, Cincinnati, Ohio, 1979.

216. C. H. Frith, S. S. Herrick, and A. J. Konvicka, "Computer-Assisted Collection and Analysis of Pathology," *J. Nat. Cancer Inst.*, **58**, 1717 (1976).

217. F. W. Lichtenberg and G. E. Devitt, "The Medical Data Base System of Owens-Corning Fiberglas Corporation," *Am. Ind. Hyg. J.*, **41**, 103 (1980).

218. V. E. Rose, "Reliability and Utilization of Occupational Disease Data," Division of Criteria Documentation and Standards Development, NIOSH, 77–189, Cincinnati, Ohio, 1977.

219. C. W. Sem-Jacobsen, "Brain/Computer Communication to Reduce Human Error: A Perspective," *Aviat. Space Environ. Med.*, **52**, 33 (1981).

220. D. H. Pearce et al., "Computer-Based System for Analysis of Respiratory Responses to Exercise," *J. Appl. Physiol.: Resp. Environ. Excercise Physiol.*, **42,** 968 (1977).

221. J. H. Cissik, T. J. Cramer, and L. L. Shelman, "Evaluation of the Cavitron Spirometric Computer for Accuracy in Clinical Screening Spirometry," *Aviat. Space Environ. Med.*, **52,** 125 (1981).

222. E. J. Engelken and J. W. Wolfe, "Analog Processing of Vestibular Nystagmus for On-Line Cross-Correlation Data Analysis," *Aviat. Space Environ. Med.*, **48,** 210 (1977).

223. E. R. Adlard, A. W. Bowen, and D. G. Salmon, "Automatic System for the High-Resolution Gas Chromatographic Analysis of Gasoline-Range Hydrocarbon Mixtures," *J. Chromatog.*, **186,** 219 (1979).

224. L. C. Dickson, et al., "Software Improvements for Gas Chromatography-Mass Spectrometry-Calculator System Used in the Analysis of Trace Organic Compounds From Environmental Samples," *J. Chromatog.*, **190,** 311 (1980).

225. D. P. Manning, "Model-Accident Statistics, Part 4," *Protection*, **16,** 34 (1979).

226. D. K. Guinan, "The Railroad Industry Hazard Information and Resource System," Control of Hazardous Material Spills. Proceedings of the 1980 National Conference on Control of Hazardous Material Spills, Louisville, Kentucky, May 13–15, 1980, pp. 350–357.

227. M. A. Ayoub, "Integrated Safety Management Information System—Part I: Design and Architecture," *J. Occup. Accidents*, **2,** 135 (1979).

228. M. A. Ayoub, "Integrated Safety Management Information System—Part II: Allocation of Resources," *J. Occup. Accidents*, **2,** 191 (1979).

229. G. L. Horton, J. R. Lowe, and C. N. Lieske, "Cholinesterase Inhibition Studies by Stopped-Flow Instrumentation and Automated Data Processing," *Anal. Biochem.*, **78,** 213 (1977).

230. A. F. Levy, "Scheduling of Toxicology Protocol Studies," *Comput. Biomed. Res.*, **10,** 139 (1977).

231. L. Claxton, and R. Baxter, "Computer Assisted Bacterial Test for Mutagenesis," *Mutation Res.*, **53,** 345 (1978).

232. R. L. Henry and C. R. Johnson, "Pathology Data Quality Assurance and Data Retrieval at the National Center for Toxicological Research," *J. Environ. Pathol. Toxicol.*, **3,** 169 (1980).

234. W. K. Chu, D. E. Raeside, and G. D. Adams, "A Computerized Radiation Safety System," *Health Phys.*, **32,** 116 (1977).

235. I. D. Brown, A. J. Hull, and A. C. Cox, "Digit Matrices for Data Energy and Retrieval: Studies of Scanning Direction and Colour Contrast in a Telephone Switchboard Application," *Ergonomics*, **22,** 1217 (1979).

CHAPTER EIGHT

Statistical Design and Data Analysis Requirements

NELSON A. LEIDEL, Sc.D., and
KENNETH A. BUSCH

1 INTRODUCTION

Industrial hygienists and allied professionals can derive substantial professional benefits from use of study design and data analysis methodologies that are based in the theory of mathematical statistics and theory of probability. Section 2 of this chapter discusses some major areas of industrial hygiene practice where statistical methods play an important role. The need for statistical experimental design of both experimental and observational studies is discussed in Section 2.1. Brief discussions appear in Section 2.2 concerning statistical methods used for occupational epidemiological studies, and in Section 2.3 concerning estimating possible threshold levels and low risk levels for occupational exposures. Finally, Section 2.4 introduces the area of application for statistics which is of primary interest and receives most attention in this chapter, namely estimation of occupational exposures to airborne contaminants. Nine possible objectives of occupational exposure estimation are discussed which have their own special requirements for study design strategies.

The study designs and data analysis methods for exposure estimation have come to be broadly called *sampling strategies* (1). These sampling strategies are plans of action based on statistical theory used to determine a logical, efficient framework for application of general scientific methodology and professional judgment.

Section 3 presents basic statistical theory which applies to occupational exposure data. Distributional models are given which identify the contributions of various sources of variation to the overall (net) random error in occupational exposure estimates. The National Institute for Occupational Safety and Health (NIOSH) nomenclature for exposure data is first given in Section 3.1. Then a model for the contributions of the various components of variation to the net random error in occupational exposure measurements (due to the measurement procedure used) is given in Section 3.2. Section 3.3 extends the model to include random and systematic variations in true exposure levels (over times, locations, or workers doing similar work).

Section 3 also includes information on the mathematical characteristics of basic distributional models. This is the starting point for deriving sampling distributions of industrial hygiene exposure data taken by various sampling strategies. General properties of the normal distribution model are given in Section 3.4, and of the lognormal distribution (both 2-parameter and 3-parameter) in Section 3.5. Section 3.6 discusses the adequacy of the normal and lognormal distributions as models for types of continuous variable data (specifically occupational exposure measurements). These data models are then used to make special interest applications of statistical theory to study design (Section 5) and to data analysis (Section 6).

Section 4 discusses the basic principles of statistical experimental design and data analysis which apply to all industrial hygiene surveys, evaluations, or studies. The first Section 4.1 presents general study design principles for experimental studies which are designed to estimate means, variances, tolerance limits, or proportions for a single population, and for those studies seeking to compare these parameters between two study groups (e.g., between an exposed "treatment" group and an unexposed "control" group). The next Section 4.2 discusses general principles of data analysis. Particular emphasis is given to the necessity for appropriate selection, estimation, and verification of a distributional model for the study data.

Section 5 gives particular study designs for use in collecting data to estimate individual occupational exposures and distributions of exposures. It first discusses the important concept of a *target population* of workers. Section 5.1 describes in detail another important concept, the *determinant variables* affecting occupational exposure levels of the target population. Section 5.2 discusses *exposure measurement* strategies selected to measure a short or long period time-weighted average (TWA) exposure of an individual worker on a given day. Both practical and statistical considerations are discussed for long-term and short-term exposure estimates. Lastly, *exposure monitoring* strategies are presented in Section 5.3 for measuring multiple exposures (e.g., multiple workers on a single day, a single worker on multiple days, or multiple workers on multiple days). Eight possible elements of monitoring programs are discussed in detail, with examples given of *exposure screening* and *exposure distribution* types of monitoring programs.

Lastly, Section 6 gives applied methods for formal statistical analysis of occupational exposure data generated by the study designs discussed in Sections 4 and 5. The first portion of Section 6 (6.1 through 6.5) covers methods for computing confidence intervals for individual true worker exposures which have been estimated using exposure measurement strategies. In addition, statistical hypothesis (significance) tests are presented for classifying individual exposure estimates relative to an exposure control limit. The second portion of Section 6 (6.6 through 6.10) covers inferential methods for computing tolerance limits, tolerance intervals, and point estimates of exposure distribution fractiles. The third portion of Section 6 (6.11 through 6.13) presents graphical techniques for plotting lognormally distributed data and their associated tolerance limits.

2 GENERAL AREAS OF APPLICATION FOR STATISTICS IN INDUSTRIAL HYGIENE

Statistical theory is applicable in both *experimental* studies (i.e., studies with planned intervention on determinant factors suspected of altering the phenomenon under study) and in *observational* surveys and studies (i.e., studies with no deliberate human intervention). With proper statistical design and data analysis, both experimental and observational studies can validly and reliably identify causes of occupational health problems, screen workplaces for excessive exposure conditions, estimate worker exposure levels, evaluate the effectiveness of engineering controls, and evaluate the protection levels afforded by personal protective equipment. Of course, depending on factors such as relative cost, conditions amenable to experimentation, availability of relevant laboratory models, and interpretability of available field data, one or the other of these two general types of studies is usually a clear choice for any given research objective. However, experimentation generally has several fundamental advantages over observational surveys and studies. These are discussed in Section 2.1 in the context of health effects studies related to workplace contaminant exposures.

2.1 Need for Statistical Experimental Design

Typical industrial hygiene *experimental* studies involve estimating distributions of worker exposures for selected conditions, determining the efficacy of exposure control systems, estimating the accuracy and variability of measurement methods, etc. Experiments may be done to study the effects of determinant variables on worker exposure levels, on the efficacy of control measures, or on exposure measurement procedures. In these experiments, the experimental unit is usually a set of controlled conditions under which an exposure level (or other physical or biological response) is measured as the response variable. If a factor is "controlled" through deliberate selection of particular existing conditions, one has a *quasi-experimental* study in a field setting. One has an *observational* study when conditions must be taken as found in a field setting (i.e., without opportunity to experiment with preselected, controlled levels for the determinant variables).

An inherent problem for industrial hygienists is that "safe" exposure levels for many substances are unknown and must be estimated using the best available evidence (2). To do this, pertinent exposure level and health effects data are collected and statistically analyzed. One approach is to expose suitable animal species and observe biological effects that may occur. A second important approach to estimation of acceptable exposure levels, the epidemiological study on exposed workers, is discussed in a limited manner in this section, but references to more comprehensive presentations are supplied.

For either type of occupational health study, experimental or observational, the resulting data will have stochastic components (i.e., random or chance variations) that cannot be ignored when evaluating the data. Different types of biological data may have fundamentally different statistical properties. Population health effects may be measurable in:

1.   The average amount of change in a quantitative biological parameter measured on a continuous scale. Examples of continuous variable measurements are lung volume, heart rate, and body weight.
2.   The presence or absence of a qualitative biological abnormality. An example would be a pathological condition such as a tumor.
3.   Values that exist only at a limited number of discontinuous (i.e., discrete) points on an ordered scale. For example, severity of lung histopathology has been graded on a 6-point rating scale.

The first and second types of data are the most frequently encountered and they will be discussed in Section 4 in relation to principles of study design and data analysis.

The objectives of an occupational health study should be the primary determinant of its statistical design, *not* expedient considerations of "available" specialized experimental facilities or presence of "experts" on a given professional staff. Availability of specialized research staff and personal interests of these researchers can be powerful incentives for inappropriately designing a research study.

It is the responsibility of the researcher, not the statistician, to assure that determinant factors such as the species used as animal models, exposure techniques, biological parameters that can be accurately measured, and exposure measurement procedures are given consideration and are appropriately incorporated into the experimental protocol. The statistician then addresses such tactical problems as sample size, allocation of subjects to exposure groups, schedules for sampling, and data analysis techniques. Proper consideration and selection of study design and data analysis parameters and methodologies will assure the researcher that the study will have adequate (but not excessive) statistical power and technical capability to detect the anticipated effects that are the focus of the research.

Sometimes technically appropriate experimental facilities or situations are available, but the feasible sample sizes are only marginal insofar as the production of definitive results is concerned. Nevertheless, an expedient decision may be made to proceed with the study under the misguided rationale that "some information is better than none." This incorrect and wasteful practice often results in studies that are predestined to be inconclusive and possibly misleading to subsequent investigators. The potential waste of time and resources can be minimized by first securing a statistician's evaluation of (or better yet, assistance with) the study design and data analysis plan. The statistician can usually warn the researcher if the planned study has low statistical power or for other reasons (e.g., bias due to confounded factors) lacks the statistical capability to detect the desired size of determinant factor effects, if indeed they exist.

If some species of experimental animal could serve as a perfect biological model for health effects on humans caused by workplace exposure, then there would be no question that controlled animal exposure studies would yield better information than observational studies of humans. The deficiencies of human observational studies are analogous to those which would occur in an animal study for which subjects were constrained within a measured but uncontrolled laboratory environment for 8 hours each day, but allowed to roam freely through the streets and alleys outside their laboratory, eating and breathing whatever they encountered, during the other 16 hours.

Unfortunately, the perfect biological model for human health effects does not exist, but the advantages of the carefully controlled animal exposure experiment are several. First, in an experimental animal study we can *control* the primary study factor(s) (e.g., exposure levels, exposure duration, exposure schedule) and determine its direct effect, at differing levels of interest, on one or more response variables. In observational studies of humans, the primary study factor(s) can be observed, but not controlled, and toxicological and mathematical extrapolation of observed results to other levels of interest must be performed.

Second, with animal studies there is more freedom to design experimental studies for complete elimination of bias (we shall put aside for now the interspecies extrapolation bias which may exist) and for high sensitivity to small effects. Given large enough sample sizes, an experimental study can be designed to have suitably high statistical power (probability) of detecting a small effect (if detecting this small effect is worth the required time and resources). Even a much larger observational study may not be able to detect a statistically significant, moderate-sized exposure effect, since there are unknown, or uncontrolled and unmeasured (even unmeasurable) secondary factors which operate within the workplaces and home environments to modify or distort the workers' biological responses to workplace exposures. Here, making a simple comparison of response variable values (the effect) between exposed and unexposed groups will yield an imprecise estimate of the exposure effect (i.e., a wide confidence interval for the true magnitude of the effect) and probably will also yield a biased (substantially inaccurate) estimate of the effect. Bias can occur because of unequal levels of the secondary factors in the control and exposed groups, so that we observe the joint (net) effect due to both the primary factor and to unintended differences in levels of the secondary factors. This joint effect is said to consist of *confounded effects* (i.e., effects of two or more factors confused with each other).

In observational studies for which levels of some secondary factors are known, methods of formal statistical analysis can be useful to adjust for the part of the bias due to confounded effects of these factors. For secondary factors which have effects that are additive to each other and to the effect of the primary factor, several methods for bias adjustment are available. These include *covariance analysis* and *analysis of variance (ANOVA) for randomized blocks* for continuous response variables. For discrete response variables there are procedures such as the *chi-square test for matched data (Cochran's Q test)* and *two-way analysis of variance (ANOVA) of ranks.*

In observational studies no statistical adjustment can be made for any bias due to unknown secondary factors, or due to covariates for which the levels were not measured. However, in experimental studies this dilemma of unknown or uncorrectable bias need never occur if the experiments are statistically well designed. Bias can be prevented in experiments by using statistical design features such as selection of subjects at random from the same pool for assignment to control and test groups, or restricted random selection with matching on a secondary factor or within blocks (e.g., ages, sexes, weight ranges). Randomizing within blocks not only prevents bias, but also improves precision. Similar blocking can be employed in observational studies during the data analysis, but only regarding the known secondary factors. It is *random assignment* of subjects to the exposure groups which is the best protection against bias due to effects of *unknown* secondary factors. Random assignment can be used in experimental studies, but not in observational studies for which there will always be some bias (one hopes it will be small) due to unequal representation of secondary factors (i.e., those not taken account of in the data analysis) in the exposed and unexposed groups.

In spite of the inherent limitations and deficiencies of observational studies on workers, they are an invaluable source of information on health effects due to work-related diseases. To some research scientists, the fact that the subjects of observational studies are workplace-exposed humans is an overriding consideration which transcends the statistical advantages and experimental flexibility afforded by animal exposure studies. Effects observed in workplace observational studies can be strongly suggestive of effects due to the primary factor(s) studied, but the results cannot be considered definitive by themselves. Replication of the result found for a given toxic agent, preferably in other studies of different worker target populations, for several occupational settings and cultural groups, affords much higher credibility and substantiation to the initial findings. This is true because the universal presence of a similar observed effect in different occupational settings (which involve different secondary factors) would tend to implicate the primary factor as the cause, not the variety of secondary factors.

On the other hand, a single controlled animal experiment, which has been toxicologically and statistically well-designed (with balance and randomization used at all phases of the work) and competently conducted, theoretically can yield a definitive result (within the limits of random error, which can be governed by its experimental design). Even so, any experimental scientist feels better after the initial findings have been replicated in an independently conducted study. Also, the criticism can be made, "What good is a 'definitive' result for animals if it does not apply to humans?" An extrapolation from animals to humans can only be defended on the basis of the similarities of their appropriate biological mechanisms and structures, supported presumptively by empirical species similarities previously observed. A good statistical experimental design does not assure validity of the interspecies extrapolation; however, it does assure validity of the findings for the animal species used.

Finally, the observational studies in workplaces can be an economical source of estimates of chronic effects due to long-term, low-level workplace exposures (although the estimates are often somewhat imprecise and ill-defined). Observational studies use existing data (although these data sometimes are not readily accessible), whereas an animal chronic exposure study requires at least several months to several years to conduct.

2.2 Epidemiological Studies

In the occupational epidemiological study an attempt is made to associate an observed incidence or severity of adverse health effects in groups of human workers with factors such as industry, job type, or some measure of exposure to potentially toxic materials. The last type of study may attempt to estimate a *dose-response* relationship between level of exposure to the material and prevalence or incidence of health effect being studied. However, the type of relationships utilized for occupational settings generally are more correctly referred to as *exposure-response* functions, since the effective dose at the critical site(s) within each individual worker's body is never really known (3). Ulfvarson (4) has noted:

The concept of exposure of an employee to a substance in the work environment may denote at least two things. It may indicate the dose of the substance absorbed in the body. It may also merely indicate the presence of the employee in an environment in which there is a more or less well determined concentration of the substance, from which an uptake of the substance is deduced.

For practical reasons, the latter concept of exposure is currently used by industrial hygienists.

Additionally, the *exposure* variable utilized may be either an average exposure level over some long period (such as a worker's total working lifetime) or a time-integrated (cumulative) exposure [e.g., ppm-years, (fibers/cu cm)-years, (mg/cu m)-years]. Estimation of these exposure-response relationships is desirable since they can be used in the process of selecting exposure control limits for workers in occupational environments (5, 6). Recognizing the statistical nature of the problem, Roach (7) first suggested that an occupational hazard (silica dust exposure leading to silicosis) should be analyzed as an exposure-response function similar to the dosage-response function utilized in toxicological research.

A major obstacle in estimation of exposure-response relationships is accurate and sufficiently precise estimation of worker exposure levels and their relation to doses delivered to the body. Gamble and Spirtas (8) have presented a systematic approach for using occupational titles to classify the "effect" and "exposure" of workers in a retrospective exposure-response estimation. Esmen (9) has proposed a process for reconstruction of the integrated exposure of one or more agents over a reasonably long period of time. He presents the basis for a model and a simplified procedure for what he calls a *retrospective industrial hygiene survey*. Roach (10, 11) has proposed sampling strategies that consider the biological half-times of the measured substances.

Combined with the substantial advantage of having human workers as subjects, occupational observational studies generally carry the frustrating disadvantages of lack of control of the exposure conditions, uncontrolled effects of the exposures outside the occupational environment, and interactions between exposure effects and demographic and socioeconomic factors. Statistically it should be possible to estimate a set of exposure-response relationships for any work-related disease given that a reasonably homogeneous and large enough group of workers is available. As a practical matter, it is difficult to obtain homogeneous and unbiased worker populations of adequate size for which one has adequate, accurate exposure information (12). These design problems for observational studies were discussed at length in Section 2.1. As a result, there have been a minimal number of exposure-response functions estimated in the last 30 years for occupational populations. These include functions estimated by Hatch for silicosis in miners and other dusty trades (13), Lundin et al. for respiratory cancer in uranium miners (14), Berry et al. for asbestosis in asbestos textile workers (15), and Dement et al. for lung cancer and other nonmalignant respiratory diseases in chrysotile asbestos textile workers (16). The Berry et al. article has an appendix with a particularly good discussion of measures of exposure and occupational "dose-response" relationships.

Better control over, and accurate measurement of, short-term human exposures can be obtained in experimental clinical studies of volunteer subjects. Of course an overriding necessity is safety of the human subjects, and this usually precludes use of exposure levels high enough, or exposures long enough, to produce the chronic toxicity that can occur in the workplace.

Ulfvarson (4) has reviewed and extensively discussed the limitations to the use of typical worker exposure data in epidemiologic studies. He concluded that there are a considerable number of possible biases in the estimation of uptake of a substance into the bodies of a group of workers, when body uptake is uncritically derived from typical sources of airborne exposure levels. He defined a positive bias as where the uptake is overestimated in comparison to the true uptake. He also rated the validity of his conclusions regarding the probable sign of the bias as follows: 3 = self evident, 2 = a conclusion with some reservation, and 1 = an educated guess. Some of the possible biases listed by Ulfvarson (with his own ratings {in braces} of the validity of their sign) follow:

1. Positive bias {2} due to the use of a measurement strategy intended to collect a sufficient mass of contaminant for the purpose of exceeding the minimum detectable level for the analytical method.
2. Positive bias {3} due to the use of a "worst case" biased measurement strategy.
3. Positive or negative bias due to the use of daily TWA exposure results that do not represent the unsampled workers they are applied to.
4. Positive bias {2} due to the use of data from area (static) sampling devices that were deliberately located to yield high results such as would be obtained with a source sampling strategy.
5. Negative bias {2} due to failure to obtain repeat measurements when the first result demonstrates compliance, but may have been unusually low.
6. Positive bias {2} due to use of exposure data from establishments that do not represent the unsampled establishments they are applied to.
7. Positive and negative biases due to seasonal variations in exposure levels.
8. Positive bias {3} due to rotation of workers to unexposed work areas, when the exposure measurements are only taken during the work operation.
9. Positive bias {3} due to the use of effective respirators by workers.
10. Negative bias {2} due to existence of unfavorable exposure patterns creating a "resonance" between intermittent airborne levels and levels in the body.
11. Negative bias {2} due to increased lung ventilation resulting from hard physical labor.

Ulfvarson (4) recommends that an epidemiologist use the list of possible biases as a checklist and try to find out the premises of the available data and thus the most probable sign of bias.

General statistical methodology for epidemiological studies is given in texts such as those by MacMahon and Pugh (17), Mausner and Bahn (18), Friedman (19), Lilienfeld and Lilienfeld (20), Schlesselman (21), and Rothman and Boice (22). Only recently have textbooks in epidemiology appeared that deal specifically with studies of occupational groups. One is Monson (23) and another is

Chiazze et al. (24). In addition, state-of-the-art general methodology for epidemiological studies in occupational populations is available in studies reported in journals such as *American Journal of Epidemiology, Archives of Environmental Health, British Journal of Industrial Medicine,* and *Journal of Occupational Health.* Both theoretical and practical problems of performing occupational health field studies are being solved, thanks to the extensive field experience that investigators have accrued in occupational observational studies. The March 1976 issue of *Journal of Occupational Medicine* deals entirely with occupational epidemiology, including an article by Enterline (25) warning of the pitfalls in epidemiological research. Potential problems in occupational studies include inaccurate or uninterpretable cause of death statements on death certificates, improper control groups, lack of reliable and accurate quantitative exposure estimates, overlapping exposure and follow-up periods, and competing (but sometimes unknown) causes of death.

The research needs in epidemiology have been summarized by an authoritative Second Task Force appointed by Department of Health, Education, and Welfare. In Chapter 15 of their 1977 report (26), a subtask force chaired by Dr. Brian MacMahon made an assessment of the state-of-the-art of epidemiology and other statistical methods used in environmental health studies. Among the specific recommendations were the following:

1. For clinical environmental research, guidelines are needed for the protection of human subjects and special attention should be given to statistical design and analysis of such studies.
2. For epidemiological studies, better exposure data are needed. Health professionals should help make decisions about what environmental data are collected by governmental agencies. Also, routine surveillance is needed of disease incidence in occupational groups along with surveillance of exposure levels. Animal studies should be used to identify biochemical or physiological early indicator effects of serious chronic disease in worker groups believed to be exposed to potentially hazardous agents. In all areas of environmental health research, more powerful statistical techniques are needed in areas of multivariate analysis, time series and sequential analysis, and nonparametric methods. Better dose-response models for mixtures of toxic agents are needed as are better models for animal-to-human extrapolation (particularly for carcinogenesis).

Another useful reference to more specific methodological techniques is the "Steelworker Series" of ten epidemiological reports by J. W. Lloyd and C. K. Redmond, published between 1969 (27) and 1978 (28) in *Journal of Occupational Medicine.* These reports are considered by some to constitute the evolution to the present state-of-the-art of modern methodology for epidemiological studies of an occupational population. In a 1975 paper Kupper et al. (29) discuss methods for selecting suitable samples of industrial worker groups and valid control groups. This subject is covered in more detail in Chapter 4.

2.3 Threshold Levels and Low Risk Levels

Ideally, the industrial hygienist would like to be able to assess the net risk of an adverse health effect as a function of a worker's past, present, and future exposures and in relation to age, race, sex, and other individual susceptibility factors. If such comprehensive toxicological knowledge were available in relation to past exposures, appropriate limits on future exposures of an individual could be recommended to control the chances of the worker experiencing adverse chronic health effects. However, achieving such a high level of toxicological understanding is not realistic—one would have to know all the "exposure-time-response" relationships for the toxic material of interest (i.e., the relationships between level of exposure, length of exposure, pattern of intermittent exposures, and the incidence and severity of health effects that occur to some or all an exposed population). However, in most cases the industrial hygienist has only an approximate estimate of some "low risk" exposure level at which the incidence of an adverse health effect appears to be low or absent in one or more working populations. Better yet, if possible, one would like to know a "safe" level of exposure (i.e., threshold level) below which all adverse health effects other than minor and transient ones are absent in almost all workers. The existence of such thresholds is a controversial question which we will not attempt to answer here. Some references to discussions of the question of existence of thresholds are (30) through (40).

Statistical models have been developed that can aid us in extrapolating to "acceptably low risk" exposure levels from higher risk dose-response data. Since a low- or minimal-risk exposure level always exists (regardless of whether a true threshold exists), a low-risk level of exposure can usually be selected which is "sufficiently safe." Note that the determination of "sufficiently safe" involves cost-benefit considerations of a political nature and is not solely a scientific determination. Hartley and Sielken (41) have reviewed the technical aspects of statistically estimating "safe doses" in carcinogenesis experiments. They leave the definition of an "acceptable" increase in the risk of carcinogenesis to the regulatory agencies. Acceptable levels of risk that have been suggested are 10^{-8} (1 in 100 million) and 10^{-6} (1 in 1 million). Either of these low-risk exposure levels is effectively impossible to determine by direct experimentation with animals because of the enormous sample sizes that would be required to distinguish such minuscule tumor incidence differences. The problem becomes particularly acute when there is a "normal" or "background" (control) tumor incidence. Therefore various mathematical models have been proposed for extrapolation to low-risk exposure levels using a mathematical model for the exposure-response curve, which has been fitted to higher-level exposure-response data. Among these mathematical modeling procedures are the Mantel-Bryan procedure (42, 43) based on extrapolation using a conservative slope on logarithmic normal probability graph paper. Other models are reviewed by Armitage (44). Some of the models are more flexible than others. For example,

Hartley and Sielken (45, 46) use a polynomial instead of a straight line to extrapolate. Other models are derived from assumed biological mechanisms. Crump et al. (47) assume a multistage biological mechanism for development of cancer. The Crump model is fitted to simple incidence versus dose data, whereas the Hartley-Sielken model can be fitted to time-to-tumor data. Recently Krewski and Brown (48) provided a comprehensive list of references for carcinogenic risk assessment, which were grouped by carcinogen bioassay, carcinogenicity screening, quantitative risk assessment, and regulatory considerations.

It is difficult to determine exposure-time-response relationships from limited and unstructured epidemiological data such as are usually available for humans reflecting a disease (adverse health effect) state that can be attributed to working environments. Nonetheless, exposure control limits can be estimated from carefully conducted, extensive epidemiological studies of workers. For example, the lower limit of a range of average personal exposure levels for a *large* group of workers in a given plant could be taken as a conservative control limit, if the "health profile" of these workers is found to be similar to that of a cohort group of unexposed but otherwise similar individuals. The condition that the unexposed cohort group be "otherwise similar" is often difficult to attain; relevance of this necessary condition to validity of the estimated health effect has been previously discussed in Section 2.1. But in any case, this approach is considerably less desirable than use of the sought after, but generally unattainable, exposure-response curves. With the latter approach the exposure experience of the entire exposed work population related to the occurrence of some adverse health effect is used to estimate an appropriate low-risk exposure control limit. With the former approach, which usually is based on only a small fraction of the total exposed group, an exposure control limit is usually set on the basis of the estimated lower range of exposures (see above) in the highest exposure group that did not show the health effect. However, such failure to find the health effect in a small sample of workers may have limited statistical significance. That is, it is possible to fail to observe a health effect in a small random sample of workers even though they had been selected from a larger group containing a proportion who would show the adverse health effect. Thus this procedure may yield an estimated control limit that may be set too high for the workforce. Also, the susceptibility characteristics of a small sample of workers, who work and live in close proximity, may be different from the rest of the workforce and other exposed groups.

2.4 Objectives of Occupational Exposure Estimation

Statistical methodologies can make substantial contributions toward achieving the three goals vital to the objective of effective worker protection: *recognition*, *evaluation*, and *control* of chemical and physical stresses to workers. Industrial hygienists are called on to examine the work environment and recognize workplace stresses and factors which have the potential for adversely affecting worker health. Then they must evaluate the magnitude of the stresses and interpret the results to be able to give an expert opinion regarding the general healthfulness of the workplace, either for short periods or for a lifetime of worker exposure. Finally, there must be a determination of the need for, or effectiveness of, control measures to minimize adverse health effects of workplace exposures.

The practical aspects of achieving these three goals require answers to two initial questions:

1. Are exposure measurements necessary?
2. If so, what type of measurements are needed in relation to our reasons for taking the exposure measurements?

Attempting to answer these two broad questions will require answering additional questions such as the following that will influence the choice of study design strategy:

1. Do we need only rough estimates or precise estimates of exposure levels?
2. Do we need only "worst case" estimates of the higher exposure levels or do we need estimates of exposure distributions?
3. What substances will we be measuring?
4. Which workers will be sampled?
5. How many samples will we be taking?
6. When will we take the samples?
7. At what locations will we take the samples?
8. What actions can be taken based on results of the data analysis?

Regarding our data analysis strategies we must ask:
How will we analyze our data and reach decisions regarding our research hypotheses or decide if the exposure levels are acceptable?

Answers to the preceding questions must be arrived at by first clearly defining the *objective(s)* of the worker exposure estimation. Clear definition of the objective(s) will facilitate the formulation of appropriate study designs and data analyses. Typically the objective(s) of worker exposure estimation will be one or more of the following nine:

1. *Hazard recognition.* Hazard recognition is the identification of toxic substances present in the workplace which are used in such a fashion that airborne levels create a possible health hazard. It involves the identification of the toxic substances present in the workplace, the job activities and work operations that involve their use, and determining if the toxic materials are intermittently or routinely released into the workplace air. The existing control measures must be identified, that might reduce the airborne levels of contaminants. It must also be determined if hazardous conditions could occur during irregular episodes or during an accident. These include toxic or harmful substances, explosive concentrations, or insufficient oxygen.

2. *Hazard evaluation.* This objective is similar to but wider in scope than *hazard recognition.* Where airborne toxic chemicals arise from industrial operations, it must be determined if it is possible for a hazardous condition to exist. This can involve an evaluation of the severity of health hazard(s) from airborne toxic chemicals, explosive concentrations, or insufficient oxygen due to industrial operations. It may involve determining if worker complaints or health problems could be due to hazardous exposure levels in the workplace.

Hazard evaluation can include an estimation of the distribution of worker exposure levels. Specifically, the shape, location, and dispersion of possible exposure distributions could be estimated. Percentiles of exposure distributions (with associated tolerance limits) could be estimated as a function of average exposure level. The important determinant factors affecting worker exposure levels might be identified. These are the qualitative or quantitative variables, factors, or parameters that influence or affect worker exposure levels. Periods of abnormally high exposure during the workshift should be looked for.

3. *Control method evaluation.* For existing control measures, their adequacy and the probability of unacceptable worker exposures would be determined. With newly installed controls, their adequacy and control effectiveness could be determined and compared to the previous exposure situation. The important design and operating parameters affecting the efficacy of control methods can be identified and evaluated.

4. *Exposure screening program.* This is a limited exposure monitoring program, which is designed to identify target populations of workers with other than acceptable exposure level distributions (i.e., those groups of workers with substantial proportions of exposures exceeding an exposure control limit or regulatory standard) for additional exposure monitoring. The program uses an Action Level as a screening cutoff to select appropriate target populations for inclusion in a limited exposure surveillance program (i.e., minimal periodic exposure measurements on only a few workers) or a more extensive *exposure distribution monitoring program.* An *exposure screening program* uses minimal resources with consideration for reasonable protection for the target population of workers. This type of a limited program was recommended by NIOSH for use in *regulatory monitoring programs* (1).

5. *Exposure distribution monitoring program.* This is a more extensive monitoring program designed to quantify exposure distributions of target populations over an initial temporal base period (e.g., on a day, over several weeks, months, or years). The initial estimates of exposure distributions would then be periodically updated with routine monitoring of the target populations. This objective involves the appropriate definition of target populations and identification of determinant variables affecting the exposure distributions.

6. *Regulatory monitoring program.* This objective necessitates establishing an exposure monitoring program that meets or exceeds the requirements of the applicable regulatory agency (e.g., Occupational Safety and Health Administration (OSHA), Mine Safety and Health Administration (MSHA)). The primary goal of this objective is "acceptable quality maintenance" regarding worker exposure levels (i.e., to ensure that all worker exposure levels meet applicable

permissible exposure limits (PELs) set by the regulatory agency). This program may involve elements from both an *exposure screening program* and an *exposure distribution monitoring program.*

7. *Epidemiological studies.* Some epidemiological studies attempt to associate the observed incidence or severity of adverse health effects in target populations with levels of exposure to potentially toxic materials. Demonstration of increasing response associated with increased exposure gives strong support to a hypothesis of disease causation due to exposure to the contaminant investigated. This objective generally requires valid and reasonably precise estimation of exposure levels for individual workers during all periods of their working lives. In exploratory studies, sometimes group-average or group-minimum exposures must be used as rough approximations to individual worker exposure levels.

8. *Measurement method comparison.* Often in industrial hygiene we desire to demonstrate the equivalency of a new exposure measurement method to an existing "standard" method. This approach is less desirable than defining a performance standard for monitoring a workplace contaminant, then performing a *measurement method validation* for a proposed new method (objective (9) below). However a methods comparison may be necessary if the "control" method has been used in past epidemiological studies or is the basis of a regulatory exposure level (e.g., Mine Research Establishment (MRE) horizontal elutriator for coal dust, vertical elutriator for cotton dust, USPHS/NIOSH method for asbestos fibers). This objective involves determining if the new method can be used as an adequately precise predictor of results that would be obtained if the standard method were used.

9. *Measurement method validation.* To properly evaluate workplace exposure results, it is desirable to estimate the accuracy of the measurement method and also identify the causes and magnitude of the components of random error in the method. Then the total precision error repeatability (variability in replicate measurements on the same sample by the same analyst) and reproducibility (variability between laboratories) for the method can be estimated. The "detection limits" for the method should be estimated. The first of these is the *limit of detection* (LOD), which is the minimum detectable level that can be differentiated from background. The LOD value should be reported for all observed values at or below the LOD (i.e., "zero" values should not be reported). The second type of "detection limit" is the *limit of quantitation* (LOQ), which is the minimum value for a method for which quantitative estimates with acceptable uncertainty can be obtained. The LOQ is greater than the LOD and is the minimum measurement level for which an uncertainty estimate (e.g., confidence interval, confidence limits) should be reported. Sometimes the accuracy and precision estimates for the method will be compared to a performance standard to determine the acceptability of a measurement procedure (e.g., the "± 25% accuracy at a 95% confidence level" criteria often used by NIOSH and OSHA (1)).

Some of the possible sampling strategies (primary and secondary study designs and data analysis strategies) suitable for the preceding nine objectives

Table 8.1 Sampling Strategies

Objectives	A	B	C	D	E	F	G	H	I	J	K
	Study Design Strategies							Data Analysis Strategies			
1. Hazard recognition	●	○							○	●	
2. Hazard evaluation	○		●						●	●	
3. Control method evaluation			○		○	○				○	○
4. Exposure screening programs				●		●	●	○	●	●	
5. Exposure distribution monitoring programs					●			●	○	○	
6. Regulatory monitoring programs				●	○		●	●	●	●	
7. Epidemiological studies					●	●			○	●	●
8. Measurement methods comparison					○	○				●	●
9. Measurement method validation									○	●	

Sampling Strategies (column key)

Study Design Strategies
A. Workplace observations
B. Estimate possible airborne levels from material usage rates
C. Worst case exposure measurements
D. Limited monitoring program to identify higher risk target populations for additional monitoring
E. Quantify exposure distributions of target populations
F. Quantify determinant variables for exposure level distributions
G. Use Action Level as a screening cut off to identify target populations for additional monitoring

Data Analysis Strategies
H. Control chart for daily TWA exposure levels and distribution parameters
I. Probability charts
J. Inferential statistical methods:
   Hypothesis testing
   Confidence limits and intervals
   Tolerance limits and intervals
K. Linear regression analysis:
   Simple (2 variables)
   Multiple (3 or more variables)

of occupational exposure estimation are presented in Table 8.1. Specific inferential statistical methods related to individual exposure estimates, such as confidence limits, confidence intervals, and hypothesis testing, will be presented in Section 6. Section 6 also presents inferential methods related to distributions of exposure estimates, such as tolerance limits and tolerance intervals, point estimates of distributional fractiles, and the use of logarithmic probability and semilogarithmic graph paper to display exposure distributions.

3 STATISTICAL THEORY, SOURCES OF VARIATION, AND DISTRIBUTIONAL MODELS FOR OCCUPATIONAL EXPOSURE DATA

To plan a study to evaluate occupational exposures and properly analyze the resulting worker exposure data, one must first understand how various types of errors can affect individual exposure data values. Evaluating the errors which affect an individual exposure measurement is analogous to evaluating how the sizes of individual trees of the same type and age vary randomly within sections of a forest. Adequate distributional models for errors in occupational exposure results and the sources of variation for single exposure data values will be discussed in Sections 3.2 and 3.3. Next, one must understand the patterns that groups of data will occur in. For example, the location parameters (mean values) of the within-group distributions may differ between groups due to assignable causes. This is analogous to interpreting differences in average tree sizes for entire forests due to such factors as soil fertility, rainfall, pollution, etc. Models for describing the behavior of such data families will be discussed in Sections 3.4 through 3.6.

3.1 Nomenclature for Exposure Data

Exposure data can be reported as single measurements or other estimates. An *exposure measurement* is the measured airborne concentration of a material in a single air sample taken near a worker.

Note: The use of the word sample *in both the statistical sense and the industrial hygiene physical sense in this chapter can present a source of confusion. Conceptually, a statistical sample consists of one or more items selected from a parent population, each of which has some characteristic measured. However, in the physical sense an industrial hygiene sample or exposure sample is an* exposure measurement *determined from a measured amount of an airborne material collected on a physical device (e.g., filter, charcoal tube, passive dosimeter). Industrial hygiene sampling is usually performed by drawing a measured volume of air through a filter, sorbent tube, impingement device, or other instrument to trap and collect the airborne contaminant. Passive dosimeters rely on diffusion to move the contaminant to the collecting media. In the sense of this chapter, an occupational exposure sample and accompanying act of sampling combine both the concept of a statistical sample (i.e., one result among many which could occur under the same conditions) and the* physical sample *that is chemically analyzed or interpreted.*

An *exposure estimate* is an estimate of a workplace exposure over a specified time period, which is calculated from one or more *exposure measurements*. An exposure estimate may be for a few seconds or represent a period from minutes to years. In the latter case it is known as a *time-weighted average (TWA)* exposure, which is a time-integral of the instantaneous exposure (i.e., the cumulative concentration) divided by the length of time for the exposure period. If cumulative exposure could be estimated from a single air sample, it would be the quotient of the weight of the material in the air sample divided by volume of air sample during the measurement period. Frequently the duration of TWA exposure estimates will be about 15 minutes or less to estimate short-term acute exposure risk, 8 hours to estimate a workday exposure risk, and 40 hours or longer to estimate prolonged chronic exposure risk. If it is appropriate from a toxicological standpoint, longer time-averaging periods (a month or longer) can be used to estimate chronic exposure risk. Certain nomenclature has been developed by Leidel et al. (1) to describe several different types of TWA exposure estimates. These are illustrated in Figure 8.1 and include:

1. *Full period single sample estimate.* A single exposure measurement taken for the full duration of the desired time-averaging period (e.g., 40 hours for a workweek TWA, 8 hours for an 8-hour workday TWA, or 15 minutes for a 15-minute TWA).

2. *Full period consecutive samples estimate.* The time-weighted average (TWA) of a continuous series of exposure measurements (equal or unequal time duration) obtained for the full duration of the desired time-averaging period.

3. *Partial period consecutive samples estimate.* The time-weighted average (TWA) of a series (continuous or noncontinuous) of exposure measurements (equal or unequal time duration) obtained for a total duration less than the desired time-averaging period. For an 8-hour TWA exposure estimate, this would mean that exposure samples assumed to represent the entire 8-hour exposure would be selected to cover about 4 to less than 8 hours. Several samples totaling less than 4 hours (e.g., eight 15-min samples) can be better described as grab samples for the purposes of statistical analysis.

4. *Grab samples estimate.* The average of several short-period samples taken during random intervals of the time-averaging period. Sometimes it is not feasible, due to technical limitations in measurement methods (e.g., direct reading instruments, some colorimetric detector tubes), to obtain a type (1) or type (2) exposure estimate. In such situations, grab samples may be taken during several short intervals (e.g., seconds, several minutes to less than about 30 minutes) of the desired larger time-averaging period such as 15 minutes or 8 hours.

Adequate distributional models for describing variability of the preceding four exposure estimates will be presented in Section 3.6.

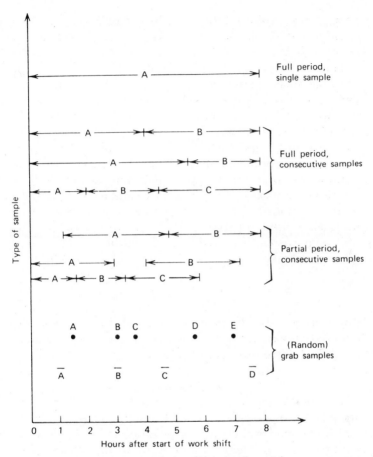

Figure 8.1 Types of exposure measurement that could be taken for an 8-hour average exposure standard. From Leidel et al. (1).

3.2 Net Error Model for Exposure Measurements

Suppose a worker's exposure to a workplace contaminant is to be measured. Assume that an appropriate measurement method is available, which *on the average* can give valid (i.e., representative and accurate) determinations of airborne concentrations. The sampling equipment and laboratory instruments must be properly calibrated to reduce systematic errors or biases. This does not imply that every sample will give the correct answer (i.e., the true value of the airborne concentration at the time and place where the sample is taken). To the contrary, every sample result will differ from the respective true average exposure that existed during the time period of the sample. "Exposure" is used here synonymously with "concentration" because it is assumed that no difference exists between the *true concentration* measured and the *true exposure* intended to be measured (e.g., as the concentration in a "breathing zone"). The discrepancies between the reported results and the unknown true exposures are termed *random errors* because they are assumed to vary in magnitude and direction in a random manner from sample to sample. Random errors, within limits, are inherent to any measurement method and equipment. The presence of random error does not imply that the method has been improperly used (i.e., that mistakes have been made). Of course, a discrepancy outside the usual range of variability (an excessively imprecise result) for the method could indicate that a mistake has been made and such data might be discarded, especially if the suspect result can be associated with an identifiable irregularity that occurred during the sampling procedure or in the analysis of the sample.

To systematically approach the statistical treatment of random errors, we use a mathematical statistical model. The true average concentration at the spatial location and temporal period of the exposure sample is denoted by the symbol μ and the particular reported result from the sample is denoted by the symbol X. Thus the total error of a single sample ϵ_T is given by

$$\epsilon_T = X - \mu \tag{1}$$

The total (net) error ϵ_T is the algebraic sum of independent measurement errors, which are typically due to the component sampling and analytical steps in the measurement procedure. For example:

$$\epsilon_T = \epsilon_S + \epsilon_A \tag{2}$$

where ϵ_S is a positive or negative random sampling error and ϵ_A is a positive or negative random analytical error. For *independent errors,* the size and sign of the analytical error does not depend on the size or sign of the sampling error. All ϵ's have the same units of concentration as the reported result X (e.g., ppm or mg/cu m).

Thus any exposure estimate X that is both an industrial hygiene and statistical *sample* can be represented as the algebraic sum of the true concentration μ and the net sampling and analytical error for the particular sample:

$$X = \mu + \epsilon_S + \epsilon_A. \tag{3}$$

If multiple samples could be taken at *exactly the same point in space and over the same time period* and *if the true value μ were identical for all samples,* they would be true *replicate samples.* Note that in actual industrial hygiene sampling it may be difficult to obtain duplicate samples that are true replicates. Nevertheless, a given sample must be thought of as a sample from the hypothetical population of all replicate samples which might have been obtained under exactly the same exposure conditions. Seim and Dickeson (49) have suggested a device for collecting actual replicate samples in the workplace.

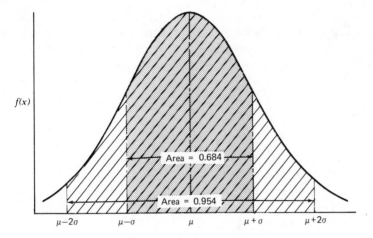

Figure 8.2 Normal distribution curve.

Figure 8.2 presents a horizontal concentration scale on which the point μ represents the true concentration. Concentration results X for replicate samples occur within intervals above and below μ with predictable *relative frequencies* (or probabilities) that are proportional to corresponding areas under the curve (the *sample distribution* curve) appearing above the concentration scale. Ordinates of the sample distribution curve *do not* give probabilities of corresponding sample results. Rather, it is the *area* under the sample distribution curve between two values of X that is proportional to the relative frequency (proportion) of replicate samples that would occur in that interval. For this area to represent a probability, which is a proportion between 0 (impossibility) and 1 (certainty), a distribution curve is *standardized* such that the total area under the curve is exactly unity (1.0).

3.3 Sources of Variation in Exposure Results

Routinely a population of exposure sample results from a given sampling strategy (e.g., grab samples for a worker during a workshift, 8-hour TWA exposures for several workers on a workshift, a series of 8-hour TWA exposures for a worker on several workshifts) will exhibit variability (i.e., scatter or dispersion). An important part of the interpretation of such results is an analysis of the pattern of variation, or distribution of the data. From the sample results one usually attempts to draw inferences about the population distribution of exposure levels (i.e., about the pattern of all levels which occur for the same conditions under which the sample results were obtained). When analyzing sample data, it is important to understand the sources of variation in exposure sample results that combine to create observed total (net) variability (due to total errors measured as differences between measurement results and true exposure levels). The sizes of these variations are a function of both the exposure levels and the measurement method. Both *random* and *systematic errors,* can affect both the *exposure levels* being measured (Sections 3.3.1 and 3.3.3) and the *measurement procedure* used to obtain the sample results (Sections 3.3.2 and 3.3.4).

3.3.1   Random Variation in Workplace Exposure Levels

An elementary mathematical model for random errors leading to net random variation in replicate sample results was discussed in the Section 3.2. Our recognition of random influences on the measurement process can be extended to a recognition of other random influences which affect the true value of what is measured. This extension of the model will enable us to better understand the sources of variation in exposure results at different sites and times. Random changes in the determinant variables affecting the workplace exposure levels can lead to random variation in the exposure results. Exposure *determinant variables* are qualitative or quantitative variables, factors, or parameters that influence or affect true airborne exposure levels. The types of determinant variables are discussed in detail in Section 5. For now, note that random variation in determinant variables can result in:

1.   *Intra*day (within a day) exposure level fluctuations in the workplace environment.
2.   *Inter*day (between days) exposure level fluctuations in the workplace environment.
3.   Variation in exposures of different workers within a job group or occupational category

It is important to realize that random variation in exposure levels and in subsequent exposure measurements can be accounted for (but not prevented) by appropriate statistical procedures. Generally the magnitude of random exposure variations over space, time, or workers cannot be quantified or predicted before making the exposure measurements, since often it is not possible to predict how many determinant variables will affect workplace exposure levels.

3.3.2   Random Variation in the Measurement Procedure

In the previous section we discussed sources of random variation in workplace true exposure levels. Random errors also occur in the exposure level measurement procedure and these physical variations lead to random variations in corresponding exposure measurements data. Examples of possible sources of random physical errors in the *process* of exposure measurement are:

1. Random changes in pump flowrate (or mass flowrate with passive dosimeters) during sample collection.
2. Random changes in collection efficiency of the sampling device.
3. Random changes in desorption efficiency of the samples during analysis.

It is important to realize that random variation in multiple results during an hour or during a day, due to fluctuations in real workplace exposure levels, will usually exceed measurement procedure variation by a substantial amount (often by factors of 10 or 20). Thus the predominant component of variation in these results will be due to the considerable variation in what is being measured. However, in contrast to random variation in exposure levels, the relatively smaller random measurement errors can generally be quantified before making the measurements (i.e., ranges of error can be estimated probabilistically from methods evaluation experiments performed before the exposure measurements). The effects of the known distribution of random measurement errors on exposure results can be minimizd by the application of sampling strategies and programs based on statistical principles.

3.3.3   Systematic Variation in Worker Exposure Levels

In contrast to random variations, *systematic variations* in either workplace exposure levels or systematic errors in the measurement procedure cannot be predicted with statistical methodologies based in probability theory. Instead, the study design must anticipate and make provision for systematic errors. During a data analysis performed to compare exposure results between two groups, the comparison should be made within appropriate subgroups which are homogeneous in all other respects, or the measurements should be corrected for possible systematic errors due to extraneous factors *before* any statistical analyses are performed.

Systematic biases or shifts in the determinant variables affecting the workplace exposure levels will lead to systematic shifts in the exposure results. For example, some systematic shifts in determinant variables, and their consequences, are:

1. Changes in a worker's exposure situation (such as several different jobs or operations during a workshift or over several days), which can result in intraday or interday shifts in worker exposure.
2. Production or process changes, which can cause shifts in worker exposure levels (intraday or interday).
3. Control procedure or control system changes, which can cause shifts in worker exposure levels (intraday or interday).

3.3.4   Systematic Errors in the Measurement Procedure

Besides the random errors in a measurement procedure, there can also be *systematic errors* or *biases* which occur during the measurement procedure, which lead to systematic errors in exposure results. Examples are:

1. Mistakes in pump calibration and drops in pump battery voltage leading to systematic errors in air flowrate.
2. Use of the sampling device at temperature or altitude conditions substantially different than the calibration conditions (see Technical Appendix G of Reference 11).
3. Physical or chemical interferences during sample collection.
4. Sample degradation during storage before analysis.
5. *Intra*laboratory errors due to chemical or optical interferences, improper procedures, mistakes in analytical instumentation calibration, or failure to properly follow the steps of an analytical procedure.
6. *Inter*laboratory differences due to use of different equipment, or different training of personnel.

Systematic measurement errors may be identified and their effects minimized with the use of quality assurance programs.

In the statistical sense, a substantial systematic shift or error (a step change during a series of measurements) in either the measurement process or the exposure levels being measured creates a different population with another location (central tendency) on the exposure level scale. If the systematic shift(s) goes undetected, the resulting two (or several) "side-by-side" sample populations can be mistakenly analyzed as a single distribution. *The inferential statistical procedures presented in this chapter will not detect and do not allow for the analysis of highly inaccurate results caused by systematic errors or shifts.* Unfortunately, systematic errors in a measurement procedure or systematic shifts in exposure levels sometimes go undetected and introduce considerably larger variation into the exposure results than would be caused by the usual random errors. This can lead to reporting inferences from the sample results that have erroneously higher uncertainty (less precision) than is stated, if precision is calculated from the known amount of random variability which has previously existed in similar individual sample distributions. Vague or uncertain inferences generally have little value; thus it becomes important to identify potential systematic errors and take steps in the study design to eliminate them, or correct them, if possible,when the results are statistically analyzed.

3.3.5   Location of the Measurement Device in Relation to the Worker

A most important goal of personal exposure measuring is to obtain *valid* estimates of the concentrations breathed by workers. A valid exposure estimate is one that measures what it is purported to measure. More specifically, criteria for evaluating the validity of worker exposure estimates include:

1. *Relevance.* Is the air concentration in the sample equivalent to the concentration breathed by the worker(s) of interest?

2. *Calibration.* Are the exposure measurements unbiased estimates of the true concentration sampled (i.e., are measurements accurate on the average)?

3. *Precision.* Was adequate exposure information obtained to derive a sufficiently precise exposure estimate for the worker(s) of interest (either on a given day or over some longer period such as several years of employment)?

Concerning relevance of an exposure estimate to a worker's actual exposure, exposure measurements should be taken in the worker's *breathing zone* (i.e., air that would most nearly represent that inhaled by the worker). There are three basic classes of exposure measurement techniques:

1. *Personal.* The measurement device is directly attached to the worker and worn continuously during all work and rest operations. Thus the device collects air from the "breathing zone" of the worker.

2. *Breathing zone.* The measurement device is held by a second person who attempts to sample the air in the "breathing zone" of the worker.

3. *General air.* The measurement device is placed in a fixed location in the work area. This technique is also called *area sampling.*

If measurements taken by the *general air* technique are to be used to estimate worker exposures, then it is necessary to demonstrate that they are valid personal exposures. Normally this is difficult to do. Refer to Technical Appendix C of Reference 1 for a discussion of the subject.

3.4 The Normal Distribution Model

3.4.1 Descriptive Parameters

A utilitarian mathematical model for the frequency distribution of some types of continuous-variable occupational health data is the *normal distribution*. This model has a simple formula which can be used to describe and compute the normal distribution curve. In routine practice the formula is rarely directly applied, since tables of the distribution are readily available. The formula, which relates ordinates $f(X)$ of the curve to values of the variable X, is called a *distribution function* or simply *distribution*. [Note. This terminology is not universally used. Our definition of the term *distribution function* is according to Hald (50), but others refer to $f(X)$ as the *frequency function* or *probability density function*.] The $f(X)$ illustrated in Figure 8.2 is of a particular type known as the normal curve. This distribution function is represented by the special notation $N(X; \mu, \sigma^2)$. That is, for the normal distribution model, the ordinate or height of the probability density curve is given by:

$$f(X) = N(X; \mu, \sigma^2) = \frac{1}{\sigma\sqrt{2\pi}} \exp\left(\frac{-\frac{1}{2}(X - \mu)^2}{\sigma^2}\right) \tag{4}$$

Two constants, or parameters, completely describe the normal distribution: μ, its mean, and σ, its standard deviation. Thus the notation $N(X; \mu, \sigma^2)$ is statistical shorthand for "the distribution of a variable X which has the true mean μ and variance σ^2." All normal curves have the same general appearance— a bell-shaped curve that is symmetrical about its mean. The true mean μ, also called the *expected value* of the random variable X, denoted $E(X)$, is the weighted average value of all values of the distribution. The weighting function is the distribution function [i.e., each X weighted by its probability density $f(X)$].

Mathematically, the mean $E(X)$ of any distribution function, say $f(X)$, is the center of gravity of the corresponding distribution curve, defined by:

$$E(X) = \int_{-\infty}^{+\infty} Xf(X) \, dX \tag{5}$$

For the mean of the normal distribution the general $f(X)$ in Equation 5 is replaced by $N(X; \mu, \sigma^2)$ so that

$$E(X) = \int_{-\infty}^{+\infty} \frac{X}{\sigma\sqrt{2\pi}} \exp\left(\frac{-\frac{1}{2}(X - \mu)^2}{\sigma^2}\right) dX = \mu \tag{6}$$

The integration in Equation 6 is not obvious and the details of its evaluation are not presented here. The point to note is that for the normal distribution, the parameter μ in its formula is the mean of the distribution.

Similarly, it can be shown that for the normal distribution, the weighted average value of squared deviations $(X - \mu)^2$ is σ^2, that is,

$$E(X - \mu)^2 = \int_{-\infty}^{+\infty} \frac{(X - \mu)^2}{\sigma\sqrt{2\pi}} \exp\left(\frac{-\frac{1}{2}(X - \mu)^2}{\sigma^2}\right) dX = \sigma^2 \tag{7}$$

For any distribution the mean square of deviations from the mean, denoted by $E[X - E(X)]^2$, is known as the *variance* of X. The *variance* is the square of the standard deviation. For the normal distribution, the variance is equal to its second parameter σ^2, so that the standard deviation is σ. The *mode* of any distribution is the point on the X-scale at which the maximum of the distribution function occurs. The *median* is the middle X-value, that is, the value exceeded by 50 percent of the area under the distribution curve. Note that hereafter we may refer merely to "proportion of the distribution," which should be understood to mean "proportion of the area under the distribution curve." Also, since the total area under any distribution curve is unity (1.0), the "proportion of the area between two X-values" can also be called the "area between two X-values." For the normal distribution the mode, median, and mean are equal to one another.

The two parameters of a normal distribution completely determine its location (central tendency) and shape (dispersion or variability). The location parameter is the mean μ, which is the center point of the curve. The variability parameter (or *measure of dispersion*) is the standard deviation σ, which indicates the dispersion of the X-values about their mean. Table 8.2 gives some examples of relationships between the mean, standard deviation, and proportions of the total distribution that lie within various intervals containing the mean. The first and fifth of these proportions are depicted graphically in Figure 8.2 as shaded and cross-hatched areas, respectively, under a normal curve. See Section 6.9 for a procedure to calculate intervals of a normally distributed variable X (with known parameters) which contain designated proportions of the distribution. Such values can be expressed by *fractile* terminology (this is Hald's terminology (50), some others use *quantile*). The *fractile* X_P is the value of X which has proportion P of the distribution $f(X)$ at or below it.

3.4.2 Coefficient of Variation

Random error in exposure concentration measurements, due to errors in exposure measurement procedures during sampling (e.g., elapsed time, air flowrate) and during subsequent chemical analyses, are generally proportional to the level of airborne concentration measured. Therefore, it is appropriate to express the magnitudes of these errors as fractions of the concentration levels. In this way, the measurement variability of an exposure measurement procedure can be adequately expressed as a constant value which is independent of the concentration measured. A measure of this proportional variability called the *coefficient of variation* (CV) is defined by: $CV = \{E[X - E(X)]^2\}^{1/2}/[E(X)]$. Chemists know it as the *relative standard deviation* (RSD) or (s_r). For the normal distribution, $CV = \sigma/\mu$. If a measurement procedure is composed of two or more independent steps (e.g., obtaining the sample, subsequent laboratory analysis), it can be shown (see Equation 2) that the net error ($\epsilon_T = \epsilon_S + \epsilon_A$)

Table 8.2 Areas Under the Normal Curve

X-Interval	Data Within Interval (%)
$\mu - \sigma$ to $\mu + \sigma$	68.4
$\mu - 1.645\sigma$ to $\mu + 1.645\sigma$	90.0
$\mu - 1.96\sigma$ to $\mu + 1.96\sigma$	95.0
$-\infty$ to $\mu + 1.645\sigma$	95.0
$\mu - 2\sigma$ to $\mu + 2\sigma$	95.4
$\mu - 1.96\sigma$ to ∞	97.5
$\mu - 2.576$ to $\mu + 2.576\sigma$	99.0
$\mu - 3\sigma$ to $\mu + 3\sigma$	99.7

for the combined steps of the procedure has the following *total coefficient of variation*:

$$CV_T = [CV_S^2 + CV_A^2]^{1/2} = \sigma_T/\mu \tag{8}$$

where the subscript S denotes the sampling step and the subscript A denotes the analytical step. It is important to realize that the CVs are not directly additive; instead, the CV_T increases as the square root of the sum of the squares of the component CVs.

The total relative standard deviation CV_T generally can be treated as constant within the range of concentrations at which the measurement method is routinely applied. At a given concentration μ within the application range, the standard deviation of the measurement error is given by

$$\sigma_T = (\mu)(CV_T) \tag{9}$$

where

$$\sigma_T = [\sigma_S^2 + \sigma_A^2]^{1/2} \tag{10}$$

Note that a random variable which is assumed to be normally distributed can theoretically attain negative values, whereas airborne concentration levels are equal to or greater than zero. Nevertheless the normal distribution model is usually an adequate approximation to the sampling distribution of replicate samples. Most measurement methods used in industrial hygiene have net random errors whose standard deviation is small compared to the true mean airborne concentration. Thus the portion of the normal distribution model lying left of zero has negligible area and the model adequately predicts the distribution of (positive) replicate measurements.

3.5 The Log Normal Distribution Model

A second mathematical model of great utility for several types of industrial hygiene data is the logarithmic normal or *log normal distribution*. The general properties of the several types of log normal distributions have been extensively discussed by Aitchison and Brown (51). Section 3.5.1 will discuss a variate whose logarithm is distributed according to normal law (i.e., the case of a *2-parameter log normal*). This is the simplest case because it involves primarily an interplay of the mathematical properties of the logarithmic function and the well-known statistical properties of the normal distribution, which were discussed in the previous section. In Section 3.5.2, the definition and scope of the log normal distribution will be extended with the use of a third parameter to shift the origin of the distribution's measurement scale.

3.5.1   Two-Parameter Log Normal

The *2-parameter* lognormal curve illustrated in Figure 8.3 is one of a family of such curves that have the general formula:

$$f(X) = \frac{1}{X(\ln \sigma_g)\sqrt{2\pi}} \exp\left[\frac{-\frac{1}{2}(\ln X - \ln \mu_g)^2}{\ln^2 \sigma_g}\right] \tag{11}$$

where $0 < X < \infty$.

Equation 11 is called the *log normal distribution function*. Its general structure is similar to Equation 4 for the normal distribution function. The relationship between these two distributions is that for a random variable X, which is log normally distributed, the values $\ln X$ or $\log X$ will be normally distributed. As with the normal distribution, the basic 2-parameter log normal distribution is fully described by only two parameters. However, the log normal parameters are known as the *geometric mean* (GM) and *geometric standard deviation* (GSD). The true geometric mean, μ_g is defined by

$$E(\ln X) = \int_{-\infty}^{+\infty} \ln X \, N(\ln X; \ln \mu_g, \ln^2 \sigma_g) \, d \ln X = \ln \mu_g \tag{12}$$

The true geometric standard deviation σ_g is defined by

$$E(\ln X - \ln \mu_g)^2 = \int_{-\infty}^{+\infty} (\ln X - \ln \mu_g)^2 N(\ln X; \ln \mu_g, \ln^2 \sigma_g) \, d \ln X$$
$$= \ln^2 \sigma_g \tag{13}$$

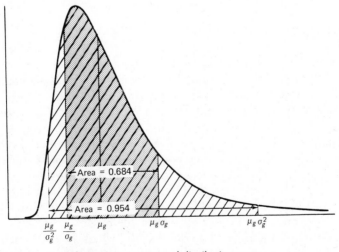

Figure 8.3   Log normal distribution curve.

In these equations, the notation used is analogous to the $N(X; \mu, \sigma^2)$ notation defined in Equation 4 for a normally distributed variable X with mean μ and variance σ^2. Thus μ_g and σ_g are antilogs to the base e of the mean and standard deviation, respectively, of the natural logarithmic transform of X. The interpretation of μ_g and σ_g parameters for a 2-parameter log normal distribution differs somewhat from the interpretation of μ and σ for a normal distribution. The similarity is that the geometric mean is the location parameter, and the geometric standard deviation is the variability, or dispersion, parameter. Note that the distribution function for the 2-parameter log normal model originates at the origin of zero and does not and cannot exist in the region below zero.

In Table 8.2 for a normal distribution, multiples of σ were added to and subtracted from μ to obtain intervals of X that contain specified proportions of the distribution. The factors that multiply σ are standard normal deviates (i.e., values of a normally distributed variable with mean zero and variance one), which are known as Z-*values*. Values of Z are listed in tables of the standard normal distribution available in statistical texts and other scientific reference books. A given Z-value, denoted Z_P, corresponds to a probability P that a randomly selected value of X will be within the interval $(\mu - Z_P\sigma)$ to $(\mu + Z_P\sigma)$. For example, Table 8.2 shows that $Z_{.684} = 1.000$, $Z_{.90} = 1.645$, and $Z_{.95} = 1.960$. Corresponding intervals for a log normal distribution are of the form $\mu_g / \sigma_g^{Z_P}$ to $\mu_g \sigma_g^{Z_P}$. Table 8.3 gives examples of 2-parameter log normal intervals corresponding to the intervals for a normal distribution in Table 8.2. The first and fifth of these intervals are depicted graphically in Figure 8.3 as shaded and cross-hatched areas under a 2-parameter log normal curve. See Section 6.10 for a procedure to calculate intervals of a log normally distributed variable X (with known parameters) which contain designated proportions of the distribution. Such intervals are bounded by *fractiles* X_P (see the definition given in Section 3.4.1).

3.5.2 Three-Parameter Log Normal

The utility of the basic 2-parameter log normal model can be considerably expanded by the introduction of a third parameter, which is a change-of-origin parameter. A simple displacement of a random variate X, which is *not* log normally distributed, can sometimes be made to define a transformed variate $X^* = (X - k)$, which *is* log normally distributed. If the range of X is $k < X < \infty$, the range of X^* will be $0 < X^* < \infty$. The 2-parameter model can be thought of as a special case of a 3-parameter log normal model, for which $k = 0$. This third parameter for the log normal model can sometimes be selected as the lower bound to the known range of values of the original variate X and can be thought of as the threshold of the 3-parameter log normal distribution, just as zero is the threshold for the basic 2-parameter model.

The 3-parameter log normal model is useful when the data, such as exposure results, seem to show more skewness to the right than would be expected for a log normal distribution which is now located close to the zero origin. Such a

Table 8.3 Areas Under the Log Normal Curve

X-Interval	Data Within Interval (%)
μ_g/σ_g to $\mu_g\sigma_g$	68.4
$\mu_g/\sigma_g^{1.645}$ to $\mu_g\sigma_g^{1.645}$	90.0
$\mu_g/\sigma_g^{1.96}$ to $\mu_g\sigma_g^{1.96}$	95.0
0 to $\mu_g\sigma_g^{1.645}$	95.0
μ_g/σ_g^2 to $\mu_g\sigma_g^2$	95.4
$\mu_g/\sigma_g^{1.96}$ to ∞	97.5
$\mu_g/\sigma_g^{2.576}$ to $\mu_g\sigma_g^{2.576}$	99.0
μ_g/σ_g^3 to $\mu_g\sigma_g^3$	99.7

distribution could result if there were log normal random additive variations in exposure levels combined with a fixed background exposure level. An appropriate constant k is subtracted from each data value to create transformed data values which are then analyzed using techniques appropriate for log normally distributed data. Then as desired, estimated parameters of the transformed distribution, along with appropriate confidence limits, are calculated. Examples are the GM of X^* and its confidence limits, or tolerance limits for the random variable X^*. The constant k is then added to all calculated values of X^* to estimate corresponding values for X relevant to the original data distribution. Details for appropriate estimation of k are presented in Section 6.12, Step 5 of the Solution.

3.6 Adequate Distributional Models for Exposure Results

To design efficient and statistically powerful studies, make rational decisions in hypothesis tests, and make valid inferences regarding expected limits on true occupational exposures, it is necessary to use *adequate* distributional models of exposures for the target populations the samples represent. Adequate distributional models are the keystones to the parametric statistical methodologies that will be presented in Sections 5 and 6.

The adequacy of a model is dependent on its ability to serve as a workable forecaster of the unsampled portions of the parent population. Moroney (52) has noted,

Probably there never is a mathematical function which fits a practical case absolutely perfectly. Nor is it at all necessary that there should be. What we seek is not a *perfect* description of a distribution but an *adequate* one; that is to say, one which is good enough for the purpose we have in view.

Wilkins (53) has remarked that statistical tests for goodness-of-fit have the characteristic that the test will reject any practical data set if there is a sufficient number of samples. The critical value for the test statistic (based on the permissible departure of the observed data distribution from the chosen distributional model) can even be smaller than the precision provided by the measurement methodology used to obtain the data. Even though we might observe a "statistically significant difference in fit," the proposed distributional model may still be adequate for our needs. One must also be cautious regarding hypothesis test outcomes of "no statistically significant difference in fit." These outcomes may be due to low statistical power for the tests resulting from small sample sizes.

3.6.1 Applications for the Normal Distribution

The *normal distribution* is usually an adequate model for the following populations of industrial hygiene results:

1. *Populations of replicate analyses* performed on an industrial hygiene sample (e.g., aerosol filter or charcoal tube). *Replicate analyses* of a given sample are defined to be repeated analyses with variability equal to that which would exist in analyses of physically different samples, if these could have been obtained *without* sampling errors of exactly the same concentration in exactly the same setting. In other words, replicate analyses are those with variability which reflects the total random error of the analytical procedure, not just components of error due to some (but not all) steps in the analysis. For an unbiased analytical method (i.e., without systematic error), the expected value of truly replicate analyses of a given physical sample is not the true time-integrated concentration the sample was obtained from. Rather, the expected value of replicate analyses is an air concentration equivalent to the amount actually present in the sample (i.e., the expected value includes the random sampling error for that sample). The sample coefficient of variation (CV_A) computed from replicate analyses is a measure of dispersion for the analytical procedure, and is usually taken to be an approximation to the true CV_A measuring the proportional error due to the analytical step of the exposure measurement procedure. Note that the total (net) error of the measurement procedure is $CV_T = (CV_A^2 + CV_S^2)^{1/2}$, where CV_S is the coefficient of variation for *sampling errors* (i.e., those introduced by the physical sampling portion of the measurement procedure during which the contaminated air is moved onto or through the sampling media).

Section 6.6 presents the computation of tolerance limits for a variable which is normally distributed with unknown parameters. Based on results for n replicate samples, this procedure can be used to compute a tolerance interval that, we can be 95 percent confident, will contain at least 95 percent of analytical results for the same concentration by the same method under the same conditions. The procedures in Sections 6.12 and 6.13 detailing the use of logarithmic probability paper to compute and display tolerance limits can be modified, where a normal distribution is expected, by substituting normal probability paper (where a linear scale is used for the original data instead of a logarithmic scale).

2. *Populations of replicate measurements* of calibrated test concentrations. The arithmetic mean of the replicate measurements is the best estimate of a calibrated test concentration. The sample coefficient of variation determined from a set of *replicate measurements* is an estimate of the measurement procedure's total CV_T (combined CV for the sampling and analytical portions of the method). Refer to Section 3.4.2 and Step 1 of this section.

Industrial hygiene researchers will usually obtain multiple measurements taken simultaneously at sampling locations in a small spatial volume (e.g., a sphere less than 30 cm in diameter), when attempting to estimate the variability (CV) of a measurement method. Unfortunately, unless the sampled workplace atmosphere is truly homogeneous, the sample results may lead to a variability estimate for the measurement method that is erroneously high. It is usually assumed that each of the measurements is a sample of the same true concentration, but this may not be a valid assumption and the researcher must demonstrate that the sample environment is truly homogeneous.

It is important to note that normality of replicate exposure measurements (or at least approximate normality that is sufficient to meet the requirements of any inferential statistical methodology one desires to use) is to be expected from theoretical considerations. The total net error of any particular exposure measurement is the net error resulting from many random incremental positive and negative additive physical influences during the sampling and analytical stages of a measurement. Determinant variables that can lead to positive and negative errors in measurements include unavoidable technician variations, small environmental variations (e.g., humidity, temperature), and functional variations in component parts of the sampling and analytical equipment (e.g., voltage, pump flowrate, operating temperature). Insofar as these various sources of random errors operate independently, their net influences tend to make the net error follow a normal probability density curve. The proof that this is true would be similar to the proof of the central limit theorem from mathematical statistics.

3. *Populations of replicate full period single sample estimates* of exposure. The justification for using the normal distribution to model the random errors of replicate exposure measurements has been discussed previously in part (2) of this section. A single full period sample estimate can be considered a sample of one from a hypothetical parent population of all replicate exposure measurements which could have been obtained at exactly the same point in space and over the same time period. Even though the sample is continuously collecting airborne material from constantly varying levels (e.g., from a log normal distribution of true levels that generally have substantial variation), it is assumed that on the average such samples faithfully and accurately integrate all the instantaneous concentration levels. The result of this integration is the mass of formerly airborne material collected on the sample. Thus the single

full period measurement is the time-integrated exposure estimate for the duration of the sample. A previously well-determined total coefficient of variation CV_T for the measurement procedure is used as the measure of dispersion for the population of possible replicate samples from which the one at hand is considered to be a random sample. This known CV_T is used for inferential decision-making and confidence interval calculations for this type of exposure estimate.

Section 6.1 presents applied statistical procedures for computing confidence limits (Section 6.1.1) and classifying exposures relative to an exposure control limit (Section 6.1.2) for the case of an exposure estimate based on a full period single sample.

4. *Populations of full period consecutive sample estimates* of exposure. The justification for using the normal distribution to model the random errors of averages of sets of consecutive samples taken during the time-averaging period of the exposure estimate is an extension of the preceding discussion for the full period single-sample estimate. It is assumed that the random proportional errors of the consecutive measurements are independent and have the same known total coefficient of variation for the measurement method.

Sections 6.2 and 6.3 present applied statistical procedures for computing confidence limits on the true exposure (Sections 6.2.1 and 6.3.1) and for classifying exposures relative to an exposure control limit (Sections 6.2.2 and 6.3.2) when exposure estimates are based on the average of full period consecutive samples. The confidence interval and decision-making computations are presented for two different types of situations. *Uniform exposure* methods are given in Section 6.2 to be used when one believes that all consecutively sampled periods had equal exposure concentrations. Conservative *nonuniform exposure* methods are given in Section 6.3, to be used if one believes the periods have had substantially different exposure concentrations.

3.6.2   Applications for the Log Normal Distribution

The *log normal distribution* (either 2-parameter or 3-parameter) is usually an adequate model for four general types of populations discussed in this section. However, be alert that a population of industrial hygiene data may be a composite of several different distributions. For example, a substantial portion of the data may occur at zero concentration with the remainder occurring in a 2-parameter log normal distribution and one or more 3-parameter log normal distributions.

1. *Populations of true exposure levels* at different times during periods of hours to years. The justification for using the log normal distribution to model workplace exposure levels at different times for a given worker, or averages for a *target population* of workers at different times (see Section 5), has been presented in Technical Appendix M of (1). Conditions conducive to (but not all necessary for) the occurrence of log normal distributions are found in

populations of workplace exposure levels. These conditions include:

a. Physical causes of variability tend to cause the same percentagewise changes in concentration, irrespective of whatever concentration is present.
b. The true exposure levels cover a wide range of values, often several orders of magnitude. The variation of the true exposure levels is of the order of the size of the exposure levels.
c. The true exposure levels lie close to a physical limit (zero concentration).
d. A finite probability exists of unusually large values (or data "spikes") occurring.

Section 6.7 presents an applied statistical procedure for the computation of tolerance limits for samples from a log normally distributed population with unknown parameters. This procedure is useful for computing from a sample of exposure levels, a tolerance interval that we can be 95 percent confident will contain at least 95 percent of the population of true exposure levels. Section 6.8 details the computation of a point estimate and confidence limits for the proportion of a log normally distributed population that exceeds a specified value (such as an exposure control limit). Section 6.11 suggests the use of semilogarithmic graph paper for plotting variables that are log normally distributed in time, which frequently is the case for true exposure levels at different times during a period. Sections 6.12 and 6.13 present procedures for using logarithmic probability paper to estimate the parameters of logarithmic normal distributions and for displaying tolerance limits for estimated log normal distributions.

2. *Grab sample populations* of intraday exposure measurements made on noncoincident short-term samples, where the duration of each sample is short compared to the total interval the samples were obtained from (e.g., less than about 5 percent of the interval). Grab samples reflect the log normal intraday distribution of true exposure levels that is sampled. Grab sample populations also have a component of normally distributed measurement process (sampling and analysis) error, but this component generally is negligible compared to the log normal variation of the true concentrations.

For evaluation of a worker's individual health risk, an inference must be made concerning the relation between the TWA exposure control limit and the true arithmetic mean of the entire population of grab samples from which the few samples at hand were selected at random. If the true exposures occurring during the interval could be considered uniform (effectively equal), then the sample arithmetic mean of the grab samples would be an adequate estimate of the TWA exposure. If there are subintervals of respectively stable but different exposures, then a sample arithmetic mean for each period of equal exposure would have to be computed from the grab samples of each period and a TWA estimate computed from the series of arithmetic means. Confidence limits for such estimates of TWA exposures would be based on the normal distribution, since the only errors are those of the measurement procedure.

However, if there were general log normal variability among the total set of grab samples intervals making up the period of the standard, an estimate of the TWA exposure would require computing a sample geometric mean and converting it to an estimate of the TWA exposure, but this is not as precise a procedure. The correction factor for converting a *GM* estimate to a *TWA exposure* estimate is a function of the *GSD* and the sample geometric standard deviation for an interval of log normal exposure would have to be computed to estimate the variability of the worker's exposure during that interval. This complex estimation technique introduces considerable sampling error into the TWA exposure estimate.

Section 6.7 presents an applied statistical procedure for the computation of tolerance limits for a log normally distributed population with unknown parameters. This procedure is useful for computing tolerance intervals that we can be 95 percent confident will contain 95 percent of the true short-term exposure levels during a given day for which grab samples were taken. Also, Section 6.11 suggests the use of semilogarithmic graph paper for plotting variables that are log normally distributed in time, which typically is the case for grab sample measurements on a given day. Sections 6.12 and 6.13 present procedures for using logarithmic probability paper to estimate the parameters of log normal distributions and for displaying corresponding tolerance limits.

3. *Populations of daily 8-hour TWA exposure estimates* for a worker. For evaluation of a worker's individual health risk due to a chronic exposure, the long-term average of daily exposures could be estimated by the arithmetic mean of a random sample of daily TWA estimates, *if* the daily exposures could be considered uniform. If not, then a sample arithmetic mean for each group of daily TWAs from each multiday period of equal exposure would have to be computed and a long-term TWA estimate computed from the series of arithmetic means. In either of these cases (where uniform exposures on each day can be assumed), confidence limits for the long-term TWA exposure would be based on the normal distribution, since the only errors are in the measurement procedure.

However , for the case of day-to-day log normal random variability (without multiday periods of equal exposures), the sample geometric mean of a random sample of daily exposures would need to be computed and converted to an estimate of the long-term arithmetic mean exposure using a function of the sample geometric standard deviation as a correction factor. This complex estimation procedure for the arithmetic mean of a log normal distribution has a relative large variance and should be avoided if possible. An alternative, for large samples (e.g., at least 30 days), is to compute the sample arithmetic mean and compute its confidence limits under normal distribution assumptions.

The justification for this is that sample means based on many random samples are approximately normally distributed even though the single samples are not normally distributed. A detailed discussion of this point is given as a technical note at the end of this part.

With populations of daily 8-hr TWA exposure estimates, measures of central tendency can be misleading regarding work health risk. For example, one may report that a river has an average depth of two feet, thus inferring that it is safe to wade in. But people can drown in those parts of the river that are more than five feet deep. Analogously, it is important to consider the upper tail (higher values) of any exposure distribution. Tolerance limits, which provide an indication of the potential upper levels of exposure distributions or an indication of the potential widths of exposure distributions, often are the relevant statistical values instead of estimates and confidence limits for central tendency parameters (arithmetic means and medians). Often investigators compute the latter merely because elementary statistical texts contain equations for estimates of and confidence limits on central tendency parameters, while the texts fail to discuss the concept and use of tolerance limits.

Section 6.7 presents an applied statistical procedure for the computation of tolerance limits for a variable which is log normally distributed with unknown parameters. This procedure is useful for computing tolerance intervals that we can be 95 percent confident will contain at least 95 percent of daily 8-hour TWA exposures. The tolerance limits should be separately estimated for each interval of anticipated different interday exposure distribution. Section 6.8 details the computation of point estimate and confidence limits for the proportion of a log normally distributed population that exceeds a specified value (such as an exposure control limit). Section 6.11 suggests the use of semilogarithmic graph paper for plotting variables that are log normally distributed in time, which frequently is the case for true exposure levels on different days during a reference period. Sections 6.12 and 6.13 present procedures for using logarithmic probability paper to estimate the parameters of log normal distributions and for displaying tolerance limits for the variable which is log normally distributed (i.e., for daily TWA exposure levels).

As a technical note, the reader should be cautioned that it is possible to erroneously apply the Central Limit Theorem of mathematical statistics and conclude that daily TWA exposure estimates, which are calculated from log normally distributed intraday exposure levels (e.g., from grab samples), should have an (interday) normal distribution. To the contrary, the theorem merely implies that the means of n *identically distributed* (e.g., log normally) independent random variables will be approximately normally distributed regardless of the distribution of the individual variables. But note that the sample means (daily TWAs) of the individual variables (all possible instantaneous exposure values over some *multi*day period) *are not* the means of independent random samples obtained from the same log normal distribution. Different log normal distributions of intraday exposure levels exist for the various days, and *another* log normal distribution exists for interday variability of the daily TWAs. Each "mean" (TWA) is then merely a *single* sample from the interday log normal distribution of daily TWAs.

An appropriate application of the Central Limit Theorem would be to a multiday exposure average computed from a log normal population of daily TWA exposure estimates. Each daily TWA would constitute a single sample.

If n randomly selected TWAs were drawn from the log normal interday population of TWAs and a sample mean calculated, then the distribution of such multiday exposure averages (each estimated from n samples) would be approximately normally distributed. The approximation improves as the sample size n increases.

4. *Populations of daily 8-hour TWA exposures* for a group of workers having similar expected exposures (e.g., in the same exposure environment, from the same job type or occupational group).

For evaluation of individual worker exposure and health risk, neither the arithmetic mean nor the geometric mean of any such population of multiple exposures of a group of workers is an appropriate parameter, unless the distribution has negligible variation. The difficulty is that a particular worker's individual distribution of exposures may consistently lie in the high (or low) tail of a multiday, multiworker distribution because the true multiday exposure average of that worker may be substantially different from the central tendency of the multiworker exposure distribution. Geometric standard deviations would be the appropriate measure of dispersion for the separate interday and interworker components of the total variation of daily exposures.

Both types of exposure distributions (multiple work shifts for a given worker and multiple workers for a given work shift) are usually approximately log normal. However, to be able to use a *single* log normal distribution for exposures of different workers on different work shifts (or days), the following conditions must apply. The *between days for a given worker* and *between workers on a given day* random variations must be independent. Such independence can be assured by randomly selecting the worker-day combinations for exposure measuring. A suitable procedure to do this would be to randomly select the days for sampling and then randomly select a different worker for measurement on each selected day. The reason this selection procedure is required in order that the resulting data will follow a single log normal distribution is that, if the same group of workers were measured on each of the same several days, exposures would be intercorrelated and would not constitute a simple random sample from the same log normal distribution. The method of analysis of variance (of a logarithmic transform of exposure results) would then have to be used to separate the total variation into components due to worker-to-worker (on the same day) and day-to-day (for the same worker) log normal variations. This technique is so complex that its complete exposition is inappropriate in this chapter. If cross-classified exposure data must be analyzed to determine tolerance limits, a professional statistician's assistance most likely will be needed. Additional discussion is given in Section 6.8 concerning an example computation of log normal tolerance limits determined from exposures for randomly selected worker-day combinations.

3.7 Decision Values for the Unknown Mean of a Normal Distribution with Known Coefficient of Variation

Frequently n independently collected consecutive samples $X_1, X_2, \ldots, X_n$ are obtained, which collectively span the period of a worker's time-weighted average exposure (e.g., 40-hour, 8-hour, 15-minute TWA). Such samples are termed *full-period consecutive samples* and the average of the n measurements is used to estimate the TWA exposure. Assume that these measurements have net random errors which are normally and independently distributed with the same total coefficient of variation CV_T. Then the mean $\bar{X}$ of the n measurements will be normally distributed. This normal distribution of $\bar{X}$-*values* has the following mean and variance:

$$\mu_{\bar{X}} = E(\bar{X})$$

$$= \left(\frac{1}{n}\right) (E(X_1) + E(X_2) + \cdots + E(X_n))$$

$$= \left(\frac{1}{n}\right) (\mu_1 + \mu_2 + \cdots + \mu_n) \tag{14}$$

$$\sigma_{\bar{X}}^2 = \left(\frac{1}{n^2}\right) (CV_T^2)(\mu_1^2 + \mu_2^2 + \cdots + \mu_n^2) \tag{15}$$

If a worker's workshift exposure were *uniform* (i.e., all consecutively sampled periods having effectively equal true average concentrations), there would be $\mu_i = \mu$ for $i = 1, 2, \ldots, n$, and the TWA measurement mean exposure $\bar{X}$ could then be treated as a random sample from a normal distribution with mean μ and variance $\sigma^2/n = (1/n)(CV_T^2)(\mu^2)$. (*Note. Substantially different exposure situations during a work shift generally result in a* nonuniform 8-hour TWA exposure.) In 95 percent of such uniform mean measurements, the average $\bar{X}$ would be within an interval $\mu \pm [(1.96)(CV_T)(\mu)/n^{1/2}]$. Equivalently, *two-sided* intervals $\bar{X} \pm [(1.96)(CV_T)(\mu)/n^{1/2}]$ would contain μ for 95 percent of the $\bar{X}$-values. Note that the latter probability intervals are centered about the randomly varying TWA measurement means $\bar{X}$.

To create a decision-making test (i.e., statistical significance test of a null hypothesis of compliance of a reported TWA exposure mean for a worker with an exposure control limit or exposure *standard* denoted ECL), assume that the worker's true TWA exposure level μ is equal to the value ECL. Under this null hypothesis, *decision intervals* surrounding the ECL can be computed that would contain the TWA exposure mean $(\bar{X})$ in *at least* 95 percent of similar cases. Such a decision interval would be of the form:

$$ECL \pm \frac{(1.96)(CV_T)(ECL)}{n^{1/2}} \tag{16}$$

The lower bound will be termed the *lower decision value (LDV)* and the upper bound will be termed the *upper decision value (UDV)*. Similar *open decision intervals,* that are upper-bounded only (by UDV, i.e., *one-sided decision intervals*) can be computed that would contain the TWA measurement mean $\bar{X}$ for *at least* 95 percent of the similar cases, given that $\mu \leq ECL$. Such a one-sided decision interval, $\bar{X} \leq [ECL + (1.645)(CV_T)(ECL)/n^{1/2}] = UDV$, is open on the left side and its upper bound will be denoted as the *Upper Decision Value (UDV)* for $\overline{X}$, since it will only be exceeded by 5 percent or less of $\overline{X}$ values, when the null hypothesis, H_0: $\mu \leq ECL$, is true. Therefore, in case $\overline{X} > UDV$, the hypothesis H_0 is considered unlikely to be true, since this occurrence would be infrequent under H_0, and an alternative hypothesis, H_1: $\mu > ECL$, is accepted because H_1 gives the observed $\overline{X}$ a more reasonable probability of having occurred by chance. If we desire to calculate a one-sided 95 percent *lower confidence limit* (*LCL*) for μ, we could rearrange the probability statement:

$$P\{\overline{X} \leq [\mu + (1.645)(\mu)(CV_T)/n^{1/2}]\} = 0.95$$

and obtain the *LCL* as

$$\mu \geq \frac{\overline{X}}{\{1 + (1.645)(CV_T)/n^{1/2}\}}.$$

The analogous one-sided 95 percent *upper confidence limit* (*UCL*) for μ is

$$\mu \leq \frac{\overline{X}}{\{1 - (1.645)(CV_T)/n^{1/2}\}}.$$

To obtain two-sided 95 percent confidence limits, use the above formulae with 1.96 substituted in place of 1.645. These formulae give exact 95 percent confidence limits in cases where CV_T is known and there are uniform exposures in n equal-duration consecutive sampling periods. For the general case of unequal sampling durations and nonuniform exposures, approximate confidence limits can be obtained using methods given in Sections 6.2.1 and 6.3.1.

Specific applied methods with examples will be presented in Sections 6.1 through 6.3, which apply the *LDV* and similar *Upper Decision Value (UDV)* concepts to decision-making regarding compliance or noncompliance of a particular TWA exposure estimate with an exposure control level or standard. Specific cases of confidence limits applications are also discussed with examples in Section 6.1 through 6.3.

4 PRINCIPLES OF EXPERIMENTAL DESIGN AND DATA ANALYSIS

One important goal of research is to make inferences about some population, or draw other general conclusions, based on results of a sample survey or experimental data. Often an investigator will seek the assistance of a statistician in analyzing the results from a research study. Unfortunately, sometimes the results presented for analysis are not only fragmentary, but incoherent, so that next to nothing can be done with them except perhaps compute some trivial descriptive statistics. This does not have to happen. Research dollars do not have to be wasted on unproductive studies. Adherence to statistical principles of study design and related data analysis can produce substantially better results.

A study should be initiated, conducted, and the results evaluated only if the investigator has a clear purpose in mind and a clear idea about the precise way the results will be analyzed to yield the desired information. Far too often studies are conducted in the blithe and uncritical belief that a subsequent "statistical analysis" will yield something useful, especially when a statistician is engaged to "juggle the data."

The methodological tools of statistics cannot extract information or inferences which are not inherent to the data. The use of any statistical technique to analyze study results requires asking certain questions concerning the hypotheses to be tested, the parameters to be estimated, the study design used, and the circumstances in which the data were collected (54). Statistical distributional assumptions are required for application of most methods of statistical analysis and these assumptions depend partly on the circumstances and pattern of the experimentation or data collection. Besides using the right research tools, we must also use appropriate experimental designs to have a good chance of detecting changes or effects of practical significance (i.e., those effects large enough to be of interest). The effects of interest are often small enough so that they might be obscured by experimental error, or hidden by other confounded effects, unless special attention were given to designing an experiment with sufficient statistical power.

There is a solution to these problems. The investigator must know enough of statistical principles and techniques to be able to recognize when advice is needed from a statistician *before* a study is initiated. Review of, and adherence to, the principles of study design and data analysis discussed in this section will yield more productive investigations and research studies. Many of these principles are common sense, but unfortunately, common sense is frequently uncommon. Altman (55) believes that the general standard of statistics in medical journals is poor. The situation is not any better for industrial hygiene journals. Of course uniform guidelines for statistical design cannot be precisely applied in every study. Section 2 discussed how *observational* and *experimental* studies are the two major classes of research studies. Special considerations may be involved in subclasses of these study types such as exploratory, methodological, and pilot or preliminary studies, where the primary purpose is to test feasibility or to evaluate alternative approaches or techniques.

4.1 Study Design Principles and Implementation Guidelines

The following guidelines are based on those suggested by Crow et al. (56), Green (57), and Soule (58).

4.1.1 Establish the Study Objectives

Clearly establish the purpose and scope of the study. How, by whom, and for what purpose will the results be used? There should be a complete, clear, and concise statement of the objectives for the study. State how the anticipated results will specifically be used to meet the objectives. Are the results intended to be definitive, or will this be a pilot test or feasibility study? Classify the study as descriptive or analytic (i.e., designed to test a specific research hypothesis). Any statistical review should, in part, concern itself with how well the study design can meet the stated objectives. The study results will be only as relevant and productive as the initial conception of the research problem.

State the study conditions, or parameters used to represent the conditions, to which the results will be applied. Provide a clear and complete definition of the study *target population* to which inferences will be made based on results obtained from a *sample*. The *target population* is a subset of the general population that is both subject to the study exposure(s) and at risk of the development of the occupational disease or adverse health effect(s). Any given *sample* is a member of a *sample population* consisting of all samples which could have been selected from the *target population*. Also, include ranges of the determinant variables and specifications for reporting units (e.g., individuals, job types, establishments, industries).

4.1.2 Formulate the Design for a Preliminary Study

Examine the precision afforded by different study sizes with consideration for the benefits of a statistically powerful study versus the disadvantages of a less powerful study, such as weak or equivocal results with appreciable risk of wrong decisions and resultant limited conclusions. Plan to take replicate samples within each "treatment" (i.e., within each combination of time, location, and any other determinate variable). Differences between "treatments" can only be demonstrated by comparison to variability within treatments. Attempt to obtain an equal number of randomly selected replicate samples for each combination of determinant variables. Taking measurements in "representative" or "typical" situations *does not* constitute random sampling, although random sampling would usually tend to be both of these. The element of random sampling is essential to methods of statistical inference which are based in the mathematical theory of probability. To test whether a treatment (exposure) has an effect, one would usually attempt to collect samples both where the condition is present and where it is absent, but all other determinant variables are the same. An effect can *best* be demonstrated by comparison with a proper control.

4.1.3 Review the Design with All Collaborators

Discuss the design with collaborators, reach an understanding, and keep notes about what decisions hinge on each outcome. Collaborators should anticipate and discuss all determinant variables that might affect the results. Review the study design in sufficient detail to discover any procedures that might lead to bias in the results. Obtain a pertinent peer review of the study design and use the comments as if the reviewers were collaborators on the study. Review and discuss the *robustness* of the chosen statistical methods (i.e., the tendency for inferences based on the methods to remain valid despite a violation of one or more assumptions underlying the theoretical development of the method). Will the study results be examined to detect serious violations of the methodology assumptions? Discuss the objectives, goals, and study design with representatives of management and labor, as appropriate to their respective interests. Review how the results should be reported and to whom.

4.1.4 Conduct the Preliminary Study

Those who skip a preliminary study due to "not enough time" usually end up wasting time by attempting to analyze a study with trivial or equivocal results. This is the opportunity to test the feasibility of the study design before substantial resources are committed to the research. Obtain sufficient data to provide adequate estimates of variance components to be used as a basis to develop an efficient design for the more definitive main study to follow. Obtain adequate information to evaluate the adequacy of the measurement equipment, personnel training and readiness, options for statistical analysis of data, and to indicate the ranges of the study variables. Verify that the chosen exposure measurement method is adequate and appropriate for the entire range of study conditions and determinant variables anticipated. Experience from past similar studies can sometimes be used as a substitute for a preliminary study or pretest. However, a careful critique of the relevancy of past studies should be made before using the data in the study design.

4.1.5 Complete the Study Design

Use the variability estimates from the preliminary study to estimate the power of the study if appropriate. If possible, specify the basis for the sample size calculation in a statement such as

To detect a true difference of ___ (units or percent) between group A and group B, with ___ percent statistical power and a ___ percent probability of making a type-I error, ___ trials are needed.

Present the design in clear terms to assure that its provisions can be followed without confusion. Include the intended data analysis methods as part of the design, including validity checks on the governing statistical assumptions. Review the principles of data analysis given in Section 4.2. Consider the necessity for a data transformation (see Section 4.2.1).

4.1.6 Conduct the Study

During the study, maintain communication among all collaborators, so that problems and intermediate results may be evaluated and dealt with in keeping with the study objectives, and the related study design previously agreed on.

If measurements are to be taken in the plant, advise representatives of management and labor in advance and during your sampling.

4.1.7 Conduct the Data Analysis

Using the principles presented in Section 4.2, follow the data analysis methods previously selected. These presumably match the statistical design and provide powerful tests of the desired hypotheses. Stay with the results from these methods. Unexpected or undesirable results *are not* valid reasons to hunt for a "better" statistical methodology. Remember that overinterpretation is an effort to compensate for underplanning.

4.1.8 Prepare a Report

Report the results in relation to the original study objectives and goals. Discuss both statistical and practical significance of the results. Present conclusions indicated and supported by the results. State necessary limitations on the inferences in your discussion and conclusions. Present summary data, statistics, and results in clear graphs and tables. In general, graphs are superior to tables for portraying trends, correlations, scatter, outliers, etc. If the results suggest a need for further studies, outline the course that such studies should take.

Just as exposure measurements should be accurate and precise estimates of worker exposure, so should reports be accurate and precise communications of your study objectives, methods, results, and conclusions. Use of guides such as those by Bates (59), Crews (60), and the CBE Style Manual Committee (61) can contribute greatly to the quality of your written reports.

4.1.9 Implement Appropriate Follow-up

Discuss study results with appropriate representatives of management and labor so that corrective action to reduce health hazards can be implemented, if necessary, along with additional exposure monitoring, biological monitoring, or medical surveillance programs as required. Are evaluations necessary for air pollution, water pollution, hazardous waste disposal, or safety?

4.2 Principles of Data Analysis

4.2.1 Choose a Distributional Model

Choose a distributional model (see Section 3.6) for the target population which the sample data represent. Note that a data *transformation* may be necessary (i.e., converting the data into such form that they follow a common distribution with known properties and readily available analytical methodology). Data transformations may be needed for other reasons than just giving the transformed results a convenient distributional form. Murphy (62) has discussed six objectives of transformation:

1. Normalization,
2. Stabilization of the variance,
3. To make the effects linear and additive,
4. To make the mean a good measurement of "the typical value,"
5. To linearize a relationship between two or more variables,
6. Remodeling the distribution into a more familiar one.

4.2.2 Review Statistical Assumptions

Review assumptions of statistical methods to be used. If a distributional model can be assumed, proceed to Section 4.2.6. The parametric methods outlined therein provide one with the benefits of lower sample sizes (hence lower costs), for the same statistical power (or with moderate gains in power for the same sample size), when compared to the more robust nonparametric methods (i.e., those "distribution free" methods which do not depend on a specific probability model with one or more parameters for the distributional form of the parent population). However, if the sample data are not *adequately* fitted by the assumed distribution (see Section 3.6) or if other assumptions (e.g., independence) of the inferential methods are suspect, then one has unknown risks of incorrect decisions, inaccurate confidence intervals, or inaccurate tolerance limits. The sizes of such inaccuracies depend on the robustness of the methods used. If sample size is insufficient (less than about 10) to qualitatively examine the data (see next section) and test the distributional model, proceed to Section 4.2.6. Then one may be in the position of having to use parametric methods without even weak verification of assumptions. A larger sample size (e.g., 30 or greater) is needed to empirically estimate distributional fractiles from ranked sample data without assuming any mathematical model for the distributional form (see Section 4.2.5).

4.2.3 Qualitatively Examine the Data

Plot the grouped sample data as a histogram or individually on appropriate probability paper (e.g., normal, log normal, Weibull) to qualitatively examine data regarding the distributional assumptions and investigate unusual patterns in the data (see Section 6.12). Data that are log normally distributed in time can be plotted on semilogarithmic graph paper so that the data plot is symmetrical about the distribution's geometric mean. This is a way of qualitatively examining for trends in the data over time (see Section 6.11).

4.2.4 Estimate Sample Distributional Parameters

From the sample data, calculate estimates for the parameters of the assumed distributional model (e.g., arithmetic mean and standard deviation for a normal distribution (see Section 3.4.1), geometric mean and geometric standard deviation for a 2-parameter log normal distribution (see Section 3.5.1), and if necessary the third parameter (for origin translocation) of a 3-parameter log normal distribution (see Section 3.5.2).

4.2.5   Verify Distributional Model

To qualitatively verify the distributional model, using the sample estimates of parameters from the previous step, plot the estimated target population distribution on the appropriate probability paper and compare to the actual sample data distribution. Mage (63) has suggested an objective method for testing normal distributional assumptions using probability paper.

   With larger sample sizes (at least 30) a histogram of the sample data can be compared to the shape of the fitted distribution function. To quantitatively verify the distributional model, the classic chi-square test can be applied to the histogram's interval frequencies. This goodness-of-fit test usually is the first to come to mind, but it generally requires substantially larger sample sizes than occur in industrial hygiene. Lilliefors (64) has adapted the Kolmogorov-Smirnov test when the mean and variance of the distributional model are unknown (as is typically the case with occupational exposure data). This modified goodness-of-fit test can be used with small sample sizes (e.g., 10 to 20), for which the accuracy of the chi-square test would be questionable. Also, it is claimed to be a more powerful test then the chi-square test for any sample size. Iman (65) has provided graphs for use with the Lilliefors test.

   If the fit is judged inadequate, return to Section 4.2.1 and choose another data transform or distributional model.

4.2.6   Apply Chosen Statistical Methodology to the Analysis of Sample Data

If the fit of the distributional model is judged adequate (or assumed to be adequate in which case one has been able to delete the steps following Section 4.2.2), or the statistical methodology is sufficiently robust for its intended application, proceed with the data analysis. For these types of data, distributional forms have been investigated using prior data collected for that purpose and these forms can usually be assumed to apply to similar samples taken thereafter. Calculate appropriate confidence intervals and tolerance limits, perform hypothesis tests etc., based on the assumed distributional model. If the data or data transforms do not adequately fit available models, certain nonparametric (distribution-free) methods are available. However, it is a problem that nonparametric methods generally require sample sizes larger than occur with sample sets used for exposure estimation.

4.2.7   Report Results and Interpretation

If a report or journal article is written, it should include a statement of the statistical experimental design and related protocol for statistical analysis of data, in sufficient detail so that a reader with statistical expertise would be able to duplicate the results if supplied with the investigator's raw data. If a published "canned" computer program was used, a precise reference to it would usually preclude the need for presenting additional computational details. If the statistical analysis had been performed by a consultant or collaborator, their assistance would usually be needed in the writing, or at least in the editing, of the final report.

5 STUDY DESIGNS FOR ESTIMATING INDIVIDUAL OCCUPATIONAL EXPOSURES AND DISTRIBUTIONS OF EXPOSURES

Study design considerations and sample size estimation techniques for obtaining estimates of occupational exposures and their distributions will be discussed from two perspectives in this section. The first perspective presented in Section 5.2 discusses exposure *measurement strategies* for making individual *exposure estimates* (see Section 3.1). The discussion concerns sampling individual workers on a single day.

The second perspective, presented in Section 5.3, covers *monitoring strategies* for obtaining exposure estimates for target populations. *Monitoring* will be defined as a series of steps necessary for estimating multiple exposures of a target population. The required steps generally include selecting a sample of workers from the target population for exposure measuring, selecting a measurement strategy, obtaining exposure measurements for the sample of workers, calculating the exposure estimates, possibly estimating an exposure distribution, and then reporting the results. A *target population* is usually defined in numbers of workers, averaging time for the exposure measurements, time period to be represented by the exposure distribution, and the ranges of determinant variables affecting the exposure levels the target population is exposed to. The number of workers can be as few as one worker and the time period could be a part of a single day. The determinant variables will be discussed in section 5.1. Some examples of target populations for several combinations of temporal periods and numbers of workers include:

1. Daily 8-hour exposure estimates for a single worker over several days to years.
2. 8-hour exposure estimates for several workers on a single day.
3. Daily 8-hour exposure estimates for many workers over several days to years.

The elementary case of a single worker on a single day would be considered under Section 5.2. For this elementary case there is but one exposure, but the parent population is considered to be all measurements (with their errors) that could have been made of that one exposure. The other two cases would be considered under Section 5.3, Exposure Monitoring Strategies, and here the parent populations consist of multiple daily exposure and/or multiple workers exposed, as well as the populations of exposure measurements of these workers at these times.

5.1 Determinant Variables Affecting Occupational Exposure Levels

Both measurement strategies and monitoring strategies must consider the *determinant variables* affecting the true exposure levels that will be estimated by measuring worker exposures. Determinant variables are the qualitative or quantitative factors that determine, or at least are associated with, the actual worker exposure levels. They generally can be classified as process, environmental, temporally associated, behavioral, and incidental. Note that a failure to identify and consider significant determinant variables generally has more deleterious consequences for a study than identifying and investigating too many determinant variables, including some that have little effect on exposure levels. **One cannot make reliable inferences from exposure results beyond the number and range of determinant variables represented by the sampling distribution**.

The following groups of typical determinant variables are presented as examples only and are not inclusive. The significant determinant variables for any given group of workers and time period must be identified, from experience with similar situations if possible, or using a research study if necessary.

5.1.1 Process Factors

These are factors related to primary contaminant levels or to the control of emissions and/or exposure levels:

1. Process type and operation. See p. 27 of (1), Burgess (66), Cralley and Cralley (67).
2. Chemical composition of material used in operation.
3. Physical state and properties of material used (e.g., vapor pressure, size distributions of particulates and aerosols).
4. Rate of operation (e.g., mass or volumetric rate, revolutions per minute, linear rate, items in a given time period).
5. Energy conditions of operation (e.g., temperature, pressure).
6. Degree of process automation.
7. Emissions from adjacent operations.
8. Airflow patterns around workers (e.g., from exhaust ventilation, from adjacent operations).
9. Heating and ventilation airflows.
10. Exposure control methods (e.g., local exhaust ventilation, respirators).

5.1.2 Environmental Determinant Variables

These are environmental factors, variations of which can modify exposure levels:

1. Meteorological conditions.
2. Age, size, and physical layout of plant.
3. Job category (e.g., responsibilities, work operations, work areas, time spent at each), see Corn (68).

5.1.3 Temporally Associated Determinant Variables

Time is an independent variable correlated with worker exposure levels. Time cannot be a direct cause of changes in workplace exposure levels, but may be useful to predict exposure levels which follow time cycles or have systematic time trends or have autocorrelation between present and past levels. Time series models are useful when the causative determinant variables are either unknown or unmeasured.

1. Contaminant build-up in the workplace air from morning to afternoon.
2. Exponential clearance due to air flushing and dilution during nonworking hours.
3. Cyclical process operations.
4. Work shift.
5. Season of year.
6. Year or decade.

5.1.4 Behavioral Determinant Variables

Behavioral factors affect work habits which in turn affect exposure levels:

1. Worker job practices, movements, habits.
2. Worker training.
3. Worker attitudes.
4. Management and supervisory attitudes.
5. Presence of exposure measurement equipment, industrial hygiene personnel, or supervisory personnel.

5.1.5 Incidental Determinant Variables

Irregular changes in exposure levels can occur due to episodes, incidents, accidents, and otherwise unintended happenings such as (69):

1. Spills due to falls, punctures, tears, corrosion, etc.
2. Equipment maintenance or lack thereof.
3. Failure of process equipment prone to corrosion and leakage (e.g., pump packings, tank vents).

4. Interruption of utilities to process equipment or exposure control systems.

5. Interaction due to accidental mixing or simultaneous release of two vessels' contents.

6. Interruption or increase in flow of one or more process streams.

7. Vessel failure.

8. Accidental overpressurization, overheating, or overcooling of process equipment.

9. Sudden plant flooding, violent storms, or earthquakes.

10. Operator errors or instrument failure.

5.2 Exposure Measurement Strategies

Exposure measurement strategies deal with the considerations necessary to measure individual worker exposures on a given day to obtain short or long period TWA exposure estimates. *Monitoring* strategies for measuring multiple exposures (e.g., multiple workers on a single day or a single worker on multiple days) are presented in Section 5.3.

5.2.1 Practical Considerations

The adequate and preferable strategies for a given measurement situation are governed by both practical and statistical considerations. An adequate measurement strategy is one which is good enough for the purpose we have in mind. Thus one should clearly identify the objective of the intended exposure estimation (see Section 2.4) and the required precision of the estimates. A measurement strategy can then be selected which meets this precision requirement.

Most of the following discussion will concern measurements obtained for estimating individual worker exposures. Generally these estimates are then compared to exposure control limits to determine the relative hazard for individual workers or the acceptability of the workplace esposure levels (e.g., in relation to OSHA PELs or ACGIH TLVs®). It should be noted that there are other specialized purposes that airborne contaminant measurements may be obtained for. These include:

1. *Source sampling* at potentially hazardous operations. Hubiak et al. (70) have discussed the utility of short-term source sampling for identifying work operations that need additional engineering effort to reduce worker exposures. Sometimes these can be considered *worst case* exposure levels, which are used in screening strategies for hazard evaluations.

2. *Evaluating work practices* to determine exposure variability and detect hazardous levels due to inappropriate work practices. This technique can also

assist in differentiating between exposure levels due to work practices and those due to inadequate engineering controls. With direct-reading measurements it is possible to make on-the-spot recommendations for improving work practices and recognize the appropriate direction for engineering control research.

3. *Worker training* to improve the effectiveness of work practices and engineering controls. Selected workers can perform the work operation while others observe the work practices and note the resulting exposure levels. The goal is to have all workers approach the results of the best or "cleanest" worker. Showing workers their exposure profile recorded on a strip chart will help them to better understand what a change in work practices or use of engineering controls can achieve.

4. *Evaluating individual job tasks* for their relative contribution to a worker's 8-hour TWA exposure. This may lead to developing appropriate administrative controls, improved work practices, engineering controls, or identify the need for personal respiratory controls.

5. *Continuous monitoring* to detect extraordinary exposure levels, so that a warning can be provided before a serious hazard develops. Generally, fixed sampling systems are used with a central analyzer. These systems may sample from multiple points in the workplace or in the return air of recirculation systems. Holcomb and Scholz (71) have reported an evaluation of typical continuous monitoring equipment.

6. *Screening measurements* for qualitative detection of airborne hazards during emergencies and uncontrolled releases. A screening strategy may also be used for quantitative determination of airborne hazards for hazard evaluation and spot checking of exposure levels. Typically, detector tubes and direct-reading meters would be used for molecular size contaminants and aerosol monitors would be used for particulates. Schneider (72) and Leichnitz (73) have discussed the use of colorimetric detector tubes for screening. King et al. (74) have described a simultaneous direct-reading indicator tube system for rapid qualitative measurements.

Some of the nonstatistical considerations affecting possible measurement strategies are as follows:

1. Amount of information available regarding the nature and concentration of airborne contaminants to be measured and possible interfering chemicals.

2. Availability and cost of sampling equipment (e.g., pumps, filters, detector tubes, direct reading meters, passive dosimeters).

3. Availability and cost of sample analytical facilities (e.g., for filters, charcoal tubes, dosimeters).

4. Availability and cost of personnel to take the measurements.

5. Location of work operations and workers to be sampled.

6. The need for obtaining results immediately, within a day or two, or after several weeks.

5.2.2   Statistical Considerations for Long-Term Exposure Estimates

Generally the statistical design of exposure measurement strategies is concerned with reducing the random uncertainty of the calculated exposure estimates and increasing the power of hypothesis tests for compliance or noncompliance with an exposure control limit. The imprecision of a long-term (e.g., 8 hours) exposure estimate is governed by two classes of factors:

1.   Random variation in the measurement procedure (i.e., the precision of the method, see Sections 3.3.2, 3.4.2, and 3.6.1), or the random variation in the true exposure levels during the estimation period (see Sections 3.3.1 and 3.6.2).

2.   Sample size (i.e., number of measurements obtained during the time-averaging period for the TWA estimate) or the duration of each measurement.

3.   Sampling period selection (i.e., whether random or systematic sampling is used to select subintervals during the TWA period of interest, or a cumulative sample taken over the entire period).

It should be realized that there is no "best" measurement strategy for all situations. However, some strategies are more desirable than others. The following discussion points out the statistical considerations for the four different types of TWA exposure estimates (see Section 3.1). Remember that the word *period* refers to the duration of the desired time-averaging period (e.g., 15 minutes for a 15-minute TWA estimate, 8 hours for an 8-hour TWA estimate).

The *full period consecutive samples estimate* is the "best" strategy in that it yields an estimate with the least uncertainty (closest confidence limits). This is because the uncertainty of the exposure estimate is a function of the measurement method error (see Section 3.2) and is independent of the substantial variability in the actual exposure levels measured (see Section 3.6.1). There are moderate statistical benefits to be gained (Technical Appendix E of (1)) from increased sample sizes (e.g., eight 1-hour samples versus four 2-hour samples), but with the substantially increased analytical costs per exposure estimate the practical benefits are negligible. Figure 8.4 illustrates the effects of increased sample size. Generally two consecutive samples for the full time-averaging period provide sufficient precision for most exposure estimation purposes.

The *full period single sample estimate* (e.g., one 8-hour measurement) is the "second-best" strategy, if an appropriate measurement method is available. An exposure estimate calculated from one 8-hour measurement effectively has nearly as good precision as an estimate computed from two 4-hour measurements, since both strategies employ full-period sampling. The disadvantage of a single measurement is that a bias or mistake in the measurement is difficult to detect. Also, substantial differences in exposure levels during the measurement period are not revealed, but this feature is also an advantage in that temporal variability has been physically "integrated out" (i.e., by the cumulative sampling procedure itself).

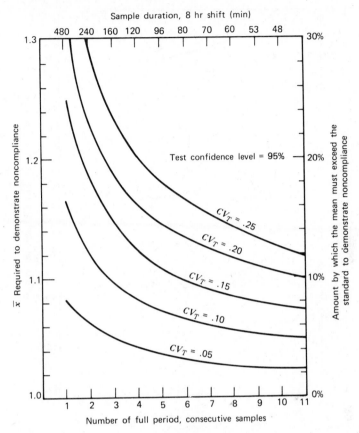

Figure 8.4 Effect of full period, consecutive sample size on noncompliance demonstration when test power is 50 percent; CV_T is coefficient of variation of sampling/analytical methods.

The *partial period consecutive samples estimate* is substantially less desirable than the preceding two estimates. The major problem created by this strategy is the unknown exposure levels during the unsampled portion of the TWA period. Strictly speaking, the measurement samples are representative only of the period of the actual sampling. If one desires to estimate an 8-hour TWA from a sample or samples spanning only 5 hours, then a problem is created of assuring that the 5-hour period results represent the entire 8-hour period. Reliable knowledge or professional judgment may sometimes be used to extrapolate the 5-hour TWA to an 8-hour estimate. This should be done only after considering the effect on the 8-hour TWA estimate that any substantial increase or decrease in actual exposure levels during the unsampled period would have. Figure 8.5 illustrates the effect of a conservative assumption of zero exposure, for the unsampled portion of the TWA period, on the value required to demonstrate noncompliance at a statistical test power of 50 percent.

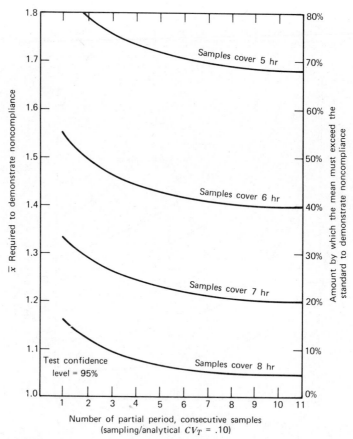

Figure 8.5 Effect of partial period, consecutive sample size and total time covered by all samples on noncompliance demonstration when test power is 50 percent.

The *grab samples estimate* is the least preferable strategy for estimating an 8-hour TWA exposure. This exposure estimate has substantially larger uncertainty than the first two types of estimates. This is because the uncertainty of a grab samples estimate is dominated by the considerable variability of the exposure levels measured, which is substantially larger than the measurement method variability. Regarding sample size, Figure 8.6 shows that the optimum number of grab samples for an exposure estimate is between 8 and 11, taken at random intervals during the TWA period. However, this applies only if the worker's exposure levels are adequately uniform during the TWA period. If the worker is at several work locations or operations during the TWA period, then at least 8 to 11 grab sample measurements should be obtained during *each* period of anticipated uniform exposure that substantially contributes to the TWA exposure. If one has to take fewer than 8 to 11 measurements during each uniform exposure period, then allocate the total number of measurements in proportion to the duration of each period. That is, take more measurements during the longer periods of anticipated uniform exposure.

If grab samples are taken, the duration of each measurement need be only long enough to collect sufficient mass of contaminant to reach the minimum level of detection for the analytical method. That is, any increase in sample duration beyond the minimum time to collect a sufficient mass of contaminant is unnecessary and unproductive. A 40-minute grab sample effectively is no better than a 10-minute one.

For grab samples it is desirable to choose the sampling periods in a statistically random fashion. The accuracy of the probability levels for the statistical methodologies for testing hypotheses of compliance or noncompliance presented in Section 6 depend on implied assumptions regarding the log normality and independence of the sample results that are averaged. These assumptions are not unduly restrictive if precautions are taken to avoid bias when selecting the sampling times during the period of the exposure estimate.

For grab sampling, a TWA exposure estimate represents a period longer than the measured interval, but an unbiased estimate of the true average can be ensured by taking samples at random intervals. It is valid to sample at equal intervals if the series is known to be stationary with contaminant levels varying randomly about a constant mean, and if exposure fluctuations are of short duration compared to the length of the sampled interval. However, if exposure

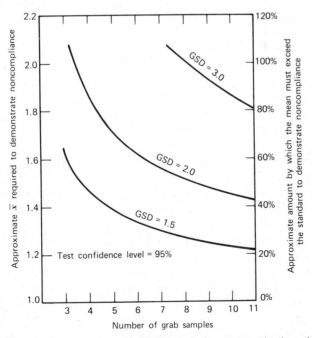

Figure 8.6 Effect of grab sample size on noncompliance demonstration. The three data geometric standard deviations (GSD) reflect the amount of intraday variation in the environment.

means and associated confidence intervals were to be calculated from samples taken at equally spaced intervals, biased results could occur if exposure cycles in the operation were in phase with the sampled intervals. The important benefits of random sampling are that subsequent results are unbiased even if cycles and trends occurred during the period of the exposure estimate.

The word *random* refers to the method used for selecting the sample. A *random sample* is one chosen in a manner such that each possible sample has a fixed and determinate probability of selection. A practical way of defining a random exposure measurement is one obtained such that any portion of the TWA exposure estimate period has the same chance of being sampled as any other. Ordinary haphazard or seemingly purposeless choice is generally insufficient to guarantee true randomness. Devices such as random number tables or random numbers generated by computer programs can be used to remove subjective biases inherent in personal choice. Technical Appendix F of (1) details a formal statistical method for choosing random sampling periods.

5.2.3 Statistical Considerations for Short-Term Exposure Estimates

Short-term exposure estimates (e.g., 15 minutes or less) generally are obtained only for determination of peak exposures during short periods. These estimates may be used for comparison with *ceiling* exposure control limits designed to prevent acute health effects. Short-term samples taken to measure short-term exposures are statistically analyzed in a manner similar to short-term samples taken to measure long-term exposures. However, two important differences should be noted.

The first difference is that the measurements taken for estimation of peak exposures are best taken in a *nonrandom* fashion. That is, all available knowledge relating to the exposure level determinant variables such as work area, worker, work practices, and type of operation should be utilized to obtain samples during periods of maximum expected exposure.

The second difference is that measurements obtained for short-term estimates are generally taken for a much shorter period than short-term samples taken for estimating 8-hour TWA exposures. Each short-term measurement usually consists of a single instrument reading (if a direct-reading device is available) or a 5- to 15-minute sample if a minimum mass of material needs to be collected (e.g., on a charcoal tube or filter). A series of samples spanning 15 minutes could also be taken and the measurements averaged.

Leidel et al. (1) recommend that a minimum of three short-term exposure estimates be obtained on any given work-shift for a worker and the highest of all estimates be used as an estimate of the worker's peak exposure for that work-shift. This recommended minimum number of estimates is not based on statistical considerations, but on practicality. Taking at least three samples increases the probability of detecting the highest exposure and facilitates the detection of gross mistakes or biased measurements. However, usually only the highest value (not the average of the three or more) would be compared to a ceiling exposure limit. If measurements are obtained for evaluation with a short-term TWA limit, such as a *Threshold Limits Value—Short Term Exposure Limit (TLV®-STEL)* of the American Conference of Governmental Industrial Hygienists (75), the total sampling time should equal the time-averaging period for the limit, which is typically 15 minutes. Thus, for some colorimetric detector tubes, it might be necessary to take several consecutive samples and average the results.

Although short-term measurements taken for estimation of peak exposures are usually best taken in a nonrandom (biased) fashion, random sampling may be useful in some work situations where the exposure levels appear uniform during a work shift. Professional judgment may be unable to identify particular periods for which exposure is higher than usual. For this case, a statistical procedure is given below which can be used as a peak exposure detection strategy. The sample size recommendations given are based on combinatorial probability formulas detailed in Technical Appendix A of Reference 1.

Purpose.   Provide a sample size to assure (i.e., have a probability of 90 or 95 percent) that at least one randomly sampled period will be from the higher exposures present during the work shift or other total duration examined (i.e., from the highest 10 or 20 percent of the work shift exposure distribution).

Assumptions.   No limiting assumptions are required. The derivation of this method is based on the hypergeometric sampling distribution (i.e., sampling from a finite population without replacement). There is no assumed mathematical model for the distribution of exposure levels present in the finite-size population of possible measured periods; therefore, these sample sizes may be larger than would be needed for situations where the parametric distributional form of the exposures is confidently known. However, this nonparametric procedure is useful when the form of the exposure distribution is irregular or unknown. Table 8.4, 8.5, and 8.6 give the required sample sizes for 32, 48, and 96 possible sampling periods in the time-averaging exposure period of interest.

Table 8.4   Required Sample Size for Detecting at Least One of the Higher Exposures Among 32 Periods in the Total Duration (15-Minute Periods in an 8-Hour Work Shift)

At Least One Period From	Confidence Level	Random Sample at Least
Top 20%	0.90	10 periods
Top 20%	0.95	12 periods
Top 10%	0.90	17 periods
Top 10%	0.95	20 periods

Table 8.5   Required Sample Size for Detecting at Least One of the Higher Exposures Among 48 Periods in the Total Duration (10-Minute Periods in an 8-Hour Work Shift)

At Least One Period From	Confidence Level	Random Sample at Least
Top 20%	0.90	10 periods
Top 20%	0.95	13 periods
Top 10%	0.90	21 periods
Top 10%	0.95	25 periods

Example.   For a target population of 32 consecutive 15-minute periods in an 8-hour work shift for a worker, estimate an appropriate sample size such that there is a 90 percent confidence that at least one sampled period will be from those periods with the highest 20 percent of exposures occurring during the work shift.

Solution.   There are 32 discrete, nonoverlapping, 15-min periods in an 8-hour work shift. Table 8.4 indicates that a random sample of 10 of the 32 15-minute periods will have 90 percent probability of containing one or more of the 6 periods during which the 20 percent highest exposures will occur. The number 6 is the largest integer representing 20 percent or less of 32.

Where the short-term time-averaging period is 10 minutes, there would be 48 such periods in an 8-hour work shift and the sample sizes in Table 8.5 would be appropriate. For example, to have 90 percent probability that at least one sampled period is among the 4 which have 10 percent or less of the highest exposures, 21 of the 48 periods should be selected at random and sampled.

Less than 10-minute time-averaged measurements may sometimes be obtained, as with a 3-minute colorimetric tube or spot readings with a direct-reading meter. Then the sample sizes in Table 8.6 are appropriate.

Table 8.6   Required Sample Size for Detecting at Least One of the Higher Exposures Among 96 or More Periods in the Total Duration (Periods Less Than 5 Minutes Each in an 8-Hour Work Shift)

At Least One Period From	Confidence Level	Random Sample at Least
Top 20%	0.90	10 periods
Top 20%	0.95	13 periods
Top 10%	0.90	21 periods
Top 10%	0.95	27 periods

5.3 Exposure Monitoring Strategies

The exposure *monitoring* strategies presented in this section are guidelines for measuring multiple exposures (e.g., multiple workers on a single day, a single worker on multiple days, multiple workers on multiple days). These strategies should be used with the exposure *measurement* strategies (for measuring an individual's exposure on a single occasion) treated in the previous section. Monitoring strategies for monitoring programs should always consider eight elements:

1. Need for exposure estimates.
2. Airborne chemical(s) to be measured.
3. Strategy for initial monitoring.
4. Criteria for decisionmaking.
5. Strategy for periodic monitoring for continuing hazard evaluation.
6. Occasions requiring extraordinary monitoring.
7. Criteria for termination of monitoring.
8. Procedure for follow-up.

Details of these elements will be presented in the following sections.

Section 2.4 listed two major types of monitoring programs as possible objectives of exposure estimation. The first type is an *exposure screening program*, which is a *limited* exposure monitoring program designed to identify target populations of workers with other than acceptable exposure distributions, for follow-up periodic monitoring. The program uses minimal resources consistent with reasonable protection for workers. This type of program uses an Action Level as a screening cutoff to identify appropriate target populations for inclusion in a limited exposure surveillance program or a more extensive *exposure distribution monitoring program*. The latter program is a more extensive one intended to quantify exposure distributions of target populations. Generally this is first done for an initial base period; then the initial estimates of the exposure distributions are periodically updated with more current estimates from routine exposure monitoring. Eight possible elements for these two types of monitoring programs will be detailed in the following sections.

There are several commonalities between exposure screening and exposure distribution monitoring programs. Guidelines noting the similarities and differences in the two types of programs will be presented in the following sections.

5.3.1   Need for Exposure Measurements

Both exposure screening and exposure distribution monitoring programs need to begin by determining the need for exposure estimates. Desirable predecessors to these programs, which may negate the need for exposure measurements, include:

1.   Conducting a workplace materials survey to determine if potentially harmful materials are being used in the workplace (see p. 21 of Reference 1).

2.   Conducting a walk-through survey to identify process operations that may be potentially hazardous and to determine if workers may be exposed to hazardous airborne concentrations of materials released into the workplace (see p. 24 of Reference 1).

3.   Estimating airborne concentrations based on the amount of material released into the workplace air. This may be useful for contaminants that are low to moderate hazards (see pp. 28–30 of Reference 1). However, this technique usually requires a substantial safety factor to account for uncertainty, which may limit its usefulness.

4.   Source sampling at potentially hazardous operations. This may be useful for screening strategies, such as may be used for hazard evaluations. The usual assumption is that the resulting exposure estimates are worst case exposure levels and all worker exposures will be less than the values found at the source(s) of the contaminant (see Section 5.2.1 above and pp. 24 and 27 of Reference 1). This assumption may be invalid if there are more than one or two contaminant sources.

5.   Preparing a written determination of the need (or lack of need) for exposure measurements. The written determination would consider:

> a.   Any information, observation, or calculation which might indicate worker exposures.
> b.   Any measurement taken.
> c.   Any worker remarks of symptoms which may be due to exposure to workplace materials or operations.
> d.   Any possible changes in production, process, or controls which could result in hazardous increases in airborne levels of contaminants or render control procedures inadequate

5.3.2   Airborne Chemical(s) to Be Measured

Both exposure screening and distribution monitoring programs need next to determine the airborne materials to be measured. This is best done by considering the information acquired from the various steps in the previous program element. It may be necessary to first do *screening measurements* [see Section 5.2.1(6)], if prior information is unavailable and qualitative detection is promptly required (e.g., emergencies and uncontrolled releases).

5.3.3   Strategy for Initial Monitoring

Initial monitoring is the first monitoring program element where an exposure screening program differs considerably from an exposure distribution monitoring program. For an *exposure screening program*, the objective of initial monitoring is to selectively obtain exposure estimates only for "maximum risk"

STATISTICAL DESIGN AND DATA ANALYSIS REQUIREMENTS

workers. These can be defined as those workers "believed to have the greatest exposure."

For exposure screening monitoring, the most efficient approach to sampling is a nonrandom selection of the highest risk workers. The selection process must use competent professional judgment which relies on experience and knowledge of the exposure level determinant factors pertinent to the target population. Some factors to consider in selecting the maximum risk worker are given on pp. 33–34 of (1). Related determinant variables affecting the exposure levels of the target populations were discussed in Section 5.1. The important point with this approach is to sample only those workers whose exposures represent the higher exposures of the target population. However, this approach is subject to frailities of professional judgment which could lead to erroneous conclusions regarding the exposure levels for the highest risk workers.

For an exposure screening program, if maximum risk workers cannot be identified for each operation or target population with reasonable confidence, a second approach is *random sampling*. The same sample size theory applies here as was discussed in Section 5.2.3 for random sampling of a homogeneous risk target population, based on the combinatorial probability formulas detailed in Technical Appendix A of Reference 1. This approach is less efficient than the first, which relies on professional judgment to select a nonrandom sample. However, the results are independent of any mistakes that might occur in professional judgment.

Purpose.   Provide a large enough sample size to assure (i.e., have a probability of at least 90 or at least 95 percent) that at least one randomly sampled worker will have a high exposure relative to most other workers in the target population (i.e., be from the highest 10 or 20 percent of the specified target population exposure distribution).

Assumptions.   The sample of workers to be measured is assumed to be randomly chosen from the specified target population. The derivation of this method is based on the theory of random sampling without replacement (see Section 5.2.3). There is no assumed exposure distribution for the target population. Thus these sample sizes may be inefficient for some situations where the exposure distribution is confidently known, such as 2 or 3-parameter log normal. However, the recommendations are robust.

Example.   Estimate an appropriate sample size for a 26-worker target population such that there is at least a 90 percent probability that the sample will include at least one higher risk worker from the top 10 or less percent of the 26-worker exposure distribution.

Solution.   Table 8.7 indicates that for the $N = 26$ workers, a random sample $n = (N - 8) = 18$ will have 90 percent probability of containing one or more of the $N_o = 2$ workers in 26 who represent no more than 10 percent of the

Table 8.7   Necessary Sample Size (n) for at Least 90% Probability (P) That One or More in the Sample Will Be from the Highest 10% or Less (N_0 Items in a Population of Size N)

N—Size of Target Population	n—Sample Size	N_o—Number of High Values
10–19	$(N - 1) = $ 9 to 18	1
20	$(N - 6) = 14$	2
21–24	$(N - 7) = $ 14 to 17	2
25–27	$(N - 8) = $ 17 to 19	2
28–29	$(N - 9) = $ 19 to 20	2
30–31	$(N - 14) = $ 16 to 17	3
32–33	$(N - 15) = $ 17 to 18	3
34–35	$(N - 16) = $ 18 to 19	3
36–37	$(N - 17) = $ 19 to 20	3
38–39	$(N - 18) = $ 20 to 21	3
40–41	$(N - 23) = $ 17 to 18	4
42–43	$(N - 24) = $ 18 to 19	4
44–45	$(N - 25) = $ 19 to 20	4
46	$(N - 26) = 20$	4
47–48	$(N - 27) = $ 20 to 21	4
49	$(N - 28) = 21$	4
50	$(N - 32) = 18$	5

highest exposures (7.7 percent in this example). The value 2 for N_o, the number of high values in the population, is the largest integer representing no more than 10 percent of $N = 26$.

Comments.   At least 18 of the 26 workers in the target population need to be randomly sampled to be 90 percent sure to "catch" one worker from the two that constitute no more than the upper 10 percent of the target population. With this small target population (and extremely small number of highest exposure workers) it is necessary to measure almost 70 percent of the total. This method is inefficient for small target populations, but works considerably better with larger ones. For a target population of 100 workers, the necessary sample would be about 20, or only 20 percent of the total, to have 90 percent probability of sampling one worker from the 10 workers that constitute the highest 10 percent of the exposure distribution. Necessary sample size tables for other upper tail percentages of exposure distributions or other confidence levels can be computed from the material in Technical Appendix A of Reference 1.

Since the target population is finite (26 in this example), only discrete probability levels (P) are attainable with the sampling plan. The sample size (n) is chosen to have 90 percent or greater probability ($P \geq 0.90$) that the (n) items in the sample will include one or more of the N_o highest items. Integer numbers (N_o) of highest values cannot be chosen to represent *exactly* 10 percent of the population size (N) unless $N = 10, 20, 30$, and so on. Therefore, N_o is chosen to be the largest integer for which $N_o \leq [(0.10)\ (N)]$.

Compared to an exposure screening program, the sampling procedure for an *exposure distribution monitoring program* is considerably different. An objective of the latter program is to estimate the exposure distribution of the target population over an initial base period, which may range from one day to several years. The necessary sample size for initial monitoring is influenced by the desired precision of the exposure distribution estimate for the target population.

One approach to exposure distribution estimation is the log normal probability plot technique presented in Technical Appendix I of Reference 1. An example of this approach is given in Paik et al. (76, 77). For this method a *minimum* sample of about 6 to 10 exposure estimates generally is required. That is, one should randomly sample at least 6 to 10 8-hour TWA exposures for the defined target population over the desired temporal base period. Note that this minimum sample will only provide the roughest estimate of the exposure distribution. It will tell us almost nothing about the goodness-of-fit to a particular distributional model, such as log normal. Data presented by Daniel and Wood (78) and discussed in Appendix I of Reference 1 indicates that considerably larger samples such as 30 to 60 are necessary to obtain an estimate with low uncertainty of the central 80 percent of the target population exposure distribution, but unusual behavior in the 10-percent upper and lower tails still cannot be confidently determined. However, as mentioned in Section 3.6, an exposure distribution estimate with moderate uncertainty may suffice because it is an *adequate* estimate. The adequacy of any distributional estimate is dependent on its ability to serve as a workable forecaster of the unsampled portions of the exposure distribution for the target population. That is, the estimate may be good enough for our purpose. This reinforces the point that the purposes of exposure measurements and monitoring need to be clearly defined *before* the measurements are obtained.

Besides providing information on the range and frequency of exposures, the exposure distribution monitoring strategy has another important advantage. The initial monitoring also can yield an estimate of the variability parameter of the distribution to be used as an indicator to judge if the target population was adequately defined. Remember that a target population of exposures generally is defined in numbers and types of workers, temporal period to be covered, and ranges of determinant variables affecting the exposure levels to which the population is exposed. If the variability of the sample exposure distribution exceeds a geometric standard deviation (*GSD*) of about 3.0, then the suitability of the factors used to define the target population should be examined (this approximate criterion is based on the professional judgment of one of the authors, NAL). Other factors may be needed to define more limited population groupings to achieve less variability for the exposure distribution. This process may need to be repeated several times until one achieves an exposure distribution variability adequate for the objective of the exposure monitoring.

If too many workers from populations with several different exposure distributions having substantially different median exposure levels are pooled into one target population, then a pooled exposure distribution of excessive variability can result. A highly disperse distribution can be unsatisfactory because the exposure estimates for individual workers will have a substantial amount of uncertainty. For example, for a 2-parameter log normal distribution with an arithmetic mean of 100 ppm and a *GSD* of 2.5 (*GM* = 65.7 ppm), the 5th percentile worker exposure is about 15 ppm and the 95th percentile worker exposure would be about 300 ppm! Note that 1 in 10 of the worker exposures would lie either below 15 ppm or above 300 ppm. For this exposure distribution, the best one could say for any individual worker is that there is a 90 percent probability of the worker's exposure lying between 15 and 300 ppm for the sampled temporal period. However, this highly imprecise estimate might be adequate if the exposure control level was substantially higher than the 95th percentile exposure level, such as 1000 ppm.

5.3.4 Criteria for Decision-Making

Decision-making is a monitoring program element where the objectives of an exposure screening program and an exposure distribution monitoring program are similar, but the techniques used differ considerably because the available information is different. The two types are discussed under separate headings below.

In this section, criteria for decision-making about *exposure distributions* cannot be discussed in a definitive framework of "legal" and "illegal" (or other black and white definitions of acceptable or unacceptable) parameters (or other statistical measures) of exposure distributions. Generally, well-accepted definitions are unavailable regarding parameters such as percentages of days workers are (or possibly are) exposed above an 8-hour TWA exposure control limit, when the long-term (multiple day) mean is below the control limit. Therefore, we will not attempt to provide explicit definitions of *acceptable* and *unacceptable* *exposure distributions*. These definitions must be derived for individual work exposure situations in the context of the particular toxicological, regulatory, and administrative considerations of a given occupational exposure environment. The technical discussions of this section are intended to assist the competent professional in estimating and interpreting worker exposure distributions. The statistical tools for making controlled-risk decisions and judgments about the acceptability of an exposure distribution, based on an estimate of that distribution, are to be found in Section 6. In particular, procedures are presented in sections 6.7 and 6.13 for computing and displaying tolerance limit for a log normal distribution of exposures, and Section 6.8 for computing a point estimate and confidence limits for the percentage of log normally distributed values exceeding an exposure control limit.

Decision-Making in an Exposure Screening Program. The main objective of exposure screening is to identify target populations with *unacceptable* exposure distributions for follow-up exposure monitoring or other actions (e.g., worker medical surveillance, worker training programs, engineering exposure controls, personal exposure controls, or other specialized measurement actions such as noted in Section 5.2.1). The decision-making technique used is to compare exposure estimates with an appropriate Action Level, which serves as a screening cutoff value. A secondary objective would be to subclassify the *unacceptable* exposure distributions as those requiring follow-up monitoring at a normal frequency (at least every two months) or at an increased frequency (at least monthly). The second screening cutoff value for these decisions is the exposure control limit chosen for adequate worker protection. For target populations judged to have *acceptable* exposure distributions, generally the indicated action for an exposure screening program would be a redetermination of need for exposure measurements (see Section 5.3.1) each time there is a change in production, process, or control measures which could result in a substantial increase in exposure levels.

In the original Action Level concept as developed by the National Institute for Occupational Safety and Health (NIOSH) and recommended to the Occupational Safety and Health Administration (OSHA) for regulatory purposes (79) (also see pp. 10–11 and Technical Appendix L of Reference 1), only single day exposure estimates for the maximum risk workers from each examined target population were to be compared to an Action Level set at 50 percent of the regulatory exposure standard for the particular substance. Single day exposure estimates would be used to reach decisions regarding the acceptability of possible exposure levels on unmeasured days and were to be the sole basis for deciding whether further exposure monitoring or other actions should be performed for the target population's workers.

Note that a required assumption for the application of an Action Level is that the companion Exposure Control Limit (*ECL*) is sufficiently protective of workers. That is, the use of an Action Level to reduce employer costs of exposure monitoring, medical surveillance, worker training, and so on, presumes that exposures below the *ECL* create minimal or acceptable risks to workers. The rationale of using an Action Level is to screen worker exposures so that there is only a low probability that even a minimal proportion of daily exposures will exceed the *ECL* linked to the Action Level. Thus it is implicit that exposures below the *ECL* lead to minimal or acceptable risks to workers. If this is not the case, then it is not appropriate to use an Action Level as a justification or screening value to reduce or eliminate exposure monitoring, medical surveillance, and so forth.

It is also important to note that the *50 Percent Action Level* should be thought of as a *regulatory Action Level*, since:

1. It is a compromise value developed for simplicity of application,
2. It was a value intended to reduce employer exposure monitoring burdens, and
3. It incorporated feasibility considerations for regulatory monitoring requirements.

During the development of the 50 Percent Action Level for regulatory purposes, the premise used was that "the employer should try to limit to 5 percent probability, that no more than 5 percent (or greater) of an employee's actual (true) daily exposure averages exceed the standard" (75). This is not meant to imply that a work situation with as many as 5 percent of overexposure days is acceptable to OSHA. The stated premise is part of the "risk tradeoff" which is characteristic of statistical decision theory. Measurements subject to random errors are used to make decisions about overexposure and the limit of 5 percent overexposure days cited above is merely the minimum of hypothetical overexposure incidences which, if they occurred, would be discovered at least 95 percent of the time they were measured.

The Leidel et al. report (79) demonstrated that an Action Level appropriately computed to achieve the stated goal of a low 5 percent probability level is a *variable value that is a function of the interday variability of the true daily TWA exposure levels*. Regarding the interday variability expressed as a *geometric standard deviation* (*GSD*) of a 2-parameter log normal distribution, they noted

Higher GSDs require lower fractional action levels. A GSD of 2.0 requires an action level as low as 0.115 of the standard! (79)

However, it was decided by NIOSH to recommend to OSHA a single Action Level computed as 50 percent of an exposure standard. This decision was intended to increase the probability of employer compliance with regulatory monitoring requirements as a result of the simplicity of applying a single Action Level, lower implementation costs, and substantially greater feasibility. The alternative would have been recommending a procedure that required an employer to obtain an estimate of the interday exposure level variability for each exposure situation before computing an Action Level.

The decision to recommend a single Action Level (50 percent) keyed to a single *GSD* of 1.22 did lead to a regulatory screening strategy with some limitations. The potential problems and performance characteristics of the "bare bones" regulatory screening strategy (1) have been comprehensively analyzed by Tuggle (80). Tuggle noted,

The intent of the NIOSH decision scheme is to give indications about periodic monitoring: whether to terminate (TERM), initiate/increase (INCR), or continue (CONT) exposure measurements.

Tuggle concluded the following:

For environments with a very low fraction of exposures above the PEL, the NIOSH
scheme renders the correct, TERM decision with a high probability—for all variabilities.
For low variability environments, the relative probability of a TERM decision decreases
sharply (the relative probability of an INCR decision increases sharply) as the fraction
of exposures above the PEL increases. In other words, at low variability, the NIOSH
scheme is very "sensitive" to an increasing fraction of overexposures. However, as
exposure variability increases, this sensitivity falls off, and the NIOSH decision scheme
assumes an increasing probability of incorrect, TERM decisions for high exposure risk
environments for which monitoring should definitely be continued or even increased.
For example, consider environments with moderate-to-high variability and with 25
percent of all exposures above the PEL—one exposure in four an overexposure: Figure
8.3 shows that the NIOSH decision scheme produces incorrect, TERM decisions for
such environments, from about 20 percent to well over 50 percent of the time.

Tuggle (80) noted the following limitations of a simplistic regulatory screening
strategy and suggested some desirable characteristics for an augmented exposure screening program:

1.   The regulatory screening stretegy bases a decision on, at most, only two
measurements: the last one and possibly the one preceding. An augmented
strategy should use *all* the data collected.
2.   The regulatory screening strategy considers only a qualitative aspect of
the exposure estimate, i.e., whether the exposure is below the 50 percent Action
Level or above the exposure standard. An augmented strategy should also
consider the quantitative aspect of *how much* an exposure estimate is below the
Action Level or above the exposure standard.
3.   The regulatory screening strategy does not determine or account for the
range of exposure level variabilities actually encountered. An augmented
strategy should, since exposure variability (*GSD*) is a factor equally as important
as median exposure (*GM*) in evaluating the occurrence of overexposures.
4.   Lastly, the regulatory screening strategy produces incorrect decisions to
terminate exposure sampling, particularly with substantial proportions of
overexposures and moderate to high variability exposure level environments.
An augmented strategy should limit such incorrect decisions to an acceptably
low probability, which is independent of exposure variability.

Tuggle (80) suggested a simple modification to the regulatory screening
strategy (1) to reduce incorrect decisions to terminate exposure monitoring that
result from analyzing initial monitoring results. He recommended that the first
decision criterion require, in all instances, two (instead of one) consecutive
exposure estimates below the 50 percent Action Level to terminate routine
exposure monitoring. He demonstrated that this would substantially reduce
the probability of incorrect decisions to terminate monitoring. Tuggle (81)
subsequently recommended an augmented monitoring strategy based on 1-

sided tolerance limits on a distribution of daily exposure estimates assumed to follow a 2-parameter log normal distribution.

It should be noted that the NIOSH "employee exposure determination and decision strategy" recommended by NIOSH to OSHA for incorporation in regulations (Figure 1.1 of Reference 1) does maintain the objective of a 5 percent limit on the risk of making incorrect decisions of noncompliance during the periods and for the workers that were actually measured for TWA exposures. However, the NIOSH regulatory screening strategy does not achieve 95 percent probability of terminating a process which has 5 percent or more overexposure days, unless the GSD of the process is at or below the value 1.22 from which a 50 percent Action Level is derived.

Decision-Making in an Exposure Distribution Monitoring Program. Compared to an exposure screening program, the decision-making process for evaluating initial monitoring results from an *exposure distribution monitoring program* is considerably different. However, the objectives of the two monitoring programs are similar.

For an *exposure distribution monitoring program*, the objective of the decision-making element is to judge the exposure distributions or target populations as *acceptable, marginal,* or *unacceptable*. An *acceptable exposure distribution* is one for which we are confident that almost all the exposures are less than the desired exposure control limit. A *marginal exposure distribution* has a moderate proportion of exposures (or part of a tolerance limit line of Section 6.13, if one desires a conservative approach) exceeding the desired exposure control limit. An *unacceptable exposure distribution* has at least a substantial proportion of worker exposures (or a substantial portion of a Section 6.13 tolerance limit line and a moderate proportion of the exposure distribution, if one desires a conservative approach) exceeding the desired exposure control limit. Note that these terms have not been defined in purely quantitative terms. Their use also connotes qualitative considerations requiring the exercise of competent professional judgment, including individual aspects of the target population, the chemical(s) creating the exposures, possible exposure levels, and determinant variables for the exposure levels.

If the exposure distribution for a target population is judged *acceptable*, generally the follow-up action would be limited to routine exposure monitoring at an appropriate future time to monitor and revise the initial or previous estimate of the exposure distribution for the target population. If the judgment for the exposure distribution is *marginal*, then the indicated action generally would be more definitive exposure monitoring, which could include the specialized measurement actions noted in Section 5.2.1. This additional monitoring should identify the marginally and unacceptably exposed workers and could assist in implementing overexposure prevention measures. Lastly, if the exposure distribution is judged *unacceptable*, the indicated action would follow the same course specified for a target population identified with an Action Level screening approach as unacceptably exposed.

Two statistical techniques which are useful in an evaluation of exposure distribution estimates for target populations of workers are probability plotting (a graphical technique) and distribution fitting to estimate one or more quantiles (a formal technique based in statistical estimation theory). The first utilizes the type of log normal probability plot presented in Technical Appendix I of Reference 1. The exposure estimates would be plotted on log normal probability paper along with a line representing the desired exposure control limit. A rough estimate of the overexposure "risk" in percentages of overexposed workers or workdays can be obtained from such probability plots. The estimate is given by the indicated probability at the intersection of the exposure estimate distribution plot and the exposure control limit line. Then a *qualitative* judgment would be made regarding the acceptability of the exposure distribution. The judgment should consider the overexposure risk estimate, slope of the exposure distribution, experience with similar exposure situations, health effects of the contaminant, and knowledge of exposure level determinant factors pertinent to the target population. Note that this technique should be used with caution and decisions made with some degree of conservatism (i.e., err in the direction of *marginal* or *unacceptable exposure distribution* decisions). Experience and competent professional judgment are needed for its proper application.

The second evaluation technique of formal point estimation complements the first, which is graphical. The two techniques can be used independently, but the authors recommend they both be used for evaluation of any exposure distribution data set. The quantitative inferential approach is to calculate the best estimate (point estimate) for the proportion of the exposure distribution exceeding some desired value (e.g., exposure control limit, exposure standard). In addition, confidence limits for the true point estimate should also be calculated to indicate the uncertainty of the estimate due to limited sample sizes. The confidence limits can be either 1- or 2-sided and typically will be calculated at the 95 percent confidence level. The details for this technique are presented in Section 6.8.

5.3.5 Strategy for Periodic Monitoring for Continuing Hazard Evaluation

Periodic follow-up exposure monitoring applies both to an exposure screening program and to an exposure distribution monitoring program, and has a similar objective for both programs. That objective is that individual workers or target populations of workers with other than acceptable (i.e., *marginal* or *unacceptable*, see Section 5.3.4) exposure distributions be periodically monitored often enough to detect *unacceptable* exposure situations. In addition, an exposure distribution monitoring program should periodically monitor at a frequency adequate to identify substantial changes in the target population's exposure distributions.

Remember that for an *exposure screening program* the set of possible decisions suggested (Section 5.3.4) regarding the target population's exposure distribution are *acceptable, unacceptable* with "normal" frequency follow-up monitoring, and *unacceptable* with "increased frequency" follow-up monitoring.

For regulatory purposes, NIOSH recommended to OSHA that minimum requirements for "normal" frequency of monitoring for an *exposure screening program* be set at 2 to 3 months. It should be noted that feasibility and simplicity of application considerations played an important role in this recommendation. Lynch et al. have commented for NIOSH (82):

Four such exposure measurements during the year are considered minimal to detect any significant fluctuations in the average level of environmental contaminants. Employee exposure measurements four times per year per exposed employee are a reasonable minimal burden on the employer. The employer should not interpret this to be an absolute minimum that would always be appropriate to determine each employee's exposure. There is information that a waiting time between sampling events (for typical data) of over 1 month results in a relatively low level of confidence in the exposure estimates. However, a maximum waiting time of 3 months between measurements should protect each employee and give some idea of variation without putting an inordinate burden on the employer.

For an "increased frequency" of regulatory follow-up monitoring, NIOSH recommended a minimum frequency of once a month. When unexpectedly or unusually high exposure results are obtained by either initial monitoring or periodic monitoring, more intensive monitoring is needed to quickly identify increasing exposure trends that may lead to hazardous exposures.

For other than regulatory monitoring programs, the selection of an adequate frequency for periodic exposure monitoring should be based on experience with similar exposure situations, knowledge of the exposure level determinant factors pertinent to the target population, and use of competent professional judgment. Guidelines for the factors to be considered include:

1. Nature and degree of health hazard from the exposure situation, evaluated at all possible exposure levels, even those that might be unexpected.

2. Ratio of previous exposure levels to the desired exposure control limit and possible trends in this ratio.

3. Degree of variability seen in previous exposure distributions, variability between the distributions, and trends over time. These can be studied by plotting the sample data sets from several longitudinal points in time on semilogarithmic paper (see Section 6.11). Another complementary approach would be the plotting of several sample data distributions on the same log normal probability graph.

4. Determinant variables affecting the target population's exposure levels (see Section 5.1).

5. Reliability of decision-making techniques used and quantity and quality of exposure estimates available. Exposure screening program decision techniques yield decisions of less reliability than the quantitative techniques available for exposure distribution estimates.

5.3.6 Occasions Requiring Extraordinary Monitoring

With either an exposure screening program or an exposure distribution monitoring program, additional exposure monitoring is indicated whenever a change occurs in production, process, personnel, exposure controls or any other determinant variable that could lead to substantial increases in worker exposure levels.

5.3.7 Criteria for Termination of Monitoring

This program element concerns a specialized type of decision-making. As with monitoring of exposures for the purpose of decision-making about exposure distributions, the objectives of termination of routine periodic monitoring in an exposure screening program and exposure distribution monitoring program are similar, but the techniques used and available information differ considerably. In some situations it may be desirable to conserve exposure measurement resources by terminating routine monitoring of target populations for which one is confident that the exposure distributions have been consistently *acceptable*. As with selecting an adequate frequency of periodic monitoring (section 5.3.5), decisions to terminate monitoring should be based on experience with similar exposure situations, knowledge of the exposure level determinant factors pertinent to the target population, availability of monitoring resources, and availability of competent professional judgment. Use the factors given in Section 5.3.5 when considering termination of routine monitoring.

For regulatory purposes in an *exposure screening program*, NIOSH recommended to OSHA that regulations allow cessation of monitoring for a worker if two consecutive exposure estimates, taken at least one week apart, were both less than the 50 percent Action Level. See Section 5.3.4 for a discussion of an analysis by Tuggle (80) of this regulatory termination rule and his suggested alternative approach for situations where a 2-parameter log normal distribution of exposures can be assumed.

For *exposure distribution monitoring programs*, decisions to terminate monitoring should also be based on experience with similar exposure situations, knowledge of the exposure level determinant factors pertinent to the target population, availability of monitoring resources, and use of competent professional judgment. Also use the factors given in Section 5.3.5 when considering termination of routine monitoring.

There is an additional quantitative tool available when considering termination of an exposure distribution monitoring program. Tuggle (81) has recommended the use of 1-sided tolerance limits if one can assume that all available exposure estimates are from a single 2-parameter log normal distribution. The computed 1-sided tolerance limit for the target population would be compared to the selected exposure control limit to assist in deciding if monitoring should be continued. All available exposure estimates would be considered sample data and utilized for computation of a tolerance limit.

However, if the exposure data are not stable over time (i.e., if differing distributions of exposures exist over time for the same target population of workers), competent professional judgment may be used to censor some of the earlier exposure estimates. If this is not done, the computed tolerance limit will probably be larger than it should be, which would cause errors in the decision-making leading to continuing monitoring. Thus if all the data are used, instead of a smaller sample based on the more recent exposure data, the decision-making would tend to be conservative (i.e., favor continuation of monitoring).

5.3.8 Procedures for Follow-up

This program element should always be used for all types of monitoring programs. One should discuss the planning, conduct, and results of each program element with appropriate representatives of management and labor. If necessary, implement corrective action to reduce potential or identified health hazards. Write periodic evaluation reports as appropriate. Determine if additional exposure monitoring, biological monitoring, or medical surveillance programs are necessary. Determine if evaluations in proximity to the workplace are necessary regarding possible environmental air or water pollution, hazardous waste disposal practices, or safety practices.

6 APPLIED METHODS FOR ANALYSIS OF OCCUPATIONAL EXPOSURE DATA

This last section of Chapter 8 is devoted to the applied statistical methods which we have earlier in the chapter suggested be used for analysis of occupational exposure data. Data to be analyzed by these methods can be generated by the study designs discussed in Sections 4 and 5. The first portion of this section (Sections 6.1 through 6.5) covers methods for computing 1-sided and 2-sided confidence limits and confidence intervals for the true exposures of individual workers during the same periods as their exposure measurements. In addition, classification tests for individual exposure estimates, relative to an exposure control limit, are presented. The second portion of Section 6 (Sections 6.6 through 6.10) covers inferential methods for exposure distributions such as computing tolerance limits, tolerance intervals, and point estimates of the proportions of values in distributions which exceed a chosen limit, along with associated confidence limits. The third portion of this section (Sections 6.11 through 6.13) presents graphical techniques for plotting log normally distributed distributions and their associated tolerance limits.

The methodologies for the exposure classification tests (statistical hypothesis tests) given in Sections 6.1 through 6.5 were originally develleoped by NIOSH for OSHA in support of their regulatory enforcement procedures. The purpose of the methodologies is to limit the risk of making unjustified noncompliance decisions (i.e., decisions not supported by sufficient accuracy in the exposure measurement data) regarding regulatory exposure limits. The necessity for development of such procedures is a reflection of the nature of the legal system used in the United States. Unfortunately, the presentation of the statistical hypothesis tests may give the impression that industrial hygienists should view any ECL, ACGIH TLV®, or legal standard as a definitive boundary between "safe" and "dangerous" exposure levels for all workers. This is not the case. The hypothesis tests for individual worker exposure estimates presented in this section should receive routine application only in legal proceedings. That is, an employer should not attempt to judge the acceptability of workplace exposures by statistically comparing only a few exposure measurements with some exposure limit.

It is recommended that the computation of confidence limits be the preferred procedure for calculating the uncertainty of individual true worker exposures. The primary purpose of computing confidence limits should be the estimation of the uncertainty of individual worker exposure estimates due to the random errors of the exposure measurement procedure used. It is important to realize that the confidence limits do not reflect the substantial interday and interworker exposure level variability. When attempting to judge the acceptability of exposures in a workplace, an industrial hygienist should if possible evaluate the multiday and multiworker exposure distributions (see Section 5.3.4). Of course, an employer continually has the legal obligation to maintain exposure levels below applicable regulatory exposure limits on *all* workdays.

6.1 Full Period Single Sample Estimate

The methods in this section are applicable to exposure estimates calculated from a single exposure measurement taken for the full duration of the desired time-averaging period (e.g., 40 hours for a workweek TWA, 8 hours for an 8-hour workday TWA, 15 minutes for a 15-minute TWA).

6.1.1 Confidence Limits for a True TWA Exposure

Purpose.   Compute 1-sided or 2-sided confidence limits for a true worker exposure, based on a single exposure measurement.

Assumptions.   The random errors in full period single sample estimates X are assumed to be independent and normally distributed with zero mean [see Section 3.6.1(3)]. This assumption implies that the arithmetic mean of many replicate exposure measurements (i.e., taken at exactly the same point in space and over the same time period) would be equal to the true exposure average. This true mean or "expected value" is unknown, but the total coefficient of variation for the exposure measurement procedure (denoted by CV_T, see Section 3.4.2) is assumed known. Adequate confidence limits can be computed for the true concentration based on a standard deviation estimated from the product of CV_T and the full period exposure estimate X. The accuracy of the computed confidence limit(s) will be a function of the accuracy of the coefficient of variation used.

Example.   An exposure measurement procedure with a CV_T of 0.09 was used to obtain an 8-hour TWA exposure estimate for a worker. A single 8-hour measurement yielded an estimate $X = 0.20$ mg/cu m. Compute *2-sided 95 percent confidence limits* for the true worker exposure based on the exposure estimate X.

Solution.

1.   Compute the *2-sided 95 percent upper* and *lower confidence limits* ($UCL_{2,.95}$ and $LCL_{2,.95}$) for the true worker exposure μ from:

$$UCL_{2,.95} = X/[1 - (1.96)(CV_T)]$$

$$LCL_{2,.95} = X/[1 + (1.96)(CV_T)]$$

In this example:

$$UCL_{2,.95} = 0.20/[1 - (1.96)(0.09)]$$

$$= 0.20/0.824$$

$$= 0.24 \text{ mg/cu m}$$

$$LCL_{2,.95} = 0.20/[1 + (1.96)(0.09)]$$

$$= 0.20/1.176$$

$$= 0.17 \text{ mg/cu m}$$

Thus, in this example, the *2-sided confidence interval* at the 95 percent confidence level for the true worker exposure which was estimated by the single sample exposure measurement of 0.20 mg/cu m is from $LCL_{2,.95} = 0.17$ mg/cu m to $UCL_{2,.95} = 0.24$ mg/cu m. That is, we can be 95 percent confident that the *true exposure* is between 0.17 and 0.24 mg/cu m.

2.   To compute either *1-sided 95 percent confidence limit* ($UCL_{1,.95}$ or $LCL_{1,.95}$) for the true exposure, use the formula for the corresponding 2-sided limit with a Z-value of 1.645 substituted for the multiplier 1.96. (The multiplier 1.96 corresponds to 2.5 percent of the area in a standard normal distribution allocated to each tail, whereas 1.645 corresponds to the total 5 percent allocated to one tail.)

Comments.   Note that, for any given estimate and associated confidence interval, the interval either does or does not contain the true value. However, we can say, in a special sense, that there is a 95 percent *probability* of the true exposure being bounded by the computed confidence interval. That is, if repeated samples were taken, and confidence limits calculated for each sample, 95 percent of the confidence limits would enclose the true worker exposure.

Also, the indicated uncertainty in the exposure estimate reflected in the width of the confidence interval (-15 to $+21$ percent of the estimate) *does not* incorporate uncertainty due to *inter*day exposure level variability [see Sections 3.3.1 and 3.6.2(3)]. The confidence interval pertains *only* to the worker's true exposure on the particular day and during the period of that day, which was actually measured.

6.1.2   Classification of Exposure (Hypothesis Testing)

Purpose.   Classify an exposure estimate based on a single exposure measurement as *noncompliance exposure, possible overexposure,* or *compliance exposure* regarding an *exposure control limit (ECL)*.

Assumptions.   The same as used for Section 6.1.1.

Example.   A charcoal tube and personal pump were used to sample a worker's exposure to alpha-chloroacetophenone. The analytical laboratory reported an exposure estimate X of 0.040 ppm and stated that the measurement procedure had a CV_T of 9 percent. The applicable ECL was 0.05 ppm. Classify the exposure estimate regarding this ECL.

Solution.

1.   For an **employer**, if $X > ECL$, no statistical significance test (of the null hypothesis of noncompliance against an alternative of compliance) need be made because the estimate itself exceeds the ECL and it would exceed the LDV (*Lower Decision Value*) by even more. However, for $X \le ECL$, it is necessary to compute the 1-sided LDV to do a hypothesis test for compliance:

$$LDV_{5\%} = ECL - (1.645)(CV_T)(ECL)$$
$$= (0.05 \text{ ppm}) - (1.645)(0.09)(0.05 \text{ ppm})$$
$$= (0.05 \text{ ppm})(1 - 0.148)$$
$$= 0.043 \text{ ppm}.$$

Then the exposure estimate is classified according to the following criteria:

   a.   If $X \le LDV_{5\%}$, classify as **Compliance Exposure**, (i.e., reject the null hypothesis and accept the alternative hypothesis) or
   b.   If $X > LDV_{5\%}$, classify as **Possible Overexposure**, (i.e., do not reject the null hypothesis).

Since the exposure measurement $X = 0.040$ ppm is less than the $LDV_{5\%}$ of 0.043 ppm, the employer can decide that the measured worker had a Compliance Exposure on the day and during the period of the measurement. The decision is made with 5 percent *or less* probability of being wrong (i.e., 0.05 or less probability of a Type 1 error). The subscript of 5% identifies the test's *level of significance* (also called the *size of the significance test*).

2.   For a **compliance officer**, if $X \leq ECL$, no statistical test of the null hypothesis of compliance against an alternative of noncompliance need be made because the estimate is already less than the ECL and it would be lower than the UDV (*Upper Decision Value*) by even more. However, for $X > ECL$, it is necessary to compute the 1-sided UDV test statistic for noncompliance:

$$UDV_{5\%} = ECL + (1.645)(CV_T)(ECL)$$

Then the exposure estimate is classified according to the following criteria:

a.   If $X > UDV_{5\%}$, classify as **Noncompliance Exposure**, (i.e., reject the null hypothesis and accept the alternative hypothesis), or
b.   If $X \leq UDV_{5\%}$, classify as **Possible Overexposure**, (i.e., do not reject the null hypothesis).

Comments.   The outcomes of these hypothesis tests are valid only for the particular worker's exposure on the day and during the period of the exposure measurement.

The use of CV_T in the classification formulae is equivalent to calculating the standard deviation of X (the single exposure measurement) as $[(CV_T)(ECL)]$ instead of $[(CV_T)(\mu)]$, where μ is the true exposure for which X is the estimate. The use of a *1-sided, 5 percent significance level, Lower Decision Value* or a *1-sided, 5 percent significance level, Upper Decision Value* ($LDV_{5\%}$ or $UDV_{5\%}$, see Section 3.7) computed in this manner is correct, since the classification rule selected for use as a *compliance officer's test for noncompliance* is algebraically equivalent to a significance test of the null hypothesis of compliance (i.e., of H_0: $\mu \leq ECL$). Similarly, the classification rule selected for use as the *employer's test for compliance* is equivalent to a significance test of the null hypothesis of noncompliance (i.e., of H_0: $\mu > ECL$). In both cases, setting $\mu = ECL$ to compute the decision value serves to keep the size of the test at or below 0.05 (≤ 5 percent probability of incorrectly rejecting the null hypothesis).

The rationale for the hypothesis tests is:

1.   For a test at the 5 percent level of significance, calculate a *1-sided Upper Decision Value* ($UDV_{5\%}$) or *Lower Decision Value* ($LDV_{5\%}$). These decision values are in fact *critical values* for the measurement X, under the null hypothesis that the true exposure is equal to the ECL. If the null hypothesis were true, the exposure estimate would not exceed the UDV (or be less than the LDV for the complementary hypothesis test) more than 5 percent of the time for the same true exposure and measurement conditions.

2.   For a test of noncompliance, if the exposure estimate exceeds the *1-sided Upper Decision Value* ($UDV_{5\%}$), reject the null hypothesis of compliance and decide for noncompliance. Similarly, for a test for compliance, if the exposure

estimate is less than the *1-sided Lower Decision Value* ($LDV_{5\%}$), reject the null hypothesis of noncompliance and decide for compliance.

6.2 Full Period Consecutive Samples Estimate for Uniform Exposure

The methods in this section are for exposure estimates calculated from a series of consecutive exposure measurements (equal or unequal time duration) which collectively span the full duration of the desired time-averaging period.

6.2.1   Confidence Limits for a True TWA Exposure

Purpose.   Compute 1-sided or 2-sided confidence limits for a true TWA exposure, based on multiple exposure measurements in a uniform exposure situation.

Assumptions.   The random errors for measurements X_i are assumed to be normally and independently distributed [see Section 3.6.1(3)]. The true worker exposure during any sampled sub-period within the total period of the TWA estimate (i.e., true arithmetic mean of the hypothetical parent population of replicate exposure measurements at exactly the same point in space during the same time period T_i) is unknown, but is assumed to be about the same for all periods sampled (uniform exposure environment). The coefficient of variation of replicate measurements made by the exposure measurement procedure is assumed known and equal (see Section 3.4.2). The accuracy of the computed confidence limit(s) will be a function of the accuracy of the coefficient of variation used.

It is assumed here that all sampled periods have equal true average concentrations; if it is expected that the samples have significantly different values because of different exposure situations during the work shift, then the conservative procedure in Section 6.3.1 can be used. Where exposures are highly variable between the sampling periods over the duration of the TWA estimate, the use of the formulae in this section will underestimate the random sampling error in the TWA exposure estimate, thus underestimating the width of the confidence interval.

Example.   An exposure measurement procedure with a CV_T of 0.08 was used to obtain an 8-hour TWA exposure estimate. A personal pump and three charcoal tubes were used to consecutively measure a worker's approximately uniform exposure to isoamyl alcohol. The analytical lab reported the following exposure estimates for the three tubes: $X_1 = 90$ ppm over $T_1 = 150$ min, $X_2 = 140$ ppm over $T_2 = 100$ min, and $X_3 = 110$ ppm over $T_3 = 230$ min. The *ECL* chosen by the industrial hygienist was 100 ppm. The CV_T for the measurement procedure is 0.08 (8 percent). Compute the *1-sided 95 percent upper confidence limit* for the true 8-hour TWA exposure. The estimated value of the TWA exposure is 110 ppm.

Solution.

1. If the sample durations $(T_1, T_2, \ldots, T_n)$ are approximately equal, the *1-sided 95 percent upper confidence limit* for the true TWA exposure can be computed from the short equation:

$$UCL_{1,.95} = \frac{\text{TWA}}{\{1 - [(1.645)(CV_T)]/n^{1/2}\}} \, .$$

In this example the sample durations are not approximately equal, but for illustrative purposes and as a first approximation:

$$UCL_{1,.95} = (110 \text{ ppm})/\{1 - [(1.645)(0.08)/(3)^{1/2}\} = (110 \text{ ppm})/(1 - 0.076)$$

$$= (110 \text{ ppm})/(0.924) = 119 \text{ ppm}$$

Thus an approximate 1-sided 95 percent $UCL_{1,.95}$ for the true TWA exposure is 119 ppm based on the TWA estimate of 110 ppm.

2. Since the sample durations *are not equal,* compute the more exact *1-sided 95 percent upper confidence limit* for the true TWA from the longer equation:

$$UCL_{1,.95} = \frac{(\text{TWA})(T_1 + \cdots + T_n)}{\{(T_1 + \cdots + T_n) - [(1.645)(CV_T)(T_1^2 + \cdots T_n^2)^{1/2}]\}} \, .$$

In the example:

$$UCL_{1,.95} = \frac{(110 \text{ ppm})(150 + 100 + 230)}{\{(480) - [(1.645)(0.08)(150^2 + 100^2 + 230^2)^{1/2}]\}}$$

$$= [(110 \text{ ppm})(480)]/[(480) - (38.5)] = 120 \text{ ppm}$$

In this case, the short equation gave a confidence limit that was only slightly lower than the more accurate estimate. The $UCL_{1,.95}$ is 120 ppm based on the TWA estimate of 110 ppm. That is, we can be 95 percent confident that the *true TWA exposure* is less than 120 ppm.

3. To compute a *1-sided 95 percent lower confidence limit* ($LCL_{1,.95}$), substitute a plus sign for the minus sign in the denominator of the Step 1 or Step 2 formulae. To compute a *2-sided 95 percent confidence interval* ($LCL_{2,.95}$ to $UCL_{2,.95}$) for the true TWA exposure, substitute a Z-value of 1.96 for 1.645.

Comments.   This method is an extension of the procedure presented in Section 6.1.1 for a TWA estimate based on a single exposure measurement. Review the comments of that section.

The indicated uncertainty in the TWA exposure estimate reflected in the difference between the exposure estimate TWA of 110 ppm and the $UCL_{1,.95}$

of 120 ppm (9 percent higher than the TWA estimate) *does not* incorporate uncertainty due to *inter*day exposure level variability [see Sections 3.3.1 and 3.6.2(3)]. The 1-sided confidence limit pertains *only* to the worker's true exposure on the day actually sampled and during the total period of the three measurements.

6.2.2   Classification of Exposure (Hypothesis Testing)

Purpose.   Classify a TWA exposure estimate, which is based on multiple exposure measurements from a uniform exposure situation, as *noncompliance exposure, possible overexposure,* or *compliance exposure,* regarding an *exposure control limit (ECL).*

Assumptions.   The same as used for Section 6.2.1. The formulae of this section strictly apply only to the case of uniform exposure. Where exposures are highly variable between the sampling periods over the duration of the TWA estimate (nonuniform exposure situation), the use of the formulae in this section will underestimate the random sampling error in the TWA estimate. This misapplication would therefore increase the chance of deciding a Noncompliance Exposure (with the test for the compliance officer) or deciding a Compliance Exposure (with the test for the employer). For a nonuniform exposure situation, use the conservative methods given in Section 6.3.2.

Example.   The same as used for Section 6.2.1, except classify the TWA exposure estimate of 110 ppm relative to the *ECL* of 100 ppm.

Solution.

1.   If the sample periods $(T_1, T_2, \ldots, T_n)$ are equal (or, for a somewhat inexact test, approximately equal), classify the TWA estimate using the short equations in Step 2 (for an employer) or Step 3 (for a compliance officer). For unequal sample periods, classify using the longer equations in Step 4 (for an employer) or Step 5 (for a compliance officer).
2.   For an **employer**, if $TWA \geq ECL$, a *possible overexposure* is indicated by the *TWA* itself and no statistical test for compliance need be made. This is the case for this example. However, when there are approximately equal sample periods and $TWA < ECL$, compute the 1-sided *LDV* (*Lower Decision Value*) to do a test for compliance:

$$LDV_{5\%} = \{ECL - [(1.645)(CV_T)(ECL)/(n)^{1/2}]\}$$

Then classify the TWA exposure estimate based on the multiple measurements according to:

   a.   If $TWA \leq LDV_{5\%}$, classify as **Compliance Exposure**.
   b.   If $TWA > LDV_{5\%}$, classify as **Possible Overexposure**.

3. For a **compliance officer**, if $TWA \leq ECL$, it is superfluous to make a statistical test for noncompliance, since the TWA estimate obviously would also be less than the UDV because by definition the ECL is always less than the UDV. However, for $TWA > ECL$, and where there are approximately equal sampling periods, the compliance officer would compute the 1-sided $UDV_{5\%}$ critical value of the TWA:

$$UDV_{5\%} = \{ECL + [(1.645)(CV_T)(ECL)/(n)^{1/2}]\}$$

Then the exposure estimate would be classified according to:

    a. If $TWA > UDV_{5\%}$, classify as **Noncompliance Exposure**.
    b. If $TWA \leq UDV_{5\%}$, classify as **Possible Overexposure**.

4. For an **employer**, when the periods of the measurements are not approximately equal and $TWA < ECL$, compute the 1-sided $LDV_{5\%}$ critical value of the TWA exposure estimate:

$$LDV_{5\%} = ECL - \frac{(1.645)(CV_T)(ECL)(T_1^2 + \cdots + T_n^2)^{1/2}}{T_1 + \cdots + T_n}$$

Then use the classification criteria of Step 2 to perform a hypothesis test for compliance.

5. For a **compliance officer**, when the periods of the measurements are not approximately equal and $TWA > ECL$, compute the 1-sided $UDV_{5\%}$ to be used as the critical value of the TWA exposure estimate:

$$UDV_{5\%} = ECL + \frac{(1.645)(CV_T)(ECL)(T_1^2 + \cdots + T_n^2)^{1/2}}{T_1 + \cdots + T_n}$$

Then use the classification criteria of Step 3 to perform a hypothesis test for noncompliance. For this example:

$$UDV_{5\%} = 100 + \frac{(1.645)(0.08)(100)(150^2 + 100^2 + 230^2)^{1/2}}{150 + 100 + 230}$$

$$= 100 + 8 = 108 \text{ ppm}$$

Since the $TWA = 110$ ppm exceeds the $UDV_{5\%}$ of 108 ppm, the TWA exposure estimate of 110 ppm is classified as a *Noncompliance Exposure*. The compliance officer can state that, for the measured worker on the day and during the period of the three measurements, the true 8-hr TWA exposure was a statistically significant *Noncompliance Exposure*. The statistical decision of *noncompliance* is made with 5 percent *or less* probability of being wrong.

Comments. The same as for Section 6.1.2, which should be reviewed before the use of this section.

The outcomes of these hypothesis tests are valid *only* for the worker's exposure on the day and during the period of the exposure measurements.

In the example, the measurement results indicate a sufficiently uniform exposure over the duration of the TWA estimation period, so that the use of the formulae in this section is appropriate.

6.3 Full Period Consecutive Samples Estimate for Nonuniform Exposure

The methods in this section are a generalization of those in Section 6.2. They are for exposure estimates calculated from a series of consecutive exposure measurements (equal or unequal time duration) which collectively span the full duration of the desired TWA exposure period. The following methods are longer than those of Section 6.2, but are not limited by the assumption of nearly equal arithmetic mean exposures during the TWA period (uniform exposure environment). For highly nonuniform exposure situations, use of the less complex methods of Section 6.2 (intended for uniform exposure situations) may slightly underestimate the sampling error in the TWA estimate. However, the conservative methods of this section will usually slightly overestimate the sampling error in the TWA estimate. The $LDV_{5\%}$ of Section 6.3.2 will be slightly lower than that from 6.2.2 and the $UDV_{5\%}$ from Section 6.3.2 will be slightly higher than that of 6.2.2.

6.3.1 Confidence Limits for a True TWA Exposure

Purpose. Compute 1-sided or 2-sided confidence limits for a true TWA exposure, based on an exposure estimate calculated from multiple exposure measurements from a nonuniform exposure situation.

Assumptions. The same as used for Section 6.2.1, *except* that there is no assumption that all arithmetic mean exposures for the sampled sub-periods within the total TWA period are essentially equal.

Example. Two charcoal tubes and a personal pump were used to obtain an 8-hour TWA exposure estimate for isoamyl alcohol. The results for the two tubes were reported as: $X_1 = 30$ ppm over $T_1 = 300$ min, and $X_2 = 140$ ppm over $T_2 = 180$ min, with a CV_T of 0.08 for the measurement procedure. The *ECL* chosen by the industrial hygienist was 100 ppm. For this nonuniform exposure situation, compute the *2-sided 95 percent confidence interval* for the true 8-hour TWA exposure based on an exposure estimate of 71 ppm.

Solution.

1. If the sample duration $(T_1, T_2, \ldots, T_n)$ were approximately equal, the *2-sided 95 percent confidence interval* for the true TWA exposure could be com-

puted from the short equation:

$$LCL_{2,.95} \text{ and } UCL_{2,.95} = TWA \pm \frac{(1.96)(CV_T)(X_1^2 + \cdots + X_n^2)^{1/2}}{(n)(1 + CV_T^2)^{1/2}}$$

In this example the durations are not approximately equal, but for illustrative purposes and as a first approximation:

$$LCL_{2,.95} \text{ and } UCL_{2,.95} = 71 \pm \frac{(1.96)(0.08)(30^2 + 140^2)^{1/2}}{(2)(1 + 0.08^2)^{1/2}}$$

$$= 71 \pm 11 = 60 \text{ and } 82 \text{ ppm}$$

Thus an approximate 2-sided 95 percent *confidence interval* for the true TWA exposure is 60 to 82 ppm. The best point estimate for the true exposure is the TWA estimate of 71 ppm.

2. Since the sample durations *are not equal,* compute the more accurate *2-sided 95 percent confidence interval* for the true TWA exposure from the longer equation:

$$LCL_{2,.95} \text{ and } UCL_{2,.95} = TWA \pm \frac{(1.96)(CV_T)(T_1^2X_1^2 + \cdots + T_n^2X_n^2)^{1/2}}{(T_1 + \cdots + T_n)(1 + CV_T^2)^{1/2}}$$

In the example:

$$LCL_{2,.95} \text{ and } UCL_{2,.95} = 71 \pm \frac{(1.96)(0.08)(300^2 \cdot 30^2 + 180^2 \cdot 140^2)^{1/2}}{(300 + 180)(1 + 0.08^2)^{1/2}}$$

$$= 71 \pm 9 = 62 \text{ and } 80 \text{ ppm}$$

Thus an improved, more accurate estimate for the *2-sided 95 percent confidence interval* for the true average exposure is 62 to 80 ppm with a point estimate *(TWA)* of 71 ppm. That is, we can be 95 percent confident that the *true TWA exposure* is between 62 ppm and 80 ppm.

3. To compute either a *lower* or *upper 1-sided 95 percent confidence limit* for the true average exposure, substitute a Z value of 1.645 for 1.96 in Step 1 or Step 2.

Comments. Similar to those given in Sections 6.1.1 and 6.2.1.

6.3.2 Classification of Exposure (Hypothesis Testing)

Purpose. Classify a TWA exposure estimate, which is based on multiple exposure measurements from a nonuniform exposure situation, as *noncompliance exposure, possible overexposure,* or *compliance exposure,* relative to an *exposure control limit* (ECL).

Assumptions.   The same as used for Section 6.2.1, *except* that there is no assumption that the arithmetic mean interval exposures during the TWA period are essentially equal.

Example.   The same as used for Section 6.3.1, except classify the TWA exposure estimate of 71 ppm relative to the *ECL* of 100 ppm.

Solution.

1.   If the sample periods $(T_1, T_2, \ldots, T_n)$ are approximately equal, classify the TWA estimate using the short equation in Step 2 (for an employer) or the equation in Step 3 (for a compliance officer). For unequal sample periods, use the longer equations in Step 4 (for an employer) or Step 5 (for a compliance officer).

2.   For an **employer**, if $TWA \geq ECL$, no statistical test for compliance need be made. Since by definition the *ECL* always exceeds the *LDV*, therefore the *TWA* must also exceed the *LDV*. However, when $TWA < ECL$ and if there are approximately equal sample periods, compute the following *LDV* for use in a 1-sided hypothesis test for compliance:

$$LDV_{5\%} = ECL - \frac{[(1.645)(CV_T)(X_1^2 + \cdots + X_n^2)^{1/2}](ECL/TWA)}{(n)(1 + CV_T^2)^{1/2}}$$

For this example *the two measurement periods are not approximately equal*, but as an illustration (and as a first approximation of $LDV_{5\%}$):

$$LDV_{5\%} \cong 100 - \frac{[(1.645)(0.08)(30^2 + 140^2)^{1/2}](100/71)}{(2)(1 + 0.08^2)^{1/2}}$$

$$= 100 - 13 = 87 \text{ ppm}$$

Then classify the TWA exposure estimate based on the multiple measurements according to:

a.   If $TWA \leq LDV_{5\%}$, classify as **Compliance Exposure**.
b.   If $TWA > LDV_{5\%}$, classify as **Possible Overexposure**.

Since the *TWA* of 71 ppm is substantially less than the *approximate* $LDV_{5\%} = $ 87 ppm, the employer can state that, for the measured worker on the day and during the period of the measurement, the exposure estimate was a *Compliance Exposure*. The probability of this decision being incorrect is considerably less than 5 percent, since the *TWA* is considerably less than its critical value $(LDV_{5\%})$ for the test of compliance. However, since the example has unequal sampling periods, the $LDV_{5\%}$-equation in the following Step 4 was used to obtain a better estimate of $LDV_{5\%}$. The formula in Step 4 yields a higher critical value ($LDV_{5\%}$ = 90 ppm) for the *TWA*, but it does not change the classification decision. The

Step 4 procedure yields a more highly statistically significant decision of *Compliance Exposure* than is provided by the more conservative Step 2 test.

3.   For a **compliance officer**, if $TWA \leq ECL$ (as in this example), it is superfluous to compute a statistical test for noncompliance (see discussion in Section 6.2.2, Step 3 of the Solution). However, when there are approximately equal sample periods and $TWA > ECL$, compute the noncompliance 1-sided test statistic:

$$UDV_{5\%} = ECL + \frac{[(1.645)(CV_T)(X_1^2 + \cdots + X_n^2)^{1/2}] (ECL/TWA)}{(n)(1 + CV_T^2)^{1/2}}$$

Then classify the exposure estimate based on the multiple measurements according to:

   a.   If $TWA > UDV_{5\%}$, classify as **Noncompliance Exposure**.
   b.   If $TWA \leq UDV_{5\%}$, classify as **Possible Overexposure**.

4.   For an **employer**, when the periods of the measurements are not approximately equal and $TWA < ECL$, compute the 1-sided LDV critical value of the TWA:

$$LDV_{5\%} = ECL - \frac{[(1.645)(CV_T)(T_1^2 X_1^2 + \cdots + T_n^2 X_n^2)^{1/2}](ECL/TWA)}{(T_1 + \cdots + T_n)(1 + CV_T^2)^{1/2}}$$

Then use the classification criteria of Step 2. In this example:

$$LDV_{5\%} = 100 - \frac{[(1.645)(0.08)(300^2 \cdot 30^2 + 180^2 \cdot 140^2)^{1/2}(100/71)}{(300 + 180)(1 + 0.08^2)^{1/2}}$$

$$= 100 - 10 = 90 \text{ ppm}$$

5.   For a **compliance officer**, when the periods of the measurements are not approximately equal and $TWA > ECL$, compute the noncompliance 1-sided test statistic:

$$UDV_{5\%} = ECL + \frac{[(1.645)(CV_T)(T_1^2 X_1^2 + \cdots + T_n^2 X_n^2)^{1/2}](ECL/TWA)}{(T_1 + \cdots + T_n)(1 + CV_T^2)^{1/2}}$$

Then use the classification criteria of Step 3.

Comments.   The same as for Section 6.1.2, plus the following Statistical Note applicable to hypothesis testing for nonuniform exposures.

Statistical Note: The following notation is used. Let μ_i = true exposure during ith exposure interval of the time-averaging period, $i = 1, 2, \ldots, n$.

T_i = length of ith sampling interval

$T = T_1 + T_2 + \cdots + T_n$ = time-averaging period of the TWA and associated ECL

X_i = exposure estimate for the ith exposure interval

To test H_0: $\mu \leq ECL$ (i.e., compliance exposure) against the alternative hypothesis H_1: $\mu > ECL$ (i.e., noncompliance exposure) a critical value must be selected for the measured TWA exposure. The critical value for a test of noncompliance is intended to be the TWA which would be exceeded only 5 percent of the time when H_0 is true. The correct critical value (denoted by $UDV_{5\%}$ for the Upper Decision Value) is:

$$UDV_{5\%} = ECL + [(1.645)(CV_T)] \left[\frac{\left(\sum_{i=1}^{n} T_i^2 \mu_i^2 \right)}{(T^2)(1 + CV_T^2)} \right]^{1/2}$$

where $(1/T) \sum_{i=1}^{n} T_i\mu_i = ECL$ is a restriction on the μ_i's. A problem of this nonuniform case is that many combinations of μ_is exist which all have the desired time-weighted average (ECL) under the null hypothesis H_0. The range of choices for μ_is is discussed next and our preferred selection is given for use in the classification decision formulae given elsewhere in Section 6.3.2.

1. Maximum *statistical power* (probability of rejecting H_0 when H_1 is true) would be obtained by choosing:

$$\mu_i = \left(\frac{T}{T_i}\right) \left(\frac{ECL}{n}\right), \quad i = 1, \ldots, n$$

However, if this choice is made, the probability of falsely rejecting the null hypothesis [i.e., of falsely deciding for noncompliance (H_1) when compliance (H_0) actually exists] would be higher than the intended 5 percent. This procedure might be acceptable for the type of "screening monitoring" which is followed by confirmatory sampling. But usually this choice of μ_is under H_0 would be unsatisfactory because it gives a liberal test of noncompliance. It could also be shown to give a liberal test of compliance. It would usually not be a good choice for use by either the compliance officer or the employer.

2. Minimum *statistical power* to reject H_0 would be obtained by choosing:

$$\mu_1 = \left(\frac{T}{T_1}\right) (ECL) \quad \text{and} \quad \mu_i = 0, \quad i = 2, \ldots, n$$

The index value $i = 1$ could be assigned to any of the n intervals. All the exposure is put into this single interval and none into the other intervals. This choice provides a conservative test with *less* than a 5 percent chance of making a Type 1 decision error (falsely rejecting H_0). The advantage to making a noncompliance decision with this procedure is that the choice of μ_is would be incontrovertible. Also, if an employer could show compliance using this procedure, there would be high assurance that the ECL was not exceeded on the day of the exposure measurement. However, the conservatism of this procedure would make it undesirable for routine use by a compliance officer because it provides less statistical power to detect true overexposure situations and this could compromise worker protection.

3. On balance, it seems best to try to control the size of the hypothesis test as close to the intended 5 percent level as we can. To attempt to do this, we have substituted the following estimates of μ_is into the UDV (or LDV) formulae given earlier in this Statistical Note:

$$\hat{\mu}_i = \left(\frac{ECL}{TWA}\right)(X_i), \qquad i = 1, 2, \ldots, n$$

This procedure provides μ_is which have ECL as their time-weighted average (under H_0) and the $\hat{\mu}_i$s are in the same ratios to each other as are their sample estimates. The resulting formulae for UDV and LDV have been presented in Section 6.3.2.

6.4 Grab Samples Estimate, Small Sample Size (Less than 30 Measurements during the Time-Averaging Period)

Grab samples are samples taken during intervals that are short compared to the duration of the time-averaging period they are drawn from (e.g., intervals of seconds to about 2 minutes for a 15-minute TWA period, up to about 30 min for TWA periods of 8 hours). Grab sample measurements should be taken at random intervals during the desired time-averaging period (see Technical Appendix F of Reference 1).

Unfortunately, there are no statistical methods available to determine exact 1-sided or 2-sided confidence limits for the *arithmetic mean* of a small sample size (less than 30) grab samples exposure estimate. This is because grab sample data generally reflect a log normal distribution of true exposure levels that are sampled, see Section 3.6.2(2). Thus the statistical problem is to estimate *confidence limits* for the *true arithmetic mean* of a *log normal distribution*. It is possible to compute 1-sided or 2-sided exact confidence limits for the *true geometric mean* of a *log normal distribution* of grab sample results, but the geometric mean is not the most suitable parameter for estimation of worker exposure risk and it is an inappropriate parameter for comparison to an *exposure control limit*. Therefore, this section will present an approximate procedure for confidence limits on the true arithmetic mean exposure (Section 6.4.1). However, an exact classification procedure for grab sample estimates is available, and it will be presented in Section 6.4.2.

6.4.1. Confidence Limits for a True TWA Exposure, Small Number of Grab Samples

In their book on the log normal distribution, Aitchison and Brown (51) state that, for small sample sizes, statistical theory fails to provide a means of obtaining exact confidence limits or intervals for the arithmetic mean (μ_X) of a log normally distributed variable X. For large sample sizes, their suggestion is to compute the limits:

$$\overline{X} \pm \frac{Z_P s_X}{n^{1/2}}$$

where $Z_P = 1.96$ for 2-sided confidence limits at the $P = 0.95$ confidence level and s_X is the sample standard deviation computed from a random sample of n log normally distributed Xs (e.g., concentrations in n grab samples taken at random intervals). See Section 6.5.1 for our recommended slight modification to the Aitchison and Brown (51) large-sample ($n \geq 30$) confidence limits. We modify the formula only slightly by using Student's t statistic, $t_{P,n-1}$, in place of the standard normal deviate, Z_P.

For the small-sample case ($n < 30$), which is the subject of this section, no theory for exact confidence limits on μ_X is available. However, such theory is not needed to perform statistical tests of significance under null hypotheses of compliance ($\mu_X \leq ECL$) and noncompliance ($\mu_X > ECL$). The hypothesis tests concerning μ_X, the arithmetic mean concentration, can be carried out with the sample mean ($\overline{y}$) and sample standard deviation (s_y) of the logarithmic transformation ($y = \ln X$). A detailed explanation of these special methods for testing hypotheses about μ_X is given in Section 6.4.2. In case confidence limits on μ_X are desired in the small-sample case, an often-used approximate method is outlined here.

1. First, compute exact confidence limits on μ_y, the true mean of $y = \ln X$. These limits are given by:

$$\overline{y} - \frac{t_{P,n-1} s_y}{n^{1/2}} < \mu_y < \overline{y} + \frac{t_{P,n-1} s_y}{n^{1/2}}$$

where $t_{P,n-1}$ is the Student's t statistic for ($n - 1$) *degrees of freedom* for the P-level of significance. For example, for 2-sided 95 percent confidence limits and a sample size of 10, $t_{.95,9} = 2.262$.

2. Then, detransform each logarithmic-value confidence limit on μ_y to a corresponding value of μ_X using the relationship:

$$\mu_X = \exp(\mu_y) \cdot \exp(\sigma_y^2/2) \cong \exp(\mu_y) \cdot \exp(s_y^2/2)$$

The approximation is due to use of the sample variance, s_y^2, instead of the true variance, σ_y^2, to perform the detransformation (i.e., to convert the geometric mean of X [$GM_X = \exp(\mu_y)$] to the corresponding arithmetic mean of $X(\mu_X)$. This method is good enough for most practical industrial hygiene applications.

6.4.2 Classification of Exposure (Hypothesis Testing), Small Number of Grab Samples

Purpose. Classify an exposure estimate based on multiple grab samples as *noncompliance exposure*, *possible overexposure*, or *compliance exposure* regarding an *exposure control limit* (*ECL*).

Assumptions. *This method cannot process zero data values.* It is assumed that none of the sample results are zero. Refer to Technical Appendix I of Reference 1 for a discussion of how to treat a sample population that includes zero values.

The grab sample measurements are drawn from a single 2-parameter log normal distribution of true exposure levels. If it is suspected that the parent population of exposure levels is more accurately described by a 3-parameter log normal distribution (see Section 3.5.2), then the location constant k must be estimated (see Section 6.12, Step 5 of the Solution) and subtracted from all measurements and the *ECL* **before** the method in this section is used.

Grab sample populations also have a component of approximately normally distributed measurement procedure error, but this component is assumed negligible compared to the log normally distributed environmental variations in the true exposure levels sampled. The net effect of measurement errors is to increase slightly the variability of the log normal distribution without changing its shape appreciably.

The true arithmetic mean TWA exposure (i.e., true arithmetic mean of a 2-parameter log normal distribution of all exposure levels which occurred at the same sampling position over the duration of the TWA period) is assumed unknown and the geometric standard deviation parameter of the parent log normal exposure distribution is also assumed unknown. Both are estimated from the sample population of grab sample results.

The grab sample measurements are assumed to be a random sample of all exposure levels which occurred at the sampling position during the TWA period of interest. One should not attempt to estimate an 8-hour TWA exposure average based upon short samples selected at random from only a small portion of the TWA period (e.g., the last 2 hours). The sample periods from the TWA period should be chosen as a random sample from the entire TWA period.

The statistical theory for the method in this section is contained in Bar-Shalom et al. (83).

Example. A personal pump and 8 charcoal tubes were used to estimate a worker's 8-hour TWA exposure to ethyl alcohol. Each tube was placed on the worker for 20 minutes. The 20-minute periods were randomly selected from the 24 possible nonoverlapping periods during the work shift. The 8-hour *ECL* used by the employer is 1000 ppm. The following results were reported back from the company laboratory: X_1 = 1225 ppm, X_2 = 800 ppm, X_3 = 1120 ppm, X_4 = 1460 ppm, X_5 = 975 ppm, X_6 = 980 ppm, X_7 = 525 ppm, and X_8 = 1290 ppm. Classify the TWA exposure estimate of 1050 ppm relative to the *ECL* of 1000 ppm.

Solution.

1. Compute a standardized value (x_i) for each sample result by dividing by the *ECL* and then compute the base 10 logarithm (y_i) of each x_i. That is, y_i = $\log_{10}(X_i/ECL)$.
2. In this example:

Original Results	Standardized Results	Logs (y_i)
1225 ppm	1.225	0.0881
800	0.800	− 0.0969
1120	1.120	0.0492
1460	1.460	0.1644
975	0.975	− 0.0110
980	0.980	− 0.0088
525	0.525	− 0.2798
1290	1.290	0.1106

3. Compute the Classification Variables, $\bar{y}$, s, and n. First compute the arithmetic mean $\bar{y}$ of the logarithmic values y_i, then compute the standard deviation s of the y_i values. The number of grab samples used in the computations is n.

In this example: $\bar{y}$ = 0.0020, s = 0.1400, and n = 8. These three variables, two sample estimates of parent population parameters and the sample size, will be used in the classification procedure.

4. Using the Figure 8.7 Classification Chart, plot the classification point which has coordinates $\bar{y}$ and s. A family of curves form the boundaries of 3 classification regions. Each pair of boundaries is for a given sample size n. Pairs of classification boundaries for odd values of n ranging from n = 3 to 25 are provided (all even sample sizes except for 4 will have to be interpolated).

5. Classify the TWA exposure:
    a. If the classification point lies on or above the upper curve of the pair corresponding to the number of measurements n (i.e., the n-curve pair), then classify as **Noncompliance Exposure**.
    b. If the classification point lies below the lower curve of the n-curve pair, then classify as **Compliance Exposure**.
    c. If the classification point is between the two curves for sample size n, then classify as **Possible Overexposure**.

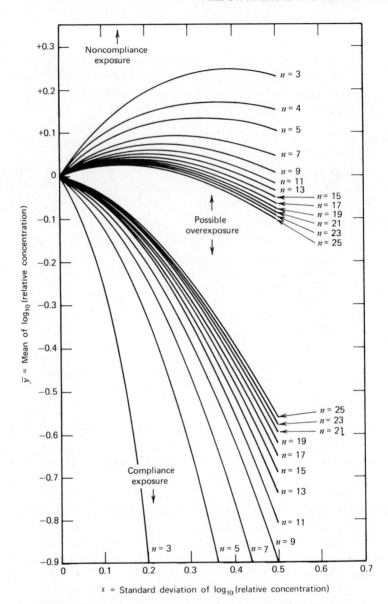

Figure 8.7 Grab sample measurement average classification chart.

d. If s exceeds 0.5 (greater than the range of the classification chart), a possible cause is that one or more of the measurements is relatively distant from the main portion of the sample distribution (i.e., an outlier which could be a mistake). Another explanation is that two or more substantially different exposure level distributions are being mistakenly analyzed as a single distribution (see Section 3.3.4).

6.   In this example the classification point lies between the upper interpolated $n = 8$ curve and the lower interpolated $n = 8$ curve. Thus the TWA exposure of 1050 ppm, estimated from the 8 grab sample measurements, is classified as a **Possible Overexposure** relative to the ECL of 1000 ppm. That is, the estimate is not precise enough to be able to confidently say that the true exposure was not in compliance.

Comments.   The ratios (X_i/ECL) computed in Step 1 of the Solution are standardized exposure values which make the concentrations of contaminant independent of the ECL (in concentration units). This enables us to use the same Decision Chart for any concentration level. All values $x_i = (X_i/ECL)$ are comparable to a single scale of compliance with an ECL of unity for x_i. That is, the ECL for the transformed variable x will always be unity, which corresponds to zero on the $y = \ln(x)$ ordinate scale of the Figure 8.7 Decision Chart.

6.5   Grab Samples Estimate of the TWA Exposure, Large Sample Size (30 or More Measurements during the Time-Averaging Period)

Usually one collects far fewer than 30 measurements during an 8-hour TWA period, or 15-minute TWA period, because of the cost of each measurement (even for inexpensive ones such as with colorimetric tubes) and limited availability of personnel to take the measurements. However, if one has a direct reading instrument available for the contaminant of interest, then it is feasible to obtain more than 30 samples during the desired TWA period. If the larger number of samples can be taken at random intervals, this strategy is preferable to the *small sample size* approach (less than 30) discussed in Section 6.4, since for larger sample sizes the uncertainty regarding the true TWA exposure is considerably less. Additionally, for sample sizes of 30 or more, the statistical analysis is less complex because the distribution of the average of the exposure measurements is adequately described by a normal distribution. This section will present methods for computing both confidence limits and hypothesis tests for TWA exposure estimates calculated from large sample sizes. One does not have to calculate the logarithms of the standardized measurements (as was necessary in Section 6.4 for TWA estimates based on small sample sizes). The hypothesis tests are less complex and this method can process zero data values, unlike the procedure in Section 6.4.

6.5.1   Confidence Limits for a True TWA Exposure, ≥ 30 Grab Samples

Purpose.   Compute 1-sided or 2-sided confidence limits for the TWA exposure, based on 30 or more grab samples.

Assumptions.   This procedure is robust regarding the actual distribution of exposure levels occurring during the TWA period of the grab sample measurements. Both the true exposure level (i.e., arithmetic mean of the distribution of all exposure levels which occurred at the sampling position over the duration of the TWA period) and the standard deviation of the parent exposure distribution are assumed unknown. Both are estimated from the results obtained from grab samples.

The grab sample measurements are assumed to be a random sample of all exposure levels which occurred at the sampling position during the TWA period of interest. One should not attempt to estimate an 8-hour TWA exposure average based on short samples selected at random from only a small portion of the TWA period (e.g., the last 2 hours). The sample periods from the TWA period should be chosen as a random sample from the entire TWA period.

Example. A direct-reading ozone meter with strip chart recorder was used to continually measure a worker's exposure to ozone. The following 35 measurement values were read off the strip chart record at 35 times randomly selected within the 8-hr period. All values are given in ppm.

$$
\begin{array}{ccccccc}
0.084 & 0.062 & 0.127 & 0.057 & 0.101 & 0.072 & 0.077 \\
0.0145 & 0.084 & 0.101 & 0.105 & 0.125 & 0.076 & 0.043 \\
0.079 & 0.078 & 0.067 & 0.073 & 0.069 & 0.084 & 0.061 \\
0.066 & 0.085 & 0.080 & 0.071 & 0.103 & 0.075 & 0.070 \\
0.048 & 0.092 & 0.066 & 0.109 & 0.110 & 0.057 & 0.107
\end{array}
$$

Compute the *2-sided 95 percent confidence limits* for the true TWA exposure based on the sample mean estimate of 0.0794 ppm ozone.

Solution.

1. Compute the arithmetic mean $\overline{X}$ and the standard deviation s of the $n = 35$ measurements X_i. Here, $\overline{X} = 0.0794$ ppm, $s = 0.0233$ ppm.
2. Compute the *2-sided 95 percent confidence limits* $(LCL_{2,.95})$ and $(UCL_{2,.95})$ for the true TWA exposure, based on the TWA exposure estimate $(\overline{X})$ from the equations:

$$
LCL_{2,.95} = \overline{X} - \frac{(t_{.95,n-1})(s)}{n^{1/2}}
$$

$$
UCL_{2,.95} = \overline{X} + \frac{(t_{.95,n-1})(s)}{n^{1/2}}
$$

Note that the factor $t_{.95,34} = 2.032$ will be different for sample sizes other than 35. See the Comments section that follows the Solution.
3. In this example:

$$LCL_{2,.95} = 0.0794 - \frac{(2.032)\,(0.0233)}{(35)^{1/2}}$$

$$= \{0.0794 - [0.0080]\} = 0.071 \text{ ppm, and}$$

$$UCL_{2,.95} = 0.0794 + \frac{(2.032)(0.0233)}{(35)^{1/2}}$$

$$= \{0.0794 + [0.0080]\} = 0.087 \text{ ppm.}$$

Thus the *2-sided 95 percent confidence interval* for the true TWA exposure based on an exposure estimate of 0.079 ppm obtained from 35 samples is 0.071 to 0.087 ppm. That is, we can be 95 percent confident that the *true TWA exposure* is between 0.071 and 0.087 ppm.

Comments.   The multiplier 2.032 in the equations for $LCL_{2,.95}$ and $UCL_{2,.95}$ is taken from the Student's t table for $(n - 1) = 34$ *degrees of freedom*. As an approximation to Student's t, the normal distribution value of 1.96 can be used as a multiplier for the larger n-values appropriate for this procedure $(n \geq 30)$.
    The indicated uncertainty in the TWA exposure estimate reflected in the width of the confidence interval (± 10 percent of the TWA estimate) *does not* incorporate uncertainty due to *interday* exposure level variability [see Sections 3.3.1 and 3.6.2(3)]. The confidence interval width reflects only the *intraday* variability of the exposure levels occurring over the period of the grab sample measurements. The computed confidence interval is valid *only* for the worker's TWA exposure on the day and during the period of the grab samples, at the same sampling position.

6.5.2   Classification of Exposure (Hypothesis Testing), $\geq$ 30 Grab Samples

Purpose.   Classify a TWA exposure estimate based on 30 or more grab sample measurements as *noncompliance exposure, possible overexposure,* or *compliance exposure* relative to an *exposure control limit (ECL)*.

Assumptions.   The same as used for Section 6.5.1

Example.   The same as used for Section 6.5.1, except classify the TWA exposure estimate relative to an *ECL* of 0.1 ppm.

Solution.

    1.   As in the Section 6.5.1, compute the arithmetic mean $\overline{X}$ (the estimated *TWA*) and standard deviation s of the $n = 35$ measurements X_i. Here, $\overline{X} = 0.0794$ ppm, $s = 0.0233$ ppm.

2.   For an **employer**, if $\overline{X} > ECL$, no statistical test for compliance need be made, since the exposure estimate exceeds the *ECL. If* $\overline{X} \leq ECL$, compute the 1-sided Lower Decision Value for comparison with the sample mean:

$$LDV_{5\%} = ECL - \frac{(1.691) \ (s)}{n^{1/2}}$$

$$= 0.1000 - \frac{(1.691)(0.0233)}{(35)^{1/2}}$$

$$= 0.093 \text{ ppm}$$

Note that the multiplier 1.645 from the normal distribution could have been used as an approximation to the correct value of 1.691 from the Student's t table corresponding to $(n - 1)$ 34 degrees of freedom.

Then classify the exposure estimate based on the single measurement according to:

    a.   If $\overline{X} \leq LDV_{5\%}$, classify as **Compliance Exposure**.
    b.   If $\overline{X} > LDV_{5\%}$, classify as **Possible Overexposure**.

Since the $\overline{X} = 0.079$ ppm is less than the $LDV_{5\%}$ of 0.093 ppm, the employer can state that for the measured worker, on the day and during the period of the measurement, the exposure estimate was a statistically significant *Compliance Exposure* at the 5 percent significance level. To check for even higher degrees of statistical significance, $LDV_{1\%}$ and $LDV_{0.1\%}$ could also be computed by using $t_{.99,n-1}$ and $t_{.999,n-1}$, respectively, as substitutes for $t_{.95,n-1}$ in the equation for *LDV*.

3.   For a **compliance officer** in this example, *TWA < ECL*, so no statistical test for noncompliance need be made, since the TWA estimate is less than the *ECL* and hence certainly less than the *UDV*, For *TWA > ECL*, the compliance officer would compute the noncompliance 1-sided *Upper Decision Value* for comparison with the sample mean:

$$UDV_{5\%} = \{ECL + [(t_{.95,n-1}]) \ (s)/(n)^{1/2}]\}$$

where $t_{.95,n-1}$ is the 95th percentile (1-sided) of Student's t distribution with $(n - 1)$ *degrees of freedom*. Note that 1.645 can be used as an approximation for $t_{.95,n-1}$, since $n \geq 30$ when this procedure is used. Then classify the TWA exposure extimate according to:

    a.   If $\overline{X} > UDV_{5\%}$, classify as **Noncompliance Exposure**.
    b.   If $\overline{X} \leq UDV_{5\%}$, classify as **Possible Overexposure**.

Comments.   The outcomes of these hypotheses tests are valid only for the worker's exposure on the day and during the total period grab sampled at random intervals, at the same sampling position.

6.6. Tolerance Limits for a Normally Distributed Variable with Unknown Parameters

Purpose.   Based on a random sample of n observations, compute 1-sided or 2-sided tolerance limits which have a desired degree of confidence (probability) that the tolerance interval they bound will contain a specific proportion of the normal distribution represented by the random sample.

Assumptions.   The parent population for which the tolerance limit statement will be made is assumed to be adequately described by a normal distribution.

Both the true mean and true standard deviation parameters of the parent normal distribution are assumed to be unknown. These two distributional parameters are estimated by sample values, $\overline{X}$ and s, computed from a random sample of n observations.

A tolerance limit can be thought of as a confidence limit for a designated percentile of the parent distribution of single observations. However, note that this procedure for estimating tolerance limits is less "robust" (against lack of normality of the parent population) than procedures for estimating confidence limits on the arithmetic mean distributional parameter. This is true because sample means tend to have a "bell-shaped" (nearly normal) distribution even if the single observations were not from a normal distribution (at least for moderate-to-large sample sizes). Thus, one must be cautious and avoid over-interpretation of tolerance limits. That is, tolerance limits should not be used for "fineline" decision-making. A useful rule-of-thumb is that they probably should not be reported to more than two significant figures. The primary use of these methods should be to obtain an indication of the potential upper levels of a parent distribution (e.g., by computation of a *1-sided upper tolerance limit* for the upper 95th percentile) or an indication of the potential width of the parent distribution (e.g., by computation of a *2-sided tolerance interval* for the central 90 percent).

Example.   An exposure measurement procedure was used to repeatedly measure a calibrated reference concentration of 100 ppm. The six replicate measurements were: 95.7, 90.9, 109.4, 107.6, 101.1, and 84.7 pm. Compute a *2-sided, 95 percent confidence level, tolerance interval* for the central 95 percent of the parent distribution of possible measurements (of a "true" concentration) of 100 ppm.

Solution.

1.   The mean will not be assumed known since any measurement method systematic error could cause the true mean of measurements to differ from 100 ppm. Compute the arithmetic mean $\overline{X}$ and standard deviation s of the $n = 6$ measurements X_i. Here, $\overline{X} = 98.2$ ppm and $s = 9.63$ ppm.

2.   Compute the *2-sided 95 percent confidence level upper and lower tolerance*

limits (*UTL, LTL*) for the central 95 percent of the parent population from the equations:

$$UTL_{2,\gamma,P} = \overline{X} + (K)(s), \quad LTL_{2,\gamma,P} = \overline{X} - (K)(s),$$

where K is obtained from appropriate tables for 2-sided tolerance limits, as a function of (a) sample size ($n = 6$ in this example), (b) confidence level ($\gamma = 0.95$), and (c) proportion of the population ($P = 0.95$).

3.   In this example K = 4.414 and the calculations are:

$$UTL_{2,.95,.95} = 98.2 + (4.414)\,(9.63)$$

$$= 98.2 + 42.5$$

$$= 141 \text{ ppm}$$

$$LTL_{2,.95,.95} = 98.2 - (4.414)\,(9.63)$$

$$= 98.2 - 42.5$$

$$= 56 \text{ ppm}$$

4.   Thus the *2-sided tolerance interval* (*95 percent confidence level, central 95 percent of the parent population*) is 56 to 141 ppm. That is, based on the frugal sample of six replicate measurements of a 100 ppm reference concentration, the best we can say about the measurement procedure is that we can be 95 percent confident that 95 percent of future results with this procedure at 100 ppm will lie between 56 and 141 ppm (under conditions similar to those during which the replicate samples were obtained).

Comments.   Tolerance limits can be thought of as γ-level "confidence limits" for a fractile interval which contains the designated proportion P of the parent distribution.

Note that we *cannot* say that there is a 95 percent *probability* of 95 percent of the parent distribution lying in the interval 56 to 141 ppm. For any given computed tolerance interval such as this one, the interval either does or does not contain the stated percentage of the parent population. However, since 95 percent of such tolerance intervals would each contain at least 95 percent of the parent population, chances are 19 to 1 that this particular tolerance interval does in fact have 95 percent or more of the distribution.

Frequently researchers or writers will erroneously state that 95 percent of their population of sample results will lie in the interval ($\overline{X} - 2s$) to ($\overline{X} + 2s$). (Generally they will use 2.000 as an approximation to the more exact value of 1.960 for the standard normal deviate.) This type of inferential statement is true *only* for large (i.e., in the hundreds) sample sizes. Typically, researchers only have small sample sizes (e.g., less than 30) and the K-factors for the appropriate 2-sided tolerance limit computations are substantially greater than 2. Thus it is possible to substantially underestimate the width of a parent distribution from small sample sizes if the proper tolerance interval computations are not performed. If we had naively estimated the central 95 percent of the results as bounded by $(\overline{X} \pm 2s)$, then we would have substantially underpredicted the possible range of future measurement method results as about 79 to 117 ppm.

The range of the six results was about 91 to 109 ppm, yet the 2-sided tolerance interval for 95 percent of the results was 56 to 141 ppm. This wide tolerance interval is partially due to the frugal sample size used for its computation. However, it also demonstrates how poor an indicator the range of a small sample can be for the span of a normal distribution.

6.7 Tolerance Limits for a Log Normally Distributed Variable with Unknown Parameters

Purpose.   From n samples taken at random, compute 1-sided or 2-sided tolerance limits, which have a desired level of confidence (γ) that the tolerance interval they bound will contain a desired proportion (P) of the log normal distribution from which the sample was taken.

Assumptions.   There is a value of the response variable for each member of a large parent population of "sampling units" to which the tolerance limits statement will apply. Examples of sampling units could be replicate workers (same sampling period and same type of work) or repeated exposure periods for a given worker. The corresponding population of exposure measurements is assumed to be adequately described by a single 2-parameter log normal distribution. If it is suspected that the parent population's exposure measurements are more accurately described by a 3-parameter log normal distribution (see Section 3.5.2), then the location constant k for the parent distribution must be estimated (see Section 6.12, Step 5 of the Solution) and subtracted from all sample values *before* the method in this section is used.

Both the geometric mean and geometric standard deviation of the parent population distribution are unspecified and assumed unknown. Both distributional parameters are to be estimated from response variable measurements made on a random sample of n sampling units selected from the parent population.

This method cannot process zero data values. It is assumed that none of the sample results are zero (or less than zero). Refer to Technical Appendix I of (1) for a discussion of how to treat a sample population that includes zero values.

Note that this procedure for estimating tolerance limits is less robust than procedures for estimating confidence limits for the geometric mean distributional parameter. The reason for this is explained in Section 6.6 relevant to tolerance limits for a normally distributed variable. We would expect that similar comments would apply here, since tolerance limits for a log normally distributed variable X are merely a detransformation of tolerance limits for the normally distributed variable $y = \ln X$. Thus, robustness which exists for the sample mean $(\bar{y})$ of the log transformation also exists for its detransformation $[GM_X = \text{antilog } (\bar{y})]$.

Example.   Five workers were randomly selected from a target population of workers. Each selected worker's 8-hour TWA exposure was measured on a different workday, randomly selected within a 6-month period. The results were: 11.1, 10.6, 21.4, 3.9, and 4.9 ppm. Compute the *1-sided 95 percent confidence level upper tolerance limit* for the lower 95 percent of a parent distribution of TWA exposures, which is defined by randomly selected workers measured on randomly selected days.

Solution.   Assume that general (i.e., group average) exposure levels (X) for the work environment are log normally distributed among workdays and that percentage differences among individual workers' exposures are similar on each workday. Workers' mean exposure levels are also log normally distributed. Then, the net variability due to both sources of variability (i.e., due to days and workers) is also log normally distributed. Under this model, the following analysis is appropriate.

1.   Compute the base 10 logarithm, $y_i = \log X_i$, of each sample value. Here the five logarithmic values are 1.045, 1.025, 1.330, 0.591, and 0.690.
2.   Compute the sample mean $\bar{y}$ and standard deviation s of the $n = 5$ logarithmic values. Here, $\bar{y} = 0.936$, and $s = 0.298$. As supplementary information, one can compute the sample *geometric mean* of X (GM) by taking the antilog$_{10}$ of $\bar{y}$ ($10^{0.936} = 8.63$ ppm) and the sample *geometric standard deviation* (GSD) by taking the antilog$_{10}$ of s ($10^{0.298} = 1.98$).
3.   Compute the *1-sided, 95 percent confidence level, upper tolerance limit* $(UTL_{1,.95,.95})$ for the lower 95 percent of the defined population of worker/day combinations exposure levels (or compute $LTL_{1,.95,.95}$, the lower 95 percent tolerance limit for the upper 95 percent of the distribution). Use the equations:

$$UTL_{1,.95,.95} = \text{antilog}_{10} \left[\bar{y} + (K)(s) \right]$$

$$LTL_{1,.95,.95} = \text{antilog}_{10} \left[\bar{y} - (K)(s) \right]$$

where K is obtained from appropriate tables of factors for 1-sided tolerance limits.
4.   In this example K = 4.202 and the calculations are:

$$UTL_{1,.95,.95} = \text{antilog}_{10} \left[0.936 + (4.202)(0.298) \right]$$

$$= \text{antilog}_{10} \left[2.188 \right] = 10^{2.188}$$

$$= 154 \text{ ppm}$$

5.   Thus the *1-sided upper tolerance limit (95 percent confidence level, lower 95 percent of the defined population)* is 154 ppm. That is, based on the frugal sample

of 5 exposure estimates over a 6-month period for the target population of workers, the best we can say is that we are 95 percent confident that 95 percent of the daily TWA exposures over the 6-month period were below 154 ppm.

Comments.   This type of 1-sided tolerance limit can be thought of as a "confidence limit" for the value of a log normally distributed random variable which is exceeded by *specified proportion* of the population of values. A companion technique is presented in Section 6.8 for computing a point estimate (and associated confidence limits) for the proportion of a log normal distribution exceeding a *specified value* of the random variable.

Note that we *cannot* say that there is a 95 percent *probability* that 95 percent of the parent distribution was below 154 ppm. For any given computed 1-sided tolerance limit such as this one, the limit either does or does not bound the stated percentage of the parent population. However, since 95 percent of such tolerance intervals would each contain at least 95 percent of the parent population, chances are 19 to 1 that this particular tolerance interval does in fact have 95 percent or more of the distribution.

The range of the five results was 4 to 21 ppm, yet the 1-sided upper tolerance limit for the lower 95 percent of the results was 154 ppm. The high $UTL_{1,.95,.95}$ is primarily due to the frugal sample size used for its computation. Note that the tolerance limit K-factor, which is 4.202 in this example, would be only about half as large if n were 50 instead of the 5 in this example. This demonstrates how poor an indicator the range of a small sample is for the span of a log normal distribution, since the distribution is skewed toward higher values.

To be able to partition the total variability (GSD) into components due to exposure variability between workers and exposure variability between days, it would be necessary to measure exposures of several workers on each of several days and examine the data by an appropriate analysis of variance method (ANOVA), using a logarithmic transformation of the exposure concentrations as the response variable.

Finally, note that in some real exposure settings, there may be "interaction" (i.e., lack of independence) between interworker exposure variations and interday exposure variations. In such cases, the interworker exposure distributions would have to be determined separately for each workday. For the solution to the example, it was assumed that equal percentage differences among individual workers' true exposures exist on each workday, which is equivalent to assuming "no interaction" on the scale of the logarithmic transformation.

6.8   Point Estimate and Confidence Limits for the Proportion of a Log Normally Distributed Population (Unknown Parameters) Exceeding a Specific Value

Purpose.   Compute a point estimate for the proportion P of a 2-parameter log normal distribution exceeding a specified value, when given a random sample from the distribution.

Assumptions.   Values of the response variable (e.g., single or multiple worker 8-hr TWA exposure estimates) for the parent population are assumed to be adequately described by a single 2-parameter log normal distribution. If it is suspected that the parent population is more accurately described by a 3-parameter log normal (see Section 3.5.2), then the location constant k for the parent distribution must be estimated (see Section 6.12, Step 5 of the Solution) and subtracted from all sample values *before* the method in this section is used.

Both the geometric mean and geometric standard deviation of the parent population distribution are unspecified and assumed unknown. Both distributional parameters are to be estimated from response variable measurements made on a random sample of n sampling units selected from the parent population.

This procedure cannot process zero data values. It is assumed that none of the sample results are zero (or less than zero). Refer to Technical Appendix I of Reference 1 for a discussion of how to treat a sample population that includes zero values. Note that this procedure for computing a point estimate of the proportion of a log normal distribution exceeding a specified value is less robust than procedures for estimating the geometric mean distributional parameter. One must be cautious and avoid overinterpretation of these point estimates and their confidence limits. A point estimate of distributional tail area is sensitive to departures from the assumption of log normality and these estimates should not be used for "fineline" decision-making. A useful rule-of-thumb is that estimated tail areas (probabilities) should not be reported to more than two decimal places. The primary use of these methods should be to obtain an indication of the potential frequency of high exposure levels from a parent log normal distribution.

The procedure assumes that the exposure estimates are from a "stable" parent distribution. Experience, professional judgment, and knowledge of the exposure level determinant factors must be relied on here for assurance of the validity of this assumption. Only current sample data that represent a "stable" exposure situation should be used in the following computations. One way of assuring that this condition is met is to plot the sample data on semilogarithmic paper (see Section 6.11). If the data are judged as trending upward (or downward) with time, then this procedure *should not be used* because an erroneous point estimate could result. Only if the long-term exposure plot appears "level" (after measurement errors have been smoothed out) should one use this procedure.

Example.   Initial monitoring of a target population of 35 workers exposed to chromic acid mist and chromates was done by sampling one worker on each of six days. A different worker was selected at random each day for exposure measurement. The monitoring yielded the following six 8-hour TWA exposure estimates for six different workers: 0.105, 0.052, 0.082, 0.051, 0.180, and 0.062 mg/cu m. Compute a point estimate P of the proportion of 8-hour TWA exposures experienced by the 35 workers that have exceeded an exposure control limit (*ECL*) of 0.10 mg/cu m and the associated 1-sided 95 percent upper confidence limit ($UCL_{1,.95}$) for this proportion.

Solution. As in the tolerance limit example of Section 6.7, we will assume that true exposures for randomly selected worker-day combinations are affected by two types of independently distributed, log normally distributed, random variations: (a) log normally distributed daily geometric means (over all workers in the target population) and (b) log normally distributed, multiplicative (proportional) factors which could be used to adjust daily means for ratios between consistently different exposure of individual workers. To be able to use a *single* log normal distribution as an appropriate model for the type of exposure monitoring data collected in these two examples, it is essential to do the exposure monitoring accordingly to a particular scheme whereby each randomly selected worker is sampled on only a *single* randomly selected day, and only *one* worker is measured each day. If a given worker were sampled repeatedly, the exposure measurements would be intercorrelated (as opposed to independent) samples and a more complex approach (i.e., ANOVA) would be needed in order to properly identify the day-to-day and worker-to-worker components of total variability.

1. Compute the *sample geometric mean (GM)* and *geometric standard deviation (GSD)* for the sample of *n* results. For the *n* = 6 8-hr TWA exposure estimates, the *sample GM* = 0.080 mg/cu m and the *sample GSD* = 1.627.

2. Compute $g(ECL) = [\log(ECL) - \log(GM)]/\log(GSD) = 0.458$.

3. From a table of areas (*P*) under the standard normal curve from *Z* to plus infinity, obtain the *P*-value for *Z* = *g(ECL)*. For *g(ECL)* = 0.458, *P* = 0.32. This is the point estimate of *P*, which is the integral of the standard normal curve from *g(ECL)* to plus infinity. Note that this quantitative point estimate of the probability *P* should be about the same as an estimate obtained with an approximate graphical plotting technique described in Section 6.12 (i.e., a probability estimate which is the indicated probability at the intersection of an exposure distribution line and an exposure control limit line).

4. To compute confidence limit(s) on the point estimate *P*, first decide whether 1-sided or 2-sided limits are desired and select the confidence level for the computation. Then from a table of *Z* values, obtain the related *Z* value for the desired confidence level. For example, to compute a 1-sided 95 percent $UCL_{1,.95}$ (or 1-sided 95 percent lower confidence limit, $LCL_{1,.95}$) use a *Z* value of 1.645 (or −1.645). For 2-sided 95 percent limits use ±1.960, and for 2-sided 99 percent limits use ±2.576.

5. The computation of the confidence limit(s) for the true value of the point estimate *P* involves the solution to a quadratic equation. The two needed quadratic roots are given by:

$$U = [-b \pm (b^2 - 4ac)^{1/2}]/(2a),$$

where

$$a = \left[\left(\frac{1}{2n-3} \right) - \left(\frac{1}{Z^2} \right) \right], \quad b = \left[\left(\frac{2g}{Z^2} \right) \left(\frac{2n-3}{2n-2} \right)^{1/2} \right],$$

and

$$c = \left[\left(\frac{1}{n} \right) - \left(\frac{g^2}{Z^2} \right) \left(\frac{2n-3}{2n-2} \right) \right]$$

The two values of U are standard normal variables (i.e., Z values) corresponding to the lower and upper confidence limits for the true value of P. The use of the larger value from the U equation leads to a 1-sided or 2-sided (depending on the Z value selected) LCL on the area to the right of ECL (i.e., an LCL on P). Use of the smaller value from the U equation leads to a UCL for P.

In this example, we will select $Z = +1.645$ to obtain a 1-sided 95 percent upper confidence limit. Since $n = 6$ and $g = 0.458$ in this example, the 3 intermediate functional variables for U are $a = -0.258$, $b = 0.321$, and $c = 0.0969$. The resulting two values for U are -0.251 and $+1.494$. The smaller value (-0.251) corresponds to a probability of $P = 0.599$ or about 0.60, which is the $UCL_{1,.95}$ for P.

6. We now have computed the point estimate of 0.32 for P (proportion of overexposures) and a 1-sided 95 percent upper confidence limit of 0.60 for the true proportion. Given the assumptions of this inferential method, our best estimate is that 32 percent of the 8-hour TWA exposures experienced by the 35 workers (on the 45 days of exposure which were sampled at random) exceeded the exposure control limit of 0.10 mg/cu m. However, this point estimate for proportion of worker-days of overexposure was based on a frugal sample of only six exposure estimates (one worker on each of six days) and the 1-sided 95 percent upper confidence limit for this estimate is 60 percent. Thus we should state that the true (actual) proportion of worker-days exceeding the exposure limit could have been as high as 60 percent.

Comments. A companion technique for computing a 1-sided tolerance limit (e.g., the value which has 95 percent confidence of cutting off at least the lower 95 percent of the distribution) was presented in Section 6.7.

If one had good reason to assume that a particular worker's exposure distribution over some long period (e.g., 2 months) is the same as the target population's distribution, then one could infer that any given worker was overexposed on 32 percent of the exposure days, but could have been overexposed as many as 60 percent of the exposure days. However, one must recognize that individual worker overexposure risks (resulting from individual exposure distributions) can be "masked" by the exposure distribution for the target population. For example, the lower portion of the target population exposure distribution might be created by workers with consistently low exposures, perhaps due to better work practices. Then these "lower exposure tail" workers would have individual overexposure risks *lower* than the 32 percent estimated for the group as a whole. Correspondingly, the upper portion of the group's exposure distribution might be due to workers with consistently high exposures, perhaps due to "dirtier" job locations. These "upper exposure tail" workers would have individual overexposure risks *higher* than the 32 percent estimated for the entire group.

6.9 Fractile Intervals of a Normally Distributed Population with Known Parameters

Purpose. Compute fractile intervals for a population adequately described by a specific normal distribution.

Assumptions. The population for which the fractile interval statement will be made is assumed to be adequately described by a normal distribution. Both the arithmetic mean and standard deviation of the population distribution are assumed known so that neither distributional parameter will be estimated from sample data. Note that these assumptions are rarely met in practice, but this procedure is presented for illustrative purposes as a teaching aid to assist in understanding the normal distribution model. The normal tolerance limits procedure of Section 6.6 should be used where the distributional parameters are estimated from typical industrial hygiene sample sizes (e.g., less than 50). If there is any question whether one has a sufficient sample size to compute fractile intervals with this procedure, use the normal tolerance limit procedure.

Example. Suppose that on a particular day, a worker's true 8-hour TWA exposure is known to be 25 ppm. On that day, if the worker's exposure were measured using a procedure with a total coefficient of variation of 10 percent ($CV_T = 0.10$), we would expect replicate measurements to follow a normal distribution with mean $\mu = 25$ ppm and standard deviation $\sigma = (\mu)(CV_T) = 2.5$ ppm (see Sections 3.4 and 3.6.1). Compute the lower and upper population values bounding the central 95 percent of the normal distribution of possible measurement values. These boundaries are the 2.5 percent and 97.5 percent fractiles of the normal distribution. There is a 95 percent probability that any single measurement will lie within the interval.

Solution.

1. Table 8.2 indicates that the X-interval for the central 95 percent of a normal distribution is from ($\mu - 1.96\sigma$) to ($\mu + 1.96\sigma$).
2. In this example the central 95 percent interval is bounded by 25 ppm $\pm$ (1.96)(2.5 ppm) = 25 ppm $\pm$ 4.9 ppm, which is 20.1 to 29.9 ppm.
3. Thus, the central interval enclosing 95 percent of the possible measure-

ments, assuming a normal distribution with mean value equal to the true TWA exposure of 25 ppm and CV_T of 10 percent, is 20.1 to 29.9 ppm. We can also say that there is a 95 percent probability that a single measurement will lie within the interval 20.1 to 29.9 ppm.

4. If we are interested in the lower 95th percentile, Table 8.2 indicates that 95 percent of the measurements will be below $(\mu + 1.645\sigma)$. In this example, the lower 95 percent fractile is bounded by $[25 \text{ ppm} + (1.645)(2.5 \text{ ppm})] = [25 + 4.1] \text{ ppm} = 29.1 \text{ ppm}$.

5. Thus the lower 95 percent of the normal distribution of possible measurements of a true TWA exposure of 25 ppm consists of all measurements at or below 29.1 ppm. We can also say that there is a 95 percent probability that a single measurement of a true TWA of 25 ppm will be *less than* 29.1 ppm.

6.10   Fractile Intervals of a Log Normally Distributed Population with Known Parameters

Purpose.   Compute fractile intervals for a population adequately described by a specified log normal distribution.

Assumptions.   The population for which the fractile interval statement will be made is assumed to be adequately described by a 2-parameter log normal distribution.

Both the geometric mean and geometric standard deviation of the population distribution are assumed known so that neither distributional parameter will be estimated from sample data. Note that these assumptions are rarely met in practice, but this procedure is presented for illustrative purposes as a teaching aid to assist in understanding the log normal distribution model. The 2-parameter log normal tolerance limits procedure of Section 6.7 should be used where the distributional parameters are estimated from typical industrial hygiene samples sizes (e.g., less than fifty).

Example.   Suppose that the interday variability of a worker's 8-hour TWA exposures is adequately described by a known 2-parameter log normal distribution with geometric mean (μ_g) of 20 ppm and geometric standard deviation (σ_g) of 1.9 (see Sections 3.5 and 3.6.2). Note that the geometric standard deviation does not have concentration units attached to it (unlike an arithmetic standard deviation, which has the same units as the corresponding arithmetic mean). Compute the lower and upper daily exposure values bounding the central 90 percent of the daily exposure distribution. These boundaries are the 5 percent and 95 percent fractiles of the 2-parameter log normal distribution.

Solution.

1. Table 8.3 indicates that the X-interval for the central 90 percent of a 2-parameter log normal distribution lies in the proportional interval $(\mu_g)/(\sigma_g^{1.645})$ to $(\mu_g)(\sigma_g^{1.645})$.

2. In this example the central 90 percent interval lies between (20 ppm)/ $(1.9^{1.645})$ and (20 ppm) $(1.9^{1.645})$, which is 7.0 to 57.5 ppm.

3. Thus the central 90 percent interval, for the 2-parameter log normal distribution of 8-hour TWA exposures with a true geometric standard deviation of 1.9, is about 7 to 57 ppm. That is, 90 percent of the daily TWAs would be within the interval 7 to 57 ppm. We can also say that there is a 90 percent probability that a single daily exposure will lie within the interval 7 to 57 ppm. Note that this central interval of the 2-parameter log normal distribution was computed to be *balanced* in the sense that it has equal percentages (5 percent) in each of the left and right tails of the distribution, which lie outside the central interval.

4. If we are interested in the lower 95th percentile, Table 8.3 indicates that 95 percent of daily TWA exposures will be below (μ_g) $(\sigma_g^{1.645})$ In this example, this lower 95th percentile is equal to (20 ppm) $(1.9^{1.645})$, which is 57.5 ppm.

5. Thus the lower 95 percent of the 2-parameter log normal distribution consists of the daily 8-hour TWA exposures at or below 57.5 pm. That is, 95 percent of the daily TWAs will be at or below 57.5 ppm. We can also say that there is a 95 percent probability that a single daily exposure will be *less than* 57.5 ppm.

6.11 Use of Semilogarithmic Graph Paper to Make Time Plots of Variables Believed To Be Log Normally Distributed in Time

Purpose. Plot data that is believed to be log normally distributed in time on semilog graph paper to check for possible cycles or trends with time.

Assumptions. This procedure is useful to qualitatively look for exposure trends or cycles in time, when the data distribution is skewed right (higher values). No quantitative inferences are made with this procedure. To produce a nearly symmetrical data plot for log normally distributed data, the midpoint of the logarithmic scale should be given a value close to the geometric mean of the sample data. Data approximately described by a 2-parameter log normal model (see Section 3.6.2) should show symmetrical random variability around the geometric mean (i.e., with no apparent trend or other systematic pattern in time).

Note that zero data values *cannot* be displayed with this procedure.

Example. A worker's exposure to chromic acid mist and chromates was measured on 7 February. The 8-hour TWA exposure estimate of 0.105 mg/cu m was judged a *possible overexposure* regarding the company's *exposure control limit (ECL)* of 0.1 mg/cu m.

Before making major capital improvements to the local exhaust ventilation, the company industrial hygienist decided to examine the interday variability of 8-hour TWA exposures for the worker on 5 other days in February and March. The measurement results in mg/cu m in chronological order are: 7 Feb., 0.105;

15 Feb., 0.052; 20 Feb., 0.082; 28 Feb., 0.051; 14 Mar., 0.180; and 25 Mar., 0.062. Plot the data to yield a symmetrical distribution in time.

Solution.

1.   Use semilogarithmic graph paper with the number of cycles and scaling appropriate for the sample data. Semilog paper has a logarithmic scale for one variable and a linear scale for the second variable. In this procedure, the exposure data is plotted on the logarithmic scale and time is plotted on the linear scale. Almost 60 types of semilog graph papers are presented by Craver (84) including a 2-cycle by 36 divisions (3 years divided into months), 2-cycle by 52 divisions (1 year divided into weeks), 1-cycle by 60 divisions (5 years divided into months), and 1-cycle by 366 divisions (1 year divided into days).

2.   In this example the range of the exposure measurements covers two decades (0.051 to 0.180 mg/cu m falls within the interval 0.01 to 1.0 mg/cu m), so that 2-cycle semilog paper would be appropriate. Also, the time variable covers 46 days (almost 8 weeks), but the time values are not evenly spaced so that about 50 or more divisions would be appropriate on the linear scale (for time in days). If the time values were spaced exactly one or more weeks apart (i.e., at multiples of seven days), then a linear scale with about 8 or more major divisions would be appropriate.

Comments.   If one desires to apply quality assurance procedures to log normally distributed data, review the article by Morrison (85).

6.12   Use of Logarithmic Probability Graph Paper to Estimate a Logarithmic Normal Distribution

Purpose.   Graphically estimate a 2-parameter or 3-parameter log normal distribution by plotting the sample cumulative distribution as a straight line on logarithmic normal probability paper.

Assumptions.   It is assumed that the sample data are drawn from a single 2-parameter log normal distribution. The distribution will then plot linearly, aside from the expected deviations accountable to random sampling variations. The procedure will also explain how to qualitatively detect 3-parameter log normal distributions as a characteristic type of nonlinearity in the plot and transform the sample results so they plot linearly (see Step 5 of the Solution). The plotting procedure will explain how to qualitatively detect multimodal distributions (such as mixtures of two or more log normal distributions). See Step 6 of the Solution.

Neither the true geometric mean nor true geometric standard deviation need be known for this plotting procedure. *This method cannot process zero or negative data values.* It is assumed that all sample results are nonzero. Refer to

Technical Appendix I of Reference 1 for a discsussion of how to treat a sample population that includes zero values.

The sample results are assumed to represent the parent distribution one desires to estimate.

Example.   The same as used in Section 6.11. Plot the data to estimate the long-term log normal distribution of daily exposures which would exist if the conditions of the 2-months representative period of the measurements persisted.

Solution.

1.   Only the basic aspects of probability plotting will be presented in this solution. Other details for this procedure are presented in Technical Appendix I of Reference 1. Use logarithmic probability paper with the number of cycles appropriate for the range of sample data. Logarithmic probability paper has one logarithmic scale for the measured variable and the other scale is a cumulative *probability* scale. The same configuration of plotted points would exist if the probability scale were replaced by one which is linear in the *probit* of the *cumulative probability*. The *probit* is a transformed variable equal to 5 plus the Z value (standard normal deviate) corresponding to the cumulative area under a normal distribution which is equal to the probability to be plotted. If the cumulative percentage of a normal distribution were plotted on a linear scale instead of on the special probability scale, it would form an S-shaped curve called an *ogive*. However, when the *ogive* curve is plotted on a *probability* scale, a linear function (straight line) results. Therefore, in this procedure, the values of the exposure are plotted on the logarithmic scale and the expected cumulative percentages determined from positions (ranks) in the ordered (ranked) data are plotted on the probability scale. The latter expected values are given as plotting positions in Table I-1 of Reference 1. Craver (84) presents 4 types of normal probability paper (with different systems for numbering the linear scale for the random variable such as exposure concentration) and 3 types of log normal probability paper (1, 2, and 3-cycle logarithmic scales for the random variable such as exposure concentration).

2.   In the example data given in Section 6.11, the range of the exposure measurements cover two decades (0.01 to 1.0 mg/cu m), so that 2-cycle logarithmic probability paper would be appropriate for estimation of a log normal distribution as a model for the data.

3.   Rank the sample data from lowest exposure result to the highest exposure value and obtain the expected cumulative percentage (plotting position) for each of the $n = 6$ values from Table I-1 of Reference 1. For this example the plotting coordinates for each of the six coordinate pairs (measurement in mg/cu m, cumulative percentage for plotting location on the probability scale) are: (0.051, 10.3%), (0.052, 26.0%), (0.062, 42.0%), (0.082, 58.0%), (0.105, 74.0%), and (0.180, 89.7%).

4. Plot a point for each sample value at the plotting coordinates. One can also plot the individual uncertainties for each measurement by first calculating individual confidence limits for each measurement with a procedure selected from Sections 6.1.1, 6.2.1, 6.3.1, or 6.5.1. This will qualitatively aid one in comparing the amount of measurement procedure uncertainty to the environmental variability of the exposure levels measured (see Section 3.3).

5. If the plotted distribution has a substantial "hockey stick" appearance, with a flattening of the curve (approaching zero slope) at substantial portions of lower cumulative probability (the lowest 20 or more percent of the sample), then it might be indicative of a 3-parameter log normal distribution (see Section 3.5.2). Such a distribution can result if there are log normal random variations which are added to a constant background level of the same contaminant. Such a data plot can be linearized by estimating the third parameter k and subtracting this constant "background level" from each measurement value before plotting. An adequate k can be estimated from the initial "hockey stick" plot by noting the value that is asymptotically approached by the "blade" of the hockey stick. That is, estimate k from the concentration value that the measurements appear to converge to at the lowest cumulative probabilities. A detailed example is given on pages 103–104 of Reference 1.

6. If the data plot appears to have one or more "dog legs" or kinks in the central region of the plot, then it may be that the plotted distribution is a mixture of two or more individual log normal distributions (multimodal data). One should then attempt to classify the sample data into two or more appropriate log normal distributions by individually examining the determinant variables for each of the sample data. Additional qualitative interpretations of log normal probability plots are given in Table I-3 on page 102 of Reference 1.

6.13 Use of Logarithmic Probability Graph Paper to Display Tolerance Limits for an Estimated Log Normal Distribution

Purpose. Graphically display 1-sided or 2-sided tolerance limits for the 75th to the 95th percentiles of a log normal distribution estimate by plotting the estimate distribution and associated tolerance limits on log normal probability paper.

Assumptions. The same as used for Section 6.7.

Example. The same as used for Section 6.12. Compute the *1-sided, 95 percent confidence level, upper tolerance limits* for the lower 75th, 90th, and 95th percentiles of the parent distribution of daily TWA exposures for the worker exposed to chromic acid mist and chromates over the 2-month period; then plot the three points on the same log normal probability plot used to display the log normal distribution estimate obtained in Section 6.12.

Solution.

1. Compute the base 10 logarithm of each sample value. Here, the six logarithmic values in chronological order from 7 February to 25 March are: -0.979, -1.284, -1.086, -1.292, -0.745, and -1.208.

2. Compute the arithmetic mean $\bar{y}$ and standard deviation s of the $n = 6$ logarithmic values. Here, $\bar{y} = -1.099$ and $s = 0.211$. As supplementary information, one can compute the *sample geometric mean (GM)* by taking the antilog$_{10}$ of $\bar{y}$ ($10^{-1.099} = 0.080$ mg/cu m), and the *sample geometric standard deviation (GSD)* by taking the antilog$_{10}$ of s ($10^{0.211} = 1.63$).

3. Compute the three *1-sided upper tolerance limits* $(UTL_{1,confidence\ level,percentile})$ for the parent population from the three equations given below, for the particular case of $n = 6$:

$$UTL_{1,.95,.75} = \text{antilog}_{10}\,[\bar{y} + (1.895)(s)]$$

$$UTL_{1,.95,.90} = \text{antilog}_{10}\,[\bar{y} + (3.006)(s)]$$

$$UTL_{1,.95,.95} = \text{antilog}_{10}\,[\bar{y} + (3.707)(s)]$$

where each tolerance limit factor K was obtained from appropriate tables for 1-sided tolerance limits, such as Table A-7 (p. T-15) of Natrella (86).

4. In this example the calculations are (all with 95 percent confidence):

$$UTL_{1,.95,.75} = \text{antilog}_{10}\,[-1.099 + (1.895)(0.211)]$$
$$= 0.20 \text{ mg/cu m for the } 75th \text{ percentile}$$
$$UTL_{1,.95,.90} = \text{antilog}_{10}\,[-1.099 + (3.006)(0.211)]$$
$$= 0.34 \text{ mg/cu m for the } 90th \text{ percentile}$$
$$UTL_{1,.95,.95} = \text{antilog}_{10}\,[-1.099 + (3.707)(0.211)]$$
$$= 0.48 \text{ mg/cu m for the } 95th \text{ percentile}$$

5. Thus the coordinates for three *1-sided upper tolerance limits (95 percent confidence level)* are (0.20 mg/cu m, 75%), (0.34 mg/cu m, 90%), and (0.48 mg/cu m, 95%). That is, based on the frugal sample of six daily exposures over a 2-month representative period for the worker, the best we can say is that we are 95 percent confident that under these conditions 75 percent of daily exposures would be below 0.20 mg/cu m, 90 percent would be below 0.34 mg/cm m, and 95 percent would be below 0.48 mg/cu m. The three tolerance limit points could then be plotted on the same log normal probability plot used to display the estimated log normal distribution obtained in Section 6.12.

Comments. A tolerance limit can be thought of as a "confidence limit" for the designated fractile of a population. Thus a "tolerance line" created by connecting the three tolerance limit points can be thought of as an approximate "confidence line" for the estimated log normal population distribution.

The confidence limits for the proportion of a log normal distribution exceeding a specified value of the variable (e.g., an exposure control limit), which can be computed with the procedures in Section 6.8, can also be plotted on the same log normal probability plot. These limits would be plotted perpendicular to the probability scale, whereas the tolerance limits would be plotted parallel to the probability scale.

REFERENCES

1. N. A. Leidel, K. A. Busch, and J. R. Lynch, National Institute for Occupational Safety and Health, *Occupational Exposure Sampling Strategy Manual*, U.S. Department of Health, Education, and Welfare (NIOSH) Publication 77-173, Cincinnati, Ohio, 1977.

2. W. W. Lowrance, *Of Acceptable Risk—Science and the Determination of Safety*, W. Kaufmann, Inc., Los Altos, Calif., 1976.

3. T. F. Hatch, "Significant Dimensions of the Dose-Response Relationship," *Arch. Environ. Health*, **16**, 571–578 (1968).

4. U. Ulfvarson, "Limitations to the Use of Employee Exposure Data on Air Contaminants in Epidemiologic Studies," *Int. Arch. Occup. Environ. Health*, **52**, 285–300 (1983).

5. T. F. Hatch, "Permissible Levels of Exposure to Hazardous Agents in Industry," *J. Occup. Med.*, **14**, 134–137 (1972).

6. T. F. Hatch, "Criteria for Hazardous Exposure Limits," *Arch. Environ. Health*, **27**, 231–235 (1973).

7. S. A. Roach, "A Method of Relating the Incidence of Pneumoconiosis to Airborne Dust Exposure," *Brit. J. Industrial Med.*, **10**, 220 (1953).

8. J. Gamble and R. Spirtas, "Job Classification and Utilization of Complete Work Histories in Occupational Epidemiology," *J. Occup. Med.*, **18**, 399–404 (1976).

9. N. Esmen, "Retrospective Industrial Hygiene Surveys," *Am. Ind. Hyg. Assoc. J.*, **40**, 58–65 (1979).

10. S. A. Roach, "A More Rational Basis for Air Sampling Programs," *Am. Ind. Hyg. Assoc. J.*, **27**, 1–12 (1966).

11. S. A. Roach, "A Most Rational Basis for Air Sampling Programmes," *Ann. Occup. Hyg.*, **20**, 65–84 (1977).

12. H. Buchwald, "The Elusive Miasma: Problems of Abnormal Responses to Atmospheric Contaminants," *Ann. Occup. Hyg.*, **15**, 379–391 (1972).

13. T. H. Hatch, "Permissible Dustiness," *Am. Industr. Hyg. Assoc. Quar.*, **16**, 30–35 (1955).

14. F. E. Lundin, J. K. Wagoner, and V. E. Archer, *Radon Daughter Exposure and Respiratory Cancer Quantitative and Temporal Aspects*, NIOSH-NIEHS Joint Monograph No. 1, U.S. Department of Health, Education, and Welfare, 1977.

15. G. Berry, J. C. Gilson, S. Holmes, H. C. Lewinsohn, and S. A. Roach, "Asbestosis: A Study of Dose-Response Relationships in an Asbestos Textile Factory," *Brit. J. Ind. Med.*, **36**, 98–112 (1979).

16. J. M. Dement, R. L. Harris, M. J. Symons, and C. Shy, "Estimates of Dose-Response for Respiratory Cancer Among Chrysotile Asbestos Textile Workers," *Ann. Occup. Hyg.*, **26**, 869–887 (1982).

17. B. MacMahon and T. F. Pugh, *Epidemiology Principles and Methods*, Little Brown, Boston, 1970.

18. J. S. Mausner and A. K. Bahn, *Epidemiology, An Introductory Text*, Saunders, Philadelphia, 1974.

19. G. D. Friedman, *Primer of Epidemiology*, McGraw-Hill, New York, 1974.

20. A. M. Lilienfeld and D. E. Lilienfeld, *Foundations of Epidemiology*, 2nd ed., Oxford University Press, New York, 1980.

21. J. J. Schlesselman, *Case-Control Studies: Design, Conduct, Analysis*, Oxford University Press, New York, 1982.

22. K. J. Rothman and John D. Boice, Jr., *Epidemiologic Analysis with a Programmable Calculator*, Epidemiology Resources, Inc., Boston, 1982.

23. R. R. Monson, *Occupational Epidemiology*, CRC Press, Boca Raton, Fl., 1980.

24. L. Chiazze, Jr., F. E. Lundin, and D. Watkins, Eds., *Methods and Issues in Occupational and Environmental Epidemiology*, Ann Arbor Science Publishers, Ann Arbor, Mich., 1983.

25. P. E. Enterline, "Pitfalls in Epidemiological Research," *J. Occup. Med.*, **18,** 150–156 (1976).

26. *Advances in Health Survey Research Methods: Proceedings of a National Invitational Conference*, sponsored by the National Center for Health Services Research, U.S. Department of Health, Education and Welfare (HRA) Publication No. 77-3154, 1977.

27. J. W. Lloyd and A. Ciocca, "Long-Term Mortality Study of Steelworkers: I. Methodology," *J. Occup. Med.*, **11,** 299–310 (1969).

28. J. F. Collins and C. K. Redmond, "The Use of Retirees to Evaluate Occupational Hazards. II. Comparison of Cause Specific Mortality by Work Area," *J. Occup. Med.*, **20,** 260–266 (1978).

29. L. L. Kupper, A. J. McMichael, and R. Spirtas, "A Hybrid Epidemiologic Study Design Useful in Estimating Relative Risk," *J. Am. Stat. Assoc.*, **70,** 524–528 (1975).

30. H. B. Elkins, "The Case for Maximum Allowable Concentrations," *Ind. Hyg. Quar.*, **9,** 22–25 (1948).

31. "Threshold Limits—A Panel Discussion," *Am. Ind. Hyg. Assoc. Quar.*, **16,** 27–39 (1955).

32. H. E. Stokinger, "Standards for Safeguarding the Health of the Industrial Worker," *Public Health Reports*, **70,** 1–11 (1955).

33. H. F. Smyth, Jr., "Improved Communication—Hygienic Standards for Daily Inhalation," *Am. Ind. Hyg. Assoc. Quar.*, **17,** 129–185 (1956).

34. H. F. Smyth, Jr., "A Toxicologist's View of Threshold Limits," *Am. Ind. Hyg. Assoc. J.*, **23,** 37–44 (1962).

35. H. E. Stokinger, "Threshold Limits and Maximal Acceptable Concentrations: Their Definition and Interpretation, 1961," *Am. Ind. Hyg. Assoc. J.*, **23,** 45–47 (1962).

36. T. F. Hatch, "Thresholds: Do They Exist?," *Arch. Environ. Health*, **22,** 687–689 (1971).

37. E. Bingham, "Thresholds in Cancer Induction," *Arch. Environ. Health*, **22,** 692–695 (1971).

38. E. R. Hermann, "Thresholds in Biophysical Systems," *Arch. Environ. Health.*, **22,** 699–706 (1971).

39. H. E. Stokinger, "Concepts of Thresholds in Standards Setting," *Arch. Environ. Health*, **25,** 153–157 (1972).

40. H. F. Thomas, "Some Observations on Occupational Hygiene Standards," *Ann. Occup. Hyg.*, **22,** 389–397 (1979).

41. H. O. Hartley and R. L. Sielken, Jr., "Estimation of 'Safe' Doses and Carcinogenic Experiments," *Biometrics*, **33,** 1–30 (1977).

42. N. Mantel and W. R. Bryan, "'Safety' Carcinogenic Agents," *J. Nat. Cancer Inst.*, **27,** 455–470 (1961).

43. N. Mantel, N. R. Bohidar, C. C. Brown, J. L. Ciminera, and J. W. Tukey, "An Improved 'Mantel-Bryan' Procedure for 'Safety Testing' of Carcinogens," *Cancer Res.*, **35,** 865–872 (1975).

44. P. Armitage, "The Assessment of Low-Dose Carcinogenicity," *Biometrics Supplement: Current Topics in Biostatistics and Epidemiology*, 119–129 (March 1982).

45. H. O. Hartley and R. L. Sielken, Jr., *A Non-parametric for "Safety" Testing of Carcinogenic Agent*,

Food and Drug Administration Technical Report 1, Institute of Statistics, Texas A & M University, College Station, Tx., 1975.

46. H. O. Hartley and R. L. Sielken, Jr., *A Non-parametric for "Safety" Testing of Carcinogenic Agent*, Food and Drug Administration Technical Report 2, Institute of Statistics, Texas A & M University, College Station, Tx., 1975.

47. K. S. Crump, H. A. Guess, and K. L. Deal, "Confidence Intervals and Tests of Hypotheses Concerning Dose-Response Relations Inferred from Animal Carcinogenicity Data," *Biometrics*, **33**, 437–451 (1977).

48. D. Krewski and C. Brown, "Carcinogenic Risk Assessment: A Guide to the Literature," *Biometrics*, **37**, 353–366 (1981).

49. H. J. Seim and J. A. Dickeson, "A Device for Collecting Replicate Samples in the Workplace," *Am. Ind. Hyg. Assoc. J.*, **44**, 562–566 (1983).

50. A. Hald, *Statistical Theory with Engineering Applications*, John Wiley and Sons, Inc., New York, 1952, p. 91.

51. J. Aitchison and J. A. C. Brown, *The Lognormal Distribution*, Cambridge at the University Press, Cambridge, Great Britain, 1957.

52. M. J. Moroney, *Facts from Figures*, Penguin Books, Baltimore, 1951, p. 261.

53. P. E. Wilkins, "Log Normal Distribution," *J. Air Pollution Control Assoc.*, **26**, 935 (1976).

54. D. J. Finney, "The Questioning Statistician," *Statistics in Medicine*, **1**, 5–13 (1982).

55. D. G. Altman, "Statistics in Medical Journals," *Statistics in Medicine*, **1**, 59–71 (1982).

56. E. L. Crow, F. A. Davis, and M. W. Maxfield, *Statistics Manual*, Dover Publications, Inc., New York, 1960.

57. R. H. Green, *Sampling Design and Statistical Methods for Environmental Biologists*, John Wiley and Sons, New York, 1979.

58. R. D. Soule, "An Industrial Hygiene Survey Checklist," in National Institute for Occupational Safety and Health, *The Industrial Environment—Its Evaluation and Control*, Department of Health, Education, and Welfare, 1973.

59. J. D. Bates, *Writing With Precision—How to Write So That You Cannot Possibly Be Misunderstood*, Acropolis Books Ltd., Washington, D.C., 1978.

60. F. Crews, *The Random House Handbook*, Random House, New York, 1974.

61. CBE Style Manual Committee, *CBE Style Manual*, Fifth Edition, Council of Biological Editors, Inc., Bethesda, Md., 1983.

62. E. A. Murphy, *Biostatistics in Medicine*, The Johns Hopkins University Press, Baltimore, 1982, Chap. 4.

63. D. T. Mage, "An Objective Graphical Method for Testing Normal Distributional Assumptions Using Probability Plots." *Am. Statistician*, **36**, 116–120 (1982).

64. H. W. Lilliefors, "On the Kolmogorov-Smirnov Test for Normality With Mean and Variance Unknown," *J. Am. Stat. Assoc.*, **62**, 399–402 (1967).

65. R. I. Iman, "Graphs for Use With the Lilliefors Test for Normal and Exponential Distributions," *Am. Statistician*, **36**, 109–112 (1982).

66. W. A. Burgess, *Recognition of Health Hazards in Industry*, John Wiley and Sons, New York, 1981.

67. L. V. Cralley and L. J. Cralley, Eds., *Industrial Hygiene Aspects of Plant Operations, Vol. 1, Process Flows*, Macmillan, New York, 1982.

68. M. Corn and N. A. Esmen, "Workplace Exposure Zones for Classification of Employee Exposures to Physical and Chemical Agents," *Am. Ind. Hyg. Assoc. J.*, **40**, 47–57 (1979).

69. B. B. Crocker, "Preventing Hazardous Pollution During Plant Catastrophes," *Chemical Engineering*, 97 (May 4, 1970).

70. R. J. Hubiak, F. H. Fuller, G. N. VanderWerff, and M. Ott, "Improving Work Practices and Engineering Controls," *Occ. Health and Safety*, **50**, 10–18 (1981).

71. M. L. Holcomb and R. C. Scholz, *Evaluation of Air Cleaning and Monitoring Equipment Used in Recirculation Systems*, National Institute for Occupational Safety and Health, Publication DHHS (NIOSH) 81-113, 1981.

72. D. Schneider, "The Draeger Gas Detection Kit," *Draeger Review*, **46**, 5–12 (1980).

73. K. Leichnitz, "Qualitative Detection of Substances by Means of Draeger Detector Tube Polytest and Draeger Detector Tube Ethyl Acetate 200/a," *Draeger Review*, **46**, 13–21 (1980).

74. M. V. King, P. M. Eller, and R. J. Costello, "A Qualitative Sampling Device for Use at Hazardous Waste Sites," *Am. Ind. Hyg. Assoc. J.*, **44**, 615–618 (1983).

75. ACGIH, *Threshold Limit Values for Chemical Substances and Physical Agents in the Work Environment with Intended Changes for 1983–84*, American Conference of Governmental Industrial Hygienists, Cincinnati, 1983.

76. N. W. Paik, R. J. Walcott, and P. A. Brogan, "Worker Exposure to Asbestos During Removal of Sprayed Material and Renovation Activity in Buildings Containing Sprayed Material," *Am. Ind. Hyg. Assoc. J.*, **44**, 428–432 (1983).

77. Errata, *Am. Ind. Hyg. Assoc. J.*, **44**, 697 (1983).

78. C. Daniel and F. S. Wood, *Fitting Equations to Data*, Wiley-Interscience, New York, 1971, Appendix 3A.

79. N. A. Leidel, K. A. Busch, and W. E. Crouse, National Institute for Occupational Safety and Health, *Exposure Measurement Action Level and Occupational Environmental Variability*, U.S. Department of Health, Education, and Welfare (NIOSH) Publication 76-131, Cincinnati, Ohio, 1975.

80. R. M. Tuggle, "The NIOSH Decision Scheme," *Am. Ind. Hyg. Assoc. J.*, **42**, 493–498 (1981).

81. R. M. Tuggle, "Assessment of Occupational Exposure Using One-Sided Tolerance Limits," *Am. Ind. Hyg. Assoc. J.*, **43**, 338–346 (1982).

82. J. R. Lynch, N. A. Leidel, R. A. Nelson, and R. F. Boggs, National Institute for Occupational Safety and Health, *The Standards Completion Program Draft Technical Standards Analysis and Decision Logics*, National Technical Information Service Publication PB 282 989, Springfield, Va, 1978.

83. Y. Bar-Shalom, D. Budenaers, R. Schainker, and A. Segall, National Institute for Occupational Safety and Health, *Handbook of Statistical Tests for Evaluating Employee Exposure to Air Contaminants*, U.S. Department of Health, Education, and Welfare (NIOSH) Publication 75-147, Cincinnati, Ohio, 1975.

84. J. S. Craver, *Graph Paper from Your Copier*, H. P. Books, Tucson, Ariz., 1980.

85. J. Morrison, "The Lognormal Distribution in Quality Control," *Applied Statistics*, **7**, 160–172 (1958).

86. M. G. Natrella, *Experimental Statistics*, National Bureau of Standards Handbook 91, Superintendent of Documents, U.S. Government Printing Office, Washington, D.C., 1963.

CHAPTER NINE

Analytical
Measurements

PETER M. ELLER, Ph.D.

1 INTRODUCTION

Analytical chemistry is important to the industrial hygienist in several areas: personal monitoring, source sampling and analysis, and area measurements. In most cases the chemical properties of the analyte must be considered to make the proper choice of sampling device and to recover the analyte quantitatively. Thus collection efficiency, stability of the sample, possible interferences, and desorption efficiency may be functions of the chemical interactions between analyte and sampling device. This chapter discusses selectively, but not exhaustively, the analytical techniques employed in industrial hygiene and their relationships to sampling systems.

The decisions relating to sampling and analytical methods to be used in industrial hygiene measurements are frequently complex. The methods may be thought of as one of the more recent fields of application of analytical chemistry (1), with its own unique set of limiting parameters. In the case of highly toxic substances, including carcinogens, sensitivity of the analytical method is commonly the limiting factor in determining the typically low permissible levels. In most cases, however, sensitivity is a secondary consideration. For example, capacity of the solid sorbent tube is the limiting factor for a number of substances with permissible levels of several hundred milligrams per cubic meter. Other factors, including specificity (Section 4.1), cost, and analytical turnaround time may be overriding. The increasing use of direct-reading methods, both instrumental and those based on color change reactions, is testimony to the need for faster feedback of analytical results.

2 VALIDITY OF THE MEASUREMENT

2.1 Accuracy and Precision

The value of the analytical result to the industrial hygiene program and the relevance to the individual worker are ultimately dependent on the accuracy of the result. This chapter deals with contributions of the sampling and analytical method to accuracy; other considerations such as sampling strategies and personal versus area sampling are discussed elsewhere in this volume.

The accuracy of an individual measurement is determined by the occurrence of two types of error. These are determinate errors, also called bias, and indeterminate errors, or random errors. Examples of determinate errors are improperly calibrated sampling pumps and desorption efficiencies that are less than quantitative. Indeterminate errors are typified by instrumental noise or interanalyst variations and are frequently the limiting factor in defining detection limits.

Contamination, both of the sampling device in the field and of the sample by reagents during analysis, is one of the most common sources of error. The positive biases introduced by the various sources of contamination are best eliminated by efforts to reduce the sources of contamination and the analysis of appropriate blanks. In practice, a distinction is made between several types of blanks, based on their history and use. Field blanks are defined as sampling devices which are treated exactly as samples, except that no air is drawn through them (2). That is, field blanks are taken to the sampling site, opened, resealed, and shipped to the laboratory along with the air samples. The function of field blanks is to detect contamination which may have occurred in the process of sampling or shipment. Media blanks, on the other hand, are sampling devices which are identical to those used for taking samples, but which are not taken to the field. Media blanks are used for several purposes: as substrate for desorption efficiency or analytical method recovery studies, as the starting point for calibration standards, or as a measure of background contamination to be subtracted from samples. Finally, reagent blanks are used in their conventional role as measures of contamination contributed by the reagents used to process the samples in the analysis.

Another source of random error arises in the analysis of samples which lie outside the optimum working range of the analytical method. For most methods the relative standard deviation (s_r) increases rapidly with decreasing sample size toward the lower end of the working range. For methods such as spectrophotometry, an increase in s_r with increasing analyte concentration is also seen above the upper limit of the working range. The best precision is obtained with sample sizes or analyte concentrations that are within an optimum range.

Losses, through chemical degradation on the sampling device, or in the analytical process, lead to negative biases that may be minimized by prompt analysis and by the practice of subjecting known quantities of analyte to the same processing steps as the samples.

Table 9.1 summarizes these and other sources of error.

2.2 Reference Materials

The use of reference materials of known composition to calibrate analytical procedures is an important step in methods development or quality control. When evaluating the results obtained by a candidate procedure using a reference material, both precision and accuracy are important. The critical question is, "Does the candidate procedure give a result that falls within the range certified?" Thus the procedure will be categorized as one of three types: inaccurate (poor precision, possibly with bias), biased (good precision, but with bias), and accurate

Table 9.1   Some Common Errors in Sampling and Analysis

Source of Error	Direction	Remedy
Sampling		
Contamination	±	Analyze blank sampling device
Flowrate uncertainty	±	Calibrate sampling train
Sample too small for precise analysis	±	Take larger sample
Interfering substances	±	Take also bulk, area, or rafter samples
Loss of analyte	−	Avoid high temperatures, long storage of sample
Low collection efficiency	−	Decrease sampling time, temperature, flowrate; use fresh collection reagents
Analysis		
Contamination	+	Use reagent blank
High recovery with standard additions	+	Correct for nonlinear calibration curve
Biased analytical method	±	Calibrate with reference material or method
Matrix effects	±	Match sample and standard matrices
Interferences	±	Analyze bulk, area, rafter samples; apply correction for matrix
Too little or too much sample	±	Work in linear portion of calibration curve
Low desorption efficiency	−	Determine recovery with spiked samples
Loss during sample processing	−	Carry standards through same processing

(good precision, with bias absent). When precision is poor, it is not correct to classify the procedure as accurate (3). These definitions are illustrated by three hypothetical candidate methods using a reference material with certified value 100 ± 5 (Table 9.2 and Figure 9.1). Method A is inaccurate because the estimate it provides extends outside the certified range of 95 to 105. It makes no difference that the mean value falls within the certified range; the entire set of values expressed by

$$\frac{\overline{X}_A + ts_A}{(N)^{1/2}} \tag{1}$$

must fall within the certified range. The t-value, 2.23 in this example, is selected to include a desired proportion of the values obtained by method A (95 percent of the values in this example). Method B is inaccurate but precise; systematic errors are present and must be identified. The amount of bias that must be removed is the difference between the mean obtained $\overline{X}_B$ and the nearest boundary of the certified range. In this case the difference is 107.59 minus 105, or 2.59. Method C qualifies as a reference method because at the 95 percent confidence level it gives a mean value that falls entirely within the certified range. In practice, a larger number of determinations than 10 would be obtained. A detailed explanation of the steps needed to establish reference methods is given elsewhwere (4).

The U.S. National Bureau of Standards offers a large number of reference

Table 9.2 Evaluation of Hypothetical Methods Using a Reference Material with Certified Value 100 ± 5

	Method		
	A	B	C
Results	99.0	103.9	93.9
	97.3	111.7	101.7
	84.8	106.0	96.0
	94.7	108.1	98.1
	100.6	107.9	97.9
	106.6	109.8	99.8
	111.3	107.3	97.3
	91.0	106.5	96.5
	103.2	109.5	99.5
	117.8	105.2	95.2
Mean, $\overline{X}$	100.63	107.59	97.59
Standard deviation, s	9.68	2.33	2.33
$ts/(N)^{1/2}$	6.83	1.64	1.64
95% confidence limits of mean	93.80 – 107.43	105.95 – 109.23	95.95 – 99.23

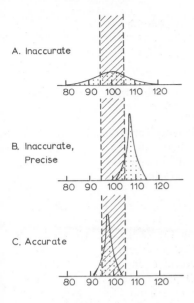

A. Inaccurate

80 90 100 110 120

B. Inaccurate, Precise

80 90 100 110 120

C. Accurate

80 90 100 110 120

Figure 9.1 Evaluation of three methods using a reference material with certified value of ±5.

materials (5), some of which are directly applicable to industrial hygiene. Table 9.3 lists these; several gases, organic solvents, metals, and quartz are included. A need exists for additional reference materials, particularly biological standards [e.g., lead in blood (6)]. The application of reference materials to proficiency testing of laboratories is covered in Section 2.5.

2.3 Validation of Sampling and Analytical Methods

After selection of candidate sampling and analytical methods, further validation under conditions appropriate to industrial hygiene use is usually desirable.

One set of validation criteria has been applied successfully to more than 200 personal sampling and analytical methods using laboratory-generated atmospheres (7, 8). The basic accuracy requirement for validation was that the overall sampling and analytical method must be capable of giving a value within ±25 percent of the true air concentration at the standard set by the Occupational Safety and Health Administration (OSHA), at the 95 percent confidence level. This requires that the overall coefficient of variation be 12.8 percent or less for a method shown to contain no systematic errors. If the method contains a bias, precision requirements are more stringent. The maximum allowable bias correction, aside from a correction for desorption efficiency, was set at 10 percent. The minimum acceptable desorption efficiency was set at 75 percent. The criterion for sample stability was that no more than a 10 percent change in analyte concentration should occur after 7 days storage at room temperature. The methods were to be suitable for sampling periods of one hour or longer, except where shorter term standards apply. Detailed instructions for sampling and analysis are given, as are data relating to accuracy, precision, working range, and possible interferences. The methods have been revised recently (8).

Table 9.3   Reference Materials Useful for Industrial Hygiene (5)

Standard Reference Material No.	Name	Substance Certified
1579	Powdered lead-based paint	Pb
1580	Organics in shale oil	Polynuclear aromatic hydrocarbons
1625–27	Sulfur dioxide permeation tube	Sulfur dioxide
1629a	Nitrogen dioxide permeation device	Nitrogen dioxide
1632a	Trace elements in coal	As, Cd, Cr, Cu, Fe, Hg, Mn, Ni, Se, Th, U, V, Zn
1648	Urban particulate matter	As, Cd, Cr, Cu, Fe, Ni, Pb, U, Zn
1677–81	Carbon monoxide in nitrogen	Carbon monoxide
1683–87	Nitric oxide in nitrogen	Nitric oxide
1878	Respirable alpha quartz	Quartz
2203	Potassium fluoride	Fluoride ion
2670	Freeze-dried urine	As, Ca, Cd, Cl, Cu, Mg, Pb
2671a	Freeze-dried urine	Fluoride ion
2672a	Freeze-dried urine	Mercury
2674	Lead on filter media (Hi-Vol)	Pb
2676b	Metals on filter media	Cd, Mn, Pb, Zn
2679	Quartz on filter media	Quartz

These validation criteria were met with the following protocol (criteria added later in the program are marked with an asterisk):

1.   Analysis of six replicate standards each at 2, 1, and $\frac{1}{2}$ the amount equivalent to the OSHA standard.

2.   Analysis of six samples collected from laboratory-generated atmospheres at 2, 1, and $\frac{1}{2}$ times the OSHA standard, with verification of the generated air concentrations by an independent method.

3.   For solid sorbents, measurement of desorption efficiency for six replicates each at 2, 1, and $\frac{1}{2}$ times the OSHA standard.

4.   For solid sorbents, measurement of breakthrough capacity at twice the OSHA standard. For moisture-sensitive sorbents, the measurement is made with at least 80 percent relative humidity.*

5.   Storage of six samples collected at the OSHA standard for 7 days at room temperature before analysis.*

2.4   Classification of Sampling and Analytical Methods

To describe the degree of confidence that can be expected in a method, it is desirable to develop a descriptive classification scheme. Of several that have been proposed, one denotes a "definitive" method as one that is directly related to fundamental (e.g., SI) units (3). An example of such a method is isotope dilution mass spectrometry. This scheme further defines a "reference" method as one that has been extensively collaboratively tested by the scientific community.

Another classification system (9) consists of five categories or classes: E (proposed), D (operational), C (tentative), B (recommended), and A (accepted). In this system the determining factors are the degrees of intra- and interlaboratory evaluation of the method. To move from E to D requires the successful use of the method on at least 15 field samples. Class C is reserved for methods in general use by other laboratories, but not evaluated for a particular industrial hygiene application. Validation of a method using standards and generated samples to characterize any biases and to obtain separate estimates of sampling and analytical errors is required for a class B method. Successful field and collaborative testing upgrades the class B method to class A.

2.5 Proficiency Testing

The goal of proficiency testing is to measure the relative abilities of laboratories, using methods of their choice, to obtain accurate results. Because proficiency samples may receive special treatment, and because they are prepared in relatively uncomplicated matrices, proficiency testing is not necessarily a measure of the accuracy of the day-to-day work of the laboratory, however. Also, it is not well suited for methods development.

For a successful test, several requirements must be met: (1) samples must be uniform and prepared reproducibly, (2) samples must have adequate shelf life, and (3) reliable analytical methods must be available. If separate estimates of intra- and interlaboratory variation are desired, each laboratory must analyze two or more samples of a given analyte (although not necessarily duplicate samples). Because the resulting data appear to be log normally distributed (10), the use of geometric, rather than arithmetic, means has been found to be useful in their interpretation. Two statistical measures of performance may be defined: the mean ratio and the error ratio. The mean ratio, a measure of the accuracy of each laboratory, is defined by

$$M = \frac{(X_1 X_2 X_3 \cdots X_n)^{1/n}}{G} \tag{2}$$

where $X_1, X_2, \ldots, X_n$ are the results obtained by the laboratory in question on its n samples of a given analyte and G is the grand geometric mean, excluding outliers, of the results obtained on this analyte by all laboratories. The value of M that signifies perfect accuracy is 1.00, and limits can be established, based on the data of a given testing session, which signify excessive deviation (bias) relative to the other participating laboratories.

The error ratio, or residuals, gives a measure of intralaboratory variation.

That is, it is a measure of the ability of a given laboratory to obtain consistent results. A residual is calculated for each analytical result submitted by each laboratory and is defined by

$$R_i = \frac{\dfrac{X_i}{(X_1 X_2 X_3 \cdots X_n)^{1/n}}}{\dfrac{F_i}{(F_1 F_2 F_3 \cdots F_n)^{1/n}}} \tag{3}$$

where X_i is one of the values $X_1, X_2, X_3, \ldots, X_n$ and F_i is the geometric mean result obtained on the sample (excluding residual outliers) by all laboratories. Perfect precision is indicated if the values of R_i are 1.00. As with the mean ratio, statistical limits can be calculated to determine whether a given R value is abnormal. For a given testing session, the product $R_1 R_2 R_3 \cdots R_n$ is 1.0 for each analyte.

The performance of a laboratory is indicated by both its mean ratio and its residuals. Four general cases exist:

1. *Mean ratio and all residuals within limits.* Both accuracy and precision are acceptable. The closer the ratios to unity, the better the performance.

2. *Mean ratio within limits, and one or more residuals out of limits.* In this case interlaboratory variation is acceptable but precision of the intralaboratory results is not. If one of the residuals is grossly different from the others, a calculation error, contamination, or other mistake may be the cause. For example, failure to correct for a nonlinear working curve whose slope decreases at high concentrations may lead to abnormally low residuals for the high concentration samples.

3. *Mean ratio out of limits, all residuals within limits.* A bias common to all the analyses is a possible cause. Calibrations, standards, and calculations should be checked.

4. *Mean ratio and one or more residuals out of limits.* This unacceptable inter- and intralaboratory variation may be due to any of the above causes.

As an example, the Proficiency Analytical Testing (PAT) program of the National Institute for Occupational Safety and Health (NIOSH), begun in 1972, sends quarterly samples to more than 300 participating laboratories (10). The samples include spiked filters containing cadmium, lead, and zinc (as the nitrates), chrysotile asbestos (in alumina matrix), quartz (with sodium silicate), and organic solvents (toluene, benzene, carbon tetrachloride, chloroform, ethylene dichloride, *p*-dioxane, trichloroethylene, and xylene) on charcoal tubes. Each set contains four samples, at different concentrations, of each analyte. Two limits are calculated and used for judging the performance. The outer limit defines a region above and below $M = 1$ or $R = 1$ that contains, at the 99 percent confidence level, at least 98 percent of the analytical results. The inner limits, slightly closer to $M = 1$ or $R = 1$, are the limits beyond which it can be stated with 99 percent confidence that a result belongs to the outer two percent of the distribution. An investigation into possible sources of error is recommended whenever (1) a mean ratio or residual exceeds an outer limit, (2) results on two consecutive rounds exceed an inner limit, or (3) results in six consecutive rounds give mean ratios on the same side of $M = 1$.

Because a variety of analytical methods may be used for a given analyte in a given PAT round, interlaboratory variance is higher than in a collaborative test. Typical values of s_r are 4 to 7 percent for the metals, 20 to 40 percent for silica, 30 to 60 percent for asbestos, and 6 to 20 percent for the organics. In all categories, the precision has improved with time as the participating laboratories have become more proficient in the analyses and as methods have been made more precise. For example, PAT s_r values for asbestos determinations declined by one-fourth to one-third in the period 1977 to 1982 (10).

Planned additions to the PAT sample types include several ketones, and a coal mine dust matrix for the silica (10).

3   SAMPLE COLLECTION AND PROCESSING

3.1   Personal Sampling Devices

Sampling is an integral part of the industrial hygiene measurement. The contribution of the sampling process to overall relative standard deviation is frequently larger than that due to the analytical method (7). The ideal personal sampling device is small and lightweight, in addition to having the sample collection characteristics discussed in Section 2.3. Thus it can be attached to the worker's clothing in the breathing zone and used to sample for extended periods of an hour or more to determine the time-weighted average concentration to which the worker is exposed. Flowrates through the sampler are set to assure efficient collection, and the minimum and maximum sample sizes are dictated by analytical sensitivity and sampler capacity, respectively.

For sampling particulates, 37-mm diameter filters in closed-face cassettes to avoid contamination and at flowrates of 1.5 to 3 liters/min are commonly used. A 13-mm diameter filter with inlet restricted to 1.2 mm, operated at 0.2 liters/min sampling flow rate, provides inlet and filter face velocities similar to a conventional 37-mm filter cassette operated at 2 liters/min. This modified filter cassette has been used to precede solid sorbent tubes for the efficient collection of pesticides (11) and arsenic trioxide vapor (12). A number of different filter materials are used, as shown by the applications summarized in Table 9.4. Considerations in the selection of a filter for a particular application are usually related to the processing involved in the recovery and determination of the analyte. Glass fiber filters are inert to all but the most vigorous chemical treatment (involving hydrofluoric and phosphoric acids) and provide relatively small pressure drops during sampling, however they may contain sufficient

Table 9.4 Some Personal Sample Collection Methods

Collection Device	Example	Reference[a]
Filters		
Cellulose ester membrane	Asbestos	7400
Cellulose ester membrane	Lead & compounds	7082
Polyvinyl chloride	Chromium (VI)	7600
Polyvinyl chloride + cyclone	Silica, crystalline	7500
Polytetrafluoroethylene	Paraquat	5003
Glass fiber	Carbaryl	5006
Cellulose ester membrane + sodium carbonate on backup pad	Arsenic trioxide	7901
Solid sorbents		
Activated coconut charcoal	Halogenated H/C	1003
Activated charcoal (2 tubes)	Vinyl chloride	1007
Silica gel	Methanol	2000
Chromosorb P (R) + Ag	Mercury	6000
Chromosorb 102 (R) + 2-(benzylamino) ethanol	Formaldehyde	2502
Activated charcoal + sodium sulfate drying tube	Carbon disulfide	1600
Ambersorb XE-347 (R)	2-Butanone	2500
Tenax-GC (R)	Nitroglycerin	2507
Liquids		
Bubbler with 1-(2-methoxyphenyl)piperazine	Isocyanate group	5505
Bubbler with 0.1N NaOH	Phenol	3502
Combination		
Glass fiber filter + Silica gel	Ethylene glycol	5500
Glass fiber filter + bubbler (isooctane)	Aldrin (R)	5502
Glass fiber filter + Florisil (R)	Polychorobiphenyls	5503
Polytetrafluoroethylene filter + XAD-2 (R)	PNAs	5506
Passive collection		
Palmes tube	Nitrogen oxides	6700

[a] Numbers refer to methods found in Reference 8.

trace metal contamination to produce unacceptably high blank values in some cases (13). Teflon filters also have a high degree of chemical inertness and provide a hydrophobic substrate as well.

For applications in which the sample is wet or dry ashed, or in which it is desired to dissolve the filter in concentrated nitric acid, acetic acid, sodium hydroxide, acetone, or dioxane, cellulose acetate or mixed cellulose ester membrane filters may be used. Polycarbonate filters provide a microscopically smooth surface that is compatible with photomicrography, and they are soluble in strong bases and dioxane, while being resistant to acids. Silver membrane filters have found application for the sampling and analysis by X-ray diffraction, without further sample processing, of crystalline species. Polyvinyl chloride (PVC) and PVC-acrylonitrile filters are relatively unaffected by relative humidity of the air sampled and thus find application where filters must be weighed; in addition, the latter type of filter is transparent in much of the infrared region; however, this property has not yet been utilized in industrial hygiene analysis.

Personal sampling devices for gases may take one of four forms: whole-air "bag" samplers, reactive solid sorbents, reactive liquids, and continuous monitors. Most whole-air samplers are useful for relatively short periods only and are effectively "grab" samples. Such samples may be useful for some applications (e.g., determination of peak, or ceiling exposures) but are less accurate for estimation of time-weighted average concentrations than are integrated samples of longer duration (Chapter 11). A recent development, a pocket-size whole-gas sampler, has been shown to eliminate this disadvantage for several gases; furthermore, the requirement for a pump is eliminated (14). The device consists of an evacuated chamber with internal volume approximately $100 \, cm^3$ connected to a sampling port containing a flow-limiting orifice. By choice of the critical orifice size, sample times of 8 hr or longer may be obtained. Connection of the sampler to a gas chromatograph by means of a simple gas handling system with provision for measurement of pressure allows for relatively easy analysis.

The second category of samplers for gases is reactive solid sorbents, including tubes filled with granular materials such as activated charcoal, silica gel, porous polymers, or other materials. The attraction between solid sorbent and gas may arise from relatively weak physical forces (e.g., collection of organic solvents on activated charcoal) or from chemical reaction involving electron transfer between gas and sorbent (e.g., sulfur dioxide on cuprous oxide) (15). Analysis may be direct in some cases such as the direct reading indicator tubes; in others desorption and analysis are performed in a laboratory. Figure 9.2 shows a three-section silica gel tube that has been applied to the sampling of aromatic amines (16). The analyst can use a relatively small (150 mg) collection section, or a larger (850 mg) section, which allows flexibility for various combinations of air concentration, relative humidity, and sampling time. As solid sorbents are studied for collection of an increasing number of substances, more sorbents are found to be useful, including a variety of porous polymers that previously were used as chromatographic columns (17). An example is the use of Tenax GC, Porapak Q, or the Century Chromasorb series for the sampling of elemental phosphorus vapor in air (8, 18).

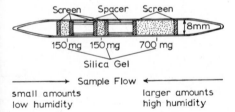

Figure 9.2 Sampling tube for aromatic amines (7).

Another approach that is simple and elegant is the use of a solid sorbent device in a passive mode. In the past few years, a number of passive monitor designs have been marketed commercially. These include several devices based on activated charcoal for collection of organic gases by diffusion and adsorption. Other devices employ color-producing reagents for reaction and field reading of the cumulative concentrations of carbon monoxide, hydrogen sulfide, sulfur dioxide and a few other reactive gases. Except in a relatively few cases, such as the Palmes tube for nitrogen oxides (19–23) and the Reiszner and West sulfur dioxide badge (24), the design parameters of the passive monitors are proprietary. In these cases, performance criteria are essential for evaluation of the accuracy of the devices. NIOSH has recently developed a thorough set of performance specifications covering analytical recovery, sampling rate, capacity, stability during both exposure and storage, overall precision and accuracy, temperature, and the effects of the factors shown in Table 9.5 (25).

Reactive liquids constitute a less favorable type of sampling medium from the point of view of the industrial hygienist because the handling and use of liquids in bubblers or impingers during sampling and shipping may lead to spillage, with resultant hazards and loss of sample. The mode of action of the liquid collection medium may be one of simple solvation (e.g., collection of tetrachloronaphthalene in isooctane), or it may involve a chemical reaction that fixes analyte (e.g., ozone + alkaline potassium iodide = molecular iodine).

Finally, sampling of gases is accomplished in some cases by continuous monitoring systems based on detection of a property such as thermal conductivity, electrical conductivity of liquid solution, or infrared absorption. These systems are usually more suited to area or source sampling than to personal sampling because of restricted portability and large physical size.

3.2 Air Sampling Parameters

The ideal sample collection device has high collection efficiency, high breakthrough capacity, and high desorption efficiency. Collection efficiency is defined as the ratio of quantity of analyte collected to quantity sampled. Breakthrough capacity is the maximum quantity of analyte collected before 5 percent of the

Table 9.5 Passive Monitor Performance Specifications

Factor	Suggested "No Effect" Range
Concentration	0.1 to 2× PEL
Exposure time	0.067 to 1.33× monitor capacity
Face velocity	10 to 150 cm/sec
Humidity	10 to 80% relative humidity
Interferent	Up to 1× PEL
Orientation	All angles to air flow

influent appears on a backup sampling stage; it should correspond to an air volume that is at least 50 percent larger than the maximum air volume to be sampled (7). Breakthrough capacity may be a function of sampling rate, air concentration, and relative humidity. For collection of toluene on activated coconut charcoal, breakthrough capacity was seen to vary from 30 mg of toluene/100 mg of charcoal at 2040 μg of toluene/liter of air (15 liters sampled) to 19 mg of toluene/100 mg of charcoal at 45 μg of toluene/liter of air (420 liters sampled) at room temperature and less than 7 percent relative humidity (26). When the relative humidity was increased to 80 percent, breakthrough volume and capacity decreased by about 50 percent. Capacity is also a function of sampling rate, as was demonstrated for vinyl chloride on activated coconut charcoal (27). Air concentrations of 500 μg of vinyl chloride per liter were sampled; the breakthrough volumes for 100 mg of sorbent were 0.9 liter at 1.0 liter/min, 2.4 liters at 0.2 liter/min, and 5.2 liters at 0.05 liter/min.

Desorption efficiency, a term usually applied only to solid sorbents, is a measure of the quantity of analyte that can be recovered from the collection device during analysis. Measurement of this parameter is done by spiking unexposed sorbent with known quantities of analyte, allowing sufficient time for equilibration, and determining the percentage of recovery. Desorption efficiency may be a function of age of the sample and should be determined for realistic storage conditions. Some typical desorption efficiencies measured for coconut charcoal are: cyclohexanone, 47 percent; methyl ethyl ketone, 62 percent; 1,1,2,2-tetrachloroethane, 68 percent; styrene, 80 percent; methyliso-butyl ketone, 89 percent; toluene, 98 percent; and benzene, cyclohexane, and trichloroethylene, 100 percent (15). In a collaborative test of seven organic solvents collected on charcoal tubes, an overall average desorption efficiency of 96 percent was observed for each of the following: benzene, carbon tetrachloride, chloroform, ethylene dichloride, trichloroethylene, and m-xylene. For dioxane the average desorption efficiency was 91 percent. The variation among laboratories was significant. For example, the range of values for benzene was 0.87 to 1.01 (28).

3.3 Sample Processing

Preparation of air filters, solid sorbents, or biological samples for analysis is an important part of the overall sampling and analysis procedure. Where appropriate, it is desirable to carry standards through all or part of the processing. Table 9.6 illustrates some sample treatment applications, including the use of various solvents, wet ashing, and thermal desorption. Not shown is the technique of low temperature, oxygen plasma ashing, which has found application for the determination of volatile elements in biological and other samples (29). The conditions under which a particular solvent is effective depend on the solute, the particle size, the temperature, and the degree of agitation. For some analytes, the use of an ultrasonic bath during desorption is required (e.g., amines on silica gel).

Preparation of air filters or biological samples for determination of metals by elemental analysis usually requires destruction of part or all of the organic matter present. Selection of an appropriate ashing technique depends on the risk of losing the analyte through volatilization, retention in the ashing vessel, or conversion to a chemical form that is unreactive or unavailable to the analytical method (30). Mercury, for example, is volatile under many wet-ashing and dry-ashing conditions. In the absence of chloride ions, however, samples containing mercury may be ashed with minimal losses using mixtures of nitric, sulfuric, and perchloric acids (30).

Volatilization losses of other elements may occur under oxidizing conditions (Rh, Os), reducing conditions (Se, Te), or in the presence of chloride ion in acidic solutions. The last category is the largest, including losses of antimony, arsenic, chromium, germanium, lead, and zinc. The chlorides of these elements are the volatile species, and the source of chloride may be the sample matrix or reduction of perchloric acid used for ashing. Retention of the analyte may

Table 9.6   Sample Treatment Methods Used in Industrial Hygiene

Method	Example	Reference[a]
Solvent desorption		
Acetone	Cresol	2001
Acetonitrile	PNAs	5506
Carbon disulfide	Hydrocarbons	1500
Carbon disulfide	Stoddard solvent	1550
Carbonate-hydroxide or dilute sulfuric acid	Chromium/Cr (VI)	7600
Ethanol	Nitroglycerin	2507
Formic acid	Acetic acid	1603
Hexane	Polychorobiphenyls	5503
Isooctane	Aldrin (R)	5502
Methanol	2,4,-D	5001
Triethanolamine	Nitrogen oxides	6700
Water	Methanol	2000
Water	Tungsten (soluble)	7074
Thermal desorption	Mercury	6000
Wet ashing		
Nitric acid + hydrogen peroxide	Arsenic compounds	7901
Nitric acid + hydrogen fluoride	Tungsten (insoluble)	7074
Nitric + perchloric acids	Elements for ICP	7300
Phosphoric acid	Silica (color)	7601
Dry ash (furnace or LTA)	Silica (XRD)	7501
Fusion in borate-carbonate	Fluoride	7902
Dissolve & resuspend in THF	Lead sulfide	7505
Clear in acetone, fix	Asbestos	7400
No treatment	Welding fume (XRF)	7200

[a] Numbers refer to methods found in Reference 8.

Table 9.7 **Methods for the Destruction of Organic Matter**

Method	Reagents Used	Temperature (°C)	Comments	Reference
Dry ashing	None	500–700	Slow, less supervision, for larger samples, loss of volatile elements	30
Wet ashing	Nitric, sulfuric, and perchloric acid	100–140	May also use hydrogen peroxide or potassium permanganate	30
	$(CH_3)_4NOH$	25–60	For tissue	32
Low temperature ashing	Oxygen plasma	25–100	Small samples only, little volatilization	29

occur by precipitation (e.g., loss of lead as the sulfate when sulfuric acid is used) or by reaction with the ashing vessel (e.g., formation of glasses with silica containers, or reduction of the element and diffusion as occurs with copper in silica). Finally, losses may be due to conversion of the element to a chemical form that is inert toward the analytical method. An example is the antimony (IV) compound formed in the absence of a strong oxidizing agent; the element in this form does not form the desired complex with Rhodamine B (31). Table 9.7 compares the common ashing techniques.

4 SAMPLE ANALYSIS

4.1 General Considerations

Analytical methods have undergone rather dramatic changes in the past decade. The classical "wet" techniques such as acid-base or oxidation-reduction titrations, or gravimetric determinations have given way to instrumental methods for many analyses. Characteristics of the instrumental methods are less dependence on the skill of the analyst, complex electronic equipment, and dramatically lowered detection limits. One of the major dividends of these methods is that specific compounds, rather than only elemental composition, can be determined. For example, chromatography in its various forms, X-ray diffraction, molecular spectrophotometry, and mass spectrometry all provide a great deal of information about the molecular species present. The importance of this to industrial hygiene analysis can be appreciated by examining a list of toxic substances (33). Approximately 95 percent of the substances are specific compounds rather than classes of compounds based on elemental analysis (Table 9.8). It is not meant to imply that the burden of speciation falls entirely on the analytical

Table 9.8 Specificity Required in Industrial Hygiene Measurements[a]

Type of Analysis	Number	Examples
Specific compound	503	Ammonia, benzene, lead arsenate
Elemental analysis only	18	Beryllium, cadmium, fluorine
Elemental analysis, soluble/insoluble forms	10	Tungsten, zinc
Elemental analysis, different oxidation states	1	Chromium

[a] Adapted from Reference 33.

method, however. In many cases (e.g., separation of gases from particulates) the sampling method is an integral part of the speciation process. For example, the separation of several species of organic and inorganic arsenic has been based on a combination of physical and chemical properties (12, 34, 35).

In certain situations it is desirable to determine more than one substance in a given sample (e.g., several metals in welding fume; mixed organic solvents). The (usually) small samples obtained in personal sampling limit the total amount of information obtainable from any one sample, however. In the development of future sampling and analytical procedures, two avenues appear promising: increased analytical sensitivity, and the use of methods with ability to determine more than one substance at a time. Examples of the first kind are provided by the development of atomic absorption spectrophotometry and gas chromatography. The early atomic absorption instruments, in the 1960s, used flame atomization. In more recent developments detection limits have been lowered by two or more orders of magnitude by various microsampling systems, including hydride generation and nonflame atomization. The development of new and more sensitive detectors for gas chromatography has also led to lower detection limits (36). Thus samples can be diluted to larger volumes and the capability for multiple determinations on a single sample is enhanced. Some techniques with capability to determine multiple substances simultaneously are emission spectroscopy, polarography, anodic stripping voltammetry, neutron activation analysis, X-ray fluorescence, and the various forms of chromatography.

4.2 Analytical Methods

The choice of which analytical method to use is frequently influenced by the sensitivities of the approaches available. For example, a number of methods have been applied to the determination of trace metals. As Figure 9.3 depicts, the range of sensititives is wide, covering more than 5 orders of magnitude between neutron activation analysis, one of the most sensitive methods for some elements, to potentiometry with specific ion electrodes. Improvements to several classical methods have made them considerably more sensitive. One of the most dramatic examples is the development of inductively coupled plasma techniques; improvements of 3 or more orders of magnitude in detection limits have resulted compared to older emission techniques (37).

In the field of industrial hygiene heavy reliance has been placed on two techniques: gas chromatography and atomic absorption spectrophotometry. An estimate of their importance is obtained from recent methods development work in the NIOSH-OSHA Standards Completion Program (7). As summarized in Table 9.9, 191 of 215 methods use one of the two techniques, with gas chromatography accounting for 165 of these. This does not constitute a claim that the two methods dwarf others in importance; it does indicate that they are extremely versatile however. Table 9.10 gives some applications of analytical methods to approximately 20 substances. The working ranges listed in the table refer to the total amount of substance in the air sample. Because the aliquot size depends on the method, varying from the entire sample for methods such as anodic stripping voltammetry to less than 1 percent of the sample in gas chromatography, instrumental detection limits must be adjusted for total sample size. For example, desorption of solid sorbents for gas chromatographic analysis requires a minimum of 1 to 2 ml of solvent even though the aliquot to the chromatograph is a few microliters. Similarly the minimum volume to which an ashed filter can be diluted accurately is 5 to 10 ml, while flame atomic absorption requires 1 to 3 ml per determination, and graphite furnace atomic absorption uses only 5 to 50 μl per aliquot analyzed. In the determination of arsenic by flame atomic absorption spectrophotometry, the instrumental detection limit of 0.1 μg/ml translates to 1 μg of arsenic per 10 ml of sample. For the more sensitive hydride generation mode (see Figure 9.4) the instrumental detection limit is approximately 0.05 μg of arsenic (8) and the entire sample can be used for a single determination. The instrumental sensitivity of graphite

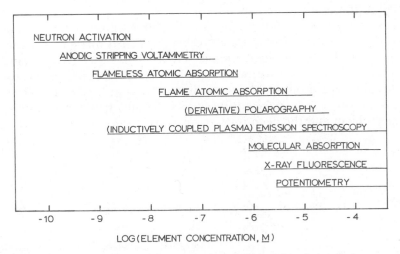

Figure 9.3 Approximate working ranges of some elemental analysis techniques.

Table 9.9   Analytical Methods in NIOSH-OSHA Standards Completion (7)

Analytical Method		Number of Methods
Gas chromatography	Flame ionization	144
	Electrolytic conductivity	8
	Flame photometric	6
	Electron capture	5
	Alkali flame ionization	2
Atomic absorption spectrophotometry		
Flame		23
Flameless		3
UV/visible spectrophotometry		13
Specific ion electrode		4
Combustible gas meter		2
Gravimetric		2
X-Ray diffraction		2
Fluorescence		1
High pressure liquid chromatography		1

furnace atomic absorption spectrophotometry is 100 pg of arsenic per determination; with the limitation of sample preparation, this equals 0.05 μg of arsenic per 10 ml of sample. Figure 9.5 illustrates similar interpretation of analytical methods for the determination of lead. The regions of applicability are shown for the lead specific ion electrode, anodic stripping voltammetry (ASV), flame atomic absorption spectrophotometry, and other AAS variations, including tantalum sampling boat (38), Delves sampling cup (39), and two varieties of graphite atomizer (40, 41).

Interaction of the sample treatment procedures with variations in analytical methods has resulted in a large number of published methods for the determination of lead in air or blood samples (Figure 9.6). Other methods, including ASV, are not shown but are nonetheless important and in general use. The determination of lead in blood is complicated by matrix effects in most of the methods shown, which makes differences in results between methods a perplexing problem and the need for a standard reference material more urgent.

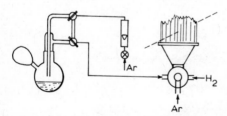

Figure 9.4   Apparatus for hydride generation: atomic absorption spectrophotometry.

4.3 Emerging Techniques

4.3.1 Inductively Coupled Plasma–Optical Emission Spectroscopy (ICP–OES)

The capability for simultaneous, multielement determinations over large dynamic ranges has made ICP–OES a promising technique (37, 42). Minimal matrix effects are seen, and the determination of elements in 25- to 50-μl volumes of biological fluids has been demonstrated (43). Some possible appli-

Table 9.10 Analytical Methods Used in Industrial Hygiene

Method	Example	Sample Working Range (μg)	Reference[a]
Chromatography			
Gas (flame ionization)	Chloroform	375–11,000	1003
	Stoddard solvent	900–27,000	1550
Gas (electron capture)	Polychlorobiphenyls	0.4–2.5	5503
Gas (conductivity)	Aldrin (R)	5–135	5502
Gas (flame photometric)	Phosdrin (R)	5–50	2503
HPLC/UV	Hippuric acid (urine)	200–1000	8301
Ion exchange	Inorganic acids	3–500	7903
Spectrophotometry			
UV/visible	Nitrogen oxides	0.5–18	6700
	Silica, crystalline	20–2000	7601
Infrared	Silica, crystalline	10–160	7602
Atomic absorption (flame)	Lead in blood	10–250	8003
Atomic absorption (graphite)	Arsenic	0.3–13	7901
Atomic absorption (hydride)	Arsenic	0.02–3	7900
Atomic absorption (cold vapor)	Mercury	0.001–2.5	6000
Emission (ICP)	Lead, other elements	2.5–1000	7300
X-Ray diffraction	Silica, crystalline	20–2000	7500
	Lead sulfide	30–2000	7505
X-Ray fluorescence	Welding/brazing fume	20–1000	7200
Electrochemistry			
Polarography	Formaldehyde	30–135	3501
Specific ion electrode	Fluoride	50–2000	7902
Microscopy	Asbestos	40,000–500,000 (fibers)	7400
Mass (weighing)	Nuisance dust	300–2000	0500

[a] Numbers refer to methods found in Reference 8.

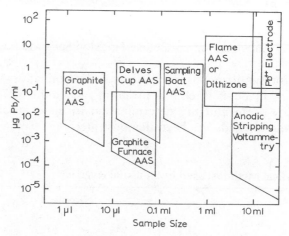

Figure 9.5 Working ranges for analytical methods for lead.

cations include the determination of more than 45 elements with good detection limits (< 0.03 μg/ml), including elements difficult to determine by other methods such as boron, phosphorus, hafnium, uranium, tungsten, yttrium, zirconium, and the rare earths.

4.3.2 Electron Spectroscopy for Chemical Analysis (ESCA)

ESCA is a technique for the examination of surfaces to a depth of only a few Ångstroms. Since the energies of photoelectrons from the target atoms are measured, their binding energies can be calculated. This gives the analyst information about the valence states of the target atoms as well as a semiquantitative, multielement analysis. Through the use of ESCA, ambient air pollution particulates were studied and found to contain sulfur tentatively identified as

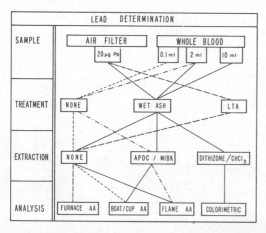

Figure 9.6 Sample treatment procedures used in some analytical methods for lead.

sulfate, sulfite, and sulfide ions, in addition to neutral sulfur and surface-bonded sulfur dioxide and sulfur trioxide (44, 45). Figure 9.7 presents an ESCA spectrum obtained on an air particulate sample (Nuclepore filter) from a smelter. The presence of lead, tin, and sulfur is indicated; the sulfur was present as sulfate in this sample (46). ESCA and other surface techniques, although the analyses are expensive and rather slow, have been used increasingly for microanalysis of air particulates (47). ESCA has been especially useful for identifying the oxidation states of sulfur and nitrogen in air pollution samples (48).

4.3.3 X-Ray Fluorescence (XRF)

Advantages of XRF for environmental samples are simultaneous multielement analyses, with no sample treatment necessary in most cases (49, 50). The analysis is nondestructive, and individual particles of diameter approximately 1 μm or larger can be analyzed. Figure 9.8 is a scanning electron photomicrograph of air particulates sampled on a Nuclepore filter at a sulfite pulp mill employing calcium bisulfite. Microanalysis of several of the particles by XRF determined that the major elements present were sulfur, calcium, and iron. Smaller amounts of phosphorus, silicon, aluminum, and magnesium were also found.

4.3.4 Ion Chromatography (IC)

A gap in inorganic and organic analysis, the rapid quantitation of anions and cations, has some promise of aid from the recent development of ion chromatography. Essentially a liquid chromatograph using pellicular ion exchange analytical columns and an electrical conductivity detector, the equipment elutes aqueous samples with carbonate-bicarbonate or other buffers for anion analysis, or acidic eluents for cation analysis (51). Applications have been demonstrated

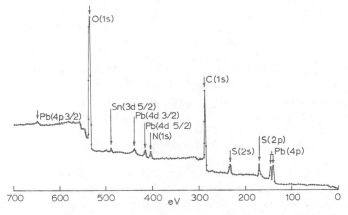

Figure 9.7 ESCA spectrum for air particulates form a copper smelter. From Eller (46).

Figure 9.8 Scanning electron micrograph of air particulates on 0.4 μm Nuclepore filter (4600 ×).

for the determination of sulfates and nitrates in ambient aerosols (52), and for electrolytes, including ammonium ion, in serum, urine, and other biological fluids (53). Conditions for the determination of a number of amines and low molecular weight organic acids have also been established (51). A chromatogram showing separation of seven anions in less than 15 min appears in Figure 9.9 (46). The recent introduction of electrochemical detection has added a number of electroactive species to the capabilities of ion chromatography, including sugar alcohols, mono- and disaccharides, sulfide, cyanide, hypochlorite, and phenols (54, 55).

4.3.5 Derivative Spectroscopy

Using wavelength modulation, derivative ultraviolet or visible spectra are obtained, providing a sensitive means of detection of gases (56). Detection limits for some gases are nitric oxide, 5 ppb; nitrogen dioxide, 40 ppb; ozone, 40 ppb; benzene, 25 ppb; formaldehyde, 200 ppb; and mercury vapor, 0.5 ppb.

4.3.6 Gas Chromatographic Developments

The power of gas chromatography in industrial hygiene and environmental analyses has been considerably enhanced in the past decade by the growing use of capillary columns (57, 58). These columns, typically 3 to 20 meters long and 0.2 to 0.4 mm internal diameter, give greatly improved speed, resolution, and sensitivity compared to packed columns. Some examples of the capabilities of capillary columns are separation of 13 hydrocarbons in 18 seconds, and resolution of over 100 components of a polycyclic aromatic hydrocarbon fraction extracted from airborne particulate (58).

The introduction of new detectors, and the use of two detectors simultaneously has also been of significant importance to gas chromatography. In addition to the widely used flame ionization detector, the thermionic, nitrogen and phosphorus flame photometric, electrical conductivity, UV absorbance, and photoionization detectors are in common use (55). As an example of the use of multiple detectors, atmospheric hydrocarbons were detected at the 1 to 10 picogram level and simultaneously classified according to degree of saturation by the use of tandem photoionization and flame ionization detectors (59).

In combination with other analytical techniques as detectors, gas chromatography comes to realize its full potential. For example, mass spectrometry contributes very powerful qualitative information in the application of gas chromatography-mass spectrometry (GC-MS). Thus, separation and identification of polynuclear aromatic hydrocarbons (PNA) in diesel exhaust is possible (60). The next section illustrates another "hyphenated" technique involving gas chromatography, GC-FTIR.

4.3.7   Fourier Transform Infrared Spectroscopy (FTIR)

FTIR is emerging as an extremely powerful analytical tool in environmental studies. Mounted on a mobile source, FTIR has been used by the U.S. EPA for remote and cross-stack measurement of stack gas concentrations (61). On a

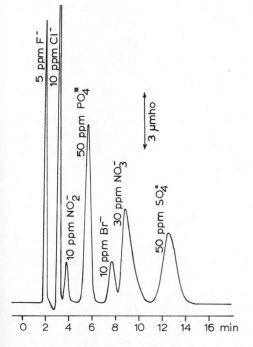

Figure 9.9 Chromatogram of anions; eluent: 0.003 M HCO_3^-, 0.0024 M CO_3^-. From Eller (46).

smaller scale, FTIR should prove no less useful in industrial hygiene. Powerful computer-based spectral search systems are available for the commercially available FTIR instruments (62, 63). With capability for extremely fast (1 sec) spectral scans, FTIR has been used successfully as a detector, in as many as five spectral windows simultaneously, for capillary column gas chromatography (64). Thus, GC-FTIR offers an attractive complement to the powerful GC-MS techniques already used to identify complex mixtures in industrial hygiene samples.

5 SAMPLING AND ANALYTICAL GAPS

A number of factors relating to sampling, analysis, or the analyte itself may hinder the development of validated sampling and analytical methods. This section discusses some of these factors, along with examples, with the hope that future research will fill these gaps.

5.1 Sampling-Related Problems

Three problems related to sampling are poor collection efficiency (low capacity), poor sample storage characteristics, and low desorption efficiencies. An example of the first type is the low capacity of conventional charcoal tubes for some of the chlorofluoromethanes, ethyl chloride, methyl acetylene, and stibine. Alternate solid sorbents should be investigated. The investigation of properties of existing and new sorbents and their application to sampling in industrial hygiene is a fertile field (65). Substances that tend to deteriorate over a period of several days when sorbed on charcoal tubes are methyl formate, nitromethane, nitroethane, 2-nitropropane, crotonaldehyde, and diglycidyl ether. Desorption efficiencies below 75 percent are characteristic of ethanolamine, *n*-butylamine, diisopropylamine, and other amines on silica gel for freshly spiked samples. Also, chloroacetaldehyde, diphenyl, furfural, and 1-nitropropane show low desorption efficiencies from activated charcoal.

Thorough theoretical and experimental evaluation of passive monitor designs is urgently needed in view of the rapid proliferation of these devices. A number of such evaluations and protocols have been published (e.g., 25, 66, 67), but additional work is needed to define more clearly the effects of interferences, mixtures, and sampling parameters.

5.2 Analysis-Related Problems

In some cases the analytical method is the limiting factor because it is lacking in specificity, sensitivity, or precision. Adaptation of a fluoride-specific electrode to the determination of perchloryl fluoride and chlorine trifluoride is unsuccessful because of interference by other fluorine compounds such as oxygen difluoride, and chlorine monofluoride. Also, oxidizing agents such as chlorine, bromine, nitrites, and chloramine interfere with colorimetric determinations for iodine (68) and chlorine dioxide (69).

For personal samples, gas chromatography is not sensitive enough in the cases of acrylamide, dinitrotoluene, nitrogen trifluoride, and perchloromethyl-mercaptan, while atomic absorption spectrophotometry lacks sensitivity for hafnium, osmium, and tantalum.

Poor analytical precision is characteristic of the particle-count methods that have been used for graphite, mica, portland cement, soapstone, and talc. Also showing poor reproducibility are gas chromatography for dibutyl phosphate, maleic anhydride, acetic acid, methylamine, and Arachlor 1242, and spectro-photometric methods for anisidine (72), *p*-phenylenediamine (70), and oxalic acid (71).

Sampling and analytical methods are also needed for low levels of polychlorinated dioxins (73).

5.3 Problems Related to the Analyte

Some analytes, such as boron trifluoride, *tert*-butyl chromate, decaborane, diborane, and sulfur monochloride, are very reactive in the presence of oxygen or water vapor, therefore very difficult to collect. Possible solutions include direct measurement in air by spectral means or specific, reactive sorbents.

REFERENCES

1. R. G. Melcher, *Anal. Chem.,* **55,** 40R (1983).
2. M. L. Bolyard and D. L. Smith, "Part I.C. in NIOSH Manual of Analytical Methods," 3rd ed., U.S. Dept. of Health and Human Services Publ. (NIOSH) 84-100 (1984).
3. U.S. National Bureau of Standards, *Accuracy in Trace Analysis: Sampling, Sample Handling, Analysis. Proceedings of the Seventh Materials Research Symposium,* Superintendent of Documents (No. C13.10:422), Washington, D.C., 1976.
4. J. P. Cali, G. N. Bowers, Jr., and D. S. Young, *Clin. Chem.,* **19,** 1208 (1973).
5. U.S. National Bureau of Standards, *NBS Standard Reference Materials Catalog,* 1981–83 ed., NBS Special Publication 260, Washington, D.C., 1981.
6. P. M. Eller and J. C. Haartz, *Am. Ind. Hyg. Assoc. J.,* **38,** 116 (1977).
7. D. G. Taylor, R. E. Kupel, and J. M. Bryant, "Documentation of the National Institute for Occupational Safety and Health Validation Tests," Department of Health, Education and Welfare Publication (NIOSH) 77-185, Washington, D.C., 1977.
8. *NIOSH Manual of Analytical Methods,* 3rd ed., U.S. Department of Health and Human Services Publ. (NIOSH) 84-100 (1984).
9. J. V. Crable and R. G. Smith, *Am. Ind. Hyg. Assoc. J.,* **36,** 149 (1975).
10. P. Schlecht, National Institute for Occupational Safety and Health, personal communication.
11. R. H. Hill, Jr. and J. E. Arnold, *Arch. Environ. Contam. Toxicol.,* **8,** 621 (1979).
12. R. J. Costello, P. M. Eller, and R. D. Hull, *Am. Ind. Hyg. Assoc. J.,* **44,** 21 (1983).
13. P. M. Eller, J. C. Haartz, R. D. Hull, and B. Frumer (unpublished data).
14. F. W. Williams, J. P. Stone, and H. G. Eaton, *Anal. Chem.,* **48,** 442 (1976).

15. E. V. Ballou, Ed., *Second National Institute for Occupational Safety and Health Solid Sorbents Round-table*, Department of Health, Education and Welfare Publication (NIOSH) 76-193, Washington, D.C., 1976, p. 3.

16. G. O. Wood and R. G. Anderson, *Am. Ind. Hyg. Assoc. J.*, **36**, 538 (1975).

17. L. D. Butler and M. F. Burke, *J. Chromatogr. Sci.*, **14**, 117 (1976).

18. H. K. Dillon, W. J. Barrett, and P. M. Eller, *Am. Ind. Hyg. Assoc. J.*, **39**, 608 (1978).

19. E. D. Palmes, A. F. Gunnison, J. DiMattio, and C. Tomczyk, *Am. Ind. Hyg. Assoc. J.*, **37**, 570 (1976).

20. E. D. Palmes and C. Tomczyk, *Am. Ind. Hyg. Assoc. J.*, **40**, 588 (1979).

21. B. C. Cadoff and J. Hodgeson, *Anal. Chem.*, **55**, 2083 (1983).

22. E. D. Palmes and A. F. Gunnison, *Am. Ind. Hyg. Assoc. J.*, **34**, 78 (1973).

23. W. Jones, E. D. Palmes, C. Tomczyk, and M. Millson, *Am. Ind. Hyg. Assoc. J.*, **40**, 437 (1979).

24. K. D. Reiszner and P. W. West, *Environ. Sci. Technol.*, **7**, 726 (1973).

25. R. D. Hull and M. E. Cassinelli, Tentative Laboratory Performance Specifications, Testing Protocol, and Evaluation Criteria for Passive Samplers (unpublished, NIOSH, September 1982).

26. E. V. Ballou, Ed., *Second National Institute for Occupational Safety and Health Solid Sorbents Round-table*, Department of Health, Education and Welfare Publication (NIOSH) 76-193, Washington, D.C., 1976, p. 55.

27. R. H. Hill, Jr., C. S. McCammon, A. T. Saalwaechter, A. W. Teass, and W. J. Woodfin, *Anal. Chem.*, **48**, 1395 (1976).

28. R. L. Larkin, J. V. Crable, L. R. Catlett, and J. J. Seymour, *Am. Ind. Hyg. Assoc. J.*, **38**, 543 (1977).

29. T. H. Lockwood and L. P. Limtiaco, *Am. Ind. Hyg. Assoc. J.*, **36**, 57 (1975).

30. T. T. Gorsuch, in Reference 3, p. 491.

31. A. A. Al-Sibbai and A. G. Fogg, *Analyst*, **98**, 732 (1973).

32. L. Murthy, E. E. Menden, P. M. Eller, and H. G. Petering, *Anal. Biochem.*, **53**, 365 (1973).

33. American Conference of Governmental Industrial Hygienists, *Threshold Limit Values*, ACGIH, Cincinnati, Ohio, 1983.

34. R. S. Braman, p. 1 in D. F. S. Natusch and P. K. Hopke, Ed., *Analytical Aspects of Environmental Chemistry*, John Wiley & Sons, New York (1983).

35. E. A. Crecelius and R. W. Sanders, *Anal. Chem.*, **52**, 1310 (1980).

36. C. H. Hartman, *Anal. Chem.*, **43**, 113A (1971).

37. V. A. Fassel and R. N. Kniseley, *Anal. Chem.*, **46**, 1110A (1974).

38. H. L. Kahn and J. S. Sebestyen, *At. Absorpt. Newsl.*, **9**, 33 (1970).

39. M. M. Joselow and J. D. Bogden, *At. Absorpt. Newsl.*, **11**, 99 (1972).

40. J. F. Lech, D. Siemer, and R. Woodriff, *Environ. Sci. Technol.*, **8**, 840 (1974).

41. J. A. Ealy, N. E. Bolton, R. J. McElheny, and R. W. Morrow, *Am. Ind. Hyg. Assoc. J.*, **35**, 566 (1974).

42. NIOSH Manual of Analytical Methods, 3rd ed., op. cit., Method 7300.

43. R. N. Kniseley, V. A. Fassel, and C. C. Butler, *Clin. Chem.*, **19**, 807 (1973).

44. N. L. Craig, A. B. Harker, and T. Novakov, *Atmos. Environ.*, **8**, 15 (1974).

45. T. Novakov, P. K. Mueller, A. E. Alcocer, and J. W. Otvos, *J. Colloid Interface Sci.*, **39**, 225 (1972).

46. P. M. Eller (unpublished results).

47. R. W. Linton, D. T. Harvey, and G. E. Cabaniss, in D. F. S. Natusch and P. K. Hopke, Ed., op. cit., p. 137.

48. T. Novakov, S. G. Chang, R. L. Dod, and L. Gundel, ibid., p. 191.

49. J. R. Rhodes, *Am. Lab.,* July 1973.

50. T. G. Dzubay, Ed., *X-Ray Fluorescence Analysis of Environmental Samples,* Ann Arbor Science Publishers, Ann Arbor, Mich., 1977.

51. H. Small, T. S. Stevens, and W. C. Bauman, *Anal. Chem.,* **47,** 1801 (1975).

52. J. Mulik, R. Puckett, D. Williams, and E. Sawicki, *Anal. Lett.,* **9,** 653 (1976).

53. C. Anderson, *Clin. Chem.,* **22,** 1424 (1976).

54. Technical Note 11, Dionex Corp., Sunnyvale, CA (1983).

55. J. S. Fritz, D. T. Gjerde, and C. Pohlandt, *Ion Chromatography,* Dr. Alfred Huthig Verlag (Heidelberg, 1982).

56. R. N. Hager, Jr., *Anal. Chem.,* **45,** 1131A (1973).

57. G. Becher, A. Bjorseth, and B. Olufsen, ibid., p. 369.

58. M. Novotny, in D. F. S. Natusch and P. K. Hopke, Ed., op. cit., p. 61.

59. W. Nutmagul, D. R. Cronn, and H. H. Hill, Jr., *Anal. Chem.,* **55,** 2160 (1983).

60. D. R. Choudhary and B. Bush, in G. Choudhary, Ed., *Chemical Hazards in the Workplace,* ACS Symposium Series 149, American Chemical Society (Washington, D.C., 1981), p. 357.

61. W. F. Herget, *Applied Optics,* **21,** 635 (1982).

62. H. B. Woodruff and G. B. Smith, *Anal. Chem.,* **52,** 2321 (1980).

63. S. R. Lowry and D. A. Huppler, *Anal. Chem.,* **53,** 889 (1981).

64. P. R. Griffiths, J. A. de Haseth, and L. V. Azarraga, *Anal. Chem.,* **55,** 1361A (1983).

65. E. C. Gunderson and E. L. Fernandez, p. 179 in G. Choudhary, Ed., op. cit.

66. D. W. Gosselink, D. L. Braun, H. E. Mullins, S. T. Rodriguez, and F. W. Snowden, ibid., p. 196.

67. R. S. Stricoff and C. Summers, ibid., p. 210.

68. J. K. Johannesson, *Anal. Chem.,* **28,** 1475 (1956).

69. American Public Health Association, *Standard Methods for the Examination of Water and Wastewater,* 13th ed., APHA, New York, 1971.

70. J. T. Steward, T. D. Shaw, and A. B. Ray, *Anal. Chem.,* **41,** 360 (1969).

71. *National Institute for Occupational Safety and Health Manual of Analytical Methods,* 2nd ed., Department of Health, Education and Welfare Publication (NIOSH) 77-157-A,B,C, Washington, D.C., 1977, P&CAM 142.

72. J. Bergerman and J. S. Elliot, *Anal. Chem.,* **27,** 1014 (1955).

73. C. Rappe and H. R. Buser, p. 319 in G. Choudhary, Ed., op. cit.

CHAPTER TEN

The Emission Inventory

ROBERT L. HARRIS, JR., Ph.D., and
EARL W. ARP, JR., Ph.D.

1 INTRODUCTION

An emission inventory for an industrial or commercial enterprise is a compilation of information from which one can calculate or estimate the rates (quantity per unit time) at which pollutants are released to the environment. For purposes of this chapter, only the emissions that contaminate workroom or community air are considered; emissions to surface or ground water, to soil, or to other environmental receptors are treated only as they may, in turn, result directly in emissions to workplace air or community air.

An emission inventory may be simple or complex. A rudimentary inventory may consist of source location, date, identification of process, a qualitative listing of materials used, and an index of size (e.g., annual production rate) for the subject enterprise. Such an inventory, along with emission factors generated by studies of other similar processes, will permit the making of an estimate of annual emissions. A comprehensive emission inventory, on the other hand, may contain sufficient detail to permit quantitation of emissions, including temporal variations, for a number of specific materials from each point of release in a complex industrial process.

An inventory of emissions, along with various other kinds of companion information, discussed later in this chapter, permits the making of estimates of the nature, and sometimes the intensity, of exposures to airborne agents in workplaces or in the community. The level of detail needed and achievable for the inventory depends both on the purposes for which it is to be used and the data sources, or data generating efforts, that can be utilized.

2 ELEMENTS OF AN EMISSION INVENTORY

The compilation of emission inventories is a well-established, specifically identifiable activity in the field of air pollution control. Practices and procedures have been highly developed and descriptions of the technology are available (1, 2). Emission inventory has been practiced in industrial hygiene for many years but has not been identified as a categorical work area in this field to the extent that it has been in the field of air pollution control. Although the types of data used and the techniques for obtaining them vary somewhat, the same basic elements appear in emission inventories in both fields of work.

2.1 Identification of Agents

Recognition, evaluation, and control of hazards are the three basic steps in the practice of industrial hygiene and community air pollution control. An emission inventory, regardless of whether it is specifically identified as such, is necessary in all three steps and is particularly important in the first two, recognition and evaluation. It is clear that if a hazard is to be dealt with, it must first be recognized; this recognition, and the identification of an emission, is a rudimentary emission inventory. Evaluation requires more than identification. In situations involving release of chemical agents to the air, evaluation may include obtaining additional information such as quantity, character, and temporal variations of emissions, all of which are part of the emission inventory. The design and implementation of emission control requires detailed information about the emission source that goes beyond the level of detail usually required for an emission inventory.

The federal Toxic Substances Control Act of 1976 (3), among other things, provides for the collection of information regarding commercially produced chemicals. Such information will permit preliminary assessment of potential exposures and possible effects on health and the environment. Implementation of the act will require identification, by process and location, of many chemical agents in industry and commerce. The notification, reporting, and recordkeeping provisions of the act will facilitate the agent identification component of a comprehensive, plantwide, emission inventory.

Not all materials handled in industry and commerce are hazardous. More than 4 million distinct chemical compounds have been identified, and the number is increasing at the rate of about 6000 per week (4). Of the millions of compounds that exist, some 63,000 are thought to be in common use (4). Some toxic dose information, based on experimental animal work or other observations, is available on about 59,000 compounds (5). Some of these are relatively nontoxic, others are not in common use, and for many the toxicity information is fragmentary. Probably fewer than 1000 compounds have been identified with occupational health or community air pollution problems sufficiently to permit development of workplace or air quality standards. The identification, by means of an emission inventory, of materials that are released from a process or operation, however, is a fundamental step in the recognition of those that represent a potential hazard in the workplace or the community.

For manufacturing processes preliminary identification of potentially hazardous agents often can be based upon the identification of process raw materials, intermediate and by-product materials, and process end products. In commercial enterprises the identification of materials that are handled permits the singling out of those that may represent potential hazards. For combustion sources information on the composition of the fuel used, and the type of combustion equipment, permits qualitative identification of pollutant components.

The evaluation phase of an industrial hygiene or community air pollution problem requires, in addition to identification of the agents of concern, a number of other kinds of information that can be obtained in an emission inventory. Among the most fundamental of these is the identification and description of the site or location at which the contaminant is released to the air.

2.2 Identification of Emission Sites

The most cursory emission inventory may identify the site or location of an emission source only as a particular plant or commerical establishment. Such location information is generally useful only for preliminary surveys of community air pollution or for indicating the need for more thorough workplace exposure evaluation. Any emission inventory use other than agent identification alone requires more specific emissions location information. For example, in air pollution control the identification of specific stacks, vents, and other points of emission is necessary for diffusion modeling and for most impact evaluations and emission regulatory activities. For industrial hygiene purposes the location of a specific workplace, process point, and perhaps even a particular process equipment opening (e.g., a mixer charging port) or work practice (e.g., the handling of shipping bags after use) may be needed. Such location information is vital to the hazard assessment process. It is necessary for identifying the workers subject to exposure from the particular source and for identifying alternatives from which to select the means for control of any hazard caused by the emission.

An emission inventory that identifies both the materials emitted from a process and the specific locations at which these materials are released can serve as the first step, and perhaps the only step necessary, in the evaluation of a hazard and initiation of a control effort.

2.3 Time Factors in Emissions

Time resolution in emission inventories may be yearly, seasonal, monthly, weekly, daily, hourly, or even less than hourly, depending on specific needs and the availability of data. For initial surveys of community air pollution, yearly average emissions by plant site may be satisfactory. At the other extreme, the assessment of emissions to workplaces that involve cyclic or intermittent operations may require use of time intervals shorter even than one hour. For short time or intermittent operations—for example, the taking of materials samples at process sampling ports—the actual time interval of emission and the frequency with which the operation occurs should be recorded in the inventory.

In some cases the interval of record for the quantity of material released is relatively long—for example, a monthly record of solvent use—even though actual release may be cyclic or may occur over short intervals. In such cases the emission inventory record should contain a sufficiently detailed description of the process or operation to permit estimation of actual emission intervals and the quantities of materials released during these intervals.

It is important that the emission inventory record include both the date on which the inventory was done and the calendar interval for which it applies. When a change in process or operation occurs that materially affects the composition, quantity, or condition of an emission, the emission inventory record should be updated to reflect the change. In the absence of any substantial change, the inventory should be revalidated at convenient intervals, perhaps annually, or as may be required by governmental regulation. When an emission inventory record is updated or revalidated, the old record should be retained; a sequential inventory file over a long period may be invaluable in future retrospective environmental epidemiologic studies.

3 QUANTITATING EMISSIONS

Emissions can be quantitated either by direct measurement, such as source testing, or by indirect means. Indirect means include techniques such as process materials balance or the determination of an index parameter—for example, a production rate—that can be related empirically to emissions.

3.1 Source Sampling

Source sampling is ordinarily associated with measurement of air pollutant emissions. Under some circumstances the techniques can be applied to industrial hygiene investigations as well. The techniques for air pollutant source testing have been described in detail by Paulus and Thron (6); their chapter "Stack Sampling" lists 72 references. The Environmental Protection Agency (EPA) has published stepwise procedures on source sampling for particulates (7); this publication contains a number of data recording forms that are useful in sampling not only for particulates but for other agents as well.

Source sampling ordinarily consists of withdrawing a representative sample from a contaminant-bearing gas stream in a duct or stack. Analysis of the sample yields data on concentration of the contaminant in the gas stream.

Concentration data, combined with companion data on gas flowrates in the ducts or stacks, yield values for contaminant emission rates for gas streams released to the atmosphere.

The critical concern in source sampling is the representativeness of the sample. Both composition and flowrate of a contaminated gas stream may vary as the processes and operations that generate it vary. Thus representativeness of a source sample depends very much on the representativeness of processes and operations at the time of sampling.

When the contaminant is particulate, the collection of a representative sample requires isokinetic sampling and use of an unbiased sampling traverse pattern. Isokinetic sampling is performed by taking the sample at such a flowrate that the sampled gas stream enters the inlet nozzle of the sampling probe with velocity equal to that which prevails at the specific point in the cross section of the stack or duct from which the sample is being withdrawn. When the velocity of gas at the sampling point in the duct or stack is greater than that in the sampling nozzle, part of the approaching gas stream is deflected around the nozzle. Smaller particles tend to follow the deflected gas stream while larger ones, by virtue of their momentum, tend to continue their trajectories and enter the nozzle; this results in a nonrepresentative overabundance of larger particles in the sample. When the velocity of the gas stream at the sampling point in the duct or stack is lower than the velocity entering the sample nozzle, the gas stream converges into the nozzle inlet, carrying with it the smaller particles but losing some of the larger ones, which are carried past the nozzle by their momentum; the sample then is nonrepresentative because of a deficiency in larger particles.

The velocity of the gas stream in a duct or stack is not uniform throughout its cross section. For this reason, and because particles are not necessarily uniformly distributed within a duct or stack, a specific traverse pattern is ordinarily used in source sampling and the measurement of velocity for determination of flowrates. For purposes of a sampling traverse, the cross-sectional area of the duct or stack is divided into equal sized subareas; sample increments and velocity readings are taken at the centers of these subareas. Sampling time should be the same for each traverse point in a duct or stack. Isokinetic sampling and equal area traverses are discussed in detail, including descriptions of apparatus and calculation methods, in the source sampling references cited earlier (6, 7).

The sampling of gases and vapors differs from sampling of particulates in that isokinetic sampling is not required unless concentrations differ from place to place in the duct cross section. The collection apparatus and reagents used in a gas sampling train also differ from those used for particulates. Filtration, inertial size classification, and impingement with capture in liquid media, are the collection mechanisms used for particulates source sampling; liquid absorption, adsorption on solids, and freeze-out are the methods usually employed for gas and vapor sampling. Sampling methods and analytic procedures for gases and vapors are described in Chapter 17, Patty's Volume I. Sampling apparatus, collecting media, and analytic methods for a number of gases and vapors have been tabulated by Paulus and Thron (6).

As mentioned earlier, the techniques of source sampling are most often applied in air pollution emission measurements. They can, in some circumstances, be applied for inplant industrial hygiene purposes as well. When a workplace is served by general dilution ventilation in the exhaust mode, the techniques of source sampling can be applied to the exhausted airstreams to determine the rate at which contaminants are released to the workplace air. The calculation methods described in Section 5.3.1 can be used to estimate emissions when the concentrations of contaminant in the supply and exhaust air, and the ventilation rate are known with the system at equilibrium. In applying the equations to exhaust air streams $K = 1$, and the contaminant generation rate is the sum of the generation rates calculated for all exhaust discharges. The procedure is most applicable when general dilution ventilation in the exhaust mode is the sole, and controlling, ventilation regimen for the workplace, that is, when there is no mechanical local exhaust ventilation and when there is no local exhaust component to natural ventilation. If local exhaust ventilation is used in the workplace, its influence as general dilution ventilation must be taken into account when using this technique to estimate workplace emissions.

Source sampling techniques may also be used to quantitate emissions from individual points of release in a workplace. The emissions may be captured using a temporary exhaust ventilation setup, and the amount of material released may be determined by sampling from that exhaust stream. Application of this technique to a single source or emission point in a space that contains several sources of the air contaminant requires either elimination of the influence of the other sources or correction for them. A mechanical arrangement can be provided to supply contaminant-free outside air to the test source. The exhaust stream from the test source will then contain only contaminant from that source and will be unbiased by contaminant from other sources. Alternatively, monitoring of the concentration of contaminant in room air that supplies the source test exhaust system permits correction for other sources; emissions from the test source can be determined by the difference in contaminant concentration in supply and exhaust air of the test system.

3.2 Materials Balance

In some cases knowledge of processes and operations permits determination of the amount of material released to the air of a workplace without emission measurements. If it is known that a gas or vapor is generated by chemical reaction or otherwise, and is released to workplace air in proportion to the use of a raw material or a production rate, that index of generation can be used to determine the release rate. Examples include the generation of products of combustion by unvented open flames, as is the case with direct fired unit heaters. Here fuel composition and use rate are indices of contaminant emissions. Uses of volatile solvents in which the solvents do not become part of a product, but evaporate completely into the workplace air, are common in industry. Solvent use rate, in such cases, is also an emission rate. When exhaust ventilation is applied to some operations in a workplace and not to others, distinction must be made between that portion of the material which is captured by exhaust ventilation and that which is released in the occupied workplace; only the portion that is released directly to the workplace air is used to estimate workplace exposures. For estimating community air pollution emission rates, the total amount of volatile material that evaporates into the atmosphere is taken as the emission rate regardless of whether the material is released to workroom air or through exhaust ventilation systems.

The American Petroleum Institute (API) has reported mathematical relationships that describe evaporation losses of petroleum products from tanks during loading and unloading (8). The materials balance concepts of the API procedure can be applied to estimating vapor emissions from the loading of volatile liquids into vessels that are vented to workroom air. The mass of vapor expelled by displacement when a volatile liquid is transferred into a vessel is

$$M = 1.37 \ VSP_v \frac{mw}{T} \tag{1}$$

where M = mass of vapor expelled (lb)
   V = volume of liquid transferred to the vessel (ft^3)
   S = fraction of vapor saturation of expelled air
   P_v = true vapor pressure of the liquid (atm)
   mw = molecular weight of the vapor
   T = temperature of the tank vapor space (°R)

Except for S, the fraction of vapor saturation, the various parameters of Equation 1 are ordinarily known or can be measured easily. For splash filling of a vessel that was initially vapor free, or for the refilling of a vessel from which the same liquid has just been withdrawn, the value of S can ordinarily be taken as 1 (8).

Equation 1 may overestimate the mass of vapor emission if the vessel walls are substantially colder than the volatile liquid. The true vapor pressure depends on the temperature of the liquid; in the case of the cold vessel, however, some vapor may condense on the inner wall surfaces and fail to escape through the vent into the workroom air.

When complete evaporation of a volatile material does not take place, some index other than total use is needed for quantitating emissions. In the simple case, when the amount used and the amount remaining can be determined, the amount released as vapor can be obtained by difference. When this is not possible, more sophisticated means such as exhaust air sampling must be employed.

With complete evaporation of a mixture of volatile materials, the quantity of each component that vaporizes is simply the quantity of that material in the mixture. Partial evaporation of a mixture, however, does not necessarily yield vapor quantity of each component in proportion to the quantity of that component in the liquid mixture. According to Raoult's law, the equilibrium partial pressure of each component of a perfect solution is the product of the vapor pressure of the pure liquid and its mole fraction in the solution:

$$p_n = P_n x_n \qquad (2)$$

where p_n = partial pressure of component n
    P_n = vapor pressure of pure liquid n
    x_n = mole fraction of component n in the liquid mixture

Thus vapor yielded by partial evaporation from a mixture of volatile materials is richer in the more volatile and leaner in the less volatile components than is the original liquid solution. The use of Raoult's law permits estimation of emission rates of components of a solution when partial evaporation takes place. The composition of the parent solution in each case must be known; values for vapor pressures of pure liquids can be found in chemical handbooks. When a substantial fraction of a liquid mixture evaporates, the change in its composition as the fractions of more volatile components decrease should be taken into account in applying Raoult's law.

Raoult's law should be used with caution in estimating emissions from partial evaporation of mixtures; not all mixtures behave as perfect solutions. Elkins, Comproni, and Pagnotto measured benzene vapor yielded by partial evaporation of mixtures of benzene with various aliphatic hydrocarbons, chlorinated hydrocarbons, and common esters, as well as partial evaporation of naphthas containing benzene (9). Most measurements for all four types of mixture showed greater concentrations of benzene vapor in air than were predicted by Raoult's law. Of five tests with naphtha-based rubber cements, one yielded measured values of benzene concentration in air in agreement with calculated values, the other four showed measured benzene concentrations in air to be 3 to 10 times greater than those calculated using Raoult's law.

Substantial deviation from Raoult's law is not always the case, however, even with benzene. Runion compared measured and calculated concentrations in air of benzene in vapor mixtures yielded by evaporation from a number of motor gasolines and found excellent agreement (10).

In the absence of other more certain means, Raoult's law can be used to estimate emissions generated by partial evaporation of mixtures that approximate ideal solutions or mixtures in which the solution is nearly pure in one component. The applicability of Raoult's law to the mixtures being assessed should be validated, or quantitative measurements of emissions should be done, if accurate emission values are needed.

For dilute solutions the partial pressure of the component present in lower concentration is given by Henry's law, expressed as follows:

$$p_n = H_n x_n \tag{3}$$

where p_n = partial pressure of component n
H_n = Henry's law constant
x_n = mole fraction of component n in the liquid mixture

Henry's law is also applicable to the solubility of a gas in dilute liquid solution, and solubilities of gases in liquid may be expressed in terms of Henry's law constants. These constants and the applicable concentration range for valid use of Henry's law can be determined only empirically.

Application of materials balance concepts for determination of emissions other than those for combustion, chemical reaction, or evaporation of volatile materials, ordinarily requires engineering analysis on a case-by-case basis.

3.3 Emission Factors

The need exists for emissions estimates for large numbers of sources in community air pollution studies, but the impracticability of source-by-source emissions tests has led to the development of emission factors. An emission factor is a pollutant emission rate for a particular type of emission source expressed as a quantity of pollutant released per unit of activity of that type source. The unit of activity chosen in each case is one that can be determined and can be related quantitatively to emissions; it may be ton of product, million Btu of heat produced, mile of vehicle travel, or other such index unit. Emission factors represent typical emissions from a class of sources and ordinarily cannot be applied with confidence to individual sources. In the absence of other information, however, emission factors for a particular source type can give useful insights into the character and general levels of emissions from individual sources of that type.

The most reliable emission factors are those based on a combination of emission measurements, process data, and engineering analysis for a large number of sources. Those that do not have a theoretical basis and are derived from only one type of data, or from data from only a few sources, should be used with caution. In some cases sufficient knowledge or information is available to permit development of empirical or analytic relationships between emission rate and some process parameter such as material composition or stream temperature. Such factors are generally the most reliable of all and can even be applied to individual sources with reasonable confidence.

Several thousands of individual air pollutant emission factors for a large number of source types have been tabulated and reported by the EPA's Office of Air and Waste Management (11). Process descriptions and emission control practices, along with typical collection performance for various types of control, are presented for most of the source types covered. Table 10.1 lists major source types for which emission factors appear in the current publication. In a separate document the EPA has published emission factors for arsenic, asbestos, beryllium, cadmium, manganese, mercury, nickel, and vanadium for processes involving these materials (12). Emission factors for hydrocarbons, carbon monoxide, and oxides of nitrogen from mobile sources have been published by the EPA Office of Mobile Source Air Pollution Control (13).

Emission factors in the EPA tabulation generally apply to identifiable point sources in processes or operations from which pollutants are released to the atmosphere through vents or stacks. As such they have limited applicability to in-plant industrial hygiene assessments. They do, however, identify some of the air contaminants that are generated by these processes; the same contaminants represent potential in-plant exposures.

3.4 Fugitive Sources

Contaminant emissions from point sources such as tank vents or transfer points, and from processes and operations that clearly involve release of a process material—for example, release of volatile components in cementing or painting operations—are ordinarily capable of identification and quantitation. There are other types of source not so easily accommodated and sometimes neglected in emission inventories. Such sources, often called fugitive sources, may be intermittent, temporary, or unpredictable; many are unrecognized or ignored in emission inventories. Fugitive sources and the generation of secondary pollutants deserve attention in emission inventories, however, even though all of them may not be capable of quantitation or prediction.

Fugitive sources that are of consequence primarily in the field of air pollution, and for which emission factors have been developed, include unpaved roads, agricultural tilling, aggregate handling and storage piles, heavy construction operations, and paved roads (11). Other sources that are consequential from the standpoints of both air pollution and industrial hygiene include the following:

1. Urban fires.
2. Industrial process fires.
3. Materials spills (accidents, equipment failure or malfunction).
4. Sample collection and analysis.
5. Process leaks (flanges and piping, valves, packing glands, conveyors, pumps, compressors, tanks and bins, etc.).
6. Relief valves and control device bypasses.
7. Maintenance activities (tank and vessel cleaning, cleaning a filter, replacing piping, pumps, etc.).
8. Emissions from waste streams and reemissions of collected materials.
9. Secondary reactions (nonproduct process reactions, extraprocess reactions).

Table 10.1   Major Source Types for Which Air Pollution Emissions Factors Have Been Adopted (11)

1. External combustion sources
    Bituminous coal combustion
    Anthracite coal combustion
    Fuel oil combustion
    Natural gas combustion
    Liquefied petroleum gas consumption
    Wood waste combustion in boilers
    Lignite combustion
    Bagasse combustion in sugar mills
    Residential fireplaces
    Wood stoves
    Waste oil disposal
2. Solid waste disposal
    Refuse incineration
    Automobile body incineration
    Conical burners
    Open burning
    Sewage sludge incineration
3. Internal combustion engine sources
    Highway vehicles
    Off-highway mobile sources
    Off-highway stationary sources
4. Evaporation loss sources
    Dry cleaning
    Surface coating
        Nonindustrial surface coating
        Industrial surface coating
    Storage of organic liquids
        Fixed roof tanks
        External floating roof tanks
        Internal floating roof tanks
        Pressure tanks
        Variable vapor space tanks
        Transportation and marketing of petroleum liquids
        Cutback asphalt, emulsified asphalt, and asphalt cement
        Solvent degreasing
        Waste solvent reclamation
        Tank and drum cleaning
        Graphic arts
        Commercial/consumer solvent use
        Textile fabric printing
5. Chemical process industry
    Adipic acid
    Synthetic ammonia
    Carbon black
    Charcoal
    Chlor-alkali

547

Table 10.1 (*Continued*)

Explosives
Hydrochloric acid
Hydrofluoric acid
Nitric acid
Paint and varnish
Phosphoric acid
Phthalic anhydride
Plastics
Printing ink
Soap and detergents
Sodium carbonate
Sulfuric acid
   Elemental sulfur-burning plants
   Spent-acid and hydrogen-sulfide-burning plants
   Sulfide ores and smelter gas plants
   Sulfur recovery
   Synthetic fibers
   Synthetic rubber
   Terephthalic acid
   Lead alkyl
   Pharmaceuticals production
   Maleic anhydride
6. Food and agricultural industry
   Alfalfa dehydrating
   Coffee roasting
   Cotton ginning
   Feed and grain mills and elevators
   Fermenting
   Fish processing
   Meat smokehouses
   Ammonium nitrate fertilizers
   Orchard heaters
   Phosphate fertilizers
     Normal superphosphates
     Triple superphosphates
     Ammonium phosphates
   Starch manufacturing
   Sugar cane processing
   Bread baking
   Urea
   Beef cattle feed lots
   Defoliation and harvesting of cotton
   Harvesting of grain
   Ammonium sulfate
7. Metallurgical industry
   Primary aluminum production
   Coke manufacturing
   Primary copper smelting

Table 10.1 (*Continued*)

Ferroalloy production
Iron and steel production
Primary lead smelting
Zinc smelting
Secondary aluminum operations
Secondary copper smelting and alloying
Grey iron foundries
Secondary lead smelting
Secondary magnesium smelting
Steel foundries
Secondary zinc processing
Storage battery production
Lead oxide and pigment production
Miscellaneous lead products
Leadbearing ore crushing and grinding
8. Mineral products industry
Asphaltic concrete plants
Asphalt roofing
Bricks and related clay products
Calcium carbide manufacturing
Castable refractories
Portland cement manufacturing
Ceramic clay manufacturing
Clay and fly-ash sintering
Coal cleaning
Concrete batching
Glass fiber manufacturing
Frit manufacturing
Glass manufacturing
Gypsum manufacturing
Lime manufacturing
Mineral wool manufacturing
Perlite manufacturing
Phosphate rock processing
Sand and gravel processing
Stone quarrying and processing
Coal conversion
Taconite ore processing
Metallic minerals processing
Western surface coal mining
9. Petroleum industry
Petroleum refining
Crude oil distillation
Converting
Treating
Blending
Miscellaneous operations
Natural gas processing

Table 10.1 *(Continued)*

10. Wood products industry
      Chemical wood pulping
         Kraft pulping
         Acid sulfite pulping
         Neutral sulfite semichemical (NSSC) pulping
      Pulpboard
      Plywood veneer and layout operations
      Woodworking waste collection operations
11. Miscellaneous sources
      Forest wildfires
      Fugitive dust sources
         Unpaved roads (dirt and gravel)
         Agricultural tilling
         Aggregate handling and storage piles
         Heavy construction operations
         Paved roads
         Industrial paved roads
      Explosives detonation

Urban structural fires are intermittent phenomena predictable only in the aggregate and not as individual events; they are considered to be emission sources primarily in the air pollution sense. An industrial hygiene consequence of urban fires, however, is the exposure to toxic materials of firefighters and others who may be involved in rescue or control activities. A number of toxic atmospheric contaminants are generated by structural fires; these tend to vary from one fire to another. Of concern in all structural fires, however, is emission of carbon monoxide. Burgess et al. (14) have described exposures of emergency personnel to carbon monoxide emissions in real fire situations; the maximum sustained air concentrations of carbon monoxide to which these persons were exposed was about 2 percent. Materials balance using pyrolysis and oxidation processes offer one means of estimating the quantities of pollutants that may be generated in any particular case.

Estimates of emissions from industrial fires require individual analysis. The quantities and characteristics of combustion products depend on the nature of the materials that burn and the circumstances of combustion (e.g., whether open or confined). Again, consideration of pyrolysis and oxidation phenomena may permit estimation of the nature and quantities of air contaminants generated by any particular event. The likelihood and consequences of accidental fires is an appropriate consideration in industrial emission inventories.

Materials spills in manufacturing, transporting, and uses of industrial materials are not infrequent occurrences. Studies of processes and operations with specific attention to spills can reveal the frequencies and magnitudes of spills if they are usual occurrences (15). Such spills can then become part of an emissions inventory. When spills are not a usual occurrence, an emission inventory can do little more than trigger consideration of the possibilities and consequences of such events. Guidance for the protection of workers when a spill of a hazardous substance occurs has been developed as a joint project of the National Institute for Occupational Safety and Health (NIOSH), the Occupational Safety and Health Administration, the U.S. Coast Guard, and the U.S. Environmental Protection Agency, and published in 1983 by NIOSH (16).

Process leaks are sometimes of major consequence from the standpoint of industrial hygiene. Control of process leaks, for example, has been a major factor in achieving acceptable working conditions in vinyl chloride polymerization plants. Process leaks can be identified by inspection or instrumental methods; timely repair may obviate quantitation for an emission inventory. Should quantitation be necessary, the techniques of source testing or materials balance may suffice.

As acceptable exposure levels become lower, attention to detail in emission evaluation takes on added importance. Routine tasks such as process sample collection may present a potential for excessive exposure under the traditional procedure of sampling at open manhole covers, open sample containers, and unconfined sample streams. Bell addressed this problem for vinyl chloride sampling and suggested techniques that may find application in other industries, particularly petrochemical operations (17).

Release of materials to community air or to a workplace through process relief valves occurs from time to time. The frequency of operation of these devices and magnitudes of releases, obtained from plant records or other sources, can be used for emission inventory purposes. Recognition of the existence of relief values or control device bypasses in a process is important in an emission inventory.

Kletz has stated that many of the foregoing sources of fugitive emissions can be controlled with emergency isolation valves (18). In addition to suggesting a method of control, this author also presents a leakage profile for both an olefin plant and an aromatic plant that could serve as the basis for developing a routine checklist at similar operations.

Volatile air contaminants may be released to the atmosphere from process sewers, drainage ditches, and/or collecting ponds. Such releases are of concern from the standpoint of both air pollution and industrial hygiene. Two examples are offered. Consider a case in which a gravity separator is used in an enclosed benzene recovery system to separate the organic and aqueous streams. Water from this separation may be reused for other purposes elsewhere in the plant. Even though benzene is only slightly soluble in water, such water reuse can result in measurable concentrations of benzene in the air at workplaces in the plant where benzene is not used. Consider another case in which alkaline scrubbing water is used to remove fluorides from a process waste gas stream. Discharge of the scrubbing water to a waste pond in which the pH is low permits release to the atmosphere of the collected fluorides. The use of the air pollution control scrubber in such a case only relocates the site of fluoride emission from the process stack to the waste pond. Air sampling and materials balance techniques are means for assessing emissions from waste streams and reemissions of collected materials.

Descriptions of processes and products do not always reveal whether by-products are formed or whether any by-products and their parent chemicals are stable throughout the manufacturing process and in the environment. This issue has been discussed, and an approach to assessing the significance of by-products and secondary reactions has been presented (19). The likelihood and consequences of hydrolysis, pyrolysis, oxidation, and other reactions are factors in the assessment. The identification of processes and principal materials is the first step in exploring for secondary pollutant emissions in an emission inventory. Literature review and perhaps laboratory exercises may yield indicators of the presence of secondary contaminants; air sampling may be required for validation.

Although it is not always possible to make quantitative, or perhaps even qualitative, assessment of fugitive sources and secondary pollutants, the possibilities of their occurrence and the opportunities for intervention to protect workers and the community merit scrutiny in emission inventories.

4 IN-PLANT EMISSION ESTIMATES FROM RECORDS AND REPORTS

In the absence of in-plant measurements of emissions, estimates derived from secondary sources of information can be used for inventory purposes. In some cases data are available from similar processes and estimates by analogy may prove sufficient for a given purpose. In other cases unique features of a process or operation may make estimates by analogy inappropriate. An alternative to measurement data or analogy makes use of records and reports whose basic purposes were other than environmental but whose content may be extracted, combined, or otherwise manipulated into a usable estimate of emissions and potential exposure. Vital information may be recorded in a variety of business documents including engineering, accounting, production, quality control, personnel, and governmental records. Although these records represent a rich source of information for emission inventory purposes, they can serve other purposes in the practice of industrial hygiene and occupational health research as well. Some of these purposes are mentioned from time to time in the following discussions of specific types of records.

4.1 Research and Development Records

Research and development units often issue a variety of reports to producing units as aids to bringing a new product on-stream, to ensure a required level of uniformity in a given item produced by different plants, or to make changes or improvement in operating equipment, materials, or techniques. Valuable insight with respect to materials and work practices employed in the past may be gleaned from several of these records. Among such records are the following:

1. *Product specification.* Final product and intermediate component data included in a product specification often give dimensions, weight, materials of construction, processing aids, processing conditions, tools and fabricating equipment, and special notes.
2. *Standard operating procedure.* This record may exist under a variety of names; SOP, Standard Practice, Uniform Methods, and (Company Name) Manual of Operations appear among the myriad of titles. All, however, share fairly specific operational instructions, usually in a step-by-step approach by processing sequence. Vital data from this record can include:

    a.   Fabrication equipment.
    b.   Authorized materials of construction.
    c.   Acceptable processing aids.
    d.   Processing conditions.
    e.   Alternate materials for cyclic operations (e.g., summer versus winter stocks).
    f.   Change notices affecting materials or techniques.
    g.   Instructions regarding protective equipment.

3. *Formulations and raw materials.* Detailed data concerning the qualitative aspects of potential exposure can be derived from study of information describing the raw materials employed in the process.

    a.   Raw material specifications may contain composition data, restrictions on contaminants, and acceptance testing schedules.
    b.   A listing of authorized vendors can offer clues to potential contaminants based on knowledge of a vendor's source of supply or processing methods.
    c.   Formulation records frequently contain the composition of mixtures as well as the methods, equipment, quantity, and conditions of processing.

4.2 Analytical Laboratory Records

Quality control considerations often dictate that incoming raw materials be monitored for selected chemical and physical properties, that supplies from prospective vendors be subjected to acceptance testing, and that production quality be assayed by sampling at intermediate stages of the process. A single journal entry of analytical results may include a wealth of intelligence applicable to an emissions inventory. Analytical laboratory records include the following:

1. *Raw material test results.* Recorded in either the analyst's journal or on prepared forms will be entries such as:

    a.   Analytical results vis-à-vis specifications.
    b.   Date of analysis, and perhaps date of receipt.

c. Quantity of the shipment.
d. Vendor.
e. Method of analysis.
f. Analyst.

2. *Authorized vendors.* Quality control laboratories often maintain vendor-supplied information regarding purchased materials. Material data sheets, product specifications, production methods, and quality control statements all offer bits of useful data.

4.3 Process and Production Records

Records generated by production units relate to such matters as scheduling, quality control, cost control, and to a lesser extent, training exercises. Depending on the product and process, production-type records can be of considerable value in generating an emissions inventory.

1. *Flow chart.* A detailed diagram of a process is particularly useful in identifying points with potential for release of air contaminants.
2. *Operating conditions.* Time, temperature, and pressure data are useful in assessing possible release of reaction products, by-products, degradation products, and unreacted raw materials.
3. *Scheduling.* Shift, cyclic, and seasonal variations may influence the emissions of contaminants. Such variations can be ascertained through production scheduling records. Included under this heading are records of batch versus continuous production modes. Gantt chart records offer a particularly attractive systematic source of scheduled activities normally both precise and detailed.

4.4 Plant Engineering Records

Virtually all industrial organizations include an engineering component responsible for the physical plant and supporting facilities. This responsibility results in a plant archive that is particularly useful in emissions inventory activity and in research that requires reconstruction of past environmental conditions.

1. *Plant layout.* The plant floor plan, coupled with elevation drawings, affords a visualization at the process, materials flow, and occupied areas, and possibly insights relating to contaminant generation and control. Furthermore, many engineering drawings are cross-referenced to other drawings of equipment, emissions control features, and adjacent work areas. Useful characteristics of plant engineering files are:

a. Processing equipment type, extent, and location appear on layouts.
b. Department floor space showing geographic boundaries and physical barriers appear on some plates.

c.   Engineering controls, particularly ventilation systems, are depicted in mechanical equipment layouts, which normally show the location, site, type, and rating of air-moving equipment, as well as a reference to the detailed plate covering that system.

d.   Dates of change or modification are often included on drawings, either by direct entry or by reference to a new set of plates.

2.   *Equipment specifications.*   Files on processing equipment installed by or under the supervision of plant engineers often include specifications that supply clues to potential problems such as process emissions, sound power rating and directivity data, or emission control features incorporated in the equipment.

3.   *Project records.*   Each major project of construction, modification, or installation normally proceeds as an integral project with one individual assigned as coordinator. Usually this project officer accumulates records that include data on:

a.   Equipment specifications.
b.   Performance checks and acceptance test results.
c.   Prime contractors and subcontractors.

4.   *Material balance.*   A detailed accounting of materials, products, and side streams, either measured or theoretical, serves as a basis for a quantitative estimate of emissions.

5.   *Emission estimates for regulatory agencies.*   For a number of years regulatory agencies have required an accounting of process losses for air and water pollution surveys, for permit applications, and for other regulatory purposes. These documents provide a record of:

a.   Process description.
b.   Effluent discharge.
c.   Contaminant(s) identification.
d.   Concentration parameters.
e.   Controls in effect.

4.5   Industrial Engineering Records

The practice of industrial engineering includes the analysis of work procedures to improve either work conditions or productivity or both. Especially for the period since World War II, industrial engineering records contain information on work methods and standards that are useful for current emission inventories and for estimating past emissions and exposures that may have resulted from them. Such records include:

1.   *Job descriptions.*

a.   Description numbers, department listings, and simple descriptive job titles can provide the link between the tasks performed and the work history recorded in an individual's personnel file.

b.   Job location within a plant, thus the potential for various exposures, may be ascertained from the geographic area, processing equipment, or department listing contained in a job description.

c.   The listing in a job description of tools, equipment, and processing aids can help establish the nature of emissions and characteristics of potential exposures (e.g., inhalation, absorption).

d.   Task evaluations often include judgments concerning such health-related items as work conditions, safety requirements, hazards, skills, effort, and responsibility.

2.   *Time and motion studies.*   Task breakdowns are useful to establish contact with materials of interest, and the fraction of time spent on certain tasks coupled with cycle time can afford an estimate of the extent of exposure.

3.   *Process analysis.*   Evaluations of an entire process focus on tasks performed by all the members of a crew rather than those of individual workers. Particularly useful in process analysis reports are listings of fractions of a workday spent at different work stations by members of the operating crew, including break time and sequencing of tasks.

4.6   Accounting Records

Accounting records cover virtually all phases of an operation for both outside reporting and internal control purposes. Major areas of interest for emission inventory purposes include records on incoming materials, internal use of these raw materials, and plant production factors (20).

1.   *Purchasing.*   Raw materials for both production and support operations, as well as processing equipment, are ordinarily obtained from outside the company, and an acquisition record is generated upon receipt. Even when raw materials are drawn from captive sources, records of resource depletion may be available.

   a.   *Purchasing.*   Materials, quantities, and supplier may be available through the ledger, often in the accounts payable portion.

   b.   *Fixed assets.*   Capital equipment is normally amortized by a recognized accounting procedure, the evaluation of which can be employed to determine the date of acquisition. Alternately many organizations maintain a property book that contains the same information.

2.   *In-plant issues.*

   a.   *Raw material consumption.*   Often these data can be developed through accumulation of issues to departments, cost centers, or plants through collected expense or work-in-process accounts.

   b.   *Cost accounting.*   Material variance accounts may serve as a surrogate for material balance data if the latter are unavailable. The absolute variance value between actual and standard material usage may be of marginal value, but changes in this quantity can be important in establishing trends.

c. *Stores' accounts.* Nonproduction supplies are often issued to maintenance and other support departments through a "general store" account, with the receipt listing the item and quantities issued.

3. *Inventory.* Inventory control may be maintained through stock records that list material, vendor, date and quantity received, date and quantity of issue, department to which issued, and location and quantity on hand.

4. *Production.* Finished goods and work-in-process production accounts normally include an identification of the work schedule, days operated, and average production level.

4.7 Personnel Records

Personnel records, per se, provide some insights into the nature and locations of contaminant releases applicable to emission inventories. In addition they are vital to other industrial hygiene occupational health research activities. In any study of causal association between conditions of work and health experience, the work experience of each individual is an integral component. Thus accurate and reliable personnel-type records are vital. Organizational units normally developing useful records include the personnel, payroll, industrial relations, safety, and medical departments.

1. *Personnel records.* The file on a particular employee usually lists department and job assignments chronologically and gives a brief description of tasks, limitations, and periods of absence. Earnings records sometimes provide a clue to potential exposure through a listing of special pay or rates for hazardous tasks, and exposure duration or regimens by shift differentials or overtime pay.

2. *Industrial relations.* Seniority listings and work force distribution charts are useful in establishing the locations of jobs and personnel currently or at various times in the past.

3. *Safety and medical.* Insurance carrier survey reports, fire and explosion inspection reports, sickness/accident/compensation records, administrative control documents that limit time at a particular task or require worker rotation, along with written SOPs for activities such as entering tanks and vessels and other hazardous tasks, represent a record of emission sites and past working conditions, however subjective, suitable to at least rank-order jobs by exposure potential to certain agents.

4. *Union records.* Seniority listings and union membership rolls can help to identify the working populations of past years, and records of negotiated agreements between labor and management often contain provisions for work schedules, protective clothing and equipment, provisions for personal hygiene, clothing changes, and other health-related clauses that help to describe conditions of work.

4.8   Government Records and Reports

Data so broad as to characterize an entire industry, or so specific as to deal with a single chemical compound, are routinely and systematically collected, assembled, and published by various government agencies. Most of these records are available to the enterprising investigator, usually at little or no cost. A few of these record types are mentioned here as illustrations.

1. *Nationwide statistics.* The U.S. Tariff Commission publishes an annual listing of Production and Sales of Synthetic Organic Chemicals (21). It includes listings of raw materials, the production chemicals by use category (surface-active agents, etc.), and a directory of manufacturers of each material. Time series analyses of the entries can indicate the dates of introduction, extent of use, and dates of decline of a given material.

2. *Industrywide data.* Government directed or controlled operations such as government owned-contractor operated plants compile data for required reports, many of which find their way into the public domain. For example, information on virtually all facets of the synthetic rubber industry during the 1940–1945 period are available through the various reports of the Rubber Reserve Company (22, 23). Included are flow charts, raw materials, formulations, plant capacities, and locations.

3. *Specific products.* On June 15, 1844, Charles Goodyear received U.S. Patent 3633 for a compound consisting of gum elastic, sulfur, and white lead (24). Thus the approximate period of time at which rubber compounders could be considered as potentially exposed to white lead as a rubber accelerator is established. Roughly 4 million U.S. patent numbers have accrued since the mid-nineteenth century, doubtless including many materials, processes, and devices of interest to health investigators.

4.9   Record Location

Within most corporate industrial organizations, records of major additions or changes, as well as events affecting the entire operating units, are likely to be available at a central location, such as the division or corporate offices. Examples of these types of records are overall production levels, descriptions of uniform practices, and product specifications. More specific records, or those dealing with minor plant alterations, are ordinarily available only at the local plant level. Current records are ordinarily found in the departments identified in Sections 4.1 through 4.7 of this chapter; historical records may be found in those departments or in a central plant records storage facility. At the plant level project officers often retain personal copies of papers related to their projects; if a plant record has been discarded, such personal records may provide valuable historical data.

4.10   Output from Records

Measures derivable from secondary sources of data are limited by the nature and detail of the record content. In general, the more distant the time of interest, the less complete will be the record, and the lower will be the confidence in its accuracy. When data are sparse and only plantwide data are available, estimated measures such as gallons per cubic foot of plant building volume per month, or pounds used per employee per day, may be the limit of detail, but this may serve for plant-to-plant contrasts. More detailed data such as quantities of materials used in specific departments and descriptions of controls for specific work areas permit more detailed estimates. Material generation rates coupled with general area ventilation rates may enable one to estimate an area average concentration. Such data may be particularly useful for within-plant contrasts or for plant-to-plant studies of particular areas or processes.

Records pertaining to the detailed nature of tasks, such as standard operating procedures or job descriptions, make possible a ranking of jobs—thus individuals—by potential for exposure. In this fashion personnel with similar exposure histories can be grouped for comparison of health experience with those sharing different common exposure profiles.

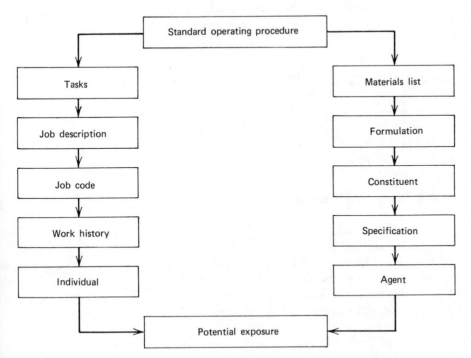

Figure 10.1   Synthesis of potential exposures associated with a manufacturing step for which a standard operating procedure is available.

As an example of the use of such information, let us offer an illustration of the type of emissions output that has been accomplished for retrospective research purposes through the use of secondary sources of information.

In Figure 10.1 the standard operating procedure is the pivotal document that identifies both tasks and materials for a step in manufacturing. A *formulation* record, or series of records, for each material used in that particular manufacturing step identifies each raw material *constituent* of each material. Examination of the raw material *specification* for each constituent then permits identification of *agents* (e.g., coal tar naphtha) that may be emitted in the workplace, often with additional compositional detail (e.g., aromatic content or benzene content). Because formulations, constituents, and specifications change from time to time, the listing of potential emissions thus identified must show inclusive dates on an agent by agent basis.

The *tasks* portion of the SOP includes a *job description*, which in turn has a specific *job code*. The job code is listed in the personnel record of each person who has held that particular job. Examination of the personnel records of the plant's current and past workers identifies complete *work history*, including dates and duration of each job code assignment for each *individual* who has held the specific job code of interest.

In this way the potential for contact with a particular chemical agent by an identified individual, including dates and duration of the contact, can be developed, even for events that occurred in the distant past.

5 USES OF EMISSION INVENTORIES

Emission inventories are fundamental to the identification, evaluation, and control of industrial hygiene and community air pollution hazards associated with chemical agents. Not all uses for inventory information can be mentioned here, but a few are discussed.

5.1 Regulatory Requirements

Various governmental regulatory programs in occupational safety and health, air pollution, and toxic substances control have reporting requirements that involve emission inventory information. Regulations regarding the nature and content of these reports, the agents covered, and the types of establishment with reporting obligations change periodically. Persons responsible for industrial or commercial enterprises that involve emissions to community air, or the handling or production of agents that may be covered in occupational health or toxic substances regulatory programs, should keep abreast of the specific requirements imposed by state and federal regulations. Emission inventory programs can then be so designed and operated that the emissions data necessary to satisfy reporting requirements are obtained.

5.2 Community Air Pollution Dispersion Estimates

Data on emissions of a pollutant to community air and companion meteorological data permit estimation and prediction of ground level concentrations of that pollutant at any desired location in the community. Methods and procedures for manual calculations to estimate the impact of individual sources (25), and a basic computer program for calculating concentrations resulting from emissions from large numbers of point and area sources (26) have been reported. The mathematical relationships and computational techniques are too extensive to be repeated here. Application is best undertaken with the workbook or guide in hand. Suffice it to say that both methods can be used by engineers and industrial hygienists when the necessary data on emissions, site locations, and meteorology are available.

5.3 Workplace Exposure Estimates

Qualitative estimates of the potential for emissions and the consequent exposures of workers can be made when agents have been identified and conditions of their use are known. Even the most rudimentary emissions inventory—the identification of agent, process, and location of use—permits this kind of assessment. When a very large number of materials are used at a single facility, and perhaps some of them are used only infrequently or in small quantity, the maintenance of a file on potential emissions and exposures may be a major undertaking. An operating program in which the industrial hygiene department of a large research facility systematically obtains descriptions of all uses of the many hundreds of chemical agents available for use at the facility has been described (27). The system includes a hazard rating for each agent based on its toxicity, flammability, reactivity, and special properties. Use information does not identify emissions per se, but retrieval of use information for a particular agent permits rapid subjective assessment of its emissions potential throughout the facility.

Quantitative information on emissions, either measured, calculated, or estimated, can be combined with data on workplace ventilation to estimate workroom concentrations of an agent. Such estimates may occasionally be of use in validating sampling data, but they are more likely to have value in such applications as assessing growth or decay of concentrations in nonequilibrium situations, in predicting the consequences of new processes or changes in operations, or in estimating retrospectively the levels of a contaminant to which particular groups of workers may have been exposed. This latter use is of particular interest in retrospective epidemiological studies of worker populations.

5.3.1 Gases and Vapors

Emission rates of gases and vapors are used in the mathematical relationships that yield values for concentration of contaminants in workroom air. In addition to emission rates, one must know physical dimensions or volume of the workplace and the ventilation rate. The mathematical relationships were derived using the concept of materials balance (28). The rate of change in the quantity of contaminant in the air of a workplace is described by Equation 4. All emissions of a contaminant to workroom air comprise a contaminant generation rate, G.

$$V\frac{dx}{dt} = G + QX_s - QX \tag{4}$$

Integration of Equation 4 yields:

$$X_t = X_i \exp\left(\frac{-Qt}{v}\right) + \left(X_s + \frac{G}{Q}\right)\left[1 - \exp\left(\frac{-Qt}{v}\right)\right] \tag{5}$$

where V = volume of the ventilated workplace (m^3)
G = rate of contaminant generation as volume per unit time of contaminant gas or vapor (m^3/min)
Q = ventilation rate, volume/unit time of air plus contaminant (for practical purposes the ventilation rate of air may be used in dilution ventilation of occupied space)(m^3/min)
t = interval of time elapsed since conditions represented by X_i and t = 0 (min)
X = fraction of contaminant in workplace air (10^{-6} ppm)
X_i = fraction of contaminant present in workplace air at time t = 0
X_t = fraction of contaminant present in workplace air at time t
X_s = fraction of contaminant in incoming dilution air.

Equation 5 is applicable only over intervals in which G and Q are constant. Should either G or Q change, a new value of X_i and a new t_0 must be established. At equilibrium, when t is great, Equation 5 simplifies to:

$$X = X_s + \frac{G}{Q} \tag{6}$$

or with equilibrium and clean supply air:

$$X = \frac{G}{Q} \tag{7}$$

When the concentration of the contaminant in the workplace air increases with time, the maximum concentration for the time interval t is X_t; for decay situations the maximum concentration for the interval t is X_i. Exposure guides and standards may list both maximum allowable short-term concentrations and allowable average concentrations. The average concentration over time t, also derivable from materials balance, is:

$$X_{av} = X_s + \frac{G}{Q} + \left(\frac{V}{Qt}\right)\left(X_i + X_s - \frac{G}{Q}\right)\left[1 - \exp\left(\frac{-Qt}{v}\right)\right] \qquad (8)$$

As is the case with Equation 5, at equilibrium, when t is great, Equation 8 simplifies to Equation 6.

Equation 5 may be used to calculate the concentration of gas or vapor remaining in a vessel or tank at any time after purging ventilation or contaminant introduction has begun. It may also be used in the following form to estimate the interval of time required for a particular ventilation rate to cause a particular concentration X_t to be reached:

$$t = \frac{V}{Q}\ln\left(\frac{G + QX_s - QX_i}{G + QX_s - QX_t}\right) \qquad (9)$$

In some repetitive operations the generation rate for a contaminant may vary in a cyclical manner, being at one rate for one interval of time and at a different rate for another interval. If, in such a case, the greater generation rate is identified as G, the lower rate as G' and the time intervals t and t' are companion to G and G', respectively, when the cycle has equilibrated so that the workroom concentration pattern has become repetitive, the maximum concentration that occurs during the cycle is represented by:

$$X_{max} = \frac{\left(X_s + \frac{G}{Q}\right)\left[1 - \exp\left(\frac{Qt}{v}\right)\right] - \left(X_s + \frac{G'}{Q}\right)\left[1 - \exp\left(\frac{-Qt'}{v}\right)\right]}{\exp\left(\frac{-Qt'}{v}\right) - \exp\left(\frac{Qt}{v}\right)} \qquad (10)$$

For the same cyclic operation after equilibrium has been reached, the average concentration over a large number of cycles is:

$$X_{av} = X_s + \left(\frac{Gt + G't'}{Q(t + t')}\right) \qquad (11)$$

In the simple case when there is no generation during the time interval t' the term G'/Q becomes zero in Equations 10 and 11.

In an excellent analysis of dilution ventilation relationships Roach has shown that the variance of emission σ_G^2 is inversely proportional to the length of the time interval t over which emissions take place, and that the variance in concentration σ_x^2 is then inversely proportional to the product of ventilation rate Q and the square of the room volume V^2 (29). Thus for a given average emission rate, the smaller the workspace and the smaller the ventilation rate, the greater will be the fluctuations in concentration, while the greater the volume of the space and the greater the ventilation rate, the smaller will be the fluctuations.

In real situations instantaneous perfect mixing does not take place. Usually when dilution ventilation is applied, the contaminant is emitted at locations that are in the occupied portion of a workroom; not all the ventilation air mixes with the contaminant while it is in this occupied space. An empirical factor K may be used in calculations of estimated exposures to account for departures from perfect mixing. The definition of K is:

$$K = \frac{Q_{actual}}{Q_{effective}} \tag{12}$$

where Q_{actual} = ventilation rate, volume/unit time, as measured or obtained from engineering records (m^3/min)

$Q_{effective}$ = ventilation rate, volume/unit time, that is effective in determining the concentration of contaminant to which people are exposed in a workplace (m^3/min)

The value of $Q_{effective}$ is the ventilation rate Q to be used in Equations 5 through 11 for calculation of exposure estimates. The choice of a value for K is a matter of judgment for an investigator; it is ordinarily in the range 3 to 10 (30). Calculations using engineering records for ventilation rates, actual emission rates, and a number of concentration measurements of three different contaminants in each of two actual dilution ventilation situations, have yielded K values ranging from 2.2 to 5.0 with a mean value of 3.25 (31). In these two dilution ventilation situations, solvent vapors were being emitted in a number of locations in the workrooms at elevations that were approximately the same as breathing zone and sampling elevations. They may be considered to represent generally good mixing of emissions and ventilation air in actual application of dilution ventilation in industrial situations. The observed values are consistent with the lower values of the range of typical K values; the greater the departure from efficient mixing of emissions and ventilation air, the greater should be the value of K.

5.3.2   Particulates

Caution must be exercised in the application of dilution ventilation relationships, as described for gases and vapors, to particulate contaminants that are released to workplaces. Gravitational settling may be an appreciable factor in the decay of air concentrations of particulates. The smaller the particles, the greater is the likelihood that they will remain airborne and the more applicable are dilution ventilation relationships. When emissions of submicrometer sized particles—for example, most components of smoke and metal oxide fumes—are known, concentrations can be approximated by use of the dilution ventilation equations for gases and vapors.

Dusts with a wide range of particle sizes do not necessarily behave in air in the same way as fumes. Estimates of the relative concentrations of respirable dusts may be made with dilution ventilation equations when particulate emissions data are sufficient for quantitation of the respirable fraction. Size characteristics that define respirable particulates are described in Chapters 6 and 7 of Volume I.

In assessing exposures to particulates, particles larger than those considered respirable should not be ignored. Of these larger particles, those that are inhaled and deposited in the nasopharynx and tracheobronchial system before they can reach the pulmonary spaces of the lungs, represent ingestion exposures. Airborne concentrations of such particles generally decay at appreciable rates because of gravitational settling; dilution ventilation relationships do not apply to them.

As particle size increases, the influence of ventilation on concentration, therefore on the magnitude of exposure, decreases. Thus for large particles, those well in the nonrespirable range, an emission rate per se is likely to be a better relative index of potential exposures than is a calculated concentration based on both emission and ventilation rates.

6   RECORDS RETENTION PROGRAM

Emissions records may be generated for a variety of purposes. Some records are for purposes that can be served in a relatively short time; then the record may be considered a candidate for discarding. Other emission records may be generated specifically for purposes such as health research, which require long retention. Emission records made for various short- or long-term purposes may have secondary use in long-term community air pollution or occupational health research. Whether they represent typical operations or unusual events, such records may be invaluable in the future for reconstructing conditions of work to compare with long-term health experience of persons who engaged in that work. Since not all specific agents of future interest are identifiable in advance, it is prudent to retain emission records for as many agents as possible.

Emission records, or records from which emission estimates can be made, such as those identified in Section 4 of this chapter, should be retained for a long time. A single specific age at which a record no longer has value cannot be stated. Guidance on retention duration may be found in federal regulations, which require that records of medical examinations of workers exposed to asbestos be kept for at least 20 years (32). For research purposes exposure records have value equal to or greater than that of medical examination records. Because long latent periods are associated with health effects of some agents, early exposure records may be invaluable in the search for causal associations. Exposure records, or emission records that identify contact with agents or from which exposure estimates can be made, should be retained long enough to be used in studies of the mortality experiences of the worker populations to which they apply.

The need for long-term retention of environmental records is recognized in the recordkeeping provisions of an Occupational Safety and Health Administration regulation of toxic and hazardous substances (33). Time periods for which employee exposure records must be retained vary with the substance being regulated. For example, records of exposures to asbestos must be retained for at least 20 years, those for vinyl chloride must be retained for not less than 30 years, and those for coke oven emissions must be retained for the duration of employment plus 20 years, or for 40 years, whichever is longer. Emission data are pertinent to these records. Such retention is appropriate for all records of emissions of air pollutants or of chemical agents released in places of work.

The responsibility for retention of emission records for any establishment should be clearly defined. This responsibility may well be placed with the same organizational unit that has responsibility for retention of air monitoring records and/or health experience records. Records that serve as secondary sources of health related information, such as those identified in Section 4 of this chapter, may be kept in different locations. A checklist of all such records that are needed for emission records purposes should be assembled and kept up to date for each establishment and retained by the unit having responsibility for emissions records. Responsibility for retention of each of these secondary records should be clearly defined and this responsibility, along with location of each record set, should appear on the records checklist.

REFERENCES

1. U.S. Environmental Protection Agency, *A Guide for Compiling a Comprehensive Emission Inventory*, APTD 1135, EPA, Research Triangle Park, N.C., 1973.
2. J. R. Hammerle, in: *Air Pollution*, Vol. 3, 3rd ed., A. C. Stern, Ed., Academic Press, New York, 1976, Ch. 17, pp. 717–784.
3. Toxic Substances Control Act, PL 94-469, 15 USC 2607, 1976.
4. T. H. Maugh, *Science*, **199**, 162 (1978).
5. U.S. Department of Health and Human Services, *Registry of Toxic Effects of Chemical Substances*, 1981–82 Edition, DHHS (NIOSH), Pub. No. 83-107, USGPO, Washington, D.C., 1983.
6. H. J. Paulus and R. W. Thron, in: *Air Pollution*, Vol. 3, 3rd ed., A. C. Stern, Ed., Academic Press, New York, 1976, Ch. 14, pp. 525–587.
7. U.S. Environmental Protection Agency, *Administrative and Technical Aspects of Source Sampling for Particulates*, EPA 450/3-74-047, EPA, Research Triangle Park, N.C., 1974.
8. American Petroleum Institute, *Evaporative Loss from Tank Cars, Tank Trucks, and Marine Vessels*, API Bulletin 2514, API, Washington, D.C., 1959.
9. H. B. Elkins, E. M. Comproni, and L. D. Pagnotto, *AIHAJ*, **24**, 99 (1973).
10. H. E. Runion, *AIHAJ*, **36**, 338 (1975).
11. U.S. Environmental Protection Agency, *Compilation of Air Pollutant Emission Factors*, A.P. 42, 3rd ed. (through Supplement No. 14), EPA, Research Triangle Park, N.C., 1983.
12. D. Anderson, *Emission Factors for Trace Substances*, EPA-450/2-73-001, U.S. Environmental Protection Agency, Research Triangle Park, N.C., 1973.

13. U.S. Environmental Protection Agency, *Mobile Source Emission Factors*, EPA-400/9-78-005, Office of Mobile Source Air Pollution Control, EPA, 2565 Plymouth Rd., Ann Arbor, Mich., 48105, 1978.

14. W. A. Burgess, et al., *AIHAJ*, **38,** 18 (1977).

15. M. Smith, *Investigations of Passenger Car Refueling Losses*, Scott Research Laboratories, Inc., San Bernardino, Calif., September 1972, National Technical Information Service, PB-212 592.

16. D. R. Streng, et al., *Hazardous Waste Sites and Hazardous Substances Emergencies*, DHHS (NIOSH) Pub. No. 83-100, PHS, CDC, NIOSH, Cincinnati, Ohio, 1983.

17. Z. Bell, J. Laflenn, R. Lynch, and G. Work, *Chem. Eng. Prog.*, **71**:9, 45 (1975).

18. T. Kletz, *Chem. Eng. Prog.*, **71**:9 (1975).

19. E. Sowinski and I. H. Suffett, *AIHAJ*, **38,** 353 (1977).

20. R. N. Anthony, *Management Accounting, Text and Cases*, 4th ed., Irwin, Homewood, Ill., 1970.

21. U.S. Tariff Commission (1958), *Synthetic Organic Chemicals, United States Production and Sales, 1958*, Report 205, TC1.9-205, 2nd Series, Government Printing Office, Washington, D.C.

22. Rubber Reserve Company, *Report on the Rubber Program 1940–1945*, February 24, 1975.

23. G. S. Whitby, Editor-in-Chief, *Synthetic Rubber*, Wiley, New York, 1954.

24. G. D. Babcock, *History of the United States Rubber Company, A Case Study in Corporate Management*, Indiana Business Report 39, Graduate School of Business, Indiana University, 1966.

25. D. B. Turner, *Workbook of Atmospheric Dispersion Estimates*, U.S. Public Health Service Publication 999-AP-26, National Air Pollution Control Administration, Cincinnati, Ohio, 1969.

26. A. D. Busse and J. R. Zimmerman, *User's Guide for the Climatological Dispersion Model*, EPA-R4-024, U.S. Environmental Protection Agency, Research Triangle Park, N.C., 1973.

27. W. E. Porter, C. L. Hunt, and N. E. Bolton, *AIHAJ*, **38,** 51 (1977).

28. R. L. Harris, Jr., *Industrial Hygiene Engineering Training Course*, U.S. Public Health Service Occupational Health Field Headquarters, Cincinnati, Ohio, 1960.

29. S. A. Roach, *Ann. Occup. Hyg.*, **20,** 65 (1977).

30. *Industrial Ventilation, A Manual of Recommended Practice*, 18th ed., Committee on Industrial Ventilations, Lansing, Mich., 1984.

31. J. C. Baker, Jr., *Testing a Model for Predicting Solvent Vapor Concentrations in an Industrial Environment*, M.S.E.E. Technical Report, Department of E.S.E., University of North Carolina, Chapel Hill, N.C., 1977.

32. *General Industry Safety and Health Standards*, OSHA 2206 (29 CFR 1910), part 1910. 1001, Department of Labor, Washington, D.C., 1976.

33. Occupational Safety and Health Standards, Subpart Z—Toxic and Hazardous Substances, 29CFR 1910 (with amendments through January 21, 1983, 48FR2768).

CHAPTER ELEVEN

Measurement of Worker Exposure

JEREMIAH R. LYNCH

1 INTRODUCTION

This chapter explains why measurements of air contaminants are made, discusses the options available in terms of number, time, and location, and relates these options to the criteria that govern their selection and the consequences of various choices.

A person at work may be exposed to a certain number of potentially harmful agents for as long as a working lifetime, upward of 40 years in some cases. These agents occur in mixtures, and the concentration varies with time. Exposure may occur continuously or at regular intervals or in altogether irregular spurts. As a result of exposure to these agents, the worker is being dosed, and depending on the magnitude of the dose, some harmful effect may occur. All measurements in industrial hygiene ultimately relate to the dose received by the worker and the harm it might do.

Early investigators of the exposure of workers to toxic chemicals encountered conditions so obviously unhealthy as evidenced by the existence of frank disease that quantitative measurements of the work environment to estimate the dose received by the afflicted were not needed to establish cause-and-effect relationships. At the same time the ability of these early industrial hygienists to make measurements was severely limited because suitable sampling equipment did not exist and existing analytical methods were insensitive. Pumps were driven by hand, equipment was large and heavy, filters changed weight at random, gases were collected in fragile glass vessels, absorbing solutions spilled or were sucked into pumps, and laboratory instrument sophistication was bounded by an optical spectrometer. To collect and analyze only a few short period samples required several days of work and the probability of failure due to one of many possible equipment defects or other mishaps was high. Consequently few measurements were made and much judgment was applied to maximize the representativeness of the measurements, or even as a substitute for measurement.

Changes in working conditions, in technology, and in society have caused the old methods of measurement to be reexamined.

1.   With few exceptions workplace exposure to toxic chemicals is not grossly excessive but is close to what is commonly accepted as a safe level.

2.   As a consequence of the reduction of exposure, frank occupational disease is rarely seen and much of the disease now present results from multiple factors of which occupation is only one.

3.   Workers are demanding to know how much toxic chemical exposure they are receiving, and this results in a need even to document the negative.

4.   Technology has provided enormously improved sampling equipment that is rugged and flexible. This equipment, used with analytical instruments of great selectivity and sensitivity, has largely replaced the old wet chemical methods.

As a consequence of these changes in the workplace and advances in technology, it is now both necessary and possible to examine in far more detail the way in which workers are exposed to harmful chemicals. Personal sampling pumps permit collection of contaminants in the breathing zone of a mobile worker. Pump-collector combinations are available for long and short sampling periods. Systems that do not require the continual attention of the sample taker permit the simultaneous collection of multiple samples. Automated sampling and analytical systems collect data day after day. Sorbent-gas chromatograph techniques permit the simultaneous sampling and analyses of mixtures and, when coupled with mass spectrometers, identify obscure unknowns. Sensitivities have improved to the degree that tens and hundreds of ubiquitous trace materials begin to be noticeable.

At the same time the demands placed on our information gathering systems are more acute. Now we must answer not only the question: "Is exposure to this agent likely to harm anyone?" We are being called on to provide data for a great many other purposes. Worker exposure must be documented to comply with the law (1). Employees are demanding to be told what they are breathing, even in the absence of hazard (2). Epidemiologists need data on substances not thought to be hazardous, to relate to possible future outbreaks of disease. Design engineers need contaminant release data to relate to control options. Process operators want continuous assurance that contaminant levels are within normal bounds. Management information systems that issue status reports when queried require monitoring data inputs. Data needs are so pervasive that a tendency to monitor for the sake of monitoring develops.

2 OBJECTIVES OF EXPOSURE MEASUREMENT

The central question that must be asked before measuring exposure is: "What use will be made of the data?" That is, what questions will the data answer, or what external information need will be satisfied? The collection of data should be looked on as part of a decision-making process. If no decision is to be made, or if nothing is to be done differently either in the short or long run as a result of the data collected, regardless of the result, then why collect the data? In some situations a correct control decision may be perfectly clear without any measurements, although measurements may serve to reinforce the decision or to convince others. In other cases it is difficult to see how any obtainable data will aid in making decisions, or that the cost of obtaining the data needed exceeds the cost of making the wrong decision. To make decisions under uncertainty, as is usually required in industrial hygiene, the techniques of decision analysis (3) are useful. These techniques also permit the calculation of the value of information which can then be compared with the cost. While the cost of information for an identified decision is important, it is also useful to consider what other questions will need to be answered or what other information will be needed. The resources available for the measurement of exposure are usually so limited that data must serve several purposes. Some of those purposes are discussed in the following sections.

2.1 Hazard Recognition

As a starting point for a complete assessment of the risk to health posed by an occupational environment, it is necessary to know the substances to which workers are exposed. Systematic recognition of all possible hazards requires inventories of materials brought into the workplace, descriptions of production processes, and identification of any new substances introduced. However, all of these sources of information may not be enough to identify substances present as trace contaminants or substances generated by production processes, either inadvertently or as unknown by-products. To complete the identification of all substances present, before going to the next step of evaluating exposure and risk, it may be necessary to make some substance recognition measurements. Since these measurements are not intended to evaluate exposure, they may be area rather than personal samples and would typically be large-volume samples for maximum sensitivity.

2.2 Exposure Evaluation

The most common reason for measuring worker exposure to a toxic chemical is to evaluate the health significance of that exposure. These evaluations are usually made by comparing the result with some reference level. Traditionally, the Threshold Limit Values (TLVs) for Airborne Contaminants of the American Conference of Governmental Industrial Hygienists (ACGIH) (4) have been used to represent safe levels or "conditions under which it is believed that nearly all workers may be repeatedly exposed day-after-day without adverse effect." More recently, OSHA was given the responsibility for establishing permissible exposure limits (PELS) in the workplace. TLVs have served as the basis for most of the OSHA's PELs. Unfortunately, published TLVs cover only a small fraction of the chemicals that occur in industrial workplaces, albeit the most common ones. Where there is exposure to a substance for which there is no TLV or legal exposure limit, it may be necessary to develop a supplemental standard for use in a particular plant or company. Standards for substances whose toxicology is not well known are generally set to avoid acute effects in man or animals and, often by analogy to other better-documented substances, at a level low enough to make chronic effects unlikely. When very little is known about a substance, it may only be possible to estimate a lower level at which it is reasonably certain that no adverse effects occur and an upper level at which adverse effects are likely. The width of the gap between these two levels is a zone of uncertainty that needs to be considered in evaluating the results of exposure measurements.

2.3 Control Effectiveness

When changes in equipment or processes are made that affect the release of substances that are contributing to worker exposure, measurements of the magnitude of that change may be needed. These measurements provided empirical data on control effectiveness to confirm design expectations or to use as a basis for the design of other modifications. In the simplest case, before-and-after measurements are made when a new control, such as a local exhaust hood, is installed on a contaminant release point. From the results of these measurements, it is possible to predict the reduction in worker exposure, which can be confirmed by subsequent exposure measurement.

Unfortunately, the situation is rarely that simple. Most worker exposures are caused by multiple release points, creating a work environment of complex spatial and temporal concentration variations through which the worker moves in a not-altogether-predictable manner. Furthermore, interaction between several release points or other factors in the environment may confound the results. A needed improvement such as a new exhaust hood may seem to be without effect because the building is air starved, or a poor hood may seem to function well because of exceptional general ventilation. The time of contaminant release may depend on obscure and uncontrollable process operation factors. As with measurements made for other purposes, control evaluation studies must be carefully designed and are likely to consist of a series of factorial measurements analyzed by statistical methods.

2.4 Methods Research

Industrial hygiene methods development research hypotheses often take the form "will sampling and analytical method A give the same result as method

B." If we are unable to reject this hypothesis in a carefully designed experiment, then we accept that methods A and B are equivalent within our limits of error and given the bounds of the experimental conditions (5–8).

Method equivalence research usually requires extensive laboratory work, but in most cases field testing is necessary because completely realistic environments with interferences usually cannot be generated in laboratory chambers, and the difficulties of making field measurements introduce errors that may affect one method more than another. For these reasons most practicing industrial hygienists tend to distrust assertions of equivalence that are not backed up by field data. To be credible, experiments of this kind should clearly define the range of concentrations and conditions over which the equivalence has been tested. Personal versus area equivalence of coal mine dust measurements made in long-wall mines should not be assumed to hold in room-and-pillar mines. Manual versus automated asbestos counting relationships based on chrysotile do not apply when counting amosite fibers.

Enough data should be collected not only to determine whether the methods are correlated, but also to determine the ability of a measurement by one method to define the confidence limits on a prediction of the result that would have been obtained by the other method (Figure 11.1). Often a high correlation coefficient is obtained when many pairs of measurements have been made, indicating that the two methods are certainly related; yet the scatter is such

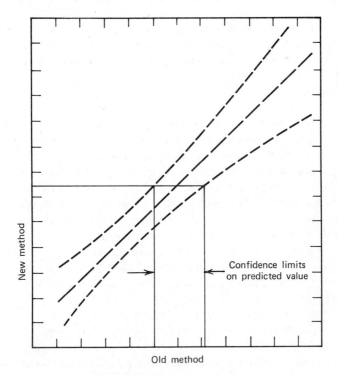

Figure 11.1 Methods comparison.

that one method may only be used as a predictor of the other method to within an order of magnitude. The design of experiments for the purpose of measuring method equivalence is discussed in Chapter 8. Considerations such as the environmental variability related to location, time, and numbers, which are described below, should be taken into account, to maximize the range of conditions over which the equivalence is evaluated without introducing so much error that the relationship has no predictive ability.

2.5 Leak Detection

Quite apart from measurements made for the purpose of evaluating intentional changes in control systems are measurements to detect leaks or other inadvertent loss of control. In addition to leaks and other "fugitive" emissions, which are a constant concern in chemical plants, local exhaust systems can fail because of plugged ducts or fan failure. Patterns of contaminant generation may change because of loss of temperature control in a vessel or tank, dullness of a chisel in a mine or quarry, or bacterial contamination of a cutting oil. These events cannot always be detected by changes in process parameters, and it may be unacceptable to wait until they show up in exposure measurements because the margin for error in a control system designed for a very strict standard, such as vinyl chloride, may be too small, or as in the case of hydrogen sulfide, because the consequences of overexposure, even for a short period, are too serious.

Automatic leak detection systems may be installed where it is important to instantly detect any leak, or leaks may be detected by periodic manual surveys which check spot concentrations near pump and valve seals, flanges, etc.

2.6 Epidemiology

Occupational epidemiology is concerned with the relationships between occupational factors and disease trends. The kinds of conclusions the epidemiologist may draw from a study depend on the kind of data used in the study. If the epidemiologist only knows the industry or place of employment of the individuals in the study group, then the conclusions must be in the form that a disease excess may be associated with work in that industry or establishment. This is a scientifically sound conclusion, but it does not lead the managers and industrial hygienists responsible for worker health to a chemical agent, if any, which may be causing the disease. To reveal such useful causal relationships the epidemiologist needs to know to what substances each worker was exposed. With data on the degree of exposure it is possible for the epidemiologist to detect dose–response relationships which can aid in confirming a causal relationship between an agent and a disease. If the degree of exposure is accurately known over time, a dose–response relationship which can be used to estimate safe levels of exposure may be calculated.

Ideally, the epidemiologist would like to have measurements of the exposure of all workers to all substances from the beginning of employment. In the National Coal Board study of coal workers pneumoconiosis in the United Kindgom, exposure and health status were measured over a long enough period so that the results could be used as a basis for the present coal dust standard in the United States. Major prospective studies such as this, however, are less common than retrospective studies which reconstruct exposure by making use of whatever measurements are available. Most measurements are made for the problem solving purposes and are not representative of the exposure of the whole population of workers. Since an industrial hygienist can often anticipate the epidemiological need for estimates of exposure of all employees over time, it may be desirable to collect some additional samples specifically for epidemiology. These would be chosen to represent the exposure of all groups of workers, including those not at risk of overexposure by current standards, and to represent exposure to substances not presently known to present a special risk. Obviously, the development of such a program is complicated by our inability to predict what workers or substances will be of interest in future research. It may be that a few "fingerprint" samples, analyzed by such detailed methods as capillary GC with mass spectroscopy, will be the best choice to generally characterize the kinds of exposures that are occurring.

In deciding to make measurements for use in future epidemiologic studies it should be known if the data will meet the need. As pointed out above, the usual industrial hygiene problem-solving measurements may not be a useful representation of exposure to relate to health effects in population. Close, early cooperation with epidemiologist is needed to avoid expensive data-collection programs which fulfill no need.

2.7 Illness Investigations

When an employee has a frank occupational disease, such as lead poisoning, confirmed by both physical findings and analysis of biological materials, or is known to have been overexposed based on analysis of biological materials, the industrial hygienist must determine the cause of the overexposure. When the conditions that led to the overexposure still exist they may be evaluated by measurements made after the event. It is also possible to evaluate overexposure that resulted from past episodes which were not observed and evaluated when they occurred. A history of past exposure opportunities can be constructed and used to estimate the dose which led to the present case. In some cases it may be necessary to reenact or simulate an event to measure what may have happened—being careful, of course, to ensure that all participants are protected. Exposure, obviously, need not always be by inhalation and may include off-the-job activities.

A much more difficult investigation is the search, in an occupational setting, for the cause of an outbreak of illness or complaints of illness that may or may not be of occupational origin, or even if related to occupation may result from factors other than exposure to toxic substances. Investigation of possible chemical exposure, however unlikely, can be extremely complex, since a release from ususual sources (off-gassing of plastics, air-conditioning system contamination) that may have occurred in the past must be considered.

2.8 Legal Requirements

Section 6b7 of the Occupational Safety and Health Act provides for " . . . monitoring or measuring employee exposure at such locations and intervals, and in such a manner as may be necessary for the protection of the employees" (9). Under this Act, OSHA has responsibility for establishing exposure limits (PELs) in the working environment. Responsible and effective implementation of this congressional intent requires that regulatory language be devised to both discriminate between employers, who may have a hazard present and those who almost certainly do not, and to do so by a scheme that is easy to understand and implement. The ideal regulation should very clearly not apply to the vast majority of establishments, which have no conceivable hazard resulting from the substance being regulated, and should apply requirements of increasing strictness as the significance of the hazard in an establishment increases, ultimately calling for measurement of sufficient frequency to ensure that the potential for harm is fully assessed in the few establishments where exposures are great enough to create significant risk.

This sorting of workplaces by level of risk can be done by prescribing a series of thresholds or triggers that lead to increasingly stringent requirements for a decreasing number of employers. First, all employers who do not have the substance present in the workplace should not be required to monitor. Although the "presence" of a material seems a simple enough criterion that everyone would interpret in the same way, the extreme bounds of interpretation, which are of concern in legal arguments, include the presence of as little as a few molecules of a gas or a single asbestos fiber. As analytical techniques become more sensitive, almost everything is to be found almost everywhere, at least at the level of a few molecules. What is needed in a regulation is an exclusion, such as a percentage concentration in a liquid, below which the substance is not considered to be present.

For employees who have a substance present in a workplace above the excluded level, the next step should be to determine whether there is any possibility that the substance is released into the workplace such that workers may be exposed. The setting of this threshold must reflect considertion of the conditions under which the substance is present and the consequences of release. Thus nuclear reactor decay products are continuously monitored against the possibility of leaks even when they are hermetically sealed. Such high toxicity materials are not released into the workplace except under very rare emergency circumstances. On the other hand, cadmium released into the workplace as a result of silver soldering should be monitored, but it is not necessary to measure exposure to cadmium where cadmium-coated auto parts are stored. Since no simple "potential for release" trigger has yet been devised, there is some regulatory error (employers included who should not have been, and vice versa) at this decision point.

A further step is needed, therefore, before a full monitoring program with its consequent expense is mandatory. One step is to use a small number of measurements of the exposure of the maximum risk employee under conditions when the exposure is likely to be the greatest. If the results of these measurements are sufficiently below the exposure limit to permit confidence about the result (10), we could say that no significant exposure is occurring and no more would be required of this employer than if the substance were not released in the first place. This initial measurement scheme, however, is not appropriate where there is such a massive presence of a highly toxic substance that continued vigilance must be maintained against the possibility of leaks or other inadvertent releases.

In cases where it has been established, by means of an initial measurement or data from other sources such as prior measurements of worker illness, that significant exposure is occurring, possibly over the limit on occasion, a regular program of periodic monitoring should be provided. The frequency of monitoring should relate to the level of exposure and should consider trends between measurements that might lead to conditions with unacceptable consequences.

Schemes for the logical stepwise analysis of data to arrive at decisions regarding monitoring and compliance have been developed by the U.S. National Institute for Occupational Safety and Health (NIOSH) (11), European Council of Chemical Manufacturers' Federations (CEFIC) (12), and the West German Federal Ministry of Labor (BMA) (13). The CEFIC occupational exposure analysis flow chart is shown in Figure 11.2. At the start a chemical inventory of products, by-products, intermediates, and impurities is assembled and annotated within data about hazards, limit values, regulations, and standards. The hazard (potential for exposure) is assessed based on the process, equipment, material volume, temperature, pressure, ventilation, work practices, and precautions. This information is analyzed to determine where and when substances may be released into the workplace and what exposures are possible as a result. The exposure status of workers in work areas identified by this analysis is then initially assessed using such a priori information as earlier measurements or computed concentrations based on comparable installations, work processes, materials, and working conditions. A compliance evaluation based on this computation is now made if possible; if not, exposure measurements are made of the maximum risk (highest exposed) employee for the job function under study. If the results are out of compliance, exposure reduction measures are taken and the process repeated. If conditions are in compliance but greater than a Decision Level (DL) then an occupational exposure monitoring protocol is developed and implemented. If exposures are below the DL the process stops. The DL, which is expressed as a fraction of the occupational exposure limit, is based on judgment. In general, it would not be greater than 0.5, would

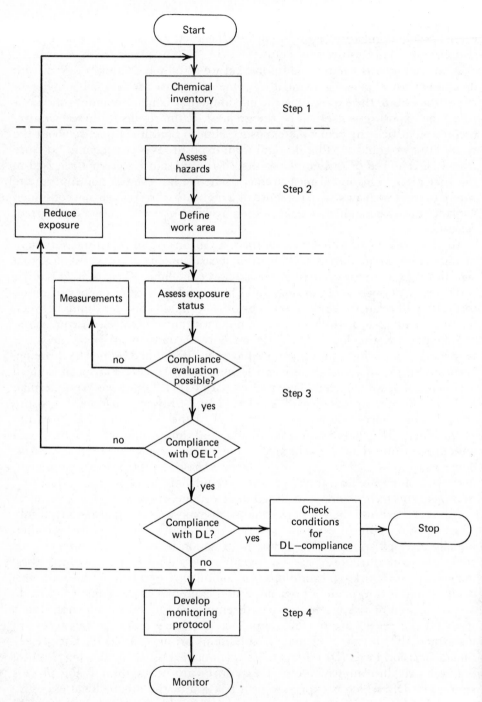

Figure 11.2 CEFIG occupational exposure analysis chart. The Occupational Exposure Level (OEL) and Decision Level (DL) used depend on the substance being evaluated.

usually be 0.25, and may be as low as 0.1 for special circumstances such as carcinogens. Unlike the OSHA Action Level, the DL is used only for monitoring decisions, not for decisions involving training, medical examinations, etc.

The above scheme is one approach to the general problem of designing employee exposure monitoring programs which fit the need. Several investigators (14–19) have commented on the limitations of various other schemes and have proposed alternatives.

3 SOURCES OF WORKER EXPOSURE

The core concern of industrial hygiene is the prevention of disease arising out of the workplace. Toxic substances cause disease when some amount, or dose, enters on or in the body. Workers have contact with, or are exposed to, toxic substances by inhaling them with the air they breathe and by several other routes. To accurately appraise the total dose being received by a worker, thus to predict biological effect, it is necessary to understand how exposure occurs.

3.1 Production Operations

Industry is generally thought of in terms of continuous repetitive operations that generate air contaminants to which workers are more or less continuously exposed. Paint is sprayed on parts passing continuously in front of a worker. Dust is generated by a foundry shakeout on a continuous casting line every few seconds. Fumes seep steadily from cracks in aluminum smelting pot enclosures. Welders join structural members on a production basis, with only short breaks between welds. Operators watch controls in the midst of a chemical plant that allows the steady release of fugitive streams. In some case, such as a grinder cleaning sand from a casting, the concentration of the contaminant is closely related to the work performed and is probably higher in the workers breathing zone than it would be several feet away. Other workers, such as dorfers and creelers in a cotton spinning mill, are exposed to dust released from hundreds of bobbins, and their own activities, short of leaving the workplace, have little effect on their exposure. All these continuous exposures are actual rather than potential and present the least difficulty in evaluation.

3.2 Episodes

Much of the exposure workers receive occurs as a result of events or episodes that occur intermittently. Glue is mixed on Wednesday. The shaker mechanism breaks and a mechanic must enter the baghouse. Coke strainers preceding a pump need to be dumped when the pressure drop becomes excessive. A drum falls off a pallet and ruptures on the floor. Samples of product are taken every two hours. A pressure relief valve opens. A packing gland bursts. A reactor vessel cover is driven from its hinges by overpressure in the vessel.

These exposure events can be periodic or they can occur at irregular intervals. They may be planned and predictable or altogether unanticipated. Some events are frequent and result in small exposures, whereas others may be catastrophic events causing massive exposures and even fatalities. As a class, episodic exposure events result in a significant fraction of the exposure burden of many workers and may not be ignored. Their evaluation, however, is extremely complex, and often only broad estimates of the probability of an unlikely event and the consequent risk may be available.

3.3 Noninhalation Exposure

When relating exposure to biological effects, either in the case of an epidemiologic study or because of an outbreak of illness that might be related to occupation, the total dose is the relevant quantity, not merely the dose received by inhalation. Amounts of a substance entering the body by any route may contribute to the total dose.

Many substances, especially fat-soluble hydrocarbons and other solvents, can enter the body and cause systemic damage directly through the skin when the skin has become wet with the substance by splashing, immersion of hands or limbs, or exposure to a mist or liquid aerosol. Some substances, such as amines and nitriles, pass through the skin so rapidly that the rate at which they enter the body is like that of substances inhaled or injested. The prevention of skin contact to phenol is as important as preventing inhalation of airborne concentrations. A few drops of dimethyl formamide on the skin can contribute a body burden similar to inhaling air at the TLV.

For some substances with low vapor pressure, like benzidine, skin absorption is the most important risk. Benzene, on the other hand, though absorbed through the skin, is absorbed at such a low rate that skin contact probably contributes little to the body burden. To judge the degree to which skin absorption is contributing to exposure, it is necessary to consider both the rate of absorption and the degree of contact. Clothing wet with a substance that remains on the worker for prolonged periods provides the maximum contact short of immersion. On the other hand, the poultice effect does not occur on wet unclothed skin; thus evaporation can take place and the result is less severe. Contact with mist that does not fully wet the clothes or body but merely dampens them is not as severe as being splashed with the bulk liquid. Theoretically gases and vapors may be absorped through the skin, but the rate would be extremely small except under extreme conditions. Protective clothing, which is impervious to the substance, will reduce absorption to nil on protected areas (20). However a leaky glove that has become filled with a solvent is providing contact with the hand equivalent to immersion. Barrier creams are often used to prevent dermatitis but may not always prevent skin absorption (21). Although skin contact and absorption must be considered as contributing to the dose for many materials, few quantitative data on rate of absorption are available, and these are in a form that is difficult to apply in an industrial setting (22, 23). Furthermore, such factors as part of the skin exposed, sweating, and the presence of abrasions or cuts can cause order of magnitude differences (24).

While on the subject of skin absorption, it should be noted that toxic substances may pass through the skin by intentional or unintentional injection. Opportunities for all kinds of materials to enter the body by injection connected with drug abuse or therapeutic accident are obvious. Bulk liquids may also break through the skin and enter the bloodstream without the aid of a needle when driven into the body as high velocity projectiles released from high pressure sources. Airless paint spray and hydraulic systems (25) often use pressures in this range, and such pressures often occur inside pipes and vessels in chemical plants. Inadvertent cracking of a flange under pressure can cause the traumatic introduction of a toxic substance into the body. Solid particles may also enter this way and if soluble or radioactive, they may cause damage beyond the initial injury.

3.4 Ingestion

Although ingestion is an uncommon mode of exposure for most gases and vapors (26), it cannot be ignored in the case of certain metals such as lead. Indeed, for workers exposed in pigment manufacture and use, ingested lead, either coughed up and swallowed or taken on food, though less well adsorbed than inhaled lead, constitutes a significant fraction of the total burden. Spot test (27) and tests with tracers have shown that materials present in a workplace tend to be widely dispersed over surfaces, thus support the essential rule prohibiting eating and smoking where toxic substances are present.

3.5 Nonoccupational Exposure

In addition to the exposure to toxic substances a worker receives at work, an increment of dose may also be delivered during nonworking hours. Air (28), water (29), and food, all contain small amounts of toxic substances that may also be found in the occupational environment. As a rule the nonoccupational dose is at least an order of magnitude below the occupational dose, just as community air standards are much lower than TLVs, and many substances present in industry are very rare in the community. Yet in certain cases these pollutants have an impact. The consequence of arsenic exposure resulting from copper smelting in northern Chile is difficult to assess, since arsenic poisoning there is endemic as a result of naturally occurring drinking water contamination. Chronic bronchitis resulting from air pollution in the industrial midlands of England is confounded with the lung diseases of coal miners. The cardiovascular consequences of urban carbon monoxide exposure may be not unlike those of marginal industrial exposures.

In addition to these somewhat involuntary exposures to toxic substances, many workers have hobbies or other leisure activities (30). Acoustic traumata from loud music and target practice are well known. Lead exposure on police and presumably private firing ranges is significant. Epoxies are used in home workshops, and garden chemicals contain a variety of economic poisons. Although no data come to hand, leisure activity exposure to toxic substances rivaling high but permissible work exposure must be rare. As a result of toxic substance legislation and product liability questions, many toxic substances are no longer sold as consumer products. Benzene, carbon tetrachloride, lead paint, and asbestos have been disappearing from retail stores. Soon it will be a rare event indeed to find a known carcinogen or a serious toxin in any consumer product. Furthermore, although a worker may apply far more diligence to a hobby than to the job, hobby activity tends to be intermittent, with long lapses. Serious acute exposures are not unlikely, but chronic disease is rare if not unknown. Thus although a supplementary increment of dose may be inferred from an off-the-job activity known to involve the toxic substance in question, it should not be assumed to be the primary source unless the assumption is supported by measurements.

Intentional exposure to solvents for their narcotic effects is well documented (31–36). Glue sniffing, gasoline sniffing, ingestion of denatured alcohol, methanol, and even turpentine have been reported. This category of chemical abuse results in exposure and often damage far beyond that normally encountered in industry. The industrial hygienist must be alert to the possibility that a case of disease may be related to intentional addictive exposure.

3.6 Environmental Variability

An important factor in the design of any measurement scheme is the degree of variability in the system being observed. This variability has a primary effect on the number of samples to be taken and the accuracy of the results that can be expected. When the system being observed is the exposure of a worker to a toxic substance in a workplace, variability tends to be quite high. During the course of a day there are minute-to-minute varitaions and daily averages vary from day to day.

A typical recording of actual intraday environmental fluctuations appears in Figure 11.3. Highly variable environmental data of this kind, which are truncated at zero, generally have been found to be best described by the log normal rather than the normal distrubution (Figure 11.4). This 2-parameter distribution (37) is described by the geometric mean (GM) and the geometric standard deviation (GSD), which is the antilog of the standard deviation of the log-transformed data (11). A rough equivalence between GSDs and the more familiar coefficient of variation (Table 11.1) is valid up to a GSD of about 1.4. Figure 11.5 illustrates the consequence of various values in terms of spread of data. Models and data derived from community air pollution measurements (38–41) have been useful in studying the in-plant micro environment. In general, studies of occupational environmental variability (42, 43) have confirmed the usefulness of the log normal description of data and have yielded

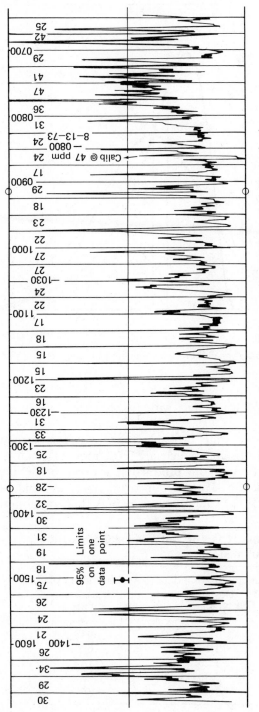

Figure 11.3 Actual industrial hygiene data showing intraday environmental fluctuations. Range of carbon monoxide data on chart is 0 to 100 ppm.

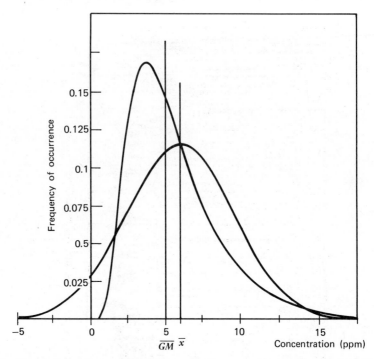

Figure 11.4 Log normal and normal distributions with the same arithmetic ($\bar{x}$) and standard deviation: $\overline{GM}$ = geometric mean.

GSDs in the range of 1.5 to 1.7, with few less than 1.1 and as many as 10 percent exceeding 2.3.

One can speculate on the probable causes for this variability in worker exposure. Fugitive emissions, which are like frequent small accidents rather than main consequences of the production process, occur almost randomly. Production rates change. Patterns of overlapping multiple operations shift irregularly. Distribution of contaminants by bulk flow, random turbulence, and

Table 11.1 Log and Arithmetic Standard Equivalence

Geometric Standard Deviation (GSD)	Coefficient of Variation (CV)
1.05	.049
1.10	.096
1.20	.18
1.30	.27
1.40	.35

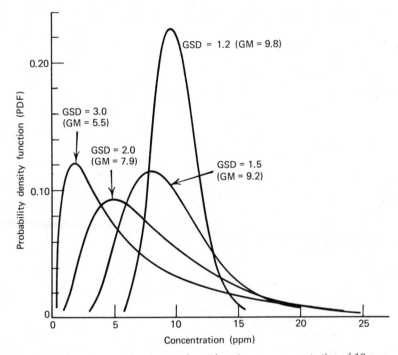

Figure 11.5 Log normal distributions for arithmetic mean concentration of 10 ppm.

diffusion is uneven in both time and space. Through all this our target system, the worker, moves in a manner that is not altogether predictable. These and other uncertainties are the probable causes of the variability typically observed, but the situation is far too complex to pinpoint each individual source of variation and its consequences.

Sampling schemes devised to fit this situation may deal with the variability from all sources as a single pool and derive whatever accuracy is required by increasing the number of samples. Alternatively one can postulate that a large part of the variability is due to some observable factor or factors and by means of a factorial design account for this portion of the variance, leaving only the residual to be dealt with as error. No hard and fast rules can be made regarding the choice except that it seems logical to expect the factors identified to account for a statistically significant fraction of the variance (F test) if it is to be worthwhile to sample and analyze the data in this manner. Shift, season of the year, wind velocity, and rate of production are some of the factors affecting worker exposure that may be worth singling out to ensure that part of the gross variance can be assigned to an accountable cause. Even if successful, it is likely that the residual or error variance will still be large, much larger than the variance due to measurement method inaccuracy, and will exert a major influence on our decisions.

4 CHARACTERISTICS OF EXPOSURE AGENTS

The physical and chemical properties of a substance are important because of the effect they have on exposure measurement quite apart from their effect on the magnitude of the exposure and its consequences.

The most important properties of any substance that affect sampling are those that determine whether it can be collected on a sampling medium, and treated or removed in a manner that permits analysis. Many vapors are rapidly adsorbed and desorbed from one or another of a variety of solid sorbents, but some high boiling materials are very difficult to desorb and some gases do not adsorb well. Similarly, most dust particles are collected efficiently by membrane filters with 0.8 μm and even larger pore sizes, however fresh fumes may pass through and hot particles can burn holes. Beyond these apparent physical and chemical dependencies of exposure measurement methods, there are some less apparent complications that frequently occur.

4.1 Vapor Pressure

Since liquids with high vapor pressure tend to evaporate completely when they are aerosolized, it is rarely necessary to measure them as mists. Liquids with very low vapor pressure may be present only as mists and must be measured with methods appropriate for aerosols. The situation can be very complicated for a liquid of intermediate volatility, high enough to produce a saturated vapor 10 percent or more of the TLV, yet low enough that mists will not quickly evaporate. If samples are collected on filters, only the mist will be caught and part of this liquid may be lost by evaporation into the air flowing through the filter during sampling, particularly if samples are collected over long periods. This effect undoubtedly results in a significant loss of the lower molecular weight two- and three-ring aromatic compounds present in coke oven effluents, when sampled by the usual filter methods.

When mixed mist-vapor atmospheres are sampled with a solid sorbent, the vapor is likely to be caught and retained efficiently, but the charcoal granules and associated support plugs do not constitute an efficient particulate filter (44). As a consequence, compound devices with a filter preceding a sorbent tube have been developed. Obviously the filter must come first, since what escapes the filter will be caught by the charcoal, but not vice versa. Despite severe sampling rate limitations, these systems are satisfactory when the sum of aerosol and vapor concentrations is to be related to a standard or effect. If, however, the aerosol and vapor mist are to be considered separately, perhaps because of different deposition sites or rates of adsorption, the single sum of the two concentrations is not enough. Furthermore, it cannot be assumed that the material collected on the filter is all the aerosol, since part of the liquid evaporates and is recollected on the charcoal. Since the reverse cannot be true, in cases when the vapor is more hazardous, this assumption is "safe" (i.e., protective). Size selective presamplers that collect part of the aerosol and none of the vapor may approximate the respiratory deposition-adsorption differential and so deliver a sample weighted toward the vapor to a biologically appropriate degree. However, even here the possibility of evaporation from the cyclone, elutriator, impactor, or other component must be acknowledged. Additional research is needed in this area.

4.2 Reactivity

Most toxic substances are stable in air at the concentrations of interest under the usual ranges of temperature and pressure encountered in inhabited places to the degree that they may be sampled, transported to the laboratory, and analyzed without significant change or loss. Some few substances are in a transient state of reaction during the critical period when the worker is exposed. In the spraying of polyurethane foam, isocyanates present in aerosol droplets are still reacting with other components of the polymer while the aerosol is being inhaled, thus must be collected in a reagent that halts the reaction and yields a product that can be related to the amount of isocyanate that was there. Similarly solid sorbents such as charcoal can be coated with a reagent to react with the substance collected to yield a new compound; which will be retained and can be analyzed.

Unwanted reactions can also occur on the collecting media. For example, a substance that will hydrolyze may do so if brought in contact with water on the solid sorbent, particularly if a sorbent like silica gel, which takes up water well, is used. Substances that may coexist in air and not react or react only slowly because of dilution may react rapidly when conentrated on the surface of a collection medium.

4.3 Particle Size

The route of entry, site of deposition, and mode of action of an aerosol all depend on some measure of the size of particle, usually but not always the aerodynamic equivalent diameter. Exposure measurement methods must discriminate among different size particles to sort out those of greater or lesser biologic effect. The most common instance of such size selection is the exclusion of particles not capable of penetrating into the terminal alveoli when measuring exposure to pneumoconiosis-producing dust. Such "respirable mass sampling" by cyclone, horizontal elutriator, or various impactors is discussed at length elsewhere in this series and is not covered in detail here. However penetration into the smallest parts of the lung is not the only division point in size selective sampling. Cotton dust particles probably do not produce the chest tightness response by penetration into the deep lung. Rather, histamine may be released in large airways (bronchioles), which can be reached by larger particles; thus a different size selective criterion (50 percent at 15 μm) and device (vertical elutriator) are used (45). Even larger particles may enter the body by ingestion if they are caught in the ciliated portion of the bronchus, coughed up, and swallowed.

At some size particles are so large they cease to be "inhalable," thus should be excluded from exposure samples (46). It is difficult to select a criterion on which to base size selectors for "inhalable dust" samplers. Particles having falling (setting) speeds greater than the upflow velocity into the nose could be said to be "noninhalable" except that some individuals breathe through their mouths. These particles turn out to be quite large, thus have falling speeds that cause them to be removed from all but the most turbulent or recently generated dust clouds. Because of their size, of course, they represent a mass far out of proportion to their number. By using the general tendency of open face filter samples to undersample large particles when pointed down (47), it is possible to use this vertical elutriation effect to discriminate against noninhalable particles in most cases.

4.4 Exposure Indices

Most toxic substance encountered in the workplace are clearly defined chemical compounds that can be measured with as much specificity as we please. Benzene need not be confused with other compounds, and other closely related compounds (e.g., ethyl benzene) have different toxicology. For some toxic substances we tend to think of an element (lead) in any one of a number of possible compounds. Thus specificity is defined in terms of the element rather than the compound. Often, however, the situation becomes more complex because the compound containing the element of interest has a significant effect on its uptake, metabolism, toxicity, or excretion (48). Although the influence of the chemical structure containing the element probably varies even among similar compounds, for simplicity the compounds are usually grouped as organic/inorganic (lead, mercury), soluble/insoluble (nickel, silver), or by valence (chromium).

The problem of differentiating the several classes of compounds of a toxic element in a mixed atmosphere adds complexity to sampling method selection, and it is sometimes necessary to make, and clearly state alongside the results, certain simplifying assumptions. It is commonly assumed when measuring lead exposure in gasoline blending, for example, that all the lead measured is organic. Similarly when measuring the more toxic soluble form of an element, the "safe" assumption may be made that all the element present was soluble.

The greatest complexity occurs when toxicity is based on the effects of a class of compounds or of a material of a certain physical description. Some polynuclear aromatic hydrocarbons (PNA) are carcinogens of varying potency, and they usually exist in mixtures with other PNAs and with compounds (activators, promotors, inhibitors) that modify their activity. Analysis of each individual compound is very difficult and when done does not yield a clear answer, since given the complexity of the mixture of biologically active agents and their interactions, a calculated equivalent dose would have little accuracy.

In these instances it is common to measure some quantity related to the active agents and to base the TLV on that index. For PNAs a TLV has been based on the total weight of benzene- or hexane-soluble airborne material. Alternatives include the single carcinogenic PNA benzo[a]pyrene, the sum of a subset of six carcinogenic PNAs (Table 11.2) or 14 or more individual PNAs (49).

Asbestos is another toxic substance for which the parameter of greatest biological relevance is difficult to define (50). In the early studies when measurements were made with an impinger, few fibers were seen and consequently a count of all particles present was used as an index of overall dustness. More recently the TLV has been based on counts of fibers longer than 5 μm as seen with a light microscope. Since most fibers present are usually shorter than 5 μm and too thin to be visible under a light microscope, it has been suggested that counts of all fibers seen by an electron microscope would provide the most meaningful estimate of risk. Long fibers may be more dangerous, however, short fibers will dominate the count; thus some adjustment may be necessary.

Byssinosis appears to be caused not by cotton itself but by inhalation of cotton plant debris dust baled with the cotton. The total dust airborne in a cotton textile mill is mostly lint (45), and indices have aimed at excluding these "inert" cellulose fibers by collecting a "lintfree" fraction by use of a screen or vertical elutriator. More relevant indices could include plant debris only or the specific biologically active agents if known.

Indices are used where a group of compounds interact to produce a biological effect, or where the active agent is unknown or unmeasurable. In choosing an index we try to maximize biological relevance with a method of measurement that is practical to use. On the one hand very simple parameters like gross dust or total count are easy to use but include much irrelevant material. On the other hand counting fibers by scanning electron microscope or detailed analyses of individual PNAs are difficult, expensive, and not likely to be undertaken frequently. In making the choice it is important to remember that the primary objective is the protection of the worker. Very exact and highly relevant methods, though scientifically satisfying, may be so tedious that very few samples are taken, and because of the larger variability of worker exposure, the true accuracy of the exposure estimate is lower than it would have been if many samples had been taken be a less specific method.

When the overall contaminant level has been reduced and with it the level

Table 11.2   Carcinogenic PNA Subset

Benz[a]anthracene
Benzo[b]fluoranthene
Benzo[j]fluoranthene
Benzo[a]pyrene
Benzo[e]pyrene
Benzo[k]fluoranthene

of the biologically active agent, the level of all correlated indices will be lower. The danger of less relevant, more indirect indices is that serious systematic bias may occur, particularly when an index from the workplace where the health effect relationship was estimated is used in other quite different workplaces. The carcinogenic risk of roofers using asphalt is far lower than for coke oven workers at the same level of exposure as measured by benzene solubles. Byssinosis patterns may be different in mills that garner old rags or process linters and in raw cotton card rooms. In using a measurement that is not perfectly specific, it must be remembered that the result obtained has a less than direct connection with the biological process, and the stronger the effect of extraneous factors, the more care is needed in interpreting the result.

4.5 Mixtures

Industrial workplaces rarely contain only one airborne contaminant, although it is uncommon for there to be several toxic substances each at or near its TLV. The measurement problems caused by the presence of gases, vapors, or dust—some in even higher concentrations than those of primary interest—are discussed below. The question of what to measure when a worker is simultaneously exposed to several agents involves biology and can be answered only by considering the mode of action of the substance in the body. Although possible interactions of substances are extremely complex, it has been the custom to accept the simplifying assumption that "In the absence of information to the contrary, the effects of the different hazards should be considered as additive" (4), and to sum the concentrations C_n of each substance as a fraction of its limit T_n.

$$\frac{C_1}{T_1} + \frac{C_2}{T_2} + \cdots + \frac{C_n}{T_n}$$

When there is "good reason to believe that the chief effects of the different harmful substances are not in fact additive, but independent, as when purely local effects on different organs of the body are produced by the various components of the mixture," exposure is judged by comparing the concentrations to each TLV. Strictly interpreted, the words "purely local effects" would place very few substances in the "independent" class, but it is common practice not to adjust TLVs for mixtures of substances that affect very different organs, even though systemic rather than local. Thus the TLV for dust in a mine is not usually reduced when carbon monoxide is present. Other interpretations are possible, and in general it should be recognized that disease from any cause lessens resistance to all other insults.

Hydrocarbons used as fuels or solvents are usually a mixture of a large number of individual aliphatic compounds and their isomers, often so numerous that analyzing for each individual compound and comparing the result with an individual limit is impractical. Indeed, since TLVs exist for only a few of the compounds present, limits must be stated for the mixture by considering the compounds as a class. Gasoline, for example, may contain aliphatic and aromatic hydrocarbons and additives such as tetraethyl lead. In view of the large differences in toxicity of the several substances and variations in content, calculation of a "TLV" for gasoline is a complex matter (51,52). One difficulty in expressing a TLV for any vapor mixture is that it has been customary to state TLVs for gases or vapors in parts per million. Analytical results, typically from gas chromatographs, emerge initially as the weight in milligrams of each fraction present, from which the concentration, in milligrams per cubic meter, can be calculated. To convert this to parts per million, it is necessary to know the mole weight, which will not be known for unidentified homologues or for the mixture as a whole. For this reason it is preferable to state TLVs for vapor mixtures as a weight concentration (mg/m^3) rather than as a volume fraction (ppm).

Certain combinations of substances present a far more complex situation than can be described by either independent or simple interaction. Benzo[a]pyrene and particulates, with and without sulfur dioxide, carbon monoxide and hydrogen cyanide, ozone, and oxides of nitrogen, and some other mixtures result in complex interactions that may cause effects beyond those predicted by the merely additive case.

4.6 Period of Standard

The concentration of industrial air contaminants varies with time, and recordings such as that in Figure 11.3 are typical. In general, where the concentration is varying with time, the height of the maxima and depths of the minimum are greater as the period of the measurement is shortened. If it were possible to make truly "instantaneous" or zero-time measurements, the peaks and troughs would be very great indeed. Real measurements using continuous reading instruments do not show quite such wide extremes, because the response time of the instrument causes some averaging, thus prevents true zero-time measurements. Even so, instruments with short response times, such as those with solid state sensors, show wide variation from the average, and even those with relatively long response times such as a beta adsorption type of dust monitor, still reveal peaks and valleys that are more than double or less than half the average. As longer and longer period measurements are made, the extremes regress toward the average and obviously, if we define our average over an 8 hr period, a single integrated sample over that period would show no extremes above or below the average. However, even daily averages have highs and lows compared to monthly or yearly averages in all but perfectly nonvarying environments, which do not occur in the real world.

Given the variance of the universe of instantaneous concentrations and the probability distribution of the concentration over one averaging time, it should be possible to determine the probability distribution of the concentration over any other averaging time. The mathematics of this conversion have not yet been developed, but the relationship of averaging time and environmental variance has the consequence that measurements made over different integrating times have a relationship to each other that is a function of the variance.

For example, if a given value of the concentration had a 50 percent (or 90 percent) chance of occurring over one averaging time, there would be a unique higher and lower pair of values that had an equal chance of occurring over a shorter averaging time. Since not all values of environmental variance are equally likely but rather tend to be in the range of GSDs of 1.5 to 3 or 4, it is possible to say in some cases that certain values for different averaging times are inconsistent with each other. Thus it is very unlikely that any environment would be so variable that the probability of exceeding the 8 hr average of 2 µg of beryllium per cubic meter would be the same as exceeding the 25 µg limit for a 15 min averaging time. To state this another way, if the probability of the 2 µg, 8 hr limit being exceeded is 50 percent, the probability of exceeding the 25 µg, 15 min limit is very much less than 50 percent.

In another example, it is very unlikely that an industrial environment would be so constant that the 100 ppm, 8 hr limit for ethyl benzene could be exceeded without exceeding the 125 ppm, 15 min short-term exposure limit (STEL). The effect of setting short-term limits close to long-term limits is to force the effective long-term limit down. Thus to avoid exceeding a 125 ppm, 15 min STEL, it will probably be necessary to achieve an 8 hr average value of much less than 100 ppm of ethyl benzene perhaps even lower than 50 ppm. There may be valid toxicological reasons (53,54) for setting a short-term limit based on, for example, acute irritation, and a TWA that is aimed at preventing some chronic effect; however, it should be recognized that the two are not independent, and when they are set outside the range of approximately equal likelihood, holding concentrations below the limit for one averaging time means holding them far below the limit for the other averaging time: thus one limit is in effect forcing the other.

In terms of sampling strategy, the significance of limits for different averaging times that are statistically inconsistent is that since one has a relatively greater likelihood of being exceeded than the other, regardless of the absolute likelihood of either, there is an opportunity to devise schemes that emphasize measurements to detect the likely event and use these measurements and knowledge of variance derived from them to draw inferences about the less likely event.

5   SAMPLING STRATEGY

Thus far this chapter has discussed the reasons for making measurements of air contaminants in the industrial environment, the nature and variability of worker exposure, and the physical and biological characteristics of toxic substances as they relate to measurement. The next subject is the factors involved in developing a plan of action or strategy for making some set of measurements in the industrial environment that will accomplish the desired purpose effectively and efficiently. A starting point in this problem-solving process must be a clear understanding of the purpose of the measurement. Rarely are data collected purely for their own sake. Even when data are collected because of a demand by others, such as the government or employees, the use of the data should be considered.

What questions will be answered by the data? What decisions could depend on those answers? For example, do we want to know whether these workers are overexposed? And if so, will a decision to take control action follow? Is the control likely to be a minor change in a work practice or an expensive engineering modification? Or is it intended to combine the data with those of health effects to answer the question: What level of exposure is safe? (55). Will it then be decided to modify the TLV or establish a new permissible exposure limit (PEL) by regulation? Or are the numbers to be assembled to answer the question: What level of control is presently being achieved in industry? And may this answer lead to new decision regarding what control is feasible? Last, will workers use the result to find out whether their health is at risk and, as a consequence, decide to change jobs or seek changes in the conditions of work? Often there are many questions that need answers and thus data are collected for multiple purposes. As the discussion that follows indicates, the purpose of the data determines the design of the measurement scheme. All too often data intended for multiple purposes turn out not to be suitable for any purpose. Thus it is usually necessary to focus on the prime need to be sure the strategy will meet this requirement; then, if possible, minor adjustment or additions can be made to meet other needs, if this can be done without losing the main purpose.

The optimum sampling strategy is that which so advantageously combines the choice of method and sampling scheme with respect to sampling location, time, and frequency that we are confident the answer to the objective question is adequate for the decisions which follow.

5.1 Location

Measurements are most frequently made to estimate the risk of adverse health effects. This follows from the primary objective of industrial hygiene which is to help provide a safe and healthy workplace. The health effects of most occupational exposures result from a dose of a substance entering the body by some route. For some substances a semidirect, albeit usually imprecise, method of measuring dose is by biological monitoring. The analysis of exhaled air or of some biological fluid such as urine, blood, tears or perspiration, or of some body component, such as hair or nails, to evaluate past exposure to a chemical is covered in Chapter 3, part B "Biological Responses," Patty's Vol. III revised and also by Baselt (56), Lauwerys (57) and Zielhuis (58). These methods are useful, but due to their limitations, industrial hygienists most often estimate inhaled dose by measurement of the concentration of a substance in inhaled air. Although air samples are sometimes collected from inside respirator face pieces, it is generally not possible to sample the air being inhaled directly. Therefore, the location of the sample collector inlet in relation to the subjects nose and mouth is important. We categorize sampling methods in terms of their closeness to the subject and the point of inhalation as personal samples, breathing zone or vicinity samples, and area or general air samples. A personal sample is one that is collected by a sampling device worn on the person of the worker, which travels with the worker. Breathing zone samples are those collected in the envelope or "breathing zone" around the worker's head, which is thought, based on observation and the nature of the operation, to have approximately the same concentration of the contaminant being measured as the air breathed by the worker. Area or general air samples are the most remote and are collected in fixed locations in the workplace.

Obviously, personal samples are perferred since they most closely measure inhaled air. OSHA enforcement operations "reflect a longstanding belief that personal sampling generally provides the most accurate measure of an employee's exposure ..." (59). However, even personal samples may not sample exactly the air being breathed as indicated by the fact that even a few inches difference in the placement of the filter head of a personal sampler has been reported to make a significant difference in the concentration measured (60, 61) particularly when dust comes from point sources or is resuspended from clothing. For uniformly dispersed aerosols, however, there appears to be no bias between forehead or lapel versus nose locations. (62).

If personal sampling cannot be used, some other means of estimating exposure must be accepted. Breathing zone measurements, made by a sample collector who follows the worker, can come close to measuring exposure. However, this intrusive measurement method may influence worker behavior, and the inconvenience of the measurement will limit the number of measurements and therefore reduce accuracy, as discussed below.

When fixed station samplers are used, knowledge of the quality of the relation between their measurements and the exposure of the workers is necessary if worker exposure is to be estimated. An experimental design that collects large numbers of pairs of measurements of quantities, that are in any way related, will yield a significant correlation coefficient. The important question in the use of general air measurements is: What confidence can be placed in the estimate of worker exposure? Not only is the regression line important, but also the width of the bounds on the confidence limits of a predicted exposure value from some set of fixed station measurement as shown in Figure 11.1.

In studying the relation between area and personal data with respect to asbestos the British Occupational Hygiene Society concluded (63):

The relationship between static and personal sampling results varies according to the characteristics of the dust emission sources and the general and individual work practices adopted in a particular work area.

The present study indicates that:

(i)   When identical sampling instruments are deployed simultaneously at personal and static sampling points and the distances between them are reasonably small, at least two-thirds of the personal sampling results obtained in a given working location are higher than those obtained from static sampling.

(ii)   The differences found between the two types of result tend to be particularly great where the static sampling points are relatively remote from dust emission points, as, for example, when "background" static testing is adopted.

(iii)   In certain cases, results from personal sampling may be lower than those from static sampling, owing to factors such as the positioning of the sampling point with respect to air extraction systems.

(iv)   The correlation coefficient between the personal and static measurements is statistically signifiant but, even so, no consistent relationship of great practical utility could be found in the limited data available.

Many attempts have been made to estimate worker exposure from fixed station air sampling schemes (64–69). The static sampling arrangements ranged from manually operated fixed stations in some studies to computer based automated monitoring systems in others. In some studies the time a worker spent in an area was taken from observations of work patterns. Leidel (11) analyzed these studies and concluded that exposure estimates based on general air (area) monitoring should only be used where it can be demonstrated that general air methods can measure exposure with appropriate accuracy. Linch (70) on the other hand concludes that "only by personal monitoring could a true exposure be determined."

Although worker exposure measurements are most often used in relation to health hazards, not all measurements made for the protection of health need be measurements of exposure. When it has been established that an industrial operation does not produce unsafe conditions when it is operating within specified control limits, fixed station measurements that can detect loss of control may be the most appropriate monitoring system for workers' protection. Local increases in contaminant concentration caused by leaks, loss of cooling in a degreaser, or fan failure in a local exhaust system can be detected before important worker exposure occurs. Continuous air monitoring (CAM) equipment which detects leaks or monitors area concentrations is often used in this way (71). All such systems should be validated for their intended purpose (72).

5.2   Period

Free of all other constraints, the most biologically relevant time period over which to measure or average worker exposure should be derived from the time constants of the uptake, action, and elimination of the toxic substance in the body (73, 74). These periods range from minutes in the case of fast-acting poisons such as chlorine or hydrogen sulfide, to days or months for slow systemic poisons such as lead or quartz. In the adoption of guides and standards, this broad range has been narrowed and the periods have not always been selected based on speed of effect of half-life. For most substances a time-weighted average over the usual work shift of 8 hours has been accepted since it is long enough to average out extremes and short enough to be measured in one work day. Several systems have been proposed for adjusting limits to novel work shifts (75, 76).

Once the time period over which exposure is to be averaged has been decided for either biological or other reasons (69, 77), there are available several alternate sampling schemes to yield an estimate of the exposure over the averaging time. A single sample could be taken for the full period over which exposure is to be averaged (Figure 11.6). If such a long sample is not practical, several shorter samples can be strung together to make up a set of full period,

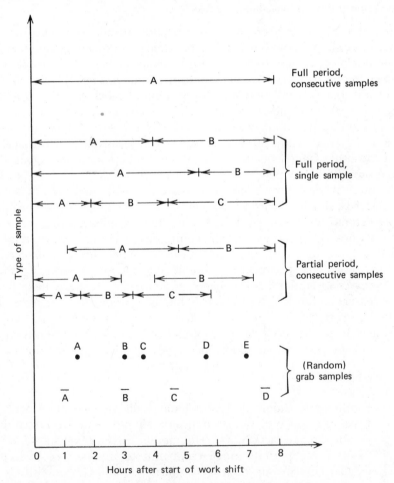

Figure 11.6 Types of exposure measurements that could be taken for an 8-hour average exposure standard.

consecutive samples. In both cases, since the full period is being measured, the only error in the estimate of the exposure for that period is the error of sampling and analytical method itself. However, when these full period measurements are used to estimate exposure over other periods not measured, the interperiod variance will contribute to the total error.

It is often difficult to begin sample collection at the beginning of a work shift, or an interruption may be necessary during the period to change samples. Several assumptions may be made with respect to the unsampled period. It may be assumed that exposure was zero during this period, in which case the estimate for the full period could be regarded as a mininum. Alternatively it could be assumed that the exposure during the unmeasured period was the same as the average over the measured period. This is the most likely assumption in the absence of information that the unsampled period was different. However, it is difficult to calculate confidence limits on the overall exposure estimate since the validity of the assumption is a factor, there is no internal estimate of environmental variance, and the statistical situation is complex.

When only very short period or grab samples can be collected, a set of such samples can be used to calculate an exposure estimate for the full period. Such samples are usually collected at random; thus each interval in the period has the same chance of being included as any other and the samples are independent. This sampling scheme of discrete measurements within a day is analogous to a set of full period samples used to draw inferences about what is happening over a large number of days. In both cases the environmental variance, which is usually large, has a major influence on the accuracy of the results.

Short-period sampling schemes can be useful with dual standards. For example, if a toxic substance has both a short term, say 15-minute limit and an 8-hour limit, 15-minute samples taken during the 8-hour period could be used to evaluate exposure against both standards. This involves some compromise, however, since samples taken to evaluate short-period exposure should be taken when exposure is likely to be at a maximum rather than at random. Statistical techniques for evaluating exposure with respect to dual standards are also available (78). As discussed earlier, when a dual standard is inconsistent, so that one limit is more likely to be exceeded than the other, sampling schemes that evaluate exposure with respect to the limit more likely to be exceeded can be used to provide some confidence about the other limit.

The traditional method of estimating full period exposure is by the calculation of the "time-weighted average." In this method the workday is divided into phases based on observable changes in the process or worker location. It is assumed that concentration patterns are varying with these changes and are homogeneous with each phase. A measurement or measurements, usually shorter than the length of the phase, are made in each phase, and the exposure estimate E is calculated according to the formula:

$$E = \frac{C_1 T_1 + C_2 T_2 + \cdots + C_n T_n}{8}$$

where C_n = concentration measured in phase n
       T_n = duration (hr) of phase n ($\Sigma T = 8$)

Figure 11.7 represents this procedure graphically. Although the exposure estimate itself is simple to make, calculation of the confidence limits on this estimate can be very complex. Each phase must be treated separately. A set of samples must be collected to determine the mean and standard deviation for the phase. These data must then be combined in a manner that weights the variance to obtain an error estimate for the whole. This complex calculation does not include, however, any consideration of the imprecision in selection of phase boundaries. Given the number of samples required in each phase to provide adequate error estimates and the lack of confidence in the end results due to the several layers of assumptions; an equal number of grab samples collected at random over the whole period may yield more accurate results for the same level of effort.

The recommendation that samples should be taken over a full shift to determine employee exposure to toxic substances has been questioned (79). It is maintained that it is possible in some instances to characterize a worker's exposure with a few short period samples. There are arguments to support both sides; ultimately, however, the issue can be decided in each individual case based on the answer to two questions: Can the risk of error in the decision to be based on these measurements be calculated from the data? Is this risk acceptable, given the consequences of the decision?

When workplace measurements are made for purposes other than the estimation of worker exposure, different considerations apply. While a single

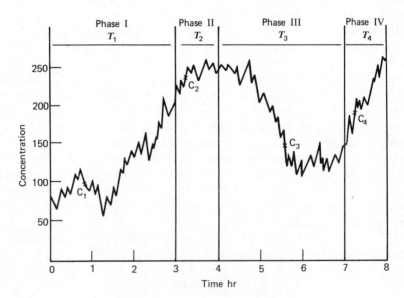

Figure 11.7 Time-weighted average.

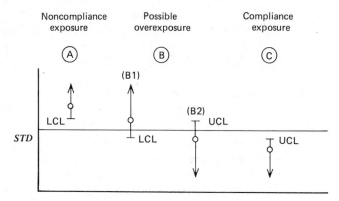

Figure 11.8 One-sided confidence limits.

8-hour sample may be an accurate measure of a worker's average exposure during that period, the exposure was probably not uniform, and the single sample gives no information on the time history of contaminant concentration. To find out when and where peaks occur, with the aim of knowing what to control, short period samples or even continuous recordings are useful. Similarly, when a control system is evaluated or sampling methods are compared, measurements need be only long enough to average out system fluctuations and provide an adequte sample for accurate analysis. As in the case of the decision on location, the purpose of the measurement is a primary consideration in the selection of a time period of a measurement.

5.3 Frequency

By increasing the number of measurements made over a period of time or in a sampling session, the magnitude of the confidence limits on the mean result can be reduced. With smaller (tighter) confidence limits it becomes easier to arrive at a decision at a chosen level of confidence or to be more confident that a decision is correct. In Figure 11.8 decisions are possible in cases A and C, but not in case B. By collecting more samples it might be possible to tighten the lower confidence limit (LCL) in case B_1, for example, permitting the conclusion that these data do in fact represent an overexposure. The choice of the number of samples to be collected rests on three factors: the magnitude of the error variance associated with the measurement, the size of a difference in results that would be considered important, and the consequence of the decision based on the result.

The error variance associated with the measurement depends in most cases on the environmental variance. An exception is the rather limited instance of evaluating the exposure of a worker over a single day by means of a full period measurement. In that case the error variance is determined by only the sampling and analytical error and confidence limits tend to be quite narrow. Usually, however, our concern is with the totality of a worker's exposure, and we wish to use the data collected to make inferences about other times not sampled. There is little choice; unless the universe of all exposure occasions is measured, we must "sample," that is, make statements about the whole based on measurement of some parts.

As discussed earlier, the universe has a large variance, quite apart from the error of the sampling and analytical method. In terms of our decision-making ability, the error of the sampling and analytical method may have very little impact. In Figure 11.9 the inner pair of curves define the decision zone for an environmental coefficient of variation equal to .60 with no sampling or analytical error; the outer zone includes a typical detector tube error having a coefficient of variation of .25. Obviously, even the relatively large error of one of the less accurate methods results in only a slight increase in the no-decision zone (80).

Figure 11.10 also illustrates the effect of sample size or our ability to arrive at a conclusion. These curves give the number of grab samples required in order to be confident that an overexposure did not occur for several typical levels of environmental variance. As can be seen, the difference between the mean and the standard necessary to achieve confidence in the conclusion decreases sharply as the number of samples used increases from 3 to 11. An important conclusion is that for a fixed sampling cost and level of effort many samples by an easy but less accurate method may yield a more accurate overall result than a few samples by a difficult but more accurate method due to the effect of increasing sample numbers on the error of the mean in highly variably environments.

In selecting a sample size it must be kept in mind that it is possible to make a difference statistically significant by increasing the number of samples even

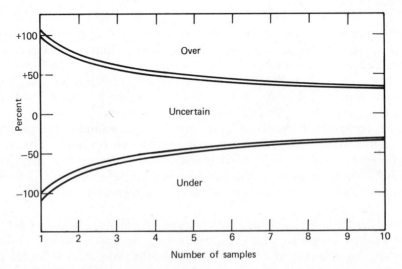

Figure 11.9 Difference between mean measured concentration and TLV required for a decision at the 95 percent level versus number of samples averaged.

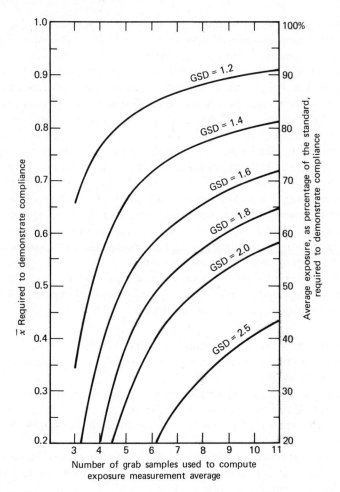

Figure 11.10 Effect of grab sample on compliance demonstration: GSD = variation of grab samples.

though the difference may be of small importance. Thus given enough samples it may be possible to show that a mean of 1.02 ppm is significantly different from a TLV of 1.0 ppm, even though the difference has no importance in terms of biological consequence. Such a statistical significance difference is not useful. Therefore, in planning our sampling strategies we should first decide how small a difference we would consider important in terms of our use of the data and then select a frequency of sampling which could prove this difference significant, if it existed, but not lesser.

The consequences of the decision made on the basis of the data collected should be the deciding factor in selecting the level of confidence at which the results will be tested. Although the common 95 percent (1 in 20) confidence level is convenient because its bounds are two standard deviations from the mean, it is arbitrary and other levels of confidence may be more appropriate in some situations. When measurements are made in a screening study to decide on the design of a larger study, it may be appropriate to be only 50 percent confident that an exposure is over some low trigger level. On the other hand, when a threat to life or a large amount of money may hang on the decision, confidence levels even beyond the three standard deviations common in quality control may be appropriate. To choose a confidence limit first consider the conseqence of being wrong and then decide on an acceptable level of risk.

Since sampling and analysis can be expensive, some thought should be given to ways of improving efficiency. Sequential sampling schemes in which the collection of a second or later group of samples is dependent on the results of some earlier set are a possibility. This common quality control approch results in infrequent sampling when far from decision points, but increases as a critical region is neared. Another means of economizing is to use a nonspecific, direct reading screening method, such as a total hydrocarbon meter, to obtain information on limiting maximum concentrations that will help to reduce the field of concern of exposure to a specific agent.

Few firm rules can be provided to aid in the selection of a sampling strategy because data can be put to such a wide variety of uses. However, the steps to be followed to arrive at a strategy can be listed.

1.   Decide on the purpose of the measurements in terms of what decisions are to be made. When there are multiple purposes, select the one or several most important for design.

2.   Consider the ways in which the nature of the environmental exposure and of the agent relates to measurement options.

3.   Identify the methods available to measure the toxic substance as it occurs in the workplace.

4.   Select an interrelated combination of sampling method, location, time, and frequency that will allow a confident decision in the event of an important difference with a minimum of effort.

6   MEASUREMENT METHOD SELECTION

The assortment of tools available to the industrial hygienist for measuring worker exposure to toxic substances is much fuller now than a decade ago, although there are still many gaps. This section describes the choices available with respect to the location at which the measurement is made, time period or averaging time of the measurement, ability to select certain size aerosols and reject others, and degree to which human involvement can be lessened by automation and computer analysis. The attributes of the various sampling methods which are important in deciding if they are appropriate for a particular sampling strategy are summarized.

6.1   Personal, Breathing Zone, and General Air Samples

Section 5 described the locations at which samples could be collected with relation to the air actually inhaled by the worker and defined the terms personal, breathing zone, and general air. Personal samplers are devices which collect a sample while being worn on the worker's clothing with the sample inlet positioned as close to the mouth as practicable, usually on the lapel. One kind of personal sampler consists of a pump on the worker's belt or pocket and a sampling head containing the sorbent tube, filter, or other collection medium, clipped to the lapel close to the nose and mouth. In recent years these pump type personal samplers have been joined by passive personal samplers or passive dosimeters. These samplers do not pump the air to the collection medium but passively allow the toxic gas or vapor to diffuse (81) onto the solid sorbent. These dosimeters or "gas badges" (82) are also worn close to the nose. As with all new methods, the accuracy and reliability of these passive dosimeters has been questioned. However, Rose and Perkins, in their 1982 review of the state of the art, concluded that passive dosimetry is as reliable as active (pump) dosimetry and is, with some exceptions, an acceptable method for monitoring gases and vapors (83). Carefully designed studies validating the accuracy of passive dosimeters of new designs or for new substances or conditions continue to be needed.

Personal samplers of both types are worn by the worker so they must be lightweight, portable, and not affected by motion or position. These restrictions tend to limit the size of sample that can be collected. In the case of passive dosimeters the geometry of the diffusion channel and the diffusion constant of the substance result in an equivalent sampling rate which is usually quite low thus requiring long sample periods. Personal sampling with wet collectors, such as impingers or bubblers, is difficult due to the danger of spillage and glassware breakage. Spill proof impingers have however been used. Gravitational size selectors such as horizontal elutriators which are affected by position have not made successful personal samplers, although inertial devices such as cyclones or impactors are satisfactory.

When limitations of weight, complexity, wet collecting media, or position prevent successful personal sampling, it is still possible to make an approximate measurement of worker exposure by collecting the sample in the "breathing zone" of the worker. This vaguely defined zone is the envelope of air surrounding the worker's head which is thought to have approximately the same concentration of the contaminant being measured as the air breathed by the worker. Breathing zone samples may be collected by a fixed sampler with the inlet near the nose of a stationary worker or, in the case of a mobile worker, by carrying the sample collecting equipment and holding the sample inlet near the worker's head, while moving around the work site with the worker. The obviously awkward and time-consuming nature of this kind of sampling limits its usefulness.

When measurements of workers exposure are not needed or indirect estimates are adequate, the concentration of a contaminant in the general air of a workplace may be measured at some fixed station. Many of the equipment limitations imposed by personal and breathing zone sampling systems do not apply to general air samplers. Portability is not critical, so electrical components may be either battery or line powered. When line power is available, powerful pumps may be used to provide enough vacuum for critical orifices to obtain precise flow control or to operate high volume vacuum sources capable of collecting very large samples. Wet devices and both horizontal and vertical elutriators are practical. Very large samples, which may be needed to obtain sufficient sensitivity for trace analyses, may be collected on heavy or bulky collecting media. Multiple samplers of different types may be arrayed close to each other to provide sampler comparison data. Most new methods of measurement of worker exposure were first tested in fixed station arrays, where they were compared with older methods before being adapted to personal sampling.

6.2 Short- and Long-Period Samplers

Available sampling methods have limited flexibility in the period of time over which the sample can or must be collected. Some methods are inherently grab samplers, although the increased interest in long period samplers has led to their adaptation. Most dectectors tubes (84) were originally intended to produce a result after a few pump strokes of up to several hundred milliliters. The interval between strokes could be lengthened, but instead of increasing the sampling period, this produces an average of several short samples taken over a longer time. Automatic systems have been developed (85) as fixed station samplers that can extend the low range of some tubes by repeated pump strokes spread out over a long period. In the continuous flow mode, short-period detector tubes have been recalibrated for use at very low flowrates over long periods, and special tubes have been developed (86, 87) as long-term samples for up to 8-hour flowrates of 5 to 50 cm^3/min. Other inherently short period methods are those that use a liquid, particularly a volatile liquid, in a bubbler or impinger. As the air passes through the sampler, the collecting medium evaporates, eventually to dryness, with loss of sample, unless terminated in time. Usually, without the addition of liquid, sampling periods of in excess of 30 min are impractical with wet devices.

Vessels that collect a whole air sample are convenient grab samplers (88), and some can be adapted for longer period sampling. Many low flowrate personal sampler pumps have an air outlet fitting that can be used to fill a bag. Allowing for possible contamination due to the air passing through the pump, long period samples may be collected. When the bag is carried in a sling on the worker's back personal samples may be collected.

Greater time flexibility is available with solid sorbents. Even though there is a fixed volume of air that can be sampled at a concentration before breakthrough occurs, the freedom to use a wide range of low flowrates permits personal and fixed sampling over periods of 8 hours or more. Other air contaminants, competing for active sorbent sites and particulates adding to the resistance to flow, limit the maximum volume of air that can be sampled in much the same way as breakthrough.

Systems that tend to be suitable for long-period samples only are usually those where the sampler can barely collect the minimum amount of material required by the sensitivity of the analytical method. A personal sample for respirable quartz using a 10 mm cyclone size selective presampler, operated at 2 liters/min or less, will sample less than a cubic meter of air in 8 hours and thus collect less than 100 ug of quartz at the TLV. Shorter samples will confront the serious sensitivity limitations of most methods for quartz (89). The same problem occurs with personal samples for beryllium and for detailed analyses for multiple compounds such as PNAs.

High volume fixed station, or even personal particulate samplers using large size selectors and filters, where necessary, allow shorter sampling periods that are still longer than "grab samples." Even when analysis is not necessary and only gross or respirable weight is being measured, analytical balance limitations prevent very short samples for particulates. There are, however, various "instantaneous" mass monitors based on beta absorption (90) or the piezoelectric effect (91) that are capable of making a measurement in as little as one minute.

6.3 Size Selection

When it is necessary in sampling for particulates to include or exclude certain size particles, limitations are created by the nature of the size selecting devices available. The unsuitability of elutriators as personal samplers due to their orientation requirements has already been mentioned. All size selectors make their stated cut only at a predetermined flowrate, which must be held constant over the period of the sample against changing filter resistance and battery conditions (92). Whereas cyclones and impactors tend to compensate for flowrate changes and elutriators compound the error, all need pumps that not only sample a known volume over the sampling period but do so at a known constant flowrate. Such pumps are usually larger and heavier than low flow pumps, which need only sample a reliably known volume, and approach the limit of practicality as personal samplers. In addition to the requirement for constant flow, pump pulsation must be damped out if it upsets the size selector.

Isokinetic conditions usually must be established when sampling for particulates in high velocity streams (93). In stacks, particles with high kinetic energy due to their weight and velocity can be improperly included or excluded from the sample under nonisokinetic conditions. However, workplace sampling is generally done at low ambient air velocities of less than 300 ft/min. Further, air velocity and direction is usually continually changing. Under these conditions isokinetic sampling of the kind used in stack sampling is not necessary or practical. However, consideration must be given to the effect of the velocity and direction of the air at the filter inlet. Open face filters may oversample when face up or undersample face down. Some undersampling may be desirable to avoid collection of large, noninhalable particles. Davies (47) has developed theoretical relationships for filter inlet performance.

6.4 Continuous Air Monitors

Leak detecting continuous air monitors (CAMs) with fixed sensors in critical locations connected to alarms and remote indicators are commonly used for carbon monoxide, hydrogen sulfide, chlorine hydrocarbons, and other acutely toxic or explosive gases. These systems may use passive sensors, they may pump the air through a detection cell located at the point of collection, or they may pump contaminated air to a remote analyzer. Sequential valving arrangements allow one remote analyzer to be coupled to many sample lines. Provision for automatic calibration and zeroing may be included.

Systems which make measurements automatically at a number of locations and gather the data at a central readout point have been installed in a number of plants. Adaptation of these systems to the estimation of worker exposure has led to the development of complex computer based monitoring systems (94) (Figure 11.11). Since the contaminant sensors do not detect the presence of a worker, some data on worker location must be added to estimate exposure. Time and motion studies that yield percentage of time in various measured locations could be used with the daily average fixed station measurement for that location to calculate a weighted average exposure. The drawbacks are that time/activity distributions exhibit considerable variation, even under routine

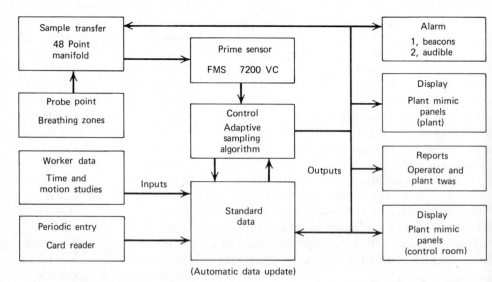

Figure 11.11 Vinyl chloride monimer (VCM) monitoring system. Courtesy of Eocom Corporation.

conditions, and the most significant exposures occur during nonroutine periods. Also an assumption is made that the concentration at a time is independent of the worker's activity.

Alternative schemes provide a device that reads a card carried by each worker to signal the computer of each entry and departure from a monitored area. Time in the area can be multiplied by the general or weighted average concentration from the sensors in that area as measured while the worker was present. This situation is analogous to estimating exposure from fixed station sampler measurements. An even more elaborate proposed system places on each worker small transmitters that are tracked automatically by a sensing network, and these detailed worker location data are combined with fixed station measurements for various locations to estimate exposure.

6.5 Mixtures

Toxic substances seldom occur by themselves. Mixtures not only cause difficulties in estimating the biological consequences of exposure, but also complicate exposure measurement. The other substances present in an air mixture need to be considered, even when only one component of the mixture is being measured and the other components are far below toxic effect levels, because of the effects of the other substances on the sampling systems. Charcoal is a very useful sampling medium because it will absorb and retain so many substances. However, because of this property it is possible for all the active sorption sites to be occupied by other materials. The substance being measured may then break through long before the recommended sample volume, based on collection of the pure substance, has been passed through the tube.

An analogous case is the measurement of a low concentration of asbestos fibers in an environment containing a great deal of other, nonfibrous dust. To collect enough fibers to give the fiber density necessary for an adequate count without counting an unreasonable number of fields would result in the collection of so many grains that the fibers would be obscured. Thus the sample volume is limited by the total dust present and low fiber density is compensated for by counting a large number of fields. Overload from other airborne substances can also result in plugging of filter samplers, particularly when liquid accumulates on the surface of membrane filters and "blinds" the pores. The use of thick depth filters of glass or cellulose fibers provides greater capacity, although at the possible sacrifice of some efficiency.

6.6 Method Selection

The purpose of this chapter is to explain the considerations which go into the choice of a sampling strategy (location, period, frequency) and the selection of a method that will permit the accomplishment of that strategy. Table 11.3 summarizes the degree to which the most common methods possess the attributes of importance in selecting various sampling strategies. The column

Table 11.3 Sampling and Analysis Method Attributes of Importance in Selecting Sampling Strategy

Method	Sampling Period	Ability to Concentrate Contaminant	Ability to Measure Mixtures	Time to Reach	Intrusiveness	Proximity to Nose and Mouth
Personal sampler/solid sorbent						
Sorption only	Medium to long	Yes	Yes—gases	After analysis	Medium	Very close
Sorption plus reaction	Medium to long	Yes	No	After analysis	Medium	Very close
Personal sampler/filter						
Gross gravimetric	Medium to long	Yes	Yes—particulate	After weighing or analysis	Medium	Very close
Respirable gravimetric	Long	Yes	Yes—particulate	After weighing or analysis	Medium	Very close
Count	Medium to long	Yes	Yes—particulate	After counting	Medium	Very close
Combination filter and sorbent	Medium to long limited	Yes	Yes	After analysis	Medium	Very close
Passive dosimeter	Long	Yes	Yes—gases	After analysis	Low	Very close
Breathing zone impinger/ bubbler						
Analysis	Medium-limited	Yes	Yes	After analysis	High	Close
Count	Medium-limited	Yes	Yes—particulate	After counting	High	Close
Detector tubes						
Grab	Short	NA	No	Immediate	High	Close
Long period	Medium to long	NA	No	Immediate	Medium	Very close

608

Gas vessels					
Rigid vessel	Short	No	Yes—gases	After analysis	High	Close
Gas bag	Short to long	No	Yes—gases	After analysis	High	Close
Evacuated/critical orifice	Medium to long	No	Yes—gases	After analysis	Medium	Very close
Direct reading portable meters					
Nonspecific (flame ion, combination gases)	Instantaneous or recorder	NA	Yes	Immediate	High	Slightly distant
Specific (carbon monoxide, hydrogen sulfide, ozone, sulfur dioxide, etc.)	Instantaneous or recorder	NA	No	Immediate	High	Slightly distant
Multiple compound (infrared, gas chromatography, etc.)	Instantaneous or recorder	Some	Yes	Almost immediate	High	Slightly distant
Mass monitor (beta absorber, piezoelectric)	Short	Yes	No	Almost immediate	High	Slightly distant
Particle counters (optical, charge)	Short	No	No	Almost immediate	High	Slightly distant
Fixed station					
High volume	Medium to long	Yes	Yes—particulate	After analysis	Low	Remote
Horizontal or vertical elutriator	Long to short	Yes	Yes—particulate	After analysis	Low	Remote
Installed monitor	Short to long	Some	No	Almost immediate	Low	Remote
Freeze trap	Medium	Yes	Yes—vapors	After analysis	Low	Remote

Table 11.3 (*Continued*)

Method	Specificity	Convenience Rating	Sample Transportability	Recheck of Analysis Possible	Accuracy
Personal sampler/solid sorbent					
Sorption only	High by analysis	High	Good	Elution—Yes; thermal des. no	Good
Sorption plus reaction	High by analysis	High	Good	Yes	Good
Personal sampler/filter					
Gross gravimetric	None for weight only—high by analysis	High	Fair	Yes	Good
Respirable gravimetric	High by analysis	Medium	Fair	Yes	Fair
Count	Fair—depends on particle identification	High	Good	Yes	Poor
Combination filter/sorbent	High by analysis	Medium	Good	Yes	Fair
Passive dosimeter	High by analysis	Very high	Good	Yes	Fair
Breathing zone impinger/ bubbler					
Analysis	High by analysis	Low	Poor	Yes	Fair
Count	Fair—depends on particle analysis	Low	Poor	Yes	Poor
Detector tubes					
Grab	Medium—some interference	High	No sample	No	Fair
Long period	Medium—some interference	High	No sample	No	Fair

Gas vessels					
Rigid vessel	High by analysis	Low	Fair	Yes	Good
Gas bag	High by analysis	Low	Fair	Yes	Good
Evacuated/critical orifice	High by analysis	High	Good	Yes	Good
Direct reading portable meters					
Nonspecific (flame ion, combination gases)	None—total of measured class	High	No sample	No	Good
Specific (carbon monoxide, hydrogen sulfide, ozone, sulfur dioxide, etc.)	Medium—some interference	High	No sample	No	Good
Multiple compound (infrared, gas chromatography, etc.)	Medium—frequent overlap	Medium	No sample	No	Fair
Mass monitor (beta absorber, piezoelectric)	Mass only	High	No sample	Not usually	Fair
Particle counters (optical, charge)	Count/size only	High	No sample	No	Fair
Fixed station					
High volume	High by analysis	Low	Fair	Yes	Good
Horizontal or vertical elutriator	High by analysis	Low	Fair	Yes	Good
Installed monitor	Medium—may be interferences	High	No sample	No	Good
Freeze trap	High by analysis	Very low	Poor	Yes	Fair

headings are explained as follows:

Method. The methods listed are sampling methods but the ratings of attributes which follow assume the usual range of analytical methods which can be applied to the size and type of sample collected.

Sampling Period. By short is meant essentially grab samples while long means approximately 8 hrs in a single sample.

Ability to Concentrate Contaminant. Sampling methods which extract the contaminant from the air and collect it in a reduced area or volume are potentially able to improve analytical sensitivity by several orders of magnitude. However, the concentrating mechanism (filtration, sorbtion) may introduce errors.

Ability to Measure Mistures. Most sampling methods provide a sample which can be analyzed for more than one gas or vapor, but usually not for both gases and vapors.

Time to Result. Certain decisions (vessel entry) must be made immediately while others can wait until after the sample is transferred to a laboratory and analyzed.

Intrusiveness. When the method requires the presence of a person to collect the sample or the wearing of a heavy or awkward sampling apparatus, this intrusion of the sampling system into the work situation may affect worker behavior and exposure.

Proximity to Nose and Mouth. As discussed earlier, locating a sampler inlet even a small distance from a worker's mouth may bias the exposure measurement and samplers remote from the worker may not be measuring the air inhaled at all.

Specificity. Some methods give only nonspecific information like total weight of all dust particles or concentration of all combustible gases while others provide a sample which can be analyzed for any specific material.

Convenience Rating. These are estimates of the amount of work or difficulty involved in collecting samples.

Sample Transportability. If the sample must be transported to a distant labortory for analysis, the ability to withstand shock, vibration, storage, and temperature and pressure changes without being altered or destroyed is important.

Recheck of Analysis Possible. Some samples may only be analyzed once while others are in a form such that rechecks, reanalyses at different conditions or analysis for other substances is possible.

Accuracy. Given all the possibilities for error from sampler calibration, sample collection, transport and analysis, an overall coefficient of variation (CV) of 10 percent is considered good. Some count methods are subject to such counter variability that poor accuracy is usual. Method inaccuracy should not be judged alone but should be seen in combination with the inaccuracy caused by environmental variability, which is usually larger, in making decisions whether a method is sufficiently accurate for a purpose.

This table shows that not all sampling strategies are possible since for some strategies the sampling and analytical method with the necessary combination of attributes may not exist. The industrial hygiene technology gaps thus revealed are fruitful areas for future research and development.

REFERENCES

1. M. Corn, *J. Am. Ind. Hyg. Assoc.*, **37,** 353–356 (1976).

2. U.S. Public Health Service–National Institute for Occupational Health and Safety, "The Right to Know: Practical Problems and Policy Issues Arising from Exposures to Hazardous Chemicals and Physical Agents in the Workplace," Department of Health, Education and Welfare, PHS– NIOSH, Cincinnati, Ohio, 1977.

3. H. Raiffa, *Decision Analysis: Introductory Lectures on Choices under Uncertainty*, Addison-Wesley Publishing Co., Reading, Mass. 1970.

4. American Conference of Governmental Industrial Hygienists, "Threshold Limit Values for Chemical Substances and Physical Agents in the Work Environment with Intended Changes for 1983–1984," ACGIH, P.O. Box 1937, Cincinnati, Ohio, 45201.

5. E. M. Thompson et al., *J. Am. Ind. Hyg. Assoc.*, **38,** 523–535 (1977).

6. H. M. Donaldson and W. T. Stringer, "Beryllium Sampling Methods," Department of Health, Education and Welfare Publication (NIOSH) 76-201, Cincinnati, Ohio, 1976, p. 21.

7. C. S. McCammon, Jr., and J. W. Woodfin, *J. Am. Ind. Hyg. Assoc.*, **38,** 378–386 (1977).

8. L. D. Horowitz, *J. Am. Ind. Hyg. Assoc.*, **37,** 227–233 (1976).

9. Occupational Safety and Health Act of 1970, PL 91-596.

10. J. R. Lynch, *Nat. Saf. News*, **113:**5, 67–72 (1976).

11. N. A. Leidel, K. A. Busch, and J. R. Lynch, *Occupational Exposure Sampling Strategy Manual*, Department of Health, Education and Welfare, Publication (NIOSH) 77-173, Cincinnati, Ohio, 1977.

12. CEFIC Report on *Occupational Exposure Limits and Monitoring Strategy*, European Council of Chemical Manufacturers Association, Brussels, 1982.

13. *Measurement and Evaluation of Concentrations of Airborne Toxic or Health Hazardous Work-Related Substances*, W. German Federal Ministry of Labor, TRgA401 Sheet 1, 1979.

14. R. P. Harvey, "Statistical Aspects and Air Sampling Strategies" in: C. F. Culles and J. G. Firth, Eds., *Detection and Measurement of Hazardous Gases*, Heinemann, London-New York, 1981, p. 147.

15. S. M. Rappaport, S. Selvin, R. C. Speer, and C. Keil, *Am. Ind. Hyg. Assoc. J.* **42,** 831 (1981).

16. S. M. Rappaport, S. Selvin, R. C. Speer, and C. Keil, *ACS Symp. Series*, **149,** 431 (1981).

17. R. M. Tuggle, *Am. Ind. Hyg. Assoc. J.*, **42,** 493 (1981).

18. R. M. Tuggle, *Am. Ind. Hyg. Assoc. J.*, **43,** 338 (1982).

19. J. C. Rock, *Am. Ind. Hyg. Assoc. J.*, **43,** 297 (1982).

20. G. R. Oxley, *Ann. Occup. Hyg.* **19,** 163–167 (1976).

21. R. R. Lauwerys et al., *J. Occup. Med.*, **20,** 17–20 (1978).

22. S. Fukabori et al., *J. Sci. Labour*, **53,** 89–95 (1976); abstracted in *Ind. Hyg. Dig.*, **41,** 16 (1977).

23. S. Fukabori et al., *J. Sci. Labour*, **52,** 67–81 (1976); abstracted in *Ind. Hyg. Dig.*, **40,** 19–20 (1976).

24. E. Cronen and R. B. Stoughton, *Arch. Dermatol.*, **36,** 265 (1962).

25. J. V. LeBlanc, *J. Occup. Med.*, **19,** 276–277 (1977).

26. M. M. Key, Ed., *Occupational Diseases, A Guide to Their Recognition*, Government Printing Office, Washington, D.C., 1977.

27. R. W. Weeks, Jr. et al., *Occup. Health Saf.* **46,** 19–23 (1977).

28. T. D. Sterling and D. M. Kobayashi, *Environ. Res.*, **3**, 1–35 (1977).

29. D. T. Wigle, *Arch. Environ. Health*, **32**, 185–190 (1977).

30. M. McCann, *Artists Beware*, Watson-Guptill, New York, 1979.

31. A. Poklis and C. D. Burkett, *Clin. Toxicol.*, **11**, 35–41 (1977).

32. J. S. Oliver, *Lancet*, **1**, 84–86 (1977).

33. R. Korobkin et al., *Arch, Neurol.*, **32**, 158–162 (1975).

34. J. W. Hayden et al., *Clin. Toxicol.* **9**, (2), 169–184 (1976).

35. R. A. Warriner III et al., *Arch. Environ. Health*, **32**, 203–205 (1977).

36. B. L. Weisenberger, *J. Occup. Med.*, **19**, 569–570 (1977).

37. J. Aitchinson and J. A. C. Brown, *The Lognormal Distribution*, University Press, Cambridge, England, 1963.

38. K. E. Bencala and J. H. Seinfeld, *Atmos. Environ.*, **10:**11, 941–950 (1976).

39. R. I. Larsen, "A Mathematical Model for Relating Air Quality Measurements to air Quality Standards," U.S. Environmental Protection Agency, Publication AP-89, Research Triangle Park, N.C., 1971.

40. Y. Kalpasanor and G. Kurchatora, *J. Air. Pollut. Control Assoc.*, **26**, 981 (1976).

41. A. C. Stern, Ed., *Air Pollution, 3rd ed., Vol. 3, Measuring, Monitoring, and Surveillance of Air Pollution*, Academic Press, New York, 1976, p. 799.

42. H. Ayer and J. Burg, "Time-Weighted Average Versus Maximum Personal Sample," paper presented at the American Industrial Hygiene Conference, Boston, 1973.

43. N. A. Leidel, K. A. Busch, and W. E. Crouse, *Exposure Measurement Action Level and Occupational Environmental Variability*, Government Printing Office, Washington, D.C., 1975.

44. C. I. Fairchild and M. I. Tillery, *J. Am. Ind. Hyg. Assoc.*, **38**, 277–283 (1977).

45. J. R. Lynch, "Air Sampling for Cotton Dust," Transactions of the National Conference on Cotton Dust and Health, University of North Carolina, Chapel Hill, 1970, p. 33.

46. J. H. Vincent, D. Mark, *Ann. Occup. Hyg.* **24**, 375 (1981).

47. C. N. Davies, *J. Appl. Phys.*, Ser. 2, **1**, 921–932 (1968).

48. B. R. Roy, *J. Am. Ind. Hyg. Assoc.*, **38**, 327–332 (1977).

49. K. A. Schute, D. J. Larsen, R. W. Hornung, and J. V. Crable, "Report on Analytical Methods used in a Coke Oven Effluent Study," National Institute for Occupational Safety and Health, 1974.

50. R. K. Zumwalde and J. M. Dement, "Review and Evaluation of Analytical Methods for Environmental Studies of Fibrous Particulate Exposures," Department of Health, Education and Welfare, Publication (NIOSH) 77-204, NIOSH, Cincinnati, Ohio, 1977, p. 66.

51. H. E. Runion, *J. Am. Ind. Hyg. Assoc.*, **38**, 391–393 (1977).

52. H. J. McDermott and S. E. Killiany, *J. Am. Ind. Hyg. Assoc.*, **39**, 110–117 (1978).

53. D. Turner, *Ann. Occup. Hyg.*, **19**, 147–152 (1976).

54. D. M. Ferguson, *Ann. Occup. Hyg.*, **19**, 275–284 (1976).

55. E. E. Campbell, *J. Am. Ind. Hyg. Assoc.*, **37**, A-4 (1976).

56. R. C. Baselt, *Biological Monitoring Methods for Industrial Chemicals*, Biomedical Publications, Davis, Calif. (1980).

57. R. R. Lauwerys, *Industrial Chemical Exposure: Guidelines for Biological Monitoring*, Biomedical Publications, Davis, Calif. (1983).

58. R. L. Zielhuis, *Scand. J. Work Environ. and Health*, **4**, 1 (1978).

59. Use of Personal Sampling Devices During Inspection, 47 Fed. Reg. 55478 (Dec. 10, 1982).

60. B. B. Chatterjee, M. K. Williams, J. Walford, and E. King, *J. Am. Ind. Hyg. Assoc.*, **30**, 643 (1969).

61. R. Butterworth, J. K. Donoghue, *Health Phys.*, **18**, 319 (1970).

62. B. S. Cohen, A. E. Chang, N. H. Harley, M. Lippmann, *Am. Ind. Hyg. Assoc. J.*, **43**, 239 (1982).

63. S. A. Roach, et al., *Ann. Occup. Hyg.*, **27**, 1 (1983).

64. A. J. Breslin, L. Ong, H. Glauberman, A. C. George, and P. LeClare, *J. Am. Ind. Hyg. Assoc.*, **28**, 56–61 (1967).

65. R. J. Sherwood *J. Am. Ind. Hyg. Assoc.*, **27**, 98–109 (1966).

66. H. F. Shulte, "Personal Sampling and Multiple Stage Sampling," paper presented at ENEA Symposium on Radiation Dose Measurements, Stockholm, Sweden, June 12–16, 1967.

67. A. L. Linch, E. G. Wiest, and M. D. Carter, *J. Am. Ind. Hyg. Assoc.*, **31**, 170–179 (1970).

68. D. B. Baretta, R. D. Stewart, and J. E. Mutcheler, *J. Am. Ind. Hyg. Assoc.*, **30**, 537–544 (1969).

69. E. J. Calabrease, *J. Am. Ind. Hyg. Assoc.*, **38**, 443–446 (1977).

70. A. L. Linch and H. V. Pfaff, *J. Am. Ind. Hyg. Assoc.*, **32**, 745–752 (1971).

71. R. A. Olson, L. W. Rampy, G. W. Engdall, "Proposed Guidelines for Classification of Continuous Air Monitoring (CAM) Equipment in Production Plants," paper presented at the Am. Ind. Hyg. Conf., Portland, Oregon (1981).

72. S. K. Norwood, G. Poetteker, R. Y. Nelson, C. N. Park, and L. B. Coyne, "Validation and Quality Assurance for Continuous Air Monitors for Determining Work Area Concentrations: A Case Study with Methylene Chloride," paper presented at the Am. Ind. Hyg. Conf., Cincinnati, Ohio (1982).

73. S. A. Roach, *J. Am. Ind. Hyg.Assoc.*, **27**, 1 (1966).

74. S. A. Roach, *Ann. Occup. Hyg.*, **20**, 65–84 (1977).

75. R. S. Brief and R. A. Scala, *J. Am. Ind. Hyg. Assoc.*, **36**, 6 (1975).

76. J. W. Mason and H. Dershin, *J. Occup. Med.*, **18**, 603–606 (1976).

77. J. L. S. Hickey and P. C. Reist, *J. Am. Ind. Hyg. Assoc.*, **38**, 613–621 (1977).

78. R. S. Brief and A. R. Jones, *J. Am. Ind. Hyg. Assoc.*, **37**, 474–478 (1976).

79. D. D. Douglas, *J. Am. Ind. Hyg. Assoc.*, **38**, A-6 (1977).

80. J. R. Lynch, "Uses and Misuses of Detector Tubes," Transactions of the 32nd Meeting of the American Conference of Governmental and Industrail Hygienists, ACGIH, Cincinnati, Ohio, 1970.

81. E. D. Palmes et al., *J. Am. Ind. Hyg. Assoc.*, **37**, 570–577 (1976).

82. F. C. Tompkins, Jr. and R. L. Goldsmith, *J. Am. Ind. Hyg. Assoc.*, **38**, 371–377 (1977).

83. V. E. Rose, J. L. Perkins, *Am. Ind. Hyg. Assoc. J*, **43**, 605 (1982).

84. American Industrial Hygiene Association, *Direct Reading Colorimetric Indicator Tubes Manual*, AIHA, Akron, Ohio, 1976.

85. K. Leichnitz, *Detector Tube Handbook, 3rd ed.*, Drägerwerk, Lübeck, Germany, 1976.

86. D. Jentysch, D. A. Fraser, *Am. Ind. Hyg. Assoc. J.*, **42**, 810 (1981).

87. K. Leichnitz, *Ann. Occup. Hyg.*, **19**, 159–161 (1976).

88. R. W. Miller et al., *J. Am. Ind. Hyg. Assoc.*, **37**, 315–319 (1976).

89. National Institute for Occupational Safety and Health *Manual of Analytical Methods, 2nd ed.*, Department of Health, Education and Welfare (NIOSH) Publication 77-157, U.S. Public Health Service-NIOSH, Cincinnati, Ohio, 1977 (3 volumes).

90. P. Lilienfeld and J. Dulchunos, *J. Am. Ind. Hyg. Assoc.*, **33**, 136 (1972).

91. G. J. Sem at al., *J. Am. Ind. Hyg. Assoc.*, **38**, 580–588 (1977).

92. D. L. Bartley and G. M. Brewer, *Am. Ind. Hyg. Assoc. J.*, **43**, 520 (1982).

93. N. A. Fuchs, *Atoms. Environ.* **9**, 697–707 (1975).

94. G. L. Baker and R. E. Reiter, *J. Am. Ind. Hyg. Assoc.*, **38**, 24–34 (1977).

CHAPTER TWELVE

Philosophy and Management of Engineering Control

KNOWLTON J. CAPLAN

1 DEFINITION

Engineering controls for industrial hygiene purposes may be defined as an installation of equipment, or other physical facilities including, if necessary, selection and arrangement of process equipment that significantly reduces personal exposure to occupational hazards.

Common examples are local exhaust systems, general ventilation systems, enclosures around noisy machines, and substitution of a nondusting mixer for a dusty mixer. Substitution of a less hazardous material in the process involves many other disciplines besides industrial hygiene, but since the proposed substitute may also require process or handling modifications, it frequently becomes involved in the question of engineering controls.

Currently the Occupational Safety and Health Administration (OSHA) defines engineering controls "by difference." OSHA defines personal protective equipment as "anything worn on the person of the worker to reduce the exposure;" administrative controls as "any adjustment of the work schedule to reduce the exposure;" and engineering controls as "all other measures to reduce the exposure." This approach has obvious flaws. For example, the third definition would place biological monitoring in the category of an engineering control, clearly not a very good fit. Future developments in industrial hygiene also may evolve measures to reduce the hazard that are not compatible with the general concept of engineering.

2 DEGREE OF ENGINEERING CONTROL REQUIRED

The determination of what degree of engineering control is required to maintain a given concentration of air contaminant is a most difficult task when approached on a theoretical or general basis. In most industries, the required degree of control has been arrived at through an evolutionary process, wherein an increasing level of control, and control of increasingly minor sources, is pursued until the objective is reached. Thus for industries and operations that are fairly common, the required level of control is based on rather wide experience; and if there is adequate cooperation and care between the hygienist and the engineer, the results can be largely on target.

For industries and operations that are relatively unique (i.e., there are only a few plants or operations) or when a threshold limit value (TLV) is to be drastically reduced from prior practice, however, the determination of adequate controls is much less certain. There are several reasons for this.

1. The properties of the materials involved and the ways in which they are handled in the manufacturing process represent a bewildering variety of combinations. In addition to the rather well-defined physical and chemical properties of the materials (such as vapor pressure), there are many other, less readily quantified variables that can lead to the generation of more or less contaminant in the air. One obvious example is the difference between a light fluffy dust and a sticky dust. The sticky dust tends to adhere to surfaces and does not become airborne easily. Thus if reasonable housekeeping is maintained, contamination of surfaces is of little import to the airborne concentrations, whereas the opposite is true if the dust is not sticky and is light and easily airborne.

2. The amount of contaminant generation from the process that would be of hygienic significance is usually so small as to be completely insignificant in the process material balance. Usually the only way in which the significance of a potential source can be quantified is by experimentally measuring the generation rate, which can be relatively simple in some few operations but is frequently difficult and expensive.

3. When contaminant concentrations are to be drastically reduced, the major sources mask and obscure the effect of the minor sources. Since as noted previously the rate of generation from sources is usually not known, one is obliged to guess just what sources require control to reach a given target level of air contamination.

Obviously the best approach in terms of securing good results with engineering controls and not wasting resources on unneeded controls is the evolutionary, stepwise method. First, the obvious sources are controlled and resulting concentrations determined. The masking effect of the major sources having been removed, the next tier of sources can be estimated and considered. This approach could well take several years of persistent and intelligent effort in the event that more than one or two rounds of application of engineering controls are required.

2.1 Hazard Classification System for Particulate

The control of airborne dust or fume is more difficult in general than is the control of gases or vapors. Dust that is not captured and removed from the environment settles on floors, work surfaces, and overhead ledges and becomes a secondary source of airborne dust, whereas this is not true of most gases and vapors. (A notable exception is mercury: this low vapor pressure material can slowly and continuously evaporate into the air from minor sources in amounts significant to the low TLV.)

Furthermore, the customary units used in reporting dust or particulate concentration versus gas and vapor concentrations (i.e., mg/m or g/m versus ppm) creates a subjective impression of concentrations which is entirely erroneous. A solvent vapor or gas that has a TLV of 1 ppm, which "sounds low" is, assuming a molecular weight of 100, actually 4080 $\mu g/m^3$ when expressed in those units.

After cautioning the reader not to treat the following as "design data" but rather as a concept level attempt at guidance, a method for estimating the level of control required for particulate matter in the form of dust and fumes can be described. The approach is to first select a "production factor" category, designated by a letter A, B, C; then to establish a hazard classification that is a combination of generation of contaminant (high, medium, or low) and the control level to be achieved. The production factor is established as follows:

Production Classes

A. Plantwide processing of hazardous material; by-products as well as products may contain hazardous material.

B. Departmental processing of hazardous material.

C. Local processing of hazardous material; or intermittent, infrequent, or very small scale; or short-term.

Generation Classes. The dust or fume generation category is selected according to the following description:

High. Vapor pressure of material as used would result in equilibrium vapor concentration over the TLV; a high degree of process heat is applied relative to melting point or boiling point; material is oxidized; violent physical dispersion occurs; high volatility rate; dry, dusty powder; fluffy, easily airborne dust; soft, friable materials; dry bulk processing.

Medium. Low vapor pressure, moderate heat relative to melting point or boiling point, less violent physical dispersion, low volatility rate; properties intermediate between "high" and "low."

Low. Negligible vapor pressure; gentle physical dispersion; hard, abrasion-resistant material; little or no oxidation; heavy or sticky dust not easily airborne.

Control Level Selection. Then the appropriate control level is selected. Since engineering controls, particularly ventilation, are physical in nature, the content of the target material in the process stream is important. For example, consider grinding on a 2 percent beryllium-copper alloy. Leaving aside all arguments over the relative toxicity of beryllium in that form, assume that the element is to be controlled to 2 $\mu g/m^3$. This means that since the dust is only about 2 percent beryllium, the total dust concentration evolved from the process can be $100/2 = 50$ times the 2 $\mu g/m^3$, or 100 $\mu g/m^3$. If on the other hand the concern is lead oxide fume, the fume will be 92.8 percent lead and the TLV for lead itself would be governing.

Hazard Class Selection. Having selected the control level for the contaminant mix, enter Table 12.1 to select a number for the hazard class. The production factor and the hazard factor are then combined to yield a control code comprised of a letter and a number, viz. B1, A3, etc. The control levels applicable to the classification derived are as follows:

Control Levels Required

Code A1. Techniques similar to and approaching those used in nuclear fuel reprocessing facilities would be required. In general, the source of contamination should be completely removed from contact with the worker. This requires complete enclosure or barrier walls, completely automated and mechanized processing, and perhaps mechanized maintenance. Even such facilities require human intervention at least for maintenance and troubleshooting, and an extraordinarily high degree of personal protection is required during such activities.

Table 12.1 Hazard Class for Dust and Fume

Control Level Target (mg/m³)	Generation Category		
	High	Medium	Low
	Hazard Factor		
<0.05	1	1	2
0.05	2	2	2
0.10	2	2	3
0.25	3	3	3
0.5	3	3	3
1.0	3	3	4
3.0	4	4	4
5.0	4	4	4
>5.0	4	4	4

Code A2. Complete control of every observable or foreseeable source; glovebox, mechanized, or remote handling in ventilated enclosures where practical; general ventilation as adjunct to local control; higher ventilation rates than "minimum standards;" scrupulous housekeeping and general cleanliness a necessary adjunct to engineering controls.

Code A3. Good control of all except minor sources; generous general ventilation; good housekeeping highly advisable; ventilation rates standard or in excess of standards.

Code A4. Standard ventilation control of significant sources.

Code B1. As in *A1*, necessary only in affected department. Segregation or isolation of affected department allows standard industrial procedures in remainder of plant. Adjunct techniques of cleanliness necessary to prevent contamination of entire plant by physical means.

Code B2. As in *B1*, necessary only in affected department. Adjunct techniques of cleanliness necessary to prevent contamination of entire plant by physical means.

Code B3. As in *A3*, applied only to affected department.

Code B4. As in *A4*, applied only to affected department.

Code C1. Total enclosure, glove-box hoods, segregation and isolation of facility.

Code C2. Good local control; ventilation rates in excess of minimum standards; extent of control depends on whether there is a "life hazard" capable of causing death or severe injury on short exposure.

Code C3. Local control of significant sources using standard ventilation techniques; if sources are sufficiently small, general ventilation may suffice; if sources are sufficiently intermittent, respiratory protection of worker may suffice.

Code C4. Local control of major sources only, using standard ventilation rates. In some cases dilution ventilation alone may suffice.

3 INTERFACE WITH THE PROCESS

Even the most simple of engineering controls work better if there is interaction between the hygienist and the plant operators, and plant engineering personnel. Even a simple exhaust hood should be designed to meet the following operating criteria.

1. It should not interfere unduly with normal operations; if some interference is required to achieve control, the acceptable design should be determined mutually with operating personnel.

2. Provisions should be made for easy opening and closing of access doors or easy removal and replacement of access panels for anticipated maintenance of enclosed equipment. "Easy removal and replacement" does not permit a multiplicity of nuts and bolts.

Such design cannot be successful if done "at a distance" but must be accomplished after adequate observation of the operations and face-to-face discussions with operators and engineers. Designs that show a simple symbol for a hood or a simple box, leaving the details to the sheet metal contractor, are almost never successful.

If a change in process or substitution of less toxic materials is a viable alternative, even greater interaction with production and engineering personnel is required. Depending on the nature of the project, research and marketing functions may also be involved.

Perhaps the most simple process change is that involving only material handling. For example, a double-cone blender or a Y-blender requires very difficult and expensive dust control if a low TLV is required. Much better dust control at lower total cost may be achieved by using a rotary blender. However, the blending characteristics of these two machines are somewhat different; before such a substitution is made, it must be determined whether the proposed new blender will in fact perform the required blending properly. A somewhat similar situation would exist if it were proposed, for example, to substitute a screw conveyor for a belt conveyor for a material that is difficult to handle.

If the proposed substitution or change in process goes deeper into the process chemistry and physics, the project is liable to assume some magnitude. In addition to the design of the process itself, the products may be changed in their characteristics and if so, there would be an impact on the market for the end products as well as on the manufacturing process.

It should be remembered that most industrial processes have a long evolutionary history. An existing process has evolved to its present state through gradual change and improvement, during which each step has been adjusted and proved in production to ensure that the system is workable. Interaction with the marketplace and knowledge of actual production costs are also necessary in such evolution. It is true, however, that the process as evolved, and with certain quality control measures, yields a product of known and acceptable properties. The evolutionary process is a combination of the theoretical and the empirical. In lay language, "if we make a certain thing in a certain way, with certain controls, it produces a product with certain known attributes and limitations." There may well be parameters involved that, although capable of measurement, have never been measured for one reason or another.

At one time, in the manufacture of natural uranium metal fuel elements for graphite-moderated reactors, the metal was produced in rather small ingots that contained various impurities. These impurities were removed by recasting the ingots in high vacuum induction furnaces. The vacuum melting boiled the impurities out of the molten uranium. The impurities included magnesium metal which, as it boiled out, swept with it the highly radioactive daughter products of natural uranium, the beta emitters UX_1 and UX_2. These materials condensed on the water-cooled lid of the vacuum furnace. Since the magnesium was pyrophoric and the UX_1 and UX_2 no longer subject to the self-shielding of the parent metal, the opening of the vacuum furnaces presented a severe fire and hygiene hazard. These problems were controlled by means of operating the top end of the furnaces in a total enclosure with robot handling of crucibles, ingots, and so on.

A new process was developed using the theoretical principles of metallurgy. It involved casting a very much larger ingot, and by controlled cooling, segregating the impurities in the top of the ingot and in the outside skin. These impurities could be removed by a simple machining operation, which was much less expensive than vacuum recasting and presented much less severe fire and hygiene problems. After machining, the interior of the ingot successfully passed the many quality control tests for the uranium metal. Since at the time a new plant was being built and an old plant abandoned, the new process was installed.

Within a few months numerous complaints came from the reactor operators. The new metal was causing an inappropriately high number of slug ruptures; that is, the aluminum jacket of the uranium metal was rupturing while in service in the reactor, releasing fission products into the cooling water, thus causing untold difficulties for the reactor operator. An immediate investigation was launched, but repeated and intensified quality control testing at the uranium plant failed to disclose the source of the problem. A statistical analysis of the batches of metal that caused slug ruptures revealed no correlation with operating parameters other than the weather. At about the same time the reactor operators reported that the cause of the slug ruptures was bubbles of hydrogen gas accumulating between the uranium and the jacket, impeding the necessary heat transfer. These two findings led to a new quality control test—the measurement of hydrogen in the uranium. Samples of metal made by the old process contained a maximum of 2 ppm hydrogen, whereas uranium made by the new process in humid weather contained 6 ppm.

This is a case in point. Uranium metal was manufactured under quality control procedures much more stringent than are employed for most commercial products; yet because hydrogen had never been a problem, the measurement of hydrogen had not been a subject of quality control. It was not even realized, before the bad experience with the new product, that the hydrogen content of the metal would be of concern. Obviously, such mistakes can be very costly. Thus major changes in industrial processes are usually slow, evolutionary, step-by-step procedures.

It not infrequently happens that the hygiene control of an existing process is very difficult and expensive, but simpler, more effective, and less expensive controls are anticipated with a new process. Assuming that hygiene control is to be accomplished for reasons of significant health hazard or perhaps because of legal enforcement actions, it seems appropriate to deduct the cost of hygiene controls for the old purpose from the projected cost of the new process. Frequently the new process will offer operating economies, but it is not at all infrequent that the capital cost of the new process cannot be amortized adequately by such advantages. In such a case the deduction of the hygiene control cost for the old process may be the deciding factor in whether the new process is implemented.

Once a new process has been developed and is being implemented, the hygienist runs the risk of getting out of touch. The new process seems to be so much better from a hygiene point of view that the problems seem minor, and perhaps complacency sets in. Furthermore, the new project is frequently so big that it is designed as a major engineering project over which the hygienist may not have adequate control. Under such circumstances, it may be a bitter disappointment to find that the new process is not in good shape, that further retrofit controls are required anyway. Very seldom does a new process accomplish the complete transition from very difficult or unsolvable hygiene problems to no problems at all. The hygienist must keep in touch with the detailed design of the new process.

4 EXPERIMENTAL SAMPLING

Since the advent of personal sampling pumps and legal enforcement of standards based on 8-hour time-weighted average exposures, most industrial hygiene sampling is oriented toward determination of the 8-hour time-weighted exposure. Although these are the most appropriate data for evaluating exposures when short-term exposure is not important, they furnish little or no information leading to the intelligent application of engineering controls. The classical case of a production worker performing essentially the same task all day long is disappearing from the industrial scene. Therefore, if hygiene controls are to be wisely applied, it is necessary to know what parts of the worker's job (i.e., which tasks) contribute what proportional amount of the day's exposure. It is also perfectly possible for significant sources of contaminant that are not directly connected with the worker's tasks to be responsible for the exposure.

The industrial hygiene of most fairly common industries has been quite well studied. In the foundry industry, for example, general process knowledge and experience may be adequate without experimental sampling. However, if the emissions of concern are not visible, not odorous (or if odor fatigue sets in), and not well-known, experimental sampling may be necessary to determine the significance of various possible sources.

The first and perhaps easiest kind of experimental sampling is task oriented. Assuming that the worker does not perform the same duties continuously during the workday, task-oriented sampling consists of taking separate samples for the different tasks in which the individual is involved to determine the exposure separately for each task. This kind of sampling imposes added strains on the logistics of sample rate, sampling time, and analytical sensitivity, to ensure the validity of the samples, and attention to these details is essential. For example, it is possible to use a given sampling filter repeatedly if the sampling time for each round of a specific task is too short to produce definitive results. It is not uncommon for such a task-oriented sampling regime to reveal that it would be more beneficial to control some part of the job that actually is at lower concentration than some other part, as is evidenced by Table 12.2.

Table 12.2 Task-Oriented Sampling

Task	GA or BZ[a]	Minutes/ day	Concentration (mg/m^3)	Minutes/day × Concentration
A. Charge pot	BZ	40	0.12	4.8
B. Unload pot	BZ	80	0.50	40.0
C. General survey	GA	250	0.16	40.0
D. Lab—sample trips	GA	20	0.05	1.0
E. Change room	GA	30	0.08	2.4
F. Pump room—repack	BZ	30	0.32	9.6
G. Lunch room	GA	30	0.07	2.1
		480		99.9

$$\text{TLV} = 0.2 \text{ mg/m}^3 \qquad \text{Wt. avg.} = \frac{99.9}{480} = 0.21 \text{ mg/m}^3$$

[a] GA = general air; BZ = breathing zone.

For the job analyzed in Table 12.2 presumably a single 8 hr sample would have shown a concentration of 0.21 mg/m³. The task-oriented sampling, however, reveals several interesting things. Column 5 shows that tasks B and C are the major contributors to the day's exposure and that a significant reduction in the concentration at either of those tasks would be adequate to bring the 8 hr exposure well below the TLV. This is true even though task C in itself is below the TLV. In addition, it shows that task F, well above the TLV concentration, is of such short duration that significant improvement in that part of the exposure would not have a large effect on the 8-hour exposure.

Sampling of a more "experimental" nature is limited only by the ingenuity of the experimenter. By means of such experimental sampling it has been shown that transportation of dry pasted battery plates by fork truck on clean floors is not, in itself, of concern at a 100 µg/m³ level of lead in air. A more complex set of experimental samples showed that in manual loading or unloading of an automatic stacking machine in a battery plant, the generation of lead dust at the rack of pasted plates is significant and requires control at the 150 µg/m³ level, even though historically control of the stacking machine alone has seemed to be adequate at the 200 µg/m³ level (1). For such experimental sampling it is necessary to follow the principles of any measurement process, and it is most difficult to measure something without disturbing that which is being measured. The general principles may be outlined as follows:

1. The operation should be as normal as can be contrived and still satisfy the requirements of the experiment.
2. The sample taking should interfere as little as possible with the operator.
3. All other contaminant-generating operations should be in normal operation or perhaps alternatively, completely shut down.

4.   If at all possible, it is advisable to estimate or measure the airflow patterns as well as the contaminant concentration, to permit the estimation of the total amount of the contaminant released.

5   EDUCATION AND PARTICIPATION OF WORKERS

Most hygiene controls, especially local exhaust ventilation hoods and enclosures, sound enclosures, and so on, require changes in the way the worker does the job. Usually the job is made somewhat more difficult. Ideally, however, good design can minimize the detractive aspects and make the hood easier rather than more difficult to live with; and sometimes the same project can include mechanical assists or other revisions that make the job easier. In any event, changes to the job are introduced.

In the absence of any participation in the design by the worker, or of any information about the proposed changes, human nature automatically generates resistance to the change. The new aspects of the job may well be resented. In addition to whatever design features prove to be less than optimum on an objective basis, the human aspects of the situation are even more important. Most of us can remember similar events from our own experience. The young fellow with the old wreck of a car, who really wants to keep it running, manages to do so with baling wire and chewing gum. The do-it-yourselfer who has created a gadget for home use will put up with major inconveniences to make it work. However the purchaser of a new car will complain bitterly of every rattle and minor inconvenience. The key is, "Does the user (worker) *want* to make it work?"

No engineering job can be perfect in its design or construction. Even though it may be almost perfect for normal operations, operations are not always normal. Accommodation must be made by the operators for unusual situations. If the design happens to be less than perfect for such an event, and if the workers are psychologically opposed to the installation in the first place, it will not be long before the cutting torches come out and the offending feature is removed, and not replaced and not re-engineered.

This destined-for-failure sequence can be minimized, perhaps completely avoided, by engaging the participation of the workers in intermediate stages of design. Operating supervisors, busy as they are with all facets of production, may not even know of some of the small problems that occur on the job. Even supervisors are people, and people tend to tell the engineer or systems designer how things "should be" rather than how they actually are.

All these problems can be minimized by observations of the job over an adequate time period, talking with the operator, asking questions, and checking the observations and conclusions with the supervisor. Later in the design effort, when preliminary drawings are available, these should be reviewed with the operators as well as the operating supervision. Many industrial organizations have frequent safety meetings or similar meetings, which usually provide a good forum for such discussions if the group is small enough. If such meetings are not routinely held, special meetings for the purpose of design review are well worthwhile.

Management sometimes fears that, in such a situation, workers will make requests or present ideas that are technically or economically unsound or cannot be accomplished within the budget for the job. It has been my experience that such questions or suggestions frequently arise, but if all suggestions are sincerely considered and at least some are adopted, no lingering resentment is detected over those that are declined or rejected for good reason. It has also been my invariable experience that operating problems are brought to light (or good ideas for improvement introduced) that had escaped the design engineer's attention and had not been mentioned by operating supervision. Some comments are as simple as noting that using an access door would be more convenient if the door opened to the right instead of to the left. Other suggestions, of course, may be much more complex.

In addition to whatever objective improvements are gained in the design of the project, the intangible gains are even more important. The workers have had a chance to get a better idea of what is forthcoming, their questions have been answered, their ideas and suggestions have been seriously considered and some of them adopted. The same meeting may also serve as a conduit for further explanations of the reason for doing the job and for answering questions concerning the hazard, therefore also assisting the motivation of the worker to "make it work."

In today's industrial social climate it is fruitless for the engineer or the supervisor to complain that the workers will not use the hygiene control properly. Effective and practical mechanisms do not exist, in general, to enforce proper utilization on a broad base, especially if there are genuine defects in the design. Therefore, the only successful alternative is to convince the workers to assist in proper utilization; and as a rule this can be achieved only by education and cooperation. It is also true that there is always a small scattering of recalcitrant individuals, for whom the standard labor relations type of disciplinary action may be required.

In short, the success of engineering controls for hygiene purposes depends on the same factors as does the success of engineering features related to production. The subject matter may not be as familiar to the operating supervisors and workers, but it certainly is not beyond comprehension. With top management backing, the same techniques of training and motivation that are applied to production will be successful if applied to hygiene.

6 DESIGN CONSIDERATIONS FOR EXHAUST VENTILATION SYSTEMS

Since ventilation is one of the major types of engineering control for hygiene problems, it serves as a good basis for discussion of some design philosophies that in principle also apply to any other kind of engineering control. This discussion assumes that the design is of good quality from a purely engineering standpoint; the remarks are addressed to the specific problems of the interface with operations represented by most engineering controls.

6.1 Operating Convenience

It is essential that interference of exhaust system hoods with normal production operations be at the minimum extent practical while obtaining the desired control objective. The philosophical aspects have been discussed previously; the design requirements for operating convenience are summarized here. The elements of the design activity necessary to obtain operating convenience involve:

1. Adequate observations of operation.
2. Consultation with operators and operating supervision.
3. Interaction with workers to offer an opportunity for understanding the purpose of the system, obtaining worker input to aid in design convenience, and involvement of the worker in the design process. Again, all normal elements of design competence are assumed to be accomplished.

6.2 Maintenance Access

The ventilation equipment itself should be arranged and designed so that maintenance is reasonable and practical, as for any other mechanical equipment.

Special considerations concerning maintenance reflect awareness that exhaust system hoods, especially if they are partial or total enclosures, necessarily interfere with the maintenance of the contaminant-generating equipment being so enclosed. (The same considerations apply in many respects to noise enclosures.) It is obvious that there must be access to the enclosed equipment so that normal maintenance and lubrication can be provided easily; what is all too often forgotten is abnormal or major maintenance access. Merely specifying a removable panel that is held in place with many nuts and bolts or other fasteners is not sufficient.

When maintenance is required on production equipment, especially breakdown maintenance rather than preventive maintenance, time is usually at a premium. Time is also important even for preventive maintenance. Thus one can rest assured that access will be obtained to the equipment requiring repair. If removal of nuts and bolts is necessary, it will be performed; in the absence of bolted panels, access will be had with a cutting torch. The problem, from the hygiene point of view, is to make possible the quick and easy replacement of the panel after maintenance has been conducted. If a cutting torch has been employed to get to the equipment, rare indeed is the equipment maintenance crew who will apply a welding torch to repair the hood. Similarly, time is seldom available to get all the nuts and bolts back in place on a bolted panel.

In addition, the multiplicity of nuts and bolts involved is in itself a problem: when the time comes to replace these fasteners, they will not all be found, much less reinstalled. It is a very unusual mechanic who will take time to go to the storeroom to get more nuts and bolts. Psychologically the offending panel is a hindrance to the accomplishment of the task at hand, rather than part of the task.

Improvement in this situation can be obtained by the application of a small set of principles.

1. A minimum number of tools should be required (preferably no tools) for any anticipated access, and at most a single tool such as a hammer or a single simple wrench.

2. A hinged door that swings completely out of the way when opened is better than a panel that must be completely removed and set aside, perhaps incurring damage or distortion in the process.

3. Any fasteners should preferably be both easy to use and rugged—a number of devices are available (various latches with handles, taper lock wedges, husky rubber tension fasteners, pins, etc.). Such fasteners should not be completely removable but should be retained on the piece, by means of intrinsic design or by means of a short length of light chain.

4. The weight and convenience of the panel should be considered and handles provided if advisable. Some closures, especially if the design is such that gravity will keep the door closed, can well be provided with a rubber flap, such as a piece of conveyor belting, so that bending or distortion of light gauge sheet metal do not cause problems.

Special provisions often can be made for items that require frequent maintenance—for example, the lubrication fittings on the idler pulleys of conveyor belts, when the conveyor belt is to be enclosed. During installation of the enclosure, the lubrication fittings can be extended, by means of a short piece of pipe through a small hole in the enclosure wall, so that nothing has to be removed for the oiler to gain access to the fitting.

An example of excellent design in this regard is the hammer mill enclosing hood illustrated in Figure 12.1. Hammer mills are a notorious source of dust. Although the dust is generated inside the machine, the joints in the machine and shaft seals do not always stay dust tight, and even if they do, the machine itself acts as an inefficient fan and creates a positive pressure on the downstream side. Thus any receiving vessels must be airtight or provided with exhaust ventilation. If provided with exhaust, it frequently happens that the ventilation control removes an excessive amount of product and accordingly the airflow is throttled, resulting in insufficient hygiene control. The principles embodied in this hood design include appropriate interface with the process in that an exhaust hood is around the source but does not actually interfere with the process containment, containing and collecting as it does only material that leaks from the process. The maintenance and operational features of the design

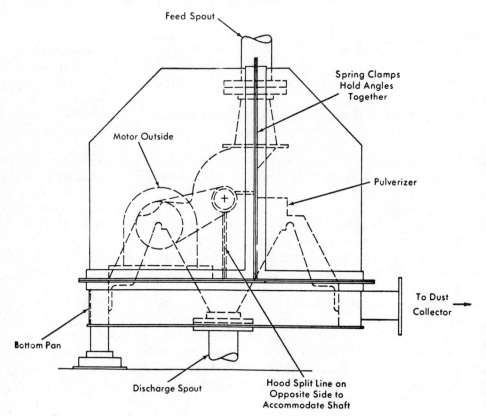

Figure 12.1 Breakaway hood design for hammer mill.

are illustrated by the following features:

1. The bottom pan is rigidly affixed to the discharge flange of the pulverizer and does not interfere with access to and maintenance of the machine itself; furthermore, a solid connection to the exhaust duct can be provided, since no movement is needed.

2. The upper part of the hood is self-supporting. When it is removed, there are no structural supports for hood panels to block access to the pulverizer.

3. The upper part of the hood is custom designed for the pulverizer in that it has a split line, offset as necessary. This permits accommodation of the feed spout and the drive shaft on the split line. Thus neither the feed spout nor the drive has to be disturbed to remove the hood.

4. The drive motor and the V-belt are outside the hood. If lubrication fittings are provided on the bearings of the pulverizer, they may be extended to the outside of the hood on the split line.

5. The fasteners that hold the hood in place are conveyor cover clips, chained to the hood itself. These can be removed by use of a screwdriver or

other similar device such as a pry bar, or by a hammer. They are replaced by a simple tap with a hammer.

6. The hood sections are shaped to have reasonable structural integrity; they will not be distorted by ordinary handling. Handles should be provided if the hood sections are so large that it would be awkward to remove the hood without them.

7. Preferably, the hammer mill discharge should be into a closed, airtight vessel.

6.3 Blast Gates and Dampers

6.3.1 General

The *Manual on Industrial Ventilation* (2) published by the American Conference of Governmental Industrial Hygienists (ACGIH) offers a brief tabulation of advantages and disadvantages of using dampers or blast gates to balance exhaust systems. These points should be amplified and their import described. Many of the opinions expressed are based on my observations over many years but are not quantified on a broad basis and perhaps are not universally true. The need for this discourse is based on an observation of a persistent trend among plant engineers and vendors of equipment to prefer the blast gate (damper) method of balancing exhaust systems because it is "easier" and "more flexible." By contrast, it is hard to find an exhaust system balanced by blast gates that is controlling adequately the problem for which it had been installed.

Balancing an exhaust system implies that the correct amount of airflow is obtained in each of the several branches of the system. There are two general ways of achieving this goal, one is to use blast gates or other types of dampers to permit field adjustment of the air flows in each branch, and the other is to design the hoods, ductwork, and fan to achieve the appropriate balance without dampers. If the contaminant to be controlled is hazardous, the maintenance of proper exhaust volumes assumes some added importance. When the contaminant is particulate matter (dust or fume), a further constraint is placed on the system in that a minimum velocity must be maintained in each branch or the ductwork will plug. Many other possible design problems arise from time to time, such as condensation in the ductwork, erosion or corrosion, and undesirable chemical reactions. This discussion, however, is limited to the problem of balancing exhaust systems handling dust or fume.

Dampers may be used for several purposes:

1. As on-off valves where only some of the branches of the system need to be active at any given time.

2. To prevent motor overload on startup, for example, in a fabric filter system with new bags.

3. When required volume in a given branch is not known with certainty and a conservative estimate may lead to excessive removal of product.

4. For balancing the system to achieve proper airflow in each branch.

Of these four reasons, the first two are obviously legitimate and do not in fact involve "balancing." Even in such circumstances, difficulty is frequently encountered when operators fail to open and shut dampers as required, and it is advisable to interconnect the damper operation with the process if a practical means of accomplishing this can be found. However, situations do arise in which the required exhaust volume is uncertain, and a conservative estimate may result in excessive volumes, which remove desired product from the process or unnecessarily overload the dust collector. A damper may be used for these situations, but there will be severe limitations in that closing the damper too far will reduce the duct velocity below the minimum; other means of compensating are available and are usually a better choice. In the third case, the blast gate would be found only in one branch of the system, not in all branches. It is the fourth case, the general use of dampers or blast gates to balance the entire system, that requires elaboration with respect to its problems. It is obvious, however, that there are occasions when blast gates are advisable, and the use of engineering judgment is to be encouraged in these cases (as opposed to siding with the use or nonuse of blast gates as an act of faith).

The *Industrial Ventilation Manual* tabulates the relative advantages of using blast gates. That list is reproduced here for convenience.

Relative Advantages of Method A and Method B

Method A: Balance without Blast Gates

1. Air volumes cannot be changed by workmen or at the whim of the operator.
2. Small degree of flexibility for future equipment changes or additions; the ductwork is "tailormade" for the job.
3. Choice of exhaust volumes for a new unknown operation may be incorrect; in such cases some ductwork revision is necessary.
4. No unusual erosion or accumulation problems.
5. Ductwork will not plug if velocities are chosen wisely.
6. Design calculation is more time-consuming than method B.
7. Total air volumes slightly greater than design air volume due to added air handled to achieve balance.
8. Poor choice of "branch of greatest resistance" will show up in design calculations.
9. Layout of system must be in complete detail, with all obstructions cleared and length of runs accurately determined. Installations must exactly follow layout.

Method B: Balance with Blast Gates

1. Air volumes may be changed relatively easily, though precautions are taken. Such changes are desirable where pickup of unnecessary quantities of material may affect process.

2.   Greater degree of flexibility for future changes or additions.
3.   Correction of improperly estimated exhaust volumes is easy, within certain ranges.
4.   Partially closed blast gates may cause erosion to slides, thereby changing degree of restriction or may cause accumulations particularly of linty material.
5.   Ductwork may plug if the blast gate adjustment has been tampered with by unauthorized persons.
6.   Design calculations are relatively brief.
7.   Balance may be achieved with design air volume.
8.   Poor choice of "branch of greatest resistance" may remain undiscovered. In such case the branch or branches of greater resistance will be "starved."
9.   Leeway is allowed for moderate variation in duct location to miss obstructions or interferences not known at time of layout.

All the points in the ACGIH list are valid, but in many cases elaboration and explanation is called for. The overall impression given by the list is improper, in that there are severe limitations to the advantages of blast gates, and a whole series of interrelated problems derives from their use.

It is perfectly true that a system can be designed and balanced with blast gates. However, that is just the beginning of the story; the following sets of conditions prevail:

1.   It is *not* less expensive to balance the system with dampers than it is to design it to be balanced.
2.   The flexibility afforded by blast gates is severely limited in one direction by the necessity for maintaining carrying velocities, and in the other direction by the available suction in the system.
3.   Changing the setting of any one blast gate affects the performance of all the other branches in the system.
4.   The purported ease of adjustment with blast gates is in itself a severe disadvantage in that the limitations are not understood by operating and maintenance personnel. The availability of blast gate "adjustments" is what leads to patchwork and butchered systems, where changes have been made with no semblance of engineering technology.
5.   Last but not least, no method of adjusting blast gates and fixing them in place has yet been developed that can withstand the onslaughts of the operator, the maintenance department, or the night shift mechanic. These people naturally enough feel that the blast gates are there to be adjusted as a solution to whatever problems may be perceived to exist at any given time.

These five problems are explained as follows.

6.3.2   Ease of Balancing

For the *designer* of a system it is easier to call for balancing by damper adjustment than it is to devise a system that is balanced. This does not mean, however, that it is easier or cheaper for the owner to have the balancing achieved by adjusting dampers. The field balancing, even for a relatively simple system, requires a large number of iterations of trial-and-error balancing of each branch, since adjustment of any damper affects each branch of the entire system. It typically takes more man-hours to achieve balance by this method than by design. Depending on the local labor situation, the owner may have to pay for a craftsman, an apprentice, and a technician—three people—to achieve the same results that could have been produced on the drawing board.

Balancing of ductwork is so extensive and expensive that a number of commercial organizations base their business on that work, and there is even a trade association, the Air Balance Council, involved in this market. It is true that the bulk of such work concerns air-conditioning and air-supply ventilation systems that are controlled by dampers, and the technical considerations for such systems differ considerably from those for exhaust systems handling dust or fume. Air-conditioning systems are usually concealed, making access to the dampers difficult, and the premises are usually used by office personnel who have little inclination to tamper with such equipment. Moreover, poor air balancing in the systems results in discomfort, and complaints are immediately forthcoming. In the case of ductwork involving excessive concentrations of harmful air contaminants, however, typically one must await industrial hygiene monitoring or OSHA inspection before the deficiency is discovered—and, more important, poor performance results in a health hazard, not merely in discomfort. All these factors taken together cause a shift in the engineering judgment involved, and the usual conclusion is that *for the air-conditioning system*, balancing by use of dampers is appropriate.

6.3.3 Flexibility

The flexibility achievable by the use of blast gates is severely limited. For an existing system, the power requirement increases with the square of the duct velocity; accordingly, it is inappropriate to design the ductwork for velocities much higher than the minimum carrying velocity actually required. This means that as dampers are closed to reduce the volumes in a given duct, the amount of such throttling that can be accomplished without reducing the duct velocity below the minimum depends entirely on which branch is being throttled. If the branch is toward the far end of the system (i.e., the branch of greatest resistance or a branch of high resistance with respect to the available static pressure), the available latitude for such throttling will be minimal. If the branch enters the system, where the static pressure in the main is greatly in excess of that required to operate that branch at design volume, there is much more latitude for throttling.

It is by no means obvious in most systems which branch is the branch of greatest resistance—it is not always the one farthest from the fan—and the subtleties of appropriate throttling of a given branch, combined with the affect of changing one damper on the rest of the system, make such adjustment a prime cause of system inadequacy.

If more air is desired in a given branch, opening the damper will work up to the available static pressure in the main where that branch joins, which again is highly variable and is just the reverse of the situation described for throttling a damper. Even worse, however, when one damper is opened, other branches in the system then draw less air than they were drawing before the adjustment.

The purported "ease of adjustment" and "flexibility" create an attitude in the minds of operating and maintenance personnel, indeed sometimes in engineering personnel, that the system is something like a utility water distribution system: "If you want more water, open the valve; and if you want less water, close the valve." In fact, in dust control systems only very minor changes can be made with dampers without engineering and redesign. The general attitude created, however, is that the flexibility exists to permit such adjustment, modifications, and additions, without recourse to engineering and design calculations. As explained earlier, such is not the case. If on the other hand the ductwork was not provided with dampers, and perforce changes would be required in duct size, orifices, and so on, recourse to engineering redesign would be unavoidable. It is true that the specific change or modification being made would cost more and take longer than would butchering the ductwork and moving the blast gates around; but in the long run the cost will be less because the performance of the system will not be ruined.

6.3.4 Fixing the Position

Many schemes have been proposed for fixing dampers and blast gates in position once a system has been balanced by their use. These include putting a bolt through the damper, clamping the damper, and tack welding the damper in position, all of them to no avail. Dust collection systems frequently receive inadequate maintenance; and any number of things can go wrong with the system. If dampers and blast gates are available, they will be the first choice for adjustment and will be the quickest available "remedy" for the problem as perceived by the operator or night shift mechanic. These people have the tools and the contacts necessary to get the inhibiting and offending lock removed, in the hope of finding a quick and easy cure for the problem. Furthermore, the locked-in-place damper is poor psychology. The "lock" itself is offensive to many of us. It further seems as though the engineer is imposing his decision regarding the proper damper position on the operator, and many fancy themselves to be amateur experts on the subject of ventilation.

In another context, it is desirable to encourage an attitude among operators and mechanics that they can and should "fine tune" their production equipment to achieve better production, higher quality and so on. This management philosophy may be appropriate if the personnel sufficiently understand the production equipment to achieve those ends and are permitted the freedom, but it should not be applied to the dust collection system and its complexities. I have never seen a blast gate system that did not generate apologies on the part of the management or the engineering personnel because it had been tampered with and was no longer in balance.

6.3.5   Why So Popular?

If balancing by dampers and blast gates is such a poor approach, why are there
so many proponents of this system? The numerous proponents include plant
engineers who are attracted to the "ease of adjustment" and "flexibility" of
such systems, as indeed they should be if such advantages were significant.
However, most plant engineers are not sufficiently experienced with these
systems to understand the pitfalls; moreover, they tend to overrate the desirable
features (flexibility, etc.).

The other two groups who usually strongly recommend the blast gate system
are the equipment vendors and the contractors, both of whom would be
expected to know their business. As a matter of fact they do know, and the
blast gate system is the one that is the best for their business. The equipment
vendors and contractors both furnish a degree of "free" engineering, which of
course is included in the price of the equipment or contract and is necessarily
limited for competitive reasons. To properly design a balanced system without
blast gates, a degree of knowledge of the operations is required that may be
beyond the expertise of the vendor or contractor. Gaining that knowledge
requires time and effort that the supplier can ill afford under the umbrella of
free engineering. Similarly, the design computations needed to balance the
system would also be effort to be charged to the free engineering; but the
equipment vendor can escape this cost almost completely by saying that a blast
gate system is better.

The same motivation exists in a somewhat different way for the contractor:
if he is required to do the balancing by blast gates, that work will add to the
value of the contract in a visible way for which he can charge without complaint
from the customer. Alternatively, if the customer is going to balance the system,
it is a cost that the contractor escapes while at the same time adding slightly to
the value of the contract by installing the blast gates or dampers.

The foregoing comments may be excessively harsh in some cases but not in
most. The worst situations occur when the architect-engineer, working at long
distance from the project, merely shows a triangle for an exhaust hood; leaving
the complete design of the hood up to the sheet metal subcontractor. At the
other end of the spectrum are the contractors who have become so sophisticated
that they not only install balanced systems without blast gates, but have developed
computer programs for the layout of elbows and fittings for any pipe diameter
without resort to shop patterns, so that they can easily and economically balance
systems by proper choice of pipe sizes.

6.3.6   Summary

For most dust and fume control systems, balancing without dampers and blast
gates is by far the preferred choice. A blast gate system that is working
adequately is seldom encountered. The advantages of blast gates, although
true, are highly exaggerated and the limitations of their use not generally

recognized. They are advantageous to the vendor or contractor, not to the owner. A major exception to this position is the type of system in which dampers or blast gates are used as on-off valves, and even there, their operation should be interlocked with the process, if practical.

6.3.7 Example

To provide an example of the limited flexibility gained by the use of a blast gate system, consider a modestly complex system, without an air cleaning device (merely to simplify the calculations required to determine what would happen if various changes were made in the system). The presence of an air cleaning device would not change the "balancing" results significantly, but changes in total system volume may change the performance of the air cleaning equipment itself.

The illustrative system (Figure 12.2) has five branches, designed as a blast gate system, with a blast gate in each branch. The depicted data are typical of such a system, showing the volume in each branch and each section of the main, and the pressure drop due to the duct and due to the blast gate adjustment "as balanced." Three cases of need for adjustment are summarized

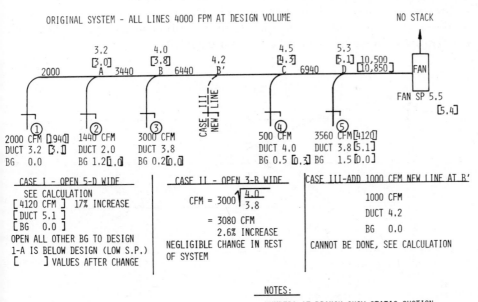

Figure 12.2 Example of blast-gate system. Numbers after "BG" (blast gate) show ΔP across blast gate; numbers after "dead" show pressure drop, numbers at branch show static suction (in. w.g.); numbers in brackets are values after change.

as follows. As in Figure 12.2, numbers in brackets are values after change and BG stands for blast gate.

Case I was chosen as a change that would be regarded as "likely to be successful" because the branch being adjusted is the one nearest the fan where the highest static pressure is available and there is a significant pressure drop across the blast gate, indicating that a great deal of upward adjustment should be available. It is desired to handle more air through branch 5D. How much more can be handled, and what will be the effect of that change on the rest of the system? By opening the blast gate wide, the entire 5.3 in. w.g. suction would be available at the branch entry, assuming that the system itself did not change, although of course it would. What the system "would be" if it did not otherwise change is called the "would be" operating situation. When *all* system effects are considered, the result is called the "actual" situation.

The example calculation shows that when all blast gates are adjusted, the "would be" system requires a fan static pressure of 5.5 in. at 11,140 cfm. Plotting this point on the fan system graph (Figure 12.3) and basing the plot of the system curve on the principle that pressure drop is proportional to the square of the volume, we have a new operating point for the system at 10,850 cfm, 5.35 in. w.g. fan static pressure. The suction at *D* is actually −5.1 in. w.g., not the −5.3 available before the change, with the result that branch 5-*D* handles 4120 cfm, a 16 percent increase over the initial design volume. This is truly a modest increase and in most cases would not be sufficient to improve materially the capture of the exhaust hood on that branch.

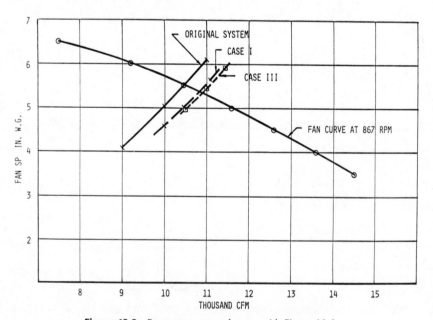

Figure 12.3   Fan system curve for use with Figure 12.2.

Case I. Open blast gate 5-*D*. Changes system. Then system "would be"

$$\text{cfm } 5\text{-}D = 3560 \sqrt{\frac{5.3}{3.8}} = 4200$$

$$\text{cfm } D\text{-}E = 6940 + 4200 = 11,140$$

$$\Delta P \; D\text{-}E = 1.2 \left(\frac{11,140}{10,500}\right)^2 = 1.35$$

Fan static pressure = 5.3 + 1.35 − 1.12 = 5.53 in. w.g.

Plot new "would be" system curve with a point at 11,140 cfm; 5.53 in. w.g. The operating point is the intersection of the fan curve and system curve. The operating point is 10,850 cfm; 5.35 in. w.g. (0.15 in. less than original). Static pressure, SP, at *E* is 6.4. $\Delta P \; D\text{-}E$ is 1.3. SP available at *D* is 6.4 − 1.3 = 5.1.

$$\text{cfm } 5D \text{ is } = 3560 \sqrt{\frac{5.1}{3.8}} = 4120 \text{ cfm (16 percent increase)}$$

Now the available suction at point *D*, as far as the rest of the system is concerned, is only 5.1 in. w.g. instead of the former 5.3.

For branches 4-*C*, 3-*B*, and 2-*A*, blast gate adjustment will maintain the designed cfm.

For branch 1-*A*, the blast gate is already wide open and the air-flow will drop from 2000 to 1940 cfm.

Thus the change is almost successful, if a 16% increase in branch 5-*D* is adequate.

Case II is an example of an adjustment that is not very likely to succeed, although an unsophisticated user of the system might attempt it. Opening the blast gate wide on branch 3-B provides only a 2.6 percent increase in volume, too small to measure, and the rest of the branches are not significantly affected. Of course more air could be sucked through branch 3-B if the other blast gates were further closed, but that would result in less than design volume and velocity in all other branches of the system.

Case III represents a not untypical modification to the system, requiring only about a 10 percent increase in volume by adding a new line. It is postulated that the new line would be desired to handle 1000 cfm. The new line is connected to the system at *B'* where the previously existing static pressure was −4.2 in. The calculations for case III show that a significant additional pressure drop in the main "would be" required to handle this amount of air. Plotting a new system curve, we see that the system operates at 10,900 cfm instead of the desired 11,500 cfm, and the fan static pressure is 5.3 in. Furthermore, this provides a volume increase of 400 cfm, not 1000. The new branch would handle about half of what was desired even with the blast gate wide open.

Branches 1, 3, and 4 will be below design because the blast gates cannot be opened enough to compensate for the lower suction at the branch entry, and branches 2 and 5 could be adjusted to original design. It is not worthwhile to work out the precise flows through each branch under these conditions, since the change would be so unsuccessful.

Case III. Add new line at B', 1000 cfm required, static pressure at this point in existing system, 4.2 in. Changes system. System "would be":

$$\text{cfm } B'\text{-}C = 6440 + 1000 = 7440 \text{ cfm}$$

$$\text{Friction } B'\text{-}C = 0.3 \left(\frac{7440}{6440}\right)^2 = 0.40 \text{ in. w.g.}$$

$$\text{New } VP = \left(\frac{7440}{6440}\right)^2 \times 1.0 = 1.33$$

$$\Delta VP = 1.33 - 1.00 = 0.33$$

$$\text{Total } B'\text{-}C = 0.40 + 0.33 = 0.7 \text{ in. w.g. (would be)}$$

Static pressure at C would be $4.5 + 0.7 = 5.0$ in. w.g.

$$\text{Friction } C\text{-}D = 0.8 \left(\frac{7940}{6940}\right)^2 = 1.05$$

Static pressure at D would be $5.0 + 1.05 = 6.05$ in w.g.

$$\text{Friction } D\text{-}E = 1.2 \left(\frac{11,500}{10,500}\right)^2 = 1.44$$

Static pressure at E would be $6.05 + 1.44 = 7.49$ in. w.g.

$$\text{Fan static pressure would be } 7.49 - 1.33 = 6.16 \text{ in.}$$

System operates at 10,900 cfm; 5.3 in. w.g. (see fan curve in Figure 12.3). Fan inlet VP is 1.08 in. w.g. SP at E is 6.4, whereas 7.5 would be required, see above.

Approximately: static pressure is 1.1 in. w.g. lower than required. Fan volume increases 400 cfm, not 1000. New branch runs between 400 and 1000 cfm (nearer 400). Branches 1, 3 and 4 are below design—blast gate cannot open enough to compensate for lower system pressure. Branches 2 and 5 can be adjusted.

The three cases above, predictable examples of change in a relatively simple system, indicate a very limited degree of flexibility. What plant foreman or even plant superintendent is going to be sufficiently sophisticated in exhaust system design, or wants to be bothered in view of the "flexibility" he has with the blast gates, to make the calculations that have been made in this example? In the real world, the blast gates would simply be changed. Then the problems would start, and depending on the circumstances, a great deal of effort might be expended in trying to correct the problems.

6.4 Recirculation of Cleaned Air

Air cleaning equipment is available with adequate efficiency to clean recirculated air of many kinds of contaminant to a safe concentration. Typical is the fabric filter type of collection equipment when applied to a dry, easily collected dust or to a fume that coagulates and filters with high efficiency. Historically, industrial hygienists in general and regulatory bodies in particular have been very reluctant to approve recirculation of air from such equipment. Although manufacturers claim and in many cases can prove that the cleaned air is cleaner than the workroom air, there always remains the vexatious problem of adequate maintenance and operation of the air cleaning equipment. Should it fail to receive the proper maintenance, as all too often happens, the recirculated air would contain dangerous or obnoxious quantities of contaminants.

The extensive history and background on this topic should be briefly reviewed. Until recently it has been a rather uniform policy of almost all official agencies and almost all professional industrial hygienists to recommend against the recirculation of air cleaned of toxic contaminants. It was recognized that the air cleaning devices, even if capable of cleaning the air to a safe degree, were not completely reliable and all too frequently received inadequate maintenance. With recirculation, therefore, there was a finite risk that the workroom air would be dosed with toxic materials by the ventilation system itself. In the days of cheap energy it was a correct decision and good policy to require that recirculation be prohibited and that the air exhausted from the building be replaced with outside air suitably heated or otherwise conditioned. It should also be remembered that historically, most proposals for such recirculation were based on the use of a single air cleaning device without any "backup" or redundancy to ensure that the recirculated air would be safe. Recirculation proposals were made on that basis because (in the days of cheap energy) the savings by recirculation would be negated if additional safeguards and monitoring devices were required for the recirculated air.

Now, however, the cost of energy is such that it becomes economically wise to spend the extra capital and maintenance cost to provide for recirculation of cleaned air *if* such a system can be designed and monitored so that it is reliable and safe.

The following discussion is based on the case of dry particulate (dust or fume) because that application is probably the easiest to design. Exhaust air containing gas-phase contaminants can also be cleaned and recirculated, using the same principles, but the implementation of the concept will probably be more difficult and costly.

The technology to permit safe recirculation of dry particulate dusts has been in existence for some time. The typical system would consist of a self-cleaning fabric filter as the first air cleaning device. For safety and reliability and more-than-adequate efficiency, the first filter would be followed by a second or backup filter, which would consist of a high efficiency filter of the high efficiency particulate air (HEPA) type. For many situations, a less efficient filter of greater dust holding capacity, namely a 90 percent AFI rated filter, would be adequate.

The second filter would not be self-cleaning but would require replacement if sufficient dust leaked through the first filter to build up resistance on the second filter. Filters of the class proposed for the second or backup filter can be tested in place to ensure that their efficiency is reliably high; there are no moving parts or other phenomena likely to destroy the efficiency of such a filter bank, once properly installed. Various other controls and devices can easily be installed to monitor the concentration of dust in the cleaned air, to automatically bypass the system in the event of failure, and so on.

Such a recirculation system, properly designed, can recover essentially 100 percent of the heat in the air exhausted from the workroom; in addition, it can recover essentially 100 percent of the energy used in moving the air and forcing it through filter resistances; and even can recapture essentially 100 percent of the wasted electrical energy represented by the inefficiency of the electric motors employed.

In typical dusty operations only a slightly greater capital investment is required to permit such recirculation than is needed to provide makeup air. Fuel costs are such that the capital cost is typically recovered in less than one year. In view of the reasonable projection that energy costs will increase faster than other costs, such economic benefits can only become more attractive.

It can be reasonably argued that a well-designed recirculation system will provide a better working environment than would a conventional exhaust and makeup air system. The dust can be removed from the air so effectively that the concentration is as low or lower than the outside air concentration in the vicinity of the plant. Furthermore, because of economic pressures, any legal requirement that recirculation cannot be practiced would result in attempts to use minimal exhaust volume to reduce the volume of makeup air required and the cost of heating it. In that event better control at the capture points would be achieved if the more generous air volumes permitted by recirculation were used.

The price of fuel is not the sole consideration. The only fuels suitable for small installations for makeup air duty are natural gas, propane or LPG, or fuel oil. For these fuels price is not the only factor because they are at times literally unavailable, and unavailability may become increasingly common in the future. The next step up would be to purchase electricity for heating, which is typically 2 to 3 times as expensive as the fuels just named, or to provide a steam plant fired by coal. The capital cost of the coal-fired steam plant would have a significant economic impact.

In recognition of the oncoming "fuel crisis" the Committee on Industrial Ventilation of the ACGIH devoted several pages of its *Industrial Ventilation Manual* (13th and 14th editions, 1974 and 1976) to a description of ways in which recirculation of air, cleaned of toxic contaminants, could be achieved; including a list of pertinent considerations to ensure that the recirculation was safe. Furthermore, the National Institute for Occupational Safety and Health (NIOSH), Engineering Branch, recognizing the same factors, has conducted studies on the same subject resulting in a report, "Guidelines for the Recirculation of Industrial Exhaust" (Contract no. 210-76-0129).

One interesting aspect of current discussions and studies of recirculation is the emphasis on reliability. There can be no argument that a high degree of reliability, especially adequate provision for safety in event of failure of the recirculation feature, is required. The only difference between an ordinary system and a recirculation system in this regard concerns the elements of the system that are necessary to prevent recycling of contaminant to the workroom in amounts or for lengths of time that would be excessive. The concern for reliability of the recirculating system seems to extend itself to the total system, to a degree inappropriate compared to concern for the reliability of the nonrecirculating system, or for adequate design in the first place. General or total system reliability should be treated equally, whether the system is recirculating or nonrecirculating.

6.5 Heat Stress Relief Versus Exhaust Control

Unless exhaust system hoods are total enclosures, they function on the principle of establishing an air velocity into the hood that is sufficient to capture and contain the contaminants. Obviously if large, high-velocity air currents are directed at the hood, as from a man cooling fan or a misplaced air supply fixture, the velocity of those currents will be high enough to completely overwhelm and spoil the capture velocities created by the hood. In some cases more air is blown into the face of the hood than is removed by the exhaust connection, thus causing the hood to "overflow" or spill contaminated air into the workroom. Almost all exhaust ventilation standards state that the absence of such disturbing high-velocity room air currents is assumed. Achieving this happy state of affairs is easier said than done.

The situation is bad enough if the heat stress is due merely to warm ambient air as from summer conditions in the typical factory. In many cases, however, there is a concentrated heat load from the process or work being done at the hood, a common example being the shakeout of hot castings.

One of the best general methods of overcoming a heat stressful situation is the provision of "spot cooling." This involves the direction of a high velocity air current of cooler outside air (or possibly artificially cooled air) directly on the worker. The airflow over the worker's body increases the rate of evaporation of sweat and increases the conduction of heat away from the body if the temperature of the airstream is lower than the temperature of the skin.

The behavior of air jets in open ambient air space, as in the workroom, is well known. Most texts or handbooks dealing with exhaust ventilation do not give data concerning such jets. Suitable information can be found in the ASHRAE *Handbook* (3). The properties of air jets must be understood by the designer and the hygienist when they confront a problem combining heat stress and contaminant control.

There is no solution to this type of problem known to be generally applicable. Each problem needs careful analysis, and frequently a compromise is the best that can be achieved. Potential solutions include possible heat stress relief by other means, such as radiant heat panels, and modification or mechanization of the process. Frequently the spot cooling air jet can be made more effective at lower velocity if the air is treated by evaporative cooling techniques. Evaporative cooling has a bad reputation because it is not effective for comfort air conditioning in any but dry desert climates. Its effectiveness for the relief of heat stress is usually not recognized, but it is surprisingly effective in typical humid midwestern climates (4).

Intelligent application of evaporative cooling principles can result in effective spot cooling using lower jet velocities, thus with less disturbance to the control of the exhaust hood. The exhaust volumes used on the hood and the corresponding capture and control velocities should be increased above the minimum standards because they are applicable only when there are no disturbing air currents. In some situations the "heat stressful" part of the work is not continuous, and a spot cooling or area cooling island of cool air can be provided at some little distance from the hot contaminant source. Thus the heat stress relief is obtained by the worker when standing back from the hot source, and the flow of cooling air does not directly interfere with the exhaust hood control.

6.6 Maintenance and Inspection

An exhaust system that has been properly designed and properly installed and is exerting adequate control must be adequately maintained. Obviously and as previously described, reasonable access must be had to the equipment itself when maintenance is required. A number of other features can be considered, however, for routine inspection of the system to determine whether maintenance is needed.

One of the easiest things to check is whether the design airflow is flowing through each hood. That, combined with a visual check of the integrity of the hood itself, is not only important but also is easy. As described in Volume I, Chapter 18, there is a pressure drop when the air flows from the exhaust hood into the connecting duct, commonly called the "throat" of the hood. The suction at this point is calculated during the design of the system so that the "as designed" value should be available. When the system is installed and operating, a measurement of the suction at the throat of the hood provides an operational "quality control" datum point that is easily checked to determine whether the hood operation is within "quality control" limits.

This verification calls for a static pressure tap. The simplest static pressure tap is a small hole—for example, $\frac{1}{16}$ to $\frac{1}{8}$ in. diameter, drilled in the side wall of the duct in the region of the hood throat (i.e., about one duct diameter downstream of the connection to the hood itself)—and otherwise suitably located to prevent distorted readings (see Volume I, Chapter 18). A simple U-tube manometer or other static pressure measuring device can be connected to this static pressure tap by holding the open end of square-cut soft rubber tubing to the hole in the duct. This requires some care, however, and human error can easily introduce a large discrepancy in readings. Therefore a slightly more sophisticated fitting is recommended.

A simple pipe T can be welded or brazed to the outside of the duct over the static pressure tap hole, and a fitting screwed into the leg of the T with a short stub of $\frac{1}{4}$-in. copper tubing, permitting the hose from the manometer to be slipped over it, assuring a reasonably airtight fit. A plug or appropriate valve can be inserted in the other arm of the T to provide an easy method for cleaning out the static pressure hole with a wire or pipe cleaner if it becomes plugged.

A manometer or other pressure gauge can be permanently installed at each static pressure tap. If the system is handling dust, corrosives, or condensables, however, this procedure is not recommended. The static pressure hole will become plugged and the gauge will give a false indication. The problem can be avoided by the use of a purge airflow as is commonly used in process instrumentation, but such an installation adds considerably to the cost and is usually not worthwhile except under unusual circumstances.

The simple pressure tap and rubber tube method assumes easy access to the static pressure tap. Lacking easy access, however, extension of copper or other suitable tubing from the hood throat to some reasonably accessible point is advisable. If the pressure tap is inaccessible, so is the clean-out plug for cleaning the static pressure hole, and an additional provision for clearing the static hole, such as a connection for blowing high pressure compressed air into the line, may be used.

Provisions for routine inspection of the airflow through hoods is required by OSHA standards. With the attribute described above (i.e., the pressure drop through the hood throat being a function of the airflow), the hood throat can be considered to be an orifice-type flow measurement device for which correct reading can be established either from the design or from the experience gained in startup.

If the static suction at the hood throat is low, one of the following causes can be found.

1. If the system has blast gates, they are out of adjustment.
2. There is a constriction (partial plugging of the ductwork, damage to the ductwork, etc.) in the pipe between the hood throat and the exhaust fan.
3. There is major leakage in the piping between the hood throat and the fan.

4.   Fan speed is low (viz., slipping belts on the fan drive).

5.   There is excessively high pressure drop through the air cleaning device on the system.

If the static suction is high, the following factors may be the cause.

1.   If the system has blast gates, they are misadjusted.

2.   Other branches in the system are plugged or constricted.

3.   There is a malfunction in the air cleaning device (e.g., broken bags in the fabric collector) causing its pressure drop to be less than normal.

More sophisticated devices may also be used if the integrity of the system is important enough to warrant the cost. One such continuous monitoring feature could be a device that indicates that the system is indeed in operation—for example, by means of a pilot light indicating whether the electrical circuit to the fan motor is on. This example also illustrates the fallacy of many such control features in that what is measured is different from the desired parameter. In this case, what is desired is knowledge of whether there is significant airflow in the system, but all the pilot light reveals is whether there is voltage on the power circuit at the point where the pilot light tap-in is made. There could easily be voltage at this point and insufficient airflow through the system; prime causes of this kind of event are badly slipping drive belts or a burned out blower motor. If the variable to be checked is the airflow in the system, an airflow switch that is actuated directly by airflow should be installed.

Other and more sophisticated monitoring devices can be used, depending on the need. Such devices may be pressure recorders or differential pressure recorders, permanently installed centerline pitot tubes with provisions for compressed air clean-out, and various electrical interlocks. Highly complex interlocks are not successful in the typical factory. The general environment is very different from that of flying an aircraft or spaceship. This matter is discussed in greater detail in Section 8 of this chapter.

6.7   Fan Noise

With surprising frequency, industrial exhaust systems that are installed to correct one hazard have the effect of generating a new hazard—excessive noise. Adequate information exists to predict the amount of noise created by a fan. This problem is unfortunately treated, to a degree, as a separate discipline and the design data are not published in the same sources as for exhaust hoods. I believe that any competent hygienist or engineer, with rudimentary training in noise technology, can also handle the average problem of fan noise. Data are available (5–8).

Fan noise may be a problem in the workroom or in the neighborhood of the plant. Some minimal attention should be paid to both possibilities when the exhaust system is being designed, and provisions should be made to prevent a hazard or a nuisance from this potential source.

7 PURCHASING SPECIFICATIONS RELATED TO INDUSTRIAL HYGIENE

Purchasing specifications can be considered in two ways: in relation to the ever-present problem of specifying the appropriate quality for a machine (regardless of whether its intended purpose is hygienic), and in relation to the special specifications associated with health and safety requirements.

A simple example will suffice to illustrate the first category without getting into the depths of engineering expertise at the detail design level. In the purchase of an exhaust fan, the primary consideration is the air volume and static pressure required. A mere statement of that requirement on a "performance specification" basis will almost guarantee an unsatisfactory installation. There are many other aspects of performance or quality that must be considered. First the type of fan must be specified, since there are several basic types, some of which may not be appropriate for the intended use. Once that is done, however, there may be several duty classes of blower that would satisfy the particular specification. For example, a class 2 blower might be adequate for the specified duty. However, there are finite limits to the speed and pressure at which a class 2 blower may be safely operated, and it is frequently advisable to buy a higher class blower to permit greater flexibility in this regard, or to permit more latitude in other subsequent potential uses of the blower.

Other questions would also arise: Is a drain plug needed? Is a clean-out door needed? Should the inlet and outlet be flanged? Should the blower be a type that permits rotation of the housing to change the direction of the discharge? Is a vibration limitation specification needed? Does the shaft require a heat slinger or a shaft seal? For any given requirement of cubic feet per minute and static pressure, the range of blowers that could be procured probably covers a cost span of 4 to 1 depending on whether a "pickup truck" or a "Euclid ore carrier" is the appropriate machine for the job. The same general considerations of detailed engineering specifications apply, of course, to all the major equipment involved in the hygiene control system.

The other category of specification tends to include such language as "to meet all applicable OSHA standards" or perhaps "to have a noise level no greater than 85 dBA at 5 ft from the machine." In this regard, the limitations of the vendor should be realized and the interaction between his equipment and the plant environment taken into account. The vendor has a moral and perhaps legal responsibility to predict the performance and characteristics of the machinery being sold. However, interactions between that machine and the location and installation of that machine are beyond the vendor's control and perhaps beyond his knowledge. The single exception to this occurs when the vendor is privy to all the conditions surrounding the proposed installation of the machine; then in fact he is functioning as a design engineer as well as an equipment vendor.

The noise specification affords a good case in point. An analogy would involve a common experience, the purchase of a light bulb. One can purchase a 100 W light bulb and be assured that its power consumption is reasonably close to 100 W. To be somewhat more technical, the wattage of the light bulb could be related to the lumen output of the bulb, and the item could be purchased on that basis. So far, we are within the duty of the vendor, that is, to predict and guarantee what the light bulb will do when appropriately furnished with 115 V, 60 Hz, AC power. The vendor of the light bulb, however, is unable to tell what level of illumination will appear on the buyer's desk because the installation environment has a high effect on the relationship between the wattage of the light bulb and the light on the desk. The light fixture, the color of the walls, the distances and geometrical relationships between the light bulb and the desk, all affect the resulting illumination level at the work surface.

A comparable situation exists with the noise from machinery. The manufacturer can and should be able to state the sound power level of the machine, that is, how much sound it generates. What the resulting sound pressure level field will be at any point in space depends on how the sound source is installed, whether it has vibration isolators, the nature and proximity of reflecting surfaces, and so on. There is no escape from the technology and engineering that needs to be applied to establish the relationship between the sound power level of the source and the sound pressure level at the receptor. If the vendor is to be burdened with this responsibility, he needs complete data concerning the installation environment, he becomes the design engineer for that portion of the project rather than solely an equipment vendor, and he should be and is (in the price of the equipment) compensated for that additional effort and responsibility.

Another problem arises in connection with the purchase or specification of innovative devices. One of the unfortunate aspects of our reasonably free market system is the appearance, when demand becomes heavy, of a number of innovative devices, many of which operate by some kind of magic that appeals to the unsophisticated purchaser. One example is the generation of "air ions" that purportedly charge the airborne dust particles and cause them to sit down on the floor and behave themselves. It is the same marketing freedom, however, that permits the easier and more rapid development of meritorious innovative devices. Fortunately most hygienists are fairly skeptical in their appreciation of new and magic gadgets. If the project is of any importance at all, it merits the expense of obtaining, from the vendor of the innovative device, a list of installations and contacts at those installations. Then the hygienist can visit these places to obtain first-hand information about the functioning of the device. Quantitative measurements of the performance are highly advisable if that can be arranged. The owners of such the innovative devices may be quite unsophisticated, and their opinion alone may or may not be valid. One instance involved the use of a new kind of dust collector, rated at 2000 cfm and 99.97 percent efficiency; when tested, the device proved to be handling 500 cfm and emitting a visible dust plume. However, the owner expressed a high degree of satisfaction with the dust collector and was preparing to buy two more.

8 MONITORS, ALARMS, AND "FAIL-SAFE" DESIGN

When engineering controls are indicated but the design problem is difficult, there is a tendency to specify the use of a monitoring device that will sound an alarm, for example, if the contaminant concentration exceeds a specified level. All too frequently, this procedure can be labeled "the designer's cop-out." Having failed to conceive a reasonably effective and reliably engineered system for control, the designer discharges his immediate responsibility by specifying a monitor hanging on the wall to give the alarm in case the control system fails to work. At the current state of the art, these devices are not highly reliable and require a considerable amount of attention if dependability is to be ensured.

A philosophical question arises: If the maintenance capability of an establishment is not adequate to keep an exhaust fan running properly, who is going to keep the delicate electronics of a monitor functioning properly? Such instruments have a tendency to suffer from zero drift and other malfunctions, including false alarms; they need to be adjusted and calibrated frequently. Typically, the occasion for the monitor sounding the alarm occurs rarely, and an unusual degree of discipline in preventive maintenance is required to assure that the monitor will be functioning correctly when it is needed. On the other hand, if it has a tendency to give false alarms, it will not be a surprise to find that it has been unplugged or turned off, although still hanging neatly on the wall, encouraging a feeling of false security. If the designer must solve problems by the use of such a monitor, the solution is valid only if arrangements are instituted for adequate maintenance and calibration on a preventive maintenance basis.

A somewhat similar situation exists in the typical factory if an intricate interlock system is designed and installed to prevent misoperation. Such interlocks work much better on paper than they do in real life. Consider a typical scenario. An intricate interlock system has been installed involving microswitches, limit switches, and pressure switches. Sooner or later one of the switches sticks or otherwise malfunctions at 3 a.m. Because of the malfunction, the process is not operable, and the foreman is pressuring the mechanic to get the machinery running again. The electrical schematics are available in two places: in the file cabinet in the locked-up maintenance office, and in the file cabinet in the locked-up engineering department. The night shift mechanic cannot figure out the interlock circuit, cannot locate the offending malfunction, and anyway replacement parts are probably not in stock in the storeroom. The solution: jumper the interlock circuit. If there was no malfunction other than in the interlock circuit, no harm is done. Less happily, the place blows up.

If interlocks are required, keep the design simple. If there is good training and attitude on the part of the workers, the opportunity for human intervention is more reliable than an intricate interlock system. The typical factory is not a space ship where the devotion of time and resources to proper automatic function is infinite, and two or three levels of redundancy are provided. Operating and maintenance personnel are accustomed to coping with malfunctions—motors that do not run, pumps that leak, valves that stick and so on. Efforts to educate, train, and motivate will be preferable to the intricate interlock approach.

If the interlock approach is chosen, it is workable only if the administrative procedures are instituted to ensure that the right information, the right parts, and the right talent are available, and the cost of such is justified. There is one alternative—also costly—which is in case of failure of the interlock system: the operation is shut down and stays down until the fault is found and corrected.

9 DEFINITION OF "FEASIBLE"

Existing OSHA standards require that to control hazardous air contaminants to or below the level of the standard, ". . . administrative or engineering controls must first be determined and implemented whenever feasible." Proposed new standards, although varying somewhat in wording, generally provide that, "engineering controls shall be instituted immediately . . . except to the extent not feasible" and even if "not sufficient to reduce exposures to or below the permissible . . . limit, they shall nonetheless be used . . . to the lowest practicable level."

Neither "feasible" nor "practicable" is defined.

Inexperienced OSHA compliance officers, motivated by natural zeal and encouraged by the language of both present and proposed standards, frequently issue weakly based citations for "failure to use feasible engineering controls." A contest results, and such contests are won by the employer more often than not. This is a waste of talent and resources and accomplishes nothing toward bettering the health of the workers.

Expert witnesses retained by OSHA sometimes recite the entire textbook list of engineering control principles as being feasible engineering controls: (a) substitution of a less hazardous substance, (b) change of process, (c) isolation, enclosure, or segregation, (d) local exhaust ventilation, (e) general or dilution ventilation—without making any realistic assessment of the practicality of applying such controls to the problem at hand, or the results that would be achieved, or the cost, or the benefits.

This "engineering controls" requirement as stated has led to many more contested citations than would have been required for enforcement of a better-defined standard. Unless a meaningful definition is provided, this situation will continue. Such excess litigation will continue to detract from the useful and constructive efforts of OSHA, thus actually diminishing the results obtained in terms of worker health.

In an effort to be constructive, a definition of feasibility is offered. It is not quantitative, but at least it includes all the basic parameters that should be considered in deciding whether an engineering control is feasible. For a specific case, the definition can be reasonably quantified, although a value judgment is still required. If such a definition were to be included in OSHA standards, it would encourage those thought processes in making an evaluation. The proposed definition is as follows:

"Feasible" means that the method or equipment is available on the market and has been used before with success in the same or closely similar applications, or that the technology exists to create the equipment and implement the method with reasonable assurance of success; that the method or equipment will result in reducing the exposure to or below the time-weighted average standard or is necessary to reduce the exposure to a level where the reasonable use of administrative controls or personal protective equipment which is not unduly onerous to the employee will adequately protect the health of the employee; that the number of employees exposed, and the frequency and severity of the exposure, are included in a cost-effectiveness consideration of the implementation of such controls; and that seriousness of the potential risk to the employee health is given due weight in all considerations.

Some application guidelines can also be furnished. To permit easier reference to the elements of the definition in analyzing a specific case, it is rewritten with parenthetical identifiers so that the text need not be frequently repeated as follows:

"Feasible" means that the following conditions are met (in the case of items 1 and 2, only one statement in the set must be met):

1a. That the method or equipment is available on the market.

1b. That it has been used before with success in the same or closely similar applications.

1c. That the technology exists to create the equipment and implement the method with reasonable assurance of success.

2a. That the method or equipment will result in reducing the exposure to or below the permissible exposure level.

2b. That the method or equipment is necessary to reduce the exposure to a level where the reasonable use of administrative controls or personal protective equipment which is not unduly onerous to the employee will adequately protect the health of the employee.

3. That the number of employees exposed and the frequency and severity of the exposure are included in a cost-effectiveness consideration of the implementation of such controls.

4. That the seriousness of the potential risk to the employee health is given due weight in all the considerations.

To be judged feasible:

1. The answer to either 1a, 1b, or 1c should be "yes."
2. The answer to either 2a or 2b should be "yes."
3. Analysis of 3 should show that the proposed control is cost-effective. (Obviously a value judgment is involved here.)
4. "Seriousness" of the hazard includes several elements. These are:

    a. The immediacy of the potential hazard to life. For example, the potential for high concentration (many multiples of permissible 8 hr TWA) of some agents such as carbon monoxide, hydrogen sulfide, or chlorine, can be fatal in a few minutes. (Obviously very serious; does not apply to TLV concentrations.)

    b. Whether a chronic effect of the hazard would be a debilitating, painful, irreversible disease, such as cancer (more serious); or would be instead reversible with little or no remaining disability, such as elevated blood lead (within limits, less serious).

    c. Whether there exist biological monitoring techniques that if implemented, would give warning before any significant health effect had occurred. Examples are testing blood for lead, sputum cytology, and audiometric tests (less serious than if no such technique is available).

    d. Whether the effect of the hazard would be a debilitating disease, such as silicosis (more serious), or whether it would result in nondebilitating diminution of performance, such as hearing loss, peripheral neuropathy, or skin sensitization (less serious).

10 UNSOLVED PROBLEMS

The textbook list of engineering controls for hygiene problems is short and simple in its classifications. Segregation, isolation, substitution, change in process—are all valid concepts. Local exhaust ventilation with its subset of principles—total enclosure, partial enclosure, get close to the source of the contaminant—embody valid concepts.

However, at the present state of the art the application of these concepts to many problems is unwise and wasteful, if included in the list of concepts is the idea that the installation should be reasonably workable and reasonably effective. The plant engineer may seem to be uncooperative and recalcitrant when approached by the hygienist to apply some reasonably clear-cut concept to a problem in the plant. The plant engineer may in some cases indeed be recalcitrant; but more likely he is merely being more skeptical and more practical. He has the problem of designing the application of the concept and living with its problems after it is installed. To the plant engineer, the project is a failure unless it is reasonably workable and reasonably effective. His career, and indeed the future of the company, cannot stand too many failures.

There are at least four categories of hygiene problems in which the application of the concept principles becomes exceedingly difficult and frequently impossible. These are moving sources of contaminant, handling of bulk materials that are heterogeneous and not susceptible to mechanized handling, manual contaminant-generating operations on very large objects, and the combination of heat stress and contaminant control hoods.

Moving contaminant sources can be controlled if they are very large—such as a shiploading gantry crane—where an entire control system can be mounted on the moving object. They are also susceptible to control if they are quite small, as in the case of handling of beakers and flasks in a laboratory fume hood. In the intermediate size range effective control of moving sources varies from the difficult and expensive to the impossible. The Hawley Trav-L-Vent concept sometimes can be applied; in other cases the kinds of motion and the available space do not permit its application. Flexible hoses and swivel joints have severe limitations.

If raw materials or by-products are to be used in amounts large enough to require bulk handling (i.e., more than can be handled in small closed containers), the nature of the materials becomes important in choosing the type of material handling that is workable. If the material is difficult to handle because it is quite heterogeneous in size and shape of particles or objects, or very sticky, the available bulk material handling equipment that is amenable to contaminant control is usually not workable. An example of difficulty in handling is provided by the recycling of scrap metals, where the feed material may vary from a fine powder, to chunks and objects measuring several inches, to dross removed from pyrometallurgical processes and consisting of granular dusty material interspersed with beads or strings of frozen metals. For such material handling problems, only the most crude equipment such as front-end loaders and dump trucks is satisfactory. Even though the dumping points for a front-end loader can be provided with ventilation control, the other activities of the front-end loader cannot be.

When manual contaminant-generating operations are conducted on very large objects, engineering controls fail, at the present state of the art. For example, a very large object can be spray painted in a spray booth under good capture and control air velocities. The ventilation typically prevents the spread of the contaminants through the rest of the plant and prevents accumulation of explosive mixtures of solvent vapor, but is not sufficient to protect the person doing the spraying from toxic pigments. It is necessary to spray large surfaces where the back spray is significant, to spray sideways or upwind to get various parts of the equipment painted, and so on.

Similar problems exist with the manual tending of large furnaces or reactors which, if they cannot be made leakproof, generate contaminants in the area where the tender is working. Any enclosure or exhaust ventilation provisions would result in the person working inside the hood, which may be advantageous for the rest of the plant but does not protect the worker.

Where work stations involve a combination of heat stress and contaminants to be controlled by local exhaust hoods, the control situation is a difficult and many times is a somewhat unsatisfactory compromise, as discussed in detail in Section 6.5.

REFERENCES

1. K. J. Caplan and G. W. Knutson, "Experimental Analysis of Lead-in-Air Sources in Lead-Acid Battery Manufacture," *Am. Ind. Hyg. Assn. J.*, **40,** 637–643 (July 1979).
2. American Conference of Governmental Industrial Hygienists, *Industrial Ventilation—A Manual of Recommended Practice*, ACGIH, Cincinnati, Ohio, current issue.
3. American Society of Heating, Refrigerating and Air Conditioning Engineers, *Handbook and Product Directory, Volume on Fundamentals*, Chap. 30, "Space Air Diffusion," ASHRAE, New York, 1977.
4. American Industrial Hygiene Association, *Heating and Cooling for Man in Industry*, 2nd ed., AIHA, Akron, Ohio, 1975.
5. American Society of Heating, Refrigerating, and Air Conditioning Engineers, *ASHRAE Handbook and Product Directory*, ASHRAE, New York, current issues.
6. Air Moving and Conditioning Association, AMCA Standard 300-67, "Test Code for Sound Rating," AMCA, 1967.
7. Air Moving and Conditioning Association Publication 303, "Application of Sound Power Level Ratings for Ducted Air Moving Devices," AMCA, 1965.
8. American Industrial Hygiene Association, *Industrial Noise Manual*, 3rd ed., AIHA, Akron, Ohio, 1975.

CHAPTER THIRTEEN

Personal Protective Equipment

**BRUCE J. HELD and
MATTHIAS J. KOTOWSKI**

1 INTRODUCTION

The use of personal protective equipment dates back to the time when mankind first put on an animal skin and found that it offered some protection against cold temperatures. Sandals to protect the feet against sharp stones on the ground have been found by archeologists in their diggings of many ancient cultures. With the development of weapons, personal protection went from passively protecting ourselves from what nature had to offer to actively thwarting off what man could do to himself to cause bodily harm. Shields to protect the body and helmets to protect the head were common in the armies of most ancient cultures, culminating in the total protection of the entire body by the chain mail or metal suits worn by knights.

In the workplace, personal protection also dates back to ancient cultures. Roman miners used pigs' bladders for breathing protection, Greek seamstresses had thimbles for finger protection when sewing, and it is easy to imagine that leather gloves and other articles of clothing have long been used for skin protection while doing various crafts that could abraid the skin.

After the industrial revolution and the rise of modern technology, the number of possible insults to the body, both internally and externally, rose at ever-increasing rates. In various jobs, every part of the body could be harmed in one way or another by physical forces, and harmful materials could enter the body through the nose, mouth, or even the intact skin. Hundreds of types of personal protective devices have been developed over the years to try to offer protection against the huge numbers of possible assaults in the workplace.

With the rise in the modern-day health and safety movement, the realization quickly came that engineering controls were far better than personal protective devices. Preventing a physical, chemical, or biological assault on the body through engineering controls or isolation give much better odds for protection than the last minute separation of the body from the assault that is offered by personal protective devices. Personal protection requires that the proper device was selected and that each user knows how to use it and, in fact, uses it. Engineering controls remove the human nature aspects of forgetfulness and willingness to take a chance because of discomfort.

However, many emergency and nonroutine situations occur in the workplace where personal protective devices are all that are available to guard the worker. These devices are discussed in this chapter. Even though this book is primarily about industrial hygiene, since health protective devices and safety protective devices overlap considerably (e.g. the same glove may prevent skin absorption as well as skin abrasion), all aspects are briefly presented.

All protective wear, for whatever purpose intended, should be routinely inspected and maintained to assure such protection. Protective wear that becomes frayed or otherwise damaged and loses its protective value should be replaced. Should breakthrough occur in the protective wear while in use, the affected skin should be immediately decontaminated.

2 HEAD AND FACE PROTECTION

2.1 Head Protection

Head protection has been in continuous use for military applications for thousands of years. It has evolved from the simple helmet designed to deflect a warrior's sword to extremely sophisticated systems to protect today's worker against a large variety of assaults: impact, penetration, acceleration, electric shock, chemical splash, fire, and radiant heat. While the user may be unimpressed by a simple hard hat, it is one of the most thoroughly engineered and tested protective devices in use today.

2.1.1 Head Protection Standards

Use of head protection is mandated by the general OSHA clause that requires the use of personal protective equipment whenever it could prevent injury or impairment (1). It is also called out in the construction standards, the maritime standards, and explicitly for several specific applications. Head protection is more narrowly specified in 29 CFR 1910.135-Occupational Head Protection.

Helmets for the protection of heads of occupational worker from impact and penetration from falling and flying objects and from limited electric shock and burn shall meet the requirements and specifications established in American National Standards Safety Requirements for Industrial Head Protection, Z89.1-1969.

This standard is, of course, the source of the OSHA requirements. A separate ANSI Standard exists for "Safety Requirements for Industrial Protective Helmets for Electrical Workers, Class B - ANSI Z89.2-1971" (3). Helmets of this type are required for work on power transmission and distribution equipment by 29 CFR 1926.951.

2.1.2   Helmet Components

Helmets consist of at least two components: the shell and the suspension system; in some cases a retention system is also provided. The shell is the outer plastic or metal covering. It provides protection against impact, penetration, electric shock, flame, etc. The suspension system typically is made of webbing, plastic headband and anchors. It not only holds the helmet in place, it also maintains a clearance of $1\frac{1}{4}$ in. between the head and the shell for ventilation and to protect against penetrating objects, and it distributes and absorbs impact shock loads. In some cases, a retention system is also provided to keep the helmet in place during adverse conditions.

2.1.3   Types of Helmets

The ANSI Standard recognizes two general types of helmets. Type 1 helmets offer the greatest protection from falling objects. They have a continuous brim at least $1\frac{1}{4}$ in. wide surrounding the entire shell. This type of hat offers protection not only for the crown, but also some limited protection to the sides and the back of the head. Type 1 hats are ideal in areas with serious hazards from falling objects and nuisance drips of low hazard chemicals. The size of the hat, however, impedes movement in close quarters, and for most applications the smaller Type 2 caps seem to be preferred.

The Type 2 cap is commonly referred to as a "hard hat." It provides the same degree of protection as the Type 1 helmet for the crown of the head, but lacks the brim to protect the shoulders and the neck. Only a peak or short visor is required by the ANSI standard. The Type 2 cap permits work in fairly close quarters, and it is the head protection chosen in most applications.

2.1.4   Classes of Head Protection

ANSI Z89.1 recognizes three different classes of protective helmets for industrial use and one for fire fighting use. The latter one has been superseded by the specifications for "Structural Fire Fighter's Helmets, NFPA/ANSI 1972–1979," (4) hence, it will not be addressed here.

All classes of helmets are available as Type 1 and Type 2 and provide similar protection against impact resistance and penetration resistance. Major differences exist in the areas of electrical protection and flammability rating.

Impact resistance is determined by dropping a $3\frac{3}{4}$ in.-diameter steel ball weighing 7.8 to 8.0 lb on the center of the helmet's shell from a height of 60 in. The specimen is mounted on a head form for this test, and the forces transmitted are measured using a Brinell penetrator assembly. Using this test, the helmet may transmit an average force of 850 lb for all classes of helmets.

Penetration resistance is determined by again mounting the helmet on a head form and dropping a 1-lb steel plumb bob on it from a height of 10 ft. The bob is sharply pointed with an included angle of 35 degrees, and it may not pierce the shell by more than $\frac{3}{8}$ in., or $\frac{7}{16}$ in. for Class C helmets.

Class A helmets must meet the above criteria, they must provide protection against voltages to 2200 V, and they must not propagage flames at a rate greater than 3 in. per min. The flame propagation rating is determined by the ASTM D635-68 method: $\frac{1}{2}$-in. wide specimens of shells are ignited in the horizontal position and the time for a 3-in. burn is recorded. The electrical insulation properties are determined by filling the inside of the helmet shell with water and then immersing the shell into a water bath. For one minute, 2200 V, AC, 60 Hz, are applied between the inner and outer water baths and the leakage current through the shell is measured. It may not exceed 3 milliamperes. An electrician wearing a Class A hard hat would not suffer any injury should he contact a 2,200 V live conductor with his hat accidentally. In addition to all of this, Class A helmets are limited to a weight of 15 oz, and they may not absorb more than 5 percent water by weight.

Class B helmets are similar to those of Class A, but they provide better protection against high voltage. A test voltage of 20,000 VAC, 60 Hz, is applied for 3 min., and the measured leakage current may not exceed 9 milliamperes. In addition, mounting holes for accessories are not permitted. To achieve this improved protection against high voltage, helmets may weigh up to 15.5 oz, and the water absorption is limited to .5 percent. (The specifications for Class B helmets are contained in the separate standard ANSI Z89.2-1971 "Safety Requirements for Industrial Protective Helmets for Electrical Workers, Class B".) (3). It may also be worth noting that Class B helmets are usually also certified as Class A helmets.

Class C helmets provide protection against impact and penetration only and have no insulation or fire protection value. All aluminum helmets, for instance, fall into this category. Care must be taken to assume that workers using Class C helmets are not exposed to electrical or fire hazards. The only advantage Class C helmets have over those of Class A or B is their somewhat lighter weight. This makes them more acceptable to many users, and user acceptance is always an important factor in selection of personal protective equipment.

2.1.5 User Comfort

Industrial head protection is of value only when it is used—hence, comfort and convenience are major considerations. Unfortunately, a hard hat without any accessories can weigh almost a pound, and in many cases some time is required to adjust to this additional weight. Fortunately, head protection has become so widely accepted that its use is often considered part of a craft or trade, and it is rare that serious objections are raised when head protection is indicated.

Nevertheless, any measures that can be taken to make head protection more acceptable to the user should be considered. In the summertime users may complain of excessive heat and irritation due to perspiration when using head protection. Actually, the surface of the scalp is shielded from the direct rays of the sun, and the clearance between the shell and the suspension provides more ventilation than is available with other sun hats, for example, baseball caps. Hence, the head is cooler than it would be without the hard hat (5). A soaked sweatband, however, can be very irritating. Where this is a problem, hard hats with replaceable sweatbands are available, and liberal access to replacement sweatbands may lessen the problem.

In the winter, hard hats can be the source of a bone chilling experience, unless a winter liner is used. Winter liners are available in a variety of styles. Some cover the skull and ears and are attached to the inside of the helmet's suspension system. For extra cold situations, such liners are available that extend far enough to also cover the neck. Other liners are attached to the brim of the hard hat and cover the forehead, ears, and nape of the neck only; a snug fit keeps the wind out. For very cold applications, the liner may also include a face covering resembling a ski mask. For added comfort, hoods of windbreakers and parkas can be worn over Type 2 caps.

Sanitary plastic liners are also available. Such liners limit scalp ventilation and render the helmet less comfortable; they are intended for use by visitors only and are not recommended for regular users.

2.1.6   Accessories

The use of a regular hard hat frequently precludes the effective concurrent use of other protective equipment—face shields, ear muffs, etc. For cases where such multiple protection is required by virtue of the assaults present, head protection is available that accepts other personal protective equipment as accessories. Ear muffs, face shields, welding shields, and flashlights all can be attached to ANSI-type helmets. One manufacturer even offers an air purifying helmet that provides head protection, a full-face shield, and a cooling stream of filtered air that flows over the worker's face.

2.1.7   Other Head Protection

In addition to the industrial head protection mentioned so far, other protectors are available for special applications. Perhaps the most frequently encountered is the bump cap. The bump cap is similar in styling to a hard hat, but fits more closely to the head; it usually has a light-weight suspension system, it may have foam pads, and it is much lighter than a hard hat. Bump caps were designed to provide some head protection for workers in confined quarters with overhead obstructions. They fulfill this task admirably. Because of their light weight they are readily accepted by the worker. This factor also leads workers occasionally to question if bump caps could not be worn in lieu of the hard hats requirement. The answer is no, since bump caps do not meet the requirements of ANSI

Z89.1 and do not provide any real protection against falling objects. However, where falling objects are not a concern, bump caps may be the protection of choice, and they should be considered seriously.

Structural fire fighters' helmets are another interesting form of head protection. As mentioned earlier, the specification in ANSI Z89.1 has been superseded by the "Standard on Structural Fire Fighters' Helmets," ANSI/NFPA 1972–1979 (4). Force transmission and electrical insulation requirements are the same as for Class A helmets. Heat and flame resistance requirements are more stringent, and penetration resistance is measured differently. In addition, the shell and suspension system assembly must be tested to assure that the head will not undergo undue acceleration when the helmet is subjected to impact. This, of course, limits the risk of concussion, and one can only hope that similar requirements will eventually be adopted for all head protection.

In many cases the main hazard to the worker's head is the danger of entangling the hair in rotating machinery. Where this hazard exists, cloth caps or hair nets that will shroud the hair completely are required. Caps with a stiff visor are usually preferred since they provide an additional warning when the worker leans into the spindle (6). Long beards present the same hazard and must be similarly protected. The question of when is hair long enough to require a hair net or cap invariably rises. A few hairs that become entangled in a spindle will just be torn from the scalp, resulting in minor injury. The hazard becomes severe, however, when a large number of hairs become entangled before the first hairs break, and this becomes more of a problem the longer the hair gets. While there is no uniform standard, hair covers are usually required when hair extends to the collar.

Finally, there is other head protection available that may be of value in special cases. Broad-brimmed sunhats may be indicated for outdoor work. There is a variety of helmets for vehicular uses and bicyclists, even riot helmets for the protection of peace officers. While not normally applicable in the industrial environment, the hygienist should be aware of such special gear and propose its use when indicated by the hazard.

2.2 Ear Protection

2.2.1 Introduction

In addition to protection of the ears from physical trauma such as blows or temperature extremes, personal protection is also available for loud noise. This type of protection is similar to respiratory protection in that an external hazard is prevented from entering the body by means of a physical barrier. Effectiveness is predicated by the ability of the barrier to keep out the hazard and the seal of the barrier to the skin of the wearer. Just as a respirator will not prevent skin absorption of skin-penetrating substances, neither will ear protective devices prevent some noise from entering the inner ear through bone conduction.

Personal ear protection from hazardous noise levels is available in two types of devices: ear plugs which fit in the ear and ear muffs which fit over the ear. Both the plugs and the muffs depend on a complete sealing of the device to the skin surface of the outer ear canal or the areas around the ear respectively. Secondly, the plugs or muffs must be able to attenuate the exterior noise to a safe level [the OSHA regulations, Part 1910.95 as amended require these devices to attenuate noise levels to at least a TWA (Time Weighed Average) of 90 dBA for everyone and 85 dBA for employees who have experienced a standard threshold shift] (7).

The attenuation rating of personal ear protective devices is provided by the manufacturer or is available in the literature. The average noise reduction rating (NRR) for several different devices using the Environmental Protection Agency's single number measure is 22–24 dB (2). This number is intended to be subtracted from the C-weighted work-place noise levels so that the employee's A-weighted exposure can be estimated. However, some investigators question these numbers and suggest that the NRR be reduced by 10 dB before subtracting it from the C-weighted sound level to take into account imperfect fits and improper or no training (8). Others stress proper audiometric testing as an early warning system to identify employees who are not wearing personal ear protectors properly (9).

2.2.2 Ear Plugs

Ear plugs are available in several types of materials, such as soft rubber or plastic, malleable wax, cotton, or impregnated cotton (10). The rubber or plastic types usually come in several different sizes to attempt to get the best fit for the wearer. Others are molded from casts or impressions made of each user's ears or molded directly in the ears. The wax and cotton types are usually disposable and are discarded after each use.

Plugs of any type should not be used until the wearer is examined by a physician to determine if ear infections or ear canal irregularities rule out their use. Fitting of permanent type plugs should only be done by physician.

2.2.3 Ear Muffs

The ear muffs are similar to earphones, but usually with a deeper cup. They include a headband that goes over the head or around the back of the neck, to allow wearing of a hat or cap. They may also be mounted directly on a hard hat, which can affect their proper sealing around the ears.

While fitting is not as hard as with plugs, several models may be necessary to get a proper seal on all users. The cup proper is usually made of a hard material such as plastic, with a soft foam rubber or liquid or grease-filled cushion to seal against the head. Ear muffs can be equipped with earphones and radios for communications in a high noise environment.

While good plugs may give slightly better attenuation at low frequencies, muffs give better attenuation at the middle frequencies (10). Plugs are convenient and inexpensive, but actual muff usage is more easily monitored by supervisors.

2.3 Respiratory Protection

2.3.1 Introduction

The philosophical basis of the use of respiratory protective devices has been that first preventive measures should be taken to keep contamination from entering the workplace and from contacting workers. Whenever possible this is accomplished through good engineering control measures. However, engineering measures cannot be used in all circumstances, and personal protective devices must be relied on in some cases.

The Respiratory Protective Devices Manual (11), the first book we know of to comprehensively cover the field of respiratory protection in industry in the United States, begins by stating:

In the control of those occupational diseases caused by breathing air contaminated with harmful dusts, fumes, mists, gases, or vapors, the primary objective should be to prevent the air from becoming contaminated. This is accomplished as far as possible by accepted engineering control measures;. . .

Similar wording is found in the American National Standard Z88.2-1969, "Practices for Respiratory Protection" (12). This standard, adopted verbatim by the Occupational Safety and Health Administration (OSHA), states that (1):

In the control of those occupational diseases caused by breathing air contaminated with harmful dusts, fogs, fumes, mists, gases, smokes, sprays, or vapors, the primary objective shall be to prevent atmospheric contamination. This shall be accomplished as far as is feasible by accepted engineering control measures (for example, enclosure or confinement of the operation, general and local ventilation, and substitution of less toxic materials). When effective engineering controls are not feasible or while they are being instituted, appropriate respirators shall be used pursuant to the following requirements.

In current and proposed health standards, OSHA further defines the use of respirators as a last line of defense for the worker in a contaminated atmosphere. The language varies from one standard to the next, but the regulatory agency basically limits respirator usage to the following conditions:

1. For routine operations while enginering controls are being instituted or evaluated.
2. When engineering controls are not technically feasible or cannot by themselves control a contaminant below an acceptable level.
3. For nonroutine operations that occur so infrequently that engineering controls would be completely impractical (e.g., certain maintenance operations).

4.   For emergency use or other unplanned events where the possibility of an overexposure to a worker exists.

The attempt to avoid the use of personal protective equipment in place of engineering controls is readily understandable for several reasons. First, the effectiveness of personal protective devices depends on the actions of many people; each and every user is responsible for knowing the limitations of the devices, using them properly, and in fact, using them at all. Engineering controls, on the other hand, can be installed and operated by a few "experts;" they can be monitored, and even can sound an alarm in the event of a failure, so that workers can be removed from the area before an overexposure occurs. If a respirator fails and the air contaminant is not readily detectable by odor or one of the other senses, however, an overexposure will not be detected until after the fact, when it may be too late.

Other undesirable features of respirators are discussed in the referenced documents. Some of these are summarized in the proposed OSHA standards for benzene (13) and cotton dust (14) in the testimony given in defense of the limitations imposed on the use of respirators in lieu of engineering controls and include:

1.   Fitting problems with persons with small faces, with persons with facial hair, such as beards or long sideburns, scars or growths which break the facepiece-to-face seal and with persons with pronounced wrinkling, a sunken nose bridge, deeply cleft chin, very narrow nose, a very wide or very narrow face, or very long or very short face.

2.   Communications problems between workers wearing respirators. Any respirator will muffle voice communications in varying degrees. With many devices, the jaw movements used for speaking can break the facepiece-to-face seal and cause inleakage of contaminated air.

3.   Vision problems can be numerous. Downward vision is obstructed with half or full-face masks, thus increasing tripping or stumbling possibilities. The inability to look straight down can easily cause serious injury. Full-face respirators will restrict the wearer to tunnel vision. Persons who must wear prescription glasses must be fitted with special glasses kits for full-face masks, as the temples of regular glasses break the facepiece-to-face seal. Some persons also have difficulties wearing glasses with half masks. The respirator either pushes the glasses above the eyes or the glasses come between the mask and the face around the nose, thus breaking the facepiece-to-face seal.

4.   Fatigue and reduced efficiency occur more quickly among workers wearing any type of covering over the face or head. This is caused by many factors, among them being increased breathing resistance, heat, some possible feelings of anxiety from being closed in, and reduced vision. On many jobs an alternative is to permit, or require 15-minute breaks every one or two hours while wearing a respirator in lieu of other rest breaks.

5.   Some persons, for reasons of health, would not be permitted to work if

they had to wear a respirator. A person with a heart problem may not be able, for example, to wear a 30-lb self-contained breathing apparatus. A person with some lung problem such as emphysema may not be able to stand the increased breathing resistance inherent in any respirator.

It is noteworthy that, in 1963, the Respiratory Protective Devices Manual (15) stated that respirators should be used only when engineering control measures are "uneconomical, inapplicable, impractical, or ineffective." A dozen years later, OSHA no longer considered the economic aspects of engineering controls in its standards. Today the belief is that human health cannot be measured in dollars and cents, and that the effectiveness of engineering controls is so superior to respirator usage that cost should not be a factor. This alone has probably caused the greatest amount of friction between industry and OSHA in the respiratory portions of the proposed standards. It is indeed difficult to define engineering feasibility solely in altruistic terms, since economics certainly plays a part at some point. Locating that point with precision, though, is the problem.

2.3.2 Basis for Performance Testing of Respiratory Protective Devices

The National Institute for Occupational Safety and Health (NIOSH) under the Department of Health, Education and Welfare has been established as the testing and approval agency for personal protective devices. At present NIOSH tests and approves only respiratory protective devices. Protective devices for the skin and eyes are not tested and approved by any agency, although it is hoped that eventually NIOSH will take on this responsibility.

Early Development of Devices.   Prior to World War II the United States was notably lacking in good respiratory protective devices. With the exception of the mining industry, which imported European devices for mine rescue work, general industry had very little in the way of protection for workers.

Industry essentially used no respirators before the twentieth century. This was probably because occupational illness was generally unidentified as such, and when it was, the worker had no legal rights or compensation available. Thus, employers did not feel obligated to offer protection.

The fire service had a more visible problem in that smoke inhalation caused an immediately identifiable effect: illness or death. Such effects led to a multitude of inventions of breathing apparatus of questionable value for the firefighter, but at least efforts were made. Unfortunately, the true worth of the device was only discovered when the "protected" firefighters died or sustained severe lung impairment. Some fire departments even budgeted for steam beer for their firefighters to be available after each firefighting episode to help "cleanse the lungs" (16). Other departments required that the firefighters have at least a 6 in. beard, which was to be dipped in water before entering a burning building. The beard was then folded up and put in the mouth; the smoke was supposed to be filtered and cooled when breathed through the growth. Wet sponges were also sometimes used to filter and cool the smoky air.

The mining industry, however, had no problem in recognizing the inadequacies of contemporary equipment for breathing protection. Mine rescue teams had to enter oxygen-deficient atmospheres for long periods of time, and only good oxygen-supplying equipment would serve. Since none was available in the United States, the mining companies looked to Europe to fulfill their needs. Devices made by Siebe Gorman in England and Dräger in Westphalia in Germany were used by the rescue teams.

The use of poison gas by Germany in World War I quickly illustrated America's woeful lag in producing adequate respiratory protection. An English device was brought to the United States, copied, and produced by the Bureau of Mines (B of M) for use by the Allied Expeditionary Forces. Gas mask development work was quickly started, resulting in devices such as the Connel, Kops, and Tissot masks.

Formation of Bureau of Mines Approval System.   After World War I, the B of M began studies of respirators for industrial uses, based on their earlier work on military devices. A method to test and approve self-contained breathing apparatus (SCBA) for mine rescue teams was developed, and the first device, the Gibbs Oxygen-Breathing Apparatus was tested and approved in 1920. The method of operation of the B of M, and now NIOSH, was and is that the respirator manufacturer first perform all required tests and submit the results; the testing agency then attempts to confirm the results.

Self-Contained Breathing Apparatus.   The first B of M approval schedule, Number 13, for SCBA, went into effect on March 5, 1919. Five revisions of that schedule were made (Schedules 13A through 13E) (15) between 1913 and 1968 before NIOSH took over the testing functions from the B of M in 1972.

Most of the B of M tests for SCBA were designed as "man tests," that is, tests that are performed while the unit is being worn by someone. The test subjects performed various exercises such as walking, running, crawling, weight lifting, and sawing wood while wearing the apparatus during a series of 15 types of tests in an irrespirable atmosphere of formaldehyde. Pulse and breathing rates of the test subjects were measured throughout the testing periods, as well as inspired air temperature and carbon dioxide content of the atmosphere within the SCBA. Certain engineering qualifications were also placed on the unit—for example, fully charged devices could weigh no more than 40 lb, and bypass valves had to be provided.

The tests, designed to provide answers to the following questions (11), are still the objectives of the NIOSH test schedule:

1.   Will the apparatus supply air or oxygen fast enough to meet the needs of the wearer?
2.   Can the wearer breathe freely at all times?

3.   Does the temperature of the inspired air remain within prescribed limits?

4.   Are vital parts of the apparatus protected so as to prevent damage or excessive wear?

5.   Is the concentration of carbon dioxide in the inspired air within prescribed limits?

6.   Does the harness hold the apparatus on the wearer's body without undue discomfort to him?

7.   For what period of time will the apparatus provide adequate respiratory protection for the wearer under the test conditions?

Gas Masks.   Two types of approval tests were necessary for the gas mask schedule. Like the SCBA schedule, man tests to determine the effectiveness of the complete system were required. In addition, tests to determine the reliability of the gas mask canisters were included.

The gas mask schedule, number 14, went into effect for approval in August 1919. Six revisions were made over the years from 1923 to 1955. Schedule 14F (17) was in effect when NIOSH took over the testing in 1972.

The canisters were machine tested against low concentrations of the contaminant of interest at low flow rates and at high concentrations at high flow rates. The low concentrations and flow rates gave information on the ability of the sorbent to hold the contaminant and also on how much of the contaminant the sorbent could hold without releasing any into the facepiece. The high concentration and flow rate tests determined whether the sorbent could react quickly enough to capture the contaminant before it could enter the facepiece. Both dry air and moist air tests were performed to determine effects on the sorbent.

The man tests were made in irrespirable atmospheres of ammonia to determine gross leakages on men with varying facial shapes and sizes, then in phosgene as an even more severe gas-tightness test. To pass these tests, the gas mask had to give complete protection to the men performing various exercises; excessive eyepiece fogging could not occur, and undue discomfort could not be experienced by any of the subjects. Gas masks were approved for use in 2 percent maximum concentrations of acid gases, organic vapors, carbon monoxide, or 3 percent ammonia.

Supplied Air Respirators.   Supplied air devices were first approved in 1932 under Schedule 19. Only two revisions were made before NIOSH took over the testing, namely, 19A in 1937 and 19B in 1955 (18). Approvals were granted for units such as hose masks with or without blowers, continuous flow or demand air lines with air supplied by a compressor or tank, and continuous flow or demand abrasive blasting respirators.

As with the SCBAs, man tests were the primary testing form to determine facepiece, helmet, or hood fit, air distribution in the respiratory inlet covering, vision clarity, safety, and comfort factors. In addition, limitations on hose length were required (75 ft for a hose mask without blower, 150 ft for a hose mask with blower, and 250 ft for all other devices in this category); inhalation and exhalation resistance limitations were set; maximum air pressures at hose inlet was established; and strength, noncollapsibility, and resistance to liquid gasoline requirements on the air hoses were set. A sandblasting man test operation involving silica dust and airflow characteristics was also required for abrasive blasting respirators.

Dust, Fume, or Mist Respirators. Respirators for protection against particulate matter, dusts, fumes, or mists, were approved under Schedule 21, which first became effective in 1934. Revisions were made in 1955 (Schedule 21A) and in 1965 (Schedule 21B) (19) before NIOSH took over the testing in 1972.

None of the dust, fume, or mist respirators approved by the B of M were for protection against dispersoids significantly more toxic than lead. The filters used in those respirators were tested against silica dust and litharge for dust respirators, and chromic acid mist and a mist formed by atomizing silica in a water suspension for mist respirators. Combinations of these tests could be used for approving respirators for protection against dusts and fumes, dusts and mists, fumes and mists, or all three, provided the respirator passed the requirements for each type. Men with varying facial sizes would also man test each respirator submitted for approval, to check on its fitting qualities.

Chemical Cartridge Respirators. The first B of M test schedule was number 23, for chemical cartridge respirators. It became effective in 1944 and was set up to approve only organic vapor chemical cartridge devices, as were the two revisions 23A and 23B (20) promulgated in 1955 and 1959, respectively. The chemical cartridge respirators were tested in a manner very similar to that employed for their gas mask counterparts, but at a much lower test concentration. Carbon tetrachloride was used as the organic vapor test agent, but the approvals limited the respirator use to concentrations of organic vapor of 0.1 percent (1000 ppm) or less. Isoamyl acetate was used as the test agent for the fitting quality tests.

Combination organic vapor and dust fume, or mist respirators were also approved, if the device passed the tests for each type of unit for which it was submitted. Paint spray respirators were also approved by testing the device against carbon tetrachloride, atomized lead paint, enamel, and lacquer.

NOISH/MSHA Approved System. In 1972, NIOSH took over the testing functions of respirators from the B of M. All testing was performed by NIOSH, the results reviewed by the B of M, and the approval was issued jointly by NIOSH/B of M. In 1973 the B of M was reorganized and the regulatory functions formed into a new agency, the Mining Enforcement and Safety Administration (MESA). Respirator approval certification was also a function of the new agency, and henceforth, the approvals were issued jointly between NIOSH/MESA. In the late 1970s, MESA was reorganized: part remained under the B of M, while part became the Mine Safety and Health Administration

(MSHA) in the Department of Labor. MSHA became the agency to jointly approve respirators, along with NIOSH.

NIOSH revised the B of M test schedules, and reissued them (21). A new schedule, Subpart M, was added for the approval of pesticide respirators.

Subpart A: General Provisions.   Subpart A states the purpose of Part II and defines words and terms used throughout. In addition, requirements were added that limited respirators used for entry or escape from hazardous mine atmospheres to those meeting Part II requirements after March 30, 1974. Subpart A also states that an "approved" respirator would no longer be approved unless it was selected, fitted, used, and maintained in accordance with provisions given in ANSI Z88.2-1969 (8).

Subpart B: Application for Approval.   The procedures for applying for approval, application contents, and the number of respirators the applicant must provide to NIOSH are outlined in Subpart B.

Subpart C: Fees.   Subpart C gives the fee schedule for respirator approvals. Fees listed on publication ranged from $500 for a nuisance dust respirator to $4100 for a type N gas mask.

Subpart D: Approval and Disapproval.   This subpart sets forth the requirements for issuing certificates of approval, for disapproving a respirator that has failed to meet the tests, for marking and labeling approved devices, for revoking an approval certificate, and for issuing a modification of a certificate of approval when the manufacturer wishes to make or change a feature of an approved device.

Subpart E: Quality Control.   Subpart E was a major addition to the B of M schedules. It sets forth detailed requirements for a quality control plan that a respirator manufacturer must prepare, submit for approval, and follow when manufacturing an approved device. Also, detailed quality control records are required, which can be reviewed from time to time by NIOSH during the period an approved device is manufactured. Failure to follow the quality control plan submitted and approved, or failure to keep proper records, can result in revocation of the approval.

Subpart F: Classification of Approved Respirators; Scope of Approval; Atmospheric Hazards; Service Time.   This subpart lists the types of respirators and includes the scope of approval. It also classifies the respirator types according to hazards in which they may be used (e.g., entry and escape, escape only, oxygen deficiency), or the contaminants against which the device protects. Service times for each respirator type must be specified by the manufacturer according to a classification set forth in Part II.

Subpart G: General Construction and Performance Requirements. Rather than repeat construction and performance requirements under each subpart for each type of respirator, Subpart G lists all those that are common to all. These requirements include good engineering, no hazards to the user by the device, pretesting by the applicant, observers permitted, and other miscellaneous items.

Subpart H: Self-Contained Breathing Apparatus. The SCBA requirements remain essentially the same as in B of M Schedule 13E. Pressure-demand apparatus was added, and the total weight of the equipment was reduced to 35 lb. Under some conditions 40 lb is still permitted. Units rated for 15 minutes or longer were permitted for entry into a hazardous area; low temperature tests were modified; and air quality standards for cylinders were modified. The man test requirements remain essentially the same.

Subpart I: Gas Masks. The test schedule for gas masks also stayed essentially the same as B of M Schedule 14F, but additions were made. A chin-style gas mask was added that contains less sorbent, and the canister is attached directly to the facepiece instead of being mounted on the chest or back and attached with a breathing hose. These devices are permitted in air contaminant concentrations up to 0.5 percent (5000 ppm). Escape gas masks consisting of a half mask or mouthpiece, a canister, and associated connections were also added.

Masks to protect against acid gas and organic vapor gas were given an added restriction: they could not be used against contaminants with poor warning properties, generally regarded as those with an odor threshold above the threshold limit value (TLV). Nitrogen dioxide was also added as a test agent for acid gas approvals.

There is now some controversy concerning the use of gas masks. NIOSH has been somewhat reluctant to approve them and some manufacturers no longer produce some types, such as type N or universal gas masks. Misuse of these devices can easily occur, since they are for use in highly dangerous concentrations of a contaminant and in emergency situations. Because air concentrations generally are not known in an emergency, the device could be incorrectly used, that is, employed as protection in concentrations beyond its capability and even in an oxygen-deficient atmosphere. Possible liability suits against manufacturers further discourage the production of such devices.

Users are generally cautioned to contact NIOSH for guidance in the use of any gas masks and are encouraged by most health and safety professionals to use an appropriate atmosphere-supplying device. The OSHA health standards likewise prohibit gas masks for use in emergency situations or in atmospheres immediately dangerous to life or health (IDLH). MESA has also withdrawn permission to use these devices, even for escape purposes.

Subpart J: Supplied-Air Respirators. This section also is very similar to the B of M counterpart, except that the air quality portions have become more restrictive. Use of the Type C supplied-air respirators (with a compressed air source) in IDLH atmospheres is prohibited.

Although Subpart J still permits the use of hose mask with blower in an IDLH atmosphere, the OSHA health standards do not, nor does the NIOSH/OSHA decision logic in use for the Standards Completion Program (22). Both types A and B supplied-air respirators generally are not permitted for use at all in the new OSHA health standards, since the disadvantages of using them coupled with the advantages of other devices now available make them obsolete.

Subpart K: Dust, Fume, and Mist Respirators.   Provisions have been made in Subpart K to approve filter-type respirators for protection against particulates significantly more toxic than lead, for protection against radon daughters, for protection against asbestos, for single-use dust respirators as protection against pneumoconiosis and fibrosis-producing dusts or dusts and mists, and for powered air-purifying respirators.

The high industrial usage of highly toxic particulates led to the need to approve respirators equipped with high efficiency particulate air (HEPA) filters, defined as being 99.97 percent efficient against airborne particles with a mass median aerodynamic diameter (MMAD) of 0.3 micron. The HEPA filter definition came about because filters were available with the 99.97 percent efficiency and a test aerosol was readily available, heated dioctyl phthalate (DOP), which produced a monodispersed aerosol of 0.3 micron MMAD. Respirators with such filters were in high demand for protection against radioactive particulates, beryllium dusts, and other highly toxic disperoids.

Advances in aerosol technology will undoubtedly force changes in the test methods for other particulate filters as well. Problems in the reproducibility of test methods using silica dust and lead fumes will be alleviated when tests using sodium chloride or some other reliable aerosol replace the older disperoid methods originated by the B of M.

Subpart L: Chemical Cartridge Respirators.   Subpart L expands the B of M chemical cartridge respirator approval system to include any gas or vapor, including acid gases, ammonia, or whatever the manufacturer stipulates. As with gas masks, organic vapors with poor warning properties are not included in the approval for organic vapor respirators. Powered air-purifying respirators are also included in the test schedule.

Subpart M: Pesticide Respirators.   This completely new test schedule replaces the listing of respirators published by the Department of Agriculture from 1950 through 1966 that could be used for protection against pesticides. The subpart entails tests using carbon tetrachloride to check the sorbent capabilities and silica dust and lead fume for testing the filter portion of the cartridge or canister. Front or back-mounted pesticide canister HEPA filters are tested with DOP.

One question that has been posed concerns the use of carbon tetrachloride as the only test for the sorbent. In studies by Nelson and Harder (23), a charcoal sorbent was found to be very effective for absorbing carbon tetrachloride but not for some other organic vapors. Thus, the vapor phase of a particular pesticide was not readily absorbed by the sorbent and if it also had poor warning properties, the user would not be aware of vapor breakthrough into the facepiece.

Research for New and Better Test Methods.   Needs for new and better respirator testing methods are recognized by most people in the industrial hygiene field and in NIOSH. Toxicology studies resulting in the lowering of TLV or time-weighted averages (TWA) are appearing in the new OSHA health standards. Consequently, respirator reliability must be increased. What may have been an acceptable leak rate for a respirator that afforded protection against a substance with a given TWA may no longer be acceptable if the TWA has been lowered appreciably.

A complaint sometimes made by respirator manufacturers is that the test and certification schedules allow no flexibility for innovations by the manufacturer. For the manufacturer to obtain NIOSH/MESA approval on a respirator, the device must be designed to meet and pass the requirements in the approval schedule. Thus it is difficult to secure approval for a new respirator concept, either because there is no approval schedule or because the concept may not meet the engineering requirements in the existing schedules, hence cannot be approved. Since OSHA requires use of approved respirators, there is no sales market for a respirator designed around a new concept.

Numerous other problems exist in connection with the NIOSH test schedules. NIOSH is aware of these and is funding research projects to try to resolve many of them. However, it may take several years for funds to be made available, research to be completed, proposed changes to be written and published in the *Federal Register*, public hearings held, differences resolved, and final changes published and made effective.

Current Research in Respiratory Face Fitting.   One major field that has been investigated over the past few years has been respirator face fitting. The Atomic Energy Commission originally funded the Los Alamos National Laboratory (LANL) in 1969 to begin investigations in facepiece fitting (24). Additional funding came from NIOSH when it was organized in 1972, to LANL for further facepiece fitting studies (25, 26) and to Webb Associates to study the facial sizes of adult workers and determine which dimensions most affected facepiece fit (27). From these studies came information leading to the development of protection factors (28) and maximum use limits (29).

The results of these studies will probably lead to a change in the NIOSH testing and certification schedules, whereby the randomly selected subjects who test facepiece fit subjectively, or qualitatively, will be replaced by a test panel of male and female subjects, selected anthropometrically to represent most of the facial sizes found in adult workers. The fitting tests themselves can also be made quantitatively to determine exact leak rates. The leak rate allowances then can be set within acceptable and measurable parameters.

Other research directed toward improving test methods for respiratory protective devices of all types include development of a sodium chloride aerosol system, a respirator exhalation valve test system, a head harness strap tension test system, a facepiece-to-face pressure measuring system (30) a gas facepiece-to-face fitting system, and a valve test unit (31).

Other NIOSH-sponsored projects at LANL to improve specific test schedules include development of fit-test procedures for powered air-purifying respirators and comparisons of sodium chloride filter test methods to silica dust and silica mist filter tests, to arrive at more reliable test methods for dust, fume, and mist respirators. Performance of single-use dust respirators is also under investigation along with other filter studies to more reliably test filter performance (26). At Lawrence Livermore National Laboratory (LLNL), an examination of chemical cartridge respirators (28) showed that many organic vapors broke through cartridges approved by NIOSH for organic vapors, in minutes or even seconds. Obviously use of the currently used carbon tetrachloride method in the approval schedules for organic vapors and pesticides must be changed and new methods developed.

Respirator Face Fitting for Miners and Firefighters. One major concern has surfaced in the SCBA test schedule (Subpart H). The schedule now in effect originated with the B of M, when there was concern about SCBA usage by miners. The fire service is now one of the major users of SCBAs in the United States, and the environments to which firefighters expose their equipment are not at all similar to those in mines. Extremes of heat may be faced, along with radiant heat, high moisture, fire decomposition products, very cold temperatures in the winter in northern cities, and high vibration from riding on speeding fire engines. Preliminary studies at LLNL (32) indicate the necessity of including ways to test these factors either in Subpart H or under a totally new test schedule for firefighters' breathing apparatus.

Along with a test schedule for the firefighters' breathing apparatus, changes in use methods and equipment used must be made. Firefighting is now considered one of the most hazardous occupations in the United States, and inhalation and cardiac diseases and deaths are major contributors to the high statistics. Some fire departments still use demand apparatus, whereas in today's technological age of synthetic materials whose decomposition products may be very highly toxic, the use of pressure demand or positive pressure equipment should be mandatory (33). Since local fire departments do not fall automatically under the regulations of OSHA, which is federal, no regulatory action may be taken (unless the state has agreed to bind itself and its municipalities to OSHA standards). It is thus incumbent on the fire service itself to correct use problems. It is also necessary for NIOSH to provide a test and certification schedule that will assure the firefighter that the equipment will not fail, no matter what the exposure environment consists of.

2.3.3 Research for Better Respiratory Protective Devices

In addition to improving the methods used in the test schedules, NIOSH is also charged with trying to develop better protective devices. Likewise, the respirator manufacturers themselves are also seeking improvements in present equipment.

Evaluation studies of existing equipment and use practices are usually conducted before any attempts are made to develop new equipment. One NIOSH-sponsored study that evaluated respirators used in paint spraying operations (34) led to several recommendations both for improvements of the respiratory equipment itself, as well as for better certification tests and methods. The published report was then made available to respirator manufacturers to encourage the development of better equipment. Similar studies were made on abrasive blasting equipment (35) and respirators used in coal mines (36, 37). Several other NIOSH-funded studies conducted at LANL on various types of particulate filters used in respirators resulted in recommendations for improved performance criteria (26, 30, 31).

Firefighters' breathing apparatus has also been studied in an effort to improve protection. Studies at LANL (38) show that the weight of personal protective equipment is the main contributor to fatigue of personnel using it. The National Aeronautics and Space Administration (NASA), in a cooperative program with the Fire Technology Division of the National Bureau of Standards and Public Technology, Inc., polled city fire departments on their needs for improved breathing apparatus and developed a prototype SCBA that was both smaller and 10 lbs lighter than currently available units (39). Several manufactures are now producing these light-weight units. The LLNL studies (32) are now uncovering problem areas where firefighters' SCBAs need to be hardened for the fire environment, particularly against high air temperatures and radiant heat. Communications systems development is another area needing attention (40).

Studies at LLNL by Nelson and Harder (23) identified the need for improvement in the sorbent capabilities of chemical cartridge respirators used for protection against organic vapors. Before these studies were conducted, it was believed that any cartridge that could effectively absorb carbon tetrachloride would be adequate for any organic vapor. The LLNL research showed that this was not true, that new sorbents, specific for certain organic vapors must be developed, and that the general all-purpose organic vapor cartridge now in use must be limited to certain substances. These limitations can also be found in the respirator selection tables used in the OSHA health standards.

Other areas that remain to be worked on include: the need for better assurance that highly toxic and carcinogenic materials cannot enter the facepiece under any conditions; methods to limit chemical cartridges and canisters from being used beyond their useful capacity (e.g., through the use of color indicators, timers, or other means); continuation studies to determine psychological and physiological problems associated with respirator usage and development of devices to counter these problems; better assurance of respirator facepiece-to-face fit each time a device is used; reproducibility studies of facepiece fit each time a mask is donned; and development of low cost, more reliable fitting methods.

2.3.4 Supplied-Air Suits

Occasionally, complete body protection is required in addition to breathing protection to prevent skin absorption. In such conditions, a supplied-air suit can provide the necessary protection. Since NIOSH does not have a test schedule for supplied-air suits, it is incumbent on the health and safety professional to determine that the suit does indeed protect the wearer adequately.

An adequate supply of air is mandatory. Too little air can cause the suit to go negative, drawing outside contaminated air inside the suit. On the other hand, too much air can cause the suit to inflate to the point where movement is difficult. Tests conducted at LANL showed that the noise levels in some suits could exceed those permitted under the OSHA standards and that the very method of donning the suit could determine whether the suit remains under positive or negative pressure while in use (38).

For highly toxic particulates, care must be taken in removing the suit to prevent the spread of contamination, both to the wearer and to the surrounding environment and unprotected personnel in the area. The quality of air supplied to the suit must be monitored in the same manner as required in the OSHA standard for supplied-air respirators. Suits should never be worn in IDLH atmospheres unless adequate precautions are taken for escape in case of failure of the air supply or a tear occurring in the suit.

2.3.5 Basis for Program Requirements

In 1959, ANSI issued the American National Standard Safety Code for Head, Eye, and Respiratory Protection, Z2.1. This standard was based on earlier work that had been published in 1938 as National Bureau of Standards Handbook H24. By 1963 it was thought advisable to separate the Z2.1 standard into separate standards for each type of personal protective equipment. The Z88 Standards Committee was organized the same year, and subcommittees were established to rewrite the respirator standards (12).

Respiratory Protection.   It was not until 1969 that the first separate respirator standards were issued by ANSI: Standard Z88.1-1969, Safety Guide for Respiratory Protection Against Radon Daughters, and Standard Z88.2-1969, Practices for Respiratory Protection. Selected portions of the latter standard were then made into law by being incorporated, verbatim, into the Code of Federal Regulations (1). Further details of respirator program requirements are spelled out in the OSHA standards for specific toxic substances.

Respirator Selection.   Proper respirator selection can be one of the most difficult jobs for the health and safety professional in a respirator program. "The selection of a proper respirator for any given situation requires consideration of the following factors: (1) nature of the hazard; (2) extent of the hazard; (3) work requirements and conditions; and (4) characteristics and limitations of available respirators" (12).

Proper respirator selection is the choice of a device that fully protects the worker from the hazards to which he or she may be exposed and will permit the worker to perform the job with the least amount of discomfort and fatigue, and the greatest reliability possible. Cost of running the entire program, not just the initial investment, must also be considered, but only to the point to which the worker's health and job performance are not affected.

Consideration of the nature and extent of the hazard is simple in some instances and difficult in others. For example, determining whether an atmosphere is or may be oxygen deficient will immediately eliminate the use of devices that do not supply oxygen to the user if the atmosphere is deficient. However, deciding whether an atmosphere is immediately dangerous to life or health may be far more difficult.

Oxygen Deficiency.   Pritchard describes the dilemma of defining "oxygen deficiency" in the NIOSH publication *A Guide to Industrial Respiratory Protection.* Definitions of oxygen deficiency range from 16.0 to 19.5 percent. In the OSHA Standards (1) it is defined as 16 percent in Part 1910.134 and as 19.5 percent in Part 1910.94. Some standards take into consideration altitude or oxygen partial pressure, others do not. The rationale for the various definitions rests primarily with the persons or person doing the defining. The 16 percent figure is one below which definite physiological reactions can be expected to occur. The 19.5 percent figure is used by others because slight symptoms such as increased heart beat and respiration rate may be observable. The physiological response, plus the occurrence of something to cause the oxygen content to fall below the normal 20.95 percent, are believed by some to be sufficient justification to set a safety factor. Pritchard's recommendation (41) for this dilemma is perhaps the best.

The important thing is the respirator wearer's safety. If the legal definition of O_2 deficiency is above the O_2 level which is considered safe for humans, following the legal definition is justified. If the O_2 deficiency level as legally defined is less than the O_2 concentration which is believed safe for human exposure, raising the minimum O_2 level above the legal limit should be considered. Although not infallible, the partial pressure oxygen (P_{O_2}) limit of 60 mm Hg in the alveolar space should be the absolute minimum to which the O_2 level should be allowed to drop. This means that (P_{O_2}) in the ambient air should not drop below about 120 mm Hg. This problem is under study, and eventually oxygen deficient atmospheres will be redefined to eliminate the present discrepancies and account for the effect of altitude.

Immediately Dangerous to Life or Health Atmospheres.   The decision of whether an atmosphere is IDLH can be difficult and the respirator selected will depend on the interpretation. ANSI Z88.2-1969 defines IDLH as including "conditions that pose an immediate threat to life or health and conditions that pose an immediate threat of severe exposure to contaminants such as radioactive materials which are likely to have adverse delayed effects on health" (12). The Joint NIOSH/OSHA Standards Completion Program Respirator Decision Logic interpreted this definition to include:

Escape without loss of life or irreversible health effects. Thirty minutes is considered the maximum permissible exposure time for escape. Severe eye or respiratory irritation or other reactions which would inhibit escape without injury. . . . Contaminant concentrations in excess of the lower flammable limit are considered to be IDLH. . . . Firefighting is defined . . . as being immediately dangerous to life. Where only acute exposure animal data is available (30 min to 4 hr exposures), the lowest exposure concentration causing death or irreversible health effects in any species is determined to be the IDLH concentration.

Several people in the respiratory protection field, including Held, have questioned whether the lower flammable limit or eye irritation should be a consideration in the selection of respirators for an IDLH atmosphere, or whether respiratory effects alone should be the governing factor.

Using the decision logic interpretation, OSHA now requires that only full-face, pressure demand SCBA or combination supplied-air respirator and auxiliary SCBA with full facepiece in positive pressure mode will be acceptable in IDLH atmospheres in all new standards promulgated or proposed.

ANSI Z88.2-1969 (and 29 CFR Part 1910.134) permitted the use of several types of respirators in IDLH atmospheres under certain conditions: gas masks, demand-flow SCBA, hose masks with blower, and air line respirators were included. Half mask facepieces were also permitted with air line respirators, provided the atmosphere in which they were being used did not cause eye irritation or injury (see "Formation of Bureau of Mines Approval System" page 665).

The decision by OSHA to prohibit the other types of devices in IDLH atmospheres was based on two factors. The first was the results of fitting studies done at the LANL (24, 28) and elsewhere, which showed that respirators with positive pressure in the facepiece performed far better to prevent inward leakage of air contaminants for many workers than did respirators with negative pressure in the facepiece. Second, it was felt that a worker in an IDLH atmosphere should be capable of escaping by himself and should not have to depend on being rescued by a second person in the event of failure of the air supply. By the process of elimination then, only the SCBA, operated in a positive pressure mode, or a combination supplied-air SCBA (for escape if the air supply fails) would be acceptable. OSHA also reasoned that half mask facepieces on these devices would not be acceptable because they are somewhat less stable and stand a greater chance of being dislodged from the face than a full facepiece, do not protect the eyes from irritants, and provide a poorer facepiece-to-face seal on a larger percentage of the working population than does the full facepiece (22).

OSHA also believes that firefighting should fall into a separate category in the health standards and permits only the use of SCBA operated in a positive pressure mode. This decision was made since a firefighter has no way of determining whether a given atmosphere is IDLH, the worst case (i.e. that IDLH conditions exist) must be assumed. Moreover, firefighters cannot be expected to drag air lines from combination supplied-air SCBA operated in a positive pressure mode around with them.

It might be noted that the ANSI Z88.2 Committee, which has rewritten the standard, did not agree with the NIOSH/OSHA respirator decision logic concerning the use of respirators in IDLH atmospheres (42). The revised standard permits the use of gas masks, combination positive or negative pressure air line devices with full facepieces and auxiliary self-contained air supply, and any self-contained breathing apparatus with any facepiece, operated in positive or negative pressure modes.

The Committee believed that if the respirator program recommended in the standard was followed, these other devices could be used safely in the IDLH atmospheres, as they had been in the past. The Committee redefined the IDLH atmosphere to "include conditions that pose an immediate hazard to life or irreversible debilitating effects on health." In light of the 1980 ANSI Z88.2 Standard, OSHA may again review the restrictions and possibly alter its position on devices permitted in IDLH atmospheres.

MSHA, which sets the safety regulations for the mining industry, does permit the use of SCBA with negative pressure in the facepiece in IDLH atmospheres (43). Mine rescue teams use closed circuit devices that can be operated for an hour or longer, a necessity in mine rescue operations. Since the currently available SCBA with positive pressure in the facepiece is available only as one hour maximum time duration, such devices are not practical for the mining industry. The same argument may also be justified for some industrial operations and for firefighting (e.g., below-deck fires on large ships or fires in highrise buildings).

Maximum Use Limits and Protection Factors. By the late 1960s it was becoming more apparent that respirators were not providing the best protection possible, even when companies that purchased them for their employees tried to establish good industrial safety programs. Overexposures to radioactive materials were detected, for example, even when workers were wearing respirators provided with good filtering materials. Researchers such as Ed Hyatt at the LANL and Bill Burgess at Harvard University were beginning to identify some of the problem areas. The National Fire Protection Association recommended that firefighters no longer use the universal canister type of gas mask, but rather SCBA to prevent fatalities from oxygen-deficient atmospheres.

With funds provided by the Atomic Energy Commission (superseded by the Nuclear Regulatory Commission), Hyatt set up a respirator research section at Los Alamos. Some of the early work measured the fitting characteristics of respirators of various makes and types. From the data obtained, fitting indices were established, and from these, more reliable protection factors (PFs) were established. The PFs were reported in several different LANL progress reports. Hyatt's final report, which appeared in 1976 (28) defined PFs as follows:

The overall protection afforded by a given respirator design may be defined in terms of its protection factor, which is defined as a ratio of the concentration of contaminant in the ambient atmosphere to that inside the facepiece under conditions of use. This definition is illustrated by the equation:

$$PF = \frac{\text{ambient airborne concentration}}{\text{concentration inside facepiece or enclosure}}$$

... When both the ambient atmospheric concentration and the contaminant TLV are known, the protection factor may be used to select a respirator so that the concentration inhaled by the wearer will not exceed the appropriate limit. . . . For example, a respirator with a PF of 10 may be selected for use where a maximum use concentration of 10X the TLV exists. . . .

The protection factors are a quantitative measurement of the leakage of an air contaminant into the facepiece of a man performing certain exercises while wearing a respirator; thus it was possible to determine what concentration of a contaminant can be present outside the respirator without the concentration inside the mask exceeding the permissible exposure level. Several hundred tests were run on different men wearing various types of respirators. The protection factors were then assigned to the various categories of devices based on the condition that at least 95 percent of the adult males tested could achieve that PF. The PFs are valid only when a good qualitative fitting program is conducted.

Protection factors used by the British are published in British Standard 4275 (44). In the United States the PF concept is used in all the later NIOSH criteria documents and OSHA standards and by the Nuclear Regulatory Commission in "Standards for Protection Against Radiation" (45). The user not only needs to have some knowledge of the PF of a given type of a respirator to make an intellignet selection of what category of device to use, but also assurance that the device fits well enough to provide that PF. This can only be done by means of a good fitting program.

Protection factors are published in ANSI Z88.2-1980 for all types of respirators (42). This standard also permits higher PFs to be used when achievable, if quantitative fitting methods as opposed to qualitative methods are used. While certainly not in their ultimate form, since new studies are still finding discrepancies, the concept is probably one of the most significant advancements in recent years for better protecting employees using respiratory protective devices.

Approved or Accepted Respirators. OSHA requires that only approved or accepted respirators be selected for use by industry (1), by reference to ANSI Z88.2-1969 (12). The ANSI Standard in turn defines approved respirators as those approved by the B of M or listed by the U.S. Department of Agriculture. Since the Department of Agriculture no longer lists respirators (for protection against pesticides) and the B of M testing and approval system was transferred to NIOSH/MSHA, the latter approvals now are required.

Accepted devices are respiratory protective ones that cannot be approved by NIOSH/MSHA "because it is outside their approval or testing authority" (12). We know of only one device that falls into this category, namely, supplied-air suits, for which there is no test schedule yet available. ANSI Z88.2-1969 (12) does state that for a device to be "accepted," the user should assure that the device is "adequate for the required service, and that quality control during manufacture can be expected." The user also should make, or have made, suitable tests of the respirator's effectiveness which, as far as feasible, simulate tests made by the official test agencies (in this case, NIOSH).

The OSHA requirement that only NIOSH/MSHA approved respirators be used is not restrictive to the user, in that all reliable respirator manufacturers have their devices so approved. The major advantage to the user is, of course, insurance that the device will operate properly within the limitations of the test procedures and as stated by the manufacturer.

Other Selection Considerations. As stated earlier, many factors must be considered carefully in respirator selection. Determination of the nature and extent of the hazard may be accomplished by monitoring and the proper respirator selected by applying the proper PF. However, working conditions may prohibit the use of the selected respirator. High heat, humidity, cold, and/or other environmental conditions may limit the device's usefulness. A full facepiece, for instance, with its inherent tunnel vision may be dangerous to the user in an area with tripping hazards.

Finally, the respirator's limitations and characteristics must be weighed with care. For example, a NIOSH/MSHA-approved organic vapor respirator can be used only for organic vapors that have good warning properties (e.g., the odor threshold is below the TLV). Nelson and Harder (23) found that many organic vapors "broke through" the organic vapor cartridge in a very short time. If the wearer could not detect the odor of the vapor at a level below the TLV, an overexposure could easily occur. A similar danger exists with respect to poorly fitting filter respirators used for protection against highly toxic or carcinogenic dusts.

All such factors and possibilities must be carefully examined by the health and safety professional when selecting any respiratory protective device.

Respirator Use. As with the matter of selection, many factors must also be considered in the use of a respirator. Too often, a respirator that will provide adequate protection is selected without consideration of whether it will be used properly.

Obviously, the wearers must be well trained, not only in the use of the device, but in the reasons for which they are required to use a respirator. The health and safety professional must periodically survey the work area and discuss problems with respirator wearers to circumvent misuse. All too frequently, users do not wear the devices, either because of lack of understanding of their necessity or because of discomfort. Discomfort from wearing a respirator may be sufficient to drive the user to gamble with his or her health rather than suffer severely in performing a job. Work times in a respirator should be well planned so that the user has ample opportunity to retreat to a clean area and remove the device. As the work area temperature and/or relative humidity increases, the period during which the user can wear the respirator without undue discomfort decreases significantly.

For air-purifying devices, an adequate supply of filters or chemical cartridges or canisters must be available. A user of a filter device must be able to change the filter when the breathing resistance increases. The health and safety professional must also calculate the length of time that a chemical cartridge or canister can last without breakthrough, provide an adequate margin of safety, and thoroughly train the user to replace the cartridge or canister at the proper time.

In summary, proper respirator use can be achieved only through teaching the wearer not only how to use the device but why it must be worn; continued follow-up by the health and safety professional is essential to be able to anticipate and alleviate misuse problems before they occur.

Respirator Program Administration. OSHA requires that a respirator program be administered by one qualified individual. When two or more people are responsible for different aspects of the program they may not be inconsistent in their administration, thus confusing the user. Even worse, each program administrator must not assume that the other is covering some aspect of the program, which could result in a major shortcoming. This does not mean to imply that the program administrator may not delegate responsibilities to other qualified individuals, which, in all likelihood is necessary in a large program. However, the ultimate responsibility for the program must be with one person only.

Standard Operating Procedure. Standard operating procedures (SOPs) are required by OSHA covering the selection, use, and care of respirators. ANSI originally made this requirement to be sure that the administrator of the program would be forced to think out all aspects and to put them into writing. Written SOPs would also serve as a checklist for the administrator and others involved in the program, could be used easily by the employees during training, and would provide a source of continuity of the program if the administrator or others were replaced.

Though not required, it is often desirable to have the SOPs approved or endorsed "in writing by the company owner, president, or other person in high authority, to give them the emphasis they deserve" (46). Periodic reviews and updates as necessary should also be made of all SOPs.

Records. OSHA requires that records be maintained on the inspection dates and findings of all respirators maintained for emergency use (emergency use respirators must be inspected monthly). This is to ensure that the equipment is operating properly and in a "go" state of readiness. OSHA also requires a record on the date of issuance of each respirator to an individual. This affords an easy way of checking whether the users are returning their equipment for maintenance, inspection, and cleaning, as required.

"Other records that are advisable are training and fitting records, procurement records, inspection records, and maintenance records. The latter can provide valuable information on common failures of a particular brand of respirator which may prove uneconomical to keep in stock" (46).

Inspections. Finally, OSHA requires appropriate surveillance of work area conditions and employee exposure and stress, along with regular inspections and evaluations to determine that the respirator program remains effective.

Surveillance of the work area conditions calls for air sampling either as required by OSHA or in accordance with good industrial hygiene practice. The results can then be used to determine whether the respirator program is adequate, whether air concentrations in the work area have increased or decreased to the extent that the respirators being used have exceeded their PFs, or whether a less expensive or more comfortable device could be substituted.

When possible, bioassay results can be used to evaluate the program's effectiveness. A rise in blood or urine sample results from employees wearing respirators can indicate misuse or lack of use of respirators.

Cleaning and Sanitizing. Cleaning and sanitizing of respirators is necessary to prevent the spread of communicable diseases, to prevent dermatitis, and to encourage worker acceptance. OSHA requires at least one daily cleaning of respirators issued for the exclusive use of a worker, but more frequent cleaning may be necessary to prevent dermatitis from contaminants lodging at the facepiece-to-face seal and being held in contact with the skin, or from skin contact with a curing agent that may be coming off from the respirator itself (particularly with new respirators). Respirators used by more than one person and emergency use devices must be cleaned and disinfected after each use to prevent the spread of communicable diseases.

ANSI Z88.2-1969 recommends that wash water and drying temperatures do not exceed 180°F (12). However, the more delicate parts of many makes of respirators can be damaged at this temperature. ANSI Z88.2-1980 states that, in the absence of specific manufacturers' directions, washing and drying temperatures should never exceed 120°F (42). Sanitizing can be accomplished either though a separate rinsing in a disinfectant solution (care must be taken not to use a disinfectant that can warp or corrode parts), or by using a cleaner-sanitizer solution for initially cleaning the device, thus eliminating a step in the cleaning-drying process.

Inspection and Maintenance.   Following cleaning, sanitizing, and drying of the respirator, the device must be thoroughly inspected for defects, or for missing, worn, or improperly operating parts. New equipment should be inspected to make sure that all parts are present and operating properly. The inspectors must make sure that the device will give the user the protection expected. For example, a missing or warped exhaust valve will allow outside contaminants into the facepiece. A warped facepiece or headstrap will affect the user's fit, thus nullifying the fitting program. An improperly seated lens will also be a leak source for contaminants into the facepiece.

For more complicated devices such as supplied-air or self-contained breathing apparatus, the inspection procedure may have to include a testing facility to ensure proper regulator operation and airflows.

Maintenance operations would naturally follow where indicated by the inspection. If parts manufactured specifically for the respirator are not used for repair, NIOSH approval is automatically voided. Furthermore, some manufacturers of SCBAs require that maintenance operations be performed only by persons certified by the respective manufacturers; failure to observe this requirement will void the NIOSH approval.

Storage.   Respirators must be stored in such a way that they are protected against dust, sunlight, heat, extreme cold, excessive moisture, or damaging chemicals. Obviously, anything that can contaminate or cause damage to the respirator must be guarded against. Storage in plastic bags may be helpful.

The respirators should not be stored together or in a tight space. Each device must be positioned so that it retains its natural configuration. Synthetic materials and even rubber will warp if stored in an unnatural shape, thus affecting the fitting characteristics of the facepiece.

Emergency use respirators must be stored at quickly accessible sites, in clearly marked compartments built for the purpose. In determining the location of the storage area, thought should be given to whether the respirators would be used for escape (in which case they must be near the workers) or for reentry (in which case they should be near the entry point, but not in an area that might itself be involved in the emergency, thus becoming inaccessible or contaminated).

Finally, care must be taken that the respirators will not become contaminated in the lunch time, break, or overnight storage areas that are chosen. Common mistakes are hanging respirators in an area with airborne contamination and placing them in lockers with contaminated work clothing.

Training and Fitting. ANSI Z88.2-1969 (12) requires that the training of respirator users include instruction in the nature of the hazard, what would happen if the respirator is not used, and why other control methods are not used. It must be explained why the particular type of respirator provides protection, and the respirator's capabilities and limitations must be outlined. Instruction and training in actual use of the respirator, classroom and field training in recognizing and coping with emergencies, and other special training must be offered as needed.

Field studies (34, 36) have shown that many users resist wearing respirators because they are uncomfortable and inconvenient. Proper training and a thorough understanding of the reasons for wearing the device can help overcome many of these objections. Motivation for proper use can be obtained only through training.

Respirator fitting studies have revealed that large facepiece-to-face seal leakages can occur. The positive and negative pressure tests described in ANSI Z88.2-1969 (12), have been proved to be inadequate and should be used only as a field check prior to each use. When performing those tests, the facepiece is pushed against the face with the hands, thus sealing off small leaks that will recur at the conclusion of the test. Furthermore, the skin is not sensitive enough to detect small leaks that may be present in a positive pressure test.

ANSI Z88.2-1980 has eliminated the positive and negative tests and recommends only qualitative or quantitative fitting tests (42). The qualitative tests, using either a stannic chloride ventilation smoke tube or isopentyl acetate, provide a more accurate fitting method. In qualitative tests some persons cannot detect either substance by odor except in large quantities, and uncooperative subjects may simply lie: either not acknowledging detection of the odor to get the test over quickly, or claiming to smell something when they do not.

The preferred fitting method is quantitative. This permits the operator to detect the leakage accurately and impartially using either dioctyl phthalate or sodium chloride. The drawbacks are the expensive instrumentation and the need for a well-qualified operator. Some private consulting firms are now offering quantitative testing, which will put this superior form of fitting within reach of smaller companies and companies with small respirator programs.

Fitting should be performed at least annually to detect changes in fitting characteristics. Changes in an employee's weight, a new set of dentures, or new blemishes or scars can cause a respirator that once fit adequately to no longer be acceptable. Refresher training should be given when the refitting work is performed.

Medical Requirements. OSHA forbids the assignment of "persons . . . to tasks requiring use of respirators unless it has been determined that they are physically able to perform the work and use the equipment. The local physician shall determine what health and physical conditions are pertinent. The respirator user's medical status should be reviewed periodically (for instance, annually)." This restriction has caused a considerable amount of confusion regarding what minimum tests should be performed, what constitutes inability to work while wearing a respirator, whether a worker's livelihood can or should be denied when a physician says the individual should not wear a respirator, and how often a physical examination must be given—the requirement states "periodically," but "annually" is suggested.

Some of the OSHA health standards and proposed health standards attempt to clarify some of these points, yet others add to the confusion.

We believe that the health and safety professional should educate the physician (especially if the company physician is not an industrial physician by training), with respect to types of respirators, positive and negative pressures in the facepiece, breathing resistances, whether the employees will work in the IDLH atmosphere, the air contaminants that are present, and the alternatives that are available. By working together, the health and safety professional and the physician usually can work out satisfactory answers to problems involving physical conditions of employees who are required to wear respirators.

Program Evaluation. Regular respirator program evaluation and appropriate surveillance of respirator users' work area conditions and exposures or stress are OSHA requirements (1). Here again, sound professional judgment must be applied by the health and safety professional to adequately perform these tasks.

Changes in operations or any other occurrences that could affect air level concentrations must be monitored to determine whether the respirators being used are still permissible in accordance with the PF tables or in compliance to applicable OSHA health standards. Visual observation on respirator usage is necessary to determine whether the training program is effective. Discussions with users often can bring to light problem areas relating to discomfort from heat factors or other conditions, which may require reexamination for appropriateness of the respirators selected.

Respirator effectiveness can be measured for substances that can be detected by bioassay. The bioassays should be performed often enough to detect overexposures before they could become serious.

Other Standards and Regulations. The respirator program administrator may be required to follow regulations other than those of OSHA. Guidance on specialized respirator programs may be obtained from other specific standards. Examples of the latter include MSHA regulations for metal, nonmetal, and coal mines (42, 43); Nuclear Regulatory Commission regulations (45); specific OSHA health standards; and other ANSI standards in the Z88 series. The peculiar nature of the hazards for which specific standards or regulations were written warrant separate presentations. General standards such as ANSI Z88.2 could not cover all the applications for which respirators might be used.

2.4 Eye and Face Protection

If head protection is perhaps the oldest form of personal protective equipment, eye protection is surely the most widely used personal protective equipment today. And for good reason: the eye is the most delicate external part of the body; it is easily scratched, bruised, pierced, or lacerated; it is readily damaged even by weak corrosive chemicals; it can be damaged not only by visible light and by ultraviolet and infrared radiation, but also by exposure to microwave radiation and ionizing radiation. If this were not enough, the eyes are invariably brought into close proximity of the hazards, if for no other reason than to see what the hazard might be.

The basic concern in eye protection programs is to shield the eyes from impact. In almost every facility this is the most frequent hazard, and it is the firm basis of ANSI Z87.1 "Practice for Occupational and Educational Eye and Face Protection" (47). Protection against chemical splashes and exposure to infrared or ultraviolet radiation is usually provided in addition to impact protection, and it is difficult to think of an industrial operation where such separate protection without impact protection could be justified. The only exception is the use of laser eye wear, which is usually a special case.

2.4.1 Impact Protection

The most widely used form of eye protection is safety glasses. While every aspect relating to safety glasses is specified in great detail by ANSI Z87.1 (47), the most worthwhile feature is the impact resistance of the lens and frame assembly.

Lenses must be between 3.0 mm and 3.8 mm thick and must be able to survive the impact of a 1 in. diameter steel ball dropped from a height of 50 in. The numbers suggest that this test derives from empirical and historical considerations, but it is a sound standard that has proved itself over time. It is not unusual to hear of individuals whose safety lenses survived the impact of metal pieces weighing several pounds that were thrown from a lathe chuck, for example.

Traditionally, lenses have been fabricated of tempered glass. Because of the prescribed thickness of the lenses, glass spectacles can become very heavy for individuals requiring substantial vision correction. Further, in cases of extreme vision correction, it may become difficult to maintain adequate thickness at the center of the lens. Fortunately, these problems can be mitigated through the use of much lighter plastic lenses, which are also permitted. Plastic lenses must also pass a penetration test in which a weighted needle is dropped on the lens.

In order to perform properly, safety lenses must be mounted in safety frames. This is at times difficult to accept for individuals who would rather have the lenses mounted in a more fashionable or lighter weight dress frame. This, however, is neither permitted by the standard, nor would it provide a reasonable degree of protection.

Safety lenses are beveled to a 113° angle at the periphery, and safety frames have a corresponding groove to accept the lens. The lens openings of the frame are also slightly smaller on the inside, making it virtually impossible for the lens to be pushed through the frame and into the eye. The mechanical strength of the remainder of the frame, in particular the hinge, is also carefully prescribed. Dress frames, by contrast, are much more fragile and are intended to hold thinner lenses with a 90° bevel. The frames can also break or deform and the lenses may fall out or break out under small fractions of the forces that safety glasses can survive.

Safety glasses are probably as strong as they need be. While it may become possible to make much stronger glasses, the value of glasses that could stop high calibre bullets is questionable. When it becomes necessary to protect against impacts that cannot be absorbed by present safety glasses, then the remainder of the head and body is in dire need of protection also.

Lateral Impact Protection.   In most industrial settings, the fragments can strike at workers' eyes from all directions, not only from the front. To prevent such lateral exposure to impact fragments, safety glasses are often equipped with side shields. Depending on the degree of hazard, they can be cup-type side shields or flat-fold. The cup-type provides more complete protection; the flat-fold side shields make the glasses easier to carry. Logic would dictate that safety glasses with cup-type side shields be worn at all times when eye protection is required, but in many cases glasses with flat folds or with no side shields are accepted. Worker acceptance is much greater for ordinary safety spectacles, and they will prevent the vast majority of eye injuries.

In settings where the worker is routinely exposed to flying particles or chips, safety spectacles are not sufficient. Eye cup goggles or chipper's goggles are designed for this purpose. They form a seal against the skin around each eye individually and protect the eyes against fragments, chips, and dust. The same degree of protection can be obtained through the use of cover goggles over safety spectacles. Cover goggles are designed to accommodate the needs of the worker who must also wear prescription safety glasses. Additional protection against fragments, chips, and dust can also be obtained through the use of face shields as described below.

2.4.2   Chemical Splash Protection

While safety glasses provide excellent impact protection, they are of limited value in chemical exposures. The protection of choice is a chemical goggle which forms a tight-fitting seal to the skin around the entire eye region. To prevent entry of chemicals through ventilation openings, all such ventilation ports are equipped with shields or hoods. A worker's eyes will be protected from direct splash or run-off from the forehead even in cases of severe exposure.

For laboratory exposures, ANSI Z87.1 (47) permits the use of safety spectacles with side shields when used in combination with face shields. Undoubtedly, this reflects the recognition that chemicals used in laboratories are usually present in small enough quantities to preclude the possibility of a drenching exposure of the head.

Face Shields.　Face shields are not intended to provide basic eye protection; their mission is to protect the forehead, face, and frontal portion of the neck. When employed to protect against chemical exposures, the worker must also wear a chemical goggle to provide the basic eye protection.

Face shields are available in a wide variety, from a minimum height of 6 in. to oversized shields that extend to cover the entire face and front of the neck and wrap around to protect the entry to the ear canal. Thickness begins from .04 in. up. Clearly the biggest and thickest face shield provides the most protection, but it also weighs the most and may restrict head movement. A realistic appraisal of the hazard is required to settle on a style that will protect the worker without presenting an unreasonable burden.

Chemical goggles and face shields are not sufficient in situations where the worker is likely to be splashed or drenched with chemicals is a possibility. Protection for those cases is described in the section on encapsulating suits.

2.4.3　Welding Exposures

Welding, cutting, brazing, and soldering operations generate copious amounts of infrared, visible, and ultraviolet radiation, as well as flying particles, sparks, and molten metal. Eye protection must provide against all of these hazards simultaneously, yet permit the worker to see the work as well as his environment.

Eye protection for such operations is provided through the use of goggles, helmets, and hand shields. The lenses in these protectors must meet the impact standards described previously, and they are provided in a variety of shade numbers. Ideally, the transmittance values for selection of appropriate filter shade numbers should be based on scientifically determined ocular TLVs. However, the present transmittances were developed empirically and have remained essentially unchanged since they were generated by Coblenz and Stair at the Bureau of Standards in 1928.

The rationale for the original specification was quoted in a letter between Astleford and Ralph Stair (48):

At that time we had no medical information, except that a lot of welders had received great injury from ultraviolet and possibly infrared. We, therefore, set the ultraviolet and infrared values as low as practicable for the manufacture of the glass and for certification by our laboratory that met the standards we had set up. In our papers and in the Federal Specifications there are suggestions that certain shades are "intended" for certain operations, but the final answer rests with the operator to choose the proper shade for "best seeing."

The "best seeing" has since been defined more closely, and Appendix A.2 of ANSI Z81 (47) provides guidelines that narrow the choice to 2 or 3 shade numbers for any specified operation.

For brazing and gas welding operations shade numbers between 3 and 8 are specified, and the protection is usually supplied in the form of goggles. The harmful radiation output of the torch is absorbed by the lenses, yet the torch illuminates the area sufficiently to permit the welder to function effectively.

The radiant output from the various arc welding processes is much higher, and the work area is illuminated only during the actual weld process. While it would be possible to provide goggles with appropriate shade numbers, the welder would not be able to see enough to strike the arc at the proper starting point. The compromise to this solution is to provide hand-held shields or welding helmets in combination with safety spectacles, usually tinted to a low shade number. The spectacles provide impact protection from flying particles and shield the eyes from other welding operations carried out in the vicinity. To weld, the electrode is placed near the position where the arc is to be struck, the helmet is lowered or the hand-held shield is raised, the arc is struck, and the welding process is viewed through a filter plate in the helmet or shield. The helmet or shield also provides protection against the sparks and molten metal particles that are emitted, and shields the face and neck from the intense radiation.

2.4.4   Ultraviolet, Visible Light, and Infrared Exposures

A wide variety of industrial processes can generate ultraviolet, visible light, and infrared radiation. Examples are furnaces, molten metal operations, arc lights, germicidal lamps, etc. Permissible exposure levels for ultraviolet radiation, and proposed levels for visible light and near-infrared radiation are detailed in the 1982 Threshold Limit Values List (49).

Selecting appropriate eye and face protection will usually require measurement of the hazard, determining an appropriate attenuation factor, and some ingenuity in locating the most suitable protective equipment. Attenuation factors for welding lenses are listed by shade number in ANSI Z87.1 (47), and welding glasses or goggles are frequently used.

In many cases, however, the hazard may be limited to a narrow band of the spectrum, and the broad-band protection of welding gear may present unnecessary and considerable visual impairment. By careful review of the transmission properties of other equipment, it may be possible to select protective gear that provides the required protection with less vision impairment. Metallized reflective coatings can provide good protection against ultraviolet and infrared radiation and may be available for insertion in welding goggles. Metallized, tinted plastic faceshields and wire screen face shields are available and will provide substantial protection from heat radiation. And finally, a variety of specialized protective equipment is now becoming available for laser users. Some of this equipment may be used beneficially in other industrial settings if it provides the desired attenuation of the proper wavelength. In almost every case, however, a thorough search of available equipment and investigation of transmission properties is necessary to assure that the TLVs are not exceeded.

2.4.5 Laser Eye Protection

Laser eye protection is the newest category of eye protection available and perhaps the most difficult to apply properly. A brief overview is presented here, but for a thorough treatment of laser safety (including eye protection) the reader is referred to the excellent work by Sliney and Wolbarsht (50).

Basic Laser Safety Strategy. In discussing laser eye protection, it is important to remember the physical properties of laser radiation—it is intense, collimated, coherent, and monochromatic. Even a relatively small laser (Class II) emits light with a far greater intensity than sunlight, albeit over a much smaller area. Since it is highly collimated, the hazard persists over great distances, often even after multiple reflections. For high powered lasers partial specular reflections from lenses and filter plates may be hazardous, as well as diffuse reflections from almost any surface. Depending on the frequency of the laser, this radiation can cause photokeratitis (in the UV region), retinal burns (in the visible and near-infrared region), corneal burns (in the far infrared region), and cataracts (in the infrared and in portions of the UV region) (50). Maximum Permissible Exposure (MPE) limits have been established for all these regions in ANSI Z136.1-1980 (51) and by the American Conference of Governmental Industrial Hygienists (49).

The preferred strategies for protection are to use lasers that do not emit radiation in excess of MPE, or to enclose lasers and beams in interlocked housings that do not emit laser radiation in excess of the MPE. These are Class I lasers and Class I laser systems. Since the MPEs are set below damage levels, no eye protection is needed.

The secondary strategy for *visible* light lasers is to limit the output to levels that do not produce eye damage within the blink response time of the eye. The rationale is that the eyelid will shut involuntarily upon exposure to the bright laser light before any damage can occur. Again, no eye protection is required. This strategy applies to Class II and Class IIIa laser systems. Class II lasers are limited to one milliwatt or less, and Class IIIa lasers are limited to an equivalent illumination level.

In most industrial applications, it is indeed possible to construct interlocked laser system enclosures that obviate the use of eye protection. However, even in these cases, the lasers need servicing and alignment during which personnel could be exposed to hazardous laser radiation. In many research and development applications such enclosures simply are not possible. In these cases, eye protection is one of the necessary strategies. In addition, it may be necessary to use minimal power levels, beam expanders, beam filters, and low-power substitute alignment lasers.

Eye Protection Selection.   The first step in selecting eye protection is to determine the frequency or frequencies at which hazardous radiation is present, and the radiant exposure that is possible at each frequency. The protection factor for eye protection is expressed as the optical density at the frequency in question, and it can be calculated by deriving the logarithm of radiant exposure divided by the MPE. Detailed guidance for this is provided in ANSI Z136.1-1980 (51).

This procedure is simple and elegant in concept but can be exceedingly tedious and time consuming in reality. Unfortunately, it is often necessary to perform these calculations to select appropriate eye protection or to specify other precautions in addition to eye protection.

Limitations of Eye Protection   Laser eye protectors are actually optical filters that selectively absorb or reflect light at the frequency in question. They can be mounted in spectacles with opaque cup-type side shields or goggles. Since it is generally not possible to manufacture filters that are specific for a very narrow frequency band, laser eyewear actually filters out large portions of the visible spectrum. The wearer sees his surroundings in a largely monochromatic manner in the remaining portion of the spectrum. This creates at least two problems.

In the first place, the wearer generally can no longer see the laser beam or its reflections. While this selective filtration is precisely what was intended, it necessitates that other means be used for locating the beam. The possibilities include fluorescing cards, electronic detectors, and "weak" eye protection, i.e., eye protection that is strong enough to protect, yet only barely so to permit some viewing of the beam. It is difficult to find filters of just the right density, but it has been done successfully in some cases. (Normally filters with optical densities far in excess of what is required are used, simply because they are more readily available.)

The second problem is frequently the inability of the wearer to distinguish colors, and some colors appear as entirely black. If the wearer is protected against the bright red helium/neon laser beam, for example, such protection also prevents seeing any red warning lights or panel indicators. Hence, it is often necessary to devise special schemes for information, warning, and operational convenience that sidestep this problem.

It is also advisable to provide bright general illumination in laser areas. This serves two purposes. When eye protection is worn, it enables the wearer to distinguish his surroundings more readily. (This can be a significant problem with some eye protectors.) When no eye protection is used, it constricts the pupils, reducing the area through which harmful radiation can penetrate the eye.

It should be noted here that the above problems are generally not true for ultraviolet and infrared lasers. Good eye protection is available that does not interfere appreciably with good vision. Further, since these laser beams are not visible in any case, there is virtually never a reason to work on such lasers without eye protection.

A second general problem area is the ability of laser eye protection to survive exposure to the laser beam. A 40-W YAG laser with a beam diameter of 1 mm may only require an optical density of 6, but the filter material may melt before the wearer knows that his eyes are in the beam path. When possible, eye protection should be selected that will be able to dissipate the laser energy without damage to the filter material. In the case of high power lasers, this may not be possible; it is then necessary to limit access to the primary beam by other means, relying on the eyewear only for protection from diffuse reflections.

Multiple Wavelengths.   In a variety of applications, multiple wavelengths are now used in the same laser system. One laser may drive a second dye laser that operates over a broad band of the spectrum. Or laser light may pass through frequency doublers and be present at two, three, and even four frequencies, covering the entire spectrum. This poses a challenging problem in selecting eye protection.

Where possible, the preferred strategy is to provide separate enclosures for the various frequencies. It is then possible to work on the segments separately, using appropriate eyewear for each. When only two frequencies are involved, it may even be possible to locate eyewear with adequate density for both frequencies when separate enclosures are not possible or not desired. However, often it is necessary to use double eyewear, for example, spectacles with filter material for one wavelength and cover goggles with a filter for the second, or goggles with two different filter plates.

When three or more frequencies are involved, eye protection usually must be custom developed. Overlaying of filters is a limited option since the filters soon combine to form an impenetrable vision barrier. Some custom protection using filters and reflective coating have been constructed, but much additional work is needed. No such eye protection is available ready-made; it must be custom designed and fabricated for each application.

In the case of tunable dye lasers, it often is possible to tune the laser over a broad portion of the spectrum. To manage the eye protection question sensibly, it generally is necessary to establish the frequencies that will actually be used in a given application and to provide protection at those frequencies only.

Damage to Laser Eye Protection.   Laser eyewear is easily damaged to the point of destruction. Any scratch in a reflective coating will render the eyewear useless, and even small scratches in plastic filter material reduce the optical density greatly. Hence, great care is required in handling, storing, and cleaning laser eyewear. Damaged eye protection must be replaced, even though the cost is substantial. Some filter materials used in laser eyewear are also bleached by light radiation and lose their protective value over time. Where such filters are used, periodic testing and/or replacement are indicated.

Laser Enhancing Goggles   Also available on the market are laser enhancing goggles. Used in surveying applications, these goggles preferentially transmit the laser light. This permits the use of lower power lasers that would otherwise be overwhelmed by other available light. They are of value in protecting the eyes of those who do not wear them, but should not be confused with protective eyewear.

3 HANDS AND ARMS

3.1 Introduction

Several types of assaults are possible on workers' hands and arms. Since most physical work requires the use of the hands, they are particularly prone to damage. Other parts of the body are usually injured due to an accident (e.g., falling object striking the head or foot), but the hands are often deliberately exposed to or immersed in a contaminant, put in a radiation beam, used to hold a part near a pinch point, cutting or abrading device, or an impacting type of machinery, or held close to extremes of temperatures such as a hot plate or cryogenic equipment.

Occasionally, double gloving may be necessary to protect the wearer, where two or more assaults on the hands may be possible. An example of this can occur when working with beryllium or a radioactive material, for which a work glove made of a heavy material is worn on the outside to prevent skin penetration by a sharp piece of contaminated metal. Rubber gloves may then be worn underneath to stop skin contamination by liquids or dusts which work their way through the more porous work glove material.

As with any piece of personal protective equipment, hand and arm protection are considered a second line of defense. The health and safety professional should always consider the use of substitution or engineering controls first. The greater the hazard to the worker of injury or ill health effects caused by exposure of the hands or arms, the greater the desire should be to devise engineering controls or substitute a less hazardous material.

3.2 Skin Contamination, Permeation, and Penetration by Hazardous Chemicals

From an industrial hygiene standpoint, the most common industrial problems involving the hands are contamination, permeation, and/or penetration of the skin. The theory behind prevention of contaminating the hands or having a hazardous material permeate or penetrate the hands is isolation or separation. Gloves (and protective sleeves, if necessary) can provide this separation, if chosen correctly.

3.2.1 Contamination

Contamination of the hands probably is the cause of most occupational dermatoses, which, in turn, outnumber all other work-incurred illnesses (52). Dermatitis often occurs directly on the hands, and contamination causing skin rash is transmitted by the hands to other parts of the body by scratching, rubbing, and touching.

Gloves or protective creams can provide separation of the hands from the dermatosis-producing chemical. Their use must, of course, be coupled with a program of employee training and personal hygiene to be effective.

The hands can also be contaminated by many other types of materials besides those that can cause dermatosis. These range from those which are nontoxic and merely soil the hands to others which are very toxic or radioactive and can, if ingested, breathed, or penetrated into the body through broken skin, cause various problems.

The use of gloves in these cases is not for separation, but to prevent the spread of contamination of these materials. Removal of the gloves will hopefully remove any contaminating materials which could be transmitted to the mouth, nose or broken skin by the hands when eating, smoking, or touching. The removal of the gloves also limits transmission of the contaminant to other clothing articles which could be carried to the worker's home and inadvertently expose other family members. The gloves will also minimize or prevent soiling of the hands with contaminants which may be difficult to remove by ordinary handwashing methods.

3.2.2 Skin Permeation

Permeation of the skin can be a pathway of entry into the body for some hazardous materials. For these chemicals, the unprotected skin does not act as a barrier but as an open pathway. Hydrogen cyanide and mercury are two examples of several such chemicals. Others can be identified in the Threshold Limit Value booklet by the word "skin" in the listing (49).

When these materials must be handled by a worker, they must be separated from any skin contact. Personal protective equipment becomes extremely important to keep this separation, and the material from which the protective equipment is made makes selection of a high priority. Obviously, the material must be nonpermeable to the chemical being handled. Care must also be taken to not only protect the skin of the hands through the proper selection of gloves, but also other parts of the body which may be exposed from splashes, carelessness in handling, etc. If more of the lower arm might be exposed than just the skin of the hand, gloves with longer cuffs and/or protective sleeves may also be necessary. If protective sleeves are a necessary part of the ensemble, the means by which an impenetrable seal is made with the gloves must be considered. This might be accomplished with an impermeable tape whereby the glove and sleeve are taped together. Whether the glove is tucked into the sleeve or the sleeve is tucked into the glove depends on which way the contamination can migrate. If the sleeve is tucked into the glove, material splashed on the sleeve could run down the sleeve into the glove, so placing the end of the glove under the sleeve may be advisable. The reverse could be true if the worker had to do much of the work when reaching overhead (e.g., working on a leaking pipe) and liquid could run down the hands to the arms.

3.2.3 Skin Penetration

Skin penetration usually consists of a sharp, hard, contaminated object such as metal, glass, or wood puncturing the skin and injecting the contaminant into the body. Besides the trauma of the injury itself, the contaminant may produce a local, complicating wound such as the granulomata of the skin produced by beryllium to injection of a systemic poison into a blood vessel, where it can be transported to a target organ.

Separation of potential sources of contaminated objects from penetrating the skin is generally achieved through the use of gloves which are thick enough and of sufficiently strong material (e.g., canvas or leather) to stop a forcible skin entry. Since the material used for strengthening gloves may be porous, it is often necessary to provide a second pair of nonpermeable gloves underneath the stronger protective ones to prevent skin contamination or permeation.

3.2.4 Personal Protective Equipment for Protecting Hands and Arms from Chemical Hazards

Gloves.   Since most physical work is done primarily by the hands, a large variety of gloves are available that are suitable for most purposes. Thus proper selection is very important to provide adequate protection to the user. The health and safety professional must consider the following criteria when selecting gloves:

1.   All hazards from which the user must be protected.
2.   The degree of manual dexterity needed to perform the job.
3.   That proper glove sizes are available to fit all the variously-shaped hands.
4.   Whether reusable or disposable gloves are preferable (and whether means for cleaning are available if reusable gloves are selected).
5.   Comfort of the user.

Very little information is available in the form of standards or recommendations for the selection of gloves. Some information is now available in the literature concerning the permeation of different types of glove materials to some organic solvents and chemicals (53–55). Glove manufacturers can often provide assistance in glove selection for protection against many chemicals.

OSHA does not cover hand or arm protection with the completeness it covers respiratory protection in 1910.134. The only specific regulations for hand or arm protection are in 1910.34 (Ventilation) with a requirement for heavy canvas or leather gloves for abrasive blasters, and for impervious gloves for open surface tank workers if they are working in a liquid other than water; in 1910.183 (Helicopters) with a requirement for ground personnel to wear protective rubber gloves if a suspended load has not had the static charge dissipated with a grounding device; and in 1910.218 (Forging Machines) with a requirement for gloves for the use of lead. All other requirements are general in nature. Information is available from NIOSH on the performance of lineman's rubber insulating gloves and on criteria for firefighters' gloves (56, 57).

The materials from which the gloves are made and the thickness of the material can vary considerably. Leather or heavy canvas or rubber gloves can provide protection against penetration of contaminated materials. When a combination of chemical and physical hazards exist, combination types of gloves are available such as cotton gloves dipped into vinyl, neoprene, etc. Cotton liners are also available to be worn under a rubber or plastic type glove. When no liner is used, contact dermatitis can occur from some glove materials with sensitized workers, so follow-up of glove usage must be made by the health and safety professional.

One of the better listings of different types of gloves available can be found in *Best's Safety Directory*, Volume I (55). Their listing includes:

Plain, reinforced or stippled fabric glove.
Rubber coated fabric glove
Leather glove
Mitten—general duty
Sleeve—general duty
Finger guard—general duty
Glove—fingerless
Glove—abrasive blasting
Glove—impact and cut hazards
Stainless steel and synthetic fiber glove
Metal-studded leather glove
Steel fingertip leather glove
Metal mesh glove
Hand pad
Gloves and mittens—fluorescent
Gloves and mittens—liquid hazards

Glove—hydrocarbon fuels, tetraethyl lead, etc.

Glove or mitten—cold work (also heated)

Clean room gloves

Glove box replacement gloves

Glove—radiation hazard

Glove or mitten—flame and heat hazard

Barrier Creams. Separation of the skin and some chemicals can be achieved through the use of a barrier cream. The cream can be applied to the hands and as far up the arms as needed, thus permitting the user complete dexterity of his or her hands without the encumbrance of wearing gloves. However, care must be taken in selecting a cream which is completely impervious to the chemical against which it is protecting. The worker must also guard against wearing the cream off and must reapply it as needed.

Generally, the protective or barrier creams and lotions are limited to certain water solutions (i.e., coolants and cutting oils), some solvents (i.e., alcohol), and some mild alkaline solutions. Their primary use is for prevention of mild dermatitis, not a harsh chemical reaction with the skin or permeation of the skin by a toxic material.

Sleeves. The same principles that apply to glove protection are used for protecting the arms with the use of sleeves. If long gauntleted gloves cannot provide protection far enough above the hands, protective sleeves can be used. However, since this section is dealing with protection from chemicals, careful consideration must be given if sleeves are deemed necessary. Getting an effective seal between the glove and sleeve can be difficult even with impervious tape. For highly hazardous materials, other means of protection besides sleeves are probably called for (i.e., whole body protection, engineering controls). Except for skin protection against penetration of a contaminated piece of material, sleeves generally are not recommended by these authors for protection against contaminating or skin permeating chemicals.

3.3 Abrasion, Cutting, and Impact

As discussed in Section 3.2, the hands are particularly subject to injury, as they are the workers' physical tools for accomplishing tasks. The hands, therefore, receive the most injuries, as evidenced by first aid reports and the number of workers with missing fingers. For protection against abrasion, cutting, and impact, safety engineers have always relied heavily on engineering controls. In fact, most wood working and metal working tools are so powerful that gloves or other hand protectors would be of little value, and the only choice is to provide engineering controls that keep the hands out of the hazard zone. But engineering controls are also preferred for lesser hazards. Examples are lifting aids for sharp-edged sheet metal pieces and special knives that preclude finger contact with the blades. Guards are even provided on X-acto knives to help protect office workers from stabbing their hands when rummaging in a desk drawer.

However, many occupations and job tasks do not lend themselves to engineering controls, yet the potential for hand injury from abrasion, cutting, and impact is high. For these jobs, many forms of hand and arm personal protective devices are available, some of which can be seen on the *Best's Safety Directory* partial listing in Section 3.2.4.

The gloves can range from just finger guards to hand pads to fingerless gloves to full gloves, depending on what part of the hand is at risk and the amount of dexterity needed. Materials from which the gloves or sleeves are made will vary, depending on whether the hazard is abrasion, cutting, impact, or a combination of any of these. The material may be canvas or leather for abrasion or minor cutting protection to chain mail or jointed metal for more severe cutting problems.

The health and safety professional can get help in proper selection from glove manufacturers, safety equipment distributors, the National Safety Council publications, the various safety journals, and the *Best's Safety Directory*.

3.4   Heat and Cold

3.4.1   Heat

Isolation of hot processes is preferred when possible, but again, often cannot be achieved. Work around ovens, furnaces, hot plates, and hot metals must often be done directly by a worker. Molten metal splash is a common occurrence in welding. Gloves or sleeves may also be required for a combination of hazards such as protection from hot chemicals, where both chemical and heat resistance is needed.

Many glove and sleeve materials are available, with asbestos or an aluminized fire-resistant fabric being most common for very hot operations. The aluminized glove reflects infrared radiation and is usually chosen when work is done near a high radiant heat source. Leather or even cotton may be used for lower temperatures, but caution should be taken that flame resistant materials be selected if the possibility of ignition exists. Gloves generally fall into four basic cuff styles (55). For one style, the cuff merely extends to the wrist. A knitwrist cuff with a tight closure around the wrist is used to prevent sparks or splatter from entering the glove, such as in welding operations. A gauntlet style extends part way up the forearm and an elbow cuff style goes all the way to the elbow.

Whenever possible, asbestos used in personal protective clothing should be in a bound state. Articles of clothing containing asbestos should be inspected routinely and used only in the manner prescribed. Any frayed or otherwise damaged protective clothing containing asbestos should be discarded to prevent broken fibers from becoming airborne.

3.4.2 Cold

Gloves for cold protection are needed in several occupations, such as outdoor construction, freezer plants, cold storage rooms, and lumbering. Insulated plastic gloves are most desirable, as they allow air circulation through minute pores, but are insulated against the cold. In cryogenics operations, gloves must be worn to protect against contact with cold pipes and other metal surfaces. Bare skin usually contains enough moisture to cause the skin to adhere to the metal via an ice bond, should the worker contact the metal at cryogenic temperatures. Once this happens, the worker has little choice but to tear himself from the metal, leaving tissue behind. In most cases, extensive tissue damage from frostbite would result if this was not done. Protection against immersion in cryogenics or extended contact with cryogens is usually not contemplated, and it is generally not possible. The common protective strategy is to provide loose fitting welding gloves, which protect against contact with cold metal, yet can be readily thrown off if they become soaked with cryogen.

For extreme, long-term exposure to cold, electronically heated gloves are available. These gloves have a rechargeable battery to provide heat throughout the glove.

3.5 Electrical Hazards

Gloves and sleeves are available for protection against electrical hazards. Rubber gloves for linemen and electrical workers generally have long cuffs and should resist 10,000 volts for three minutes (55). Manufacturers of these gloves recommend that the electrical resistance be tested monthly. The gloves are available in combination with insulation to give cold protection and with leather to prevent punctures and abrasion.

3.6 Radiation

Lead impregnated plastic and rubber gloves and sleeves are available for low level X and gamma ray protection. It is important that the health and safety professional work with the glove manufacturer and the radiation-emitting device manufacturer to be assured that the glove/sleeve provides adequate protection for the user. Again, it is much better to isolate the worker from the radiation than to use personal protective equipment.

3.7 Training

As with any personal protective equipment, it is vital that all glove users be trained. Training should include: what the hazards are, why a certain type or types of gloves were selected, what the consequences of not using the gloves are, and prompt reporting of any problems from the glove or glove usage to the health and safety professional.

4 FOOT AND LEG PROTECTION

4.1 Safety Shoes

The most likely foot injury is perhaps a crushing of the toes from a falling or rolling object. Hence, virtually all foot protection incorporates steel toe boxes. A foot protection program that stops at this point would not be complete, however, where the possibility of puncture wounds, chemical burns, hot surfaces, molten metal, etc., exists. Foot protection is available in a very large variety that addresses just about any hazard that might be encountered by a worker. The major types of protection are described here, but it is important to remember that most of the protective features are also available in combination. Hence, usually a single foot protector can be found that addresses all hazards.

4.1.1 Safety-Toe Footwear

Safety shoes and safety boots are covered by the "American National Standard for Men's Safety-Toe Footware", ANSI Z41.1 (58). While the Standard was written explicitly for men's footwear, it should be noted that women's footwear meeting this standard is also readily available now. This standard is specified by OSHA (1), and it is also specified as a prerequisite in all other ANSI standards relating to protective footwear.

ANSI Z41.1 specifies that a toe box be incorporated into the design of safety shoes to protect against impact and crushing forces. Three different classes of protection are recognized, and test measures for impact and compression testing are specified for each. The greatest protection is provided by Class-75 footwear. In closely specified tests, the shoe is subjected to a 75 foot-pound impact test and a 2,500 pound compression test. In each case the clearance between the sole and the toe cap must remain at least $\frac{1}{2}$ in. The other two classes must also maintain this clearance, but the test loading is reduced. Class-50 shoes are subjected to 50 foot-pound impact loading and 1,750 pound compression testing, and Class-30 shoes are subjected to 30 foot-pound impact and 1,000 pounds compression.

For heavy industrial and construction applications the Class-75 shoes is clearly preferred, but the other classes have applications where foot hazards are less severe. The required rating footwear can be calculated from the exposure easily.

When foot accidents do occur to workers at forces approximately equal to the rating of the shoe, the worker is likely to suffer serious contusions of the toes, but no fractures. When forces are applied far in excess of the rating, the toe box may rotate backward and cut into the foot. The backward rotation arises because the toe box does not cover the bottom of the toes, but rests on a narrow flange on the sole of the shoe. To combat this phenomenon, at least one manufacturer now offers shoes with toe boxes that have a much thinner flange at the tip than at the sides to induce a forward rotation. This offers some additional margin for safety. If the load on the toe box becomes too great, it will, of course, be crushed entirely.

Whatever model or design of safety shoe is chosen, under excessive loading it will fail, and foot injury will result. However, it is difficult to postulate a case where lesser injury would have resulted without foot protection.

4.1.2 Metatarsal Protection

Safety shoes literally protect the toe area only. Since the toe extends farthest from the leg, and since the majority of objects that could strike the foot are fairly large, they provide sufficient protection for the entire foot in most applications. But in many other applications it is entirely possible that the foot might be struck by objects in the metatarsal area, bridging over the steel toe or bypassing it all together. In those cases, metatarsal protection is required.

Metatarsal Footwear. The "American National Standard for Metatarsal Safety-Toe Footwear", ANSI Z41.2 (59), specifies protection for this hazard. Shoes meeting this standard must meet the impact and compression testing specified for safety shoes (58), and must also pass an impact test in the metatarsal area, $3\frac{1}{2}$ inches from the tip of the toe. Depending on the class of the shoe, the impact is either 75, 50, or 30 foot-pounds, and a clear opening of at least one inch must remain during the test. A compression test is not specified.

Metatarsal Protectors. A variety of add-on devices are available to provide metatarsal protection without purchasing footwear that meets ANSI Z41.2 (59). Most prominent among these are perhaps the footguards. These are metal or plastic covers that are worn over the shoes and are strapped to the heel or ankle. They cover the toe, metatarsal, and instep areas and provide excellent protection when serious foot hazards exist. McElroy (6) reports that footguards may withstand impact loading of 300 foot-pounds and more. They can be easily donned and removed, facilitating transfer from worker to worker and use as special protection for specific jobs, for example, operating a jackhammer. Use of footguards in this manner may be more readily accepted by workers than the wearing of boots with built-in metatarsal protection. Also available are metatarsal covers that rest on the steel toe and the welt of safety boots and are fastened to the shoelaces.

4.1.3 Conductive Shoes

In situations where static electricity build-up on personnel must be prevented, corrective measures must be taken. Conductive shoes are only one part of a static electricity prevention program; they may be totally useless unless all necessary other measures are also taken.

Type-1 Footwear. The "American National Standard for Conductive Safety-Toe Footwear", ANSI Z41.3-1976 (56) specifies two types of footwear. Type-1 footwear is designed to prevent static electricity build-up on personnel who work around sensitive explosive mixtures. Applications include explosives processing areas and operating rooms where flammable anesthetizers are used. At least one source also recommends their use for cleaning tanks that have contained gasoline or other volatile hydrocarbons (6).

Type-1 footwear is required to have an electrical resistance rating of between 25,000 and 500,000 ohms. To determine the resistance, the shoe is placed on a conductive surface, a $2\frac{1}{2}$-in. electrode weighing 5 pounds is placed into the shoe over the heel, and the resistance is measured (60). This resistance rating must be maintained while the shoe is used. Hence, regular cleaning, testing, and careful repair are required. If the resistance becomes too great, for example through build-up of dirt and grime, the shoe will not prevent static build-up. If the resistance drops, due to embedded metal pieces or improper repair, for example, an unreasonable electrical hazard exists.

Conductive shoes only work properly if they are used on a conductive floor. Conductive flooring is specified in ANSI/UL 779-1980 "Standard for Electrically Conductive Flooring" (61). Note that this standard specifies a maximum resistance of 1,000,000 ohms between points three feet apart, and a minimum resistance of 25,000 ohms to ground. The latter is required for protection against electrical hazards. Use of metal flooring is definitely not an acceptable substitute.

In cases where toe protection is not necessary, conductive sole shoes, conductive booties, and ankle straps are also available. These primarily have applications in hospitals and electronic micro-processor assembly operations.

Whenever conductive footwear is used, it is necessary to have programs for cleaning, testing, and repairing footwear and flooring. Further, such programs usually only make sense when other spark sources are also controlled. This means control of clothing materials to eliminate synthetics capable of static built-up, use of nonsparking tools, Class I electrical wiring, etc.

Type-2 Footwear.   ANSI Z41.3 (60) also provides specifications for Type-2 footwear. Type-2 shoes are more conductive than Type-1, and the test permits a *maximum* resistance of 10,000 ohms. To achieve this low resistance the shoes may be equipped with conductive straps that fit around the calf of the worker. Type-2 footwear is intended for use by lineman and other personnel working on high voltage equipment where the potential between persons and energized parts must be equalized. It is also of value in cases where induced voltage may be a problem.

4.1.4   Insulating Footwear

The final ANSI Standard that deals with protective shoes in the "American National Standard for Electrical Hazard Safety-Toe Footwear", ANSI Z41.4-1976 (62). Shoes meeting this standard provide electrical insulation for the worker to reduce the likelihood of serious electric shock. The sole and heel may not contain any metal parts, and a stringent resistance test is specified: the inside of the shoe is soaked for at least 5 minutes with a 1 percent sodium chloride solution to simulate the effects of a perspiring foot. The shoe is then placed on a metal electrode, and another metal electrode covering most of the insole and weighing 5 pounds is applied to the interior of the shoe. Fourteen thousand volts of 60 Hz are applied for one minute, and leakage currents must remain below 0.5 milliamperes.

Insulating shoes must be carefully inspected for imbedded tacks and metal particles, as well as wear and damage that might reduce insulating values. Excessive cracking of the uppers may well create a path for leakage current, in particular when combined with moisture and soil built-up. It should also be noted that insulating shoes are not intended to take the place of other protective equipment for work on live electrical systems: insulating mats and gloves, insulated tools, etc. They are intended to provide an extra measure of safety.

4.1.5   Puncture Resistant Footwear

Painful and serious foot injuries and troublesome infections result when the sole of the shoes are penetrated by nails and other objects. This can be a particularly serious problem in construction and demolition trades, but occurs also in other operations. To preclude this injury, safety-toe footwear is widely available with flexible steel insoles. They add little to the weight of the shoe and permit comfortable flexing of the foot in normal walking, yet they will stop or deflect nails and other objects that have penetrated the sole.

4.1.6   Safety Shoe Programs

To begin, safety shoes are truly personal items. They must be carefully fitted and tried on by the worker, just like any other shoe. Different manufacturers use different lasts in fabricating shoes, hence, individuals may have strong preferences for certain brands that were made to a last that approximates their foot. The individual may not know why he prefers a shoe—it just fits better.

Secondly, safety footwear is available in a large variety of styles. Just about any style can be obtained, from work shoe to dress shoe to sneaker to cowboy boot. Worker acceptance of safety shoes can be greatly improved if the individual is permitted to choose the shoe or boot as long as it is consistent with job requirements.

When one considers the above and the great number of different shoe sizes, it is readily apparent that the worker must obtain the shoes from a true safety shoe store. While some large employers operate successful safety shoe stores, it is generally not fruitful to carry two or three shoe styles of various sizes in the company stores. Many employers have found it more advantageous to enter into arrangements with shoemobiles (traveling shoe stores) or shoe stores located near their facility. In most communities safety shoes are also available from regular shoe stores, and some employers permit the individual to select the place of purchase as well as the style of the shoe. Whatever the case may be, a safety shoe program will only be successful when the shoes look good and feel good to the worker.

4.2 Chemical Protection

Footwear for protection from chemicals generally is contemplated in cases where minor spills may occur or where puddles may be present. Where more serious exposures are possible, boots may be used in conjunction with protective suits.

The rock-bottom minimum protection required in any given area where chemicals are handled are closed, solid shoes. This may be obvious to the safety or hygiene professional, but enforcement of such a rule may require major efforts in academic environments and in places where chemical use is minimal. The use of canvas shoes in particular should be discouraged since they tend to soak up chemicals and keep them in intimate contact with the skin.

Where very small spills and puddles of relatively benign chemicals may be encountered, adequate protection can often be obtained through properly selected safety shoes. Various manufacturers provide soles that are resistant to attack by hydrocarbons, preventing rashes caused by oil-soaked soles and providing good slip resistance.

When serious spills are possible or where routine exposures to chemicals underfoot are encountered, protective boots are normally required. Boots are available from ankle height to hip height as primary footwear, with steel toe box and steel insole or as cover boots over safety shoes, and they are obtainable in natural or synthetic rubber and in a variety of plastic materials. Most of these boots are intended for work in wet environments—outdoor work, food processing, mining, firefighting, etc. However, some of the boots are made of oil resistant or acid resistant formulations. Data pertaining to resistance to various other chemicals are generally not available for these boots, and the industrial hygienist may need to compare the formulations of boot material to glove material, or it may even be necessary to subject boots to chemical penetration tests.

The question of breakthrough and penetration is far less critical for most boot applications than for gloves. The worker may handle chemicals or chemically coated parts routinely, but exposure to the foot is usually accidental or short-term. Further, boots are much thicker than gloves, hence, even a relatively poor formulation may withstand occasional exposures satisfactorily.

As with hand and arm protection, all foot protective wear that becomes damaged and does not provide the protection intended should be replaced. Also, should the breakthrough occur in the protective wear while in use, the affected skin should be immediately decontaminated.

4.3 Heat and Cold Protection

Safety shoes and boots are available with various degrees of thermal insulation built in to meet all conceivable requirements for cold environments. Finding the protection is not a problem, rather a thorough search to locate all applications for insulated boots may be appropriate. Work in snow and slush clearly requires thermal boots, but so does any work where the feet are constantly exposed to a cold environment. Thermal boots may be required in refrigeration plants, environmental chambers, and where cold water is present, even on the hottest summer day.

Protection against heat is less frequently required, but equally important. Heat may come from hot surfaces underfoot, or the worker may be exposed to radiant heat and splash from molten substances. Walking on hot surfaces is most commonly associated with asphalt paving work, but also occurs in various other industries with hot processes. Protection is provided through wooden sandals (pavers' sandals) that fasten to the worker's shoe or through wooden soled overshoes made of fire resistant material. These permit the worker to function on fairly hot surfaces. Protective shoes are not available for work on surfaces that are hot enough to char the wood, but then work on such surfaces would approach the hazard of an entry into a fire. (Wooden-soled sandals, incidentally, can also be used for work on surfaces with broken glass or other sharp objects to prevent rapid deterioration of shoes and foot lacerations.)

Some protection against radiant heat and splashes of molten metal can also be obtained through wooden-soled asbestos overshoes and fire resistant overshoes. The best protection, however, is provided by foundry shoes. Typically they have a heat-resistant or wooden sole for protection from heat underfoot and a snug fit at the ankle and a closed construction of the uppers to prevent entry of molten substances into the shoe. Once the shoe is struck by molten metal or glass, injury can be lessened if the shoe is quickly removed to lessen exposure of the foot to the heat. Hence, foundry shoes are also constructed to permit quick removal.

4.4 Spats, Leggings, and Chaps

4.4.1 Hot Processes

For many hot processes, it is often necessary to provide protection for the lower body against radiant heat and molten material or glowing slag. Examples of such operations include foundries, molten glass work, furnace repair, torch cutting of metals, and many more. This protection is often provided through the use of spats, leggings, and chaps. These garments may be made of fire-retardant or fireproof fabric (including asbestos) or of chrome leather, and they are available with or without aluminized reflective coatings. In addition, the pieces may be reinforced with fiberboard or other material to provide impact protection.

Since there are no standards covering these garments, they may be of any design and construction the manufacturer chooses. Traditionally, chrome leather has been used, and this is still one of the more frequently used materials for moderate heat and hot metal splash. Leather is often preferred for its durability and abrasion resistance. Similar protection can also be obtained through garments made of fire-retardant fabric, fire resistant synthetics or asbestos. For protection against radiant heat, the garments are provided with an aluminized coating. For radiant heat only, light-weight construction is preferred. When protection must also be provided against molten metal, sturdy construction with additional insulation is preferred.

In the case of the foundry worker, spats cover the ankles, the instep, and the uppers of the foundry shoes. Leggings cover the legs from the knees down and extend over the spats. Chaps cover his upper legs and extend over the top of the leggings. (Aprons are frequently used instead of chaps.) All pieces are carefully layered and overlapping to shed molten metal like roof tiles shed water. Each piece is separately fastened to the leg to stay in place and to prevent gaping holes during body movement. Similar protectors may cover the arms and torso, and the face would be protected by a reflective or wire mesh faceshield over safety glasses. The "industrial knight" in full armor!

These sectional protectors provide several advantages over full protective suits. The individual pieces can be fairly heavy and rigid. Fastening each piece individually to the legs keeps them in place and permits better body movement than would be possible with one-piece trousers. An individual protector can also quickly be removed should this become necessary after a molten metal splash.

4.4.2   Other Applications

Special protectors are also available for other purposes. One of the primary applications is for impact protection in the shin area. Reinforced leggings may be used for heavy duty applications, or shin guards may be worn discreetly under the trousers by security personnel for crowd control.

For outdoor workers it may be necessary to provide protection against snake bite. Snake-proof boots are available, but adequate protection can also be obtained through the use of special leggings worn over sturdy work boots. Snake bites generally occur between the ankle and the knee, hence, there is little concern about bites to the foot. Snake-proof leggings typically are of multilayer construction, incorporating at least one layer of tightly woven wire mesh that cannot be penetrated very far by the fangs. Underlying layers prevent penetration to the skin.

Special chaps for use in conjunction with chain saw work are also coming into wide use. Made of ballistic nylon, these chaps will not prevent the saw from cutting through, but the nylon will quickly bind up the chain and keep the damage to a minimum.

5   WHOLE BODY PROTECTION

5.1   Introduction

The principles of whole body protection are the same as those of the component parts (e.g., gloves). It is even more important to consider the possibility of using engineering controls as opposed to whole body personal protective devices, as the area of exposure to the hazard is now greatly increased; hence, the possibility of an error is proportionately increased. Recognizing this possibility, there are still some situations during emergencies and nonroutine operations where whole body protection is mandatory.

5.2 Chemical/Radioactive Material Protection

Like gloves, the objective of whole body protection is to separate the person from a contaminating or hazardous material. This may be simply a suit of work clothes or coveralls or a lab coat worn over street clothes to protect the worker's personal clothing from nonhazardous dirt and grease. The protective clothing may also be used to separate the person from chemical or radioactive materials ranging from simple hazards which can cause dermatitis (i.e., fiberglass insulation) to very toxic systemic poisons (i.e., beryllium, plutonium, mercury).

Again, the principles of skin contamination, skin permeation, and skin penetration are the governing factors, and whole body protective devices must be selected according to design, material of construction, and comfort.

For contaminating materials of greater toxicity, primarily dusts and liquids, suits made of materials impervious to the contaminate are used. These may be coveralls or separate shirts and pants. Coveralls or shirts are available with hoods, which can be taped to full-face respirators, while the cuffs of the sleeves and pants are taped to gloves and booties, affording a complete sealing of the wearer's skin against the outside contaminant. Plastic, cloth, or other synthetic materials are available in both a reusable or disposable form. With suit materials which are impermeable to water, the wearer can be hosed down to effect a decontamination before removing the clothing. Caution is advised with the use of clothing made of materials impermeable to liquids in that air circulation is also restricted and body heat build-up can occur quickly, even on cool days. Hence, the length of time in which the wearer can work is limited.

When using tape to seal the cuffs, hood, and buttoning or zippered portions of a suit or shirt and pants is required, tape selection and method of taping also is important. The tape itself must be impervious to the contaminant and must have a glue which strongly adheres to the materials it must seal. The nuclear industry, in particular, has found that there are tricks in properly taping protective clothing. Radiation protection technicians must be trained to become adept in properly dressing radiation workers where anticontamination clothing is used.

For gases and vapors and very highly toxic or radioactive dusts or liquids, supplied-air suits are usually used. These suits are also useful for protection against high concentrations of skin permeating contaminants, as the suits are at positive pressure with respect to the ambient air. Thus, if there is a leak in a tape seal, the positive pressure will prevent inward leakage of contaminated air. Unfortunately, there is no NIOSH approval schedule for supplied-air suits. Users must, therefore, rely on the manufacturer to provide good merchandise. Note, however, that the American Society for Testing and Materials (ASTM) has the F-23 Chemical Protective Clothing Committee which is in the process of developing a series of standards for testing chemical protective clothing.

Because of high usage of supplied-air suits at Department of Energy (DOE) nuclear facilities, the Los Alamos National Laboratory was asked to develop a suit testing procedure (63). This procedure could be used by non-DOE facilities to test supplied-air suits. During the development of the Los Alamos test procedure, the researchers found that some suits could go negative when the wearer made certain movements, such as bending over. This bellows effect would force air out of the suit through a leak during the movement, then suck outside air back in when the wearer straightened up. They also found that the donning methods of suits and blouses and pants minimized or maximized the bellows effect. Thus, donning procedures are important in suit use.

Due to lack of standards regarding supplied-air suits, the International Organization for Standardization (IOS) is developing a specification. While only in draft form at this writing, users should find this specification useful in the near future (64).

Various specialized suits have been developed for some specific jobs. These may range from the "moon" suits used by the astronauts to the NASA-developed environmental suits such as those used by rocket fuel handlers to butyl rubber suits permitting use of self-containing breathing apparatus for cleaning up chemical spills. For those occupations with a large enough demand, the specialized suits are available commercially. For those with a smaller demand, the user must develop and have the suits made by a firm specializing in suit construction. Studies on suit permeation are available in the literature to give guidance in a few instances (65, 66).

5.3 Flame, Heat, and Hot Metal

Leather, asbestos, and aluminized fabrics are the most common types of clothing for protection against flame, heat, and hot metals. Leather is good for heat and hot metal splashes for welding operations, and also provides limited impact, UV, and IR protection. The design should be such that no fastenings can gape during body movements. There also should be no cuffs, pockets without flaps, or other projections to catch and hold hot metal (62). Specially treated asbestos clothing is available which is impervious to metal splashes up to 1650°C and aluminized fabrics for hot operations including fire fighting for up to 1090°C. (67).

The subject of clothing for firefighters is treated extensively by various publications of the National Fire Protection Association, and will not be attempted here. However, a mention of the classification of the 4 types of suits worn when fire is present may have application in the industrial setting. These are:

Farthest away—turnout clothing
Closer in—approach suits
Next to source—proximity suits
Into total flame—fire entry suits

Some precautions for heat and flame resistant clothing noted by the National Safety Council (5.5) are worth reiterating. These are: (1) to always follow manufacturer's recommendations when laundering or cleaning flame-retardant clothing; (2) avoid cleaning with organic solvents and; (3) never use chlorine bleaches with flame-retardant cotton clothing. The Council also recommends that vacuum cleaning be used to remove excess dust and dirt from any clothing, but if compressed air is used, the air pressure should not exceed 200 kPa (30 psi).

5.4 Cold

Protection from cold temperatures may be necessary in some occupations, such as cold storage workers or outdoor construction personnel. Many types of heavy garments are available, including insulated coats and underwear.

5.5 Miscellaneous

While the above categories cover the most widely used whole body personal protective clothing, there are specialized applications which should be mentioned.

5.5.1 Impact

As indicated previously, leather clothing is commonly used for light impact protection. For heavier impacts, fiberglass bats are sometimes inserted into flaps in leather aprons or leggings, while metal studded leather or metal mesh is also available. Heavy duty plastic and aluminum breast protectors and jock straps may be used for additional impact protection.

5.5.2 Radiation

Leaded rubber, plastic, and fiberglass clothing is available for low level X-ray or gamma radiation protection.

5.5.3 High Visibility

For outdoor workers who may be exposed to traffic hazards, clothing that is highly visible both day and night is available. Bright colors, usually reds or oranges, are used most often in the daytime, while fluorescent or yellow clothes are used at night.

6 SAFETY BELTS AND HARNESSES

Safety belts and harnesses are used to secure workers in high places, to arrest falls, to lower workers, and to rescue workers. These devices are generally part of a wider program to assure safety for work at high places, including barriers, railings, and safety nets. A brief review is presented here since some of the devices have applications in tank entry situations and other cases of interest to the industrial hygienist. For further details the reader is referred to the cited standards and to the review by Ellis (68).

6.1 ANSI A10.14

The principal standard governing these devices is ANSI A10.14-1975 "American National Standard Requirements for Safety Belts, Harnesses, Lanyards, Lifelines, and Droplines for Construction and Industrial Use" (69). Additional regulations are contained in the OSHA standards, in particular in the Construction Regulations. Special rules exist in some states and should also be consulted.

While the ANSI Standard recognizes four classes of belts and harnesses, the requirements for strength, testing, and tie-off are similar. All fall protection must be secured to a fixed anchorage point capable of withstanding a minimum static load of 5,400 pounds. Ellis (68) notes that this probably was specified because it is the breaking strength of $\frac{3}{4}$ in. first grade manila rope. There may be vertical droplines or horizontal lifelines attached to fixed anchorage points, provided they are capable of withstanding the same 5,400-pound loading at the center. Attached to these lines is the belt or harness, either directly or through a six-foot lanyard. Connection may be made through special hooks or rope grabbing devices that can be moved on the lifelines.

Safety belts or harnesses and lanyards are tested as a system. A 250 dummy torso is dropped through a 6-foot free fall, and the system must survive this test for three iterations without failure. Construction materials and methods for webbing, hardware and lanyards are closely specified, and tests for individual components are also detailed.

6.1.1 Class I Body Belts

Class I belts are commonly referred to as safety belts, body belts, or work belts. These belts are primarily intended for use as restraints to keep the worker from falling in the first place. The preferred body belt comes with a lanyard as an integral part of the belt; this assures that the belt/lanyard combination was tested and is safe. Other belts come with two D rings at opposing sides to facilitate connection of a lanyard.

Body belts can be used to save workers from very limited falls only. The ANSI standard specifies that they may not be used in situations where the stopping force is more than 10X gravity. Ellis (68) and McElroy (6) both advise a free fall for the worker of not more than two feet. This requires the worker to keep the lanyard properly adjusted at all times. Additional freedom of movement can be obtained through the use of Class III harnesses and/or shock absorbers.

Body belts are not suitable for raising or lowering workers or for rescue.

6.1.2 Class II Chest Harnesses

Class II chest harnesses are perhaps of the greatest interest to the industrial hygienist since they are specifically designed for use in situations where there are no vertical free fall hazards "and for retrieval purposes, such as removal of a person from a tank or a bin" (69). The Class II harness consists of a strap that fits around the rib cage and is held in place with shoulder straps. The lanyard or safety line attaches to a D ring high in the center of the back. Should the worker lose consciousness in a tank or otherwise become helpless, he can be lifted out by the harness in a fairly vertical position. If the worker needs horizontal extraction, the harness will also permit this while presenting a small cross-section, enabling the rescuers to pull the victim through close spaces and a narrow hatch.

Since the Class II harness is not intended for free fall protection, the design and the ANSI specifications do not make any provisions for this. The harness would probably survive in a free fall situation as well as a Class I belt, but it is fastened to the body in an unsuitable manner and location for this purpose. Hence, Class II harnesses must not be used where free falls are possible.

In some cases wristlets are preferred for extraction of personnel from narrow spaces as described below.

6.1.3 Class III Harnesses

Class III fall protection consists of full body harnesses, and they are intended for maximum fall protection. The webbing is strapped to the torso and the upper thighs, and the entire harness is constructed to transmit the force of the stopping motion to the thighs, the seat, the chest, and the shoulders. The lanyard attaches to a D ring high in the center of the back, in approximately the same location as for the Class II harness.

Class III harnesses are designed for free falls up to six feet, and the harness/lanyard combination must not generate stopping forces in excess of $35 \times$ gravity. Again, the worker must carefully position the lanyard to assure that a free fall of more than six feet is precluded.

6.1.4 Class IV Suspension Belts

Class IV suspension belts are intended for raising and lowering the worker and for holding him suspended while he is working. These belts are often referred to as saddle belts, boatswain's chairs, or tree trimmer belts, and are available in a variety of designs to fit the worker's preference and the needs of the job. Some are double belts, where one of the belts is used as a seat; others have wooden seats built in; still others are of a parachute harness design suitable for raising and lowering the worker vertically.

Depending on the design, Class IV harnesses may be suitable for personnel retrieval from tanks. Again, however, they are not designed to arrest falls and may not be used in such applications.

6.2   Shock Absorption and Prevention

The limiting factor in preventing injury to the worker is not really in the distance he falls but in the deceleration that the body undergoes when the fall is arrested. When reliance is placed exclusively on lanyards and belts or harnesses, this requires that the worker monitor and adjust his attachment point constantly. Two examples of other strategies are given.

The first is the self-retracting lifeline. This device is mounted some distance above the worker's head and constantly pays out or takes up the slack in the lifeline that is attached to the worker's harness. In case of fall, the rapid spooling out of the line causes a centrifugal lock to engage, and the descent is halted virtually immediately. This device gives much greater freedom of movement to the worker and minimizes impact loading. Up to 50 feet of lifeline can be dispensed from some devices. It is still important that the worker monitor his attachment point carefully—it should be as near overhead as possible. If the cable extends too much in the horizontal direction, appreciable vertical drop is possible before the centrifugal lock will engage. Further, the worker may swing in pendulum fashion and may strike a wall or obstruction—hardly better than striking the ground.

The second strategy is to employ a shock absorber. Typically this is a lanyard with a web-tearing system that activates during the fall and spreads out the deceleration in time and distance. Additional clearance is required below the worker when such lanyard is used, since the lanyard will lengthen considerably when the shock absorber is activated.

6.3   Use Consideration

From the discussion to this point, it should be clear that fall protection requires careful engineering of lifelines and anchorages as well as knowledgeable, trained users. If the worker fails to adjust the belt or harness properly, or if the lanyard is improperly positioned, the worker may be subject to serious or fatal injuries. The worker should also be instructed in inspecting his fall protection gear daily for damage and wear, and at least twice a year the harness or belt must be inspected by a thoroughly trained individual; this semiannual inspection must be recorded on a tag attached to the gear (69).

Any time belts or harnesses and lanyards are subjected to impact loading as during a free fall, they are presumed to be damaged and must be destroyed. The gear is not necessarily damaged, but it is not possible to inspect the inner portions of ropes and webbing; hence, the integrity of the system cannot be assured. This is a general problem with ropes and other woven material. The effects of shock loading simply are not that well understood.

6.4   Other Belts and Harnesses

A variety of other specialized gear is available that is not addressed by ANSI A10.14. Of particular interest to the industrial hygienist should be wristlets.

Wristlets are specifically designed to permit rescue extraction of personnel through narrow hatches and close spaces. Wristlets are attached to both wrists or lower forearms and are connected to a rope yoke. During raising or lowering of the worker the arms are fully extended above the head, and the body presents the smallest possible cross-section. This presents two advantages over the Class II harness. As already mentioned, it permits maneuvering through smaller openings. But it also puts the body in a streamlined position where it is less likely that the shoulders will be bruised on obstructions. Where obstructions exist, the unconscious victim is also less likely to hang up on obstructions that might prevent rescue.

Other fall protection equipment that the hygienist should be aware of but that is not covered here are pole and tree climbing belts, ladder safety devices [covered in ANSI A14.3 "Safety Requirements for Fixed Ladders" (70)], window cleaner belts [ANSI A39.1 "Safety Requirements for Window Cleaning" (71)], catch platforms, and safety nets [ANSI A10.11 "Safety Nets Used During Construction, Repair, and Demolition Operations" (72)].

REFERENCES

1. 29 CFR 1910 "Occupational Safety and Health Standard," Subpart I—Personal Protection Equipment.

2. "Safety Requirements for Industrial Head Protection," ANSI Z89.1-1969, American National Standards Institute, New York, 1969.

3. "Safety Requirements for Industrial Protective Helmets for Electrical Workers, Class B," ANSI Z89.2-1971, American National Standards Institute, New York, 1971.

4. "Structural Fire Fighters' Helmets," ANSI/NFPA 1972-1979, American National Standards Institute, Boston, 1979.

5. "Safety Hats," Data Sheet 561, Revision A (Extensive), National Safety Council, Chicago, 1974.

6. F. E. McElroy, Ed., *Accident Prevention Manual for Industrial Operations-Administration and Programs*, 8th ed., National Safety Council, Chicago, 1981.

7. Part 1910 (Amended), Paragraphs (C) through (P) and Appendices A through I of 29 CFR 1910.95, Federal Register, Vol. 48, No. 46, Washington, D.C., March 8, 1983.

8. E. H. Berger, "Using the NRR to Estimate the Real World Performance of Hearing Protectors," *Sound and Vibration*, January, 1983.

9. Joseph Sataloff, M.D. and Robert, T., Sataloff, M.D., "Ear Protectors vs Intense Impact Noise," *Occupational Health and Safety*, August 1983.

10. Harry F. Schulte, "Personal Protective Devices," Chapter 26, *The Industrial Environment, It's Evaluation and Control*, U.S. Government Printing Office, Washington, D.C., 1973.

11. American Industrial Hygiene Association-American Conference of Governmental Industrial Hygienists, *Respiratory Protective Devices Manual*, Braun and Brumfield, Ann Arbor, Mich., 1963.

12. American National Standards Institute, American National Standard Practices for Respiratory Protection, ANSI Z88.2-1969, ANSI, New York, 1969.

13. B. J. Held, Respirator Testimony for Proposed Benzene Standard, Occupational Safety and Health Administration public hearing proceedings, July 20, 1977.

14. B. J. Held, Respirator Testimony for Proposed Cotton Dust Standard, Occupational Safety and Health Administration public hearing proceedings, April 6, 1977.

15. Bureau of Mines, "Respiratory Protective Apparatus, Self-contained Breathing Apparatus," Schedule 13E, Title 30, Code of Federal Regulations, Part II, 1968.

16. Robert Wells (a member of the 1910 Berkeley, California, Fire Department) and Robert Foley, Napa, Calif., private correspondance.

17. U.S. Bureau of Mines, "Respiratory Protective Apparatus, Gas Masks," Schedule 14F, Title 30, Code of Federal Regulations, Part 13, 1955.

18. U.S. Bureau of Mines, "Respiratory Protective Apparatus, Supplied-air Respirators," Schedule 19B, Title 30, Code of Federal Regulations, Part 12, 1955.

19. U.S. Bureau of Mines, "Respiratory Protective Apparatus, Filter-type Dust, Fume, and Mist Respirators," Schedule 21B, Title 30, Code of Federal Regulations, Part 14, 1965.

20. U.S. Bureau of Mines, "Respiratory Protective Apparatus, Nonemergency Gas Respirators (Chemical Cartridge Respirators, Including Paint Spray Respirators)," Schedule 23B, Title 30, Code of Federal Regulations, Part 14a, 1959.

21. U.S. Bureau of Mines–National Institute for Occupational Safety and Health, "Respiratory Protective Apparatus," Title 30, Code of Federal Regulations, Part II, March 10, 1972.

22. National Institute for Occupational Health and Safety–Occupational Safety and Health Administration, *Respirator Decision Logic*, latest edition, available from NIOSH or OSHA in computer printout form.

23. G. O. Nelson and C. A. Harder, "Respirator Cartridge Efficiency Studies: V. Effect of Solvent Vapor," *Am. Ind. Hyg. Assoc. J.*, **35,** 391 (1974).

24. E. C. Hyatt et al., "Respirator Efficiency Measurement Using Quantitative DOP Man Tests," *Am. Ind. Hyg. Assoc. J.*, **33:** 10 (1972).

25. A. L. Hack et al., "Selection of Respirator Test Panels Representative of U.S. Adult Facial Dimensions," LA-5488, Los Alamos Scientific Laboratory, Los Alamos, N.M., 1973.

26. P. L. Lowry et al., "Respirator Studies for the National Institute for Occupational Safety and Health," LA-6722-PR, Los Alamos Scientific Laboratory, Los Alamos, N.M., 1977.

27. J. T. McConville, "Ethnic Variability and Respirator Sizing," American Industrial Hygiene Association Conference, Paper 128, 1973.

28. E. C. Hyatt, "Respirator Protection Factors," LA-6084-MS Los Alamos Scientific Laboratory, Los Alamos, N.M., 1977.

29. Mine Safety Appliances Co., "Key Elements of a Sound Respiratory Protection Program," Bulletin 1000-16, MSA, Pittsburgh, 1977.

30. B. J. Held et al., "Respirator Studies for the National Institute for Occupational Safety and Health," LA-5805-PR, Los Alamos Scientific Laboratory, Los Alamos, N.M., 1974.

31. D. D. Douglas et al., "Respirator Studies for the National Institute for Occupational Safety and Health," LA-6386-PR, Los Alamos Scientific Laboratory, Los Alamos, N.M., 1976.

32. B. J. Held and C. P. Richards, "Hazards Control Progress Report 53, July–December 1976, "UCRL-50007-76-2, Lawrence Livermore Laboratory, Livermore, Calif., 1977.

33. International Association of Fire Fighters, *1975 Annual Death and Injury Survey*, IAFF, Washington, D.C., 1976.

34. C. R. Toney and W. L. Barnhart, "Performance Evaluation of Respiratory Protective Equipment Used in Paint Spraying Operations," Department of Health, Education, and Welfare Publication (NIOSH) 76-177, National Institute for Occupational Safety and Health, Cincinnati, Ohio, 1976.

35. A. Blair, "Abrasive Blasting Respiratory Protection Practices," Department of Health, Education, and Welfare Publication (NIOSH), 74-104, National Institute for Occupational Safety and Health, Cincinnati, Ohio, 1976.

36. H. E. Harris and W. C. Di Sieghardt, "Factors Affecting Protection Obtained by Underground Coal Miners from Half-mask Dust Respirators," American Industrial Hygiene Association Conference, Paper 117, 1973.

37. H. E. Harris et al., "Respirator Usage and Effectiveness in Bituminous Coal Mining Operations," American Industrial Hygiene Association Conference, Paper 116, 1972.

38. T. O. Davis et al., "Respirator Studies for the ERDA Division of Safety, Standards, and Compliance," LA-6733-PR, Los Alamos Scientific Laboratory, Los Alamos, N.M., 1977.

39. National Aeronautics and Space Administration, Technology Utilization Program Report 1974, NASA, Washington, D.C., 1975.

40. B. J. Held and C. P. Richards, "Research and Development Needs in Firefighter's Breathing Protection," *Fireline Mag.* (San Francisco), April–May 1977.

41. J. A. Pritchard, *A Guide to Industrial Respiratory Protection*, Department of Health, Education and Welfare Publication (NIOSH) 76-189, National Institute for Occupational Safety and Health, Cincinnati, Ohio, 1976.

42. American National Standards Institute, American National Standard Practices for Respiratory Protection, ANSI Z88.2-1980, ANSI, New York, 1980.

43. "Mining Enforcement and Safety Administration, Metal and Nonmetal Mine Health and Safety Standards and Regulations," Title 30, Code of Federal Regulations, Parts 70 and 75, latest edition.

44. British Standards Institution, "Recommendations for the Selection, Use and Maintenance of Respiratory Protective Equipment," BS 4275, BSI, London, 1974.

45. J. L. Caplin, B. J. Held, and R. J. Catlin, *Manual of Respiratory Protection Against Airborne Radioactive Materials*, NUREG-0041, U.S. Nuclear Regulatory Commission, Washington, D.C., 1976.

46. W. E. Ruch and B. J. Held, *Respiratory Protection OSHA and the Small Businessman*, Ann Arbor Science Publishers, Ann Arbor, Mich., 1975.

47. "Practice for Occupational and Educational Eye and Face Protection," ANSI Z87.1-1979, American National Standards Institute, New York, 1979.

48. T. Dunham, "Occupational Safety Research Specifically Related to Personal Protection, A Symposium," Department of Health, Education and Welfare Publication (NIOSH) 75-143, National Institute for Occupational Safety and Health, Cincinatti, Ohio, 1975, p. 65.

49. American Conference of Governmental Industrial Hygienists, *Threshold Limit Values for Chemical Substances and Physical Agents in the Work Environment with Intended Changes for 1982*, ACGIH, Cincinnati, Ohio, 1982.

50. D. Sliney and M. Wolbarsht, *Safety with Lasers and Other Optical Sources*, Plenum Press, New York, 1980.

51. "American National Standard for the Safe Use of Lasers," ANSI Z136.1-1980, American National Standards Institute, New York, 1980.

52. Donald J. Birmingham, in G. D. and F. E. Clayton, Eds., *Patty's Industrial Hygiene and Toxicology*, 3rd rev., ed., Vol. I, John Wiley & Sons, New York, 1978.

53. Larry L. Hipp, in J. B. Olishifski, Ed., *Fundamentals of Industrial Hygiene*, 2nd ed., National Safety Council, Chicago, Ill., 1979.

54. G. O. Nelson et al., *Glove Permeation by Organic Solvents*, AIHA Journal (42) 3/81, 217 Akron, Ohio, 1981.

55. A. M. Best Company, *Best's Safety Directory*, Vol. I, 1980, A. M. Best Co., Oldwick, N.J., 1979.

56. National Institute for Occupational Safety and Health, *A Report on the Performance of Lineman's Rubber Insulating Gloves*, NIOSH 77-196, Cincinnati, Ohio, 1977.

57. National Institute for Occupational Safety and Health, *The Development of Criteria for Fire Fighters' Gloves*, NIOSH 77-134A and 77-134B, Cincinnati, Ohio, 1977.

58. "American National Standard for Men's Safety-Toe Footwear," ANSI Z41.1-1967 (R1972), American National Standards Institute, New York, 1972.

59. American National Standard for Metatarsal Safety-Toe Footwear," ANSI Z41.2-1976, American National Standards Institute, New York, 1976.

60. "American National Standard for Conductive Safety-Toe Footwear," ANSI Z41.3-1976, American National Standards Institute, New York, 1976.

61. "Standard for Electrically Conductive Floorings," ANSI/UL 779-1980, Underwriters Laboratories, Inc., Northbrook, 1980.

62. "American National Standard for Electrical Hazard Safety-Toe Footwear," ANSI Z41.4-1976, American National Standards Institute, New York, 1976.

63. Wm. H. Revoir, et al., *Specifications and Test Procedures for Airline-Type Supplied-Air Suits*, Los Alamos National Laboratory, LA-5958-MS, Los Alamos, N.M., 1975.

64. International Organization for Standardization, *Clothing for Protection Against Radioactive Contamination*, Draft Standard of Working Group 6, Geneva, Switzerland, June 1982.

65. R. W. Weeks, Jr. and M. J. McLeod, *Permeation of Protective Garment Material by Liquid Benzene*, Los Alamos National Laboratory, LA-8164-MS, Los Alamos, N.M., 1979.

66. R. W. Weeks, Jr., and M. J. McLeod, *Permeation of Protective Garment Material by Liquid Halogenated Ethanes and a Polychlorinated Biphenyl*, Los Alamos National Laboratory, LA-8572-MS, Los Alamos, N.M., 1980.

67. National Safety Council, *Dress Right for Safety, General Body Protection*, National Safety News, Chicago, Ill., March 1983.

68. J. Nigel Ellis, *The Basics of Fall Protection* in "Best's Safety Directory—1980", A.M. Best Company, Inc., Oldwick, 1979.

69. "American National Standard Requirements for Safety Belts, Harnesses, Lanyards, Lifelines, and Droplines for Construction and Industrial Use," ANSI A10.14-1975, American National Standards Institute, New York, 1975.

70. "American National Standard Safety Requirements for Fixed Ladders," ANSI A14.3-1974, American National Standards Institute, New York, 1974.

71. "American National Standard Safety Requirements for Window Cleaning," ANSI A39.1-1969, American National Standards Institute, New York, 1969.

72. "American National Standard for Safety Nets Used During Construction, Repair, and Demolition Operations," ANSI A10.11-1979, American National Standards Institute, New York, 1979.

CHAPTER FOURTEEN

Job Safety and Health Law

MARTHA HARTLE MUNSCH, J.D.

The principal legislation relating to job safety and health is the federal Occupational Safety and Health Act of 1970 (OSHA). However, certain industries and/or portions of industries are subject to regulation by federal statutes other than OSHA. In addition, OSHA does not entirely preclude regulation of job safety and health by the states or their political subdivisions. Nevertheless, OSHA is clearly the most comprehensive legislative directive relating to workplace safety and health; accordingly, this chapter and the next focus primarily on developments and requirements pursuant to this act.

1 LEGISLATIVE HISTORY AND BACKGROUND OF OSHA

The Occupational Safety and Health Act of 1970 was enacted by Congress on December 17, 1970, and became effective on April 28, 1971. It represents the first job safety and health law of nationwide scope (1). Passage of the act was preceded by a dramatic and bitter labor-management political fight. The legislative history of OSHA is summarized in *The Job Safety and Health Act of 1970* (1, pp. 13–21).

Congress enacted OSHA for the declared purpose of assuring "so far as possible every working man and woman in the Nation safe and healthful working conditions" [§ 2(b)]. The act is intended to *prevent* work-related injury, illness, and death.

1.1 Agencies Responsible for Implementing and Enforcing OSHA

The Department of Labor is responsible for implementing OSHA. On the date the act became effective, the Department of Labor created the Occupational

Safety and Health Administration (OSH Administration or OSHA) to carry out such responsibilities. The OSH Administration is headed by an assistant secretary of labor for occupational safety and health and is responsible, among other things, for promulgating rules and regulations, setting health and safety standards, evaluating and approving state plans, and overseeing enforcement of the act (2).

Section 12(a) of the act establishes the Occupational Safety and Health Review Commission (OSAHRC) as an independent agency to adjudicate enforcement actions brought by the Secretary of Labor. The commission is composed of three members appointed by the President for six-year terms. The chairman of the commission is authorized to appoint such administrative law judges as he deems necessary to assist in the work of the commission [§12(e)].

Sections 20 and 21 of the act give the Secretary of Health, Education and Welfare (HEW) [now Health and Human Services (HHS)] broad authority to conduct experimental research relating to occupational safety and health, to develop criteria for and recommend safety and health standards, and to conduct educational and training programs (3). Section 22 establishes the National Institute for Occupational Safety and Health (NIOSH) to perform the functions of the Secretary of Health and Human Services under Sections 20 and 21.

The act specifically directs NIOSH to develop criteria documents that describe safe levels of exposure to toxic materials and harmful physical agents and to forward recommended standards for such substances to the Secretary of Labor (4). The act also directs NIOSH to publish at least annually a list of all known toxic substances and the concentrations at which such toxicity is known to occur [§ 20(a)(6)].

Section 7(a) establishes a National Advisory Committee on Occupational Safety and Health (NACOSH), whose basic functions are to advise, consult with and make recommendations to the Secretary of Labor and the Secretary of Health and Human Services on matters relating to the administration of the act. NACOSH consists of 12 members who represent management, labor, occupational safety and occupational health professions, and the public. The members are appointed by the Secretary of Labor, although four members are to be designated by the Secretary of Health and Human Services.

The Secretary of Labor is authorized by Section 7(b) to appoint other advisory committees to assist him in the formulation of standards under Section 6. For example, ad hoc advisory committees have been used to assist in developing standards for exposure to asbestos and coke oven emissions. The Secretary of Labor has also appointed various standing advisory committees (5).

1.2 Scope of OSHA's Coverage

The Occupational Safety and Health Act of 1970 applies to every private employer engaged in a business affecting commerce, regardless of the number of employees. It applies with respect to employment performed in a workplace in any of the 50 states, the District of Columbia, Puerto Rico, the Virgin Islands,

American Samoa, Guam, the Trust Territory of the Pacific Islands, Wake Island, the Outer Continental Shelf Lands, Johnston Island, and the Canal Zone [§§ 3(5) and 4(a)].

The act's definition of "employer" does not include the states, political subdivisions of the states, or the United States. However, Section 19 directs the head of each federal agency to establish and maintain an effective and comprehensive occupational safety and health program that is consistent with the standards required of private employers (6).

2 REGULATION OF JOB SAFETY AND HEALTH BY FEDERAL STATUTES OTHER THAN OSHA

Section 4(b)(1) of OSHA states that nothing in the act shall apply to working conditions of employees with respect to which other federal agencies exercise statutory authority to prescribe or enforce standards or regulations affecting occupational safety or health. Thus federal agencies other than the OSH Administration that are authorized by statute to regulate employee safety and health can continue to do so after the effective date of OSHA; in fact, the *exercise* of such authority preempts OSHA from regulating with respect to such working conditions.

Although Section 4(b)(1) seems to be self-defining, it has generated a tremendous volume of litigation. Three major interpretive questions have been raised:

1. What constitutes a sufficient exercise of regulatory authority to preempt OSHA regulation?
2. Does the exercise of authority by another federal agency in substantial areas of employee safety exempt the entire industry from OSHA standards?
3. Must the other federal agency's motivation in acting have been to protect workers?

It appears to be well settled that the mere existence of statutory authority to regulate safety or health is not sufficient to oust OSHA's regulatory scheme; some exercise of that authority is necessary. Furthermore, at least three federal courts of appeals have taken the position that speculative pronouncements of proposed regulations by a federal agency are not sufficient to warrant preemption of OSHA standards. Rather, it has been held that Section 4(b)(1) requires a concrete exercise of statutory authority (7).

The same courts of appeals have also rejected the notion that the exercise of statutory authority by another federal agency creates an industrywide exemption from OSHA regulations. Rather, the courts have agreed that the term "working conditions" in Section 4(b)(1) refers to something more limited than every aspect of an entire industry. Ambiguity remains, however, with respect to the scope of the displacing effect of another agency's regulation of a working condition.

For example, in *Southern Pacific Transportation Company* (7), the Fifth Circuit explained that the term "working conditions" has a technical meaning in the language of industrial relations; it encompasses both a worker's surroundings and the hazards incident to the work. The court stated that the displacing effect of Section 4(b)(1) would depend primarily upon the agency's articulation of its regulations (8):

Section 4(b)(1) means that any FRA [Federal Railroad Administration] exercise directed at a working condition—defined either in terms of a "surrounding" or a "hazard"— displaces OSHA coverage of that working condition. Thus comprehensive FRA treatment of the general problem of railroad fire protection will displace all OSHA regulations on fire protection, even if the FRA activity does not encompass every detail of the OSHA fire protection standards, but FRA regulation of portable fire extinguishers will not displace OSHA standards on fire alarm signaling systems.

The Fourth Circuit defined "working conditions" as "the environmental area in which an employee customarily goes about his daily tasks." The court in *Southern Railway Company* (7, 539 F.2d at 339; 3 OSHC at 1943) explained that OSHA would be displaced when another federal agency had exercised its statutory authority to prescribe standards affecting occupational safety or health for such an area (9).

The courts of appeals seem to indicate, at least implicitly, that regulation of a working condition by another federal agency need not be as effective or as stringent as an OSHA standard to preempt the OSHA standard [see *Southern Pacific Transportation Company* (7), 539 F.2d at 391-92; 4 OSHC at 1695-96]. But it remains unclear whether a decision by another federal agency that a particular aspect of an industry should not be regulated at all would preempt or preclude OSHA regulation of that same aspect. Resolution of these and other issues involving the scope of the displacement effect under Section 4(b)(1) awaits future litigation or legislation.

Finally, the commission has held that to be cognizable under Section 4(b)(1), "a different statutory scheme and rules thereunder must have a policy or purpose that is consonant with that of the Occupational Safety and Health Act. That is, there must be a policy or purpose to include employees in the class of persons to be protected thereunder" (10). In *Organized Migrants in Community Action Inc. v. Brennan* (11), the United States Court of Appeals for the District of Columbia, although not deciding the issue, implicitly rejected the argument that preemption under Section 4(b)(1) exists only where the allegedly preempting statute was passed *primarily* for the protection of employees (12).

3 OVERVIEW OF FEDERAL REGULATORY SCHEMES OTHER THAN OSHA

The following material represents an overview of the major federal regulatory schemes other than OSHA that deal with or relate to job safety and health. The listing *is by no means exhaustive*, and employers are urged to consult specific statutory schemes in substantive areas relating to their respective industries.

3.1   Mine Safety and Health Legislation

Occupational safety and health matters with respect to the nation's mining industry are regulated pursuant to the Federal Mine Safety and Health Act of 1977, which became effective in March 1978 (13). The Department of Labor is responsible for enforcing that legislation. The Secretary of Labor has delegated its enforcement authority to the Mine Safety and Health Administration (MSHA).

Prior to the effective date of the 1977 act, safety and health matters with respect to the mining industry were covered by two separate statutes. The Metal and Non-Metallic Mine Safety Act of 1966 covered mines of all types other than coal mines. Safety and health matters concerning coal mines were regulated by the Coal Mine Health and Safety Act of 1969. The 1977 Mine Safety and Health Act is now the single mine safety and health law for all mining operations. The 1977 act directs the Secretary of Labor to, *inter alia*, "develop, promulgate, and revise as may be appropriate improved mandatory health or safety standards for the protection of life and prevention of injuries in coal or other mines" (14). The secretary is also empowered to enforce those standards.

3.2   Environmental Pesticide Control Act of 1972

The Federal Environmental Pesticide Control Act of 1972 (FEPCA) regulates the use of pesticides and makes misuse civilly and criminally punishable. The Court of Appeals for the District of Columbia has held that FEPCA authorizes the Environmental Protection Agency (EPA) to promulgate and enforce occupational health and safety standards with respect to farm workers' exposure to pesticides. The EPA has exercised that authority, and thus has preempted OSHA from regulating in that area [*Organized Migrants in Community Action* (11), (15)].

3.3   Federal Railroad Safety Act of 1970

The Federal Railroad Safety Act of 1970 authorizes the Federal Railroad Administration (FRA) within the Department of Transportation (DOT) to promulgate regulations for all areas of railroad safety, including employee safety. To date, however, DOT has not adopted railroad occupational safety standards for all railroad working conditions or workplaces (16). The Department of Labor (OSHA) retains jurisdiction over safety and health of railroad employees with respect to those "working conditions" for which DOT has not adopted standards (17).

3.4   Federal Aviation Act of 1958

The Federal Aviation Act of 1958, as amended, empowers the Federal Aviation Administration (FAA) within the Department of Transportation "to promote safety of flight of civil aircraft in air commerce" (18), as well as to "establish minimum safety standards for the operation of, airports that serve any scheduled or unscheduled passenger operation of air carrier aircraft designed for more than 30 passenger seats" (19). If the congressional mandate in that statute is deemed to include the safe working conditions of airline and/or airport employees, OSHA would be precluded from exercising its jurisdiction with respect to the working conditions regulated by the FAA. At least one administrative law judge has determined that the FAA's mandate encompasses the safe working conditions of airline ground crews when performing aircraft maintenance (20).

3.5 Hazardous Materials Transportation Act

The Hazardous Materials Transportation Act (HMTA) authorizes the Secretary of Transportation to issue regulations governing any safety aspect of the transportation of materials designated as hazardous by the Secretary (21). The act encompasses shipments by rail, air, water, and highway. If worker safety is deemed to be a purpose of the HMTA, safety standards promulgated by DOT under the statute would trigger a preemption of OSHA jurisdiction with respect to the working conditions covered by such standards.

3.6 Natural Gas Pipeline Safety Act of 1968

The Natural Gas Pipeline Safety Act of 1968 (NGPSA) authorizes the Secretary of Transportation to establish minimum federal safety standards for pipeline facilities and the transportation of gas in commerce. In *Texas Eastern Transmission Corp.* (22), the Occupational Safety and Health Review Commission determined that the NGPSA was intended to affect occupational safety and health. Thus employers engaged in the transmission, sale, and storage of natural gas would be exempt from OSHA with respect to working conditions covered by DOT standards promulgated under NGPSA (23).

3.7 Federal Noise Control Act of 1972

Although a health and safety standard adopted pursuant to OSHA governs the level of noise to which a worker covered by the act may be exposed in the workplace (see Section 5.3), other federal statutes deal with noise abatement and control as well (24).

The first such enactment, a 1968 amendment to the Federal Aviation Act, required the Administrator of the FAA to include aircraft noise control as a factor in granting type certificates to aircraft under the act (25). To the extent that the Administrator of the FAA denies certification to an aircraft that produces noise in excess of the standards, or prohibits the operation of a certified aircraft in a manner that violates his regulations, the general environmental noise level in workplaces covered by OSHA and located adjacent to airports and landing patterns is correspondingly reduced.

The first attempt to deal with noise on a nationwide basis, however, was the federal Noise Control Act of 1972. The control strategy of that act is generally as follows: the Administrator of the EPA is required to develop and publish criteria with regard to noise, reflecting present scientific knowledge of the effects on public health and welfare that are to be expected from different quantities or qualities of noise. The administrator then must identify products or kinds of products that in his opinion are major sources of environmental noise, and publish noise emission regulations where it is feasible to limit the amount of noise produced by such products (26).

Under the act, the administrator is further charged with publishing regulations identifying "low noise emission products." Such products, once so designated, must thereafter be purchased by federal agencies in preference to substitute products, provided the "low noise emission product" costs no more than 125 percent of the price of the substitute.

As the Administrator of the EPA identifies more and more products as major sources of noise, and subjects those products to noise emission standards adopted under the Noise Control Act of 1972, the noise levels found in workplaces covered by OSHA and in which such products are used should decrease.

3.8 Federal Toxic Substances Control Act of 1976

The Toxic Substances Control Act of 1976 establishes a broad, nationwide program for the federal regulation of the manufacture and distribution of toxic substances (27). The act divides all "chemical substances" and "mixtures" into two categories: the old and the new. The Administrator of the EPA is charged under Section 8(b) with the gargantuan task of compiling and publishing in the *Federal Register* an inventory or list of all chemical substances manufactured or processed in the United States.

Manufacturers of substances that appear on that inventory are at liberty to continue to manufacture and distribute such substances unless the administrator by rule promulgated under Section 4 of the act first requires that a designated substance be tested and data from the tests be submitted to the EPA. If EPA thereafter makes a determination under Section 6 that the continued manufacture, processing, or distribution in commerce of the substance presents an unreasonable risk of injury to health or the environment, the administrator may either prohibit altogether the manufacture of the chemical substance or may impose restrictions (limitations on the quantity manufactured, the use to which the chemical may be put, the concentrations in which it may be used, the labels and warnings that must accompany its sale, etc.).

Section 5 of the act provides a different treatment with respect to a "new chemical substance" or a "significant new use" of a substance that appears on the Section 8(b) inventory. The manufacturer is not at liberty to commence manufacture or distribution of such a "new" substance but must first submit a notice to the administrator of his intention to manufacture such a new substance or to engage in a significant new use. Then testing data that relate to the toxicity and the effect on health and on the environment of the substance must be submitted. If the administrator does not act within 90 days, the manufacturer is at liberty to proceed with manufacture or distribution. During the initial period, however, the administrator may extend his time for action an additional 90 days. If during the original period (or its extension) the administrator believes that the information available to him is inadequate to make a reasoned finding that the proposed new substance or use does not present an unreasonable risk of injury to health or the environment, he may prohibit or limit the manufacture of the substance and obtain an injunction in court for that purpose. It would appear that in the absence of testing data submitted in compliance with Section 4, this injunction against manufacture or distribution of the new substance or use would continue indefinitely. If, however, the administrator finds, based on information before him, that the proposed new substance or use does present an unreasonable risk to health and safety, the administrator must proceed by means of the provisions of Section 6 to prohibit manufacture or to impose restrictive conditions (28).

The administrator of the EPA has promulgated a myriad of regulations implementing the Toxic Substances Control Act. These regulations can be found at 40 CFR Subchapter R, Parts 702-775. Besides detailing the procedural and chemical inventory requirements of the act, the regulations set forth guidelines for the manufacture, processing and distribution of polychlorinated biphenyls and fully halogenated chlorofluoroalkanes (29). Moreover, the regulations require local education agencies to identify friable asbestos-containing material in public and private school buildings (Part 763). The regulations also impose notification requirements on any person who disposes waste material containing tetrachlorodibenzo-p-dioxin (Part 775). It is likely that the Toxic Substances Control Act will continue to be a major weapon in the federal health, safety, and environmental arsenal, and significant developments under it should be expected (30).

3.9   Federal Consumer Product Safety Act

The Consumer Product Safety Act of 1972 was drafted to apply only to "consumer products." That term is defined in the act in a manner that serves to exclude most products destined principally for use in workplaces covered by OSHA. Nevertheless, the act promises to provide increased protection of the American worker from hazardous products that by their nature are "consumer products" within the meaning of the act, yet are frequently found in the workplace (31).

The act created a Consumer Product Safety Commission and empowered that agency to promulgate "consumer product safety standards" applicable to consumer products found by the commission to present an unreasonable risk of injury. Such standards may be performance standards, or they may require that products not be sold without adequate warnings or instruction. Where no feasible safety standard that could be promulgated would eliminate an unreasonable risk of injury, the commission is empowered to ban the consumer product altogether from interstate sale or distribution.

In addition to publishing safety standards, the commission is empowered to file suit and seek the seizure of a consumer product believed to be "imminently hazardous," regardless of whether the product in question is covered by already promulgated consumer product safety standards. The commission is also authorized to find, after hearing, that a consumer product presents a "substantial product hazard." In the event of such a finding, the commission may order the manufacturer, distributor, or retailer of the product to give public notice of that finding, and to repair, replace, or refund the purchase price of the product affected (32).

3.10 Hazardous Substances Act

The federal Hazardous Substances Act provides a mechanism by means of which the Consumer Product Safety Commission may find that a substance distributed in interstate commerce is "hazardous." After such a finding the commission may either impose packaging and labeling requirements to protect public health and safety or, in the cases of hazardous substances intended for the use of children or likely to be subject to access by children, or substances intended for household use, prohibit distribution altogether ("banned hazardous substance") (33). Insofar as safety in the American workplace is concerned, the effect of the act is that hazardous substances distributed in interstate commerce and utilized by the American worker will arrive safely packaged and accompanied by appropriate warnings.

3.11 The Atomic Energy Act of 1954 and Other Statutory Sources of Radiation Control

As discussed later (Section 5.3), the Department of Labor has published occupational health and safety standards regulating exposure to ionizing (i.e., alpha, beta, gamma, X-ray, neutron, etc.) radiation and nonionizing (i.e., radiofrequency, electromagnetic) radiation. However, the primary federal law regulating human exposure to radiation is not OSHA but rather the Atomic Energy Act of 1954, as amended. Exercising power under that statute, the Nuclear Regulatory Commission (NRC) has published "Standards for Protection Against Radiation" (34).

A detailed discussion of those regulations is beyond the scope of this chapter. The operation of the standards can be summarized briefly as follows, however. Any person holding a license issued under the Atomic Energy Act of 1954 and using "licensed material" (i.e., radioactive or radiation-emitting material) may not permit the exposure of individuals within a "restricted area" (i.e., an area in which radioactive materials are being used) to greater doses of radiation than are set forth in the regulations (35).

Although the primary thrust of the NRC regulations is to control ionizing radiation within the "restricted area," the NRC has also published regulations on permissible levels of radiation in unrestricted areas, in effluents discharged into unrestricted areas, and for the disposal of radioactive materials by release into sanitary sewerage systems (36).

The Administrator of the EPA, exercising authority under the Atomic Energy Act of 1954, which he acquired by means of the Reorganization Plan No. 3 of 1970, has also promulgated regulations limiting exposure of the general population to ionizing radiation produced during the operation of nuclear power plants licensed by the NRC (37).

There are additional federal agencies empowered to set standards for ionizing radiation control within areas under their jurisdiction. The Department of Labor, for example, has promulgated regulations regarding radiation exposure in underground mines (38). The Department of Labor has also issued radiation standards for uranium mining conducted under the Walsh-Healey Public Contracts Act (39).

Radiation generated by devices and products that are not governed by the Atomic Energy Act and licensed by the NRC is regulated by the Federal Radiation Control for Health and Safety Act of 1968 (40).

3.12 Outer Continental Shelf Lands Act

The Outer Continental Shelf Lands Act authorizes the head of the department in which the Coast Guard is operating to promulgate and enforce "such reasonable regulations with respect to lights and other warning devices, safety equipment, and other matters relating to the promotion of safety of life and property" on the lands and structures referred to in the act or on the adjacent waters (41). If worker safety is deemed to be within the mandate of this statute, OSHA jurisdiction may be preempted with respect to working conditions that are governed by standards issued by the Coast Guard pursuant to this act.

4 REGULATION OF JOB SAFETY AND HEALTH BY THE STATES

One of the primary factors that induced Congress to enact the Occupational Safety and Health Act was the failure of many of the states adequately to regulate workplace safety and health (42). In passing OSHA, Congress hoped to ensure at least a minimum level of regulation of the conditions experienced by workers throughout the country.

The Occupational Safety and Health Act of 1970 preempts state regulation of job safety and health with respect to matters which OSHA regulates, even when a state has a more stringent regulation with respect to a particular hazard (43). However, OSHA does not totally ban the states from developing and enforcing occupational safety and health standards. Pursuant to Section 18(b) of the act, a state may regain jurisdiction over development and enforcement of occupational safety and health standards by submitting to the federal government an effective state occupational safety and health plan. Final approval of a state plan can lead ultimately to exclusive authority by a state over the matters included in its plan.

The process of regaining jurisdiction over the regulation of occupational safety and health begins with the submission of a plan that sets forth specific procedures for ensuring workers' safety and health. According to the regulations of the Secretary of Labor, the states can submit either of two types of plan: a complete plan or a developmental plan.

A "complete" plan (44) is a plan that, upon submission, satisfies the criteria for plan approval set forth in Section 18(c) of the act, as well as certain additional criteria outlined by the Secretary of Labor in his administrative regulations (45). Complete plans are given "initial" approval by the Secretary of Labor upon submission. For at least 3 years following the "initial" approval, the Secretary of Labor will monitor the state plan to determine whether on the basis of the actual operations of the plan, the criteria set forth in Section 18(c) are being applied. If this determination [the "Section 18(e) determination"] is favorable, the state plan will be granted "final approval" and the state will regain exclusive jurisdiction with respect to any occupational safety or health issue covered by the state plan. Federal (i.e., OSHA) standards continue to apply to hazards not covered by the state program; thus state plans need not address all hazards, yet gaps in protection are avoided.

A "developmental" plan (46) is a plan that, upon submission, does not fully meet the criteria set forth in the statute or in the regulations. A developmental plan may receive initial approval upon submission, however, if the plan contains "satisfactory assurances" that the state will take the necessary steps to bring its program into conformity within 3 years following commencement of the plan's operation.

If the developmental plan satisfies all the statutory and administrative criteria within the 3-year "developmental period," the Secretary of Labor will so certify and will initiate an evaluation of the actual operations of the state plan for purposes of making a Section 18(e) determination. The evaluation must proceed for at least one year before such a determination can be made.

Plans that have received final approval will continue to be monitored and evaluated by the Secretary of Labor pursuant to Section 18(f) of the act, which authorizes the secretary to withdraw approval if a state fails to comply substantially with any provision of the state's plan.

Although a state does not regain exclusive jurisdiction over matters contained in its plan until the plan receives final approval, a state with initial plan approval may participate in the administration and enforcement of the act prior to final approval by satisfying the following four criteria (47):

1. The state must have enacted enabling legislation conforming to that specified in OSHA and the regulations.
2. The state plan must contain standards that are found to be at least as effective as the comparable federal standards.

3. The state plan must provide for a sufficient number of qualified personnel who will enforce the standards in accordance with the state's enabling legislation.

4. The plan's provisions for review of state citations and penalties (including the appointment of the reviewing authority and the promulgation of implementing regulations) must be in effect.

If the criteria above are met, the state plan is deemed to be "operational." Thereupon the federal government enters into an operational agreement with the state whereby the state is authorized to enforce safety and health standards under the state plan (47). During this period the act permits, but does not require, the federal government to retain enforcement activity in the state (48). Thus during this period an employer could be subject to enforcement activities by both the state and federal authorities. However, the secretary's regulations provide that once a plan (either complete or developmental) becomes "operational," the state will conduct all enforcement activity, including inspections in response to employee complaints, and accordingly, the federal enforcement activity will be reduced and the emphasis will be placed on monitoring state activity (47).

As of August 1, 1983, no states had submitted "complete" plans. However, the Secretary of Labor had certified the developmental plans of the following states: Alaska, California, Hawaii, Iowa, Kentucky, Maryland, Minnesota, North Carolina, South Carolina, Tennessee, Utah, and Vermont. Twelve additional states or territories had received initial approval of developmental plans as of that date: Arizona, Connecticut, Indiana, Michigan, Nevada, New Mexico, Oregon, Puerto Rico, Virgin Islands, Virginia, Washington, and Wyoming.

Thirteen states or territories plus the District of Columbia had submitted plans and were awaiting initial approval by the Secretary of Labor: Alabama, American Samoa, Arkansas, Delaware, Florida, Guam, Idaho, Massachusetts, Missouri, Oklahoma, Rhode Island, Texas, and West Virginia.

The following 12 states submitted plans at one time but have withdrawn them: Colorado, Georgia, Illinois, Maine, Mississippi, Montana, New Hampshire, New Jersey, New York, North Dakota, Pennsylvania, and Wisconsin. Five states (Kansas, Louisiana, Nebraska, Ohio, and South Dakota) have never submitted plans to the Department of Labor (49).

Most states that are presently operating approved and/or certified plans have adopted standards that are substantially similar, if not identical, to the federal standards. At least five state plans, however, contain certain provisions that vary from the federal standards and have been approved as being "at least as effective" as the federal standards. The states are California, Hawaii, Michigan, Oregon, and Washington (49). In some instances these states have adopted standards that are more stringent than the analogous federal standards. Employers who are operating in more than one state should be aware that they may have to deal with different regulations, different enforcement procedures, and perhaps different interpretations of similar standards for purposes of complying with the applicable occupational safety and health laws (50).

4.1 State Jurisdiction in Areas Regulated by Federal Legislation Other than OSHA

If state legislation regulates a job safety or health issue that OSHA does not cover, the state may continue to enforce its relevant standards unless other applicable federal law has preempted state enforcement.

In some areas a federal regulatory scheme permits concurrent federal and state regulation of job safety and health. For example, the Federal Mine Safety and Health Act of 1977 states that no state law, that was in effect on December 30, 1969, or may become effective thereafter, shall be superseded by any provisions of the federal mine act unless the state law is in conflict with the mine act. State laws and rules that provide for standards more stringent than those of the mine act are deemed not to be in conflict (51).

On the other hand, the Federal Railroad Safety Act of 1970 has essentially preempted the states' regulation of railroad safety and health except in cases of a state having a more stringent law because of the need to eliminate or reduce an essentially local safety hazard (52).

5 EMPLOYERS' DUTIES UNDER THE OCCUPATIONAL SAFETY AND HEALTH ACT OF 1970

A private employer's primary duties under the Occupational Safety and Health Act of 1970 are found in Section 5(a), which provides that each employer:

1. Shall furnish to each of his employees employment and a place of employment which are free from recognized hazards that are causing or are likely to cause death or serious physical harm to his employees; and
2. Shall comply with occupational safety and health standards promulgated under the act.

5.1 The General Duty Clause [§ 5(a)(1)]

The essential elements of the so-called general duty clause of OSHA are the following (53): (1) the employer must render the workplace "free" of hazards that arise out of conditions of the employment, (2) the hazards must be "recognized," and (3) the hazards must be causing or likely to cause death or serious physical harm (54).

5.1.1 Failure to Render Workplace "Free" of Hazard

It is fairly well settled that Congress did not intend to make employers strictly liable for the presence of unsafe or unhealthful conditions on the job. The employer's general duty must be an achievable one (55). Thus the term "free" has been interpreted by the courts and the commission to mean something less than absolutely free of hazards. Instead, the courts and the commission have held [*National Realty and Construction Company* (55)] that the employer has a duty to render the workplace free only of hazards that are preventable.

The determination of whether a hazard is preventable generally is made in the context of an enforcement proceeding under the act when an employer asserts the inability of preventing the hazard as an affirmative defense to a proposed citation. The employer often contends that: (1) compliance is impossible because either the technology does not exist to prevent the hazard or the cost of the technology is prohibitive, (2) compliance with a standard will result in a greater hazard to employees than would noncompliance, or (3) the hazard was created by an employee's misconduct that was so unusual that the employer could not reasonably prevent the existence of the hazard.

When technology does not exist to prevent a hazard, OSHA does not require prevention by shutting down the employer's operation. Rather, the Secretary of Labor must be able to show that "demonstrably feasible" measures would have materially reduced the hazard [*National Realty and Construction Company* (55)]. Similarly, it seems that measures that, even though technologically feasible, would have been so expensive as to bankrupt the employer, are not "demonstrably feasible." (For further discussion relating to feasibility, see Section 5.2.3 of this chapter.)

To establish the "greater hazard" defense, the employer must show that: (1) the hazards of compliance are greater than the hazards of noncompliance, (2) alternative means of protecting employees are unavailable, and (3) a variance either cannot be obtained or is inappropriate (56).

In addition, the courts and the commission have recognized that certain isolated or idiosyncratic acts by an employee that were not foreseeable by the employer could result in unpreventable hazards for which the employer should not be held liable. For example, in *National Realty and Construction Company*, the Court of Appeals stated (489 F.2d at 1266; 1 OSHC at 1427):

Hazardous conduct is not preventable if it is so idiosyncratic and implausible in motive or means that conscientious experts, familiar with the industry, would not take it into account in prescribing a safety program. Nor is misconduct preventable if its elimination would require methods of hiring, training, monitoring, or sanctioning workers which are either so untested or so expensive that safety experts would substantially concur in thinking the methods infeasible.

The court in *National Realty* emphasized, however, that an employer does have a duty to attempt to prevent hazardous conduct by employees. Thus the employer must adopt demonstrably feasible measures concerning the hiring, training, supervising, and sanctioning of employees, to reduce materially the likelihood of employee misconduct (57).

5.1.2   The Hazard Must Be a "Recognized" Hazard

The general duty clause does not apply to all hazards but only to the hazards that are "recognized" as arising out of the employment. The test for determining a "recognized" hazard is whether the hazard is known by the employer or generally by the industry of which the employer is a part (58). This test involves an objective determination and does not depend on whether the employer is aware in fact of the hazard (59).

In *American Smelting and Refining Company v. OSAHRC* (60) the Eighth Circuit held that the general duty clause is not limited to recognized hazards of types detectable only by the human senses but also encompasses hazards that can be detected only by instrumentation.

5.1.3 The Hazard Must be Causing or Likely to Cause Death or Serious Physical Harm

It is not necessary that there be actual injury or death to trigger a violation of the general duty clause. The purpose of the act is to prevent accidents and injuries. Thus, violation of the general duty clause arises from the existence of a statutory hazard, not from injury in fact (61).

Proof that a hazard is "causing or likely to cause death or serious physical harm" does not require a mathematical showing of probability. Rather, if evidence is presented that a practice could eventuate in serious physical harm upon other than a freakish or utterly implausible concurrence of circumstances, the commission's determination of likelihood will probably be accorded considerable deference by the courts (62).

The term "serious physical harm" is defined neither in the act nor in the secretary's regulations, but OSHA's Field Operations Manual defines it to mean:

(i)   Permanent, prolonged, or temporary impairment of the body in which part of the body is made *functionally useless* or is *substantially reduced in efficiency* on or off the job. Injuries involving such impairment would require treatment by a medical doctor, although not all injuries which receive treatment by a medical doctor would necessarily involve such impairment. Examples of such injuries are amputations; fractures (both simple and compound); deep cuts involving significant bleeding and which require extensive suturing; disabling burns and concussions.

(ii)   Illnesses that could *shorten life* or *significantly reduce physical or mental efficiency* by inhibiting the normal function of a part of the body, even though the effects may be cured by halting exposure to the cause or by medical treatment. Examples of such illnesses are cancer, silicosis, asbestosis, poisoning, hearing impairment and visual impairment.

5.2   The Specific Duty Clause [§5(a)(2)]

Section 5(a)(2) of OSHA imposes on employers a duty to comply with the occupational safety and health standards promulgated by the Secretary of Labor. These standards constitute the employers' so-called specific duties under the act. Specific promulgated standards preempt the general duty clause, but only with respect to hazards expressly covered by the specific standards (63).

5.2.1 Processes for Promulgating Standards

The act established processes for promulgating three types of occupational safety and health standards: interim, permanent, and emergency.

Interim standards consist of standards derived from (1) established federal standards, or (2) national consensus standards that were in existence on the effective date of OSHA (64). Section 6(a) of the act directed the Secretary of Labor to publish such standards in the *Federal Register* immediately after the act became effective (i.e., April 28, 1971) or for a period of up to 2 years thereafter. These standards became effective as OSHA standards upon publication without regard to the notice, public comment, and hearing requirements of the Administrative Procedure Act (65).

The intent of the interim standards provisions was to give the secretary a mechanism by which to promulgate speedily standards with which industry was already familiar and to provide a nationwide floor of minimum health and safety standards (66). The secretary's 2-year authority to promulgate interim standards expired on April 29, 1973.

Pursuant to Section 6(b) of the act, the Secretary of Labor is authorized to adopt "permanent" occupational safety and health standards to serve the objectives of OSHA. With respect to standards relating to toxic materials or harmful physical agents, Section 6(b)(5) specifically directs the secretary to set the standard "which most adequately assures, to the extent feasible, on the basis of the best available evidence, that no employee will suffer material impairment of health or functional capacity even if such employee has regular exposure to the hazard dealt with by such standard for the period of his working life."

The promulgation of these "permanent" occupational safety and health standards requires procedures similar to informal rule-making under Section 4 of the Administrative Procedure Act. Upon determination that a rule should be issued promulgating such a standard, the secretary must first publish the proposed standard in the *Federal Register*. Publication is followed by a 30-day period during which interested persons may submit written data or comments or file written objections and requests for a public hearing on the proposed standard. If a hearing is requested, the secretary must publish in the *Federal Register* a notice specifying the standard objected to and setting a time and place for the hearing. Within 60 days after the period for filing comments, or, if a hearing has been timely requested, within 60 days of the hearing, the secretary must either issue a rule promulgating a standard or determine that no such rule should be issued (67). Once a rule is issued, the secretary may delay the effective date of the rule for a period not in excess of 90 days to enable an affected employer to learn of the rule and to familiarize itself with its requirements.

Section 6(c)(1) of OSHA authorizes the secretary to issue emergency temporary standards if he determines: (1) that employees are exposed to grave danger from exposure to substances or agents determined to be toxic or physically harmful or from new hazards, and (2) that such emergency standard is necessary to protect employees from such danger.

An emergency temporary standard may be issued without regard to the notice, public comment, and hearing provisions of the Administrative Procedure Act. It takes immediate effect upon publication in the *Federal Register*.

The key to the issuance of an emergency temporary standard is the necessity to protect employees from a grave danger, as defined in *Florida Peach Growers* (68). After issuing an emergency temporary standard, the secretary must commence the procedures for promulgation of a permanent standard, which must issue within 6 months of the emergency standard's publication in accordance with Section 6(c)(3).

5.2.2   Challenging the Validity of Standards

Any person who may be adversely affected by an OSHA standard may file a petition under Section 6(f) challenging its validity in the United States Court of Appeals in the circuit wherein such person resides or has his principal place of business. The petition may be filed at any time prior to the 60th day after the issuance of the standard. Unless otherwise ordered, the filing of a petition does not operate as a stay of the standard.

Section 6(f) of the act directs the courts to uphold the Secretary of Labor's determinations in promulgating standards if those determinations are "supported by substantial evidence in the record considered as a whole" (69). In practice, the courts have generally declined to apply a strict "substantial evidence" standard of review. Instead, the courts have chosen to apply two different standards depending on whether the agency determination to be reviewed is one of fact or policy. They have essentially taken the position that only the secretary's findings of fact should be reviewed pursuant to a substantial evidence standard while the secretary's policy determinations should be substantiated by a detailed statement of reasons, which are subject to a test of reasonableness (70). The courts have adopted this approach with respect to emergency temporary standards as well as permanent standards [e.g., in *Florida Peach Growers* (67)].

In addition to a direct petition for review under Section 6(f), a majority of the courts of appeals have held that both procedural and substantive challenges to the validity of OSHA standards may be raised in enforcement proceedings under Section 11 (71). In fact, the Third Circuit in *Atlantic & Gulf Stevedores* (71) stated that the validity of a standard may be challenged not only in a federal court of appeals as part of an appeal from an order of the commission, but in the commission proceedings themselves. The Third Circuit explained, however, that in an enforcement proceeding invalidity is an affirmative defense to a citation and the employer bears the burden of proof on the issue of the reasonableness of the adopted standard. To carry its burden the employer must produce evidence showing why the standard under review, *as applied to it*, is arbitrary, capricious, unreasonable, or contrary to law (72).

5.2.3   Economic and Technological Feasibility of Standards

In enacting OSHA, Congress did not intend to make employers strictly liable for unavoidable occupational hazards. Accordingly, feasibility of compliance is a factor the Secretary of Labor must consider in developing occupational safety and health standards. As the United States Supreme Court explained in *American Textile Mfrs. Inst. Inc. v. Donovan* (73), OSHA's legislative history makes clear that any standard that is not economically or technologically feasible would *a fortiori* not be "reasonably necessary or appropriate" as directed by Section 3(8) of the Act (74). Thus, in enacting OSHA, "Congress does not appear to have intended to protect employees by putting their employers out of business" (75).

In analyzing economic feasibility, the secretary has tried to determine whether proposed standards threaten the competitive stability of an affected industry (76). But the Supreme Court has not yet decided whether a standard that actually does threaten the long-term profitability and competitiveness of an industry would be "feasible" (77).

The Supreme Court has expressly decided, however, that in promulgating a toxic material and harmful physical agent standard under Section 6(b)(5), the secretary is *not* required to determine that the costs of the standard bear a reasonable relationship to its benefits [*American Textile Mfrs. Inst. Inc. v. Donovan* (73)]. Rather, Section 6(b)(5) directs the secretary to issue the standard that "most adequately assures . . . that no employee will suffer material impairment of health," limited only by the extent to which this is economically and technologically feasible, or, in other words, capable of being done (*Id.* at 509)(78). The Supreme Court left open the possibility, however, that cost-benefit analysis might be required with respect to standards promulgated under provisions other than Section 6(b)(5) of the act (*Id.* at 513, n. 32)(79). The Court also left open the question of whether cost-benefit balancing by the secretary might be appropriate for deciding between issuance of several standards regulating different varieties of health and safety hazards (*Id.* at 509, n. 29).

Finally, in cases of violations of standards caused by employee disobedience or idiosyncratic behavior, the decisions of the courts and the commission have been similar to those rendered under the general duty clause (80) (see also discussion in Section 5.1.1 of this chapter).

For example, in *Brennan v. OSAHRC & Hendrix (d/b/a Alsea Lumber Co.)* (81) an employer was cited for violation of OSHA standards requiring workers to wear certain personal protective equipment. The record established that the violations resulted from individual employee choices, which were contrary to the employer's instructions. The Ninth Circuit affirmed the commission's decision vacating the citations, explaining as follows (82):

The legislative history of the Act indicates an intent not to relieve the employer of the general responsibility of assuring compliance by his employees. Nothing in the Act, however, makes an employer an insurer or guarantor of employee compliance therewith at all times. The employer's duty, even that under the general duty clause, must be one which is achievable. See *National Realty, supra*. We fail to see wherein charging an employer with a nonserious violation because of an individual, single act of an employee, of *which the employer had no knowledge* and which was contrary to the employer's instructions, contributes to achievement of the cooperation [between employer and employee] sought by the Congress. Fundamental fairness would require that one charged with and penalized for violation be shown to have caused, or at least to have knowingly acquiesced in, that violation [emphasis added, footnote omitted].

Nevertheless, even though Congress did not intend the employer to be held strictly liable for violations of OSHA standards, an employer is responsible if it knew or, with the exercise of reasonable diligence, should have known of the existence of a violation. Thus in *Brennan v. Butler Lime & Cement Co. and OSAHRC* (83) the Seventh Circuit drew on general duty clause concepts from the *National Realty* case (55) and explained that a particular instance of hazardous employee conduct may be considered preventable even if no employer could have detected the conduct or its hazardous nature at the moment of its occurrence, where such conduct might have been precluded through feasible precautions concerning the hiring, training, or sanctioning of employees (84).

In *Atlantic & Gulf Stevedores, Inc.* the Third Circuit held that such feasible precautions include disciplining or dismissing workers who refuse to wear protective headgear, even where such employer action could subject the company to wildcat strikes by employees adamantly opposed to the regulation (85).

Compare *Horne Plumbing and Heating Company v. OSAHRC and Dunlop*, where the Fifth Circuit found that an employer had taken virtually every conceivable precaution to ensure compliance with the law, short of remaining at the job site and directing the employees' operations himself. The court held that the final effort of personally directing the employees was not required by the act and that such an effort would be a "wholly unnecessary, unreasonable and infeasible requirement" (86).

5.2.4 Environmental Impact of Standards

Section 102(2)(C) of the National Environmental Policy Act of 1969 (NEPA)(87) requires all federal agencies, including OSHA, to prepare a detailed environmental impact statement in connection with major federal actions significantly affecting the quality of the human environment. The Secretary of Labor has identified the promulgation, modification, or revocation of standards which will significantly affect air, water or soil quality, plant or animal life, the use of land or other aspects of the human environment as always constituting such major action requiring the preparation of an environmental impact statement. 29 CFR §11.10(a)(3). On the other hand, promulgation, modification or revocation of any safety standard, such as machine guarding requirements, safety lines, or warning signals would normally qualify for categorical exclusion from NEPA requirements because "[s]afety standards promote injury avoidance by means of mechanical applications of work practices, the effects of which do not impact on air, water or soil quality, plant or animal life, the use of land or other aspects of the human environment." 29 CFR § 11.10(a)(1). The secretary's regulations regarding the procedure for preparation and circulation of environmental impact statements can be found at 29 CFR § 11.1 *et seq.*

5.2.5 Variances from Standards

Section 6(d) of OSHA provides that any affected employer may apply to the Secretary of Labor for a variance from an OSHA standard. To obtain a variance, the employer must show by a preponderance of the evidence submitted at a hearing that the conditions, practices, means, methods, operations, or processes used or proposed to be used by him will provide his employees with employment and places of employment that are as safe and healthful as those that would prevail if he complied with the standard.

If granted, the Section 6(d) variance may nevertheless be modified or revoked on application by an employer, employees, or by the Secretary of Labor on his own motion at any time after 6 months from its issuance. Affected employees are to be given notice of each application for a variance and an opportunity to participate in a hearing (88).

The act also provides mechanisms to enable employers to obtain variances of a more temporary nature than those sought under Section 6(d). Section 6(b)6(A) provides for "temporary" variances upon application when an employer establishes, after notice to employees and a hearing, that: (1) he is unable to comply with a standard by its effective date because of unavailability of professional or technical personnel or of materials and equipment needed to come into compliance with the standard or because necessary construction or alteration of facilities cannot be completed by the effective date, or (2) he is taking all available steps to safeguard his employees against the hazards covered by the standard, and he has an effective program for coming into compliance with it.

Section 6(b)6(C) authorizes the secretary to grant a variance from any standard or portion thereof whenever he determines, or the Secretary of Health and Human Services certifies, that the variance is necessary to permit the employer to participate in an experiment approved by the Secretary of Labor or the Secretary of Health and Human Services designed to demonstrate or validate new and improved techniques to safeguard the health or safety of workers.

Finally, Section 16 permits the Secretary of Labor, after notice and an opportunity for hearing, to provide such reasonable limitations and rules and regulations allowing "reasonable variations, tolerances, and exemptions to and from any or all provisions of" the act as he may find necessary and proper to avoid serious impairment of the national defense.

5.3 Overview of Occupational Safety and Health Standards (89)

To date, the bulk of federal job safety and health standards deal with occupational safety rather than with occupational health (90).

The occupational safety standards promulgated by the Secretary of Labor pursuant to OSHA are voluminous, encompassing hundreds of pages in the Code of Federal Regulations. A comprehensive analysis of these safety standards, most of which were adopted in 1971 as interim standards, is accordingly beyond the scope of this chapter (91).

For purposes of simplification, however, OSHA's safety standards can be broken down into the following general categories: (1) requirements relating to hazardous materials (e.g., compressed gas, acetylene, hydrogen, oxygen) and related equipment, (2) requirements for personal protective equipment and first aid, (3) fire protection standards and the national electrical code, (4) design and maintenance requirements for industrial equipment and working surfaces, and (5) operational procedures and equipment utilization requirements for certain hazardous industrial operations such as welding, cutting and brazing, and materials handling. Certain industries are also subject to specialized safety standards (92).

In the Spring of 1983, the OSH Administration's director for safety standards reported that the agency was "on schedule" for issuing 16 safety standards in final rule form by the middle of 1984 (93). Those safety standards will deal with areas that include: grain handling facilities, oil and gas well drilling and servicing, marine terminals, tunnels and shafts, electrical safety in construction, underground construction, ladders and scaffolds, floor and wall openings, latch-open devices, powered platforms, walking/working surfaces, single-piece rim wheels, mechanical power transmission apparatus, and excavations, trenching and shoring (94). In addition, as of April 1983, the agency had added onto its agenda of rules the topic of hoisting or suspension of employees on lifting buckets or work platforms (95).

The agency is also proceeding with the revocation of various safety and health standards (96). The provisions to be revoked are those which include the advisory word "should" instead of the mandatory "shall." OSHA will eventually replace the revoked standards with ones worded in compulsory terms but will not do so simultaneously with the elimination of the current ones. General duty clause citations, however, will be issued concurrently for serious hazards (97).

Like the safety standards, the bulk of OSHA's health standards were promulgated in 1971 as interim standards under Section 6(a). Established federal standards set workplace exposure to approximately 400 chemical and hazardous substances. These levels were referred to as threshold limit values (TLVs) and were expressed in terms of milligrams of substance per cubic meter of air and/or parts of vapor or gas per million parts of air. TLVs are defined as representing conditions under which it is believed nearly all workers may be repeatedly exposed day after day without adverse effects. The established federal TLVs had been developed principally in 1968 by the American Conference of Governmental Industrial Hygienists (ACGIH) and were subsequently incorporated into the Walsh-Healey Act (98).

Also under Section 6(a), the secretary promulgated certain ANSI health standards as national consensus standards (99). These ANSI standards established TLVs for 22 hazardous substances (100).

The TLVs have been sharply criticized. A congressional committee (101) has pointed out that the substances on the list of TLVs represent only a small portion of the estimated 19,000 unique toxic chemicals that are found in the workplace. In addition, constant changes in technology create new hazards and increase the difficulty of maintaining an up-to-date listing of toxic substance standards. Furthermore, the TLVs were apparently developed to be used as guidelines rather than as standards. As they now stand, they do not include several key provisions that are required for standards promulgated under the permanent rule-making procedures of Section 6(b). For example, Section 6(b)(7) of the act states that standards promulgated under Section 6(b) should prescribe, among other things, suitable protective equipment, engineering controls, medical surveillance, monitoring of employee exposure to hazards (102), and the use of labels to apprise employees of hazards (103).

In addition to the TLVs, the Secretary of Labor also promulgated interim health standards relating to radiation, ventilation, and noise. A permanent standard under Section 6(b) has subsequently been promulgated with respect to noise and is discussed further in Section 5.3.11. With respect to radiation, employers are responsible for proper controls to prevent any employee from being exposed to either ionizing or electromagnetic radiation in excess of acceptable limits. The radiation standard also requires posting of radiation areas, monitoring of employee exposure to radiation, and keeping records of such monitoring (104). Ventilation standards are included for various industrial processes, such as abrasive blasting operations, grinding, polishing, buffing, and spray finishing (105).

Since the effective date of OSHA, the Secretary of Labor has promulgated several major health standards under the permanent rule-making procedures. Each of these permanent standards can be viewed as a "complete" standard since each provides not only a specific value for the level of exposure to the toxic substance but also specifications as to monitoring, engineering controls, personal protective equipment, recordkeeping, medical surveillance, and other matters. A summary of several of these major health standards follows.

5.3.1   Asbestos

On December 7, 1971, the Secretary of Labor published an emergency temporary standard that limited the 8-hour-time-weighted average of airborne concentration of asbestos to 5 fibers greater than 5 μm in length per milliliter of air (the "five-fiber standard"). On June 7, 1972, a permanent standard was promulgated that retained the five-fiber standard for 4 years, then required reduction to two fibers (the "two-fiber standard")(106).

In *Industrial Union Department, AFL–CIO v. Hodgson* (4) the unions whose members are affected by the health hazards of asbestos dust challenged the timetable established by the permanent standard for achievement of permissible levels of concentration of asbestos and also objected to other portions of the standard concerning methods of compliance, monitoring intervals and techniques, cautionary labels and notices, and medical examinations and records. The Court of Appeals upheld the standard with two exceptions. The court directed the secretary to consider the effective date for the two-fiber standard and to determine whether this date might be accelerated for all or some of the industries affected. The court also directed the secretary to review the standard's recordkeeping provision, which required only a 3-year retention period for exposure monitoring records.

In response to the court's order, OSHA initiated a new rule-making proceeding and issued a *proposed* revised standard for asbestos on September 30, 1975. The proposal, which goes beyond the issues the secretary had been directed by the court to consider, would revise the existing standard by reducing the permissible exposure level to 0.5 fiber per cubic centimeter for all employments covered by the act except the construction industry. Other modifications were suggested, as well (107).

As of September, 1983, the 1975 proposal had still not been promulgated as a final rule. In the meantime, on or about July 1, 1976, the two-fiber standard became effective in keeping with the timetable set forth in the 1972 standard. However, in the spring of 1983, OSHA revived the 1975 proposal and instituted accelerated rule-making to change the two-fiber standard to .5 μg/cm^3. In addition, in late summer 1983, OSHA was considering the promulgation of an emergency temporary standard to reduce the permissible exposure limit for asbestos to .5 μg/cm^3 (108).

5.3.2 Vinyl Chloride

On April 5, 1974, the secretary promulgated an emergency standard for vinyl chloride. One of its requirements was that no worker be exposed to concentrations of vinyl chloride in excess of 50 ppm over any 8 hr period. The prior TLV had set an exposure limit of 500 parts vinyl chloride per million parts of air (109).

On October 1, 1974, a permanent standard was promulgated that calls for a permissible exposure limit no greater than one part per million averaged over an 8 hr period. The standard also requires feasible engineering and work practice controls to reduce exposure below the permissible level wherever possible; these practices are to be supplemented by respiratory protection (110).

In *Society of the Plastics Industry* the Court of Appeals for the Second Circuit upheld the standard and it became effective in April 1975 (111).

5.3.3   Carcinogens

The Department of Labor excluded the ACGIH's list of carcinogenic chemicals from its interim standards package, which it promulgated in 1971. After approximately one year of consulting with NIOSH and receiving data and commentary from interested groups, the secretary promulgated emergency temporary standards on May 3, 1973, for a list of 14 chemicals found to be carcinogenic. Permanent standards for the 14 carcinogens were issued on January 29, 1974 (112).

In *Synthetic Organic Chemical Manufacturers Association v. Brennan* (113) the Third Circuit upheld the standards for 13 of the chemicals except for the provisions pertaining to medical examinations and to laboratory usage of said chemicals. The standard for 4,4 methylene bis(2-chloraniline)(MOCA) was remanded and as of August 1, 1983, had not been reissued by OSHA.

In January 1980, OSHA promulgated a generic carcinogen policy for identifying, classifying, and regulating workplace carcinogens (114). The status of the carcinogen policy, however, was in doubt when this chapter went to print in view of, among other things, the Supreme Court's decision in *Industrial Union Dept. v. American Petroleum Institute* (115) wherein the Supreme Court held that before the secretary can promulgate any permanent health or safety standards, the secretary must find that significant risks are present and can be eliminated or lessened by a change in practices. In fact, OSHA had announced in 1982 that it intended to completely revoke the carcinogen policy (116), but in April 1983, OSHA indicated that the policy would not be revoked in its entirety (117). Rather, a draft proposal to amend the current carcinogen policy was being considered by OSHA (118), but as of September 1, 1983, no such proposal had been issued.

5.3.4   Coke Oven Emissions

In October 1976, the Secretary of Labor issued a permanent standard regulating workers' exposure to coke oven emissions. The standard defines coke oven emissions as the benzene-soluble fraction of total particulate matter present during the destructive distillation of coal for the production of coke. It limits exposure to 150 /μg of benzene-soluble fraction of total particulate matter per cubic meter of air averaged over an 8 hr period. The standard also mandates specific engineering controls and work practices, which were to be in use as soon as possible after the standard became effective and no later than January 20, 1980 (119).

The steel industry challenged the validity of the coke oven standard, but the Third Circuit upheld the standard in *American Iron and Steel Institute v. OSHA*, 577 F.2d 825, 6 OSHC 1451 (3d Cir. 1978), *certiorari* dismissed, 448 U.S. 917 (1980).

5.3.5 Lead

The OSH Administration issued a permanent standard regulating occupational exposure to metallic lead in organic lead compounds and organic lead soaps on or about November 14, 1978. That standard sets a permissible exposure limit of 50 $\mu g/m^3$ over an 8-hour period and an action level of 30 $\mu g/m^3$ averaged over an 8-hour period (120). In addition, the standard requires, among other things, the use of respirators and protective work clothing and equipment whenever lead exposure exceeds the PEL; compliance with vigorous rules on housekeeping and hygiene; and biological monitoring and medical surveillance whenever exposure exceeds the action level for more than 30 days in a year. The standard also contains a controversial medical removal protection (MRP) provision pursuant to which certain workers must be removed from the exposed workplace without loss of earnings, benefits, or seniority for at least 18 months. The standard further requires employers to create safety and health training programs for their workers exposed to lead, to keep detailed records on environmental (workplace) monitoring, biological monitoring, and medical surveillance and to make those records available to workers and certain of their representatives as well as to the government.

Virtually every aspect of the lead standard was challenged by the industry and by organized labor. The United States Court of Appeals for the District of Columbia rejected those challenges and upheld the standard as to certain industries (121). However, with respect to 38 other industries, the court held that OSHA failed to present substantial evidence or adequate reasons to support the feasibility of the standard for those industries and thus remanded to the Secretary of Labor for reconsideration of the technological and economic feasibility of the standard as to those industries (122). The OSH Administration subsequently amended paragraph (e)(1) of the standard so that it now requires employers who cannot reach the PEL to reduce exposure only to the lowest feasible level (123).

5.3.6 Cotton Dust

On June 23, 1978, OSHA promulgated its final Cotton Dust Standard (124). That standard establishes mandatory PELs over an 8-hour period of 200 $\mu g/m^3$ for yarn manufacturing, 750 $\mu g/m^3$ for slashing and weaving operations, and 500 $\mu g/m^3$ for all other processes in the cotton industry. The standard required full compliance with the PELs through a mix of engineering and work practice controls within four years, except to the extent that employers could establish that the engineering and work practice controls were unfeasible. The standard also requires, among other things, the provision of respirators at various times, the monitoring of cotton dust exposure, medical surveillance, medical examinations, employee education and training programs, the posting of warning signs, and the transfer without loss of earnings or other rights and benefits of employees unable to wear respirators to another position, if available, having a dust level at or below the PELs.

The cotton industry challenged the validity of the standard in the United States Court of Appeals for the District of Columbia. The Court of Appeals upheld the standard in all major respects (125) and the case moved on to the Supreme Court. On June 17, 1981, the Supreme Court affirmed the Court of Appeals' decision in all respects except as to its approval of the wage guarantee provision in the standard (126). With respect to that provision, the Supreme Court held that OSHA had acted beyond its statutory authority because it failed to make the necessary determination or statement of reasons that its wage guarantee requirement is related to the achievement of a safe and healthful work environment. The Court did not decide whether OSHA had the underlying authority to require employers to guarantee employees' wage and employment benefits following a transfer.

In June 1983, OSHA issued a proposed revision of the cotton dust standard which would add an action level of one-half of the PEL, exempt nontextile industries from the standard and delete the wage retention provisions. The PELs and the requirements for engineering controls in the current standard would be maintained (127).

5.3.7 DBCP

In March 1978, OSHA promulgated a permanent standard regarding 1,2 dibromo-3-chloropropane (DBCP) (128). The standard, which became effective April 17, 1978, sets a PEL of 1 part DBCP per billion parts of air over an 8-hour period (129). Where engineering controls and work practices are not sufficient to reduce exposure to permissible limits, respirators may be used as a supplement in order to achieve the required protection. Protective clothing is required where eye or skin contact may occur. The standard does not apply to the use of DBCP as a pesticide or when it is stored, transported, or distributed in sealed containers.

5.3.8 Acrylonitrile

In September 1978, OSHA issued a permanent standard governing workplace exposure to acrylonitrile (130). The standard, which became effective November 2, 1978 (except as to training programs, which were to be set up by January 2, 1979, and engineering controls, which were to be installed by November 2, 1980), sets a permissible exposure limit of 2 parts per million parts of air over an 8-hour period. The standard also sets a ceiling limit of 10 ppm for any 15-minute period and an action level of 1 ppm. Exposure above the action level triggers periodic monitoring requirements, medical surveillance, protective clothing and equipment requirements, employee information and training, and housekeeping. Skin or eye contact with the substance is prohibited (131).

5.3.9 Inorganic Arsenic

The OSH Administration issued a permanent standard for inorganic arsenic in May 1978 (132). The standard establishes a PEL of 10 $\mu g/m^3$ and also specifies various other requirements such as employee monitoring, engineering and work practice controls, respiratory protection, and training. The standard was challenged by the industry in *ASARCO, Inc. v. OSHA*, 647 F.2d 1; 9 OSHC 1508 (9th Cir. 1981) and was remanded to OSHA by the Ninth Circuit for reconsideration in light of the Supreme Court's decision invalidating the benzene standard in *Industrial Union Dept. v. American Petroleum Institute* (115, 133).

In April, 1982, OSHA reopened the rule-making record on risk assessment issues concerning inorganic arsenic in response to the Ninth Circuit's remand order (134) and in January 1983, OSHA published a final risk assessment indicating that the 10 $\mu g/m^3$ limit in the current standard reduces the risk of lung cancer by about 98 percent from the previous TLV of 500 $\mu g/m^3$ and that such a reduction satisfies the Supreme Court's "significant harm" test (135).

5.3.10 Benzene

The Secretary of Labor announced an emergency temporary standard for workplace exposure to benzene on April 29, 1977. The standard was to have gone into effect May 21, 1977, but it was temporarily stayed on May 20, 1977, by the Court of Appeals for the Fifth Circuit pending review. A permanent standard for benzene was issued by the secretary in February 1978 (136). Both the emergency standard and the permanent standard would have limited workplace exposure to one part of benzene per million parts of air and would have required the use of engineering and work practice controls to reduce employee exposure to or below the permissible limit.

The Fifth Circuit invalidated the benzene standard on October 5, 1978 (137) and the Supreme Court affirmed (138), finding that OSHA had failed to show that the 1 ppm exposure level presented a "significant risk of material health impairment" to workers. Thus, the only regulation still in effect concerning benzene exposure as this chapter goes to press is the TLV that was promulgated in 1971 as an interim standard (139). It sets a limit of 10 parts benzene per million parts of air. However, OSHA has announced that it will proceed again to try to lower the benzene exposure limit and plans to have a final standard completed by the summer of 1984 (140).

5.3.11 Noise

The OSH Administration's existing standard for occupational exposure to noise specifies a maximum permissible noise exposure level of 90 decibels for a duration of 8 hours. Employers are required to use feasible engineering or administrative controls, or combinations of both, whenever employee exposure to noise in the workplace exceeds the permissible exposure level. Personal protective equipment may be used to supplement the engineering and administrative controls where such controls are not able to reduce the employee exposure to within the permissible limit (141).

In March 1983, OSHA issued a final hearing conservation amendment to the noise standard. Among other things, the amendment requires employers to establish a hearing conservation program pursuant to which employers must provide hearing protectors and institute exposure monitoring, audiometric testing, and training for all employees who have occupational noise exposures equal to or exceeding an 8-hour time-weighted average of 85 decibels. The hearing conservation amendment covers all employees except those engaged in construction, agriculture, or oil and gas well drilling and servicing operations (142).

The hearing conservation amendment had originally been issued in January 1981 during the last week of the Carter Administration, but the Reagan Administration postponed its effective date until August 22, 1981, at which time some of the provisions were put into effect while the more controversial provisions were stayed pending review by OSHA and the Office of Management and Budget (143). Various industry groups are continuing to challenge the validity of the hearing conservation rules (144).

5.3.12   Proposed Health Standards

The Secretary of Labor issued a proposed standard for workplace exposure to ethylene oxide (EtO) in April 1983. The proposed standard would limit workplace exposure to one part ethylene oxide per million parts of air (145). The existing OSHA standard for EtO was the Walsh-Healey standard of 50 parts per million which was adopted by OSHA as an interim standard in the early 1970s (146).

In January 1983, a federal district court in the District of Columbia had directed OSHA to issue an emergency temporary standard concerning EtO within 20 days (147). That order was stayed pending expedited appeal. On appeal, the United States Court of Appeals for the District of Columbia held, on March 15, 1983, that the district court had erred in directing OSHA to issue an emergency standard concerning EtO, but the Court of Appeals did order OSHA to issue a notice of proposed rule-making within 30 days of the date of the court's decision. Although the court dictated no fixed dates for issuance of a final rule, it did direct "OSHA to proceed on a priority, expedited basis and to issue a permanent standard as promptly as possible" (148). The proposed standard which OSHA announced in April 1983 was the first formal step in response to the Court of Appeals' order (145).

In December 1981, OSHA issued an Advance Notice of Proposed Rulemaking concerning workplace exposure to ethylene dibromide, a fruit and grain fumigant (149). As of mid-September 1983, no proposed rule had yet been issued, but a proposal had reportedly been submitted to the Office of Management and Budget for review. That proposal, if adopted, would lower the current exposure limit of 20 parts per million to 0.1 ppm averaged over 8 hours with a short term ceiling limit of 0.5 ppm (150). The state of California adopted a permanent standard in January 1982 setting the permissible exposure limit for ethylene dibromide at 130 parts per billion. The California rule, which had previously been adopted as an emergency temporary standard, has been approved by OSHA as one that is at least as effective as the federal rule (150).

In the summer of 1983, OSHA was nearing completion of a hazard communication standard which would require chemical manufacturers to assess and communicate to employees the hazards posed by the chemicals they produce (151).

In August 1982, the United Auto Workers filed suit against OSHA alleging that OSHA's denial of a petition for an emergency temporary standard on formaldehyde was capricious, arbitrary, and not in accordance with law (152). As this chapter went to press, the UAW's motion for summary judgment, asking the court to order OSHA to take immediate steps to lower the exposure limits for formaldehyde, was still pending before the court (153).

There are several proposed standards still pending on OSHA's docket which have not been the subject of any formal action for several years. For example, proposed standards regulating workplace exposure to ammonia, beryllium, sulfur dioxide, toluene, and trichloroethylene had been issued in the 1970s, but as of September 1983, no final standards under § 6(b) had been promulgated concerning these substances. Similarly, in 1977, the so-called Standards Completion Project had issued proposed standards for six ketones [2-butanone, 2-pentanone, cyclohexanone, hexone, methyl (n-amyl) and ketone] and for several other toxic substances as well. Those proposals were submitted for public comment, but no further action has been taken on them (154).

5.3.13 Mine Industry Health Standards

The Secretary of Interior had promulgated health standards pursuant to the Coal Mine Health and Safety Act of 1969 and the Metal and Non-Metallic Mine Safety Act of 1966 focusing primarily on the following matters: (1) concentration of respirable dust in the coal mine atmosphere, (2) noise, (3) airborne contaminants, (4) surface bathing facilities, (5) sanitary toilet facilities, (6) respiratory equipment, and (7) radiation (underground metal and nonmetallic mines only). According to Section 301(b)(1) of the Federal Mine Safety and Health Act of 1977 (see Section 3.1 of this chapter) such mandatory standards issued by the Secretary of Interior that were in effect on the date of enactment of the 1977 act are to remain in effect as mandatory health standards applicable to metal and nonmetallic mines and to coal or other mines, respectively, under the 1977 act until the Secretary of Labor has issued new health standards applicable to such mines or revised mandatory standards (155).

6 EMPLOYEE RIGHTS AND DUTIES UNDER JOB SAFETY AND HEALTH LAWS

6.1 Employee Duties

Section 5(b) of OSHA requires employees to "comply with occupational safety and health standards and all rules, regulations, and orders issued pursuant" to the act that are applicable to their own actions and conduct.

The act does not, however, expressly authorize the Secretary of Labor to sanction employees who disregard safety standards and other applicable orders. The United States Court of Appeals for the Third Circuit has held that although Section 5(b) would be devoid of content if not enforceable, Congress did not intend to confer on the secretary or the commission the power to sanction employees (156).

Section 110(g) of the Federal Mine Safety and Health Act of 1977 not only requires mine employees to comply with health and safety standards promulgated under that statute, it also authorizes the imposition of civil penalties on miners who "willfully violate the mandatory safety standards relating to smoking or the carrying of smoking materials, matches, or lighters."

6.2 Employee Rights

The Occupational Safety and Health Act grants numerous rights to employees and/or their authorized representatives. The most fundamental employee right is the right set forth in Section 5(a) to a safe and healthful employment and place of employment. Other significant rights of employees and/or their authorized representatives include the following:

1. The right to request a physical inspection of a workplace and to notify the Secretary of Labor of any violations that employees have reason to believe exist in the workplace [§ 8(f)].

2. The right to accompany the secretary during the physical inspection of a workplace [§ 8(e)].

3. The right to challenge the period of time fixed in a citation for abatement of a violation of the act and an opportunity to participate as a party in hearings relating to citations [§ 10(c)].

4. The right to be notified of possible imminent danger situations and the right to file an action to compel the secretary to seek relief in such situations if he has "arbitrarily and capriciously" failed to do so [§§ 13(c) and 13(d)].

5. Various rights, including the right to notice, regarding an employer's application for either a temporary or permanent variance from an OSHA standard [§ 6(b)(6) and § 6(d)].

6. The right to observe monitoring of employee exposures to potentially toxic substances or harmful physical agents, the right to records thereof, and

the right to be notified promptly of exposures to such substances in concentrations which exceed those prescribed in a standard [§ 8(c)].

7. The right to petition a court of appeals to review an OSHA standard within 60 days after its issuance [§ 6(f)].

Section 11(c) of OSHA makes it unlawful for any person to discharge or in any manner discriminate against an employee because the employee has exercised his rights under the act. This provision is designed to encourage employee participation in the enforcement of OSHA standards (157).

Employees who believe they have been discriminated against in violation of Section 11(c) must file a complaint with the Secretary of Labor within 30 days after such violation has occurred. The secretary will investigate the complaint, and if he determines that Section 11(c) has been violated, he is authorized to bring an action in federal district court for an order restraining the violation and for recovery of all appropriate relief, including rehiring or reinstatement of the employee to his former position with back pay. The statute authorizes only the Secretary of Labor to bring an action for violation of Section 11(c)(158).

The Federal Mine Safety and Health Act of 1977 also contains a broad anti-retaliation provision and grants other rights to mine employees as well (159).

An employee has no explicit right, under OSHA, to refuse a work assignment because of what he feels is a dangerous working condition. The Secretary of Labor, however, has issued an administrative regulation that interprets the act as implying such a right under certain limited circumstances (160). The United States Supreme Court has held the promulgation of that regulation to be a valid exercise of the secretary's authority under the act. *Whirlpool Corp. v. Marshall*, 445 U.S. 1; 8 OSHC 1001 (1980). The Court observed that despite the detailed statutory scheme for speedily remedying dangerous working conditions, circumstances may arise when an employee justifiably believes that the statutory scheme will not sufficiently protect him:

[S]uch a situation may arise when (1) the employee is ordered by his employer to work under conditions that the employee reasonably believes pose an imminent risk of death or serious bodily injury, and (2) the employee has reason to believe that there is not sufficient time or opportunity either to seek effective redress from his employer or to apprise OSHA of the danger (161).

In holding that the regulation conformed to the fundamental objective of the act to prevent occupational deaths and injuries, the Court observed that the regulation also served to effectuate the general duty clause of § 5(a)(1)(162).

It is fairly well settled that OSHA does not create a private right of action for damages suffered by an employee as the result of an employer's violation of the act [see, e.g., *Skidmore v. Travelers Insurance Co.* (163)]. Finally, Section 4(b)(4) states that the act does not supersede or in any manner affect any worker's compensation law.

NOTES AND REFERENCES

1. See Bureau of National Affairs (BNA), *The Job Safety and Health Act of 1970*, Washington, D.C., 1971, p. 13. The Occupational Safety and Health Act is codified in Title 29 of the United States Code (USC), sections 651–678. References to the act in this chapter and the next are to the appropriate section of the statute itself and do not include a corresponding citation to the United States Code.

2. N. Ashford, *Crisis in the Workplace*, M.I.T. Press, Cambridge, Mass., 1976, pp. 141, 236–237.

3. See American Bar Association, *Report of the Committee on Occupational Safety and Health Law*, ABA Press, Chicago, 1975, p. 107.

4. For a discussion of the weight to be accorded by the Secretary of Labor to the NIOSH recommendations, see *Industrial Union Department, AFL–CIO v. Hodgson*, 499 F.2d 467, 476–77; 1 OSHC 1631 (D.C. Cir. 1974). [References to "F.2d" designate the volume (*e.g.*, 499) and page (*e.g.*, 467) of the *Federal Reporter*, Second Series, which contains the official reported decisions of the United States Courts of Appeals as published by the West Publishing Company. References to "OSHC" designate the same decision as reported in the "Occupational Safety and Health Cases" published by the Bureau of National Affairs (BNA).

5. See Ashford, *Crisis in the Workplace* (2), pp. 249–251. Section 27(b) of the act established a National Commission on State Workmen's Compensation Laws, which was directed to study and evaluate such laws to determine whether they provide an adequate, prompt, and equitable system of compensation for injury or death arising out of or in the course of employment. The commission's tasks were completed in July 1972, and the commission was disbanded. For a description of the activities of the commission and an evaluation of its work, see Ashford, *Crisis in the Workplace*, pp. 246, 289–292.

6. Numerous proposals had been introduced in Congress throughout the 1970s to amend the scope of OSHA's coverage, but all of those bills died either in committee or on the House or Senate floor. The scope of OSHA's coverage has not been legislatively amended since OSHA's enactment.

7. *Southern Pacific Transportation Co. v. Usery and OSAHRC*, 539 F.2d 386; 4 OSHC 1693 (5th Cir. 1976), *certiorari* denied, 434 U.S. 874; 5 OSHC 1888 (1977); *Southern Railway Company v. OSAHRC and Brennan*, 539 F.2d 335; 3 OSHC 1940 (4th Cir. 1976), *certiorari* denied, 429 U.S. 999; 4 OSHC 1936 (1976); *Baltimore & Ohio Railroad Co. v. OSAHRC*, 548 F.2d 1052; 4 OSHC 1917 (D.C. Cir. 1976).

8. 539 F.2d at 391; 4 OSHC at 1696.

9. The Third Circuit Court of Appeals has also adopted this definition of "working conditions." See *Columbia Gas of Pennsylvania Inc. v. Marshall*, 636 F.2d 913; 9 OSHC 1135 (3d Cir. 1980).

10. *Fineberg Packing Co.*, 1 OSHC 1598 (Rev. Comm. 1974).

11. 520 F.2d 1161; 3 OSHC 1566, 1572 (D.C. Cir. 1975).

12. American Bar Association, *Report of the Committee on Occupational Safety and Health Law*, ABA Press, Chicago, 1976, pp. 247–248.

13. The Federal Mine Safety and Health Act of 1977 is codified at 30 USC §§801 *et seq.*

14. 30 USC §811(a). Comprehensive regulations have been promulgated by the Mine Safety and Health Administration. Those regulations appear in title 30 of the Code of Federal Regulations at Parts 1 to 100. The Code of Federal Regulations is hereinafter cited as "CFR." The mandatory health standards can be found at 30 CFR Parts 70 *et seq.* References to specific provisions of the mine safety and health legislation and standards promulgated pursuant thereto are made throughout this chapter and the next.

15. The FEPCA is codified at 7 USC §§ 136 *et seq.* It is a comprehensive revision of the Federal Insecticide, Fungicide and Rodenticide Act of 1970, 7 USC §§ 135 *et seq.* (1970). The regulations promulgated by EPA to protect farm workers from toxic exposure to pesticides are found in 40 CFR §§ 170.1 *et seq.*

16. See, for example, *PBR, Inc. v. Secretary of Labor,* 643 F.2d 890, 896; 9 OSHC 1357, 1361 (1st Cir. 1981). The Secretary of Labor has acknowledged that under the Federal Railroad Safety Act the Department of Transportation (DOT) has *authority* to regulate all areas of employee safety for the railway industry. *Southern Railway Company v. OSAHRC* (Ref. 7; 539 F.2d at 333; 3 OSHC at 1941). However the scope of DOT's statutory authority to regulate matters relating to worker health is still unsettled. *Southern Pacific Transportation Company v. Usery* (Ref. 7; 539 F.2d at 389; 4 OSHC at 1694, n.3).

The Federal Railroad Safety Act is codified at 45 USC §§ 421 *et seq.* Other federal statutes dealing with railway safety include the Safety Appliance Acts, 45 USC §§ 1-16; the Train Brakes Safety Appliance Act, 45 USC § 9; the Hours of Service Act, 45 USC §§ 61 *et seq.*; and the Rail Passenger Service Act, 45 USC §§ 501 *et seq.* Department of Transportation regulations relating to railway safety can be found in 49 CFR, Chapter II.

The Department of Transportation also has statutory authority to regulate safety in modes of transportation other than rail. For example, the Secretary of Transportation may prescribe requirements for "qualifications and maximum hours of service of employees of, and safety of operation and equipment of, a motor carrier." 49 USC § 3102 (Revised Special Pamphlet 1983). For additional areas of DOT jurisdiction, see Sections 3.4–3.6 of this chapter.

17. See Section 2 of this chapter.

18. The provisions of the statute regarding safety regulation of civil aeronautics are codified at 49 USC §§ 1421 *et seq.*

19. See 49 USC § 1432(a).

20. See decision in *Usery v. Northwest Orient Airlines, Inc.,* 5 OSHC 1617 (E.D. N.Y. 1977). ["E.D. N.Y." refers to the federal district court in the Eastern District of New York. Reported decisions of federal district courts dealing with job safety and health matters are reported in BNA's *Occupational Safety and Health Cases* (OSHC) and many are also reported in West Publishing Company's *Federal Supplement* (F. Supp.)]. The administrative law judge determined that the FAA had exercised its authority by requiring each air carrier to maintain a maintenance manual that must include all instructions and information necessary for its ground maintenance crews to perform their duties and responsibilities with a high degree of safety.

21. The HMTA is codified at 49 USC §§ 1801 *et seq.* A table of materials that have been designated as hazardous by the secretary can be found at 49 CFR Part 172. The secretary's regulations prescribe the requirements for shipping papers, package marking, labeling, and transport vehicle placarding applicable to the shipment and transportation of those hazardous materials.

22. 3 OSHC 1601 (Rev. Comm. 1975).

23. The Secretary of Transportation has exercised statutory authority under NGPSA to promulgate safety standards for employees at natural gas facilities. These regulations can be found at 49 CFR Part 192. The NGPSA is codified at 49 USC §§ 1671 *et seq.* See also *Columbia Gas of Pennsylvania, Inc. v. Marshall, supra* (9) (regulation by DOT requiring operators to take steps to minimize danger of accidental ignition of gas while employees were performing a "hot tap" on existing gas main while installing auxiliary natural gas pipeline preempted authority of OSHA over the matter).

24. There also exist thousands of state and local laws regulating noise, discussion of which is beyond the scope of this work. See *Compilation of State and Local Ordinances on Noise Control,* 115 *Cong. Rec.* 32178 (1969).

25. See 49 USC § 1431. In carrying out this task the Administrator of the FAA has adopted and published aircraft noise standards and regulations. See 14 CFR Part 36. Section 7 of the Noise Control Act of 1972, discussed below, amended the noise abatement and control provision of the Federal Aviation Act to provide generally that standards adopted with respect to aircraft noise must have the prior approval of the Administrator of the EPA. In addition, the Noise Control Act was amended by the Quiet Communities Act of 1978, P.L.

95-609, 92 Stat. 3079, to provide for, among other things, a unified effort amongst state, local, and federal authorities to develop an effective noise abatement control program with respect to aircraft noise associated with airports.

26. The Noise Control Act is codified at 42 USC §§ 4901 *et seq.* The products with which the administrator is statutorily authorized to deal must come from among four categories: construction equipment, transportation equipment (including recreational vehicles and related equipment), any motor or engines (including any equipment of which an engine or motor is an integral part), and electrical or electronic equipment.

The administrator has promulgated comprehensive regulations under the Noise Control Act which appear at 40 CFR Subchapter G, Parts 201–211. To date these regulations cover interstate rail carriers (Part 201), motor carriers engaged in interstate commerce (Part 202), procedure and criteria for determining low-noise-emission products (Part 203), construction equipment including portable air compressors (Part 204), and transportation equipment, including medium and heavy trucks (Part 205).

Pursuant to Section 8 of the Noise Control Act, the administrator has also published product noise labeling requirements which can be found at 40 CFR Part 211.

27. The Toxic Substances Control Act of 1976 is codified at 15 USC §§ 2601 *et seq.* Prior to 1976 there existed no enactment that authorized the federal government to regulate toxic substances generally. Although regulations published under OSHA did govern worker exposure to toxic substances in the workplace (see Section 5.3, this chapter), federal law lacked a general authority to prohibit or restrict the manufacture of such substances.

28. Section 7 of the act empowers the Administrator of EPA to commence a civil action in federal court for the purpose of seizing an "imminently hazardous chemical substance," defined in the act as one that "presents an imminent and unreasonable risk of serious or widespread injury to health or the environment."

29. Polychlorinated biphenyls (PCBs) are the only group of chemicals with which the administrator is statutorily obligated to deal. See Section 6(e) of the act. For regulations concerning fully halogenated chlorofluoroalkanes, see 40 CFR Part 762.

30. The regulations issued by the EPA in 1979 dealing with PCBs were judicially reviewed by the United States Court of Appeals for the District of Columbia in *Environmental Defense Fund v. EPA*, 636 F.2d 1267 (D.C. Cir. 1980). In that proceeding, the Environmental Defense Fund ("EDF") challenged: (1) the determination by the EPA that certain uses of PCBs were "totally enclosed" and hence exempt from regulation under the Toxic Substances Control Act; (2) the applicability of the regulations to materials containing concentrations of PCBs greater than 50 parts per million (50 ppm); and (3) the decision of the EPA to authorize the continued availability of 11 nontotally enclosed uses of PCBs. The Court upheld the regulations regarding the continued availability of the 11 nontotally enclosed PCB uses, but it set aside the regulations classifying certain PCB uses as "totally enclosed" and the 50 ppm cutoff figure for materials containing PCBs. In response to this decision the EPA revised and amended the Part 761 regulations on PCBs and published new regulations at 47 *Federal Register* 37342-60 (Aug. 25, 1982). (The *Federal Register* is hereinafter cited as *Fed. Reg.*)

31. The Consumer Product Safety Act, as amended, is codified at 15 USC §§ 2051 *et seq.*

32. Provision is made in the act for suit by "any person who shall sustain injury by reason of any knowing (including willful) violation of a consumer product safety rule or any other rule or order issued by the Commission" against the responsible party in a federal district court. This right to sue is in addition to existing common law, federal, and state remedies.

Comprehensive regulations promulgated by the Consumer Product Safety Commission can be found at 16 CFR Parts 1000 *et seq.*

33. The Hazardous Substances Act is codified at 15 USC §§ 1261–1274. Regulations published under this statute can be found at 16 CFR Subchapter C, Parts 1500 *et seq.*

34. The Atomic Energy Act is codified at 42 USC §§ 2011 *et seq.* The Nuclear Regulatory Commission, an independent executive commission, was created by the Energy Reorgani-

zation Act of 1974, 42 USC § 5841(a), and all licensing and related regulatory functions of the Atomic Energy Commission were then transferred to the NRC. See 42 USC § 5841(f) and (g). The NRC's "Standards for Protection Against Radiation" can be found at 10 CFR Part 20.

35. The permissible dosage per calendar quarter within such a "restricted area" is 1.25 rems to the whole body, head and trunk, active bloodforming organs, lens of the eyes, or gonads; 18.75 rems to hands and forearms, feet and ankles; 7.5 rems to the skin of the whole body. Dosage standards are also set forth for the inhalation of radioactive substances. See 10 CFR § 20.101-103. Detailed personnel monitoring and reporting requirements are also included in the regulations.

36. See 10 CFR §§ 20.105, 20.106, and 20.303.

37. See 40 CFR Part 190 *et seq.*

38. See 30 CFR Part 57 *et seq.* These regulations are revisions of regulations previously promulgated by the Secretary of Interior under the Metal and Non-Metallic Mine Safety Act which was repealed by the Federal Mine Safety and Health Amendments Act of 1977 (see Section 3.1 of this chapter).

39. The Walsh-Healey Act is codified at 41 USC §§ 35 *et seq.* The relevant regulations can be found in 41 CFR § 50-204. These standards were later promulgated by the Secretary of Labor as established federal standards under Section 6(a) of OSHA. See Section 5.2.1 of this chapter.

40. That statute amended the Public Health Service Act and is codified at 42 USC § 263 b-n. The regulations promulgated by the Food and Drug Administration under this statute can be found at 21 CFR Parts 1000-1050 (Radiological Health). See also the standard applicable to diagnostic X-ray systems at 21 CFR § 1020.30.

41. The Outer Continental Shelf Lands Act is codified at 43 USC §§ 1331 *et seq.*

42. See Ashford, *Crisis in the Workplace* (2), pp. 47–51.

43. See D. Currie, "OSHA," *Am. Bar Found. Res. J.*, 1976, 1107, 1111 (1976). Section 18(a) of OSHA explicitly directs, however, that the states may assert jurisdiction under state law with respect to occupational safety or health issues for which no federal standard is in effect. See Section 4.1 of this chapter.

44. A so-called complete plan is described in an administrative regulation issued by the Secretary of Labor and codified at 29 CFR § 1902.3.

45. Pursuant to Section 18(c), the state plan must:

   a. Designate a State agency or agencies as the agency or agencies responsible for administering the plan throughout the state.

   b. Provide for the development and enforcement of safety and health standards relating to one or more safety or health issues, which standards (and the enforcement of which standards) are or will be at least as effective in providing safe and healthful employment and places of employment as the standards promulgated under Section 6 of OSHA which relate to the same issues.

   c. Provide for a right of entry and inspection of all workplaces subject to this chapter which is at least as effective as that provided in Section 8 of OSHA and include a prohibition on advance notice of inspections.

   d. Contain satisfactory assurances that such agency or agencies have or will have the legal authority and qualified personnel necessary for the enforcement of such standards.

   e. Give satisfactory assurances that such state will devote adequate funds to the administration and enforcement of such standards.

   f. Contain satisfactory assurances that such state will, to the extent permitted by its law, establish and maintain an effective and comprehensive occupational safety and health program applicable to all employees of public agencies of the state and its political subdivisions, which program is as effective as the standards contained in an approved plan.

g.   Require employers in the state to make reports to the Secretary in the same manner and to the same extent as if the plan were not in effect.

h.   Provide that the state agency will make such reports to the Secretary in such form and containing such information as the Secretary shall from time to time require.

The additional criteria outlined in the Secretary's regulations can be found at 29 CFR §§ 1902.3 and 1902.4

46.   29 CFR § 1902.2(b).

47.   29 CFR § 1954.3, 1954.10.

48.   The commission has held that OSHA is not precluded from exercising its own enforcement authority during this period. *Par Construction Co., Inc.*, 4 OSHC 1779 (Rev. Comm. 1976); *Seaboard Coast Line Railroad Co. and Winston-Salem Southbound Railway Co.*, 3 OSHC 1767 (Rev. Comm. 1975).

49.   A chart on the status of state plans can be found in BNA, *OSHR*, Reference File at 81:1003.

50.   See P. Hamlar, "Operation and Effect of State Plans," in: *Proceedings of the American Bar Institute on Occupational Safety and Health Law*, ABA Press, Chicago, 1976, pp. 42–45.

51.   See Section 303(e) of the Federal Mine Safety and Health Act of 1977.

52.   See Section 205 of the Federal Railroad Safety Act.

53.   There is language in the legislative history of OSHA to indicate that the general duty clause merely restates the employer's common law duty to exercise reasonable care in providing a safe place for his employees to work. However, the courts have generally characterized such statements as "misleading." For example, in *REA Express v. Brennan*, 495 F.2d 822, 825; 1 OSHC 1651 (2d Cir. 1974), the Second Circuit could not "accept the proposition that common law defenses such as assumption of the risk or contributory negligence will exculpate the employer who is charged with violating the Act."

54.   At a multi-employer construction worksite, this duty may extend to hazardous conditions which the employer neither creates nor fully controls under what have become known as the *Anning-Johnson/Grossman* rules, unless the employer can show that it took realistic or reasonable measures to protect its employees or that it neither knew or reasonably could have known of the violation. *Dun-Par Engineered Form Co. v. Marshall*, 676 F.2d 1333, 1335–1336; 10 OSHC 1561, 1562 (10th Cir 1982); *Electric Smith, Inc. v. Secretary of Labor*, 666 F.2d 1267, 1268–1270; 10 OSHC 1329, 1330–1332 (9th Cir. 1982); *DeTrae Enterprises, Inc. v. Secretary of Labor*, 645 F.2d 103, 104; 9 OSHC 1425, 1426 (2d Cir. 1981); *Bratton Corp. v. OSAHRC*, 590 F.2d 273, 275; 7 OSHC 1004, 1005 (8th Cir. 1979).

55.   *National Realty and Construction Company, Inc. v. OSAHRC*, 489 F.2d 1257; 1 OSHC 1422 (D.C. Cir. 1973); *Brennan v. OSAHRC and Canrad Precision Industries*, 502 F.2d 946; 2 OSHC 1137 (3d Cir. 1974).

56.   See *True Drilling Co. v. Donovan*, 703 F.2d 1087, 1090; 11 OSHC 1310, 1311 (9th Cir. 1983); *Carlyle Compressor Co. v. OSAHRC*, 683 F.2d 673, 677; 10 OSHC 1700 (2d Cir. 1982); *PBR, Inc. v. Secretary of Labor*, 643 F.2d 890, 895; 9 OSHC 1357, 1360 (1st Cir. 1981).

57.   489 F.2d 1266–1267; 1 OSHC 1427; see also, *Capital Electric Line Builders of Kansas, Inc. v. Marshall*, 678 F.2d 128, 130; 10 OSHC 1593, 1594 (10th Cir. 1982); *H. B. Zachry Co. v. OSAHRC*, 638 F.2d 812, 818; 9 OSHC 1417, 1421–1422 (5th Cir. 1981); *General Dynamics Corp. v. OSAHRC*, 599 F.2d 453, 458; 7 OSHC 1373, 1375 (1st Cir. 1979). See also the discussion in Section 5.2.3 of this chapter.

58.   *Pratt & Whitney Aircraft v. Secretary of Labor*, 649 F.2d 96, 100; 9 OSHC 1554, 1557 (2d Cir. 1981); *Continental Oil Co. v. OSHRC*, 630 F.2d 446, 448; 8 OSHC 1980, 1981 (6th Cir. 1980), *certiorari* denied, 450 U.S. 965 (1981); *Brennan v. OSAHRC and Vy Lactos Laboratories*, 494 F.2d 460, 464; 1 OSHC 1623, 1625 (8th Cir. 1974); *National Realty* [(55) 489 F.2d at 1265 n.32; 1 OSHC at 1426].

59.   *National Realty* (55); *Pratt & Whitney Aircraft* (58).

60.   501 F.2d 504; 2 OSHC 1041 (8th Cir. 1974).

61. *Brennan v. OSAHRC and Vy Lactos Laboratories* (58), 1 OSHC at 1624; R. Morey, "The General Duty Clause of the Occupational Safety and Health Act of 1970," *Harv. Law Rev.*, 86, 988, 991 (1973). The same is true for violations of the specific duty clause.

62. *National Realty* [(55) 489 F.2d at 1265, n.33]; *Babcock & Wilcox Co. v. OSAHRC*, 622 F.2d 1160, 1165; 8 OSHC 1317, 1319 (3d Cir. 1980); *Illinois Power Co. v. OSAHRC*, 632 F.2d 25, 28; 8 OSHC 1512, 1514–1515 (7th Cir. 1980); *Titanium Metals Corp. of America v. Usery*, 579 F.2d 536, 541; 6 OSHC 1873, 1876–1878 (9th Cir. 1978). "[T]he 'likely to cause' test should be whether reasonably foreseeable circumstances could lead to the perceived hazard's resulting in serious physical harm or death—or more simply, the proper test is plausibility, not probability." Morey, "The General Duty Clause of the Occupational Safety and Health Act of 1970" (61, pp. 997–998).

63. 29 CFR § 1910.5(f).

64. An "established Federal standard" is defined in Section 3(10) of the act as "any operative occupational safety and health standard established by any agency of the United States and presently in effect, or contained in any Act of Congress in force on the date of enactment of this Act." Section 4(b)(2) of the act listed several federal statutes from which established federal standards were to be derived, including the Walsh–Healey Act, 41 USC §§ 35–45, the Service Contract Act of 1965, 41 USC §§ 351–357, and the National Foundation on Arts and Humanities Act, 20 USC §§ 951–960.

    A "national consensus" standard is defined in Section 3(9) of the act as any occupational safety and health standard, which "(1) has been adopted and promulgated by a nationally recognized standards producing organization under procedures, whereby it can be determined by the Secretary that persons interested and affected by the scope or provisions of the standard have reached substantial agreement on its adoption, (2) was formulated in a manner which afforded an opportunity for diverse views to be considered, and (3) has been designated as such a standard by the Secretary, after consultation with other appropriate Federal agencies." The principal sources for national consensus standards were the American National Standards Institute (ANSI) and the National Fire Protection Association. *American Federation of Labor v. Brennan*, 530 F.2d 109, 111 at n. 2; 3 OSHC 1820, 1821 at n.2 (3d Cir. 1975).

65. The Administrative Procedure Act was enacted by Congress in 1946 to impose some coherent system of procedural regularity on the growing regulatory bureaucracy of the federal government. It provides procedures for administrative "rule making" and administrative "adjudication," among other things. See generally H. Linde and G. Bunn, *Legislative and Administrative Processes*, Foundation Press, Mineola, N.Y., 1976, p. 814. The Administrative Procedure Act is codified in 5 USC §§ 551 *et seq.*

    The courts have held that OSHA does not have the right to change advisory national consensus standards ("should") to mandatory standards ("shall") upon adoption as OSHA standards without following formal rule-making procedures. Absent such rule-making, citations issued to employers pursuant to these standards have been vacated. *Usery v. Kennecott Copper Corp.*, 577 F.2d 1113, 1117–1118; 6 OSHC 1197, 1199 (10th Cir. 1977); *Marshall v. Pittsburgh-Des Moines Steel Co.*, 584 F.2d 638, 644; 6 OSHC 1929, 1933 (3d Cir. 1978). See also *Marshall v. Anaconda Co.*, 596 F.2d 370, 376–377; 7 OSHC 1382, 1385–1386 (9th Cir. 1979).

66. *The Job Safety and Health Act of 1970*, BNA, Washington, D.C., 1971, p. 23.

67. See *Florida Peach Growers Association v. U.S. Department of Labor*, 489 F.2d 120, 124; 1 OSHC 1472 (5th Cir. 1974) and Sections 6(b)(1) to 6(b)(4) of OSHA. In *National Congress of Hispanic American Citizens v. Usery*, 554 F.2d 1196; 5 OSHC 1255 (D.C. Cir. 1977), the Court of Appeals for the District of Columbia Circuit held that the statutory deadlines in Sections 6(b)(1) to 6(b)(4) for the promulgation of permanent standards were discretionary rather than mandatory as long as the secretary's exercise of discretion was honest and fair.

68. *Florida Peach Growers Association* (67), 489 F.2d at 124; see also *Industrial Union Dept. AFL–CIO v. American Petroleum Institute*, 448 U.S. 607, 651 n. 59; 8 OSHC 1586, 1602 (1980).

69.   The "substantial evidence" standard of judicial review is traditionally conceived of as suited
      to adjudication or formal rulemaking. OSHA, however, calls for informal rulemaking which
      under the Administrative Procedure Act generally entails judicial review pursuant to the less
      stringent "arbitrary and capricious" test. This apparent anomaly can be explained historically
      as a legislative compromise. The Senate OSHA bill called for informal rulemaking, but the
      House version specified formal rulemaking and substantial evidence review. The House
      receded on the procedure for promulgating standards, but the substantial evidence standard
      of review was adopted. *Industrial Union Department, AFL–CIO v. Hodgson* (4), 499 F.2d at 473;
      1 OSHC at 1635. For a more detailed discussion of these legislative events, see *Associated
      Industries of New York State, Inc. v. U.S. Department of Labor*, 487 F.2d 342; 1 OSHC 1340 (2d
      Cir. 1973).

70.   See B. Fellner and D. Savelson, "Review by the Commission and the Courts," in: *Proceedings
      of the American Bar Association Institute on Occupational Safety and Health Law*, ABA Press, Chicago,
      1976, pp. 113–114. This approach has been summarized as one requiring the reviewing
      court to determine whether the agency (1) acted within the scope of its authority; (2) followed
      the procedures required by statute and by its own regulations; (3) explicated the bases for
      its decision; and (4) adduced substantial evidence in the record to support its determination.
      United Steelworkers of America, AFL–CIO v. Marshall and Bingham, 647 F.2d 1189, 1206; 8 OSHC
      1810, 1816 (D.C. Cir. 1980), *certiorari* denied, 453 U.S. 913 (1981). See also *Texas Independent
      Ginners Assoc. v. Marshall*, 630 F.2d 398, 404–405; 8 OSHC 2205, 2209–2210 (5th Cir. 1980);
      American Iron and Steel Institute v. OSHA, 577 F.2d 825, 830–31; 6 OSHC 1451, 1455 (3d Cir.
      1978); *Society of the Plastics Industry, Inc. v. OSHA*, 509 F.2d 1301, 1304; 2 OSHC 1496, 1498
      (2d Cir. 1975).

71.   *Atlantic & Gulf Stevedores, Inc. v. OSAHRC*, 534 F.2d 541; 4 OSHC 1061 (3d Cir. 1976);
      Arkansas-Best Freight Systems, Inc. v. OSAHRC and Secretary of Labor, 529 F.2d 649; 3 OSHC
      1910 (8th Cir. 1976); *Deering Milliken, Inc. v. OSAHRC*, 630 F.2d 1094, 1099; 9 OSHC 1001,
      1004 (5th Cir. 1980); *Marshall v. Union Oil Co. and OSAHRC*, 616 F.2d 1113, 1117–1118; 8
      OSHC 1169, 1173 (9th Cir. 1980); *Daniel International Corp. v. OSAHRC and Secretary of Labor*,
      656 F.2d 925, 928–930; 9 OSHC 2102, 2104–2106 (4th Cir. 1981). But see *National Industrial
      Contractors v. OSAHRC*, 583 F.2d 1048, 1052–1053; 6 OSHC 1914, 1916–1917 (8th Cir.
      1978), in which the Eighth Circuit held that procedural challenges to an OSHA standard
      must be brought in a pre-enforcement proceeding pursuant to Section 6(f) within 60 days
      from the challenged standard's effective date.

72.   *Atlantic & Gulf Stevedores* (71), 534 F.2d at 550–552; 4 OSHC at 1067–1068.

73.   452 U.S. 490; 9 OSHC 1913 (1981).

74.   452 U.S. at 513, n. 31; 9 OSHC at 1922, n. 31. Section 3(8) of OSHA contains the general
      definition of an occupational safety and health standard. It provides as follows:

      The term "occupational safety and health standard" means a standard which requires
      conditions, or the adoption or use of one or more practices, means, methods, operations, or
      processes, *reasonably necessary or appropriate* to provide safe or healthful employment and
      places of employment. (Emphasis added.)

      For standards dealing with toxic materials or harmful physical agents, Section 6(b)(5) imposes
      the following additional requirements:

      The Secretary, in promulgating standards dealing with toxic materials or harmful physical
      agents under this subsection, shall set the standard which most adequately assures, *to the
      extent feasible*, on the basis of the best available evidence, that no employee will suffer material
      impairment of health or functional capacity, even if such employee has regular exposure to
      the hazard dealt with by such standard for the period of his working life. (Emphasis added.)

75.   *Industrial Union Department, AFL–CIO v. Hodgson* (4), 499 F.2d at 478; 1 OSHC at 1639, cited
      approvingly by the Supreme Court in *American Textile Mfrs. v. Donovan* (73), 452 U.S. at 513,
      n. 31; 9 OSHC at 1922, n. 31. In *Industrial Union Department*, the Court of Appeals applied

Section 6(b)(5) in a case challenging OSHA's standard for exposure to asbestos dust and held that the secretary, in promulgating the standard, could properly consider problems of both economic and technological feasibility.

76. *American Textile Mfrs. v. Donovan* (73), 452 U.S. at 530, n. 55; 9 OSHC at 1928–1929, n. 55; *United Steelworkers of America, AFL–CIO v. Marshall and Bingham* (70), 647 F.2d 1189, 1265; 8 OSHC 1810, 1864 (D.C. Cir. 1981).

77. *American Textile Mfrs. v. Donovan* (73), 452 U.S. at 530, n. 55; 9 OSHC at 1928–1929, n. 55.

78. *American Textile Mfrs. v. Donovan* involved a challenge by the textile industry to OSHA's standard governing occupational exposure to cotton dust. The industry contended, among other things, that the act required OSHA to demonstrate that its standard reflected a reasonable relationship between the costs and benefits associated with the standard. The Supreme Court rejected that argument and upheld the validity of the entire cotton dust standard except for a wage guarantee requirement for employees who are transferred to another position when they are unable to wear a respirator.

79. In *Donovan v. Castle & Cooke Foods and OSAHRC*, 692 F.2d 641; 10 OSHC 2169 (9th Cir. 1982), the United States Court of Appeals for the Ninth Circuit held that the Supreme Court's holding concerning cost/benefit analysis in the *American Textile Manufacturers* case applied only to standards promulgated under Section 6(b) of OSHA and did not apply to standards promulgated under Section 6(a), such as the noise standard at issue in *Castle & Cooke* which had been originally promulgated as an established Federal standard under the Walsh-Healey Act. But compare *Sun Ship, Inc.*, 11 OSHC 1028 (Rev. Comm. 1982), where the Review Commission applied the reasoning of *American Textile Manufacturers* to the noise standard and rejected the application of cost benefit analysis in enforcing that standard. See BNA, OSHR, Current Report for February 3, 1983, pp. 735–736.

80. Ashford, *Crisis in the Workplace* (2), p. 169.

81. 511 F.2d 1139; 2 OSHC 1646 (9th Cir. 1975).

82. 511 F.2d at 1144–1145; 2 OSHC at 1650–1651. See also *Daniel International Corp. v. OSAHRC and Secretary of Labor*, 683 F.2d 361; 10 OSHC 1890 (11th Cir. 1982).

83. 520 F.2d 1011; 3 OSHC 1461 (7th Cir. 1975).

84. *Accord, H. B. Zachry Company v. OSAHRC*, 638 F.2d 812; 9 OSHC 1417 (5th Cir. 1981) (defense of employee negligent misconduct fails because of the employer's inability to establish to the satisfaction of the fact-finder that it effectively communicated and enforced work rules which were necessary to ensure compliance with OSHA standards).

85. *Atlantic & Gulf Stevedores* (71), 534 F.2d at 555; 4 OSHC at 1068–1069. See Note, "Employee Noncompliance with OSHA Safety Standards," *Harv. Law Rev.*, **90**, 1041 (1977).

86. 528 F.2d 564, 570; 3 OSHC 2060, 2064 (5th Cir. 1976).

87. 42 USC § 4332.

88. As of June 30, 1982, OSHA had received 1,529 variance requests, of which 1,476 had been processed to completion. Commerce Clearing House (CCH), *Empl. Safety & Health Guide*, 1981–1982 Transfer Binder ¶ 12,604. Permanent variances had been granted in 113 cases as of June 30, 1982. *Id.*

89. Health and safety standards promulgated by the Secretary of Labor pursuant to OSHA can be found at 29 CFR Part 1910. A standards digest (OSHA Publication 2201) outlining the basic applicable standards is published in BNA, *OSHR*, Reference File at 31:4001.

   Federal health and safety standards for the construction industry were initially promulgated under the Contract Work Hours and Safety Standards Act, 40 USC §§ 327 *et seq.* These standards were incorporated by reference under OSHA, are enforceable under both laws, and can be found at 29 CFR Part 1926. A standards digest (OSHA Publication 2202) outlining the basic applicable construction standards is published in BNA, *OSHR*, Reference File at 31:3001. Health and safety standards for ship repairing, shipbuilding, shipbreaking, and longshoring were initially promulgated pursuant to the Longshoremen's and Harbor

Worker's Compensation Act, 33 USC §§ 901 *et seq.* These standards were incorporated by reference by OSHA, are enforceable under both laws, and can be found in 29 CFR Parts 1915–1918. Health and safety standards originally promulgated under the Walsh-Healey Public Contracts Act, the McNamara-O'Hara Service Contract Act of 1965, and the National Foundation on the Arts and Humanities Act of 1965 can be found in 41 CFR Part 50-204. The majority of these standards were also adopted as interim OSHA standards and are enforceable pursuant to the act. Standards promulgated under the aforementioned statutes will be superseded if corresponding standards that are promulgated under OSHA are determined by the Secretary of Labor to be more effective. See Section 4(b)(2) of OSHA.

Federal health and safety standards for coal mines were promulgated by the Department of Interior pursuant to the federal Coal Mine Health and Safety Act of 1969. Standards under the 1969 act were adopted without change by the Federal Mine Safety and Health Act of 1977. CCH, *Empl. Safety and Health Guide* ¶¶ 5924, 5931. Health standards for underground coal mines can be found in 30 CFR Part 70; health standards for surface work areas of underground coal mines and surface coal mines are codified in 30 CFR Part 71. Requirements for approval of coal mine dust personal sampler units designed to determine the concentrations of respirable dust in coal mine atmospheres can be found in 30 CFR Part 74; health standards for coal miners with evidence of pneumoconiosis are codified in 30 CFR Part 90. The safety standards for underground coal mines can be found in 30 CFR Part 75, and the safety standards for surface coal mines and surface work areas of underground coal mines are codified in 30 CFR Part 77.

The Secretary of Interior promulgated health and safety standards for metal and nonmetallic mines pursuant to the federal Metal and Non-Metallic Mine Safety Act of 1966. Mandatory standards adopted under the 1966 act were adopted without change by the Federal Mine Safety and Health Act of 1977. However, advisory metal and nonmetallic standards did not become mandatory standards under the 1977 act. A committee later reviewed these advisory standards and made recommendations for conversion to mandatory standards which MSHA accepted in August 1979 by converting scores of noncoal advisory standards to mandatory standards. These standards are included among the other mandatory standards for open pit mines, for sand, gravel and crushed stone operations, and for metal and nonmetallic underground mines at 30 CFR Parts 55, 56, and 57 (1982), respectively. CCH, *Empl. Safety and Health Guide* ¶¶ 5924, 5931.

90. Safety standards generally focus on the time that an employee is actually working. The harm created by a safety hazard is generally immediate and violent. An occupational health hazard, on the other hand, is slow acting, cumulative, irreversible, and complicated by nonoccupational factors. Ashford, *Crisis in the Workplace* (2), pp. 68–83.

91. For a listing of the initial package of national consensus and established federal standards, see 36 *Fed. Reg.* 10466–10714 (1971). Since the effective date of OSHA, the secretary has promulgated safety standards under the permanent rule-making procedures of Section 6(b) with respect to at least the following matters: commercial diving operations, 29 CFR §§ 1910.401–441; agricultural operations, 29 CFR § 1928; helicopter operations, 29 CFR § 1910.183; telecommunications operations, 29 CFR § 1910.268; slings, 29 CFR § 1910.184; ground-fault circuit interrupters, 29 CFR § 1910.304(b)(1); multipiece rim wheels, 29 CFR §1910.177; perimeter guards on low-pitched roofs, 29 CFR §1926.500(g); and onshore marine cargo-handling, 29 CFR Part 1917. The standard concerning ground-fault circuit interrupters was successfully challenged in *National Constructors Association v. Marshall*, 581 F.2d 960; 6 OSHC 1721 (D.C. Cir. 1978), as the Court of Appeals remanded it to cure the defect of failure to consult adequately with the Advisory Committee on Construction Safety and Health. The standard has since been reviewed by the Advisory Committee which concluded that no modification was necessary. OSHA thus reaffirmed the standard and it is presently in effect. The diving standard, which actually establishes mandatory safety *and health* requirements for commercial diving operations, was also challenged. In *Taylor Diving and Salvage Company et al. v. U. S. Department of Labor*, 599 F.2d 622; 7 OSHC 1507 (5th Cir.

1979), the Fifth Circuit invalidated the medical examination and cost allocation provisions contained in § 1910.411. That selection has since been amended.

92. Industries covered by specific OSHA regulations include pulp, paper and paperboard mills, textiles, bakery equipment, laundry machinery and operations, sawmills, pulpwood logging, telecommunications, agriculture, commercial diving, construction, ship repairing, shipbuilding, shipbreaking, and longshoring. The mining industry is subject to comprehensive safety regulations issued pursuant to the Federal Mine Safety and Health Act of 1977 (formerly the Federal Coal Mine Health and Safety Act of 1969 and the Federal Metal and Non-Metallic Mine Safety Act of 1966). See also Section 3 of this chapter.

93. BNA, *OSHR*, Current Report for March 17, 1983, p. 894.

94. See BNA, *OSHR*, Current Reports for: March 17, 1983, p. 894; March 10, 1983, pp. 828–829; and March 11, 1982, pp. 812–813. For a summary of the proposed underground construction standard, see BNA, *OSHR*, Current Report for August 11, 1983, p. 227. A final rule on marine terminals (onshore marine cargo handling) was issued on July 5, 1983. See BNA, *OSHR*, Current Report for July 7, 1983, p. 123.

95. BNA, *OSHR*, Current Report for April 14, 1983, p. 957–958. The catalyst for this addition was the death of four workers in a crane accident which occurred in March of 1983 at the Tampa, Florida Stadium.

96. BNA, *OSHR*, Current Report for March 17, 1983, p. 894.

97. BNA, *OSHR*, Current Report for March 17, 1983, p. 894.

98. See Ashford, *Crisis in the Workplace* (2), p. 154. The Secretary of Labor did not include the ACGIH's carcinogen standards in his Section 6(a) package but instead preferred to develop his own standards regarding carcinogens. Ashford, pp. 154, 247–248.

99. See 29 CFR § 1910.149.

100. The TLVs that have been developed by the ACGIH and the ANSI and promulgated by OSHA can be found at 29 CFR § 1910.1000, Tables Z-1, Z-2, and Z-3.

101. U.S. Congress, Committee on Governmental Operations, "Chemical Dangers in the Workplace," House Report 1688, 94th Congress, 2d Session, 1976, pp. 15–16. In addition, see Ashford, *Crisis in the Workplace* (2), pp. 295–296.

102. For example, OSHA's vinyl chloride standard (see Section 5.3, this chapter) requires employers to undertake a program of initial monitoring and measurement to determine whether there is any employee exposure to vinyl chloride (without regard to the use of respirators) in excess of the action level specified in the standard. If there is such exposure, the employer must establish a program for determining exposures for each such employee. The standard specifies how frequently those employees must be monitored. The standard also: (1) requires monitoring when the employer has made a change in production, process, or control that may result in an increase in release of vinyl chloride; (2) specifies the required accuracy of the method of monitoring and measurement; and (3) gives employees the right to observe the monitoring and measuring. 29 CRF § 1910.1017(d). Compare OSHA's coke oven emissions standard, which requires employers to notify each employee in writing of the exposure measurements that represent that employee's exposure and, if the exposure exceeds the permissible exposure limit, to so notify the employee and to inform the employee of the corrective action being taken to reduce exposure to or below the permissible exposure limit. 29 CFR § 1910.1029(e).

103. Labeling requirements in standards that have been promulgated under the permanent rule-making provisions of Section 6(b) generally tend to specify the language to be used in labels or other visual warning devices as well as the locations where such labels or other devices should be placed. For example, see the labeling requirements for OSHA's vinyl chloride standard, 29 CFR § 1910.1017(1).

104. The standard for ionizing radiation was substantially derived from an established federal standard and can be found at 29 CFR § 1910.96. The standard for electromagnetic

(nonionizing) radiation was derived from an ANSI standard and can be found at 29 CFR § 1910.97. See also the discussion regarding radiation in this chapter, Section 3.11.

105. The ventilation standards were derived from ANSI standards and can be found at 29 CFR § 1910.94.

106. The OSHA asbestos standard can be found at 29 CFR § 1910.1001. Asbestos standards have also been promulgated by the Secretary of Interior pursuant to the Coal Mine Health and Safety Act of 1969 (30 CFR § 71.202) and the Metal and Non-Metallic Mine Safety Act of 1966 (30 CFR §§ 55.5, 56.5, and 57.5) (now the Federal Mine Safety and Health Act of 1977; see this chapter, Section 3.1).

107. For the text of this proposal, see BNA, *OSHR*, Current Report for October 16, 1975, p. 714.

108. BNA, *OSHR*, Current Reports for: April 28, 1983, p. 992; July 7, 1983, p. 124; August 25, 1983, p. 307; and September 29, 1983, p. 403.

109. See *Society of the Plastics Industry v. OSHA* (70), 509 F.2d at 1306; 2 OSHC at p. 1500.

110. *Society of the Plastics Industry v. OSHA*, 509 F.2d at 1307; 2 OSHC at pp. 1500–1501. The standard can be found at 29 CFR § 1910.1017.

111. *Society of the Plastics Industry v. OSHA* (70), 509 F.2d at 1311; 2 OSHC at 1504.

112. See *Dry Color Manufacturers' Association, Inc. v. U.S. Department of Labor*, 486 F.2d 98; 1 OSHC 1331, 1332 (3d Cir. 1973). The 14 chemicals included in the carcinogen standard are 4-nitrobiphenyl, A-napthylamine, 4,4-methylene(bis)(2-chloraniline), methyl chloromethyl ether, 3,3-dichlorobenzidine, bis-chloromethyl ether, B-naphthylamine, benzidine, 4-aminodiphenyl, ethyleneimine, B-propiolactone, 2-acetylaminofluorene, 4-dimethylaminoazobenzene, and N-nitrosodimethylamine. The permanent standards can be found at 29 CFR §§ 1910.1003–1910.1016.

113. 506 F.2d 385; 2 OSHC 1402 (3d Cir. 1974), *certiorari* denied, 423 U.S. 830 (1975).

114. The policy is entitled "Identification, Classification and Regulation of Potential Occupational Carcinogens." Pursuant to the policy, a "potential occupational carcinogen" is defined as "any substance, or combination or mixture of substances, which causes an increased incidence of benign and/or malignant neoplasms, or a substantial decrease in the latency period between exposure and onset of neoplasms in humans or in one or more experimental mammalian species as the result of any oral, respiratory or dermal exposure, or any other exposure which results in the induction of tumors at a site other than the site of administration." Potential carcinogens will be classified as either Category I or Category II substances, depending on the nature and extent of available scientific evidence. The policy also requires OSHA to publish an annual list of candidates for regulation as suspected carcinogens and a semiannual list of those that are found to be of the highest priority. Setting priorities for regulating carcinogens is one of the stated purposes of the policy. For a summary of the policy, see BNA, *OSHR*, Current Report for January 17, 1980, pp. 763–765. The full text of the policy (which fills almost 300 pages in the Federal Register) can be found at 45 *Fed. Reg.* 5002 and in BNA, *OSHR*, Reference File, 41:7304.

115. 448 U.S. 607; 8 OSHC 1586 (1980).

116. A proposal which was to be released by February 1, 1983, was to have included a complete revocation of the old policy. See BNA, *OSHR*, Current Report for December 2, 1982, p. 539.

117. See BNA, *OSHR*, Current Report for April 28, 1983, p. 987.

118. See BNA, *OSHR*, Current Report for June 16, 1983, p. 45.

119. The entire text of the standard can be found at 29 CFR § 1910.1029.

120. The standard can be found at 29 CFR § 1910.1025. For a detailed history of the standard, see *United Steelworkers of America, AFL-CIO v. Marshall*, 647 F.2d 1189; 8 OSHC 1810 (D.C. Cir. 1980), *certiorari* denied, 453 U.S. 913 (1981).

121. *United Steelworkers of America v. Marshall* (120).

122. For a listing of the industries as to which the standard was remanded, as well as for a summary of the court's order, see *United Steelworkers of America v. Marshall* (120), 647 F.2d at 1311; 8 OSHC at 1901–1902.

123. See BNA, *OSHR*, Current Report for December 17, 1981, pp. 539–540.

124. The standard, which occupied 69 pages of the Federal Register, can be found at 29 CFR § 1910.1043. For a detailed history of the standard, see *American Textile Mfrs. Inst. v. Donovan*, 452 U.S. 490, 498–502; 9 OSHC 1913, 1916–1918 (1981).

125. *American Textile Mfrs. Inst. v. Bingham*, 617 F.2d 636; 7 OSHC 1775 (D.C. Cir. 1979).

126. *American Textile Mfrs. Inst. v. Donovan* (124), 452 U.S. at 541; 9 OSHC at 1933.

127. See 48 *Fed. Reg.* 26962-26984 and BNA, *OSHR*, Current Reports for June 9, 1983, pp. 27–28, and June 16, 1983, pp. 55–77.

128. The standard can be found at 29 CFR § 1910.1044. It supersedes the emergency temporary standard for DBCP which had been in effect since the fall of 1977.

129. The PEL specified in the emergency temporary standard was 10 parts per billion.

130. The standard can be found at 29 CFR § 1910.1045.

131. An emergency standard for acrylonitrile had been in effect from January 17, 1978 through July 1978.

132. The standard can be found at 29 CFR § 1910.1018.

133. The Ninth Circuit did stay the application of the arsenic standard to four firms which had demonstrated irreparable harm and willingness to adopt less costly methods of protecting workers during the remand period. As to all other employers, the standard was to remain in effect during the remand period. *ASARCO, Inc. v. OSHA*, 647 F.2d at 2–3; 9 OSHC at 1509.

134. BNA, *OSHR*, Current Report for April 8, 1982, p. 915.

135. BNA, *OSHR*, Current Report for January 13, 1983, pp. 643–644; 48 *Fed. Reg.* 1864–1903.

136. For a detailed history of the benzene standard, see *Industrial Union Dept. v. American Petrol. Instit.* (115), 448 U.S. at 615–630; 8 OSHC at 1588–1593.

137. 581 F.2d 493; 6 OSHC 1959.

138. *Industrial Union Dept. v. American Petroleum Instit.* (115).

139. That standard can be found at 29 CFR § 1910.1000 (Table Z-2).

140. BNA, *OSHR*, Current Reports for April 7, 1983, p. 940, and May 5, 1983, p. 1043.

141. The noise standard can be found at 29 CFR § 1910.95.

142. BNA, *OSHR*, Current Report for March 10, 1983, pp. 827–828. The full text of the amendment can be found at *Id.*, pp. 841 *et seq.*

143. For a history of the amendment as well as a discussion of the changes between the original rule and the final rule, see BNA, *OSHR*, Current Report for March 10, 1983, pp. 827–828.

144. See BNA, *OSHR*, Current Report for May 19, 1983, p. 1080.

145. BNA, *OSHR*, Current Report for April 21, 1983, p. 971.

146. The existing standard can be found at 29 CFR § 1910.1000 (Table Z-1).

147. *Public Citizen Health Research Group v. Auchter*, 554 F. Supp. 242; 11 OSHC 1049 (D.C. D.C. 1982).

148. *Auchter v. Public Citizen Health Research Group*, 702 F. 2d 1150, 1159; 11 OSHC 1209, 1215 (D.C. Cir. 1983).

149. BNA, *OSHR*, Current Report for December 24, 1981, p. 579. The existing OSHA standard for ethylene dibromide is 20 parts per million, which was adopted in the early 1970s as an interim standard.

150. BNA, *OSHR*, Current Reports for September 15, 1983, pp. 371–372, and September 22, 1983, pp. 388–389.

151. BNA, *OSHR*, Current Report for June 23, 1983, p. 93.

152. *International Union, United Automobile, Aerospace, and Agricultural Implement Workers of America v. Donovan* (No. 82–2401, D.C. D.C.).

153. BNA, *OSHR*, Current Report for August 18, 1983, p. 291. The current OSHA standard for formaldehyde is 3 parts per million, with a 5 ppm ceiling limit and a 10 ppm peak. *Id.*

154. A status chart of OSHA's proposed standards and NIOSH criteria documents can be found at BNA, *OSHR*, Reference File at 11:2151–2164. See also BNA, *OSHR*, Current Report for September 29, 1983, pp. 405–406.

155. See note 89. The standards adopted by the Secretary of Interior regarding air contaminants are the TLVs developed by the ACGIH. With respect to coal mines, the TLVs apply only to surface coal mines and to surface work areas of underground coal mines. A separate standard has been promulgated by the Secretary of Interior for asbestos.

156. *Atlantic & Gulf Stevedores* [(71) 534 F.2d at 553; 4 OSHC at 1069–1070].

157. *Dunlop v. Trumbull Asphalt Company, Inc.*, 4 OSHC 1847 (E.D. Mo. 1976).

158. See *Powell v. Globe Industries, Inc.*, 431 F. Supp. 1096; 5 OSHC 1250 (N.D. Ohio 1977). The National Labor Relations Board (NLRB) has concurrent jurisdiction over Section 11(c) cases. In 1975 the General Counsel of the NLRB and the Secretary of Labor entered into an understanding for the procedural coordination of litigation arising under Section 11(c) of OSHA and Section 8 of the National Labor Relations Act, to avoid duplicate litigation. See J. Irving, "Effect of OSHA on Industrial Relations and Collective Bargaining," in: *Proceedings of the ABA National Institute on Occupational Safety and Health Law*, ABA Press, Chicago, 1976, pp. 125–127. See also 40 Fed. Reg. 26083 (June 20, 1976).

159. The general anti-retaliation provision in the 1977 Mine Safety and Health Act is set forth in Section 105(c) of the statute. The 1977 mine act also provides for immediate inspection of a coal mine at the request of a miner [§ 103(g)], the right of employees to accompany the inspector on his walk-around inspection of the coal mine [§ 103(f)], and limited payments to miners when a safety violation closes the mine [(§ 111)]. Black lung (coal worker's pneumoconiosis) benefits are provided to totally disabled coal miners and surviving dependents of coal miners whose deaths were due to black lung disease (§§ 401 *et seq.*).

160. 29 CFR § 1977.12(b)(2).

161. 445 U.S. at 10–11; 8 OSHC at 1004.

162. With respect to work stoppages over safety disputes in the context of collective bargaining agreements and the National Labor Relations Act, see *Gateway Coal Company v. United Mine Workers*, 414 U.S. 368; 1 OSHC 1461 (1974) and *National Labor Relations Board v. Tamara Foods, Inc.*, 692 F.2d 1171 (8th Cir. 1982).

163. 483 F.2d 67; 1 OSHC 1294 (5th Cir. 1973). Also, *Taylor v. Brighton Corp.*, 616 F.2d 256; 8 OSHC 1010 (6th Cir. 1980).

CHAPTER FIFTEEN

Compliance and Projection

MARTHA HARTLE MUNSCH, J.D., and
LARAINE R. ALLEN, J.D.

1 INVESTIGATIONS AND INSPECTIONS

With the enactment of the Occupational Safety and Health Act of 1970 (OSHA), Congress authorized the Secretary of Labor to enter, inspect, and investigate places of employment to discover possible violations of the employer's general and specific duties under the act.

Section 8(a) authorizes the secretary, upon presenting appropriate credentials to the owner, operator, or agent in charge:

1. To enter without delay and at reasonable times any factory, plant, establishment, construction site, or other area, workplace or environment where work is performed by an employee or employer.
2. To inspect and investigate during regular working hours and at other reasonable times, and within reasonable limits and in a reasonable manner, any such place of employment and all pertinent conditions, structures, machines, apparatus, devices, equipment, and materials therein, and to question privately any such employer, owner, operator, agent, or employee.

The Federal Mine Safety and Health Act of 1977 directs authorized representatives of the secretary to make frequent, unannounced inspections and investigations in coal or other mines each year. The purposes of these visits include determining whether an imminent danger exists and whether there is compliance with the mandatory health and safety standards issued under that statute (1).

1.1 Inspection Procedures—Warrants

Section 8(a) of OSHA on its face gives the Secretary of Labor the unqualified right to enter and inspect, in a "reasonable manner," any place of employment upon the presentation of credentials. However, the United States Supreme Court in 1978 held Section 8(a) unconstitutional insofar as it authorized nonconsensual warrantless inspections at an employer's establishment. In *Marshall v. Barlow's, Inc.* (2), the Court held that an employer may refuse entry to an OSHA compliance officer unless a warrant is obtained, reasoning that employers have a reasonable expectation of privacy in their commercial property and are therefore guaranteed the right, under the Fourth Amendment, to be free from unreasonable official intrusions (3).

The Supreme Court explained, however, that the secretary need not make a showing of probable cause in the criminal sense to obtain a warrant. Instead, probable cause authorizing an administrative inspection may be based either upon specific evidence of a violation or upon a showing that reasonable legislative or administrative standards for conducting an inspection are satisfied with respect to a particular establishment (4). By way of illustration, the Court stated that a warrant could properly be issued upon a showing that a particular business was chosen for inspection pursuant to a general enforcement plan derived from "neutral sources" (5).

1.1.1 Probable Cause Necessary for Issuance of Warrants

Following the *Barlow's* decision, several lower federal courts have addressed the question of the specific showing of probable cause necessary to obtain an OSHA inspection warrant. The Supreme Court stated in *Barlow's* that probable cause authorizing an OSHA inspection may be based upon evidence that a specific violation exists at the establishment to be inspected. Thus, although probable cause requirements for warrant issuance may be established by OSHA's receipt of an employee complaint (6), the Seventh and Tenth Circuit Courts of Appeals have held that the nature of the violation complained of must be described in the warrant application so the magistrate issuing the warrant may make an independent determination that probable cause exists (7). In addition, the majority of federal courts addressing the issue have concluded that a warrant based on specific employee complaints is overly broad if it purports to authorize a "wall-to-wall" inspection of the entire plant. Instead, the scope of the warrant and resulting inspection must bear a reasonable relationship to the specific violations complained of (8).

There appears to be some question whether a past history of OSHA violations satisfies the requirement of probable cause for issuance of a warrant. The Second Circuit Court of Appeals has upheld the issuance of a warrant based on an employer's record of past violations (9). The Seventh Circuit has also upheld a warrant based on a showing that the employer had been cited for a violation at an old plant, and that the employer had tried to abate the violation by moving the unsafe operations to a new plant (10). The same court, however, has stated that a history of past violations, standing alone, is insufficient to establish probable cause (11).

Finally, the Supreme Court in *Barlow's* stated that an inspection warrant could be issued pursuant to an inspection plan based on "neutral criteria." Under this standard, the warrant application must describe the administrative plan being used, so the magistrate can make an independent determination that it is based on "neutral criteria" (12). Thus, warrant applications describing OSHA's general inspection program, in which employers are randomly chosen for inspection from a list of firms in industries with above average lost workday rates, have been held to satisfy probable cause (13).

In addition, the Seventh Circuit has held that an inspection warrant was properly issued on the basis of OSHA's National Emphasis Program targeting high-hazard industries, even without evidence in the warrant application showing why a particular firm within the industry was selected (14). On the other hand, district courts in Pennsylvania and New Jersey have ruled that an employer's involvement in a high-hazard industry is insufficient, by itself, to establish probable cause for issuance of a warrant. Instead, these courts have held that the warrant application must contain information from which the magistrate can conclude that the particular firm within the high-hazard industry was chosen for inspection on the basis of "neutral criteria" (15).

1.1.2 *Ex Parte* Warrants

Additional questions raised by the Supreme Court's decision in *Barlow's* involve the proper procedure for obtaining, enforcing, and contesting OSHA inspection warrants. Although the Fifth Circuit Court of Appeals has held that the United States district courts do not have jurisdiction to issue injunctions compelling employers to submit to OSHA inspections (16), it is well settled that both the district courts and the United States magistrates have the authority to issue administrative inspection warrants (17).

A more difficult question has been presented regarding OSHA's right to obtain an *ex parte* inspection warrant: *i.e.*, a warrant obtained without prior notice to the employer. In *Barlow's*, the Supreme Court suggested that the secretary could, by appropriate regulation, provide for *ex parte* warrants although the regulation then in force called instead for "compulsory process" (18). When the original regulation (19) was held not to include *ex parte* warrants (20), an "interpretive rule" was issued stating that the term "compulsory process" was intended to include *ex parte* warrants (21). In *Cerro Metal Products, Division of Marmon Group, Inc. v. Marshall* (22), the Third Circuit held the "interpretive rule" invalid because it was inconsistent both with OSHA's prior interpretations of the "compulsory process" regulation and with the Supreme Court's dictum in *Barlow's*. Thus the court held that *ex parte* warrants were unavailable to the secretary. The Seventh, Ninth, and Tenth Circuits disagreed, holding that the secretary's "interpretive rule" properly authorized *ex parte* warrants (23). OSHA subsequently promulgated a new regulation (24) which clearly states that OSHA may obtain an inspection warrant without prior notice to the employer (25).

1.1.3   Challenging Validity of Warrant

A final question that faced the federal courts as a result of the *Barlow's* decision was the proper procedure for an employer to use in challenging the validity of an OSHA inspection warrant. In general, it appears that an employer who wishes to contest a warrant has two choices: (1) he may refuse to obey the warrant and may move to quash it in district court; or (2) he may allow the inspection to proceed and challenge the warrant's validity before the Review Commission if he contests any citations issued pursuant to the inspection.

In *Babcock & Wilcox Company v. Marshall*, the Third Circuit indicated that an employer may obtain a district court hearing on a warrant's validity prior to inspection by refusing to obey the warrant (thereby risking contempt), moving to quash the warrant, and promptly appealing if the motion is denied (26). However, the Third Circuit held that if the employer obeys the warrant, he must "exhaust his administrative remedies" before the Review Commission prior to obtaining judicial review of his objections to the warrant (27).

The First, Fifth, and Eighth Circuits have agreed with the Third Circuit that the doctrine of exhaustion of remedies precludes an employer from challenging an executed warrant in federal court (28). The Seventh Circuit, on the other hand, has held that an employer may obtain district court review of a warrant even after an inspection has been completed (29).

1.2   Inspection Procedures Concerning Matters Other than Warrants

It is well settled that OSHA inspections must be made at reasonable times, in a reasonable manner, and within reasonable limits pursuant to the act (30). The act also requires the inspector to present his credentials to the employer before beginning the inspection (31). The Fifth Circuit Court of Appeals has held, however, that even if the inspector fails to present his credentials, such failure cannot operate to exclude evidence obtained in the inspection when there is no showing that the employer was prejudiced thereby in any way (32).

Section 8(e) of OSHA requires that a representative of the employer and an authorized employee representative be allowed to accompany the OSHA inspector during the physical inspection of any workplace. In *Chicago Bridge & Iron Company v. OSAHRC and Dunlop* (33), the Seventh Circuit held the dictates of Section 8(e) to be mandatory rather than merely directory. The court refused, however, to hold that the absence of a formalized offer of an opportunity to accompany the compliance officer on his inspection rendered the citations for violations observed during that inspection void *ab initio*. Rather, the court explained that when there has been substantial compliance with the mandate of the act regarding walk-around rights and the employer is unable to demonstrate that prejudice resulted from his nonparticipation in the inspection, citations issued as a result of the inspection are valid (34).

The court in *Accu-Namics* (32, at p. 833) did not reach the question of whether the language of Section 8(e) was mandatory or directory, but it did refuse to adopt a rule that would exclude all evidence obtained illegally, no matter how minor or technical the government's violation and no matter how egregious or harmful the employer's safety violation.

The Ninth and Tenth Circuits have also concluded that minor violations of Section 8(e)'s inspection procedures do not justify dismissing a citation (35) or suppressing evidence gained from an inspection (36), at least as long as there has been substantial compliance with the act and the employer's defense on the merits has not been prejudiced.

The Fourth and Eighth Circuits have gone even further, holding as a matter of law that regardless of whether there has been substantial compliance by the inspector, Section 8(e) violations do not affect the validity of citations issued or evidence obtained unless the employer can show that he has been prejudiced in preparing or presenting his defense on the merits. In *Pullman Power Products, Inc. v. Marshall and OSAHRC* (37), the Fourth Circuit Court of Appeals held that, in the absence of such prejudice, the validity of citations was not affected by an inspector's alleged failure to properly present his credentials, conduct opening and closing conferences, and provide walk-around rights. In so holding, the court declined to reach the issue of substantial compliance, stating broadly that the employer's inability to show prejudice bars its attack on the validity of the citations.

The Court of Appeals for the Eighth Circuit reached a similar conclusion in *Marshall v. Western Waterproofing Co., Inc.* (38). Although stating that the requirements of Section 8(e) are not merely directory, the court nevertheless held that, in the absence of prejudice to the employer, an inspector's failure to comply with these requirements will not justify suppression of evidence, regardless of whether there has been substantial compliance (39).

2 RECORDKEEPING AND REPORTING

Section 8(c)(1) of OSHA requires employers to make, keep, and preserve such records regarding their OSHA-related activities as the Secretary of Labor, in cooperation with the Secretary of Health, Education, and Welfare (now Health and Human Services "HHS"), may prescribe as necessary or appropriate for the enforcement of the act or for developing information on the causes and prevention of occupational accidents and illnesses. These records must also be made available to the Secretaries of Labor and/or HHS (40).

Section 8(c)(2) more specifically directs the Secretary of Labor, in cooperation with the Secretary of HHS, to issue regulations requiring employers to maintain accurate records of, and to make periodic reports on, work-related deaths, injuries, and illnesses (other than minor injuries requiring only first aid treatment and not involving medical treatment, loss of consciousness, restriction of work or motion, or transfer to another job).

Regulations promulgated by the Secretary of Labor implementing Sections 8(c)(1) and (2) and in effect as of August 1, 1983, require employers to keep the following records or the equivalent thereof:

OSHA Form 200   A log and summary of all recordable occupational injuries and illnesses.
OSHA Form 101   A supplementary record for each occupational injury or illness (41).

The regulations further require that a copy of Form 200, summarizing the year's occupational illnesses and injuries, be posted in each establishment in a conspicuous place or places where notices to employees are customarily posted.

The records, required by Forms 200 and 101, must be retained in each establishment for five years following the end of the year to which they relate. The regulations further require that within 48 hours after an on-the-job accident that is fatal to one or more employees or results in hospitalization of five or more employees, the employer must report the accident either orally or in writing to the nearest office of the Area Director of the OSH Administration.

In addition, Section 24 of the act directs the Secretary of Labor, in consultation with the Secretary of HHS, to develop and maintain a program of collection, compilation, and analysis of occupational safety and health statistics. The Secretary of Labor has given the Commissioner of the Bureau of Labor Statistics (BLS) the authority to develop and maintain such a program. This program requires employers to participate in periodic surveys of occupational injuries and illnesses. The survey form is OSHA Form 200S and an employer who receives such a form has a duty to complete and return it promptly (42).

Small employers are exempt from many, but not all, of OSHA's recordkeeping requirements. For example, employers who had no more than ten employees at any time during the calender year preceding the current one are exempt from the requirements of keeping the log and summary (OSHA Form 200) and supplementary record (OSHA Form 101) described previously (43). However, small employers are not exempted from the requirement of reporting accidents resulting in fatalities or multiple hospitalizations. In addition, small employers may be selected to participate in the BLS periodic surveys and, if selected, must maintain a log and summary on Form 200 for the survey year, and must make the required reports on Survey Form 200S.

Finally, with respect to toxic materials or harmful physical agents, the Secretary of Labor, in cooperation with the Secretary of HHS, is directed by Section 8(c)(3) of the act to issue regulations requiring employers to maintain accurate records of employee exposure to potentially toxic materials or harmful physical agents that are required to be monitored or measured under Section 6. These regulations must guarantee employees or their representatives an opportunity to observe the required monitoring or measuring, and to have access to the records of employee exposure. Further, the regulations must ensure employees and former employees access to such records as will indicate their own exposure to such substances (44).

These statutory directives have been implemented by the Secretary of Labor in a rule governing employee exposure and medical records, issued in May 1982 (45). This rule requires employers to maintain exposure and medical records pertaining to their employees' exposure to toxic substances and harmful physical agents (46). The exposure records must be made accessible to exposed and potentially exposed employees, as well as to designated employee representatives and to OSHA. Access to medical records must also be ensured for the employee and for OSHA; because of the privacy interests involved, however, an employee's medical records are open to his collective bargaining representative only with his consent (47).

3 SANCTIONS FOR VIOLATING SAFETY AND HEALTH LAWS

3.1 Citations

The Occupational Safety and Health Act authorizes the Secretary of Labor to issue citations and proposed penalties to employers who are believed to have violated the act or its implementing regulations. Section 9(a) directs the secretary to issue citations "with reasonable promptness" following an inspection or investigation. No citation may be issued after the expiration of 6 months from the occurrence of any violation [§9(c)] (48).

Each citation is to be in writing and must describe "with particularity the nature of the violation, including a reference to the provision of the chapter, standard, rule, regulation, or order alleged to have been violated" (49). Section 9(a) states that each citation must also fix a reasonable time for the abatement of the violation (50).

Section 9(b) requires employers to post each citation prominently at or near each place where a violation referred to in the citation occurred. The mechanics of how, when, where, and how long to post the citations are set forth in regulations issued by the secretary (51).

3.2 Penalties

Within a reasonable time after a citation has been issued, the secretary is directed by Section 10(a) to notify the employer by certified mail of the penalty, if any, that will be assessed for the violation. The penalty will be based at least in part on the nature of the violation (52). Violations fall into the following general categories: serious, nonserious, *de minimis*, willful, repeated, and criminal.

3.2.1   Serious Violations

Section 17(k) of OSHA defines a serious violation as follows:

[A] serious violation shall be deemed to exist in a place of employment if there is a substantial probability that death or serious physical harm could result from a condition which exists, or from one or more practices, means, methods, operations, or processes which have been adopted or are in use, in such place of employment unless the employer did not, and could not with the exercise of reasonable diligence, know of the presence of the violation.

The *probability* of an *accident* occurring need not be shown to establish that a violation is serious (53). Rather, the Ninth Circuit court ruled in *California Stevedore and Ballast* (53) that a serious violation exists if any accident that should result from a violation would have a substantial probability of resulting in death or serious physical harm (54). No actual death or physical injury is required to establish a serious violation (55).

Employer knowledge is clearly an element of a serious violation. The knowledge requirement in Section 17(k) deals with actual or constructive knowledge (56) of practices or conditions that constitute violations of the act; it is not directed to knowledge of the law (57). The burden of proof is on the secretary to prove knowledge (58) as well as the other elements of a serious violation of the act (59).

Section 17(b) provides that an employer who has received a citation for a serious violation of the act *must* be assessed a civil penalty of up to $1000 for each such violation.

3.2.2   Nonserious Violations

The original Senate version of the occupational safety and health bill treated all violations as "serious." As finally enacted, however, OSHA incorporated a House proposal for violations "determined not to be of a serious nature" (60).

The statute does not describe the elements of a nonserious violation and provides no guidelines for determining when a violation is not serious. The Fifth Circuit, however, has described nonserious violations as violations that do not create a substantial probability of serious physical harm (61). The commission has explained that serious and nonserious violations are distinguished on the basis of the seriousness of injuries that experience has shown are reasonably likely to result when an accident does arise from a particular set of circumstances (62). At least one federal court of appeals has held that employer knowledge is an element of a nonserious violation (63).

When a violation is determined not to be serious, the assessment of a penalty is discretionary rather than mandatory (64). Section 17(c) states that the employer *may* be assessed a civil penalty of up to $1000 for each nonserious violation. In the Department of Labor's appropriations for fiscal years 1977 through 1983, however, Congress exempted employers from penalties for nonserious, first-instance violations unless 10 or more such violations were uncovered during the inspection. In addition, funding measures for fiscal 1977 through 1983 have prohibited the assessment of penalties for nonserious violations against any employer with 10 or fewer employees, who has previously requested an on-site consultation and who makes good-faith efforts to correct workplace hazards (65).

Section 110(a) of the Federal Mine Safety and Health Act of 1977 does not allow discretionary penalties for so-called nonserious violations, but instead requires the Secretary of Labor to assess a civil penalty of up to $10,000 for each violation of a mandatory health or safety standard under the act.

3.2.3 De Minimis Violations

If noncompliance with an OSHA provision or standard presents no direct or immediate threat to the safety or health of employees, the violation is *de minimis* and the Secretary of Labor may issue only a notice—not a citation—to the employer (66). The notice contains no proposed penalty (67). An OSHA Program Directive issued in October 1978 explains that, in keeping with a "common sense" approach to OSHA enforcement, a violation should be considered *de minimis* if: (a) an employer complies with the intent of a standard but deviates from its particular requirements in a way that has no direct or immediate relationship to safety and health; or (b) an employer complies with a proposed amendment to a standard and the amendment provides equal or greater safety and health protection than the standard itself; or (c) an employer's workplace is "state of the art" that is, it is technically advanced beyond the requirements of a standard and provides equal or greater safety and health protection (68).

3.2.4 Willful or Repeated Violations

The Occupational Safety and Health Act provides more stringent civil penalties for employers who "willfully or repeatedly" violate the act or any regulations promulgated pursuant thereto. Willful or repeated violations are subject under Section 17(a) to penalties of up to $10,000 for each violation (69). The act contains no definition of either "willful" or "repeated" as applied to violations. Thus it is not surprising that the courts have had difficulty in agreeing on the elements of these types of violations.

In *Frank Irey Jr., Inc. v. OSAHRC* (70), the Third Circuit initially held that "[w]illfulness connotes defiance or such reckless disregard of consequences as to be equivalent to a knowing, conscious, and deliberate flaunting of the Act. Willful means more than merely voluntary action or omission—it involves an element of obstinate refusal to comply."

The majority of the circuits, as well as the Occupational Safety and Health Review Commission, have declined to follow the *Frank Irey* definition of "willfulness." The First and Fourth Circuits have interpreted a willful action as a "conscious, intentional, deliberate voluntary decision," regardless of venial motive (71).

The Review Commission has agreed that no showing of malicious intent is necessary to establish "willfulness," (72) defining a willful violation as one "committed with either an intentional disregard of, or plain indifference to, the Act's requirements" (73). The Second, Fifth, Sixth, Eighth, Ninth, and Tenth Circuits have all either adopted the Review Commission's standard or have embraced similar definitions that do not require a showing of a bad motive (74).

Thus, a conflict developed between the Third Circuit's view, as expressed in *Frank Irey*, and the majority approach. The Court of Appeals for the District of Columbia Circuit, however, characterized this conflict as more apparent than real. The court in *Cedar Construction Co. v. OSAHRC and Marshall* (75) indicated that the two approaches were likely to yield the same results in particular cases, since there is little practical difference between "obstinate refusal to comply" and "intentional disregard" of the act.

The Third Circuit agreed when it addressed the "willfulness" question again in a later decision. In *Babcock & Wilcox Co. v. OSAHRC* (76), the court explained its holding in *Frank Irey*, reasoning that (77):

[t]he supposed conflict among the circuits on this point has been generated by several courts of appeals reading into our *Irey* definition a requirement that the employer act with "bad purpose." Read in this fashion, *Irey* has not been followed by some circuits. . . . To our way of thinking, an "intentional disregard of OSHA requirements" differs little from an "obstinate refusal to comply;" nor is there in context much to distinguish "defiance" from "intentional disregard." . . . We also believe, as does the District of Columbia Circuit, that the same results would likely be reached in various cases, including the one here, regardless of the verbiage utilized. . . . It is not unusual that different words are used to describe the same basic concept.

In light of the Third Circuit's decision in *Babcock & Wilcox*, it appears there is now general agreement among the circuits that the Review Commission's "intentional disregard" standard is the correct one, and that no malicious intent need be shown to establish a willful violation (78).

The interpretation of "repeated" violation has also generated disagreement among employers, the courts and the commission. In *Bethlehem Steel Corp. v. OSAHRC and Brennan* (79), the Third Circuit held that the commission can find a repeated violation only when the evidence shows the employer consciously ignored or "flaunted" the requirements of the act and was cited for a similar violation on at least *two prior* occasions.

The word "flaunting" had been used by the Third Circuit in *Frank Irey Jr.* in determining whether a violation was properly classified as "willful." The court in *Bethlehem Steel* reasoned that a repeated violation, like a willful violation, must consist of particularly flagrant conduct. The court explained its "test" as follows (80):

The mere occurrence of a violation of a standard or regulation more than twice does not constitute that flaunting necessary to be found before a penalty can be assessed under Sec. [17(a)]. What acts constitute flaunting of the requirements of the Act must be determined, in the first instance, by the Secretary and the Commission, but they should be guided by our statements in *Frank Irey*. . . . It should be noted that Sec. [17(a)] can be applicable even if the same standard is never violated twice, if the general or specific duty clauses of Sec. [5(a)] are repeatedly violated in such a way as to demonstrate a flaunting disregard of the requirements of the Act. Among the factors the Commission should consider when determining whether a course of conduct is flaunting the requirements of the Act are the number, proximity in time, nature and extent of violations, their factual and legal relatedness, the degree of care of the employer in his efforts to prevent violations of the type involved, and the nature of the duties, standards, or regulations violated.

The court further explained that in applying the "repeatedly" portion of the act, the commission must determine that the acts themselves "flaunt" the requirements of the statute, but need not determine whether the acts were performed with an *intent* to "flaunt" the requirements of the statute. A "repeated" violation is established by proof of facts from which it can be inferred that an employer's conduct constitutes disregard of the act's requirements (81).

The majority of federal circuit courts that have considered the issue have refused to adopt the Third Circuit's interpretation of a "repeated" violation. In addition, a majority of the members of the commission as well as several state tribunals have rejected the Third Circuit's test.

The Fourth Circuit, in *George Hyman Construction Company v. OSAHRC* (82), reasoned that the requirement of flagrant misconduct to establish a repeated violation fails to "recognize a meaningful distinction between willful and repeated violations." (83) The Fifth, Sixth, Ninth, and Tenth Circuits have agreed that an employer need not have a particular state of mind or motive for "flaunting" the act, nor otherwise exhibit an aggravated form of misconduct to be guilty of a repeated violation (84). Each of these courts has also rejected the Third Circuit's conclusion that a repeated violation must be based on at least two prior violations, requiring instead only one prior and substantially similar infraction (85).

In addition, a majority of the members of the current Review Commission agree that no "flaunting" of the act need be shown to establish a repeated violation. In *Potlach Corporation* (86), then-Chairman Cleary and Commissioner Cottine joined in an opinion defining a "repeated" violation as follows (87):

A violation is repeated under section 17(a) of the Act if, at the time of the alleged repeated violation, there was a Commission final order against the same employer for a substantially similar violation.

Thus, although the circuits appear to have resolved their disagreement concerning the definition of a "willful" violation, the controversy surrounding "repeated" violations will undoubtedly continue until a definitive interpretation of that term is rendered by either the Supreme Court or Congress (88).

3.2.5   Criminal Sanctions

Job safety and health legislation also provides criminal sanctions for certain specified conduct. The most stringent criminal sanctions are set forth in the Federal Mine Safety and Health Act of 1977. Section 110(d) states that a mine operator can be subjected to a fine of up to $25,000 or a prison term of up to one year (or both) for willfully violating mandatory health or safety standards or for knowingly refusing to comply with certain orders issued under that statute.

Willful violations (89) of OSHA (or of any standard or rule promulgated pursuant thereto) that cause death to any employee can result in a fine of up to $10,000 or imprisonment for up to six months or both (90).

Criminal penalties may also be imposed for: (a) knowingly making any false statement, representation, and so forth, in any document filed pursuant to or required to be maintained by OSHA or by the Mine Safety and Health Act of 1977 (91); (b) giving advance notice of an OSHA or Mine Safety and Health inspection without the authority of the Secretary of Labor (92); (c) knowingly distributing, selling, and so on, in commerce any equipment for use in coal or other mines that is represented as complying with the Mine Safety and Health Act and does not do so (93); (d) killing an OSHA inspector or investigator on account of the performance of his duties (94).

3.2.6   Failure to Abate a Violation

The Occupational Safety and Health Act does not specify fixed periods within which violations must be remedied, but Section 9(a) does require that each citation "fix a reasonable time for the abatement of the violation" (95). An employer who fails to correct a violation within the period specified in the citation may receive an additional citation pursuant to Section 10(b) for failure to abate. Failure to abate may result in the assessment of civil penalties of not more than $1000 per day for each day the violation continues. According to Section 17(d) the abatement period does not begin to run until the date of the final order of the commission affirming the citation, as long as the review proceeding, if any, initiated by the employer was in good faith and not solely for delay or avoidance of penalties.

Notices of violations under Sections 104(a) and 104(b) of the Federal Mine Safety and Health Act of 1977 must similarly specify time periods for abatement of violations. Failure to abate under that statute can result in an order directing all persons to be withdrawn from the affected area of the mine until a representative of the secretary determines that the violation has been abated.

3.3 Contesting Citations and Penalties

Section 10(a) of OSHA and regulations promulgated thereunder provide a means for contesting citations and proposed penalties. After the employer has been notified of the penalty proposed by the secretary, the employer has 15 working days to notify the secretary that he wishes to contest the citation or the proposed assessment of penalty. A failure to notify the secretary within 15 days of intent to contest the citation or proposed penalty will render the citation or penalty "a final order of the Commission and not subject to review by any court or agency" (96).

The secretary's regulations (97) instruct the employer that "[e]very notice of intention to contest shall specify whether it is directed to the citation or to the proposed penalty, or both." Similarly, the courts have construed the OSHA enforcement scheme as mandating a distinction between contesting a citation and contesting a proposed penalty (98). Thus, in *Dan J. Sheehan* the Fifth Circuit held (99) that an employer's letter that contested the proposed penalty but failed to contest the citation (in fact, the letter affirmatively admitted the violation) constituted waiver of the employer's right to challenge the citation on appeal. However, the commission has stated that it will construe notices of contest that are limited to the penalty to include a contest of the citation as well if the cited employer indicates later that it was his intent to contest the citation (100).

If an employer files a timely notice of contest (or if within 15 working days of the issuance of a citation, a representative of his employees files a notice challenging the period of abatement specified in the citation), the secretary must immediately advise the Occupational Safety and Health Review Commission of the intent to contest. The commission then must afford an opportunity for an administrative hearing (101).

A commission hearing is conducted pursuant to the Administrative Procedure Act and is presided over by a single administrative law judge employed by the commission. After taking testimony, the judge writes an opinion, which is subject to review by the full three-member commission at its discretion (102). An aggrieved party may petition for discretionary review before the full commission (103), and any commission member may direct review of a case on his own motion (104). If no commissioner directs review, or if a timely petition for review is not filed, the administrative law judge's decision becomes a final order of the commission (105).

Section 10(a) authorizes the commission to review either the citation or the proposed penalty or both. The commission's scope of review is set forth in Section 10(c), which provides:

The Commission shall thereafter [i.e., after hearing] issue an order, based on findings of fact, affirming, modifying or vacating the Secretary's citation or proposed penalty, or directing other appropriate relief, and such order shall become final thirty days after its issuance.

Furthermore, Section 17(j) empowers the commission to assess appropriate civil penalties, giving due consideration to the size of the business of the employer being charged, the gravity of the violation, the good faith of the employer, and the history of previous violations (106). On several occasions the commission has taken the position that it may exercise its power under Section 17(j) to *increase* the secretary's proposed penalty after considering the factors outlined above. At least three courts of appeals have expressed the view that the commission may act in this manner (107).

The Ninth Circuit in *California Stevedore* (53) also sanctioned the commission's right to increase the degree of a violation from nonserious to serious. The commission has taken the view that it can reduce the degree of a violation as well (108).

The final stage of an OSHA enforcement proceeding is review in the Court of Appeals and thereafter discretionary review by the Supreme Court. Any person adversely affected or aggrieved by the commission's disposition (109) may obtain review in the Court of Appeals pursuant to Section 11(a) of the act. Section 11(b) provides that the Secretary of Labor may also obtain review or enforcement of any final order of the commission by filing a petition for such relief in the appropriate Court of Appeals. The reviewing court is bound by Section 11(a) to apply the "substantial evidence test" to the commission's findings of fact (110). The same section empowers the court to direct the commission to consider additional evidence if the evidence is material and reasonable grounds existed for a party's failure to admit it in the hearing before the commission. Regarding the penalty imposed by the commission, the reviewing court may inquire only whether the commission abused its discretion because the assessment of a penalty is not a finding of fact but rather the exercise of a discretionary grant of power (111).

3.4   Imminent Danger Situations

Section 13(a) of OSHA confers jurisdiction on the United States District Courts, upon petition of the Secretary of Labor, to restrain hazardous employment conditions or practices if they create an imminent danger of death or serious physical harm that cannot be eliminated through the act's other enforcement procedures.

As originally reported out of the House Committee, the act contained a provision that would have permitted an OSHA inspector to close down an operation for up to 72 hours without a court order if he found that an imminent danger existed. The original Senate version of the bill also contained a provision allowing an inspector to close down an operation for 72 hours, but this provision was revised so that no shutdown can occur unless a federal district judge grants an application for a temporary restraining order (112).

The Federal Mine Safety and Health Act of 1977, however, permits coal or other mine operations to be shut down without a restraining order from a court. Section 107(a) of the act provides that when a federal inspector finds that an imminent danger is present in a mine, he shall order the withdrawal of all other persons from a part or all of that mine until the imminent danger no longer exists (113).

3.5 Constitutional Challenges to the OSHA Enforcement Scheme

The citation and penalty scheme of OSHA has been subject to constitutional challenge on several fronts.

In *Atlas Roofing Company, Inc. v. OSAHRC* (114) a cited employer contended that the act was constitutionally defective because: (a) civil penalties under OSHA are really penal and call for the constitutional protections of the Sixth Amendment and Article III; (b) even if the penalties are civil, OSHA violates the Seventh Amendment because of the absence of a jury trial for fact finding; (c) the act denies the employer his right to a Fifth Amendment "prejudgment" due process hearing since commission orders are self-executing unless the employer affirmatively seeks review; and (d) the overall penalty structure of OSHA violates due process because it "chills" the employer's right to seek review of the citation and penalty.

The Fifth Circuit in *Atlas Roofing* rejected all the employer's constitutional contentions. The Supreme Court granted a petition for *certiorari* in that case, limited to the Seventh Amendment issue (115), and subsequently upheld the act's provision for imposition of civil penalties without fact finding by a jury (116).

Several other Courts of Appeals have considered similar constitutional challenges to the act's enforcement scheme and most, if not all, have likewise rejected the constitutional contentions (117).

4 THE FUTURE OF JOB SAFETY AND HEALTH LAW

According to Representative Joseph M. Gaydos (D-Pa.), chairman of the House Education and Labor Subcommittee on Safety and Health, the future of the Occupational Safety and Health Act is "stable." Rep. Gaydos predicted "little if no activity" in the House of Representatives in 1983 to change the structure of OSHA because occupational safety and health "are no longer the controversial issues they once were" (118).

Notwithstanding the present lack of legislative activity in the area of occupational safety and health, the Reagan Administration has made and continues to make an effort to redirect the focus of OSHA. This effort has drawn criticism and even anger from organized labor, whose spokesmen contend the administration is weakening OSHA standards and attempting to remold OSHA for its own purposes (119).

The present OSH Administration seems to have departed from the traditional approach of emphasizing standards enforcement and instead has emphasized voluntary compliance programs, abatement and compliance assistance through consultation services, labor/management assistance programs, and worker training and education (120). For example, on July 15, 1982, a policy became effective whereby OSHA compliance officers and inspectors are to assist employers in identifying methods for correcting possible safety and health violations (121). Moreover, in July 1982, OSHA announced a nationwide effort to encourage voluntary compliance that would focus on employers with specified safety and health programs which exceed the OSHA standards in providing a safe working environment. Participants would be exempt from OSHA's general schedule inspections list and would be given priority attention on variance applications (122).

A further example of the current administration's new direction for OSHA is the lifting of recordkeeping rules regarding workplace injuries and illnesses for employers in certain low-hazard retail trade, finance, and service industries (43). As a result of this exemption, which became effective January 1, 1983, approximately 474,000 employers in industries shown to be the least hazardous will no longer have to adhere to the recordkeeping rules.

In addition to the current administration's emphasis on voluntary compliance and education, it has also adopted a new approach to issuing standards and reviewing current regulations. Assistant Labor Secretary Thorne G. Auchter has stated that OSHA will follow a three-step evaluation process and avoid acting on an emotional basis when making decisions concerning standards and regulations (123). Pursuant to this process, the OSH Administration, first, will collect data and objectively analyze it to determine if a suspected hazard presents a significant risk. Second, the OSH Administration will decide, based on clear and objective evidence, if a proposed rule would reduce the risk. Finally, as required by President Reagan's Executive Order 12291, the OSH Administration will perform a cost/benefit analysis to aid in giving a "clear indication of an approach we should take" (124).

The cost/benefit aspect of the OSH Administration's evaluation process has been sharply criticized by organized labor. The AFL–CIO has termed cost/benefit analysis a "primitive and unreliable tool for regulations," and has argued that cost/benefit analysis should be used only as one of the many tools to determine whether a rule should be issued (125).

Despite the current emphasis on voluntary compliance, statistics released in 1983 indicate that inspections were still a major aspect of the OSH Administration's overall safety and health program (126). Approximately 63,914 inspections were conducted by the OSH Administration in 1982, an increase of nearly 15 percent over the 1981 total and the second highest total in the previous six years (127). In addition, more general schedule inspections directed toward high hazard workplaces were conducted in 1982 than in the two previous years (128). The OSH Administration acknowledged , however, that the total number of inspections for 1982, while higher than the 55,593 total of 1981, was less than the 64,876 sum for 1980, the last full year of the Carter Administration (129).

Finally, the OSH Administration announced that it would concentrate its efforts in 1983 on providing leadership and assistance in what it termed the six key areas in its overall safety and health program: internal agency management, employer/employee assistance programs, state plan programs, compliance, federal agency programs, and standards development (130). Also, in 1983 the OSH Administration was expected to begin implementing its integrated management information system, as well as to place particular emphasis on the needs of new workers in its employer/employee assistance programs, and to seek to expand its targeting mechanisms for improving the effective use of field resources (131).

In sum, although there is no doubt that OSHA is here to stay, its final form and direction is anything but certain. It seems that each succeeding presidential administration intends to place its mark on OSHA. For the immediate future, it will be interesting to see how the Reagan Administration's focus and direction for OSHA fares, particularly in light of organized labor's criticism of the Reagan policies and programs in safety and health.

NOTES AND REFERENCES

1. See Section 103(a) of the Federal Mine Safety and Health Act of 1977. For a discussion of "imminent danger," see Section 3.4 of this chapter. Underground coal or other mines must be inspected at least four times each year. Each surface coal or other mine must be inspected at least twice a year.

2. 436 U.S. 307; 6 OSHC 1571 (1978). Several state courts have likewise held the inspection provisions of their respective state occupational safety and health acts unconstitutional, or have held that inspections are permissible under those provisions only pursuant to a warrant. For example: *Woods & Rohde, Inc., d/b/a Alaska Truss & Millwork v. State*, 565 P.2d 138; 5 OSHC 1530 (Alaska Sup. Ct. 1977) (provision authorizing warrantless inspections violates state constitution); *State v. Albuquerque Publishing Co.*, 571 P.2d 117; 5 OSHC 2034 (New Mexico Sup. Ct. 1977), *certiorari* denied, 435 U.S. 956; 6 OSHC 1570 (1978) (nonconsensual inspection pursuant to state occupational safety and health act requires warrant satisfying administrative standards of probable cause); *Yocom v. Burnette Tractor Co., Inc.*, 6 OSHC 1638 (Kentucky Sup. Ct. 1978).

3. By contrast, the Supreme Court has upheld the constitutionality of warrantless inspections under the Federal Mine Safety and Health Act of 1977 on the grounds that mining has long been a pervasively regulated industry, in which a businessman can have no reasonable expectation of privacy against official inspections. In *Donovan v. Dewey*, 452 U.S. 594 (1981), the Court reasoned further that the Federal Mine Safety and Health Act inspection provisions are more narrowly drawn, and provide for less administrative discretion concerning inspections than does OSHA, thus making such inspections reasonable under the Fourth Amendment.

4. 436 U.S. at 320-21; 6 OSHC at 1575-76. Several state courts have held that this relaxed administrative standard of probable cause is sufficient to support an inspection warrant pursuant to their respective occupational safety and health acts. See, for example, *Yocom v. Burnette Tractor Co., Inc.*, (2); *State v. Keith Manufacturing Co.*, 6 OSHC 1043 (Oregon Ct. of Appeals 1977) (provision of Oregon Safe Employment Act authorizing inspection warrant based on administrative standard of probable cause is constitutional); *State v. Kokomo Tube Co.*, 426 N.E. 2d 1338; 10 OSHC 1158 (Indiana Ct. of Appeals 1981) (administrative probable cause standard articulated in *Barlow's* is applicable to inspection warrants under Indiana OSH Act). By contrast, the California Court of Appeals has held that a higher standard of

probable cause is required for inspections under Cal/OSHA because that statute authorizes extensive criminal sanctions. *Salwasser Manufacturing Co., Inc. v. Municipal Court*, 156 Cal. Rptr. 292; 7 OSHC 1492 (California Ct. of Appeals 1979) (requiring a showing of probable cause to believe that a safety violation currently exists at the workplace to be inspected).

5. 436 U.S. at 321; 6 OSHC at 1575-76. Such "neutral sources," according to the Court, could include statistics indicating "dispersion of employees in various types of industries across a given area, and the desired frequency of searches in any of the lesser divisions of the area. . ." *Id.* at 1576.

6. See *Northwest Airlines, Inc.*, 587 F.2d 12; 6 OSHC 2070 (7th Cir. 1978); *Marshall v. Chromalloy American Corp. (Gilbert & Bennett Manufacturing Co.)*, 589 F.2d 1335; 6 OSHC 2151 (7th Cir. 1979), *certiorari* denied, 444 U.S. 884; 7 OSHC 2238 (1979); *Marshall v. W and W Steel Co., Inc.*, 604 F.2d 1322; 7 OSHC 1670 (10th Cir. 1979); *Burkart Randall Div. of Textron, Inc. v. Marshall*, 625 F.2d 1313; 8 OSHC 1467 (7th Cir. 1980). On the other hand, it has been held that the occurrence of an accident on the employer's premises is insufficient evidence of a specific violation to establish probable cause for an inspection. See, for example, *Donovan v. Federal Clearing Die Casting Co.*, 655 F.2d 793; 9 OSHC 2072 (7th Cir. 1981) (newspaper reports of accident do not satisfy "specific evidence of existing violation" standard); *Marshall v. Pool Offshore Co.*, 467 F. Supp. 978; 7 OSHC 1179 (W.D. La. 1979) (while actual fatalities do not constitute evidence of specific violation, an OSHA policy of investigating all fatalities may satisfy administrative "neutral criteria" standard).

7. *Weyerhaeuser Co. v. Marshall*, 592 F.2d 373; 7 OSHC 1090 (7th Cir. 1979); *Marshall v. Horn Seed Co., Inc.*, 647 F.2d 96; 9 OSHC 1510 (10th Cir. 1981). The Seventh Circuit has concluded that, since probable cause in the strict criminal sense is not required in OSHA inspection cases, the warrant application need not establish the reliability of the complainant or the basis of his complaint. *Marshall v. Chromalloy American Corp. (Gilbert & Bennett Manufacturing Co.)* (6); *Burkart Randall Div. of Textron, Inc., v. Marshall* (6). Accord, *Establishment Inspection of BP Oil, Inc.*, 509 F. Supp. 802; 9 OSHC 1282 (E.D. Pa. 1981). The Tenth Circuit, on the other hand, held in *Horn Seed Co.* that a warrant application based on an employee complaint must contain evidence showing the complaint was actually made by a complainant who was sincere in asserting a violation existed, and who had some plausible basis for his belief.

8. *Marshall v. Central Mine Equipment Co.*, 608 F.2d 719; 7 OSHC 1907 (8th Cir. 1979); *Marshall v. North American Car Co.*, 626 F.2d 320; 8 OSHC 1722 (3d Cir. 1980); *Donovan v. Sarasota Concrete Co.*, 693 F.2d 1061; 11 OSHC 1001 (11th Cir. 1982); *Establishment Inspection of Asarco, Inc.*, 508 F. Supp. 350; 9 OSHC 1317 (N.D. Tex. 1981). The Seventh Circuit has held to the contrary, reasoning that a limitation on the scope of the inspection would enable employers to present a special "sanitized" area for inspection, while concealing real violations. See *Marshall v. Chromalloy American Corp. (Gilbert & Bennett Manufacturing Co.)* (6); *Burkart Randall Div. of Textron, Inc. v. Marshall* (6); *Marsan Co., Inc.*, 7 OSHC 1557 (N.D. Ind. 1979). The majority opinion in *Burkart Randall* suggests that "wall-to-wall" inspections are permissible in every case involving an employee complaint, but this view was not espoused by a majority of the judges deciding the case. Instead, the result in *Burkart Randall* seems to have been based on the fact that the complaints were not confined to isolated areas of the plant, so that a broad inspection was viewed as reasonable by a majority of the court. Because the *Gilbert & Bennett* and *Marsan* cases also involved complaints that could not be pinpointed to specific areas of the workplace, the Seventh Circuit decisions may indicate merely that a "wall-to-wall" inspection is permissible when it reasonably appears that the complained-of violation affects the entire plant. This view is in accord with the decisions of other circuits. For example, see *In re Establishment Inspection of Seaward International, Inc.*, 510 F. Supp. 314 (W.D. Va. 1980), *affirmed without opinion*, 644 F.2d 880 (4th Cir. 1981) (complaint involving exposure to carcinogens); *Hern Iron Works, Inc. v. Donovan*, 670 F.2d 838; 10 OSHC 1433 (9th Cir.), *certiorari* denied, 103 S. Ct. 69 (1982) (complaint pertaining to ventilation system).

9. *Marshall v. Northwest Orient Airlines, Inc.*, 574 F.2d 119; 6 OSHC 1481 (2d Cir. 1978).

10. *Pelton Casteel, Inc. v. Marshall*, 588 F.2d 1182; 6 OSHC 2137 (7th Cir. 1978).

11. *Marshall v. Chromalloy American Corp. (Gilbert & Bennett Manufacturing Co.)* (6). *Accord, Marshall v. Weyerhaeuser Co.,* 456 F. Supp. 474; 6 OSHC 1920 (D.N.J. 1978) (past violations do not establish probable cause when a follow-up inspection of those violations revealed they had been corrected).

12. *Northwest Airlines, Inc.* (6); *Marshall v. Weyerhaeuser Co.* (11).

13. *Peterson Builders, Inc.,* 525 F.Supp. 642; 10 OSHC 1169 (E.D. Wis. 1981); *Donovan V. Athenian Marble Corp.,* 10 OSHC 1450 (W.D. Okla. 1982); *Erie Bottling Corp.,* 539 F. Supp. 600; 10 OSHC 1632 (W.D. Pa. 1982); *Urick Foundry Co. v. Donovan,* 542 F. Supp. 82; 10 OSHC 1765 (W.D. Pa. 1982).

14. *Marshall v. Chromalloy American Corp. (Gilbert & Bennett Manufacturing Co.)* (6). *Accord, Marshall v. Multi-Cast Corp.,* 6 OSHC 1486 (N.D. Ohio 1978); *The Fountain Foundry Corp. v. Marshall,* 6 OSHC 1885 (S.D. Ind. 1978).

15. *Urick Foundry,* 472 F.Supp. 1193; 7 OSHC 1497 (W.D. Pa. 1979) (warrant would be sufficient if it showed, for example, that the particular firm chosen for inspection had a history of prior OSHA violations, or that it was selected at random from the list of firms in the high-hazard industry); *Marshall v. Weyerhaeuser Co.* (11) (fact that employer was member of high-hazard industry does not establish probable cause in the absence of additional facts explaining why this particular employer was selected).

16. *Marshall v. Gibson's Products, Inc. of Plano,* 584 F.2d 668; 6 OSHC 2092 (5th Cir. 1978). Based on its decision in *Gibson's Products,* the Fifth Circuit has further held that the district court has no jurisdiction to issue an injunction enforcing a warrant that has already been issued. *Marshall v. Shellcast Corp.,* 592 F.2d 1369; 7 OSHC 1239 (5th Cir. 1979). In *Marshall v. Pool Offshore Co.* (6), however, the court held that, despite *Gibson's Products,* it had jurisdiction to impose sanctions on an employer for its refusal to permit an inspection because the proceeding was for civil contempt for failure to obey a warrant rather than for an injunction to compel an inspection.

17. The Seventh Circuit has held that federal magistrates have the authority to issue OSHA inspection warrants pursuant to 28 U.S.C. § 636, which permits district judges to assign to magistrates any duties not inconsistent with the constitution or federal law. Since the *Barlow's* decision authorizes OSHA inspections pursuant to warrants, it is consistent with federal law to permit magistrates to issue such warrants. See *Marshall v. Chromalloy American Corp. (Gilbert & Bennett Manufacturing Co.)* (6); *Pelton Casteel, Inc. v. Marshall* (10). *Accord, Marshall v. Multi-Cast Corp.* (14). Other federal courts have held that the district courts and the United States magistrates have the authority to issue warrants simply because it would be inconsistent with the rationale of *Barlow's* to hold otherwise. *The Fountain Foundry Corp. v. Marshall* (14); *Marshall v. Weyerhaeuser Co.* (11); *Marshall v. Huffhines Steel Co.,* 478 F. Supp. 986; 7 OSHC 1850 (N.D. Tex. 1979).

18. 436 U.S. at 316–317; 6 OSHC at 1575.

19. The regulation permitting the secretary to obtain "compulsory process" to compel an inspection was published at 29 CFR § 1903.4.

20. See *Cerro Metal Products v. Marshall,* 467 F. Supp. 869, 872; 7 OSHC 1125, 1126–1127 (E.D. Pa. 1979), *affirmed,* 620 F.2d 964; 8 OSHC 1196 (3d Cir. 1980).

21. 43 *Fed. Reg.* 59,838 (1978); 29 CFR § 1903.4(d). See discussion in *Cerro Metal Products* (20), 7 OSHC at 1127.

22. 620 F.2d 964; 8 OSHC 1196 (3d Cir. 1980). *Accord, Marshall v. Huffhines Steel Co.,* 7 OSHC 1910 (N.D. Tex. 1979), *affirmed without opinion sub nom. Donovan v. Huffhines Steel Co.,* 9 OSHC 1762 (5th Cir. 1981).

23. *Rockford Drop Forge Co. v. Donovan,* 672 F.2d 626; 10 OSHC 1410 (7th Cir. 1982); *Stoddard Lumber Co. v. Marshall,* 627 F.2d 984; 8 OSHC 2055 (9th Cir. 1980); *Marshall v. W & W Steel Co., Inc.* (6).

24. 45 *Fed. Reg.* 65,916–65,924 (1980).

25. The new regulation, specifically held valid by the court in *Donovan v. Blue Ridge Pressure Castings, Inc.*, 543 F. Supp. 53; 10 OSHC 1217 (M.D. Pa. 1981), reads as follows [29 CFR § 1903.4(d) (1980)]:

   (d) For purposes of this section, the term compulsory process shall mean the institution of any appropriate action, including *ex parte* application for an inspection warrant or its equivalent. *Ex parte* inspection warrants shall be the preferred form of compulsory process in all circumstances where compulsory process is relied upon to seek entry to a workplace under this section.

26. 610 F.2d 1128, 1135–1136; 7 OSHC 1880, 1884 (3d Cir. 1979). An employer's right to a pre-inspection hearing in federal court on a warrant's validity was reaffirmed by the Third Circuit in *Cerro Metal Products v. Marshall* (22), holding an employer could bring an action in district court, prior to inspection, seeking to enjoin OSHA from obtaining an *ex parte* warrant. The Seventh Circuit has also held an employer may, in limited circumstances, raise a pre-inspection warrant challenge in district court. See *Blocksom & Co.*, 582 F.2d 1122; 6 OSHC 1865 (7th Cir. 1978) (employer may raise invalidity of warrant as a defense in action seeking to hold it in contempt for failure to obey warrant).

27. 610 F.2d at 1135–1137; 7 OSHC at 1883–1885. *Accord, Marshall v. Whittaker Corp.*, *Berwick Forge & Fabricating Co.*, 610 F.2d 1141; 6 OSHC 1888 (3d Cir. 1979); *Establishment Inspection of the Metal Bank of America, Inc.*, 11 OSHC 1193 (3d Cir. 1983).

28. *Quality Products, Inc.*, 592 F.2d 611; 7 OSHC 1093 (1st Cir. 1979); *Baldwin Metals Co., Inc. v. Donovan*, 642 F.2d 768; 9 OSHC 1568 (5th Cir.), *certiorari denied sub nom. Mosher Steel Co. v. Donovan*, 454 U.S. 893 (1981); *Central Mine Equipment Co.* (8). Each of these cases involved post-inspection motions to suppress evidence, but the language of the opinions did not specifically limit the exhaustion doctrine to post-inspection challenges. Therefore, it is unclear whether the First, Fifth, and Eighth Circuits would permit an employer to challenge a warrant in federal court prior to inspection, as the Third Circuit suggested in *Babcock & Wilcox*. However, a district court in the Fifth Circuit has held that the exhaustion doctrine does not bar an employer's counterclaim attacking the validity of a warrant in a contempt action by the secretary. *Marshall v. Huffhines Steel Co.* (22).

29. *Weyerhaeuser Co. v. Marshall*, 592 F.2d 373; 7 OSHC 1090 (7th Cir. 1979); *Federal Casting Div., Chromalloy American Corp. v. Donovan*, 684 F.2d 504; 10 OSHC 1801 (7th Cir. 1982) (following *Weyerhaeuser*).

30. *Dunlop v. Able Contractors, Inc.*, 4 OSHC 1110 (D. Mont. 1975), *affirmed sub nom. Marshall v. Able Contractors, Inc.*, 573 F.2d 1055; 6 OSHC 1317 (9th Cir.), *certiorari denied*, 437 U.S. 826 (1978).

31. See Section 8(a). The secretary's regulations provide in relevant part [29 CFR § 1903.7(a)]:

   At the beginning of an inspection, Compliance Safety and Health Officers shall present their credentials to the owner, operator, or agent in charge at the establishment; explain the nature and purpose of the inspection; and indicate generally the scope of the inspection and the records . . . which they wish to review.

   The secretary's regulations further provide for a conference at the conclusion of the inspection. During this conference the Compliance Safety and Health Officer advises the employer of any apparent safety or health violations disclosed by the inspection, and the employer is afforded an opportunity to bring to the attention of the officer any pertinent information regarding conditions in the workplace [29 CFR § 1903.7(e)].

32. *Accu-Namics, Inc. v. OSAHRC and Dunlop*, 515 F.2d 828; 3 OSHC 1299 (5th Cir.), *rehearing denied*, 521 F.2d 814 (5th Cir. 1975), *certiorari denied*, 425 U.S. 903 (1976). Another controversial question involving the exclusion of evidence has arisen as a result of the Supreme Court's decision in the *Barlow's* case: whether the exclusionary rule, borrowed from the field of criminal law, applies in OSHA hearings to prohibit the use of evidence obtained without a valid search warrant. The Ninth Circuit, in *Todd Shipyards Corp. v. Secretary of Labor*,

586 F.2d 683; 6 OSHC 2122 (9th Cir. 1978), suggested that the rule should not apply because OSHA hearings are civil rather than criminal in nature. The Tenth Circuit indicated the contrary in *Savina Home Industries, Inc. v. Secretary of Labor and OSAHRC*, 594 F.2d 1358; 7 OSHC 1154 (10th Cir. 1979), holding that the rule should apply in order to deter improper OSHA inspections. The Eleventh Circuit, while not deciding whether the exclusionary rule must be applied in OSHA enforcement proceedings, has stated that the OSHA Review Commission is free to apply the rule if it sees fit, and need not allow an exception permitting the use of evidence obtained by OSHA "in good faith." *Donovan v. Sarasota Concrete Co.* (8). On the other hand, the Seventh Circuit held in *Donovan v. Federal Clearing Die Casting Co. and OSAHRC*, 11 OSHC 1014 (7th Cir. 1982), that the commission must apply a "good faith" exception to the exclusionary rule where an OSHA search was made pursuant to a warrant upheld by the district court, even though the warrant was later invalidated on appeal. As these conflicting decisions indicate, the applicability of the exclusionary rule in OSHA enforcement proceedings remains an unresolved question.

33. 535 F.2d 371; 4 OSHC 1181 (7th Cir. 1976).

34. 535 F.2d at 377; 4 OSHC at 1185. The Seventh Circuit found substantial compliance in the *Chicago Bridge & Iron Co.* case because of the on-site representative of Chicago Bridge & Iron was informed of the pending inspection and was given literature setting forth the directives of the act.

35. See *Marshall v. C.F. & I. Steel Corp. and OSAHRC*, 576 F.2d 809; 6 OSHC 1543 (10th Cir. 1978), in which the inspector failed to give the employer formal notice that its facilities—in addition to those of its subcontractor—were the target of the inspection. Because the employer's personnel manager did in fact accompany the inspector on his rounds, the court held that dismissal of the citation would be a grossly excessive sanction in relation to the inspector's minor violations of Section 8(e).

36. See *Hartwell Excavating Co. v. Dunlop*, 537 F.2d 1071; 4 OSHC 1331 (9th Cir. 1976), in which the court found substantial compliance even though the employer's superintendent was not notified of the inspection until it had been partially completed, because the inspector had made an unsuccessful attempt to locate the supervisor earlier.

37. 655 F.2d 41; 9 OSHC 2075 (4th Cir. 1981).

38. 560 F.2d 947; 5 OSHC 1732 (8th Cir. 1977).

39. In *Leone v. Mobil Oil Corp.*, 523 F.2d 1153 (D.C. Cir. 1975), the District of Columbia Circuit held that neither OSHA nor the Fair Labor Standards Act of 1938, 29 U.S.C. § 203(o) requires an employer to pay wages for time spent by employees in accompanying OSHA inspectors on walk-around inspections of the employer's plant. However, in September 1977, OSHA announced an amendment of its administrative regulations (29 CFR § 1977.21) to reflect a new policy that employees should be paid by their employers for time spent on walk-arounds. The Chamber of Commerce of the United States filed suit challenging the policy, and the Court of Appeals for the District of Columbia Circuit ultimately held the regulation invalid because OSHA had failed to promulgate it properly under the notice and comment requirements of the Administrative Procedure Act. *Chamber of Commerce of United States v. OSHA*, 636 F.2d 464; 8 OSHC 1648 (D.C. Cir. 1980). The regulation was subsequently revoked (45 *Fed. Reg.* 72,118 (1980); BNA, *OSHR*, Current Report for October 30, 1980, pp. 593–594), and a new one was proposed and issued. (45 *Fed. Reg.* 75,232 (1980); 46 *Fed. Reg.* 3852 (1981); BNA, *OSHR*, Current Reports for: November 20, 1980, pp. 653–658; January 22, 1981, pp. 845–853). After the Reagan Administration took office, however, implementation of the new walk-around pay rule was delayed in response to a memorandum from the President (BNA, *OSHR*, Current Report for February 5, 1981, p. 1225), and was ultimately withdrawn (46 *Fed. Reg.* 28,842 (1981); BNA, *OSHR*, Current Report for June 4, 1981, pp. 21–24).

40. Recordkeeping and reporting requirements under the Federal Mine Safety and Health Act of 1977 are set forth in subsections 103(c)–103(e) and 103(h) of that statute.

41. These regulations are found in 29 CFR Part 1904. The secretary has defined "recordable occupational injuries or illnesses" as those that result in [29 CFR § 1904.12(c)]:

    a. Fatalities, regardless of the time between the injury and death, or the length of the illness.

    b. Lost workday cases, other than fatalities, that result in lost workdays.

    c. Nonfatal cases without lost workdays, which result in transfer to another job or termination of employment, or require medical treatment (other than first aid) or involve: loss of consciousness or restriction of work or motion. This category also includes any diagnosed occupational illnesses that are reported to the employer but are not classified as fatalities or lost workday cases.

42. 29 CFR §§ 1904.20–1904.21.

43. 29 CFR § 1904.15. As of January 1, 1983, some 474,000 employers in low-hazard industries were also exempted from keeping an injury/illness log and summary, and a supplemental record. See 29 CFR §§ 1904.12(h), 1904.16 (summarized in CCH, *Employment Safety and Health Guide*, No. 608, January 4, 1983). See also BNA, *OSHR*, Current Report for January 6, 1983, pp. 611–612. Exempted industries include automotive dealers, banks, real estate firms, and furniture stores. None of the exempt industries has been subject to a routine general schedule inspection over the past five years. Despite the new regulation, however, some of the firms may be required to participate in the BLS annual surveys and to maintain an injury/illness log for the survey year in which they are selected.

44. Subsection 103(c) of the Federal Mine Safety and Health Act of 1977 contains a similar requirement guaranteeing employees access to toxic materials exposure records.

45. 29 CFR § 1910.20. The medical records rule is augmented by the specific permanent standards issued by the Secretary of Labor for certain toxic materials and harmful physical agents. See, for example, the OSHA standard regarding employee exposure to coke oven emissions, 29 CFR § 1910.1029(m).

46. In July 1982, OSHA proposed modifications that would narrow the scope of the medical records rule. The proposed changes would limit the number of employees and the kinds of records covered by the rule, shorten the amount of time records must be retained, and increase trade secret protection. BNA, *OSHR*, Current Report for July 15, 1982, p. 147. As of July 1983, the proposed modifications had not yet been adopted.

47. Despite these privacy interests, OSHA is specifically granted access to employee medical records. 29 CFR § 1910.20(c) (10). However, additional regulations require OSHA to observe administrative procedures designed to protect the confidentiality of this information. 29 CFR Part 1913. In a case involving similar privacy concerns, the Supreme Court refused to review the Sixth Circuit's ruling that NIOSH may subpoena employee medical records in the course of a health hazard inquiry. In *General Motors Corp. v. NIOSH*, 636 F.2d 163; 9 OSHC 1139 (6th Cir. 1980), *certiorari* denied, 454 U.S. 877; 10 OSHC 1032 (1981), the court held that, as long as there is no public disclosure of this medical information, NIOSH may obtain access to it without violating the employees' privacy rights. At least one federal district court has reached a similar conclusion regarding OSHA's medical records access rule. *Louisiana Chemical Assoc. v. Bingham*, 550 F. Supp. 1136; 10 OSHC 2113 (W.D. La. 1982) (appeal pending).

48. The Review Commission has held that this 6-month limitation period does not apply if the secretary's inability to discover the violation was caused by the employer's failure to report a fatal accident as the secretary's regulations require [29 CFR § 1904.8]. The commission reasoned that allowing an employer to escape a citation because of its own failure to comply with OSHA's reporting requirements would reward it for its own wrongdoing. *Yelvington Welding Service*, 6 OSHC 2013 (Rev. Comm. 1978).

49. See Section 9(a). Interpretations of the "particularity" requirements have generally dealt with: (a) the precision of the reference to the standard allegedly violated, and (b) the

adequacy of the description of the alleged violation. For a general discussion of the particularity issue, see CCH, *Employment Safety & Health Guide*, ¶ 4107 (1980), and annotations thereto. Several federal courts have held that, to meet the particularity requirement, a citation must provide the employer with "fair notice" of the violation sufficient to enable it both to prepare its defense and to correct the cited hazard. *Whirlpool Corp. v. OSAHRC and Marshall*, 645 F.2d 1096; 9 OSHC 1362 (D.C. Cir. 1981); *Noblecraft Industries, Inc. v. Secretary of Labor and OSAHRC*, 614 F.2d 199; 7 OSHC 2059 (9th Cir. 1980); *Marshall v. B.W. Harrison Lumber Co. and OSAHRC*, 569 F.2d 1303; 6 OSHC 1446 (5th Cir. 1978).

50. The Fifth Circuit Court of Appeals has held that an employer may raise a citation's lack of particularity in a later failure-to-abate action by the secretary. In so holding, the court reasoned that a citation that is too vague to give notice of the action necessary to correct the cited hazard also makes it impossible for the Review Commission to determine whether the hazard has been abated. See *Marshall v. B.W. Harrison Lumber Co. and OSAHRC* (49).

51. American Bar Association, *Report of the Committee on Occupational Safety and Health Law*, ABA Press, Chicago, 1975, p. 57. See also CCH, *Employment Safety & Health Guide*, ¶ 4118 (1980). The secretary's regulations can be found at 29 CFR § 1903.16.

52. In determining the amount of the penalty, the secretary is also directed to consider "the size of the business of the employer being charged, the gravity of the violation, the good faith of the employer, and the history of previous violations." 29 CFR § 1903.15(b).

53. *Dorey Electric Co. v. OSAHRC*, 553 F.2d 357; 5 OSHC 1285 (4th Cir. 1977); *California Stevedore and Ballast Co. v. OSAHRC*, 517 F.2d 986; 3 OSHC 1174 (9th Cir. 1975); *Shaw Construction, Inc. v. OSAHRC*, 534 F.2d 1183, 1185 & n.4; 4 OSHC 1427, 1428 & n.4 (5th Cir. 1976). The Review Commission has held that the probability of an accident, while not necessary to establish a serious violation, is a factor to be considered in determining the gravity of the violation for penalty assessment purposes. [See the secretary's penalty assessment regulations at 29 CFR § 1903.15(b)]. Thus, an employer's accident-free history, while not a defense to a citation for a serious violation, may justify reducing the proposed penalty. *George C. Christopher & Sons, Inc.*, 10 OSHC 1436, 1446 (Rev. Comm. 1982).

54. *Accord, Usery v. Hermitage Concrete Pipe Co. and OSAHRC*, 584 F.2d 127; 6 OSHC 1886 (6th Cir. 1978); *Kent Nowlin Construction Co., Inc. v. OSAHRC*, 9 OSHC 1709 (10th Cir. 1981).

55. *Brennan v. OSAHRC and Vy Lactos Laboratories Inc.*, 494 F.2d 460; 1 OSHC 1623 (8th Cir. 1974).

56. An employer has an obligation to inspect the workplace to discover and prevent possible hazards, and its failure to do so can support a finding that it should have known of the violations it failed to discover. *Joseph J. Stolar Construction Co., Inc.*, 9 OSHC 2020 (Rev. Comm.), appeal denied, 681 F.2d 801; 10 OSHC 1936 (2d Cir. 1981). However, the existence of an effective employer inspection and maintenance program, designed to discover and remedy hazardous conditions, may preclude a finding that the employer should reasonably have known of violations it actually failed to discover. *Cullen Industries, Inc.*, 6 OSHC 2177 (Rev. Comm. J. 1978). *Cf. East Texas Motor Freight, Inc. v. OSAHRC and Donovan*, 671 F.2d 845; 10 OSHC 1457 (5th Cir. 1982) (because employer's inspection program was not effective in discovering and repairing defective machinery, employer was charged with knowledge of undiscovered defects).

57. *Mid-Plains Construction Co.*, 3 OSHC 1484 (Rev. Comm. 1975); *Southwestern Acoustics & Specialty, Inc.*, 5 OSHC 1091 (Rev. Comm. 1977).

58. *Brennan v. OSAHRC and Hendrix (d/b/a Alsea Lumber Co.)*, 511 F.2d 1139, 1144; 2 OSHC 1646, 1648–1649 (9th Cir. 1975). *Accord, Ocean Electric Corp. v. Secretary of Labor and OSAHRC*, 594 F.2d 396; 7 OSHC 1149 (4th Cir. 1979); *Diversified Industries Div., Independent Stave Co. v. OSAHRC and Marshall*, 618 F.2d 30, 31 n.8; 8 OSHC 1107, 1108 n.8 (8th Cir. 1980).

59. See, for example, *Consolidated Rail Corp.*, 10 OSHC 1564, 1568 (Rev. Comm. 1982) (appeal pending) (secretary's failure to prove the probability of serious injury in case of an accident requires a finding that the violation was nonserious).

60. Conference Report 91-1765, 1970 USC, *Congressional and Administrative News*, p. 5237; OSHA § 17(c).

61. *Ryder Truck Lines, Inc. v. Brennan*, 497 F.2d 230, 233; 2 OSHC 1075, 1077 (5th Cir. 1974).

62. *Standard Glass and Supply Co.*, 1 OSHC 1223–1224 (Rev. Comm. 1973). See also *Consolidated Rail Corp.*, 10 OSHC 1564, 1568 (Rev. Comm. 1982) (where the evidence did not show the violation created a substantial probability of serious injury in case of an accident, the citation was affirmed as nonserious rather than serious). In determining whether a violation is serious or nonserious, the commission has held that a number of nonserious violations may be grouped together to form a serious violation if the cumulative effect of the violations could result in an accident causing death or serious injury. *H.A.S. & Associates*, 4 OSHC 1894, 1897–1898 (Rev. Comm. 1976). When the violation poses a risk of illness through cumulative exposure to a toxic substance, the relevant question is whether the secretary can show a substantial probability of serious harm resulting from the degree and length of actual employee exposure. *Bethlehem Steel Corp.*, 11 OSHC 1247, 1252 (Rev. Comm. 1983); *Texaco, Inc.*, 8 OSHC 1758, 1761 (Rev. Comm. 1980).

63. *Brennan v. OSAHRC and Hendrix (d/b/a Alsea Lumber Co.)* (58); *National Steel and Shipbuilding Co. v. OSAHRC*, 607 F.2d 311, 315–316 n.6; 7 OSHC 1837, 1840 n.6 (9th Cir. 1979). See also *Dunlop v. Rockwell International*, 540 F.2d 1283, 1291; 4 OSHC 1606, 1611–1612 (6th Cir. 1976), in which the court upheld a divided Review Commission's affirmance of a nonserious, rather than a serious, violation where employer knowledge was not established. The court, in dictum, approved of the reasoning of the Ninth Circuit in the *Alsea Lumber* case. Contra, *Arkansas-Best Freight Systems, Inc. v. OSAHRC*, 529 F.2d 649, 655 n.11; 3 OSHC 1910, 1913 n.11 (8th Cir. 1976); *Brennan v. OSAHRC and Interstate Glass Co.*, 487 F.2d 438, 442 n.19; 1 OSHC 1372, 1375 n.19 (8th Cir. 1973).

64. Section 17(i) of the act requires that a penalty be assessed for violation of OSHA's posting requirements. However, the commission has stated that such a violation is nevertheless considered "nonserious." *Thunderbolt Drilling, Inc.*, 10 OSHC 1981 (Rev. Comm. 1982).

65. BNA, *OSHR*, Current Reports for: August 11, 1977, p. 348; October 19, 1978, pp. 668–669; October 18, 1979, p. 467; November 29, 1979, p. 603; October 2, 1980, p. 483; December 18, 1980, p. 761; June 11, 1981, p. 35; October 8, 1981, p. 364; November 26, 1981, p. 491; December 17, 1981, p. 541; April 8, 1982, p. 920; October 7, 1982, p. 371; December 16, 1982, p. 582. See also CCH, *Employment Safety & Health Guide*, ¶¶ 4152, 6258 (1980); ¶¶ 6260, 6270 (1981); ¶ 6272 (1983).

66. *Lee Way Motor Freight, Inc. v. Secretary of Labor*, 511 F.2d 864, 869; 2 OSHC 1609, 1612 (10th Cir. 1975). See § 9(a) of the act and the regulations promulgated thereunder. 29 CFR § 1903.14.

67. Although the secretary's regulations provide for the issuance of a notice for a *de minimis* violation, OSHA's current policy is to discuss *de minimis* violations with employers and retain records of them in the employer's case file, but not to issue formal notices of them in most cases. BNA, *OSHR*, Current Report for November 23, 1978, pp. 934–935.

68. BNA, *OSHR*, Current Report for November 23, 1978, pp. 934–35.

69. The Federal Mine Safety and Health Act of 1977 [Section 110(d)] does not provide more stringent civil penalties for willful violations of that act's safety and health standards, but does impose criminal liability on mine operators who are found guilty of such willful conduct. The act does impose [Section 110(g)] a civil penalty on miners who willfully violate the safety standards relating to smoking or the carrying of smoking materials.

70. 519 F.2d 1200, 1207; 2 OSHC 1283, 1289 (3d Cir. 1974), *affirmed on other points sub nom. Atlas Roofing Co., Inc. v. OSAHRC*, 97 S.Ct. 1261 (1977).

71. *Messina Construction Corp. v. OSAHRC*, 505 F.2d 701; 2 OSHC 1325 (1st Cir. 1974); *Intercounty Construction Co. v. OSAHRC*, 522 F.2d 777, 780; 3 OSHC 1337, 1339 (4th Cir. 1975), *certiori* denied 423 U.S. 1072, 96 S.Ct. 854, 47 L.Ed.2d 82, 3 OSHC 1879 (1976).

72. *Kent Nowlin Construction, Inc.*, 5 OSHC 1051, 1055 (Rev. Comm. 1977), *affirmed in relevant*

part 593 F.2d 368, 369; 7 OSHC 1105, 1108 (10th Cir. 1979). For an interpretation of the California Occupational Safety and Health Act's "willful" violation provision, see *Rawly's Division of Merit Ends, Inc.* (Docket No. 823-75) (summarized in BNA, *OSHR*, Current Report for February 24, 1977, p. 1234).

73. *Kus-Tum Builders, Inc.*, 10 OSHC 1128, 1131 (Rev. Comm. 1981); *A. Schonbek & Co., Inc.*, 9 OSHC 1189, 1191 (Rev. Comm. 1980), *affirmed* 646 F.2d 799; 9 OSHC 1562 (2d Cir. 1981). Several state tribunals have also adopted the "intentional disregard and indifference" test for "willfulness." See *Monadnock Fabricators, Inc.* (Docket No. RB491), a ruling of the Vermont Occupational Safety and Health Review Board (summarized in BNA, *OSHR*, Current Report for June 24, 1982, p. 96). The New Mexico Health and Safety Review Commission applied the same standard in *Environmental Improvement Division v. Stearns-Roger, Inc.* (Docket No. 82-3) (summarized in BNA, *OSHR*, Current Report for November 18, 1982, pp. 491–492).

74. *A. Schonbek & Co., Inc. v. Donovan and OSAHRC*, 646 F.2d 799, 9 OSHC 1562 (2d Cir. 1981), *affirming* 9 OSHC 1189 (Rev. Comm. 1980); *Mineral Industries & Heavy Construction Group (Brown & Root, Inc.) v. OSAHRC and Marshall*, 639 F.2d 1289, 9 OSHC 1387 (5th Cir. 1981); *Georgia Electric Co. v. Marshall and OSAHRC*, 595 F.2d 309; 7 OSHC 1343 (5th Cir. 1979); *Empire-Detroit Steel Div., Detroit Steel Corp. v. OSAHRC and Marshall*, 579 F.2d 378; 6 OSHC 1693 (6th Cir. 1978); *Western Waterproofing Co., Inc. v. Marshall and OSAHRC*, 576 F.2d 139; 6 OSHC 1550 (8th Cir.), *certiorari* denied, 439 U.S. 965 (1978); *National Steel & Shipbuilding Co. v. OSAHRC* (63); *Kent Nowlin Construction Co. v. OSAHRC and Marshall*, 593 F.2d 368; 7 OSHC 1105 (10th Cir. 1979), *affirming in relevant part* 5 OSHC 1051 (Rev. Comm. 1977).

75. 587 F.2d 1303; 6 OSHC 2010, 2012 (D.C. Cir. 1978).

76. 622 F.2d 1160; 8 OSHC 1317 (3d Cir. 1980).

77. 622 F.2d at 1167–1168; 8 OSHC at 1322 (footnotes omitted).

78. In *A. Schonbek & Co., Inc. v. Donovan and OSAHRC* (74), the Second Circuit joined the ranks of those courts adopting this view. Its citation of the Third Circuit's *Babcock & Wilcox* decision in support of its holding indicates that the split in the circuits on the "willfulness" issue no longer exists. 9 OSHC at 1563.

79. 540 F.2d 157; 4 OSHC 1451 (3d Cir. 1976).

80. 540 F.2d at 162; 4 OSHC at 1454–1455.

81. The Third Circuit has reaffirmed its *Bethlehem Steel* holding as recently as 1980. See *Jones & Laughlin Steel Corp. v. Marshall and OSAHRC*, 636 F.2d 32; 8 OSHC 2217 (3d Cir. 1980).

82. 582 F.2d 834; 6 OSHC 1855 (4th Cir. 1978).

83. 582 F.2d at 840–841; 6 OSHC at 1859.

84. See *Bunge Corp. v. Secretary of Labor and OSAHRC*, 638 F.2d 831; 9 OSHC 1312 (5th Cir. 1981); *J.L. Foti Construction Co. v. OSAHRC and Donovan*, 687 F.2d 853; 10 OSHC 1937 (6th Cir. 1982); *Todd Shipyards Corp. v. Secretary of Labor*, 586 F.2d 683; 6 OSHC 2122 (9th Cir. 1978); *Dun-Par Engineered Form Co. v. Marshall and OSAHRC*, 676 F.2d 1333; 10 OSHC 1561 (10th Cir. 1982).

85. *Id.* See also *George Hyman Construction Co.* (82).

86. 7 OSHC 1061 (Rev. Comm. 1979).

87. 7 OSHC at 1063. See also *Bethlehem Steel Corp.*, 9 OSHC 1346 (Rev. Comm. 1981), in which Chairman Cleary and Commissioner Cottine reaffirmed their holding in *Potlatch Corp.* Commissioner Barnako did not participate in this decision. Commissioner Barnako, who disagreed with the *Potlatch* majority's allocation of the burden of proof on the issue of substantially similar violations, was not reappointed to his post when his term expired in April 1981. BNA, *OSHR*, Current Report for April 30, 1981, p. 1497. Barnako was succeeded by Robert A. Rowland, who assumed the chairman's post in August of 1981. BNA, *OSHR*, Current Report for August 6, 1981, p. 187. Chairman Rowland has not yet expressed an opinion on the proper definition of a "repeated" violation.

88. Although the Fourth, Fifth, Sixth, Ninth, and Tenth Circuits have unanimously rejected the

Third Circuit's definition of a "repeated" violation, confusion still remains regarding the definition of "substantially similar" violations and the proper allocation of the burden of proving such similarity. For a discussion of these problems, see *Bunge Corp. v. Secretary of Labor and OSAHRC* (84).

Another question that has arisen regarding "repeated" violations involves the treatment of transient as opposed to fixed work sites. Chapter VIII of the *OSHA Field Operations Manual* sets forth the following guidelines for the treatment of these respective types of work sites:

> For purposes of considering whether a violation is repeated, citations issued to employers having fixed establishments (e.g., factories, terminals, stores) will be *limited to the cited establishment*. For employers engaged in businesses having no fixed establishments (construction, painting, excavation) repeated violations will be alleged based on prior violations occurring anywhere within the same State.

In *Desarrollos Metropolitanos, Inc. v. OSAHRC*, 551 F.2d 874; 5 OSHC 1135 (1st Cir. 1977), the First Circuit held that these guidelines, insofar as they distinguish between employers with fixed and transient work sites, do not violate the equal protection clause of the Federal Constitution. The court also ruled that there is no constitutional infirmity concerning the guidelines' distinction between intrastate and interstate construction work.

89. For an interpretation of "willful" in the context of Section 17(e) of OSHA, see *United States v. Dye Construction Co.*, 510 F.2d 78 (10th Cir. 1975).

90. See Section 17(e) of OSHA. More than one conviction under Section 17(e) or under Section 110(d) of the Federal Mine Safety and Health Act of 1977 can result in much greater fines and/or prison terms.

91. See Section 17(g) of OSHA and Section 110(f) of the Federal Mine Safety and Health Act of 1977.

92. See Section 17(f) of OSHA and Section 110(e) of the Federal Mine Safety and Health Act of 1977.

93. See Section 110(h) of the Federal Mine Safety and Health Act of 1977.

94. See Section 1114 of the United States Criminal Code, 18 U.S.C.A. § 1114.

95. See American Bar Association, 1975 Report of the Committee on Occupational Safety and Health Law (51), p. 68; and CCH, *Employment Safety and Health Guide*, ¶¶ 4201, 4205 (1980). In determining the reasonableness of an abatement order, the following factors are considered: the gravity of the violation, the number of employees exposed to the hazard and the extent of their exposure, and the availability and time necessary for the installation of any equipment necessary to abate the violation. See CCH, *Employment Safety and Health Guide*, ¶ 4205 (1980). Thus, in *United Parcel Service of Ohio, Inc. v. OSAHRC and Marshall*, 570 F.2d 806; 6 OSHC 1347 (8th Cir. 1978), the Eighth Circuit upheld a citation but vacated the Review Commission's abatement order requiring the employer to provide safety shoes for all its workers who handled parcels. In light of the small size of most of the parcels handled, the low incidence of injuries resulting from dropped parcels, and the small number of full-time, long-term employees exposed to the hazard, the court concluded the abatement requirements were too harsh.

96. Despite the requirements of § 10(a), the Review Commission may in limited circumstances entertain a late notice of contest pursuant to Federal Rule 60(b). This rule permits a federal court to grant relief from a final order for a number of reasons, including a party's mistake, surprise or excusable neglect, the presence of newly discovered evidence, the fraud or misconduct of an adverse party, and so on. Rule 60(b), which was held applicable to the Review Commission by the Third Circuit in *J.I. Hass Co. v. OSAHRC*, 648 F.2d 190; 9 OSHC 1712 (3d Cir. 1981), has been viewed by the Commission as affording a possible basis for considering an untimely notice of contest. *Special Coating Systems of New Mexico, Inc.*, 10 OSHC 1671 (Rev. Comm. 1982).

97. 29 CFR § 1903.17.

98. *Brennan v. OSAHRC and Bill Echols Trucking Co.*, 487 F.2d 230; 1 OSHC 1398 (5th Cir. 1973); *Dan J. Sheehan Co. v. OSAHRC and Dunlop*, 520 F.2d 1036; 3 OSHC 1573 (5th Cir. 1975), *certiorari* denied, 424 U.S. 956 (1976).

99. *Dan. J. Sheehan* (98), 520 F.2d at 1038–1039; 3 OSHC at 1575.

100. *Turnbull Millwork Co.*, 3 OSHC 1781 (Rev. Comm. 1975); *State Home Improvement Co.*, 6 OSHC 1249 (Rev. Comm. 1977). The Seventh and Eighth Circuits have expressed their approval of the Commission's *Turnbull* rule, at least in situations in which the employer is a layman proceeding *pro se*. *Penn-Dixie Steel Corp. v. OSAHRC and Dunlop*, 553 F.2d 1078; 5 OSHC 1315 (7th Cir. 1977); *Marshall v. Gil Haughan Construction Co. and OSAHRC*, 586 F.2d 1263; 6 OSHC 2067 (8th Cir. 1978). The Commission has gone further and has applied the rule even where the employer is represented by counsel. *Nilsen Smith Roofing & Sheet Metal Co.*, 4 OSHC 1765 (Rev. Comm. 1976). However, the Commission has made it clear that the same lenient policy does not apply to an employer who understands the difference between contesting a penalty and a citation, and who nevertheless limits his notice of contest to the penalty. *F.H. Sparks of Maryland, Inc.*, 6 OSHC 1356 (Rev. Comm. 1978).

101. See Section 10(c) of OSHA. The rules governing practice before the commission can be found at 29 CFR §§ 2200.1–2200.211. The rules found at §§ 2200.200–2200.211 were added in 1981 (46 *Fed. Reg.* 63,041, December 30, 1981), and permit any party to request simplified proceedings before the ALJ in cases that do not involve alleged general duty violations or alleged violations of certain enumerated standards. Procedures under the new regulations are simplified in several ways: the pleadings are limited, discovery is generally not permitted, the Federal Rules of Evidence do not apply, and interlocutory appeals are not permitted.

102. *Brennan v. OSAHRC and Interstate Glass Co.* (63), 487 F.2d at 438; 1 OSHC at 1373.

103. The commission's Rules of Procedure provide that a petition for discretionary review must state specific grounds for relief. The commission normally limits review to cases in which a party asserts that: (a) a finding of material fact is not supported by a preponderance of the evidence; (b) the administrative law judge's decision is contrary to law or to the rules or decisions of the commission; (c) a substantial question of law or policy or an abuse of discretion is involved; or (d) the administrative law judge committed a prejudicial procedural error. 29 CFR § 2200.92(b). See also M. Rothstein, *OSHA After Ten Years: A Review and Some Proposed Reforms*, 34 VAND.L.REV. 71, 119 (1981).

104. 29 CFR §§ 2200.91 and 2200.92. Any member of the commission clearly has the authority to direct review of a case on his own motion, even if no party has requested review. *GAF Corp.*, 8 OSHC 2006 (Rev. Comm. 1980). However, the commission's rules limit the grounds on which a member may independently order review. Except in extraordinary circumstances, review is limited to the issues raised by the parties before the administrative law judge, and commission members may normally direct review only when the case involves "novel questions of law or policy or questions involving conflict in Administrative Law Judges' decisions." 29 CFR § 2200.92(d). Thus, it is normally preferable for a party to obtain review by filing a petition. This is especially important in light of a recent amendment to the commission's rules, which provides that a party's failure to file a petition may foreclose later judicial review of any objections to the administrative law judge's decision. 29 CFR § 2200.91(a)(1980). See also *Keystone Roofing Co., Inc. v. OSAHRC and Dunlop*, 539 F.2d 960; 4 OSHC 1481 (3d Cir. 1976); *McGowan v. Marshall*, 604 F.2d 885; 7 OSHC 1842 (5th Cir. 1979).

105. B. Fellner and D. Savelson, "Review by the Commission and the Courts," in: *Proceedings of the American Bar Association Institute on Occupational Safety and Health Law*, ABA Press, Chicago, 1976, p. 102.

106. The Review Commission's failure to adequately consider these statutory penalty assessment factors can result in the Court of Appeals remanding the case to the commission for more particularized findings regarding the penalty. *Astra Pharmaceutical Products, Inc. v. OSAHRC and Donovan*, 681 F.2d 69; 10 OSHC 1697 (1st Cir. 1982).

107. *REA Express v. Brennan*, 495 F.2d 822; 1 OSHC 1651 (2d Cir. 1974); *Brennan v. OSAHRC*

and *Interstate Glass Co.*, 487 F.2d 438; 1 OSHC 1372 (8th Cir. 1973); *California Stevedore & Ballast Co. v. OSAHRC* (93). Reduction of the secretary's proposed penalty is also a matter within the commission's discretion. *Western Waterproofing Co., Inc. v. Marshall and OSAHRC* (74). Compare *Dale M. Madden Construction, Inc. v. Hodgson*, 502 F.2d 278; 2 OSHC 1236 (9th Cir. 1974), where the court held that the commission has no authority to modify *settlements* made by the secretary and the cited employer. See also *Marshall v. Sun Petroleum Products Co. and OSAHRC*, 622 F.2d 1176; 8 OSHC 1422 (3d Cir. 1980), *certiorari* denied, 449 U.S. 1061, holding that the commission may review and modify settlements, but only for the limited purpose of determining the reasonableness of the abatement period when it has been challenged by employees.

108. See, for example, *Dixie Roofing and Metal Co.*, 2 OSHC 1566 (Rev. Comm. 1975). See also *Consolidated Rail Corp.*, 10 OSHC 1564, 1568 (Rev. Comm. 1982), in which the commission reduced the degree of a violation from serious to nonserious without specifically discussing its authority to do so.

109. Although § 11(a) permits "any person" aggrieved by a commission order to obtain judicial review, the Third Circuit has held that the role of employees or their representatives in a proceeding initiated by the employer is strictly limited to challenging the reasonableness of the abatement period pursuant to § 10(c) of the act. In *Marshall v. Sun Petroleum Products* (107), the court indicated that this limitation should also apply to the right of employees to instigate judicial review. On the other hand, the District of Columbia Circuit has concluded that employees, although prohibited from *instituting* a commission action on matters other than the reasonableness of the abatement period, may nevertheless participate fully as parties in employer-initiated proceedings. The right of employees to participate in enforcement proceedings, the court reasoned, must include the right to appeal from an unfavorable commission decision. *Oil, Chemical, and Atomic Workers v. OSAHRC*, 671 F.2d 643; 10 OSHC 1345 (D.C. Cir. 1982), *certiorari denied sub nom. American Cyanimid Co. v. Oil, Chemical and Atomic Workers*, 103 S.Ct. 206.

110. The Sixth Circuit has defined "substantial evidence" in the context of OSHA proceedings as "such relevant evidence as a reasonable mind might accept as adequate to support a conclusion." *Martin Painting & Coating Co. v. Marshall*, 629 F.2d 437; 8 OSHC 2173, 2174 (6th Cir. 1980), *certiorari* denied, 449 U.S. 1062, quoting *Dunlop v. Rockwell International*, 540 F.2d 1283, 1287; 4 OSHC 1606, 1608 (6th Cir. 1976). While the commission's factual findings are conclusive if supported by substantial evidence, its interpretations of statutory language are legal conclusions, which are not accorded the same deference on review. *Usery v. Hermitage Concrete Pipe Co. and OSAHRC* (54). The commission's resolution of questions of credibility, on the other hand, are insulated from reversal on appeal unless they are contradicted by "uncontrovertible documentary evidence or physical facts." *Super Excavators, Inc. v. OSAHRC and Secretary of Labor*, 674 F.2d 592; 10 OSHC 1369, 1370 (7th Cir. 1981), *certiorari* denied, 102 S.Ct. 2958 (1982), quoting *International Harvester Co. v. OSAHRC*, 628 F.2d 982, 986; 8 OSHC 1780, 1783 (7th Cir. 1980).

111. *Secretary v. OSAHRC and Interstate Glass* (102), 487 F.2d at 442; 1 OSHC at 1375. The procedures for judicial review under the Federal Mine Safety and Health Act of 1977 are set forth in Section 106. For a summary of the review procedures available under the earlier law, the Coal Mine Health and Safety Act of 1969, see *National Independent Coal Operators Assoc. v. Kleppe*, 423 U.S. 388 (1976).

112. See *Usery v. Whirlpool Corp.*, 416 F. Supp. 30, 34; 4 OSHC 1391, 1392–1393 (N.D. Ohio 1976), *reversed on other points sub nom. Marshall v. Whirlpool Corp.*, 593 F.2d 715; 7 OSHC 1075 (6th Cir. 1979), *affirmed*, 445 U.S. 1 (1980). The United States Supreme Court in the *Whirlpool* case was faced with the question whether an employee is protected from retaliation by his employer if he walks off the job in an imminent danger situation. The Secretary of Labor, in 1973, promulgated a regulation that protected employees from discrimination if they refused in good faith to work under life-threatening conditions. The rule applied only if the employee reasonably believed that a real danger of death or serious injury existed, and that

the danger was too immediate to be eliminated through the act's normal enforcement channels. 29 CFR § 1977.12(b)(2). In *Whirlpool Corp.*, the Supreme Court upheld § 1977.12(b)(2) as a valid exercise of the secretary's authority under the act. Thus, employees are afforded a limited right to refuse to work in imminent danger situations. See also G. Scarzafava and F. Herrera, Jr., *Workplace Safety—The Prophylactic and Compensatory Rights of the Employee*, 13 ST. MARY'S L.J. 911, 931–933 (1982).

113. Imminent danger is defined by Section 3(j) of the Federal Mine Safety and Health Act of 1977 as "the existence of any condition or practice in a coal or other mine which could reasonably be expected to cause death or serious physical harm before such condition or practice can be abated. . . ." The definition of "imminent danger" in the Coal Mine Health and Safety Act of 1969 was almost identical to that which now appears in the 1977 act. In a case arising under the 1969 provision, the Seventh Circuit affirmed the Board of Mine Operations Appeals' holding that an "imminent danger" situation exists if, in a reasonable man's estimation, it is at least as probable as not that continuation of normal coal extraction operations in the disputed area will result in the occurrence of the feared accident or disaster before the danger is eliminated. *Freeman Coal Mining Co. v. Interior Board of Mine Operations Appeals*, 504 F.2d 741; 2 OSHC 1308 (7th Cir. 1974). *Accord, Old Ben Coal Corp. v. Interior Board of Mine Operations Appeals*, 523 F.2d 25; 3 OSHC 1270 (7th Cir. 1975); *Eastern Assoc. Coal Corp. v. Interior Board of Mine Operations Appeals*, 491 F.2d 277 (4th Cir. 1974).

114. 518 F.2d 990; 3 OSHC 1490 (5th Cir. 1975).

115. The Supreme Court also granted *certiorari* in *Frank Irey, Jr., Inc. v. OSAHRC and Brennan*, 519 F.2d 1215; 3 OSHC 1329 (3d Cir. *en banc* 1975) to review the same issue. The *Atlas Roofing* and *Frank Irey* cases were decided by the Court in consolidated proceedings.

116. 430 U.S. 442; 5 OSHC 1105 (1977).

117. See, e.g., *Dan J. Sheehan Co. v. OSAHRC* (98); *Clarkson Construction Co. v. OSAHRC and Secretary of Labor*, 531 F.2d 451; 3 OSHC 1880 (10th Cir. 1976); and *Savina Home Industries, Inc. v. Secretary of Labor and OSAHRC* (38).

118. BNA Daily Labor Reporter, April 21, 1983, at p. A-2.

119. See BNA, *OSHR*, Current Reports for: August 13, 1981, p. 203; February 26, 1981, p. 1289; October 8, 1981, pp. 368–369; October 21, 1982, p. 404; March 31, 1983, pp. 926–927.

120. See BNA, *OSHR*, Current Reports for: July 29, 1982, pp. 206–207; January 21, 1982, p. 643 (discusses OSHA's guidelines for voluntary compliance programs); July 23, 1981, p. 151 (discusses OSHA's guidelines for joint labor/management committees); August 6, 1981, pp. 187–189 (discusses OSHA's proposal to replace inspections at particular workplaces with cooperative labor/management committees—the STAR Program). For adverse labor reaction to the STAR Program, see BNA, *OSHR*, Current Report for September 10, 1981, pp. 293–294.

121. See BNA, *OSHR*, Current Report for July 1, 1982, p. 107.

122. See BNA, *OSHR*, Current Report for July 8, 1982, pp. 123–124, describing OSHA's three voluntary compliance plans: the PRAISE, STAR, and TRY Programs. The PRAISE Program grants an exemption from general schedule inspections for firms in low-hazard industries with good safety records. The STAR Program provides a similar exemption for companies with safety and health programs that exceed regulatory requirements, focusing on firms with established labor/management safety committees. Finally, the experimental TRY Program is open to employers with a low (or improving) injury record and a health and safety program involving active employee participation. As of February 1983, a total of eleven worksites had been approved for participation in these voluntary programs. BNA, *OSHR*, Current Report for February 10, 1983, p. 759. See also BNA, *OSHR*, Current Report for October 7, 1982, pp. 371–372 (describing General Motors' and United Auto Workers' establishment of a six-member advisory panel to work together in improving worker health and safety).

123. BNA, *OSHR*, Current Report for April 16, 1981, pp. 1425–1426.

124. *Id* at 1425.

125. BNA, *OSHR*, Current Report for May 7, 1981, p. 1528 (remarks of AFL–CIO Secretary/
Treasurer Thomas R. Donahue). The Reagan Administration's approach to occupational
safety and health continues to receive sometimes harsh criticism from organized labor and
other interested groups. Indeed, on March 28, 1983, nine environmental organizations urged
President Reagan to "reverse two years of efforts to dismantle crucial programs," and
contended that the Reagan Administration's handling of OSHA parallels the policies that
led to "destruction" of the Environmental Protection Agency. BNA, *OSHR*, Current Report
for March 31, 1983, pp. 926–927. Specifically, organized labor and environmental groups
have been upset by OSHA's re-evaluations of the cotton dust and lead standards and the
workplace cancer policy, cutbacks in personnel, closing of area offices, and denial of petitions
to regulate or impose stricter standards on formaldehyde, ethylene oxide, asbestos, and
benzene. *Id*. In addition, Eula Bingham, former head of the OSH Administration during
the Carter Administration, and several other former federal agency directors have announced
the creation of a "Regulatory Audit Project" to "monitor the Reagan Administration's assault
on public health and safety programs." BNA, *OSHR*, Current Report for March 10, 1983,
pp. 835–836.

126. See BNA, *OSHR*, Current Report for May 5, 1983, pp. 1043–1044.

127. *Id*.

128. In 1982, 45,038 general schedule inspections were conducted as compared with 36,942 in
1981 and 34,794 in 1980. *Id*.

129. *Id*. The average number of inspections during the Carter years was 59,822 according to
OSHA sources. Moreover, follow-up visits were down from 3,386 in 1981 to 1,426 in 1982;
this is in line with the present OSH Administration's decision to reduce this type of visit,
because more than 99 percent of follow-up inspections revealed no failure to correct
violations.

130. See BNA, *OSHR*, Current Report for October 14, 1982, p. 395. For example, in the area of
state plan programs OSHA's efforts will be directed toward revising the system for monitoring
states' performance, resolving the issues of states promulgating standards different from
federal ones, and granting final approval to fully effective state programs. In addition,
OSHA is preparing new staffing benchmarks for state plans. See BNA, *OSHR*, Current
Reports for May 5, 1983, p. 1044.

131. *Id*.

Index

Abate, health effects, 158
Abrasion, hand and arm protection, 696–697
Absorption, 130–131
  absorbed dose formula, 135
  indicators, 101
Accidents:
  and employee assistance program, 56
  prevention, 58–59
Accounting records, as source for emission
    information, 556–557
Acetaldehyde, health effects, 158
Acetic acid, health effects, 158
Acetic anhydride, health effects, 158
Acetone, health effects, 158
Acetonitrile, health effects, 158
2-Acetylaminofluorene, health effects, 158
Acetylene, health effects, 158
Acetylene dichloride, health effects, 158
Acetylene tetrabromide, health effects, 158
Acid phosphatase, biochemical marker of
    cancer, 103
Acrolein, health effects, 158
Acrylamide, health effects, 158
Acrylonitrile:
  health effects, 158
  OSHA standard, 742, 759
Action Level, 459, 461, 579
  TERM decision, 461
Address, defined, 362
Aerosol:
  absorption, 130
  microbial, modified exposure limits, 236
Air contaminants:
  accumulation, 129
  averaging time and environmental variance,
    591–592

body burden comparisons for shift work,
  129
buildup in target tissue, 202–203
community dispersion estimates, 561
determination, 454
emission factors, 545–546
exposure limits, 127
immediately dangerous to life or health
    atmosphere, 676–679
indexing of levels, history, 11–12
inhaled, pharmacokinetic behavior, 201–
  202
monitoring stations, 327–328
process sewers, drainage ditches and
    collecting ponds, 551
sampling parameters, 520–521
sensor-based data acquisition, 305
simulation of body uptake, 203
TLVs:
  basis, 128
  maximum body burden, 128–129
  unusual schedules, 128–129
workplace air, 542
Air Force, safety belt usage, 59
Alarms, 649–650
Alcohol:
  associated with smoking, 53
  employee assistance programs, 53–56
  fetal alcohol syndrome, 53
  prevalence of alcoholism, 53–54
Aldrin, health effects, 158
ALGOL, defined, 362
Algorithm, defined, 363
Allergens, modified exposure limits, 237
Allyl alcohol, health effects, 158
Allyl chloride, health effects, 158

791